D1289224

Angelica M. Stacy
Professor of Chemistry
University of California at Berkeley

with

Janice A. Coonrod
Bishop O'Dowd High School
Senior Writer and Developer

Jennifer Claesgens
Northern Arizona University
Curriculum Developer

W. H. Freeman and Company/BFW

Editors	Ladie Malek, Jeffrey Dowling
Project Administrators	Elizabeth Ball, Rachel Merton, Janis Pope
Consulting Editors	Heather Dever, Joan Lewis, Andres Marti
Editorial Advisor	Casey FitzSimons
Editorial Production Supervisor	Kristin Ferraioli
Copyeditor	Margaret Moore
Production Coordinator	Ann Rothenbuhler
Production Director	Christine Osborne
Text Designer	tani hasegawa
Compositor	Precision Graphics
Art Editors	Maya Melenchuk, Lisa Torri
Illustrators	Ken Cursoe, Greg Hargreaves, Tom Ward
Technical Artist	Precision Graphics
Photo Researcher	Laura Murray Productions
Cover Designer	Diana Ghermann

Textbook Product Manager	Tim Pope
Executive Editor	Josephine Noah
Publisher	Steven Rasmussen

This material is based upon work supported by the National Science Foundation under award number 9730634. Any opinions, findings, and conclusions or recommendations expressed in this publication are those of the author and do not necessarily reflect the views of the National Science Foundation.

W. H. Freeman and Company
41 Madison Avenue
New York, NY 10010
Houndmills, Basingstoke
RG21 6XS, England
http://www.highschool.bfwpub.com

ISBN-13: 978-1-55953-941-8
ISBN-10: 1-55953-941-0

Fourth Printing 2014
Printed in the United States of America

Acknowledgments

A number of individuals joined the project as developers for various periods of time along the way to completing this work. Thanks go to these individuals for their contributions to the unit development: Karen Chang, David Hodul, Rebecca Krystyniak, Tatiana Lim, Jennifer Loeser, Evy Kavaler, Sari Paikoff, Sally Rupert, Geoff Ruth, Nicci Nunes, Gabriela Waschewski, and Daniel Quach.

David R. Dudley contributed original ideas and sketches for some of the wonderful cartoons interspersed throughout the book. His sketches provided a rich foundation for the art manuscript.

This work would not have been possible without the thoughtful feedback and great ideas from numerous teachers who field-tested early versions of the curriculum. Thanks go to these teachers and their students: Carol de Boer, Wayne Brock, Susan Edgar-Lee, Melissa Getz, David Hodul, Richard Kassissieh, Tatiana Lim, Evy Kavaler, Geoff Ruth, Nicci Nunes, Gabriela Waschewski, and Daniel Quach.

Dr. Truman Schwartz provided a thorough and detailed review of the manuscript. We appreciate his insights and chemistry expertise.

Ladie Malek and Jeffrey Dowling served as the developmental editors for the project, giving feedback and advice.

Science Content Advisor

Dr. A. Truman Schwartz, Macalaster College (emeritus), St. Paul, MN

Teaching and Content Reviewers

Scott Balicki, Boston Latin School, Boston, MA

Greg Banks, Urban Science Academy, West Roxbury, MA

Randy Cook, Tri County High School, Howard City, MI

Thomas Holme, University of Wisconsin, Milwaukee, Milwaukee, WI

Mark Klawiter, Deerfield High School, Deerfield, WI

Carri Polizzotti, Marin Catholic High School, Larkspur, CA

Matthew Vaughn, Burlingame High School, Burlingame, CA

Rebecca Williams, Richland College, Dallas, TX

Note to Students

Welcome to *Living By Chemistry.* In this course, you will actively participate in uncovering the chemistry in the laboratory and in the world around you. Rather than simply writing "correct" answers to chemistry questions and problems, you will learn to support answers with evidence. Learning chemistry is a bit like learning a new sport or dance—you will get better with practice. The more you participate, the deeper your understanding will be.

This textbook is designed to be used for reading and reference. The readings focus on the concepts you investigate in class. There are two or three pages of reading to go along with each day's activity, followed by homework exercises. The readings provide real-life examples to further explain and clarify the chemistry concepts. After each day's lesson, the reading and exercises will help you understand and remember what you learned in class. We designed this textbook to be highly readable, and sincerely believe it will make the study of high school chemistry enjoyable for you.

Angelica M. Stacy

About the Author

Angelica M. Stacy is a Professor of Chemistry at the University of California, Berkeley. At Berkeley, she teaches introductory chemistry and does research in materials chemistry and chemistry education. She has received numerous awards and honors for her teaching. Through partnerships with high school teachers and students, Dr. Stacy has worked for many years in designing and refining the *Living By Chemistry* curriculum. She developed this curriculum to support students by offering them engaging challenges.

Contents

Content Coverage Chart

Concepts are introduced and then reinforced, often across different units. Coverage usually consists of an introduction (I), then practice (P), and finally teaching to mastery (M).

	Unit and Section								
	Alchemy					Smells			
	I	II	III	IV	V	I	II	III	IV
Content									
Atomic and Molecular Structure			I	P	P	P	P	P	M
Chemical Bonds					I	P	P	P	P
Conservation of Mass and Stoichiometry	I	P	P						
Gases and Their Properties									
Acids and Bases		I				P			
Solutions		I							
Energy/Chemical Thermodynamics									
Reaction Rates						I			
Chemical Equilibrium									
Organic Chemistry and Biochemistry						I	P	P	P
Nuclear Processes			I/P						
Investigation and Experimentation	I	P	P	P	P	P	P	P	P

Unit and Section													
Weather			Toxins					Fire				Showtime	
I	II	III	I	II	III	IV	V	I	II	III	IV	I	II
										M			
			P	P	P	P	P	P	P	P	M		
I	P	M											
						P						P	M
					P	P	P				P	P	M
	I							P	P	P	P	M	
										P		P	P
												I	P
P	P	P	P	P	P	P	P	P	P	P	P	P	M

How to Use This Book

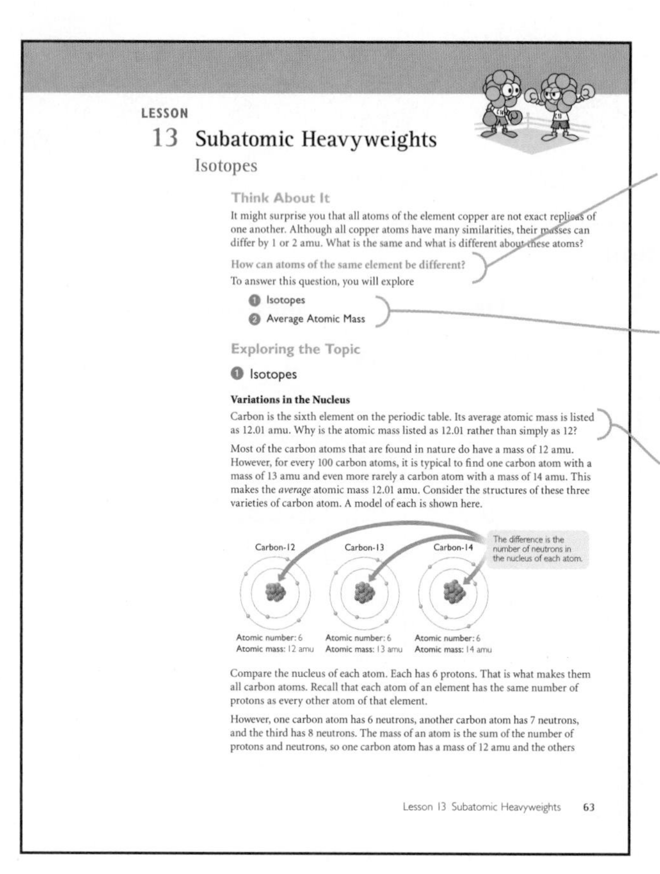

LESSON

13 Subatomic Heavyweights

Isotopes

Think About It

It might surprise you that all atoms of the element copper are not exact replicas of one another. Although all copper atoms have many similarities, their masses can differ by 1 or 2 amu. What is the same and what is different about these atoms?

How can atoms of the same element be different?

To answer this question, you will explore

1. Isotopes
2. Average Atomic Mass

Exploring the Topic

1. Isotopes

Variations in the Nucleus

Carbon is the sixth element on the periodic table. Its average atomic mass is listed as 12.01 amu. Why is the atomic mass listed as 12.01 rather than simply as 12?

Most of the carbon atoms that are found in nature do have a mass of 12 amu. However, for every 100 carbon atoms, it is typical to find one carbon atom with a mass of 13 amu and even more rarely a carbon atom with a mass of 14 amu. This makes the *average* atomic mass 12.01 amu. Consider the structures of these three varieties of carbon atom. A model of each is shown here.

Compare the nucleus of each atom. Each has 6 protons. That is what makes them all carbon atoms. Recall that each atom of an element has the same number of protons as every other atom of that element.

However, one carbon atom has 6 neutrons, another carbon atom has 7 neutrons, and the third has 8 neutrons. The mass of an atom is the sum of the number of protons and neutrons, so one carbon atom has a mass of 12 amu and the others

Lesson 13 Subatomic Heavyweights 63

- This **question** lets you know what to focus on while you're reading. This is the question that you explored in class, and the reading will tell you more about it.
- The **main topics** covered in the reading are listed here.
- The **reading** helps you review and remember what you learned in class.

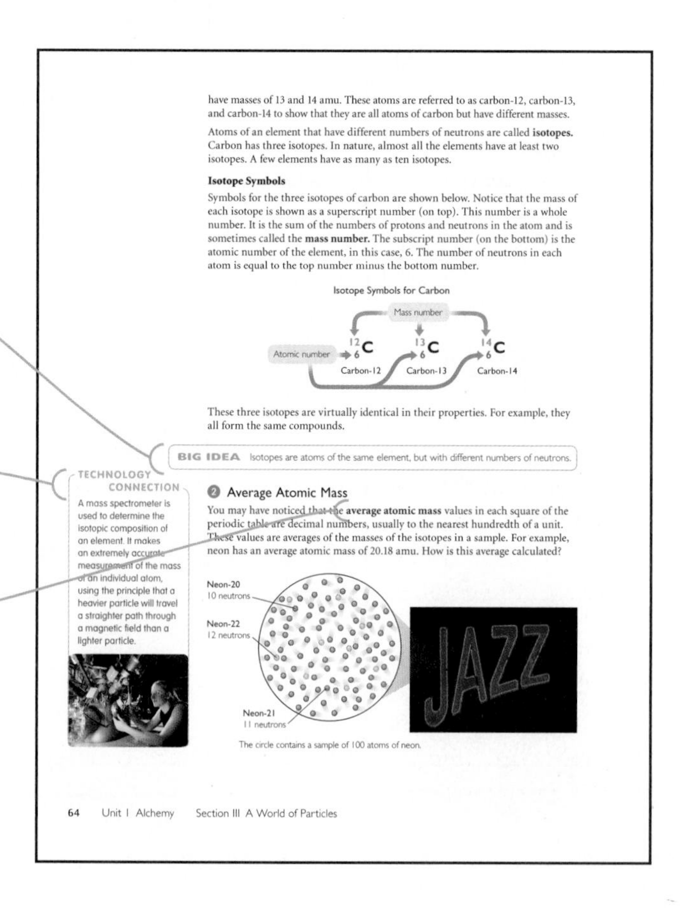

have masses of 13 and 14 amu. These atoms are referred to as carbon-12, carbon-13, and carbon-14 to show that they are all atoms of carbon but have different masses.

Atoms of an element that have different numbers of neutrons are called **isotopes.** Carbon has three isotopes. In nature, almost all the elements have at least two isotopes. A few elements have as many as ten isotopes.

Isotope Symbols

Symbols for the three isotopes of carbon are shown below. Notice that the mass of each isotope is shown as a superscript number (on top). This number is a whole number. It is the sum of the numbers of protons and neutrons in the atom and is sometimes called the **mass number.** The subscript number (on the bottom) is the atomic number of the element, in this case, 6. The number of neutrons in each atom is equal to the top number minus the bottom number.

Isotope Symbols for Carbon

These three isotopes are virtually identical in their properties. For example, they all form the same compounds.

BIG IDEA Isotopes are atoms of the same element, but with different numbers of neutrons.

TECHNOLOGY CONNECTION

A mass spectrometer is used to determine the isotopic composition of an element. It makes an extremely accurate measurement of the mass of an individual atom, using the principle that a heavier particle will travel a straighter path through a magnetic field than a lighter particle.

2. Average Atomic Mass

You may have noticed that the **average atomic mass** values in each square of the periodic table are decimal numbers, usually to the nearest hundredth of a unit. These values are averages of the masses of the isotopes in a sample. For example, neon has an average atomic mass of 20.18 amu. How is this average calculated?

The circle contains a sample of 100 atoms of neon.

64 Unit I Alchemy Section III A World of Particles

- Pay attention to **Big Ideas,** especially when you are reviewing for a quiz or exam. These are fundamental concepts of chemistry.
- **Connections** highlight how chemistry is used in other areas of study, careers, or everyday situations.
- Words in bold are **key terms.** These important chemistry vocabulary terms are listed at the end of the lesson and at the end of each section and unit. Key terms are defined in the glossary.

About 90% of all neon atoms have an atomic mass of 20 amu, 9% have an atomic mass of 22 amu, and 1% have an atomic mass of 21 amu. By considering a random sample of 100 neon atoms, you can calculate their average atomic mass like this:

$$\text{average atomic mass} = \frac{\text{total mass}}{\text{number of atoms}}$$

$$= \frac{(90)(20\ \text{amu}) + (9)(22\ \text{amu}) + (1)(21\ \text{amu})}{100}$$

$$= 20.19\ \text{amu}$$

As you can see, this number is nearly identical to 20.18, the average atomic mass listed for neon on the periodic table.

The percentage of each isotope of an element that occurs in nature is called the *natural abundance* of the isotope. For example, the natural abundance of neon-20 is about 90.48%.

Important to Know The mass of an isotope refers to the mass of a single specific atom of an element. The average atomic mass given on the periodic table is the average of the masses of all the isotopes in a large sample of that element. ◂

BIOLOGY CONNECTION

Less common isotopes are more easily detectable, so researchers often use them to trace how an element moves through a living creature or the environment. For example, using a fertilizer that is enriched in nitrogen-15 allows a scientist to track the nitrogen atoms to see how much nitrogen a plant is using.

Example 1

Isotopes of Copper

There are two different isotopes of copper. The isotope names and symbols are given here.

$^{63}_{29}Cu$ $^{65}_{29}Cu$

copper-63 copper-65

a. Explain why both symbols have 29 as the bottom number.

b. Explain how the two isotopes are different from each other.

c. Scientists have found the natural abundances of each isotope: 69% copper-63 and 31% copper-65. Explain why the average atomic mass listed on the periodic table for copper is 63.55.

Solution

a. Copper's atomic number is 29, so both isotope symbols have a subscript 29 indicating 29 protons.

b. The two isotopes have different atomic masses. Both isotopes have 29 protons, so copper-63 has 34 neutrons and copper-65 has 36 neutrons.

c. You could consider a sample of 100 atoms of copper and calculate their average mass. The average mass of the 100 atoms is determined by adding the masses of the 100 atoms and dividing this total by 100.

$$\frac{69(63\ \text{amu}) + 31(65\ \text{amu})}{100} = \frac{6362\ \text{amu}}{100} = 63.6\ \text{amu}$$

This is very close to the value found on the periodic table.

Lesson 13 Subatomic Heavyweights 65

■ The **examples** show you how to solve problems step-by-step. Try to answer the question yourself before reading the solution, to check your understanding.

■ The **lesson summary** recaps the main ideas of the lesson that answer the opening question. Reading the lesson summaries is a good way to review for a quiz or an exam.

■ Doing the **exercises** will help you check that you understood the reading and can solve problems on your own. You can check some of your answers in the Answers to Selected Exercises section in the back of the book.

■ Using **SciLinks,** you can find information online to help you learn more about chemistry topics covered in *Living By Chemistry.* Visit the website and enter the topic and code.

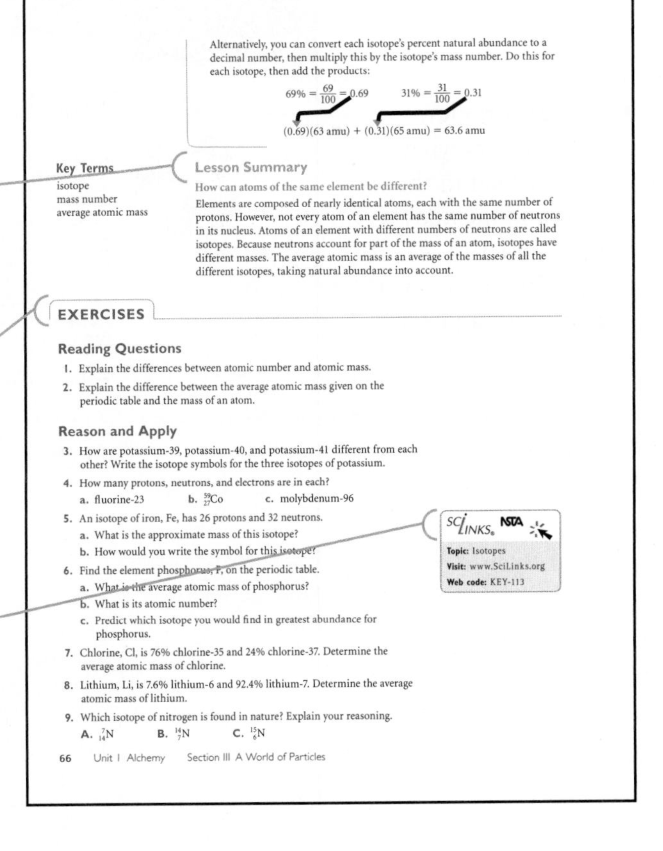

Alternatively, you can convert each isotope's percent natural abundance to a decimal number, then multiply this by the isotope's mass number. Do this for each isotope, then add the products:

$69\% = \frac{69}{100} = 0.69$ $31\% = \frac{31}{100} = 0.31$

$(0.69)(63\ \text{amu}) + (0.31)(65\ \text{amu}) = 63.6\ \text{amu}$

Key Terms
isotope
mass number
average atomic mass

Lesson Summary

How can atoms of the same element be different?

Elements are composed of nearly identical atoms, each with the same number of protons. However, not every atom of an element has the same number of neutrons in its nucleus. Atoms of an element with different numbers of neutrons are called isotopes. Because neutrons account for part of the mass of an atom, isotopes have different masses. The average atomic mass is an average of the masses of all the different isotopes, taking natural abundance into account.

EXERCISES

Reading Questions

1. Explain the differences between atomic number and atomic mass.
2. Explain the difference between the average atomic mass given on the periodic table and the mass of an atom.

Reason and Apply

3. How are potassium-39, potassium-40, and potassium-41 different from each other? Write the isotope symbols for the three isotopes of potassium.
4. How many protons, neutrons, and electrons are in each?
   a. fluorine-23 b. $^{59}_{27}Co$ c. molybdenum-96
5. An isotope of iron, Fe, has 26 protons and 32 neutrons.
   a. What is the approximate mass of this isotope?
   b. How would you write the symbol for this isotope?
6. Find the element phosphorus, P, on the periodic table.
   a. What is the average atomic mass of phosphorus?
   b. What is its atomic number?
   c. Predict which isotope you would find in greatest abundance for phosphorus.
7. Chlorine, Cl, is 76% chlorine-35 and 24% chlorine-37. Determine the average atomic mass of chlorine.
8. Lithium, Li, is 7.6% lithium-6 and 92.4% lithium-7. Determine the average atomic mass of lithium.
9. Which isotope of nitrogen is found in nature? Explain your reasoning.
   A. $^{7}_{14}N$ B. $^{14}_{7}N$ C. $^{15}_{6}N$

SciLINKS NSTA
Topic: Isotopes
Visit: www.SciLinks.org
Web code: KEY-113

66 Unit I Alchemy Section III A World of Particles

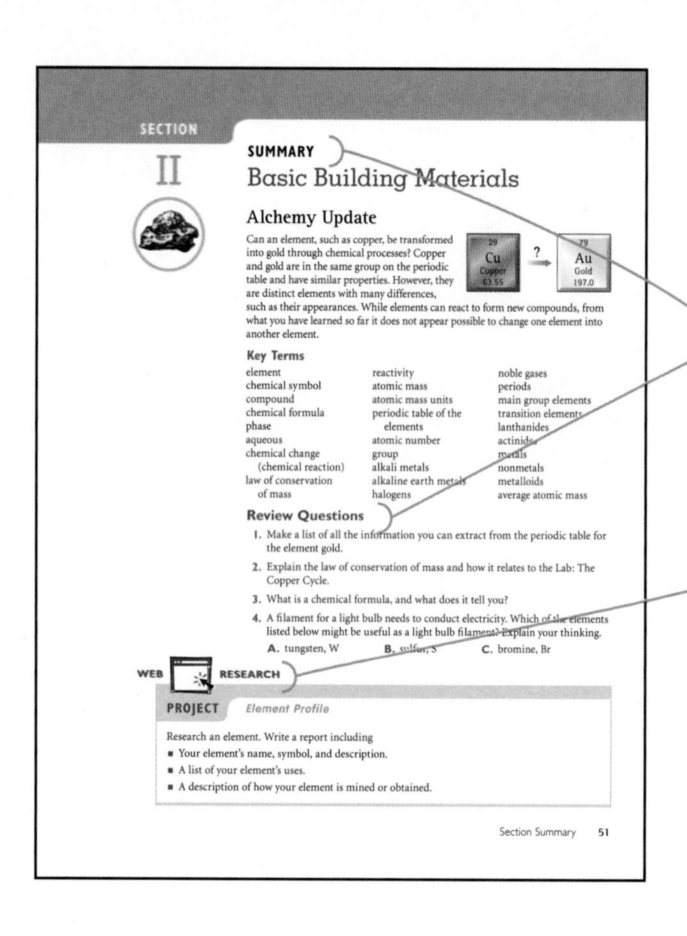
SECTION

II

SUMMARY

Basic Building Materials

Alchemy Update

Can an element, such as copper, be transformed into gold through chemical processes? Copper and gold are in the same group on the periodic table and have similar properties. However, they are distinct elements with many differences, such as their appearances. While elements can react to form new compounds, from what you have learned so far it does not appear possible to change one element into another element.

29 Cu Copper 63.55 ? 79 Au Gold 197.0

Key Terms

element	reactivity	noble gases
chemical symbol	atomic mass	periods
compound	atomic mass units	main group elements
chemical formula	periodic table of the elements	transition elements
phase	atomic number	lanthanides
aqueous	group	actinides
chemical change (chemical reaction)	alkali metals	metals
law of conservation of mass	alkaline earth metals	nonmetals
	halogens	metalloids
		average atomic mass

Review Questions

1. Make a list of all the information you can extract from the periodic table for the element gold.
2. Explain the law of conservation of mass and how it relates to the Lab: The Copper Cycle.
3. What is a chemical formula, and what does it tell you?
4. A filament for a light bulb needs to conduct electricity. Which of the elements listed below might be useful as a light bulb filament? Explain your thinking.
   A. tungsten, W B. sulfur, S C. bromine, Br

WEB RESEARCH

PROJECT Element Profile

Research an element. Write a report including

- Your element's name, symbol, and description.
- A list of your element's uses.
- A description of how your element is mined or obtained.

Section Summary 51

- The **section summary** and **review questions** help you review after each group of lessons.
- **Projects** may be assigned by your teacher to give you a chance to do some research on your own.

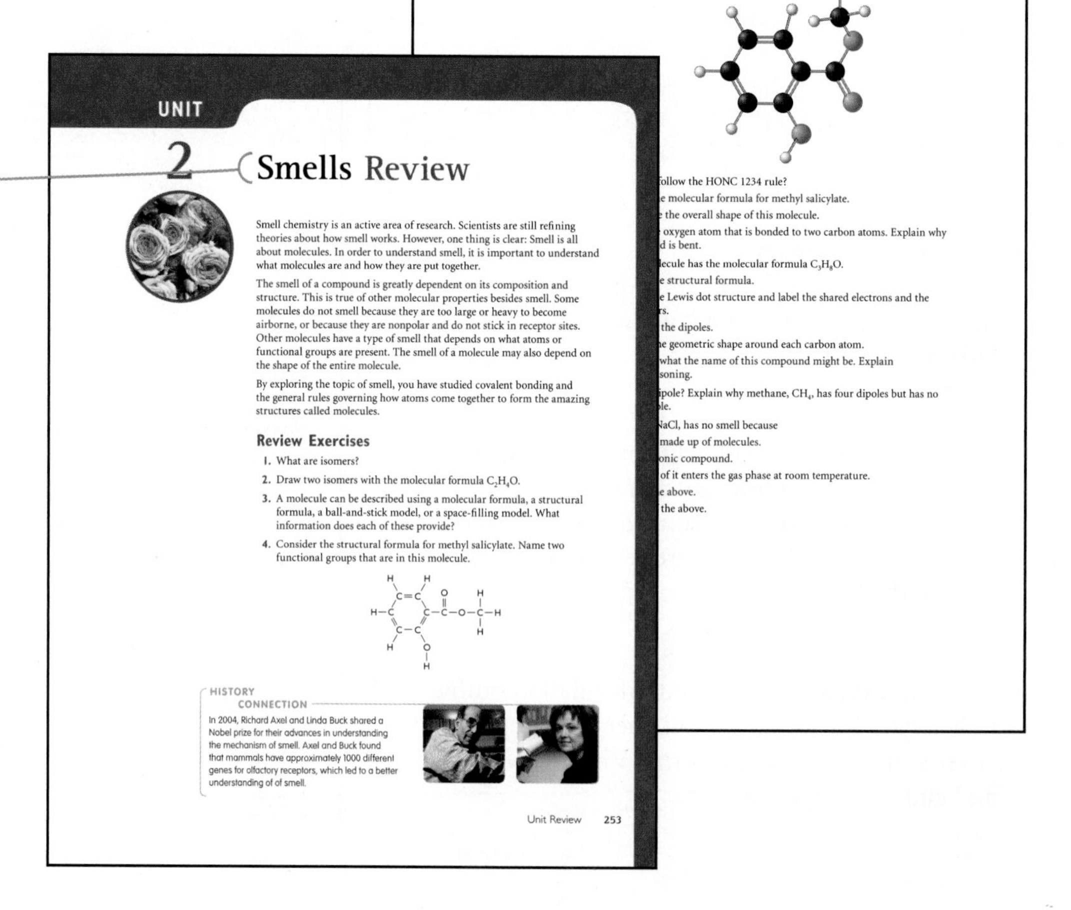
5. Here is a ball-and-stick model of methyl salicylate.

ollow the HONC 1234 rule?
e molecular formula for methyl salicylate.
the overall shape of this molecule.
oxygen atom that is bonded to two carbon atoms. Explain why
d is bent.
lecule has the molecular formula C_3H_8O.
e structural formula.
e Lewis dot structure and label the shared electrons and the
s.
the dipoles.
e geometric shape around each carbon atom.
what the name of this compound might be. Explain
soning.
pole? Explain why methane, CH_4, has four dipoles but has no
le.
NaCl, has no smell because
made up of molecules.
nic compound.
of it enters the gas phase at room temperature.
e above.
the above.

UNIT

2 Smells Review

Smell chemistry is an active area of research. Scientists are still refining theories about how smell works. However, one thing is clear: Smell is all about molecules. In order to understand smell, it is important to understand what molecules are and how they are put together.

The smell of a compound is greatly dependent on its composition and structure. This is true of other molecular properties besides smell. Some molecules do not smell because they are too large or heavy to become airborne, or because they are nonpolar and do not stick in receptor sites. Other molecules have a type of smell that depends on what atoms or functional groups are present. The smell of a molecule may also depend on the shape of the entire molecule.

By exploring the topic of smell, you have studied covalent bonding and the general rules governing how atoms come together to form the amazing structures called molecules.

Review Exercises

1. What are isomers?
2. Draw two isomers with the molecular formula C_2H_6O.
3. A molecule can be described using a molecular formula, a structural formula, a ball-and-stick model, or a space-filling model. What information does each of these provide?
4. Consider the structural formula for methyl salicylate. Name two functional groups that are in this molecule.

HISTORY CONNECTION

In 2004, Richard Axel and Linda Buck shared a Nobel prize for their advances in understanding the mechanism of smell. Axel and Buck found that mammals have approximately 1000 different genes for olfactory receptors, which led to a better understanding of of smell.

Unit Review 253

- The **unit review** summarizes what you learned in the unit and includes more practice exercises to help you review for an exam.

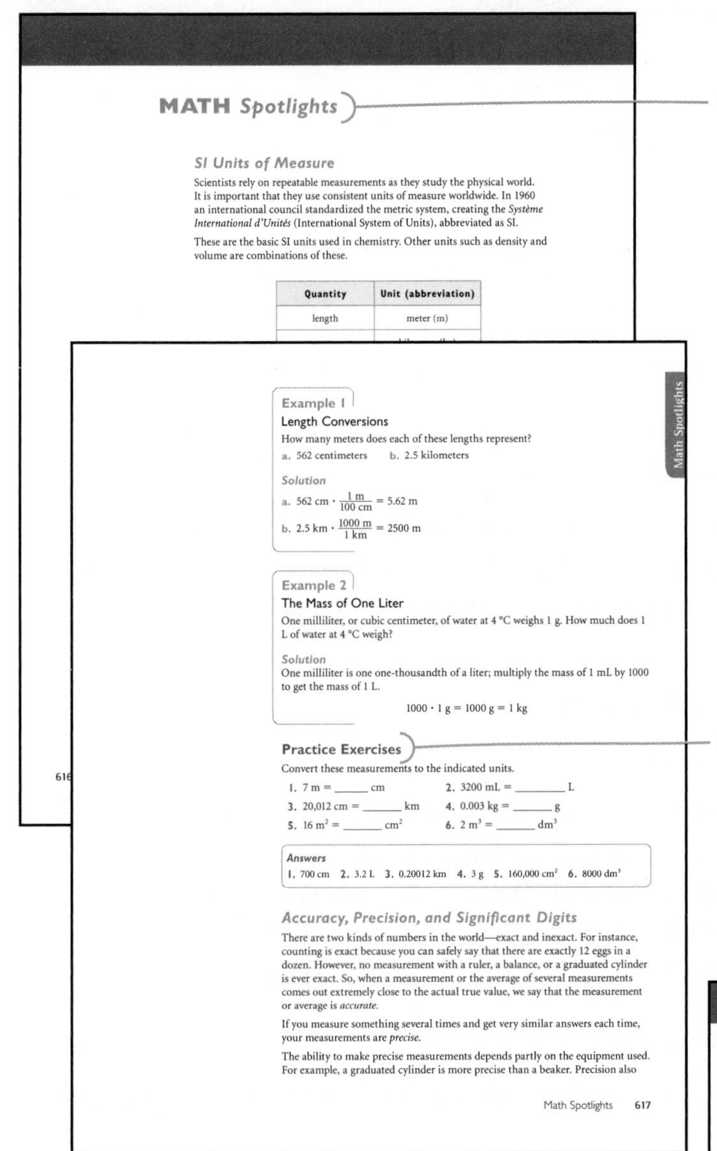

MATH *Spotlights*

SI Units of Measure

Scientists rely on repeatable measurements as they study the physical world. It is important that they use consistent units of measure worldwide. In 1960 an international council standardized the metric system, creating the *Système International d'Unités* (International System of Units), abbreviated as SI.

These are the basic SI units used in chemistry. Other units such as density and volume are combinations of these.

Quantity	Unit (abbreviation)
length	meter (m)

Example 1
Length Conversions

How many meters does each of these lengths represent?

a. 562 centimeters b. 2.5 kilometers

Solution

a. $562 \text{ cm} \cdot \frac{1 \text{ m}}{100 \text{ cm}} = 5.62 \text{ m}$

b. $2.5 \text{ km} \cdot \frac{1000 \text{ m}}{1 \text{ km}} = 2500 \text{ m}$

Example 2
The Mass of One Liter

One milliliter, or cubic centimeter, of water at 4 °C weighs 1 g. How much does 1 L of water at 4 °C weigh?

Solution

One milliliter is one one-thousandth of a liter; multiply the mass of 1 mL by 1000 to get the mass of 1 L.

$$1000 \cdot 1 \text{ g} = 1000 \text{ g} = 1 \text{ kg}$$

Practice Exercises

Convert these measurements to the indicated units.

1. $7 \text{ m} = \_\_\_\_ \text{ cm}$ 2. $3200 \text{ mL} = \_\_\_\_ \text{ L}$
3. $20{,}012 \text{ cm} = \_\_\_\_ \text{ km}$ 4. $0.003 \text{ kg} = \_\_\_\_ \text{ g}$
5. $16 \text{ m}^2 = \_\_\_\_ \text{ cm}^2$ 6. $2 \text{ m}^3 = \_\_\_\_ \text{ dm}^3$

Answers
1. 700 cm 2. 3.2 L 3. 0.20012 km 4. 3 g 5. 160,000 cm^2 6. 8000 dm^3

Accuracy, Precision, and Significant Digits

There are two kinds of numbers in the world—exact and inexact. For instance, counting is exact because you can safely say that there are exactly 12 eggs in a dozen. However, no measurement with a ruler, a balance, or a graduated cylinder is ever exact. So, when a measurement or the average of several measurements comes out extremely close to the actual true value, we say that the measurement or average is *accurate*.

If you measure something several times and get very similar answers each time, your measurements are *precise*.

The ability to make precise measurements depends partly on the equipment used. For example, a graduated cylinder is more precise than a beaker. Precision also

- When you see a note about **Math Spotlights** in a lesson, you can turn to this section in the back of the book for a quick review of that math topic.

- Do these **practice exercises** for a quick refresher. Answers are provided below so you can check your understanding.

- The **glossary** contains all the key terms, defined in English and Spanish.

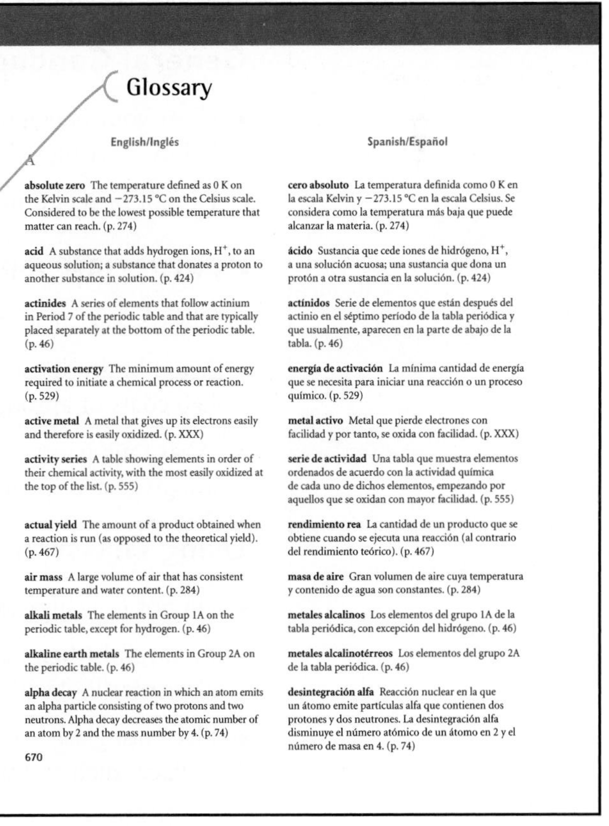

Glossary

English/Inglés	Spanish/Español
A	
absolute zero The temperature defined as 0 K on the Kelvin scale and −273.15 °C on the Celsius scale. Considered to be the lowest possible temperature that matter can reach. (p. 274)	**cero absoluto** La temperatura definida como 0 K en la escala Kelvin y −273.15 °C en la escala Celsius. Se considera como la temperatura más baja que puede alcanzar la materia. (p. 274)
acid A substance that adds hydrogen ions, H^+, to an aqueous solution; a substance that donates a proton to another substance in solution. (p. 424)	**ácido** Sustancia que cede iones de hidrógeno, H^+, a una solución acuosa; una sustancia que dona un protón a otra sustancia en la solución. (p. 424)
actinides A series of elements that follow actinium in Period 7 of the periodic table and that are typically placed separately at the bottom of the periodic table. (p. 46)	**actínidos** Serie de elementos que están después del actinio en el séptimo período de la tabla periódica y que usualmente, aparecen en la parte de abajo de la tabla. (p. 46)
activation energy The minimum amount of energy required to initiate a chemical process or reaction. (p. 529)	**energía de activación** La mínima cantidad de energía que se necesita para iniciar una reacción o un proceso químico. (p. 529)
active metal A metal that gives up its electrons easily and therefore is easily oxidized. (p. XXX)	**metal activo** Metal que pierde electrones con facilidad y por tanto, se oxida con facilidad. (p. XXX)
activity series A table showing elements in order of their chemical activity, with the most easily oxidized at the top of the list. (p. 555)	**serie de actividad** Una tabla que muestra elementos ordenados de acuerdo con la actividad química de cada uno de dichos elementos, empezando por aquellos que se oxidan con mayor facilidad. (p. 555)
actual yield The amount of a product obtained when a reaction is run (as opposed to the theoretical yield). (p. 467)	**rendimiento rea** La cantidad de un producto que se obtiene cuando se ejecuta una reacción (al contrario del rendimiento teórico). (p. 467)
air mass A large volume of air that has consistent temperature and water content. (p. 284)	**masa de aire** Gran volumen de aire cuya temperatura y contenido de agua son constantes. (p. 284)
alkali metals The elements in Group 1A on the periodic table, except for hydrogen. (p. 46)	**metales alcalinos** Los elementos del grupo 1A de la tabla periódica, con excepción del hidrógeno. (p. 46)
alkaline earth metals The elements in Group 2A on the periodic table. (p. 46)	**metales alcalinotérreos** Los elementos del grupo 2A de la tabla periódica. (p. 46)
alpha decay A nuclear reaction in which an atom emits an alpha particle consisting of two protons and two neutrons. Alpha decay decreases the atomic number of an atom by 2 and the mass number by 4. (p. 74)	**desintegración alfa** Reacción nuclear en la que un átomo emite partículas alfa que contienen dos protones y dos neutrones. La desintegración alfa disminuye el número atómico de un átomo en 2 y el número de masa en 4. (p. 74)

Laboratory Safety

Laboratory experiments are an important part of chemistry. By following these safety precautions, you can avoid danger to yourself and your classmates.

Before Working in the Lab

- Read and become familiar with the entire procedure before starting.
- Listen to instructions. When you are in doubt, ask your teacher.
- Know the location of emergency exits and escape routes.
- Learn the location and operation of all safety equipment in your laboratory, including the safety shower, eye wash, first aid kit, fire extinguisher, and fire blanket.

Emergencies and Accidents

- Immediately report any laboratory accident, however small, to your teacher.
- If you get chemicals on you, rinse the affected area with water.
- In case of chemicals on your face, wash off with plenty of water before removing your goggles. In case of chemicals in your eyes, remove contact lenses and wash eyes with water for at least 15 minutes.
- Minor skin burns should be held under cold, running water.

General Conduct

- Clear your bench top of all unnecessary materials, such as books and jackets, before starting work.
- Do not bring gum, food, or drinks into the laboratory.

Appropriate Apparel

- Always wear protective safety goggles when working in the laboratory.
- Avoid bulky, loose-fitting clothing; roll up long sleeves; and tie back loose hair. Lab coats or aprons may be required.
- Wear long pants and shoes that cover the whole foot (not sandals) so your feet are protected from accidental spills or broken glassware.

Using Glassware and Equipment

- Do not use chipped or cracked glassware or damaged equipment.
- Be careful when handling hot glassware or apparatus. Remember, hot glassware looks like cold glassware.
- Place hot glassware or apparatus (such as a crucible) on an appropriate cooling surface, such as a wire gauze.

- Never point the open end of a test tube toward yourself or anyone else.
- Never fill a pipette using mouth suction. Always use a bulb.
- Keep electrical equipment away from sinks and faucets to minimize the risk of electrical shock.

Using Chemicals

- Never taste substances in the laboratory and avoid touching them if possible.
- Check chemical labels twice to make sure you have the correct substance. Some chemical formulas and names differ by only a letter or a number.
- Read and follow all the hazard classifications shown on the label.
- Never pour anything down the drain unless instructed to do so by your teacher.
- When transferring chemicals from a common container to your own test tube or beaker, take only what you need. Do not return any extra material to the original container, because this may contaminate the original.
- Mix all substances together slowly. Add concentrated solutions to dilute solutions. When working with acids and bases, always add concentrated acids and bases to water; never add water to a concentrated acid or base as this can cause dangerous spattering.
- If you are instructed to smell something, do so by wafting (fanning) some of the vapor toward your nose. Do not place your nose near the opening of the container.

Before Leaving the Lab

- Clean your lab station and return equipment to its proper place.
- Make sure that gas lines and water faucets are shut off.
- When discarding used chemicals, carefully follow the instructions provided.

UNIT

1 Alchemy

Emperor Rudolf II (1552–1612) governed the Holy Roman Empire from 1576 to 1608. However, he often neglected his ruling duties and is better known as an alchemist and patron of the arts.

Why Alchemy?

Chemistry has some of its roots in the ancient practice of alchemy. The alchemists experimented with trying to make gold out of ordinary substances. In the process, they learned a great deal about matter and about chemistry. When you understand the nature of matter and its composition, you will be able to answer the question, "Is it possible to turn ordinary substances into gold?"

In this unit, you will learn

- what matter is composed of
- to use the language of chemistry
- to decode information contained in the periodic table
- how new substances with new properties are made
- what holds substances together

SECTION

I

Defining Matter

This photo shows gold. What makes gold different from other metals? How could you identify a metal based on its properties?

With a simple procedure, a copper penny can be made to look like gold. But is it really gold, or has something else been made? In order to tell real gold from other substances, you can compare their properties. Chemists study various properties of matter and use the results to compare and identify substances.

In this section, you will study

- the tools of chemistry
- how matter is defined
- how to measure mass, volume, and density
- how types of matter differ from one another

LESSON

1 Tools of the Trade

Lab Equipment and Safety

Think About It

A chef depends on a wide variety of gadgets and kitchenware to create delicious meals in the kitchen—from whisks and mixers, to ovens and saucepans. An auto mechanic relies on a toolbox of wrenches. In every profession it is important to have the right tool for the job. Chemists have their own special tools and equipment that allow them to study the world around them. They also have a set of guidelines for using the tools safely.

What tools and equipment do chemists use?

To answer this question, you will explore

1. The Tools Chemists Use
2. Laboratory Safety

Exploring the Topic

1 The Tools Chemists Use

Chemistry often brings to mind a laboratory filled with unusual glassware and bubbling beakers. Chemists depend on a variety of tools in their explorations. In particular, chemists need tools that allow them to measure the mass and volume of substances, mix them, heat and cool them, and observe and separate them. Take a moment to examine these illustrations to see some of the tools that are used for these purposes.

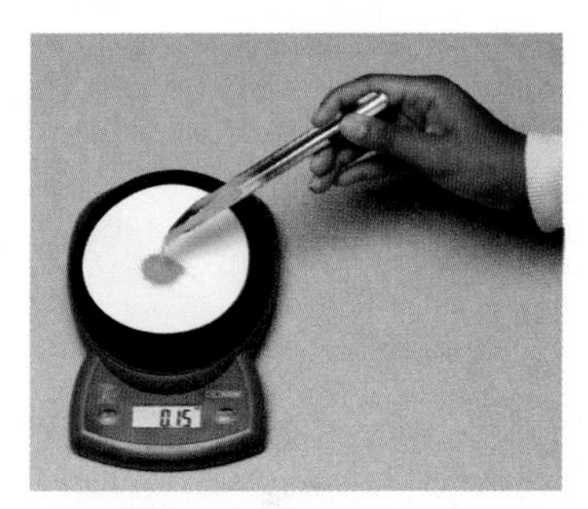

Tools for measuring mass (balance, weighing paper, spatula)

Tools for measuring volume (graduated cylinder, Erlenmeyer flask, burette)

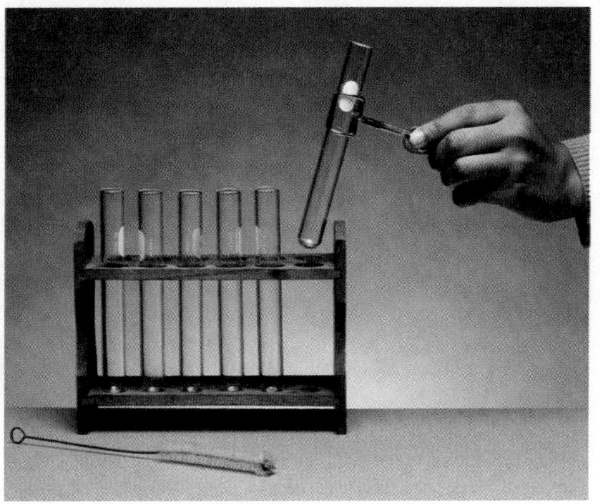

Tools for observing change (test tube holder, test tube rack, brush for cleaning)

Tools for mixing (stirring rod, beaker)

Tools for separating (funnel, filter paper, beaker, wash bottle)

Tools for heating (hot plate, beaker, boiling chips, stirring rod, thermometer)

Tools for heating (Bunsen burner, striker, ring stand, utility clamp, triangle, crucible)

Measuring accurate amounts is important to chemists. They weigh solids on electronic balances and measure volumes of liquids in special glassware. You might notice that many of the containers chemists use are made of glass. Glass is a material of choice because substances in a glass container are visible. Chemists use tempered glass containers, which can be heated over flames without shattering. Also, glass containers are relatively easy to clean and reuse. Finally, notice that chemists use ring stands and special clamps to keep glassware from toppling. Spills can be hazardous.

2 Laboratory Safety

The chemistry laboratory is a place for discovery. However, as in any workplace that uses specialized equipment, safety is always an important consideration. There are many situations in a lab that can become dangerous. Before participating in any chemistry activities, you should familiarize yourself with the safety equipment in your lab.

Take a moment to examine these illustrations. What safety equipment and precautions do you notice?

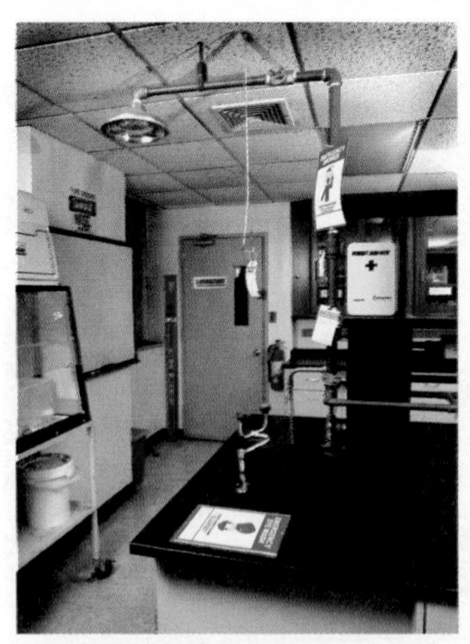

Know the location of the safety equipment and how to use it. Immediately report any laboratory accident, however small, to your teacher.

Never taste or touch chemicals. Never touch hot glassware. If you get chemicals on you, rinse with plenty of water.

CONSUMER CONNECTION

Like labware, bakeware is often made of tempered glass so that it can withstand rapid temperature changes and high temperatures without shattering.

A few do's and don'ts: When working in a lab, dress appropriately. Roll up your sleeves, tie back long hair, and wear closed-toe shoes. Be sure that you have read the instructions for the procedure carefully. Double-check that you are using the correct chemicals. Do not put chemicals back into the original bottle because you might contaminate the chemicals in the bottle. Before you begin working with chemicals or glassware, put on safety goggles. Before leaving the lab, clean your lab station and return equipment to its proper place. Your teacher will provide waste containers; never put chemicals or solutions down the drain unless instructed to do so by your teacher.

Lesson Summary

What tools and equipment do chemists use?

Chemists use their own specialized equipment in the laboratory. This equipment is designed to allow chemists to measure mass and volume, and to mix, heat, cool, observe, and separate substances. It is important to work safely and carefully. When working in a chemistry laboratory, always wear safety goggles and appropriate clothing and closed-toe shoes. Be prepared to know what to do in case of an accident.

EXERCISES

Reading Questions

1. Why are most chemistry containers made of glass?
2. Describe the appropriate clothing to wear in a chemistry lab.

Reason and Apply

3. List three things you should do before beginning any laboratory procedure.
4. Describe what you would do in the case of an accidental spill in class.
5. List three things you should do before leaving the laboratory.
6. What is a fire blanket used for? If necessary, do some research to find out.
7. What is a hood used for in the chemistry laboratory? If necessary, do some research to find out.
8. Why do chemists use clamps and ring stands?

CONSUMER CONNECTION

When using a fire extinguisher, remember the three P's: pull, point, and press. First, pull out the pin on the handle. Second, point the nozzle at the base of the fire. Third, press the trigger.

DEMO A Penny for Your Thoughts

Purpose

To observe a chemical transformation firsthand.

Materials and Safety

1. List the equipment used in the demonstration.
2. Briefly describe your observations of each substance used in the demonstration.
3. Safety is extremely important in the chemistry lab. Write three important safety considerations for this demonstration.

Procedure and Observations

Record your observations for each step of the demonstration.

1. Place a beaker containing zinc filings and sodium hydroxide on a hot plate set to 4.
2. Use tongs to pick up the penny and place it in the heated beaker.
3. While holding the beaker steady with tongs, remove the penny with the other tongs.
4. Put the hot penny in a beaker of cold water to cool and rinse it.
5. Use tongs to place the penny on the hot plate.
6. When the penny has changed color, use the tongs to place it in the beaker of cold water.

Analysis

Working with the students at your table, spend a few minutes discussing what you observed during the demonstration. Then answer the questions individually on your own paper.

7. Describe what happened to the penny during the demonstration.
8. What do you think turned the penny silver?
9. What do you think turned the penny gold?
10. **Making Sense** Do you think you made real gold? Why or why not? How could you find out?

LESSON

2 A Penny for Your Thoughts

Introduction to Chemistry

Think About It

Gold is worth a lot more than copper. If you could turn pennies into gold, you would be very rich. More than a thousand years ago, people known as alchemists tried to transform substances into other substances. In particular, some of them tried to turn ordinary metals into gold. Today, we recognize these alchemists as early chemists. In fact, the word *chemistry* is derived from the Arabic word for alchemy, *al-khemi.*

What is chemistry?

To answer this question, you will explore

1. The Roots of Chemistry
2. Chemistry: The Study of Matter and Change

Exploring the Topic

1 The Roots of Chemistry

While trying to make gold, alchemists developed some of the first laboratory tools and chemistry techniques. They classified substances into categories and experimented with mixing and heating different substances in order to create something new. When alchemists succeeded in creating a new substance, they faced the challenge of figuring out whether that new substance was really gold. Often alchemists were fooled into thinking that a substance was gold just because it looked like gold.

In class, you watched a procedure to make a "golden" penny.

During the procedure, when the silver-colored penny was heated, it turned a gold color. You came up with a **hypothesis** to explain what happened. A hypothesis is a possible explanation for an observation. It can be tested by further investigation or experimentation. Suppose your hypothesis is that the penny turned to gold during the procedure. In order to test that hypothesis, you can compare your gold-colored penny to actual gold to see if it has all the **properties,** or characteristics, of gold. You can check the penny's physical properties, such as color, hardness, weight, and the temperature at which it melts. You can also test its chemical properties, such as whether it changes when you pour acid on it or rusts over time.

Over the course of this unit, you will explore whether it is possible to turn copper or any other substance into gold. But unlike the alchemists, you will have the advantage of hundreds of years of chemistry knowledge to help you answer this question.

CONSUMER CONNECTION

Why is gold so valuable? It retains its shine and resists change, even after hundreds of years. Gold is soft and easy to fashion into beautiful jewelry. It is a vital component in computers and cellular phones. Gold is also relatively rare.

2 Chemistry: The Study of Matter and Change

Changes are constantly occurring all around you. Nails rust, colors fade, milk sours, and plants grow. You can mix ingredients and bake them in an oven to make cookies. You can bleach your hair to change its color and turn liquid water

into ice cubes. Your body can transform cheeseburgers and burritos into muscle, fat, and bone. Chemists seek to understand changes such as these.

Chemistry is the study of what substances are made of, how they behave, and how they can be transformed. It is the study of matter and how matter changes. In this first unit you will investigate matter and how it can be changed. You will learn to describe and explain what happens when matter is changed and will begin to understand what changes are possible.

Chemistry at work—an iron train rusting

Key Terms
hypothesis
property
chemistry

Lesson Summary

What is chemistry?

Chemistry is the study of the substances in the world around you. It is the study of matter and how matter can be changed. The modern study of chemistry emerged from the experimentation and effort of the alchemists. The alchemists invented useful tools and discovered many valuable laboratory techniques in their efforts to create gold out of ordinary substances.

EXERCISES

Reading Questions

1. How did the alchemists contribute to the modern study of chemistry?
2. What is chemistry?

Reason and Apply

3. **WEB RESEARCH** Use the library or the Internet to research the development of alchemy in one of these regions: China, India, the Middle East, Greece, Spain, England, or Egypt. Write a two-paragraph essay on the history of alchemy for your chosen region. Be sure to list your sources.
4. **WEB RESEARCH** Use the library or the Internet to research common uses for sodium hydroxide, which is also called "lye."
5. Write down at least ten changes that you observe in the world around you. Which changes involve chemistry? Explain your reasoning.

HISTORY CONNECTION

The ancient art of alchemy has been traced to many different cultures and areas around the world. Some alchemists sought to turn lead into gold or find a potion that would bring eternal life. This etching shows alchemists at work in the early 16th century.

LESSON

3 What's the Matter?

Defining Matter

Think About It

People tend to value gold over other substances. You don't often see someone wearing aluminum jewelry or putting coal in a high-security bank vault. What is it about gold that makes it unique? Is it possible to create gold from another substance?

This lesson begins to explore the nature of matter as the first step toward proving whether you can or cannot create gold. After all, chemistry is the study of matter and its properties.

What is matter?

To answer this question, you will explore

1. Defining Matter
2. Is It Matter?
3. Measuring Matter

Exploring the Topic

1 Defining Matter

Matter is the word chemists use to refer to all the materials and objects in the world. Your desk, this book, and the paper and ink in the book are all matter. These are all things you can see or feel. However, your senses alone are not always enough to tell you whether something is matter. For instance, you cannot see the virus that gave you a cold, but it is matter. Conversely, you can see shadows on the ground cast by the light from the sun, but they are not matter.

A gold ring, the ink in this book, and a virus each have *substance,* which means they are made out of material, or "stuff." The amount of substance, or material, in an object is called **mass.** Mass is a property of matter that can be measured. So, although the virus has very little substance, it still has mass. Another property that a gold ring, the ink in this book, and a virus have in common is that they take up space, which means they have dimensions. The amount of space something takes up is called **volume** and is also a property of matter that can be measured. So, matter is anything that has mass and volume. You can also say matter is anything that has substance and takes up space. This explains why a virus is matter but a shadow is not.

INDUSTRY CONNECTION

Gold has several special properties. It is shiny, it does not rust or tarnish, and it does not melt in a candle flame. It is smooth, bendable, easy to dent, and even a small piece of gold is surprisingly heavy.

2 Is It Matter?

It is easy to see that solids and liquids have mass and volume. When water is poured into a container, you can see how much space it takes up. When filled with water, the container has more mass. You can see this if you use a balance.

On a two-pan balance, a cup with water will be lower than an identical cup with no water because it has more mass. Thus, water is matter.

When identifying matter, gases can be misleading. For example, most of the time you do not see or feel the air and it may seem as if nothing is there. However, when you fill a balloon with carbon dioxide gas, you can see that the carbon dioxide gas inside occupies space. On a two-pan balance, the pan with a balloon filled with carbon dioxide gas will be lower than the pan with an empty balloon because the balloon filled with carbon dioxide gas has more mass. Therefore, gases do have mass and volume even though gases might be harder to detect than solids and liquids.

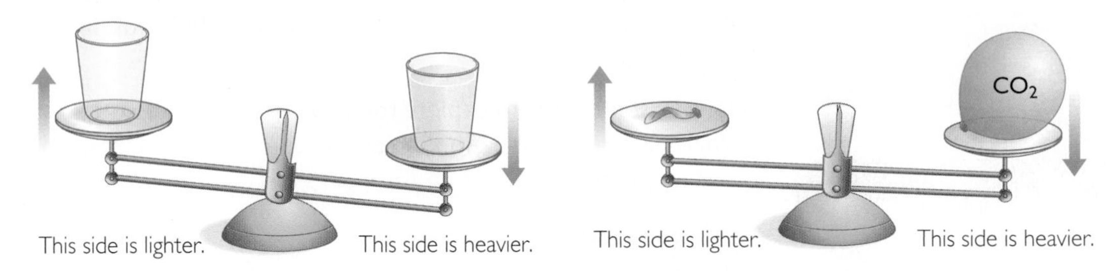

What about other things, like heat and sound? Are they matter? When you heat soup it does not gain mass, and sound may "fill the room" but it doesn't have mass or volume. Sound is the movement of air against your eardrums. Without matter sound can't exist, but sound itself is not matter. Similarly, heat can't exist without some form of matter, but heat itself is not matter. Both sound and heat are referred to as types of energy, and energy by itself is not matter. Likewise, feelings and thoughts are not matter, although you could argue that they require an interaction of matter and energy.

BIOLOGY CONNECTION

Viruses are so small that the length of a virus or other microorganism is measured in microns. A micron is equal to 0.001 millimeter. The viruses in this image have been magnified 10,000 times and colorized to make them easier to see.

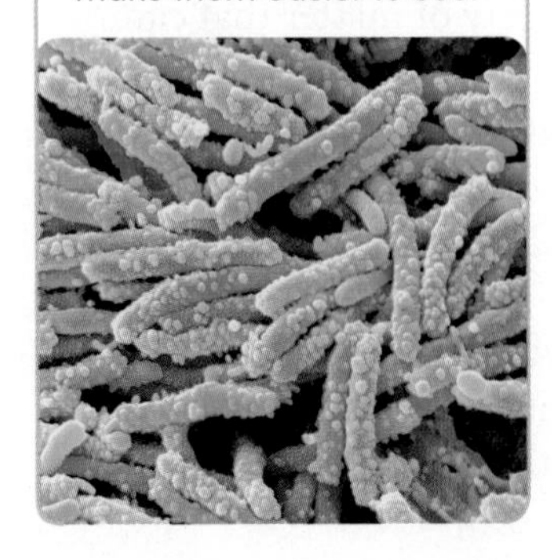

Example

Is It Matter?

Classify wind, music, and clouds. Are they matter? Explain your answer.

Solution

Wind is the movement of air. Air is matter, but the movement of air is not matter because movement has no mass or volume. Therefore, wind is not matter.

Music is sound, which is the movement of air against your eardrums. Music does not have mass or volume. Therefore, music is not matter.

Clouds are made of water droplets, which have mass and volume. Therefore, clouds are matter.

3 Measuring Matter

Measuring Mass

To find the mass of something, you weigh it. In everyday life, things are usually weighed in pounds, lb, or kilograms, kg. For example, you've probably weighed yourself and know approximately how many pounds you

This electronic balance will measure the mass to the nearest thousandth of a gram.

ASTRONOMY CONNECTION

The words *weight* and *mass* are often used interchangeably. However, the difference between weight and mass is apparent when you consider astronauts on a space walk. The astronauts weigh much less than they do on Earth, but they have not gotten skinnier. Even in space they have the same mass.

weigh. And when you go to the grocery store, you might buy a pound of butter or flour. However, chemists measure mass in kilograms, kg, and grams, g (1 kg = 1000 g). [For review of this math topic, see **MATH** ***Spotlight: SI Units of Measure*** on page 616.]

In the chemistry classroom, an electronic balance is used to measure mass. The balance is precise and shows more digits (places) than you may need. You need to decide how many places to pay attention to when using this balance. For example, when you weigh yourself, you measure to the nearest kilogram. You would state your mass as 70 kg, not 69.903 kg. But if you are measuring the mass of a penny, you may want to be more exact. The mass of the copper penny to the nearest hundredth of a gram is 3.11 g. The mass to the nearest gram is 3 g.

This beaker measures to the nearest tenth of a liter.

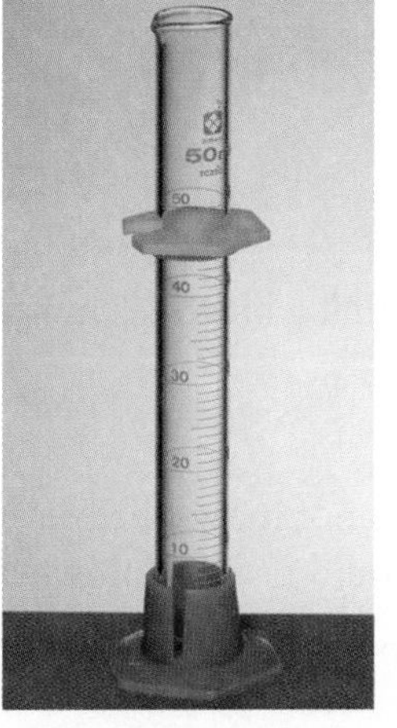

This graduated cylinder measures to the nearest milliliter.

Read the measurement at the bottom of the meniscus. Make sure your eye is at the water level.

Measuring Volume

You can measure the volume of a regularly shaped object by measuring its dimensions, such as length, width, and height, and using a mathematical formula. Volume measured this way is reported in cubic units, for example, cubic meters, m^3, or cubic centimeters, cm^3.

Chemists often measure the volume of gases and liquids in liters, L, and milliliters, mL (1 L = 1000 mL). The volume of a large soda bottle is 2 L, or 2000 mL. This is the volume of the bottle, and if the bottle is full, it is also the volume of liquid inside. Chemists use a variety of special containers for measuring volume. For approximate measurements, they use beakers. For more precise measurements, they use graduated cylinders. [For review of this math topic, see **MATH** ***Spotlight: Accuracy, Precision, and Significant Digits*** on page 617.]

You can measure the volume of a sample of water by pouring it into a graduated cylinder and reading the number of milliliters, mL, on the side. However, the surface of the water is curved because of the way it adheres, or clings, to the sides of the cylinder. The curvature of the top of a liquid in a container is called a **meniscus.** It is most accurate to read the water level at its lowest point, the bottom of the meniscus, because most of the liquid is at this level. To get an accurate measurement, you must read the graduated cylinder at eye level.

Key Terms

matter
mass
volume
meniscus

Lesson Summary

What is matter?

Chemists seek to understand matter and its properties. Matter can be defined as anything that has mass and volume. Mass is the amount of substance or "stuff" in a material or object. Solids, liquids, and gases all have mass and volume and are classified as matter. To measure the mass of something, you use a balance. Volume is the amount of space taken up by matter. To find the volume of something, you can use a container such as a beaker or graduated cylinder, or you can calculate the volume using dimensions, such as length, height, and width.

EXERCISES

Reading Questions

1. Explain the difference between mass and volume.
2. Describe how you could prove that a bicycle is matter.

Reason and Apply

3. Someone might claim that air is not matter because you can't see it. Write a paragraph showing how you would prove to this person that air is matter.
4. The Sun is considered matter, but sunlight is not considered matter. Explain why this is so.
5. Sit somewhere and observe your environment. From your observations, make a list of ten things that are matter and a list of ten things that are not matter. Explain your reasoning.
6. Give examples of five words that describe the movement of matter.

LESSON

4 Mass Communication

Mass and Volume

Think About It

Suppose you have two samples of gold, a gold ring and a gold nugget. Is there more gold in the ring or in the nugget? They feel similar in weight, and they look fairly similar in volume. While your senses can give you valuable information, they can't tell you exactly how much gold you have in each sample.

How do you determine the masses and volumes of different substances?

To answer this question, you will explore

1. Measuring Volume
2. Comparing Mass and Volume

Exploring the Topic

1 Measuring Volume

As you learned in Lesson 3, volume is a measure of size, or how much space each sample takes up.

There are two common ways of measuring the volume of solids: (1) by measuring their dimensions and using a geometric formula, and (2) by water displacement. The first method is convenient if the object has a regular shape. The second method is more convenient for irregularly shaped objects.

Using Geometric Formulas to Determine Volume

If a solid is rectangular, you can find its volume by measuring its three dimensions—length, width, and height—and multiplying these three values. Volume measured in this way is reported in *cubic* units such as cubic centimeters, cm^3; cubic meters, m^3; or cubic inches, in^3. The formula for volume is

$$V = lwh$$

Consider these two solid blocks. You can use the formula to figure out their volumes.

$$V = lwh$$
$$= 1.0 \text{ cm} \cdot 1.0 \text{ cm} \cdot 1.0 \text{ cm}$$
$$= 1.0 \text{ cm}^3$$

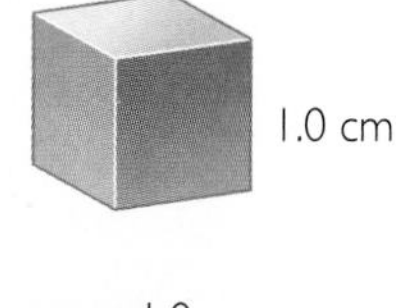

$$V = lwh$$
$$= 1.0 \text{ cm} \cdot 0.5 \text{ cm} \cdot 2.0 \text{ cm}$$
$$= 1.0 \text{ cm}^3$$

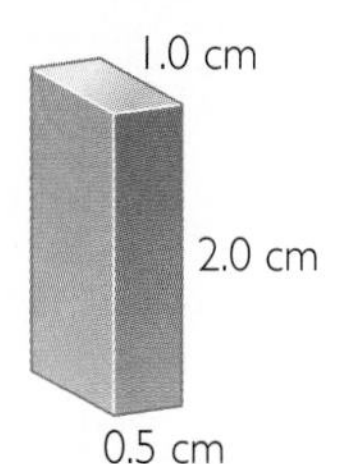

Notice that the two solids have different dimensions, but they have the same volume, 1.0 cubic centimeter.

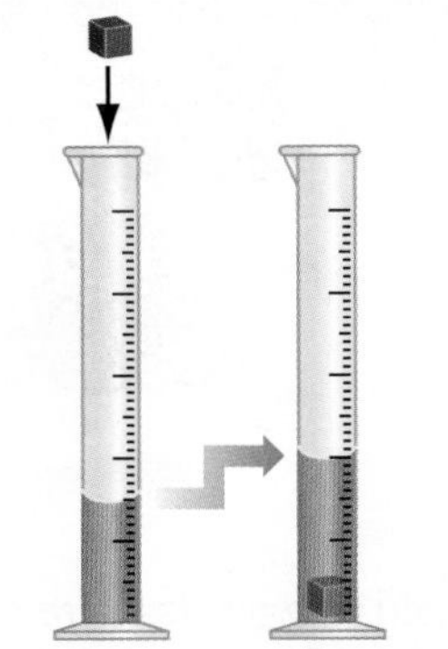

The water level rises by 5.0 mL because the cube has a volume of 5.0 cm^3.

Using Water Displacement to Determine Volume

The second method of measuring the volume of a solid object is called **water displacement.** First, pour water into a graduated cylinder. Next, add the object whose volume you are measuring. An object takes up space, so it displaces some of the water when it is placed in the graduated cylinder. This causes the water level to rise by an amount equal to the volume of the object. If you read the volume of water in the graduated cylinder before you submerge the object and again after submerging it, the volume of the object is the difference between these two volumes.

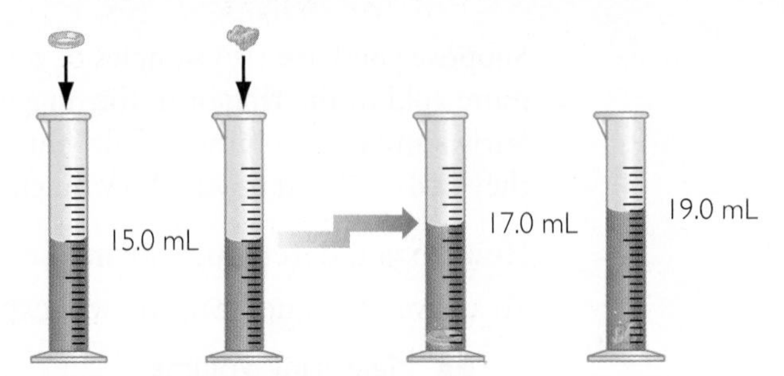

Imagine you place a gold ring and a gold nugget in graduated cylinders partly filled with water. The graduated cylinder is marked in milliliters, mL. You observe that the water level in one container rises by 2.0 mL and by 4.0 mL in the other. The ring has a volume of 2.0 mL or 2.0 cm^3 and the nugget 4.0 mL or 4.0 cm^3.

Important to Know One milliliter is exactly equal in volume to one cubic centimeter: $1\ mL = 1\ cm^3$. ◀

Of course, the water displacement method works only for solids that do not dissolve in water! The solid object also needs to sink so that it is completely submerged. If the object floats, the volume reading will not be accurate.

❷ Comparing Mass and Volume

Imagine you have two objects with the same volume that are *not* made from the same material. You can compare their masses using a balance. The masses of 1.0 cm^3 of gold and 1.0 cm^3 of plastic are given in the table below along with the masses of the same volumes of wood, glass, and copper.

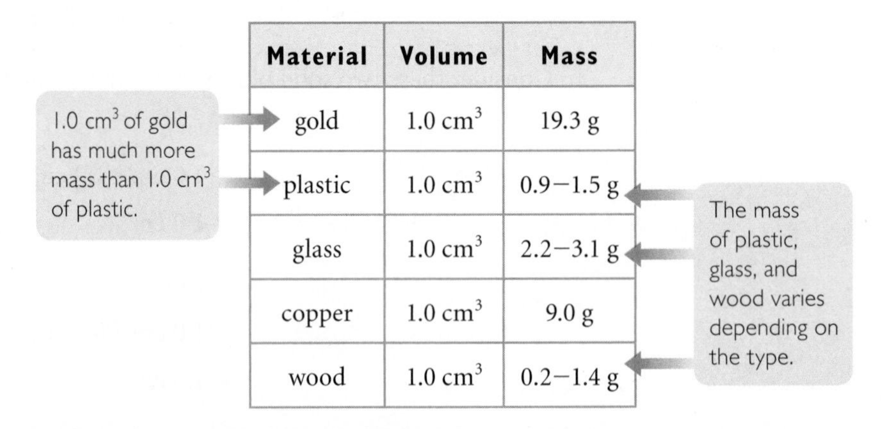

Material	Volume	Mass
gold	1.0 cm^3	19.3 g
plastic	1.0 cm^3	0.9–1.5 g
glass	1.0 cm^3	2.2–3.1 g
copper	1.0 cm^3	9.0 g
wood	1.0 cm^3	0.2–1.4 g

Notice that 1.0 cm^3 of gold is heavier than the same volume of any of the other materials. Thus, two objects can have exactly the same volume but different masses.

CONSUMER CONNECTION

There are many different types of plastic. Soft drink bottles are made of polyethylene terephthalate, also called PETE. Plastic pipes and outdoor furniture are made of polyvinyl chloride, also called PVC. Styrofoam cups are made of polystyrene.

The mass of the plastic depends on the type of plastic being considered. The mass of a 1.0 cm^3 sample of plastic varies between 0.9 g and 1.5 g. Wood and glass also show variations. For instance, if you compare a 1.0 cm^3 sample of solid wood from an oak tree with a 1.0 cm^3 sample of solid wood from a pine tree, they will have different masses. The oak sample will be considerably heavier than the pine. A 1.0 cm^3 sample of pure gold will always have a mass of 19.3 g because there is only one type of pure gold.

Example 1

Try to work out the answers yourself before reading the solutions.

Cubes of the Same Volume

Suppose you have two cubes. One cube is made of solid gold and the other cube is made of solid plastic. The sides of the cubes each measure 2.0 cm in length.

a. What is the volume of each cube?

b. How much water will each cube displace?

Solution

a. The volume of a cube is equal to length times width times height.

$$\begin{aligned} V &= lwh \\ &= 2.0\text{ cm} \cdot 2.0\text{ cm} \cdot 2.0\text{ cm} \\ &= 8.0\text{ cm}^3 \end{aligned}$$

b. Because 1 cm^3 = 1 mL, you can reason that 8.0 cm^3 = 8.0 mL. So each cube will displace the same amount of water, or 8.0 mL.

Important to Know The rise in the water level does not depend on how heavy the cube is. The water-level rise depends only on the volume of the object. ◄

Example 2

Cubes of the Same Mass

One cube is made of solid copper and another cube is made of solid glass. They have exactly the same mass. How do their volumes compare?

Solution

If you look at the table of masses on page 14, you will see that 1.0 cm^3 of copper has much more mass than 1.0 cm^3 of glass. Thus, in order for the two cubes to have the same mass, the copper cube must be much smaller than the glass cube.

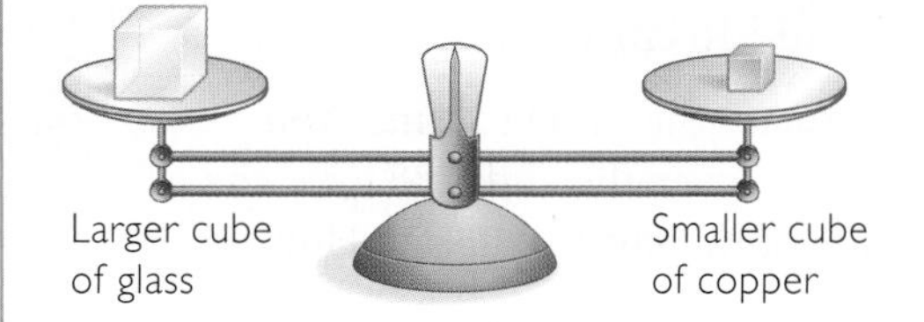

The small cube of copper exactly balances the mass of the larger cube of glass.

Key Term

water displacement

Lesson Summary

How do you determine the masses and volumes of different substances?

The volume of a solid object is determined either by water displacement or by measuring its dimensions and using a geometric formula for volume. The mass of any solid sample is found by weighing it on a balance, or scale. The relationship between mass and volume depends on the type of material being considered. Two objects may have the same volume but different masses, or two objects may have the same mass but different volumes.

EXERCISES

Reading Questions

1. Describe two ways to determine the volume of a solid object.
2. If you have two objects with equal volume, do they have the same mass? Explain your thinking.

Reason and Apply

3. Can you predict the volume of an object just by looking at it? Explain.
4. Can you predict the mass of an object just by looking at it? Explain.
5. If you stretch a rubber band, does it still have the same mass? Explain.
6. You submerge a piece of clay in water and measure the total volume. You change the shape of the clay and put it back into the same amount of water.
   a. The total volume does not change. Explain why.
   b. Does the total mass change? Explain.

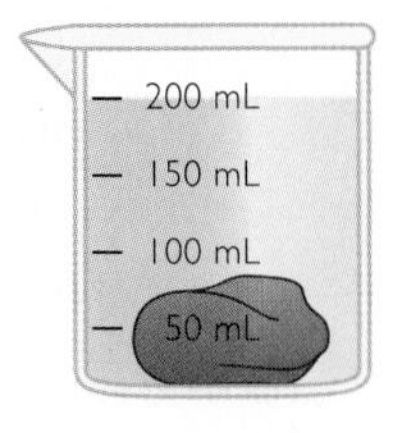

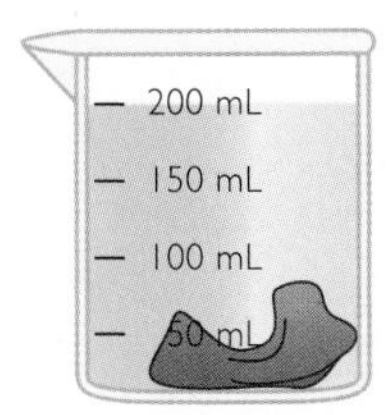

7. Describe how you might find the volume of these items:
   a. pancake mix
   b. hair gel
   c. a shoe box
   d. a penny
   e. lemonade
8. Draw two things with the same mass, but different volumes.
9. Draw two things with the same volume, but different masses.
10. Use the illustrations to complete these statements.
    a. What is the volume of the liquid inside the container?
    b. What is the volume of the rock in mL? In cm^3?
11. Suppose that you have two cubes of exactly the same volume. You weigh them and find a mass of 8.91 g for one cube and 8.88 g for the other cube even though they are made of the same material. How is this possible?

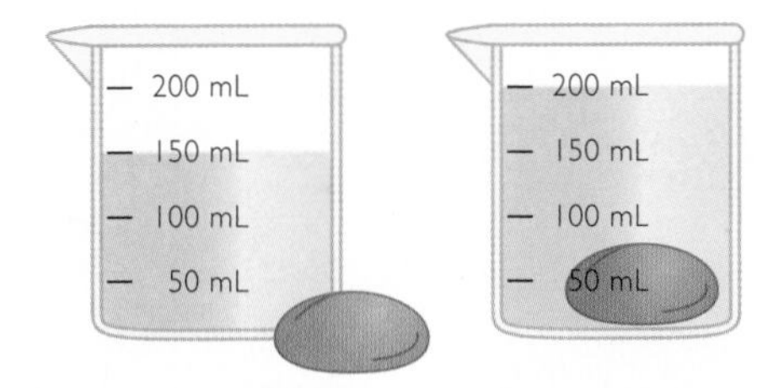

LESSON

5 All That Glitters

Density

Think About It

Gold and lead are metals that are soft and bendable. They are easy to tell apart at a glance because lead is dull gray. But suppose someone tried to trick you by coating a block of lead with a thin layer of gold. How could you prove that this block was a fake?

How can you use mass and volume to determine the identity of a substance?

To answer this question, you will explore

1. The Definition of Density
2. Calculating Density
3. Identifying Matter Using Density

Exploring the Topic

1 The Definition of Density

Which is heavier, gold or copper? It depends on the amount of each that you have. If you compare a tiny gold earring and a large copper pipe, the copper will be heavier. If you compare the same volume of each, gold is *always* heavier than copper. That is because there is more matter in 1.0 cm^3 of gold than in 1.0 cm^3 of copper. You could also say that gold is *denser* than copper. The mass of a substance per unit of volume is called its **density.**

Imagine comparing 1.0 cm^3 cubes of oak wood, water, and copper from larger samples of each material. All three cubes have the same volume, but each cube has a different mass and a different density.

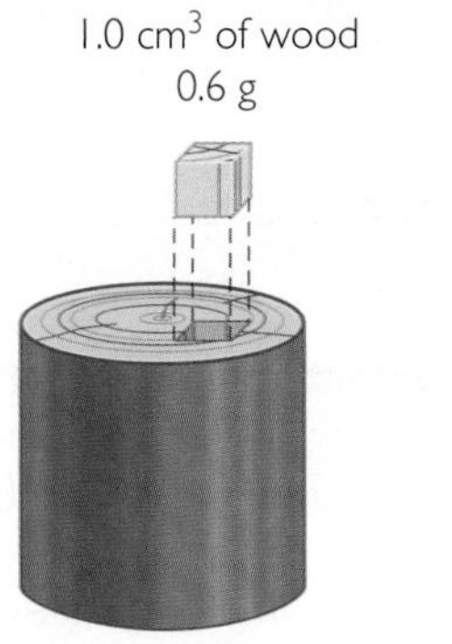

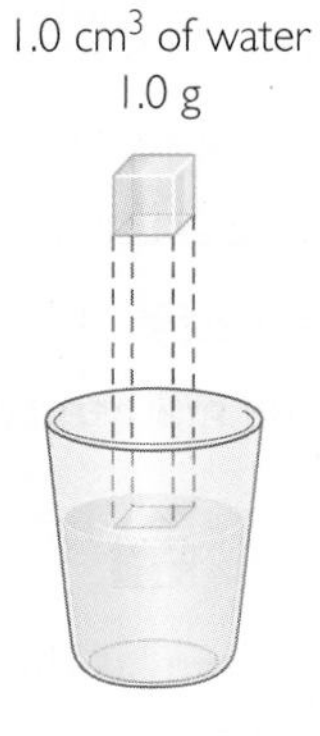

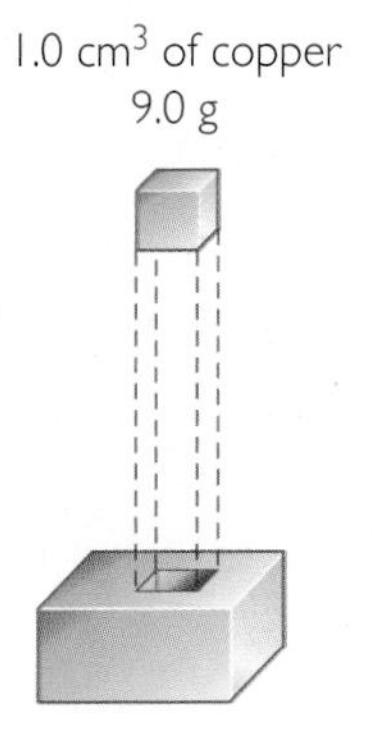

INDUSTRY CONNECTION

Most substances will expand when temperature increases and contract when temperature is decreased. For this reason, there are special joints built in between sections of sidewalks, bridges, railroad tracks, and other structures to prevent cracking or breaking when the volume of the materials changes due to temperature.

Example 1

Differences in Density

Consider the objects on balances shown in these illustrations. How can differences in density account for what you observe in these pictures?

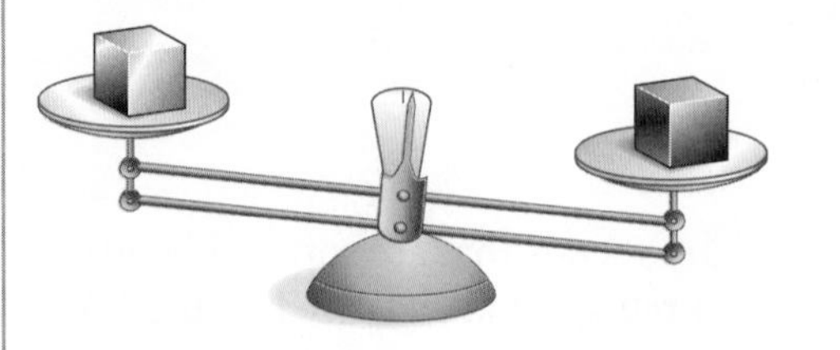

The balance is uneven. The two solid cubes have the same volume but different masses.

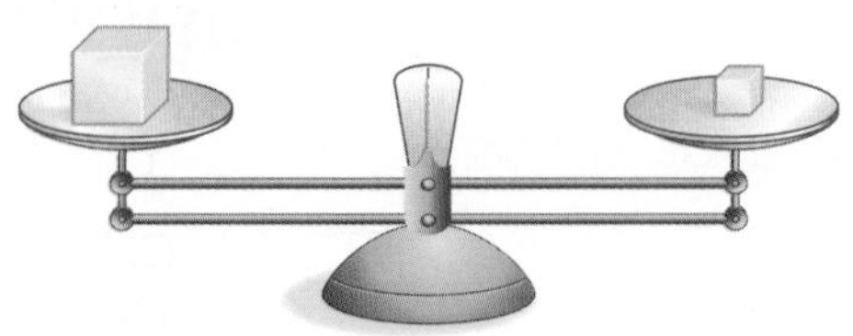

The balance is level. The two solid cubes have the same mass but different volumes.

Solution

Because the cubes on the first balance are exactly the same size, one cube must be denser than the other. So the two cubes are made of different materials.

On the second balance the larger cube must have a lower density than the smaller cube. Again, the two cubes are made of different materials.

2 Calculating Density

You can determine the density of a substance without actually cutting out a cubic centimeter sample. Suppose a large chunk of gold has a mass of 309 g and has a volume of 16.0 cm^3. Divide the mass of the gold by its volume to find its mass *per* cubic centimeter. This is equal to the density.

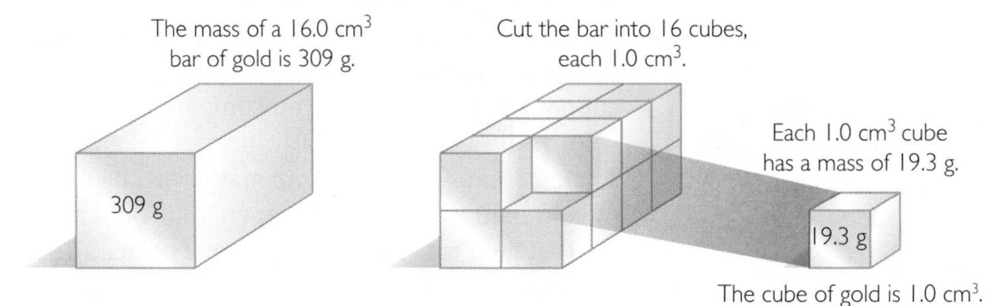

$$309 \text{ g} \div 16.0 \text{ cm}^3 = 19.3 \text{ g/cm}^3$$

The density of gold is 19.3 grams per cubic centimeter, or 19.3 g/cm^3. This answer is rounded to three significant digits. [For review of this math topic, see **MATH** ***Spotlight: Accuracy, Precision, and Significant Digits*** on page 617.]

Density

The mathematical formula for density is

$$D = \frac{m}{V}$$

where D is the density, m is the mass, and V is the volume.

Example 2

Density of a Gold Ring

Suppose you have a gold ring that weighs 7.50 g and has a volume of 0.388 mL. Does it have the same density as a big piece of gold?

Solution

Use the formula to find the density of the ring.

Start with the formula. $D = \frac{m}{V}$

Substitute the values of m and V. $D = \frac{7.50 \text{ g}}{0.388 \text{ mL}}$

Solve for D. $= 19.3 \text{ g/mL} = 19.3 \text{ g/cm}^3$

The ring has the same density as any other sample of gold, big or small. The density of a material does not depend on the shape or size of the sample.

Important to Know The density of a substance does not change with its size or its shape. A solid gold bracelet will have the same density as a solid gold brick, and the density of one penny is the same as the density of two pennies. ◀

Densities of Some Metals

Metal	Density
copper	9.0 g/cm^3
zinc	7.1 g/cm^3
gold	19.3 g/cm^3
lead	11.4 g/cm^3
aluminum	2.7 g/cm^3
brass	8.4 g/cm^3

3 Identifying Matter Using Density

Every substance has certain properties, such as color, hardness, melting point temperature, and density, that depend on the type of matter, not on the amount or size of the sample. These properties are called **intensive properties.** Intensive properties do not change if the quantity of the substance changes. Therefore, they can be used to help identify that substance.

On the other hand, **extensive properties,** such as mass or volume, do change depending on the amount of matter. Extensive properties alone can't be used to help identify a type of matter. For example, knowing that a metal earring has a mass of 3 g doesn't help you identify what the earring is made of.

Because density is an intensive property, it can be used to help identify the type of matter that an object or sample is made of. First, determine the density of the object and then compare it with known density values in a reference table like this one to help identify the type of matter.

BIG IDEA Each specific type of substance has a particular density. You can identify a substance by its intensive properties, including density.

The Fake Bar of Gold

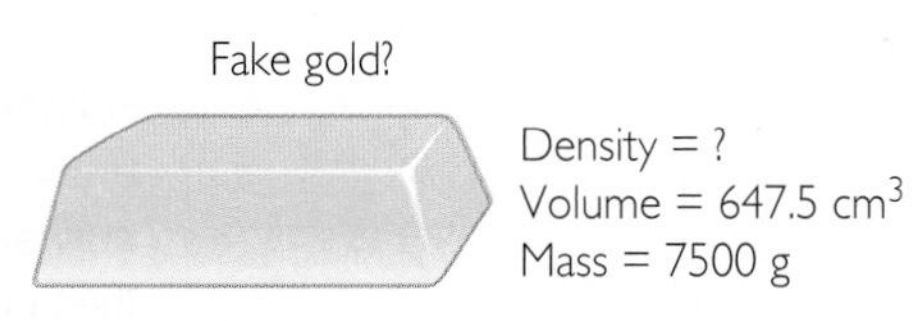

If someone tried to trick you by coating a block of lead with a thin layer of gold, how could you prove the bar is a fake?

One approach would be to scratch the bar to reveal that the inside isn't gold. Another method would be to use density to prove that the bar is or isn't gold.

If the block is pure gold, it should have a density of 19.3 g/cm^3.

Key Terms

density
intensive property
extensive property

Lesson Summary

How can you use mass and volume to determine the identity of a substance?

To identify substances, you examine their intensive properties, qualities that do not depend on size or amount. Intensive properties include color, hardness, and density. Density is the mass of a substance per unit of volume. If two substances have different densities, then they are probably made of different types of matter.

EXERCISES

Reading Questions

1. In your own words, define density.
2. Explain how density can be used to determine if the golden penny is made of solid gold.

Reason and Apply

3. How does the density of aluminum compare with the density of gold? What does this tell you about the amount of matter within each?
4. If two objects have the same mass, what must be true? Choose the correct answer(s).
   - A. They have the same volume.
   - B. They are made of the same material.
   - C. They contain the same amount of matter.
   - D. They have the same density.
5. Two objects each have a mass of 5.0 g. One has a density of 2.7 g/cm^3 and the other has a density of 8.4 g/cm^3. Which object has a larger volume? Explain your thinking.
6. A piece of metal has a volume of 30.0 cm^3 and a mass of 252 g. What is its density? What metal do you think this is?
7. A glass marble has a mass of 18.5 g and a volume of 6.45 cm^3.
   - a. Determine the density of the marble.
   - b. What is the mass of six of these marbles? What is the volume? What is the density?
   - c. How does the density of one marble compare with the density of six of the marbles?

HISTORY CONNECTION

Sacagawea was a Shoshone woman who guided Lewis and Clark in their exploration of the western United States between 1804 and 1806. The U.S. Mint began making the Sacagawea dollar in the year 2000. It is called the Sacagawea Golden Dollar, but it is made of 88.5% copper, 6% zinc, 3.5% manganese, and 2% nickel. The golden color is due to a coating of brass, which is made by combining copper and zinc.

SECTION

I

SUMMARY

Defining Matter

Key Terms

hypothesis
property
chemistry
matter
mass
volume
meniscus
water displacement
density
intensive property
extensive property

Alchemy Update

Can a copper penny be turned into gold through chemical processes?

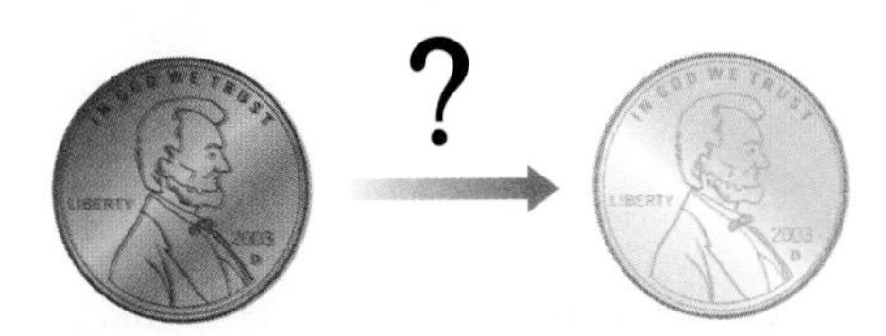

The "golden" penny looks like gold and has essentially the same volume as a true gold penny. However, the golden penny has a much lower mass than a true gold penny. This means the density of the "golden" penny is less than the density of gold. It cannot be a solid gold penny.

The concept of density has provided the evidence needed to confirm that the gold-colored penny is not actually made of gold. But if it is not gold, then what is it? Many questions remain to be answered. Is it still possible to make gold some other way? And, what is it about gold that makes it gold?

Review Exercises

1. Explain how you would determine the volume of a powdered solid, a liquid, and a rock.
2. Use your own words to define *matter.*
3. Will an object with a higher density displace more water than an object with a lower density? Explain why or why not.
4. How does the density of one penny compare with the density of two pennies?

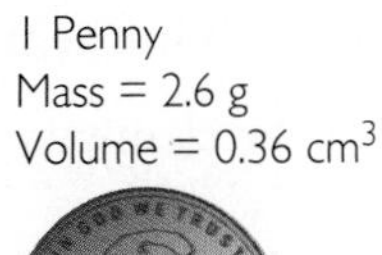

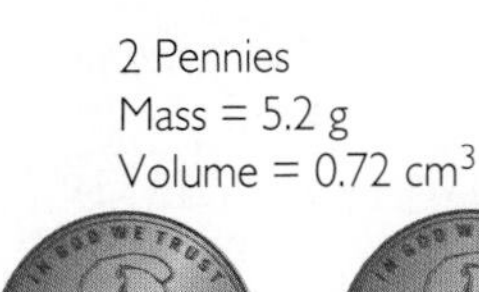

5. A small pebble breaks off of a huge boulder. The pebble has the same density as the boulder. In your own words, explain how this can be true.
6. Archeologists discover a silver crown in an ancient tomb. When they place the crown in a tub of water, it displaces 238.1 cm^3 of water. The density of silver is 10.5 g/cm^3. If the crown is really silver, what should its mass be?

SECTION

II

Basic Building Materials

5 B Boron 10.81	6 C Carbon 12.01	7 N Nitrogen 14.01	8 O Oxygen 16.00
13 Al Aluminum 26.98	14 Si Silicon 28.09	15 P Phosphorus 30.97	16 S Sulfur 32.07
31 Ga Gallium 69.72	32 Ge Germanium 72.64	33 As Arsenic 74.92	34 Se Selenium 78.96

Chemists separate substances into their simplest components in order to understand the nature of matter. A few of these components are shown here.

Matter is composed of components called elements. There are a limited number of elements in existence. Everything in our world, natural and synthetic, is made of individual elements or some combination of them. Chemical formulas indicate what element or elements a substance is made of.

In this section, you will study

- how chemists use chemical formulas to track changes in matter
- how matter is conserved in chemical processes
- the organization of the elements into the periodic table

$C_{12}H_{22}O_{11}$ NaCl

LESSON

6 A New Language

Chemical Names and Symbols

Think About It

There are two bottles on a shelf in a chemistry lab. Each contains a substance that resembles diamonds. Bottle 1 is labeled C(*s*), and Bottle 2 is labeled $ZrO_2(s)$. Does either bottle contain diamonds? Do both bottles contain diamonds?

What do chemical names and symbols tell you about matter?

To answer this question, you will explore

1. The Language of Chemistry
2. Names and Symbols
3. Physical Form

Exploring the Topic

1 The Language of Chemistry

Some chemical names are used in daily language, such as aluminum and iron, which cookware and machinery are made of, and ammonia, which is used for cleaning. Other names are used mainly by chemists, such as sodium chloride for salt and calcium carbonate for chalk.

In the illustration, the symbol C is used when the word *carbon* appears in the name, and the symbol O appears when the word *oxygen* or *oxide* is in the name.

CONSUMER CONNECTION

Neon is the gas used in neon signs. The gas is colorless but glows a bright orange red when an electric current is run through it. Neon gas is put into glass tubes that have been bent and shaped to create colorful signs.

2 Names and Symbols

Elements are the building materials of all matter. In total there are about 118 known elements. A few are shown here.

Here are some things you might notice. The **chemical symbols** for the elements consist of one or two letters. The first letter is always capitalized. If there is a second letter, it is lowercase. An element can be a solid, liquid, or gas. Some elements are metals, some are not. Sometimes the symbol for an element is an abbreviation for its name. Other times, the symbol comes from another source, such as a word in another language. For example, the symbol for iron is Fe from the Latin word *ferrum*, and the symbol for gold is Au from the Latin word *aurum*.

Elements combine in specific ratios to form **compounds.** A compound is represented by a **chemical formula.** For example, sodium chloride, or table salt, is NaCl, which tells you that salt is made of the elements sodium, Na, and chlorine, Cl, in a 1:1 ratio. The chemical formula for carbon dioxide is CO_2. The subscript number "2" in the formula indicates that the elements carbon and oxygen are combined in a 1:2 ratio. (If the subscript is 1, you normally don't write it.)

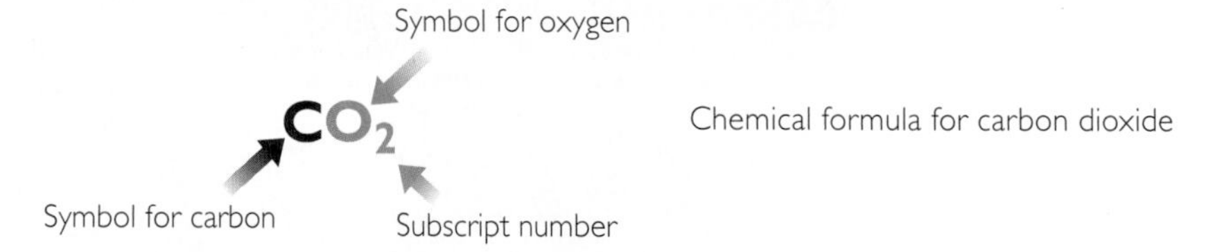

Compounds can be very different in appearance and behavior from the elements that they are composed of. For example, sodium is a shiny metal and chlorine is a gas. They are very different from sodium chloride, table salt.

BIG IDEA All matter is made up of elements or compounds, or mixtures of these.

These minerals contain copper, Cu. Notice that not one of them is copper colored.

Copper oxide, $Cu_2O(s)$

Copper sulfate, $CuSO_4(s)$

Copper sulfide, $CuS(s)$

Copper carbonate hydroxide, $Cu_2(CO_3)(OH)_2(s)$

3 Physical Form

Elements and compounds can exist as solids, liquids, or gases, and this is represented by using (*s*), (*l*), or (*g*) after the chemical formula. These are called the **phases** of matter. For example, water in the gas phase is written as $H_2O(g)$. Water is written as $H_2O(l)$ when it is in the liquid phase and $H_2O(s)$ when it is in the solid phase as ice. There is another symbol for physical form, the symbol (*aq*) for **aqueous.** A substance is aqueous when it dissolves and forms a clear mixture with water. Many familiar liquids are actually aqueous solutions, such as grape juice, vinegar, and ocean water.

Important to Know When solid sugar, $C_{12}H_{22}O_{11}(s)$, appears to "melt" in your mouth, it does not become liquid sugar, $C_{12}H_{22}O_{11}(l)$. It dissolves and becomes aqueous sugar, $C_{12}H_{22}O_{11}(aq)$. ◄

Most substances are a mixture of compounds. For example, an orange is a mixture of water, H_2O, fructose, $C_6H_{12}O_6$, citric acid, $C_6H_8O_7$, limonene, $C_{10}H_{16}$, and many other compounds. Notice that in mixtures, the components are *not* combined in a specific ratio. Some oranges are sweeter or more sour or more juicy because they have more fructose or citric acid or water than other oranges. But the ratios of the elements in compounds never change. Water is always water, H_2O.

Key Terms

element
chemical symbol
compound
chemical formula
phase
aqueous

Lesson Summary

What do chemical names and symbols tell you about matter?

All matter is composed of elements. Each element has a unique symbol. Elements combine with one another to form compounds. A chemical formula specifies which elements are present in a compound. The letters in chemical formulas are symbols for the elements, and the subscript numbers indicate the amount of each element in that substance. A chemical formula may also include the lowercase letters, (*s*), (*g*), (*l*), and (*aq*), which stand for solid, gas, liquid, and aqueous.

EXERCISES

Reading Questions

1. Describe the difference between an element and a compound.
2. What is meant by physical form?

Reason and Apply

3. How many elements are included in the chemical formula for sodium nitrate, $NaNO_3$? Name them.
4. What is the difference between NaOH(*s*) and NaOH(*aq*)?
5. You see a ring with a stone that looks like a diamond but wonder why it's so cheap. The jeweler says the stone is a type of diamond called cubic zirconia. How can chemical symbols prove that cubic zirconia is not a diamond?
6. **Lab Prep** Read the Lab: The Copper Cycle on the next page, paying special attention to the safety instructions. Describe each step that may be dangerous in this lab. Be prepared to pass a safety quiz before the lab.

CAREER CONNECTION

Jewelers can tell if a diamond is real based on properties such as hardness, density, and how it bends light.

LAB The Copper Cycle

Purpose

To find out what happens when you perform a series of reactions, starting with copper metal.

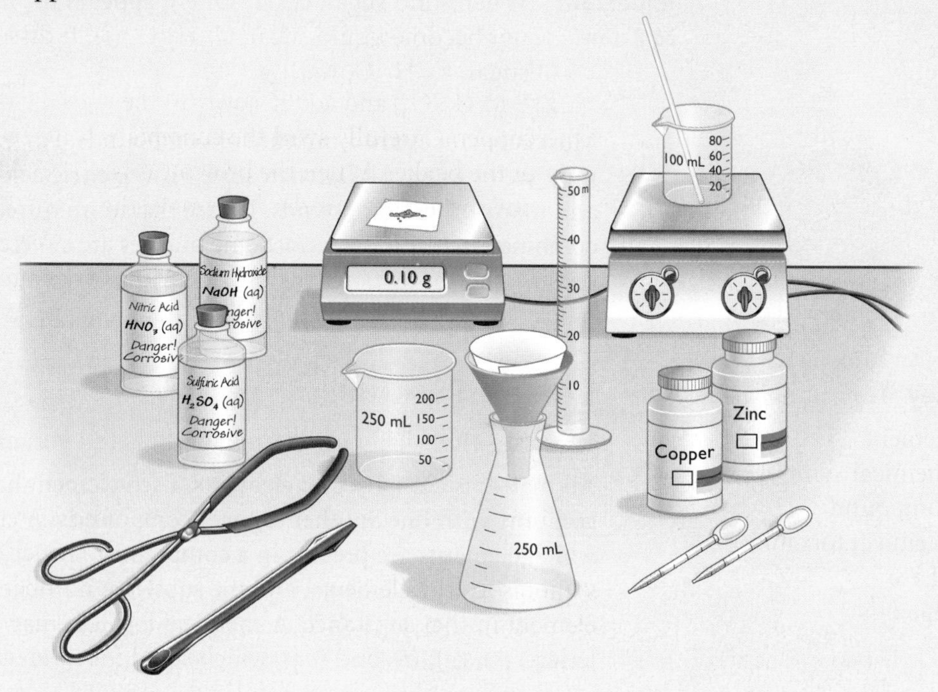

Materials

- 250 mL beaker
- 100 mL beaker
- 2 graduated pipettes
- 50 mL graduated cylinder
- funnel and filter paper
- spatula
- copper powder, 0.1 g
- 8 M nitric acid, HNO_3, 2 mL
- zinc filings, 0.1 g
- 1 M sulfuric acid, H_2SO_4, 15 mL
- 8 M sodium hydroxide, NaOH, 2 mL
- stirring rod
- hot plate
- beaker tongs
- balance

Safety Instructions

Wear safety goggles at all times.

Be very careful handling the nitric acid, sulfuric acid, and sodium hydroxide.

Be careful not to breathe the nitrogen dioxide gas. This part of the lab has to be done in a fume hood or outdoors.

Always be careful when heating chemicals.

Procedure and Observations

Make a table like the one shown. Fill in the table as you complete each step.

Procedure	Observation
1.	

1. Measure approximately 0.10 g of copper powder and place it in a 250 mL beaker. Move the beaker to a fume hood.

2. Measure 2 mL of nitric acid ($8\ M\ HNO_3$) and add it slowly to the copper. Carefully swirl the contents of the beaker. When the brown gas is no longer being produced, remove the beaker from the fume hood.

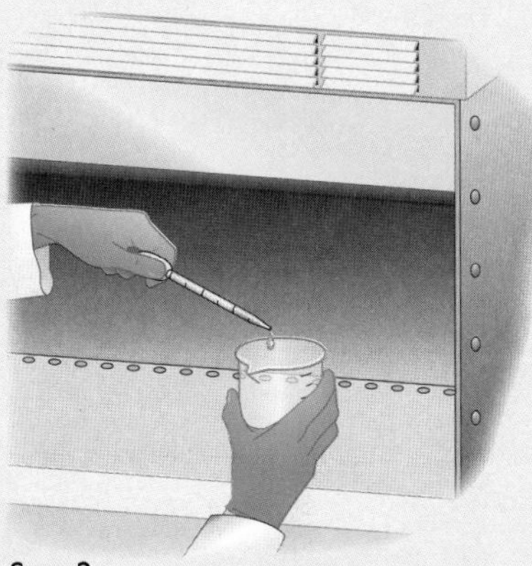

Step 2

3. Add approximately 25 mL of water to the beaker. Carefully measure 2 mL of the concentrated sodium hydroxide, NaOH. Slowly add it to the beaker. Observe the liquid closely.

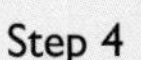

Step 4

4. Place the beaker on a hot plate and set the hot plate to medium. Stir with a glass rod and continue heating until a solid appears in the solution.

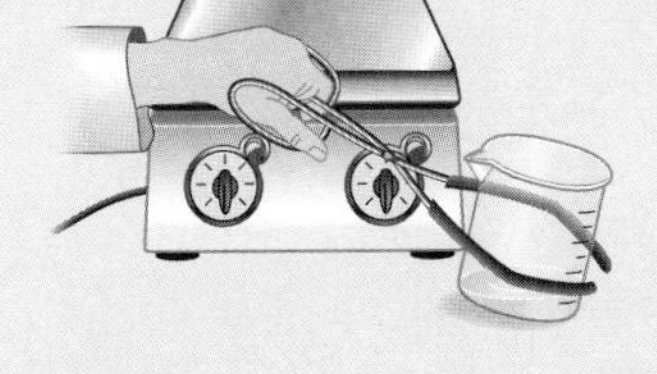

Step 5

5. Use beaker tongs to move the beaker from the hot plate to the lab table. Allow the beaker to cool before proceeding.
6. Filter the solution with a funnel and filter paper.
7. Using a small spatula, gently scrape the solid from the paper into a 100 mL beaker.

8. While stirring, slowly add 15 mL of sulfuric acid, $1\ M\ H_2SO_4$ to the 100 mL beaker.
9. Measure approximately 0.1 g of zinc filings.
10. Add the zinc to the beaker. Stir the solution until it is colorless.
11. Pour off most of the liquid into the correct waste container, making sure not to pour away any of the solid at the bottom of the container.
12. Add about 10 mL of water, swirl, and again pour off most of the liquid into the correct waste container.

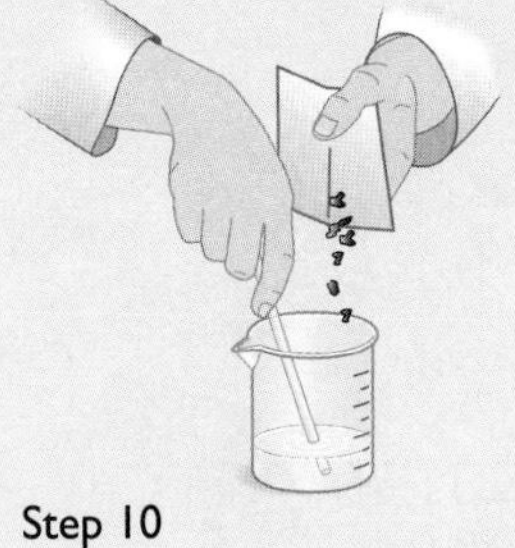

Step 10

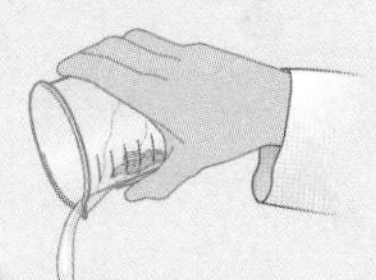

Step 11

Analysis

Write a lab report for this experiment. In your Results section, explain where you think the copper was during each step of the procedure.

LESSON

7 Now You See It

The Copper Cycle

Think About It

At this point, you are probably convinced that a copper penny cannot be transformed into real gold in a high school chemistry classroom. However, the question remains, what happened to the copper penny?

What happens to matter when it is changed?

To answer this question, you will explore

1. The Copper Cycle
2. Evidence of Chemical Change
3. Interpreting Observations

Exploring the Topic

1 The Copper Cycle

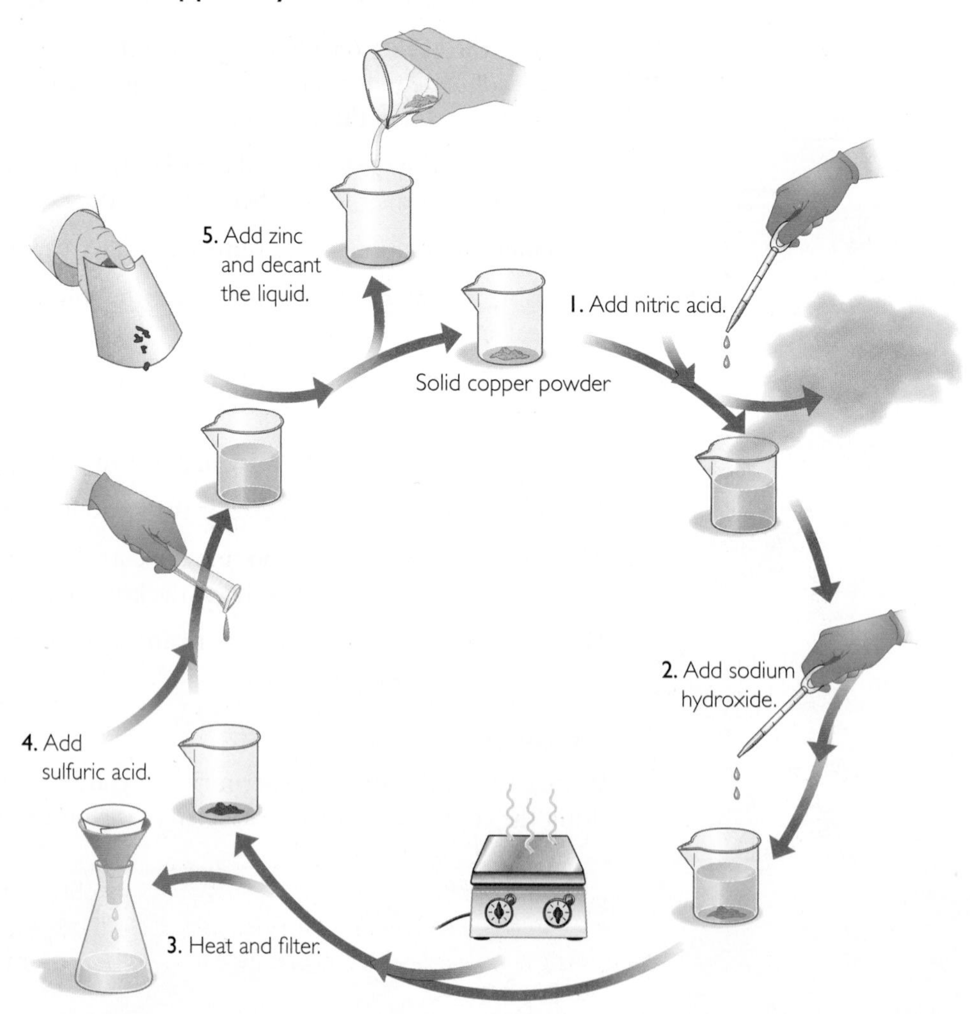

BIOLOGY CONNECTION

Octopus blood is blue-green in color. It contains the copper-rich protein hemocyanin as opposed to the iron-rich hemoglobin found in vertebrates.

In order to make sense of the copper penny experiment, it is important to learn more about how copper can be transformed. Copper itself is an element, represented by the chemical symbol Cu(*s*). As a solid powder, copper has a distinctive orange-brown appearance.

In class, you transformed copper powder through a series of chemical reactions. If all worked well, copper should have reappeared at the end of the experiment. But what happened between the beginning and the end?

The diagram on page 28 shows the steps of the copper cycle. Start at the top and follow the arrows.

❷ Evidence of Chemical Change

In each of the five steps of the copper cycle, a substance is changed into a different substance by mixing it with a liquid or by heating it. These changes are called **chemical changes** or **chemical reactions.** You can usually tell when a chemical reaction occurs because there is evidence of new substances forming. For instance, in class you observed some color changes when two of the liquids turned a bright blue color. You also observed the release of a brown gas and the formation of a dark-colored solid. These are all possible signs that new substances are forming. After a chemical reaction, you have new substances with properties that are different from the starting substances.

Copper powder

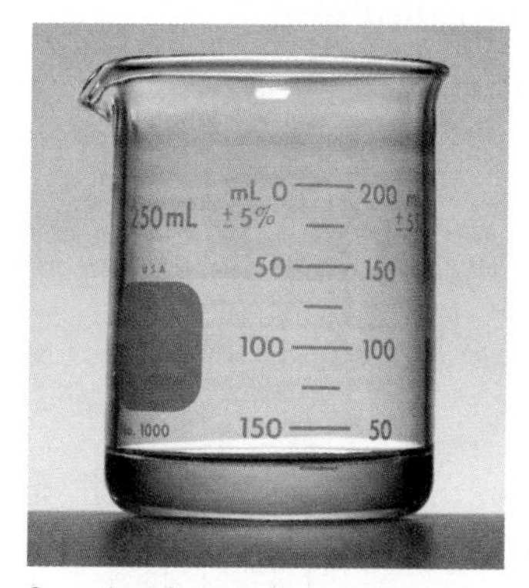

Step 1: After adding nitric acid

Step 2: After adding aqueous sodium hydroxide

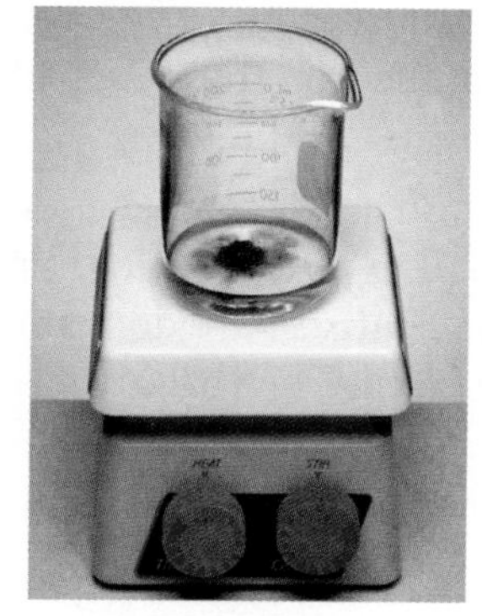

Step 3: After heating

Step 4: After adding sulfuric acid

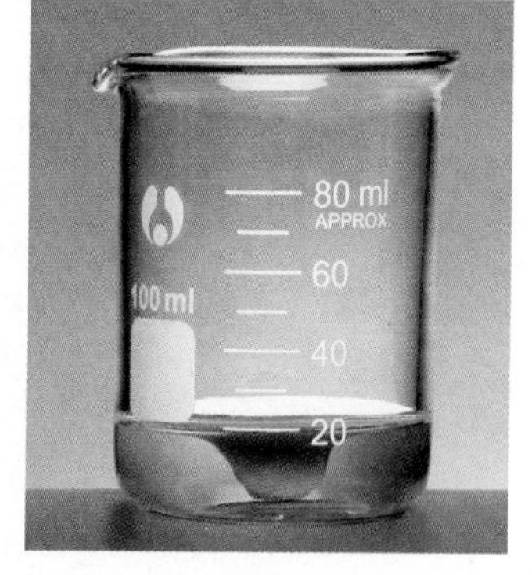

Step 5: After adding zinc

❸ Interpreting Observations

Look again at the first step in the copper cycle diagram. When nitric acid was added to the copper, a blue liquid formed and a toxic brown gas left the beaker. But where did the copper powder go? You may have wondered if the copper left the beaker

HISTORY CONNECTION

Copper is not usually found as a pure element in nature. It is found in compounds. About 7,000 years ago, people learned how to extract copper from rocks. They heated certain rocks together with charcoal in an oven, producing molten copper and carbon dioxide gas.

with the brown gas. However, copper reappeared at the end of the experiment. This is evidence that the copper doesn't leave the beaker at any point in the cycle.

Was copper put back in again? If copper was not added in again and it did not leave with the brown gas, the only other explanation is that it never left the beaker at all. This would mean that the copper is present in copper-containing compounds at each step of the cycle. Copper must be present somewhere in the blue solutions and the blue solid that were observed. Thus, the copper was not created or destroyed as a result of its chemical journey.

Over many centuries, chemists have carried out procedures like the copper cycle. In so doing, they have learned about the properties of substances and even discovered new substances. They have also learned how to predict change. Chemists know that pouring nitric acid on copper produces a clear blue liquid and a brown gas. They have even determined the identity of the blue liquid and the brown gas. Perhaps further investigation will allow you to track the changes to the copper penny as well.

Example

Rusting Nail

An iron nail that is left in contact with water and air starts to form a reddish-brown coating called rust.

a. Is this a chemical change?

b. Is rust an element, a compound, or a mixture? Explain your thinking.

c. How could you gather more evidence for your answer to part b?

Solution

Rust forms on the surface of the nail, so you can deduce that it is probably caused by exposure to water and air.

a. There is evidence of a new substance being formed on the surface of the nail. There is a color change and the rust is softer than iron; therefore, a chemical change has taken place.

b. Rust is probably a compound, a combination of the iron from the nail plus one or more elements from the air or water.

c. To gather more evidence for chemical change, you could complete several experiments. For example, you could place a nail in pure water or in pure oxygen to test what the iron is reacting with. (Scientists have done this and found that iron rusts only in the presence of oxygen. They have determined that rust is actually iron oxide.)

Lesson Summary

What happens to matter when it is changed?

When one substance is transformed, or changed, into another substance, a chemical change has occurred. A chemical change is also called a chemical reaction. When chemical reactions occur, new substances with new properties are

Key Terms

chemical change
chemical reaction

produced. Evidence of new substances, such as the production of a gas, a color change, or the formation of a solid in solution, is a sign that a chemical change has taken place. When substances undergo chemical changes, the elements themselves are not created or destroyed.

EXERCISES

Reading Questions

1. Explain what a chemical reaction is. What are some possible signs that a chemical reaction is taking place?
2. In your own words, explain how it might be possible to start with copper and end up with copper after a series of chemical reactions.

Reason and Apply

3. Look again at the copper cycle diagram. After adding zinc, the blue solution turns colorless, and copper appears as a solid. What do you think happened to the zinc? Explain your thinking.
4. Suppose you use sugar, butter, eggs, and flour to make some cookie dough. You bake the dough in the oven until the cookies are done. Do the ingredients undergo a chemical change? Give evidence to support your answer.

5. Baking soda is a white powder used for baking or cleaning. When you mix baking soda, $NaHCO_3$, with vinegar, $C_2H_4O_2$, you get a clear colorless liquid and bubbles of CO_2.

   a. Specify the phase of each of the compounds.
   b. Is this a chemical change? Give evidence.
   c. Where is the sodium, Na, before the change? After the change?
6. **WEB RESEARCH** Research how copper is extracted from a compound found in nature. Write a paragraph describing the process.

LESSON

8 What Goes Around Comes Around

Conservation of Matter

Think About It

The element copper can be mixed with other substances to make a colorful assortment of compounds. What are these compounds? How can you demonstrate that they all contain copper?

What happens to elements in a chemical change?

To answer this question, you will explore

1. Translating the Copper Cycle
2. Tracking an Element
3. Conservation of Matter

Exploring the Topic

1 Translating the Copper Cycle

Chemical names and chemical formulas are powerful tools you can use to keep track of matter. In fact, you can use them to figure out what you made at various steps in the Lab: The Copper Cycle.

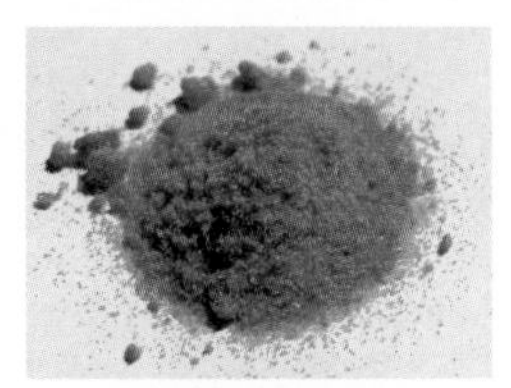

Copper nitrate, $Cu(NO_3)_2(s)$

Copper sulfate, $CuSO_4(s)$

There are several approaches you can take to figure out what was made at each step of the cycle. First, you can compare the appearance of the compounds that you obtained in each step with compounds that you have seen before. For example, in an earlier class you examined samples of copper compounds that looked like these photos.

Visual observation of compounds can give you some clues, but it is not enough to make a definite identification. For example, it is hard to tell the difference between copper sulfate and copper nitrate through observation because they are both blue. In fact, several copper compounds are blue.

Another approach is to examine the chemical names and formulas of the substances that were mixed together. The new substances are formed from parts of the starting materials. For example, when sodium *hydroxide,* NaOH, is added to the beaker in the second step of the cycle, a compound called copper *hydroxide,* $Cu(OH)_2$, is produced. Since a blue solid is formed and hydroxide, OH^-, is one of the starting ingredients, you can deduce that copper hydroxide is the product.

Adding Nitric Acid to Copper

Start at the beginning of the copper cycle and see if you can figure out what was created after the first step by translating each step into chemical symbols and formulas.

Step 1: Nitric acid is added to copper powder. A clear blue solution and a brown gas are formed.

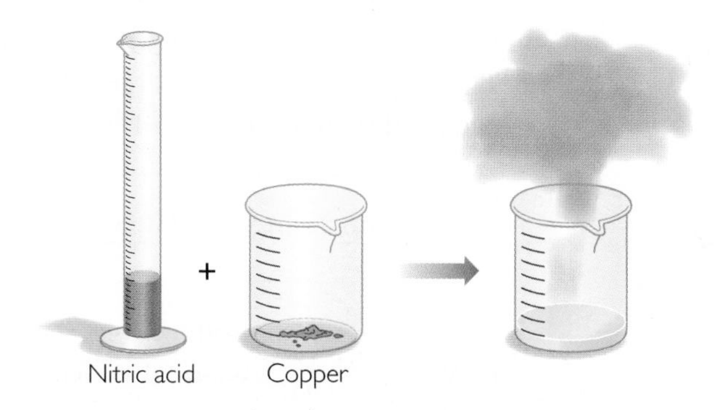

We know the chemical names that go along with the first half of this step. Nitric acid *is added to* copper powder *resulting in* a blue solution and a brown gas. Using chemical formulas, you can write this as:

$HNO_3(aq)$ is added to **$Cu(s)$** resulting in ___?___ (blue solution) and ___?___ (brown gas).

So far in class, you've seen two blue liquids. One was copper sulfate and the other was copper nitrate.

Nitric acid, $HNO_3(aq)$, was added to the copper powder, so you can deduce that the blue solution that formed was copper nitrate, $Cu(NO_3)_2(aq)$, and not copper sulfate, $CuSO_4(aq)$.

$HNO_3(aq)$ is added to **$Cu(s)$** resulting in **$Cu(NO_3)_2(aq)$** and ___?___.

How about the brown gas? The brown gas must contain some combination of H, N, O, or Cu because these are the only starting ingredients. Copper, Cu, does not form gaseous compounds, but the other three elements do combine to form several different gases. The chemical formulas and colors of these gases can be found in reference books and are listed in the table on the next page.

Aqueous copper nitrate, $Cu(NO_3)_2(aq)$

Gases Containing H, N, or O

Gas	Color
$H_2(g)$, hydrogen gas	colorless
$N_2(g)$, nitrogen gas	colorless
$O_2(g)$, oxygen gas	colorless
$H_2O(g)$, water vapor	colorless
$NH_3(g)$, ammonia gas	colorless
$NO(g)$, nitrogen oxide gas	colorless
$NO_2(g)$, nitrogen dioxide gas	**brown**

The only *brown* gas in this table is nitrogen dioxide, $NO_2(g)$. The completed chemical sentence for Step 1 is:

$HNO_3(aq)$ is added to **$Cu(s)$** resulting in **$Cu(NO_3)_2(aq)$** and **$NO_2(g)$**.

This reaction also produces one more compound: water, H_2O. You would not have noticed this because it is clear and colorless. So the final chemical sentence for Step 1 is:

$HNO_3(aq)$ is added to **$Cu(s)$** resulting in **$Cu(NO_3)_2(aq)$** and **$NO_2(g)$** and **$H_2O(l)$**.

Notice that all the elements in the starting ingredients also appear in the products. No elements are created or destroyed.

You can deduce the products for the other steps in the copper cycle in a similar way.

ENVIRONMENTAL CONNECTION

Nitrogen dioxide, $NO_2(g)$, is a part of smog. It causes the red-brown color in the skies above cities with large amounts of air pollution. Nitrogen dioxide irritates the eyes, nose, throat, and respiratory tract. Continued exposure can cause bronchitis.

❷ Tracking an Element

Once you have figured out the products of Steps 2, 3, and 4 of the copper cycle, you can track the journey of copper through the cycle. The illustration below shows the copper compounds that form at each step of the cycle.

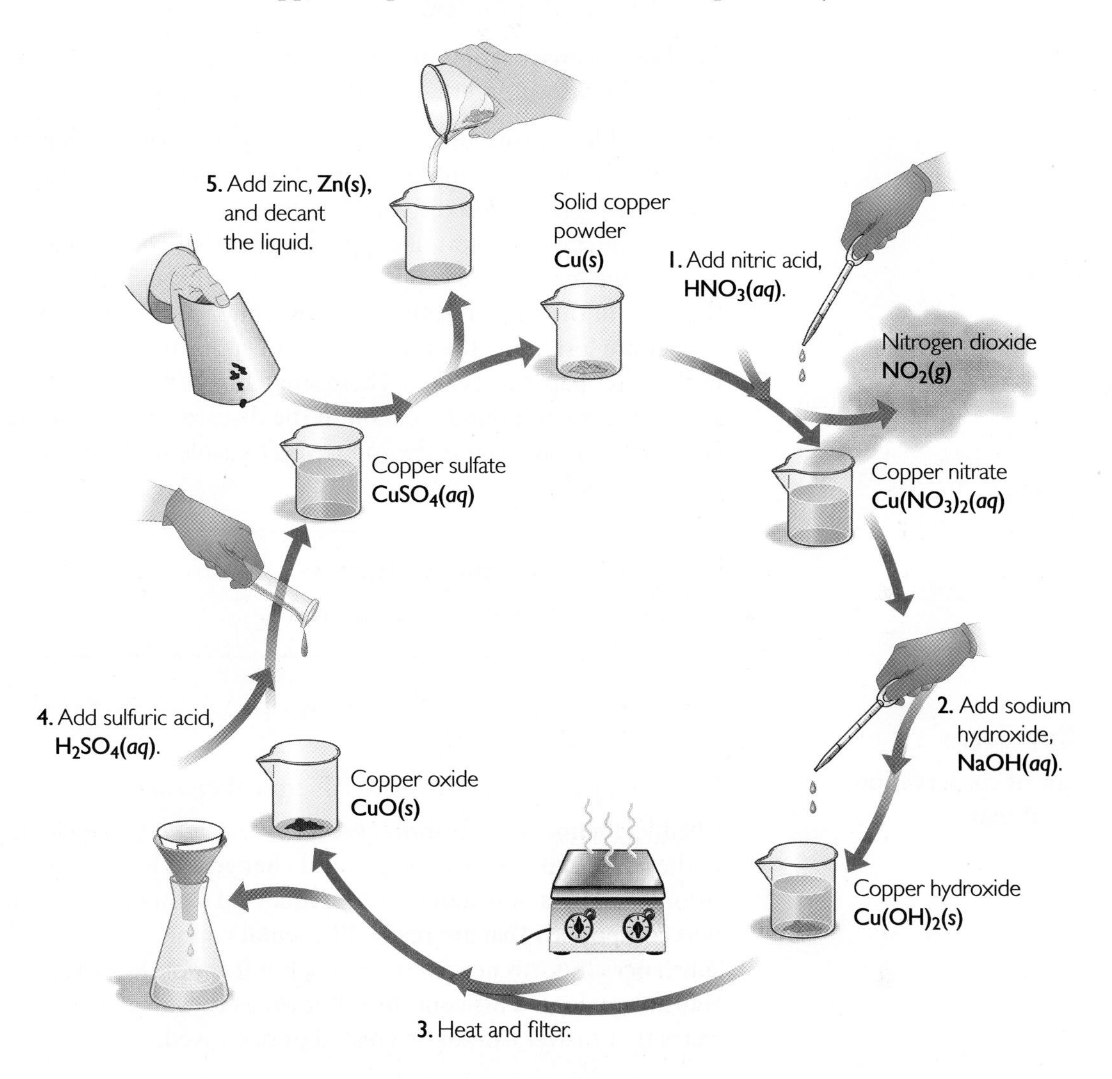

The symbol Cu is found at each stage of the cycle. Thus, the element copper is somehow combined in each of these compounds. And of course, the solid that forms at the end of the experiment is elemental copper, Cu(*s*).

What you have observed with copper is true of other elements as well. For example, nickel can be taken through a similar cycle, where various substances are added to nickel powder, Ni(*s*). Just as with the copper cycle, you end up with nickel in the end.

❸ Conservation of Matter

The copper cycle experiment brought you full circle, back to where you started. You took a sample of the element copper and added substances to it. After several steps, you ended up with copper powder once again. No matter what was done to the copper, the copper was always there in some form. In other words, it was not created or destroyed by the chemical transformations. Over many centuries, scientists have gathered evidence that matter can never be destroyed or created through chemical transformation. There is so much evidence that this is considered a scientific law.

The **law of conservation of mass** states that mass cannot be gained or lost in a chemical reaction. In other words, matter cannot be created or destroyed.

It is possible to prove that no copper was gained or lost during the copper cycle experiment by measuring the mass of the copper powder at the beginning and again at the end. If you did the experiment perfectly, you would end up with exactly the same amount of copper powder that you started with. However, in real life the mass of the copper at the end of this experiment is a bit less than the mass of the copper at the beginning due to several factors. Little amounts of copper are lost along the way, because of spills, measurement errors, and sticking to the filter paper or beaker. These small errors are hard to avoid. In addition, some copper compounds remain in the discarded solutions. Nevertheless, mass is still conserved; even if the copper isn't visible in the beaker at the end, it still exists somewhere.

Law of Conservation of Mass

Matter cannot be created or destroyed.

Key Term

law of conservation of mass

Lesson Summary

What happens to elements in a chemical change?

Chemical names and formulas are used to keep track of elements and compounds as they undergo chemical or physical changes. When elemental copper is tracked through a series of reactions, the symbol Cu shows up in the formulas of the new compounds that are made. Elemental copper can be combined with other substances to form new compounds, but it is not destroyed by the chemical transformations. This concept is known as the law of conservation of mass, which states that matter cannot be created or destroyed.

HISTORY CONNECTION

Pennies were last made of solid copper in 1836. Pennies made from 1962 to 1982 are 95% copper and 5% zinc. These pennies have a density of 8.6 g/cm^3, which is just slightly less than the density of copper, 9.0 g/cm^3. Since 1982, pennies have been made mostly of zinc with a copper coating. These pennies have a density of 7.2 g/cm^3, which is very close to the density of pure zinc, 7.1 g/cm^3.

EXERCISES

Reading Questions

1. How can chemical names and symbols help you figure out what copper compound you made in each step of the copper cycle? Give an example.
2. Explain the law conservation of mass in your own words.

Reason and Apply

3. **Lab Report** Write a lab report for the Lab: The Copper Cycle. In your report, include the title of the experiment, purpose, procedure, observations, and conclusions.

(Title)

Purpose: (Explain what you were trying to find out.)

Procedure: (List the steps you followed.)

Observations: (Describe what you observed during the experiment.)

Conclusions: (What can you conclude about what you were trying to find out? Provide evidence for your conclusions.)

Analysis: (Explain what happpened to the copper during the experiment.)

4. Explain how the copper cycle experiment supports the claim that copper is an element—a basic building block of matter.
5. Nickel sulfate, $NiSO_4(aq)$, is a green solution. Nickel chloride, $NiCl_2(aq)$, is a yellow solution. And hydrochloric acid, $HCl(aq)$, is a clear, colorless solution. If you add nickel, $Ni(s)$, to hydrochloric acid, $HCl(aq)$, what color solution do you expect to form? Explain your reasoning.
6. In the final step of the copper cycle, zinc, $Zn(s)$, is added to copper sulfate, $CuSO_4(aq)$. Elemental copper appears as a solid. Explain what you think happens to the elemental zinc.
7. Matter cannot be created or destroyed. List at least two long-term impacts that this concept has for us on this planet.

ACTIVITY Create a Table

Purpose

To create your own periodic table of the elements from data given on element cards.

Materials

- Create a Table card deck

Be

Sr

Instructions

1. Work in groups of four with one set of cards.
2. Find Be, Mg, Ca, and Sr in the deck of cards, and arrange them in a column the way Mendeleyev did. These cards are all yellow. Look for similarities and differences in these cards. Find at least one pattern or trend, and describe it to your group.
3. With your group, decide how to organize the rest of the cards into a table. Try to organize them in a way that produces as many patterns as possible.

Questions

1. What characteristics did you use to decide how to sort the cards?
2. What patterns appear in your arrangement? List at least four.
3. Where did you put H and He? What was your reasoning for their placement?
4. Did you notice any cards that didn't quite fit or that seemed out of order? Explain.
5. **Making Sense** Below are possible cards for the element germanium.
   a. Where does germanium belong in the table?
   b. Which card seems most accurate to you? What is your reasoning?

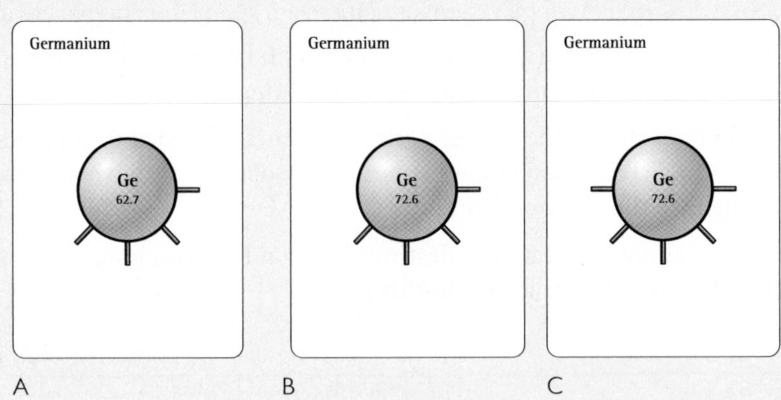

A B C

   c. What would you add to the three empty corners to complete the card?

LESSON

9 Create a Table

Properties of the Elements

Think About It

In the late 1860s, a Russian chemist and teacher named Dmitri Mendeleyev was looking for a way to organize the elements known at the time. He wanted to make it easier to remember and understand their chemical behavior. He started by placing the elements in groups based on similarities in their properties. You can understand Mendeleyev's organization of the elements by examining the information he used to sort them.

How is the periodic table organized?

To answer this question, you will explore

1. Properties of the Elements
2. A Table of Elements

Exploring the Topic

1 Properties of the Elements

Mendeleyev was intrigued by patterns in the properties of the elements. For example, tin, Sn, sodium, Na, and magnesium, Mg, are all shiny, silvery solids at room temperature.

Tin, Sn

Sodium, Na

Magnesium, Mg

However, by itself, visual appearance is not a reliable characteristic for sorting the elements into groups. For instance, there are many elements that are shiny, silvery solids at room temperature. And, while both oxygen and neon are colorless gases, they are not similar enough in their other properties to be grouped together. Mendeleyev focused on three properties in addition to appearance to sort the elements: reactivity, the formulas of chemical compounds that form when the element combines chemically, and atomic mass.

Reactivity

Mendeleyev focused on the reactivity of elements to sort them. **Reactivity** is a property that describes how easily an element will combine with other substances to form new compounds. An element that is highly reactive combines rapidly with other substances. An explosion, smoke, or a flash of light is a sign that a reaction is proceeding quickly. For instance, when the metal, sodium, Na, comes into the slightest contact with water, it reacts vigorously. Sodium even reacts with water

HISTORY CONNECTION

It took thousands of years for human beings to discover the elements that are the building blocks of matter. Ancient civilizations discovered and used seven metals: gold (Au), copper (Cu), silver (Ag), lead (Pb), tin (Sn), iron (Fe), and mercury (Hg). These seven elements are called the *metals of antiquity*. Many of the early elements were discovered because they are less reactive than other elements and therefore more likely to be found in their pure forms.

vapor in the air. Magnesium, Mg, is another metal that reacts with water. Sodium and magnesium are both shiny, silvery metals that react with water. Based on their properties, these two metals might belong in the same group.

Formulas of Compounds

Mendeleyev also paid attention to which elements combine with which, and he noted the ratios in which their atoms combine. For example, magnesium, Mg, can combine with chlorine, Cl. When it does, it forms the compound magnesium chloride, $MgCl_2$. This chemical formula indicates that atoms of magnesium and chlorine combine in a 1:2 ratio.

Sodium, Na, also reacts with chlorine, but it combines in a 1:1 ratio, forming sodium chloride, NaCl. Perhaps magnesium and sodium do not belong in the same group after all.

Examine the table, which shows some of the elements that react with chlorine and the compounds that are formed.

You can sort the elements into three groups according to the formulas of compounds with chlorine. In the Activity: Create a Table, you sorted these groups into separate columns, as Mendeleyev did.

Some Elements That React With Chlorine

Element	Symbol	Compound
magnesium	Mg	$MgCl_2$
sodium	Na	NaCl
aluminum	Al	$AlCl_3$
hydrogen	H	HCl
calcium	Ca	$CaCl_2$
indium	In	$InCl_3$
gallium	Ga	$GaCl_3$

Atomic Mass

Mendeleyev used another property, called atomic mass, to sort the elements. Each element is made of a different kind of atom. (You will study atoms in Lesson 11: Atomic Pudding.) All atoms of the same element have approximately the same mass. The mass of an atom is called its **atomic mass** and is measured in **atomic mass units,** or **amu.**

Atomic mass will be explained in more detail later in the unit. For now, simply keep in mind that each element has an average atomic mass that is expressed as a decimal number.

You can place the elements in order of their atomic masses. However, sorting the cards just by atomic mass doesn't tell you which elements are similar in their properties.

❷ A Table of Elements

Mendeleyev combined all of these sorting tactics to create his table. He put the elements with similar reactivity and chemical formulas of compounds into columns. Mendeleyev also sorted the elements in order of their atomic masses. He placed the lighter elements at the top of the columns and the heavier elements at the bottom. When he placed the columns next to each other, the atomic masses increased from left to right as well as from top to bottom.

Dmitri Mendeleyev's Periodic Table of the Elements

	Group I	Group II	Group III	Group IV	Group V	Group VI	Group VII	Group VIII
1	H = 1							
2	Li = 7	Be = 9	B = 11	C = 12	N = 14	O = 16	F = 19	
3	Na = 23	Mg = 24	Al = 27	Si = 28	P = 31	S = 32	Cl = 35	
4	K = 39	Ca = 40	_ = 44	Ti = 48	V = 51	Cr = 52	Mn = 55	Fe = 56 Co = 59 Ni = 59 Cu = 63
5	Cu = 63	Zn = 65	_ = 68	_ = 72	As = 75	Se = 78	Br = 80	
6	Rb = 85	Sr = 87	Yt = 88	Zr = 90	Nb = 94	Mo = 96	_ = 100	Ru = 104 Rh = 106 Pd = 106 Ag = 108
7	Ag = 108	Cd = 112	In = 113	Sn = 118	Sb = 122	Te = 125	I = 127	
8	Cs = 133	Ba = 137	Di = 138	Ce = 140	—	—	—	— —
9	—	—	—	—	—	—	—	—
10	—	—	Er = 178	La = 180	Ta = 182	W = 184	—	Os = 195 Ir = 197 Pt = 198 Au = 199
11	Au = 199	Hg = 200	Tl = 204	Pb = 207	Bi = 208	—	—	
12	—	—	—	Th = 231	—	U = 240	—	— — — —

Note: Mendeleyev's symbol for iodine, "J," has been changed to "I" to match modern symbols.

Mendeleyev organized the 63 elements that were known at the time into a table. In his table, the average atomic masses of the elements increase as you proceed across each row and down the table. The elements in each column have similar physical properties and reactivity, and they tend to form compounds with other elements in the same ratios. The table Mendeleyev created, with elements organized in rows and columns, became known as the **periodic table of the elements.**

The periodic table is an extremely useful organization of the elements. Mendeleyev was even able to predict the existence and properties of as-yet undiscovered elements based on gaps he located in his table.

BIG IDEA Elements are arranged on the periodic table based on similarities in their chemical and physical properties.

Key Terms

reactivity
atomic mass
atomic mass unit, amu
periodic table of the elements

Lesson Summary

How is the periodic table organized?

Dmitri Mendeleyev created one of the first organized tables of the elements. He sorted the elements based on their properties, specifically reactivity, the formulas of compounds created when elements chemically combine, and atomic mass. He placed elements in rows and columns with increasing atomic mass across a row from left to right and down a column. The elements in each column of the table have similar properties. The table Mendeleyev created came to be called the periodic table of the elements.

EXERCISES

Reading Questions

1. List three properties of the elements that are useful in sorting the elements.
2. Do you expect carbon, C, to be more similar to nitrogen, N, oxygen, O, or silicon, Si? Why?

Reason and Apply

3. **WEB RESEARCH** Use a reference book or the Internet to look up the average atomic masses and properties of silicon, Si, germanium, Ge, tin, Sn, phosphorus, P, antimony, Sb, sulfur, S, and selenium, Se.
   a. Organize these elements in rows and columns.
   b. List two properties that the elements in each column have in common.
4. **WEB RESEARCH** Use a reference book or the Internet to look up some of the properties of iron, barium, and phosphorus. Explain why nails are made of iron but they are never made of barium or phosphorus.
5. Suppose you have equal amounts of calcium, Ca, in two beakers. You react the calcium in one beaker with oxygen, O, and the other with sulfur, S. The reaction with oxygen forms the compound calcium oxide, CaO.
   a. What do you predict is the chemical formula of the compound formed from the reaction between calcium and sulfur?
   b. Which compound has more mass, the compound containing calcium and oxygen, or the compound containing calcium and sulfur? Explain your thinking.
6. **WEB RESEARCH** Use a computer to research Dmitri Mendeleyev, the Russian chemist credited with the discovery of the periodic table of the elements. Write a brief paragraph describing Mendeleyev's life and work. Include how he became a chemistry professor and how he came up with the idea for the periodic table.
7. Find at least two different versions of the periodic table and bring a copy of each to class.
   a. Write down what you think makes these two versions similar to each other.
   b. Write down what you think makes these two versions different from each other.

LESSON

10 Breaking the Code

The Periodic Table

Think About It

The elements copper, Cu, and gold, Au, share many similarities. Both are relatively unreactive elements. They are soft so it is easy to bend and shape them. They are called *coinage metals* because they have been made into coins by many cultures. Copper and gold have high values as jewelry because they remain shiny for many years. Is the similarity in their properties related to their locations on the periodic table?

What information does the periodic table reveal about the elements?

To answer this question, you will explore

1. The Modern Periodic Table
2. Trends in Properties

Exploring the Topic

1 The Modern Periodic Table

Scientists have detected around 114 different elements on the planet. Each is unique. Yet, groups of elements have similar properties. Recall from Lesson 9: Create a Table that Dmitri Mendeleyev constructed a table based on patterns in the properties of the elements. His table has been replaced over the decades with many updated versions, such as the one shown on pages 44 and 45. The modern periodic table is a storehouse of valuable information about the elements. Over time you will learn how to make use of the information that is contained there.

Element Squares

Each element has a square on the periodic table. Within each square is information about that element including its name and symbol. The whole number in each square is called the **atomic number.** Hydrogen is the first element in the table and has the atomic number 1. Helium is the second element and has the atomic number 2. Each atomic number corresponds to a different element.

The decimal number in each square on the periodic table square is the average atomic mass in amu.

Here is a square from the periodic table.

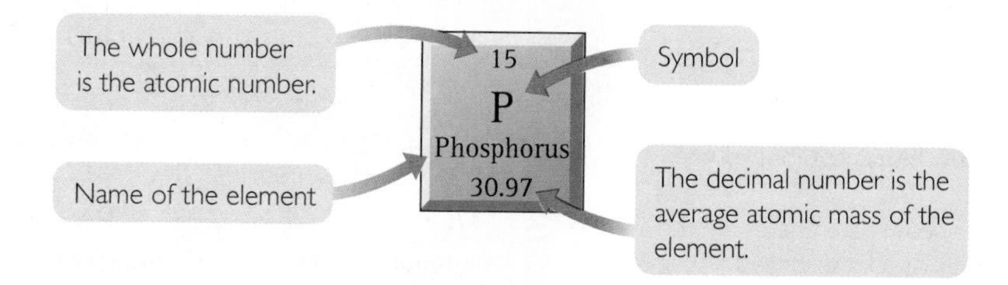

Periodic Table of the Elements

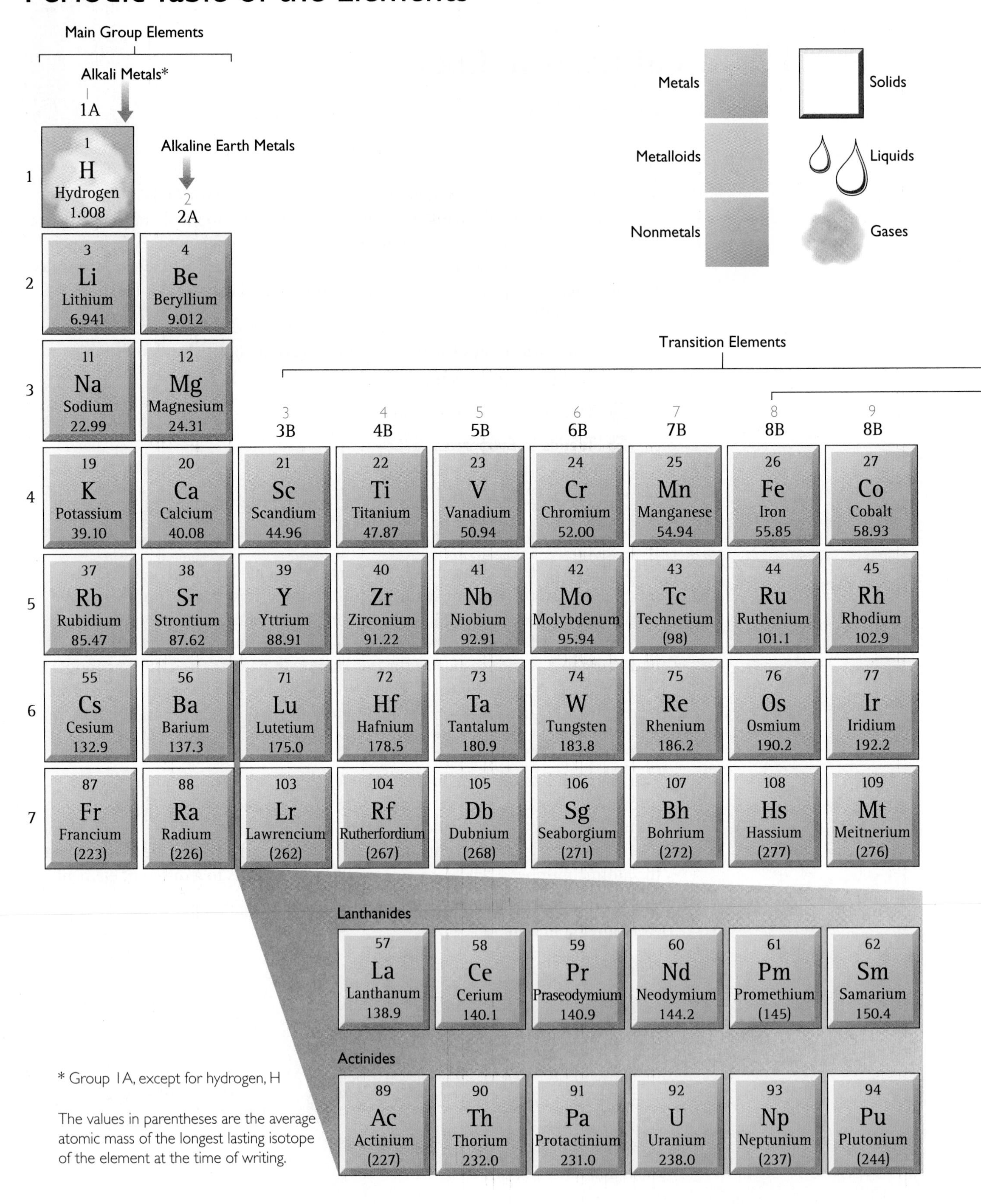

Main Group Elements

Halogens

Noble Gases

Key
1 (Atomic number) H (Symbol) Hydrogen (Name) 1.008 (Average atomic mass)

10 8B	11 1B	12 2B	13 3A	14 4A	15 5A	16 6A	17 7A	18 8A
								2 He Helium 4.003
			5 B Boron 10.81	6 C Carbon 12.01	7 N Nitrogen 14.01	8 O Oxygen 16.00	9 F Fluorine 19.00	10 Ne Neon 20.18
			13 Al Aluminum 26.98	14 Si Silicon 28.09	15 P Phosphorus 30.97	16 S Sulfur 32.07	17 Cl Chlorine 35.45	18 Ar Argon 39.95
28 Ni Nickel 58.69	29 Cu Copper 63.55	30 Zn Zinc 65.39	31 Ga Gallium 69.72	32 Ge Germanium 72.64	33 As Arsenic 74.92	34 Se Selenium 78.96	35 Br Bromine 79.90	36 Kr Krypton 83.80
46 Pd Palladium 106.4	47 Ag Silver 107.9	48 Cd Cadmium 112.4	49 In Indium 114.8	50 Sn Tin 118.7	51 Sb Antimony 121.8	52 Te Tellurium 127.6	53 I Iodine 126.9	54 Xe Xenon 131.3
78 Pt Platinum 195.1	79 Au Gold 197.0	80 Hg Mercury 200.6	81 Tl Thallium 204.4	82 Pb Lead 207.2	83 Bi Bismuth 209.0	84 Po Polonium (209)	85 At Astatine (210)	86 Rn Radon (222)
110 Ds Darmstadtium (281)	111 Rg Roentgenium (280)	112 Uub Ununbium (277)	113 Uut Ununtrium (284)	114 Uuq Ununquadium (289)	115 Uup Ununpentium (288)	116 Uuh Ununhexium (293)	117 Uus Ununseptium ( )	118 Uuo Ununoctium (294)

63 Eu Europium 152.0	64 Gd Gadolinium 157.3	65 Tb Terbium 158.9	66 Dy Dysprosium 162.5	67 Ho Holmium 164.9	68 Er Erbium 167.3	69 Tm Thulium 168.9	70 Yb Ytterbium 173.0
95 Am Americium (243)	96 Cm Curium (247)	97 Bk Berkelium (247)	98 Cf Californium (251)	99 Es Einsteinium (252)	100 Fm Fermium (257)	101 Md Mendelevium (258)	102 No Nobelium (259)

HEALTH CONNECTION

Transition metals are important to human health. For example, iron is a central component of the protein hemoglobin, which helps to transport oxygen in the blood. Chromium helps our bodies metabolize glucose, and zinc helps to protect our immune systems.

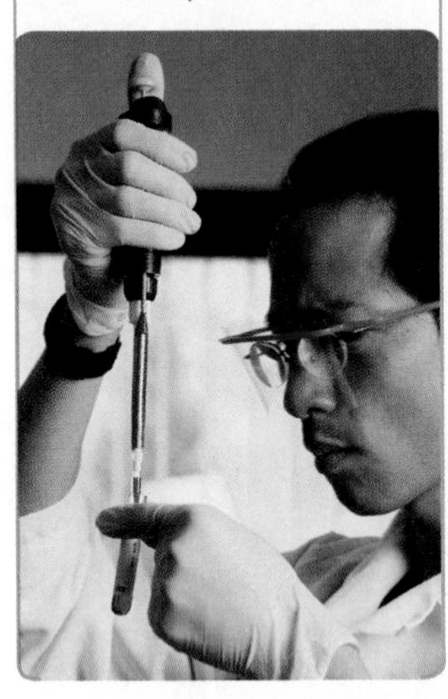

Parts of the Periodic Table

Most modern periodic tables have 18 vertical columns and 7 horizontal rows. The vertical columns are also called **groups,** or families. Hydrogen, H, is in Group 1A, along with lithium, Li, and five other elements in that column. Some of the groups have specific names as shown below.

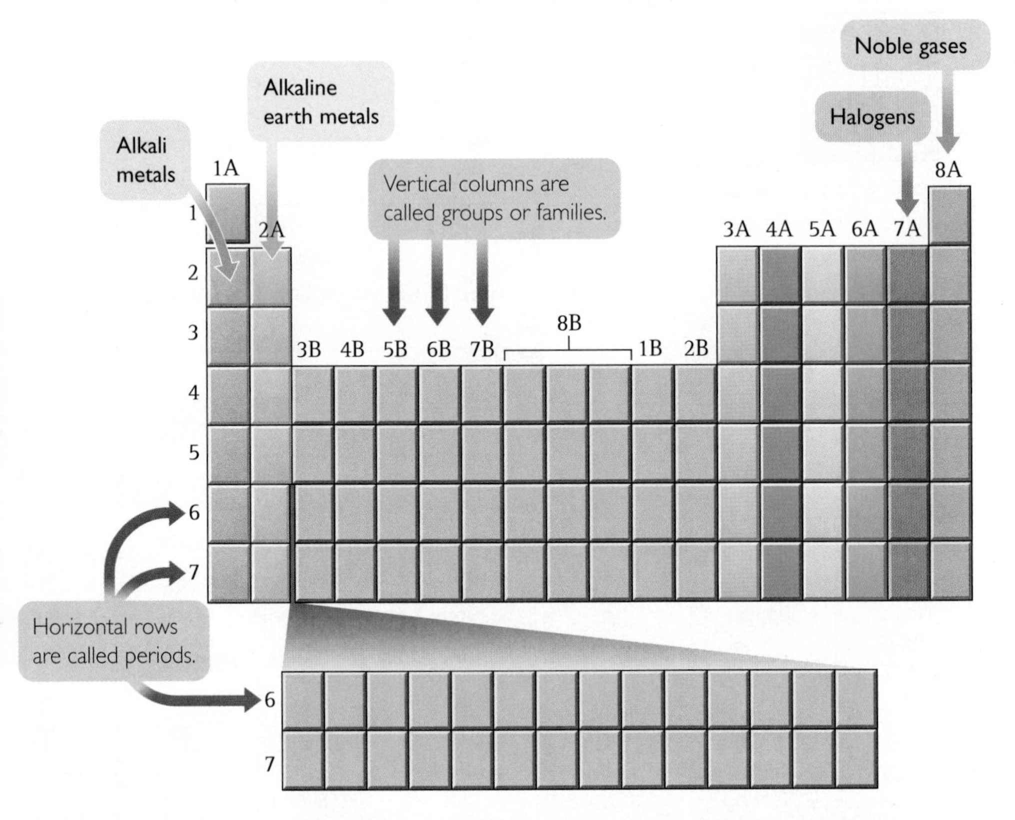

The horizontal rows of the table are called **periods** because patterns repeat periodically, or over and over again, in each row. There are only two elements in Period 1, hydrogen and helium. However, there are eight elements in Periods 2 and 3, and 18 elements in Period 4.

Chemists also have names for sections of the periodic table. Between Group 2A and Group 3A, for example, is where the transition elements fit in.

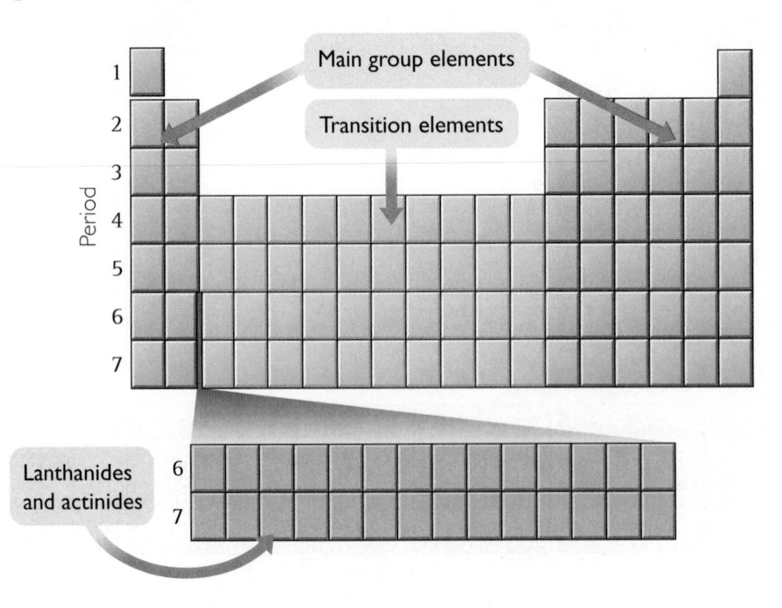

CAREER CONNECTION

Silicon Valley is a region in northern California that is known for its many high-technology and computer-based businesses. Silicon is a Group 4A element and a metalloid. It is the main component of semiconductor devices such as the microchips used in computers and electronic equipment.

In addition, there are two rows of elements usually shown at the bottom of the table. These elements are called the lanthanides and actinides. If you examine the atomic numbers of these elements, you'll see that they belong in the sixth and seventh rows. If they were included where they belong, the table would look like this.

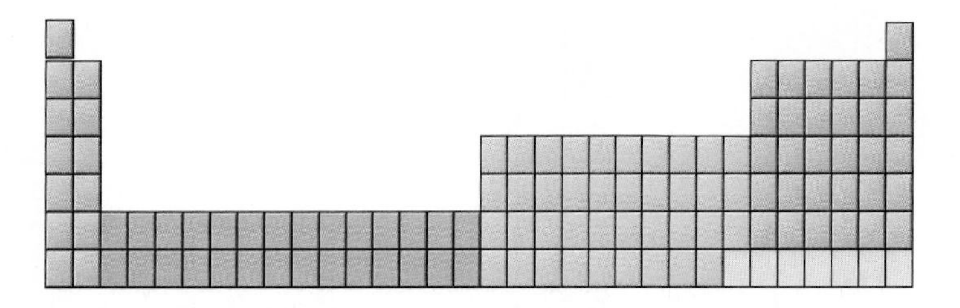

Most periodic tables show them at the bottom so everything will fit onto one page.

2 Trends in Properties

Once the elements are arranged according to their general properties, many other patterns or trends can be found. These three drawings illustrate some of the trends contained within the periodic table.

Solids, Liquids, and Gases

Most of the elements are solids at room temperature. There are several elements that are gases at room temperature, and only a few that are liquids at or near room temperature.

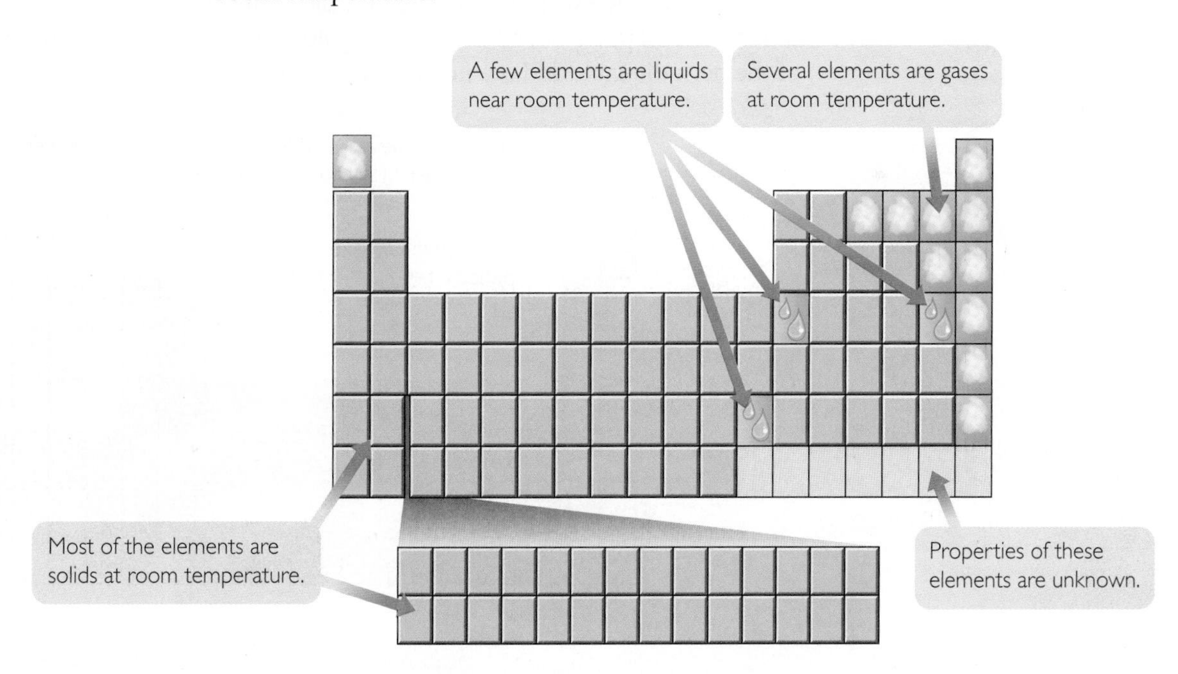

Metals, Metalloids, and Nonmetals

The majority of the elements are **metals.** On most periodic tables there is a stair-step line that divides the table. Metals are found to the left and **nonmetals** are found to the right of the stair-step line. The elements found along the stair-step line are called **metalloids.** Metalloids have properties similar to those of both metals and nonmetals.

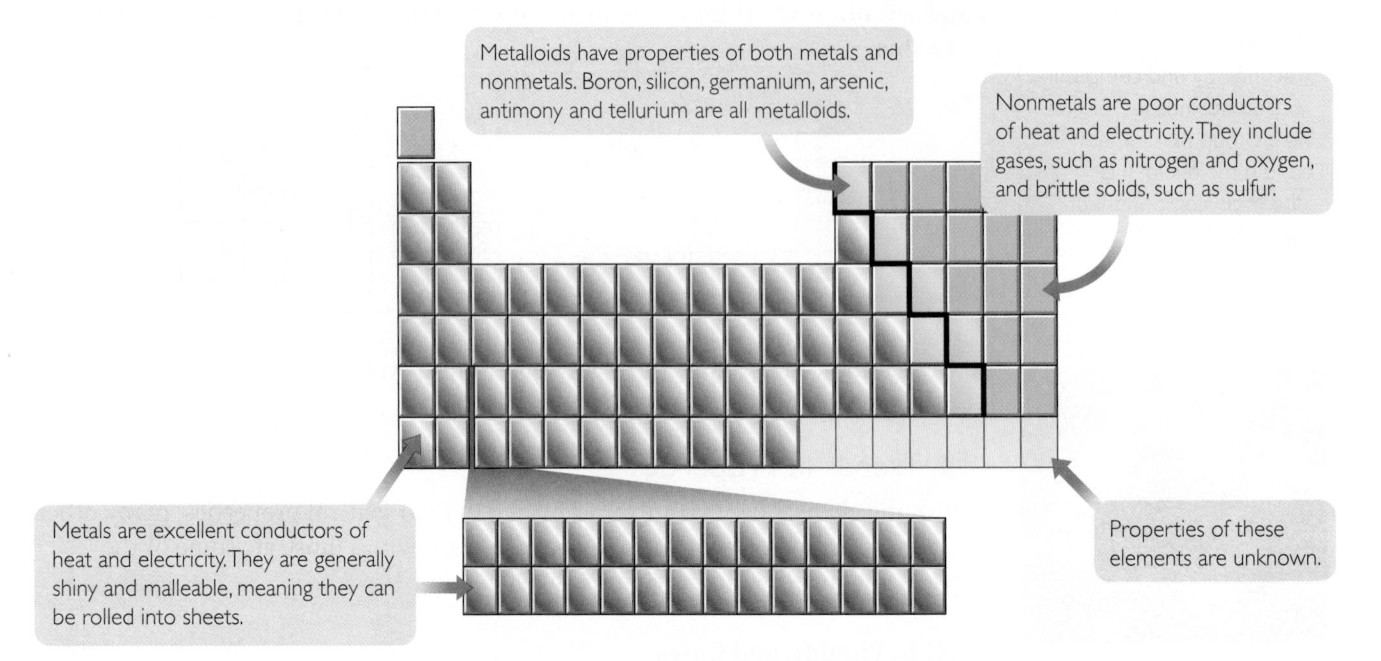

Reactivity

Elements in the lower left and upper right of the periodic table are the most reactive, with the exception of the noble gases in Group 8A, which are very unreactive. Copper, Cu, silver, Ag, and gold, Au, are metals that are in the middle of the periodic table and are not very reactive.

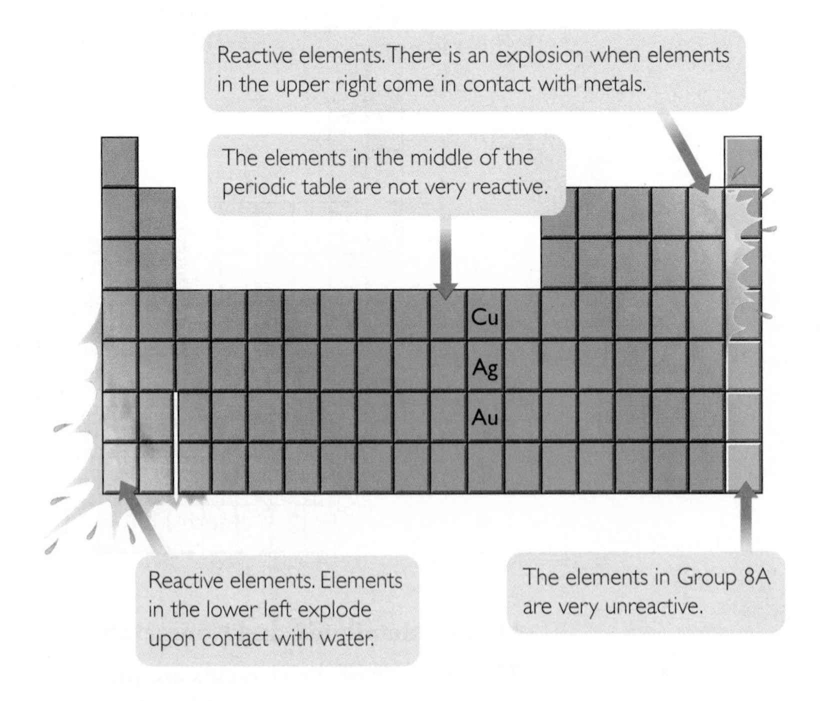

Example 1

Iodine, I

Find iodine, I, on the periodic table.

a. Find iodine's atomic number, average atomic mass, period, and group.
b. Would you expect iodine to be a solid, liquid, or gas at room temperature?
c. Is iodine a metal, metalloid, or nonmetal? How can you tell?
d. Do you expect iodine to be reactive? Explain.

Solution

Iodine is in the lower-right area of the main group elements.

a. The atomic number is 53. Average atomic mass is 126.9 amu. Iodine is in Period 5 and Group 7A, halogens.
b. Iodine is a gas at room temperature.
c. Iodine is a nonmetal, because it is to the right of the stair-step line.
d. Yes, you can expect it to be reactive, though not as reactive as elements above it in Group 7A.

Example 2

Coinage Metals

Which element would make the best coin: phosphorus, P, silver, Ag, potassium, K, or xenon, Xe? Explain your thinking.

Solution

Xenon, Xe, is a gas, so it is definitely not a candidate for making a coin. Phosphorus, P, is a nonmetal that is dull and brittle. It would be difficult to shape into a coin. Silver, Ag, is a shiny, malleable metal, so it would make the best coin. Potassium, K, is also a metal, but it is too soft and reactive, so it would not make a good coin. A good coin should not react with other substances.

Lesson Summary

What information does the periodic table reveal about the elements?

The periodic table is an organized chart of the elements. Each element square contains valuable information, including the element name, symbol, atomic number, and average atomic mass. These elements are arranged in vertical columns called groups, or families, and horizontal rows called periods. Most elements are solids and metals, except for those in the upper right of the table. The most reactive elements are located in the lower left and upper right of the table, excluding the noble gases in the last column on the right, which are unreactive.

Key Terms

atomic number
group
alkali metals
alkaline earth metals
halogens
noble gases
periods
main group elements
transition elements
lanthanides
actinides
metals
nonmetals
metalloids

EXERCISES

Reading Questions

1. Describe how reactivity changes as you go down Group 1A.
2. Choose two different properties and describe how they vary across a period.

Reason and Apply

3. You will need a handout of the periodic table.
   a. On your periodic table, clearly label the alkali metals, the alkaline earth metals, the halogens, and the noble gases (if you wish, you may color them and provide a color key at the top).
   b. Label the main group elements, the transition metals, and the lanthanides and actinides.
4. Name two elements that have properties similar to those of beryllium, Be, and have average atomic masses higher than 130.
5. Which of these elements are solids?
   A. fluorine, F **B.** titanium, Ti **C.** lead, Pb
   D. oxygen, O **E.** potassium, K **F.** silicon, Si
6. Which of these elements are nonmetals?
   A. bromine, Br **B.** carbon, C **C.** boron, B
   D. thallium, Tl **E.** phosphorus, P **F.** aluminum, Al
7. Which two of these elements are the least reactive? Explain your thinking.
   A. chlorine, Cl **B.** barium, Ba **C.** copper, Cu
   D. rubidium, Rb **E.** potassium, K **F.** mercury, Hg
8. Can you make jewelry out of each of the elements listed below? Explain your thinking.
   a. copper, Cu **b.** neon, Ne
   c. sodium, Na **d.** platinum, Pt

Topic: Periodic Table
Visit: www.SciLinks.org
Web code: KEY-110

SECTION II

SUMMARY

Basic Building Materials

Alchemy Update

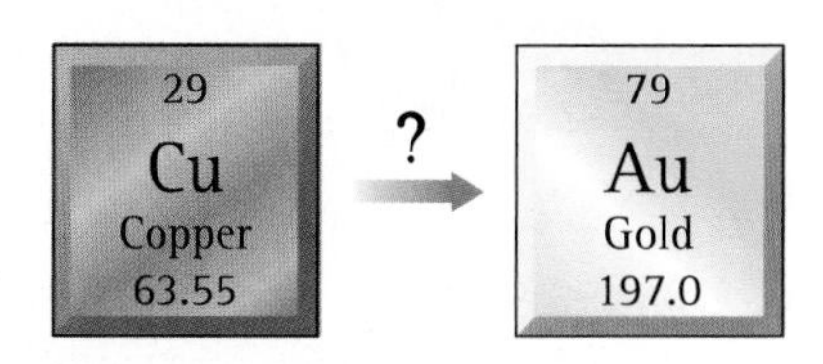

Can an element, such as copper, be transformed into gold through chemical processes? Copper and gold are in the same group on the periodic table and have similar properties. However, they are distinct elements with many differences, such as their appearances. While elements can react to form new compounds, from what you have learned so far it does not appear possible to change one element into another element.

Key Terms

element
chemical symbol
compound
chemical formula
phase
aqueous
chemical change (chemical reaction)
law of conservation of mass
reactivity
atomic mass
atomic mass units
periodic table of the elements
atomic number
group
alkali metals
alkaline earth metals
halogens
noble gases
periods
main group elements
transition elements
lanthanides
actinides
metals
nonmetals
metalloids
average atomic mass

Review Questions

1. Make a list of all the information you can extract from the periodic table for the element gold.
2. Explain the law of conservation of mass and how it relates to the Lab: The Copper Cycle.
3. What is a chemical formula, and what does it tell you?
4. A filament for a light bulb needs to conduct electricity. Which of the elements listed below might be useful as a light bulb filament? Explain your thinking.

   A. tungsten, W   **B.** sulfur, S   **C.** bromine, Br

WEB  RESEARCH

PROJECT *Element Profile*

Research an element. Write a report including

- Your element's name, symbol, and description.
- A list of your element's uses.
- A description of how your element is mined or obtained.

A World of Particles

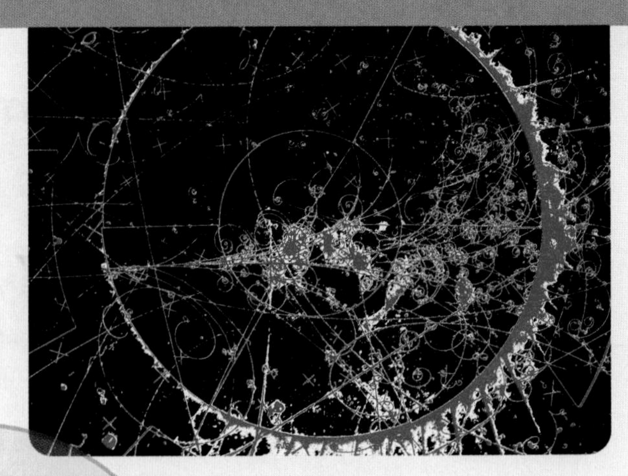

Matter is made up of atoms, and atoms are made up of even smaller, subatomic particles. This image shows the interactions of subatomic particles in a bubble chamber of liquid hydrogen.

All matter is composed of tiny particles called atoms. Based on their observations, chemists agree that atoms themselves are composed of even tinier structures, a nucleus and electrons orbiting around it. What are atoms and what does their structure tell you about matter?

In this section, you will study

- models of the atom
- how atoms differ from one another
- nuclear reactions
- how elements are created

LESSON

11 Atomic Pudding

Models of the Atom

Think About It

The drawing depicts a very tiny sample of gold taken from a gold ring.

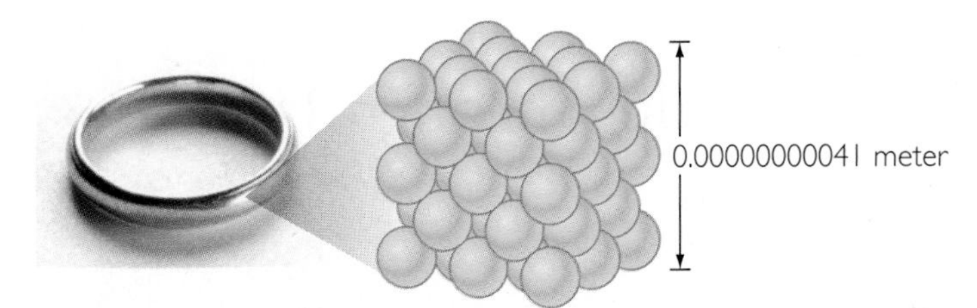

The spheres in the cube of gold are so small that they cannot be seen. What are the spheres, and what does this drawing tell you about the element gold?

How are the smallest bits of matter described?

To answer this question, you will explore

1. Atoms: Small Bits of Matter
2. Models of the Atom
3. Simple Atomic Model

Exploring the Topic

1 Atoms: Small Bits of Matter

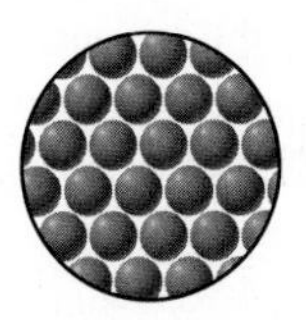

Titanium, Ti

+

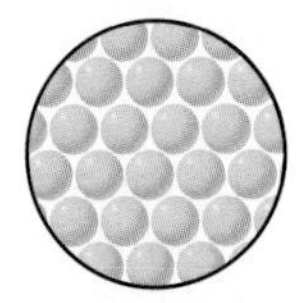

Sulfur, S

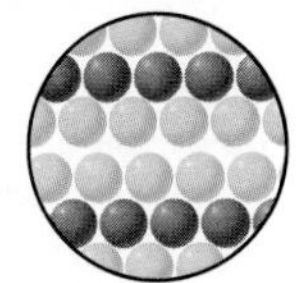

Titanium sulfide, TiS_2

Imagine you break a piece of matter in half, and then break it in half again and again. How many times can you do this? Can you keep going, getting ever smaller? Around 460 B.C.E., the Greek philosopher Democritus wondered the same thing. He thought that if he could just keep breaking matter in half he would eventually end up with the smallest bit of matter possible.

Democritus proposed that all matter was composed of tiny particles that could not be divided further. Today we use the word **atom** to describe these bits of matter. Of course, atoms are too small to actually be seen. Democritus' idea was disregarded for the next two thousand years, in part, because Democritus did not have evidence to support it.

In 1803, the British scientist John Dalton suggested that the idea of atoms could help explain why elements come together in specific ratios when they form compounds. He imagined atoms of different elements combining to form compounds in the ratios specified by the chemical formulas of the compounds. For example, to form the compound titanium sulfide, TiS_2, titanium and sulfur atoms combine in a 1:2 ratio.

Dalton had more than an idea about atoms. He conducted experiments and made observations to back up his idea. His observations provided strong evidence to support his explanation of how matter behaves.

In science, the word "theory" indicates that an explanation is supported by overwhelming evidence. The word "theory" allows room for doubt and revision, but indicates a greater degree of certainty than the word does in everyday use. The **atomic theory** states that all matter is made up of atoms. The atomic theory helps us make accurate predictions about the behavior of matter.

❷ Models of the Atom

Since Dalton's time, scientists have created many **models** to describe atoms and their parts. Models are simplified representations of something you want

The Atomic Model Through Time

EXPERIMENT 1

In 1803, John Dalton studied how elements combine chemically to form compounds. He observed that elements combine in whole-number ratios to form compounds and that matter is not created or destroyed in chemical reactions. Dalton reasoned that elements are made of tiny, indivisible spherical particles called atoms.

SOLID SPHERE MODEL

The atom is a solid sphere that cannot be divided up into smaller particles or pieces.

1803

EXPERIMENT 2

In 1897, J. J. Thomson, a British scientist, zapped atoms with electricity. He observed that negatively charged particles were removed. Thomson reasoned that atoms contain negatively charged particles, which he called electrons.

PLUM PUDDING MODEL

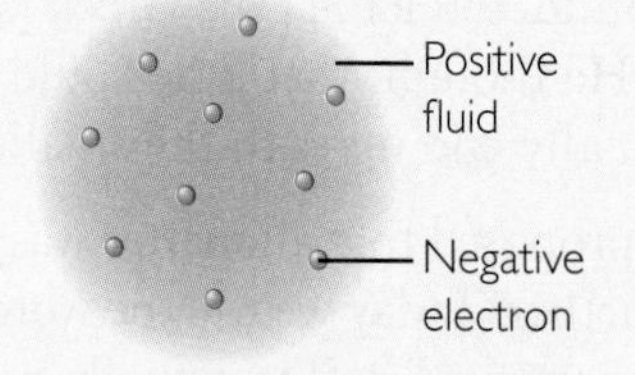

The atom can be divided into a fluid (the "pudding") and electrons. Most of the atom is made of fluid. The fluid spreads out in the atom and is positively charged. The electrons are very tiny and negatively charged.

1897

EXPERIMENT 3

In 1911, Ernest Rutherford, a New Zealand–born scientist, shot tiny positively charged particles, called alpha particles, at thin gold foil. He observed that most of the alpha particles went through the foil, but a few bounced back. Rutherford reasoned that there must be something small, massive, and positively charged in an atom, which he called the nucleus.

NUCLEAR MODEL

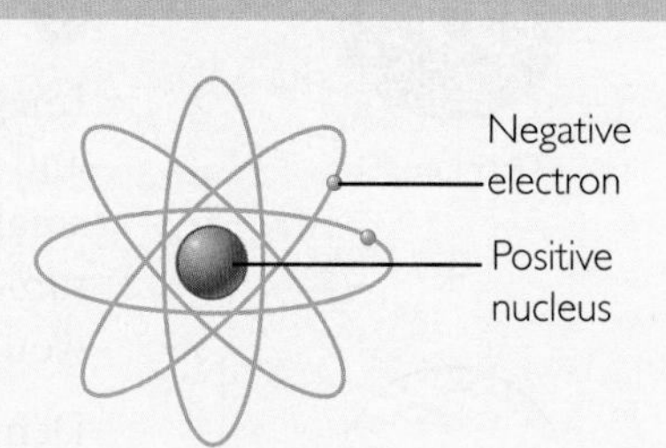

The atom can be divided into a nucleus and electrons. The nucleus occupies a small amount of space in the center of the atom. The nucleus is dense and positively charged. The electrons circle around the nucleus. The electrons are tiny and negatively charged. Most of the atom is empty space.

1911

to explain. For example, a model airplane is a small representation of a larger aircraft. Models take many forms. They can be a plan, a physical structure, a drawing, a mathematical equation or even a mental image. A model that represents the structure of an atom is called an atomic model.

Dalton pictured the atom as a hard, solid sphere. Over the next two hundred years, scientists gathered evidence to support and expand on Dalton's model of the atom. It became clear that the atom was more than just a solid sphere.

EXPERIMENT 4
In 1913, Neils Bohr, a Danish scientist, developed a model of the atom that explained the light given off when elements are exposed to flame or electric fields. He observed that only certain colors of light are given off. For example, hydrogen atoms give off red, blue-green, and blue light. Bohr reasoned that the electrons orbit around the nucleus at different distances like planets orbiting the Sun. The electrons in these orbits have different energies. When an electron falls from an outer to an inner orbit, the color of the light given off depends on the energies of the two orbits.

EXPERIMENT 5
In 1918, Rutherford made a further contribution. He found he could use alpha particles as bullets to knock off small positively charged particles, which he called protons. He reasoned that the nucleus must be a collection of protons.

EXPERIMENT 6
In 1927, Werner Heisenberg, a German scientist, proposed a cloud model of the atom. Heisenberg suggested that the location of an electron could not be specified precisely. Instead, it is only possible to talk about the probability of where an electron might be. This led to a cloud model of the atom; the electron cloud indicates where you will most likely find a single electron.

EXPERIMENT 7
In 1932, a British physicist, James Chadwick, found that the nucleus also included uncharged, or neutral, particles, which he called neutrons. He reasoned that the neutrons were important in holding the positively charged protons together.

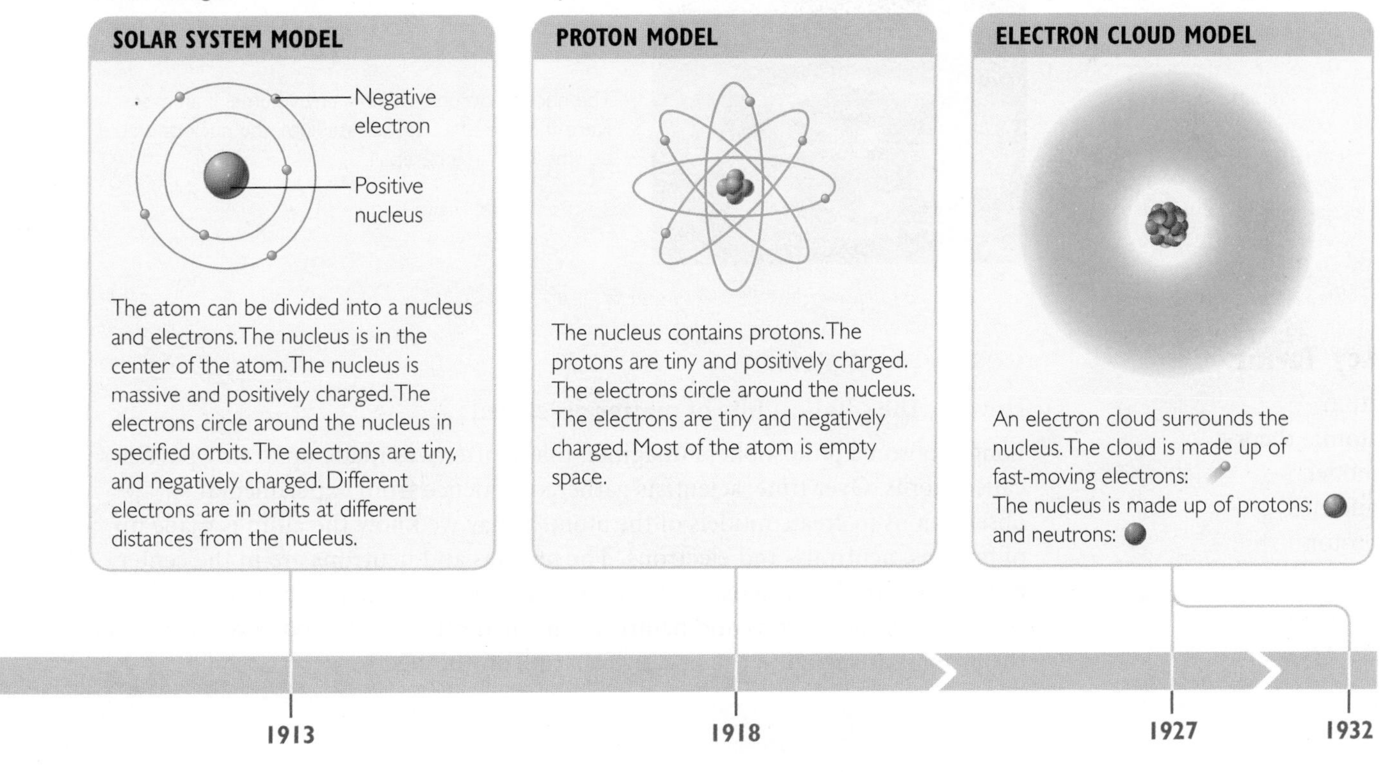

TECHNOLOGY CONNECTION

Using a scanning tunneling microscope, today it is possible to create an image of atoms. This instrument does not magnify a sample of matter like a traditional microscope. Instead, the instrument has a tiny tip that scans the surface of the sample in order to create a topographic map of the surface.

But how did scientists gather evidence about something too small to be seen? Scientists found they could learn more about atoms and their structure by shooting small pieces of matter at them or by heating them in a flame. Observations from these experiments provided evidence that helped scientists make changes and refine the model of the atom.

The model of the atom was refined and changed as new evidence was gathered. This is what science is all about—a continual process of gathering new knowledge to improve our understanding of the world.

3 Simple Atomic Model

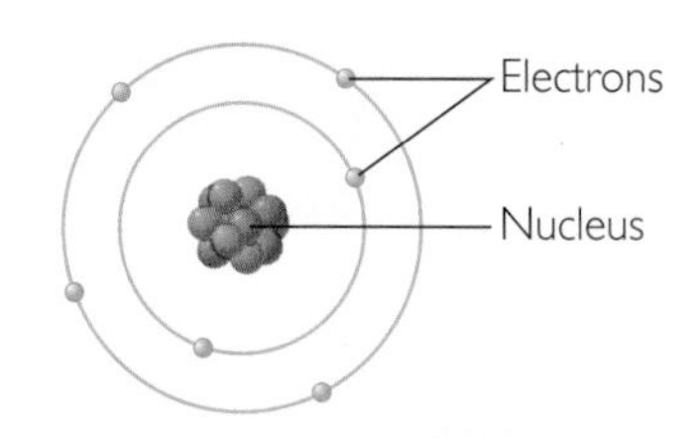

All of the models have something valuable to offer in terms of visualizing matter at an atomic level. At right is a simple atomic model of an atom. In the very center of the atom is the **nucleus.** The nucleus consists of positively charged **protons,** and **neutrons,** which have no charge. The **electrons** are even tinier than the protons and neutrons, and they orbit the nucleus. In this particular atom the electrons are located at two different distances from the nucleus.

Each electron has a charge of −1. The neutrons are neutral and thus have no charge. Each proton has a charge of +1. A neutral atom has no overall charge. It has equal numbers of positive protons and negative electrons.

BIG IDEA An atom has a nucleus made of protons and neutrons, and electrons orbiting the nucleus.

The nucleus occupies a very tiny volume. If an atom were the size of a baseball stadium, the nucleus would be smaller than a baseball.

Key Terms

atom
atomic theory
model
nucleus
proton
neutron
electron

Lesson Summary

How are the smallest bits of matter described?

Long ago, some philosophers imagined that matter was made up of tiny particles called atoms. Over time, scientists gathered evidence from experimental observations to create models of the atom. Today we know the atom is made up of protons, neutrons, and electrons. The protons and neutrons are in the center of the atom, in the nucleus. Electrons are outside the nucleus. They are much smaller than the protons and neutrons. In a neutral atom, the positive charges on the protons are equal to the negative charges on the electrons.

EXERCISES

Reading Questions

1. What evidence caused Thomson to change Dalton's solid sphere model into the plum pudding model?
2. What evidence caused Rutherford to change Thomson's plum pudding model into the nuclear model?
3. What evidence caused Bohr to change Rutherford's nuclear model into the solar system model?

Reason and Apply

4. Positive and negative charges are attracted to one another. Which of the following are attracted to a negative charge: an electron, a proton, a neutron, a nucleus, an atom? Explain your thinking.
5. Hydrogen and helium are different elements. How can you use the plum pudding model to show how atoms of the two elements might be different from one another?
6. Suppose you discovered protons shortly after Thomson discovered electrons. How would you revise the plum pudding model to include protons? Draw a picture of your revised model of the atom.
7. Draw a solar system model showing one electron, one proton, and one neutron.
8. **WEB RESEARCH** Use the Internet or other resource to find out how the size of an atom compares with the size of its nucleus. Is the diameter of an atom 10 times, 1,000 times, or 100,000 times the diameter of the nucleus?
9. The nuclear model and the solar system model both show atoms with electrons circling around the nucleus.
   a. How do these two models differ?
   b. How are these two models similar?
   c. How can you refine the solar system model so that the atoms do not look flat?
10. **WEB RESEARCH** The ancient Greeks discarded the atomic theory because there was no evidence to support it. Try to provide evidence that atoms do indeed exist. Use the Internet to help you.
11. The ancient Greeks claimed that atoms were the smallest pieces of matter. Were they correct? Explain your thinking.
12. Give an example that shows how science is a process of gathering evidence and refining models.

PHYSICS CONNECTION

When charged particles are placed near each other, they move toward or away from one another. Similar charges repel, or move away from, one another. Opposite charges attract, or move toward, one another. This photo shows iron filings that have oriented themselves around a magnet's positive and negative ends.

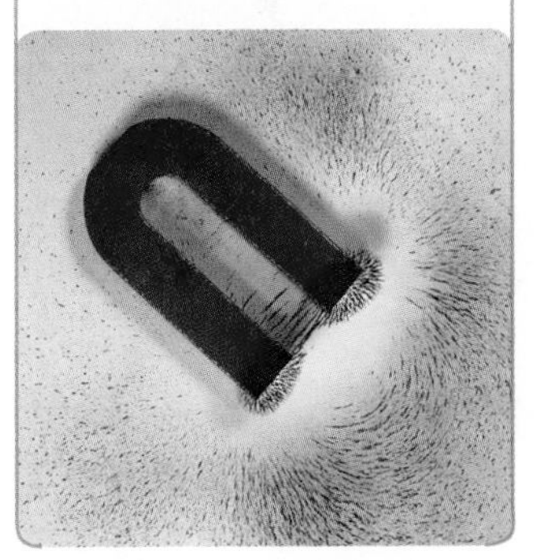

LESSON

12 Atoms By Numbers

Atomic Number and Atomic Mass

Think About It

The element copper is made up of copper atoms. Likewise, gold is made up of gold atoms. We know that copper and gold are different elements. But what makes a copper atom different from a gold atom?

How are the atoms of one element different from those of another element?

To answer this question, you will explore

1. Atomic Number
2. Atomic Mass
3. The Periodic Table and Atomic Models

Exploring the Topic

1 Atomic Number

In Lesson 9: Create a Table, you learned how Mendeleyev arranged the elements in order of increasing atomic mass. Around 1913, Henry Moseley, a British scientist, discovered an amazingly simple and important property of the elements as well. He determined that the atoms of each element differ by one proton from the atoms of the element before it on the periodic table. The first element, hydrogen, H, has one proton, the second element, helium, He, has two protons. The third element, lithium, Li, has three protons, beryllium, Be, has four, and so on.

The **atomic number** is equal to the number of protons in the nucleus of an element. The elements on the periodic table are arranged in order by their atomic numbers. So the element iron, Fe, with atomic number 26, has 26 protons in its nucleus.

BIG IDEA If you know the number of protons in an atom, you know its atomic number and what element it is.

For a neutral atom, the atomic number is also equal to the number of electrons. This is because the overall charge of a neutral atom is 0. Protons have a $+1$ charge and electrons have a -1 charge. If the overall charge on an atom is 0, then the number of protons must be equal to the number of electrons.

2 Atomic Mass

The atomic mass is the mass of a single atom. The protons and neutrons account for almost all of the mass of an atom. Electrons have a much tinier mass by comparison. The atomic mass is approximately equal to the total mass of the neutrons and protons because electrons have so little mass.

CONSUMER CONNECTION

There are about 29,400,000,000,000,000,000,000 atoms of copper in each penny, or 29,400 quintillion atoms.

Every proton in every atom, whether it is an atom of gold or an atom of oxygen, has the same mass. Scientists assign a value of one atomic mass unit, 1 amu, to the mass of a single proton. Neutrons have almost exactly the same mass as protons, so each neutron also has a mass of 1 amu. To determine the mass of a single atom, you add the number of protons and neutrons.

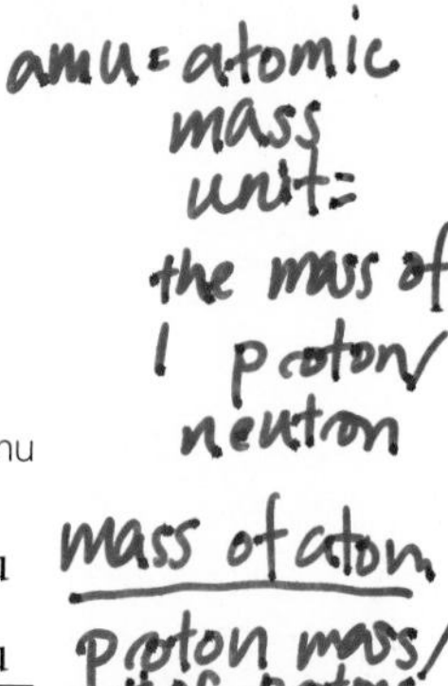

Lithium, Li

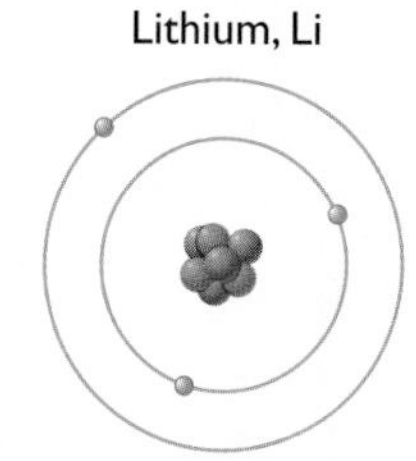

Atomic number: 3
Average atomic mass: 6.941 amu

$$\begin{array}{r} 3 \text{ protons} = 3 \text{ amu} \\ +\ 4 \text{ neutrons} = 4 \text{ amu} \\ \hline \text{atomic mass} = 7 \text{ amu} \end{array}$$

Carbon, C

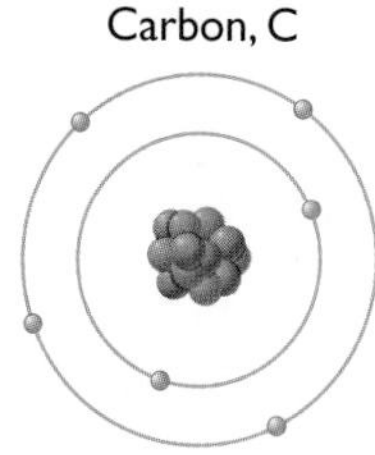

Atomic number: 6
Average atomic mass: 12.01 amu

$$\begin{array}{r} 6 \text{ protons} = 6 \text{ amu} \\ +\ 6 \text{ neutrons} = 6 \text{ amu} \\ \hline \text{atomic mass} = 12 \text{ amu} \end{array}$$

How Many Neutrons?

If you look at the atomic models, you will notice that the numbers of protons and electrons are exactly the same as the atomic number, but the number of neutrons is sometimes different from the number of protons. So, how can the periodic table tell you how many neutrons are in an atom? If you know the mass of an atom *and* you know how many protons it has, you can find out how many neutrons it has by subtracting the number of protons from the atomic mass.

Average Atomic Mass

It turns out that not every atom of an element is identical. So, the decimal number in each square of the periodic table is the average atomic mass of that element in atomic mass units, amu. This number can also be used to estimate the number of neutrons in a nucleus. Simply subtract the atomic number (number of protons) from the average atomic mass and round the result. For example, lithium has an average atomic mass of 6.941 amu, so a typical lithium atom probably has a mass of 7 amu. The atomic number of lithium is 3, so there are 3 protons. This accounts for 3 amu of the mass. The other 4 amu must be due to 4 neutrons.

In Lesson 13: Subatomic Heavyweights, you will investigate how the average atomic mass is arrived at and how atoms of an element may differ from one another.

Important to Know When considering atomic mass, it is necessary to know whether you are focusing on the mass of one particular atom or the average mass of a group of atoms. ◂

❸ The Periodic Table and Atomic Models

To draw an atomic model of a specific element, you must know the numbers of protons, neutrons, and electrons. You find this information on the periodic table.

This illustration shows you how to get information about atomic structure from each square of the periodic table to build a basic atomic model of an element.

Build an Atomic Model in Three Steps

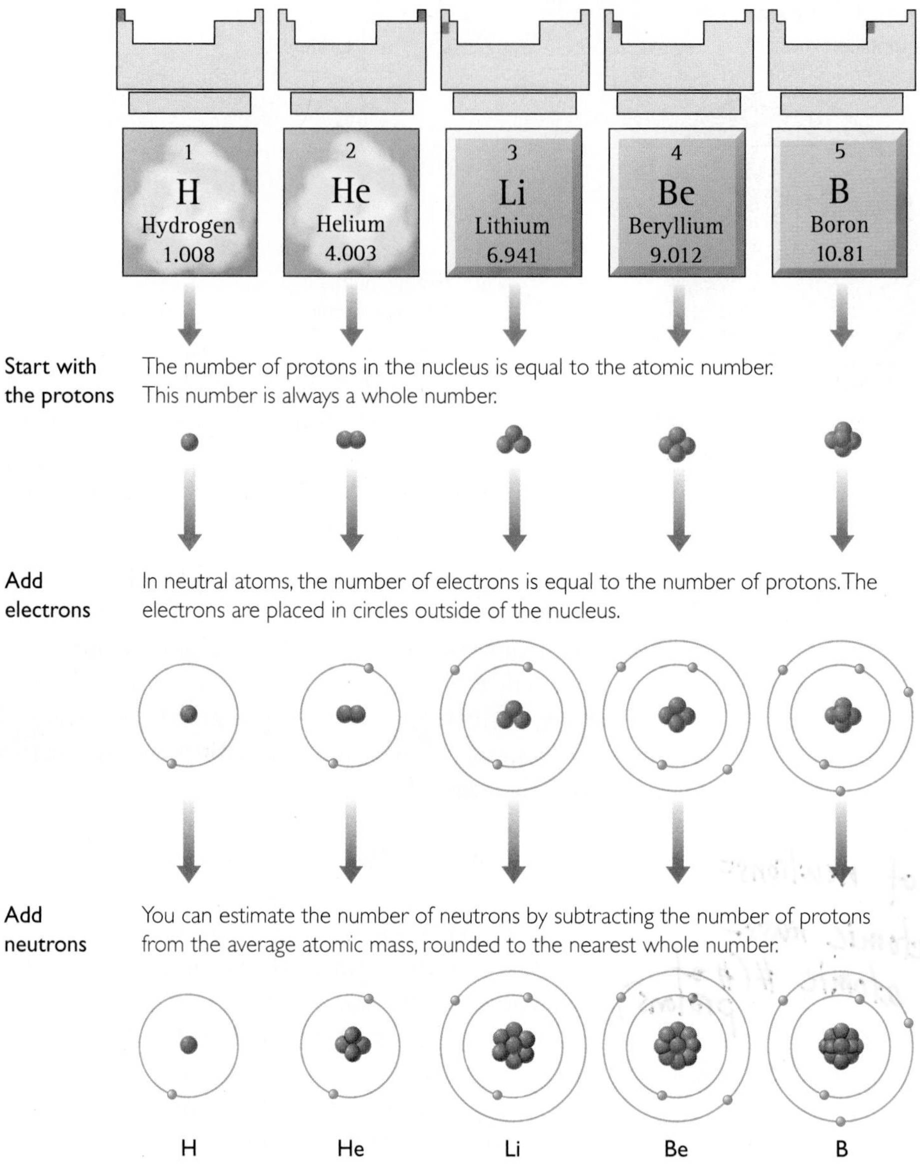

1 H Hydrogen 1.008	2 He Helium 4.003	3 Li Lithium 6.941	4 Be Beryllium 9.012	5 B Boron 10.81

Start with the protons The number of protons in the nucleus is equal to the atomic number. This number is always a whole number.

Add electrons In neutral atoms, the number of electrons is equal to the number of protons. The electrons are placed in circles outside of the nucleus.

Add neutrons You can estimate the number of neutrons by subtracting the number of protons from the average atomic mass, rounded to the nearest whole number.

H He Li Be B

HISTORY CONNECTION

Ernest Rutherford, a New Zealand–born scientist, is generally credited with the discovery of the proton. He was awarded the Nobel Prize in Chemistry in 1908, and his image appears on the New Zealand 100 dollar note.

Example

Copper and Gold Atoms

How is an atom of gold, Au, different from an atom of copper, Cu?

Solution

You can find the information you need on the periodic table. The atomic number of copper is 29. So neutral copper atoms have 29 protons and 29 electrons.

To estimate the number of neutrons in the atom, round the atomic mass to 64 and subtract the atomic number.

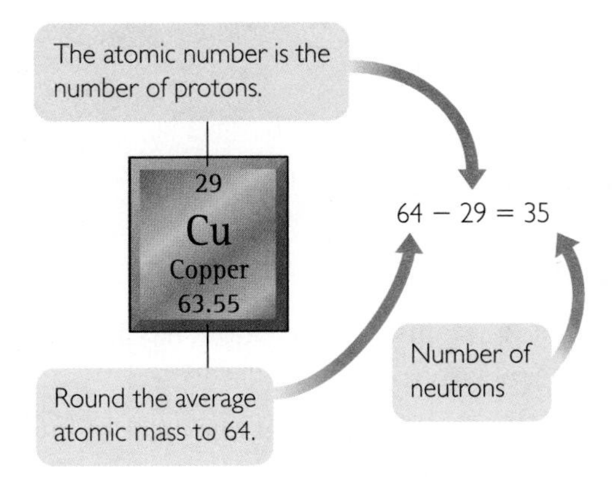

Number of protons = 29

Number of electrons = 29

Number of neutrons ≈ 35

You can follow the same steps for gold atoms.

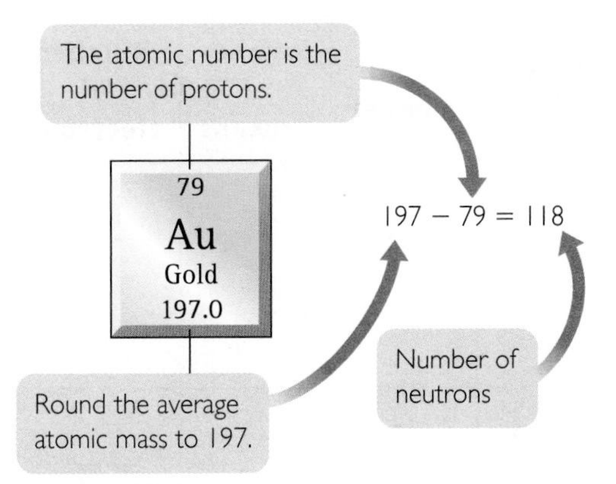

Number of protons = 79

Number of electrons = 79

Number of neutrons ≈ 118

So a gold atom has 50 more protons, 50 more electrons, and about 83 more neutrons than a copper atom.

Key Term

atomic number

Lesson Summary

How are the atoms of one element different from those of another element?

The periodic table reveals information about atomic structure. The atomic number of an element is equal to the number of protons in each of its atoms. The atomic number is also equal to the number of electrons in a neutral atom of an element. You can identify an element by the number of protons in the nucleus of an atom of the element. The protons and neutrons account for almost all the mass of an atom. Therefore, the mass of an atom in amu is approximately equal to the number of protons plus the number of neutrons. You can estimate the number of neutrons in an atom by subtracting the number of protons from the average atomic mass.

EXERCISES

Reading Questions

1. What does the atomic number tell you?
2. What does the atomic mass tell you?

Reason and Apply

3. If you have a sample of atoms and each atom has 12 protons in its nucleus, which element do you have?
4. If you want to identify an element, what one piece of information would you ask for? Explain your thinking.
5. Why does carbon, C, have a larger atomic mass than boron, B, even though they each have six neutrons?
6. Make a table like the one below. Use a periodic table to fill in the missing information.

Element	Chemical symbol	Atomic number	Number of protons	Number of electrons	Number of neutrons	Average atomic mass
nickel						
	Ne					
						24.31
		15				
				30		

7. Draw a simple atomic model for an atom of neon, Ne.
8. Place the following elements in order from lowest number of protons to highest number of protons: S, Mg, N, Na, Se, Sr. Then give the following information about a neutral atom of each: name, atomic number, number of protons, number of electrons, group number.

LESSON

13 Subatomic Heavyweights

Isotopes

Think About It

It might surprise you that all atoms of the element copper are not exact replicas of one another. Although all copper atoms have many similarities, their masses can differ by 1 or 2 amu. What is the same and what is different about these atoms?

How can atoms of the same element be different?

To answer this question, you will explore

1. Isotopes
2. Average Atomic Mass

Exploring the Topic

1 Isotopes

Variations in the Nucleus

Carbon is the sixth element on the periodic table. Its average atomic mass is listed as 12.01 amu. Why is the atomic mass listed as 12.01 rather than simply as 12?

Most of the carbon atoms that are found in nature do have a mass of 12 amu. However, for every 100 carbon atoms, it is typical to find one carbon atom with a mass of 13 amu and even more rarely a carbon atom with a mass of 14 amu. This makes the *average* atomic mass 12.01 amu. Consider the structures of these three varieties of carbon atom. A model of each is shown here.

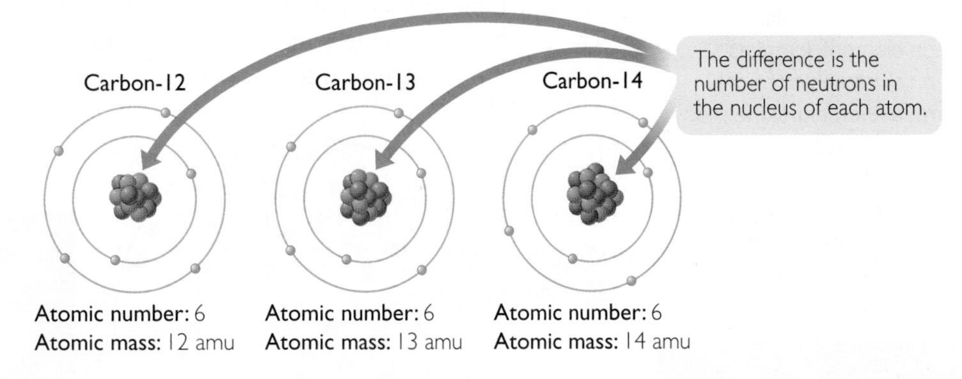

# of protons determine the element

Compare the nucleus of each atom. Each has 6 protons. That is what makes them all carbon atoms. Recall that each atom of an element has the same number of protons as every other atom of that element.

However, one carbon atom has 6 neutrons, another carbon atom has 7 neutrons, and the third has 8 neutrons. The mass of an atom is the sum of the number of protons and neutrons, so one carbon atom has a mass of 12 amu and the others have masses of 13 and 14 amu. These atoms are referred to as carbon-12, carbon-13, and carbon-14 to show that they are all atoms of carbon but have different masses.

Atoms of an element that have different numbers of neutrons are called **isotopes.** Carbon has three isotopes. In nature, almost all the elements have at least two isotopes. A few elements have as many as ten isotopes.

Isotope Symbols

Symbols for the three isotopes of carbon are shown below. Notice that the mass of each isotope is shown as a superscript number (on top). This number is a whole number. It is the sum of the numbers of protons and neutrons in the atom and is sometimes called the **mass number.** The subscript number (on the bottom) is the atomic number of the element, in this case, 6. The number of neutrons in each atom is equal to the top number minus the bottom number.

Isotope Symbols for Carbon

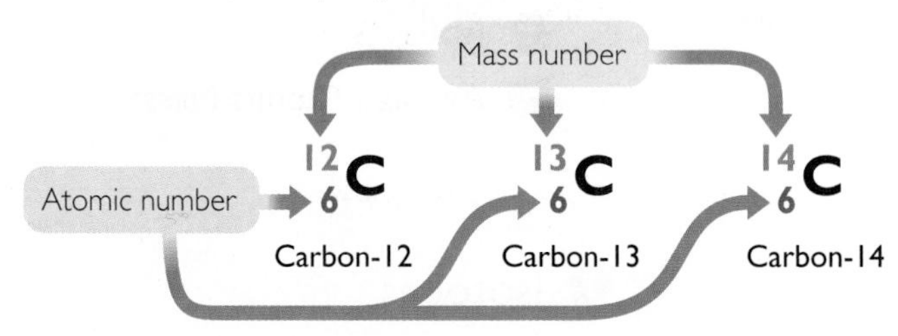

These three isotopes are virtually identical in their properties. For example, they all form the same compounds.

BIG IDEA Isotopes are atoms of the same element, but with different numbers of neutrons.

TECHNOLOGY CONNECTION

A mass spectrometer is used to determine the isotopic composition of an element. It makes an extremely accurate measurement of the mass of an individual atom, using the principle that a heavier particle will travel a straighter path through a magnetic field than a lighter particle.

2 Average Atomic Mass

You may have noticed that the **average atomic mass** values in each square of the periodic table are decimal numbers, usually to the nearest hundredth of a unit. These values are averages of the masses of the isotopes in a sample. For example, neon has an average atomic mass of 20.18 amu. How is this average calculated?

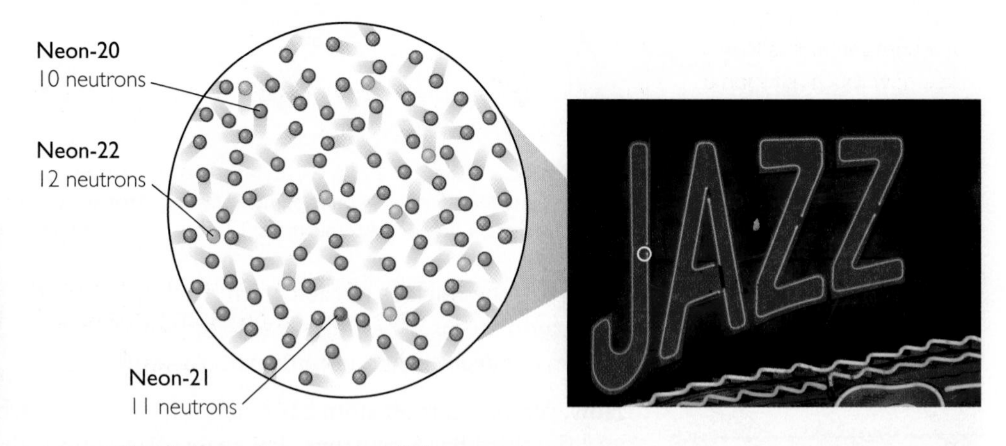

The circle contains a sample of 100 atoms of neon.

About 90% of all neon atoms have an atomic mass of 20 amu, 9% have an atomic mass of 22 amu, and 1% have an atomic mass of 21 amu. By considering a random sample of 100 neon atoms, you can calculate their average atomic mass like this:

$$\text{average atomic mass} = \frac{\text{total mass}}{\text{number of atoms}}$$

$$= \frac{(90)(20\text{ amu}) + (9)(22\text{ amu}) + (1)(21\text{ amu})}{100}$$

$$= 20.19\text{ amu}$$

As you can see, this number is nearly identical to 20.18, the average atomic mass listed for neon on the periodic table.

The percentage of each isotope of an element that occurs in nature is called the *natural abundance* of the isotope. For example, the natural abundance of neon-20 is about 90.48%.

Important to Know The mass of an isotope refers to the mass of a single specific atom of an element. The average atomic mass given on the periodic table is the average of the masses of all the isotopes in a large sample of that element. ◂

Example

Isotopes of Copper

There are two different isotopes of copper. The isotope names and symbols are given here.

$^{63}_{29}\text{Cu}$ $^{65}_{29}\text{Cu}$

copper-63 copper-65

a. Explain why both symbols have 29 as the bottom number.

b. Explain how the two isotopes are different from each other.

c. Scientists have found the natural abundances of each isotope: 69% copper-63 and 31% copper-65. Explain why the average atomic mass listed on the periodic table for copper is 63.55.

Solution

a. Copper's atomic number is 29, so both isotope symbols have a subscript 29 indicating 29 protons.

b. The two isotopes have different atomic masses. Both isotopes have 29 protons, so copper-63 has 34 neutrons and copper-65 has 36 neutrons.

c. You could consider a sample of 100 atoms of copper and calculate their average mass. The average mass of the 100 atoms is determined by adding the masses of the 100 atoms and dividing this total by 100.

$$\frac{69(63\text{ amu}) + 31(65\text{ amu})}{100} = \frac{6362\text{ amu}}{100} = 63.6\text{ amu}$$

This is very close to the value found on the periodic table.

BIOLOGY CONNECTION

Less common isotopes are more easily detectable, so researchers often use them to trace how an element moves through a living creature or the environment. For example, using a fertilizer that is enriched in nitrogen-15 allows a scientist to track the nitrogen atoms to see how much nitrogen a plant is using.

Alternatively, you can convert each isotope's percent natural abundance to a decimal number, then multiply this by the isotope's mass number. Do this for each isotope, then add the products:

$$69\% = \frac{69}{100} = 0.69 \qquad 31\% = \frac{31}{100} = 0.31$$

$$(0.69)(63 \text{ amu}) + (0.31)(65 \text{ amu}) = 63.6 \text{ amu}$$

Key Terms

isotope
mass number
average atomic mass

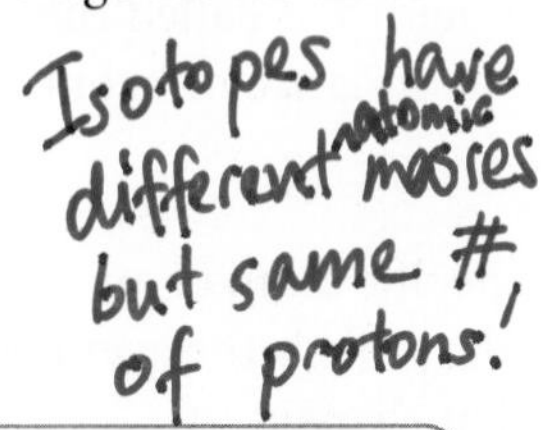

Lesson Summary

How can atoms of the same element be different?

Elements are composed of nearly identical atoms, each with the same number of protons. However, not every atom of an element has the same number of neutrons in its nucleus. Atoms of an element with different numbers of neutrons are called isotopes. Because neutrons account for part of the mass of an atom, isotopes have different masses. The average atomic mass is an average of the masses of all the different isotopes, taking natural abundance into account.

EXERCISES

Reading Questions

1. Explain the differences between atomic number and atomic mass.
2. Explain the difference between the average atomic mass given on the periodic table and the mass of an atom.

Reason and Apply

3. How are potassium-39, potassium-40, and potassium-41 different from each other? Write the isotope symbols for the three isotopes of potassium.
4. How many protons, neutrons, and electrons are in each?
   a. fluorine-23 **b.** $^{59}_{27}Co$ **c.** molybdenum-96
5. An isotope of iron, Fe, has 26 protons and 32 neutrons.
   a. What is the approximate mass of this isotope?
   b. How would you write the symbol for this isotope?
6. Find the element phosphorus, P, on the periodic table.
   a. What is the average atomic mass of phosphorus?
   b. What is its atomic number?
   c. Predict which isotope you would find in greatest abundance for phosphorus.
7. Chlorine, Cl, is 76% chlorine-35 and 24% chlorine-37. Determine the average atomic mass of chlorine.
8. Lithium, Li, is 7.6% lithium-6 and 92.4% lithium-7. Determine the average atomic mass of lithium.
9. Which isotope of nitrogen is found in nature? Explain your reasoning.
   A. $^{7}_{14}N$ **B.** $^{14}_{7}N$ **C.** $^{15}_{6}N$

Topic: Isotopes
Visit: www.SciLinks.org
Web code: KEY-113

LESSON

14 Isotopia

Stable and Radioactive Isotopes

Atomic # | Synthetic Elements
- 99 · Einsteinium (Es)
- 100 · Fermium (Fm)
- 101 · Mendelevium (Md)
- 102 · Nobelium (No)
- 103 · Lawrencium (Lr)
- 104 · Rutherfordium (Rf)
- 112 · Copernicium (Cn)
- 107 · Bohrium (Bh)

Think About It

Some isotopes are found in nature and others exist only because they were created in the laboratory or under unusual conditions. If you went digging in a copper mine and analyzed the samples, you would find copper atoms with either 34 or 36 neutrons. But do any other isotopes of copper exist?

What types of isotopes do the various elements have?

To answer this question, you will explore

1. Naturally Occurring Isotopes
2. Stable and Radioactive Isotopes

Exploring the Topic

1 Naturally Occurring Isotopes

When an atom or element can be found somewhere on Earth, it is called *naturally occurring*. Most chemists agree that there are about 92 naturally occurring elements on the periodic table. The rest of the elements on the table have existed only as a result of human activity or unusual circumstances.

The element beryllium, Be, has only one naturally occurring isotope. This means all beryllium atoms have four protons and five neutrons. The element helium, He, has two naturally occurring isotopes. Its atoms have two protons and either one or two neutrons. Helium and beryllium are shown plotted on a graph of neutrons versus protons. Notice that helium-4 lies on the diagonal line, because its neutrons and protons are equal in number.

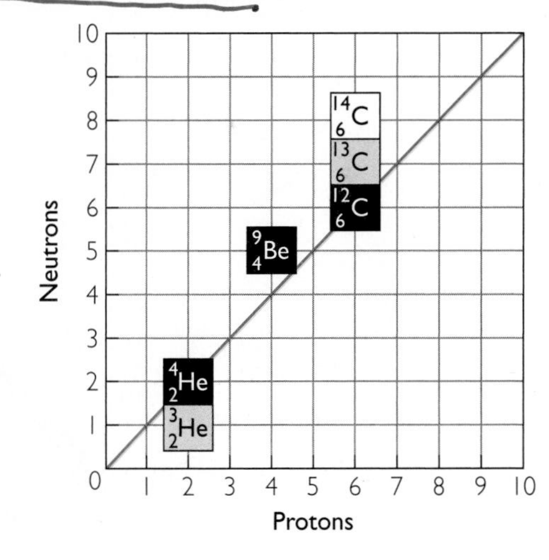

Now consider an element like carbon, C, with more than one isotope. Its isotopes, carbon-12, carbon-13, and carbon-14, are also on the graph. Notice that carbon's isotopes line up vertically. This is because they all have 6 protons. One of carbon's isotopes, with 6 protons and 6 neutrons, lies on the diagonal. The other two isotopes have more neutrons than protons in their nuclei.

Isotopes of the First 95 Elements

The graph on the next page has been expanded to include *all* of the isotopes of the first 95 elements. As before, each square plotted on the graph represents a different isotope. Take a few minutes to study the graph.

> **TECHNOLOGY CONNECTION**
>
> Technetium, Tc, and promethium, Pm, are not found in nature. They are human-made isotopes that are very unstable.

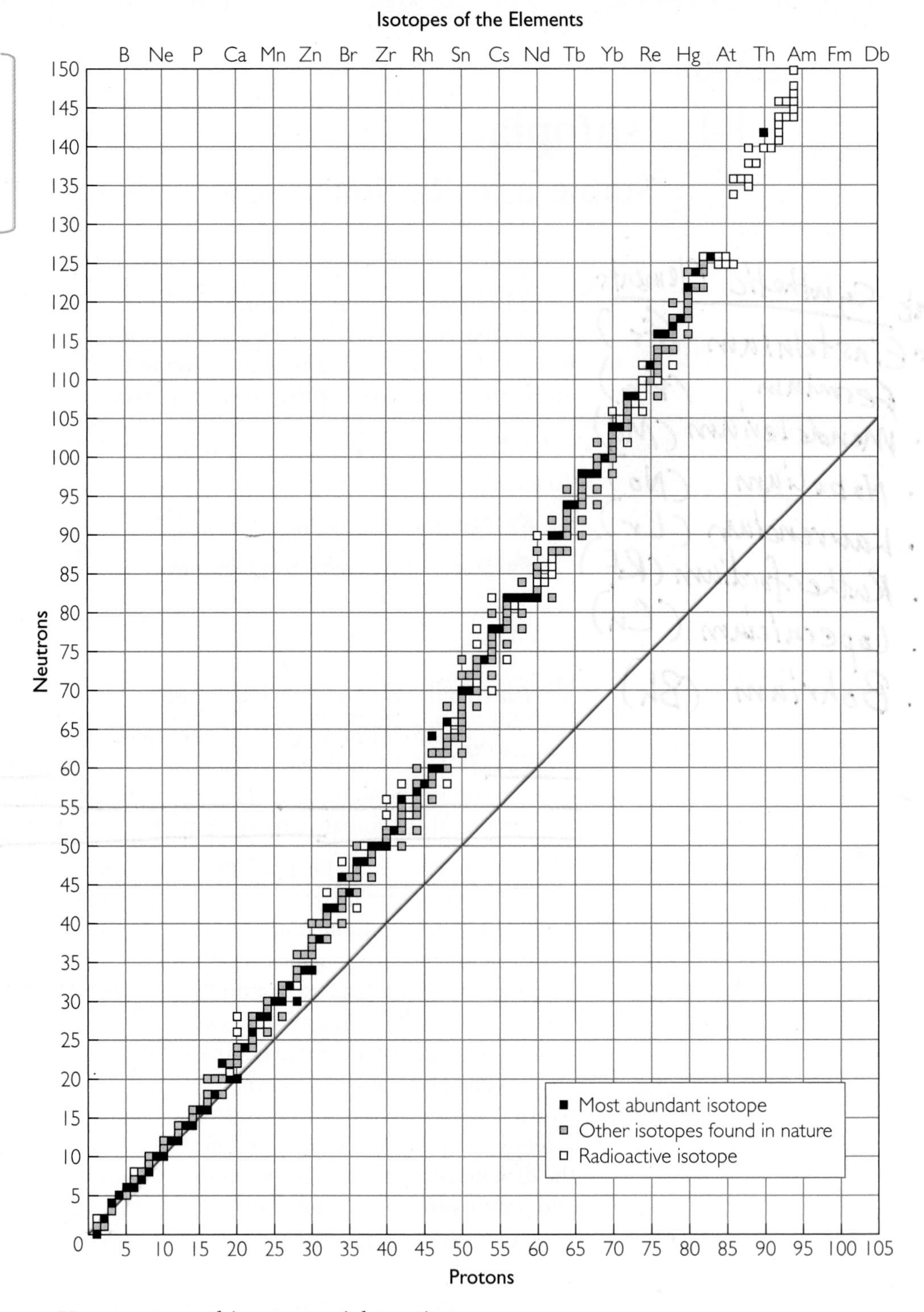

Here are some things you might notice:

- Some elements have only one isotope. However, *most* elements have more than one isotope.
- The squares that lie on the diagonal line represent isotopes with equal numbers of protons and neutrons.
- Except in hydrogen-1 and helium-3 atoms, the number of neutrons in an atom's nucleus is always equal to or greater than the number of protons. So most of the points lie on or above the diagonal line.

TECHNOLOGY CONNECTION

Rubidium-strontium dating is a method used to determine the age of geological and lunar rock samples. It is based on the fact that rubidium-87 decays over time to become strontium-87.

- Radioactive isotopes are not common. (You will learn more about radioactivity later in this lesson.)
- No element has more than 10 isotopes. Most elements have between 1 and 6 isotopes.
- The number of neutrons is roughly equal to the number of protons for atoms up to atomic number 20.
- The collection of plotted points curves up, away from the diagonal line. Beyond atomic number 20, elements have considerably more neutrons than protons.
- The majority of radioactive isotopes are elements with atomic numbers above 80.
- The elements with even numbers of protons have more isotopes than elements with odd numbers of protons. Even numbers of neutrons are also more common than odd numbers.

2 Stable and Radioactive Isotopes

When atoms are not stable, they emit small particles. This means that small bits of the nucleus come flying out of the atom. These less stable isotopes are called **radioactive isotopes.** They decay over time as particles are spontaneously emitted from the nucleus in a process called *radioactive decay.* You will learn more about radioactive decay in upcoming lessons.

As you saw in the graph, hydrogen, H, argon, Ar, and cerium, Ce, are examples of elements that have radioactive isotopes shown with white squares. There are only a few white squares scattered throughout the graph until you reach polonium, Po, element number 84. However, technetium, Tc, promethium, Pm, and every element from polonium and beyond, have *only* radioactive isotopes.

Stability Starts in the Nucleus

A stable isotope has a stable nucleus. Stability is related to the balance between the number of protons and the number of neutrons in the nucleus. An atom that doesn't have enough neutrons will disintegrate. So will an atom with too many neutrons. The larger the atom, the more neutrons it takes to make a stable nucleus. For example, isotopes of element number 74, tungsten, W, require at least 110 neutrons to be stable.

Example

Hafnium-144

Do you expect the isotope hafnium-144 to exist in nature? Explain your thinking.

Solution

The task here is to find out if there is a square on the isotope graph that corresponds to hafnium-144. The way to do this is to figure out how many protons and neutrons are in hafnium-144. Use the periodic table to find the atomic number of hafnium, Hf. The atomic number is 72, so a hafnium atom has 72 protons. An isotope of hafnium with a mass of 144 has $144 - 72 = 72$ neutrons. There is no square on the graph corresponding to 72 protons and 72 neutrons. So a hafnium atom with only 72 neutrons does not exist in nature.

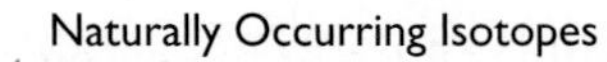

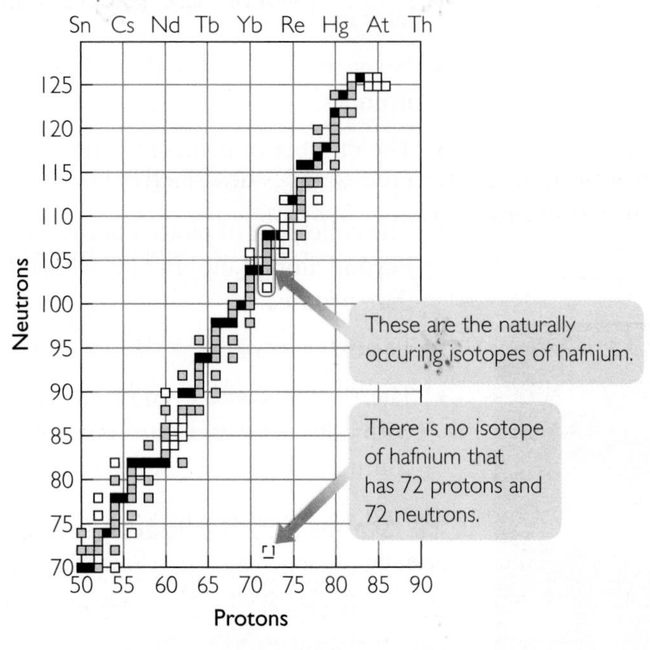

Key Term

radioactive isotope

Lesson Summary

What types of isotopes do the various elements have?

Most elements have more than one naturally occurring isotope. Most of these isotopes are stable. Stable isotopes have stable nuclei, with just the right balance of protons and neutrons. In atoms with atomic numbers up to 20, the number of neutrons is roughly equal to the number of protons. Atoms with atomic numbers between 20 and 84 require progressively more neutrons in the nucleus to attain stability. Unstable isotopes are called radioactive isotopes. The nucleus of an unstable isotope will decay and emit radioactive particles. All elements beyond atomic number 83 are unstable, and therefore, all their isotopes are radioactive.

EXERCISES

Reading Questions

1. Explain the relationship between the words *atom* and *element.*
2. Explain the relationship between the words *atom* and *isotope.*

Reason and Apply

3. Find these elements on the isotope graph on page 68. How many stable isotopes does each element have?

   oxygen, O   neodymium, Nd   copper, Cu   tin, Sn

4. Use the isotope graph to determine which of these isotopes would be found in nature.

$^{24}_{12}Mg$ $^{81}_{35}Br$ $^{152}_{60}Nd$ $^{195}_{78}Pt$ $^{238}_{92}U$

5. What does the diagonal line on the graph represent?

6. Use the isotope graph to find the isotopes described below. In each case, give the isotope name and the isotope symbol.
   a. Find three isotopes with equal numbers of neutrons and protons.
   b. Find two isotopes with the same number of protons.
   c. Find two isotopes with the same number of neutrons.
   d. Find two isotopes with the same atomic mass units.

7. Is an atom with a nucleus of 31 protons and 31 neutrons a stable isotope? Why or why not?

8. Draw basic atomic models for all the stable isotopes of oxygen, O.

9. Name five elements on the periodic table that have only radioactive isotopes. Determine the neutron-to-proton ratio for each of their isotopes, in decimal form.

10. How many protons could a stable atom with 90 neutrons have? Which elements would these be?

11. Would an atom with 60 protons and a mass number of 155 be stable?

12. Which of the following isotopes are likely to be found in nature? Identify the element by name if it is an isotope that might be found in nature.

$^{162}_{63}\underline{?}$ $^{75}_{33}\underline{?}$ $^{112}_{56}\underline{?}$ $^{260}_{88}\underline{?}$

13. The diagonal line on the isotope graph on page 68 represents nucleii with
   A. the same number of protons.
   B. the same number of neutrons.
   C. the same number of protons as neutrons.
   D. the same number of protons plus neutrons.

14. In general, elements with an even atomic number
   A. have only one isotope.
   B. have more isotopes than those with an odd atomic number.
   C. have the same number of protons as neutrons.
   D. are halogens.

ACTIVITY Nuclear Quest

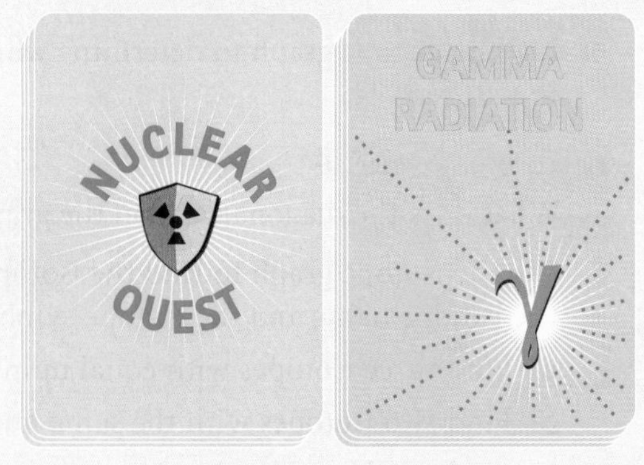

Sort cards into two piles.

Purpose

To explore nuclear chemistry.

Instructions

Play Nuclear Quest. You will need a game board, a pair of dice, Nuclear Quest cards, Gamma Radiation cards, and a game piece for each player.

Goal

Each player rolls the dice to move along the periodic table, then draws a card from the Nuclear Quest pile and does what the card says. The goal of the game is to discover element 112 and name it. This is accomplished by using nuclear chemistry to proceed through the periodic table.

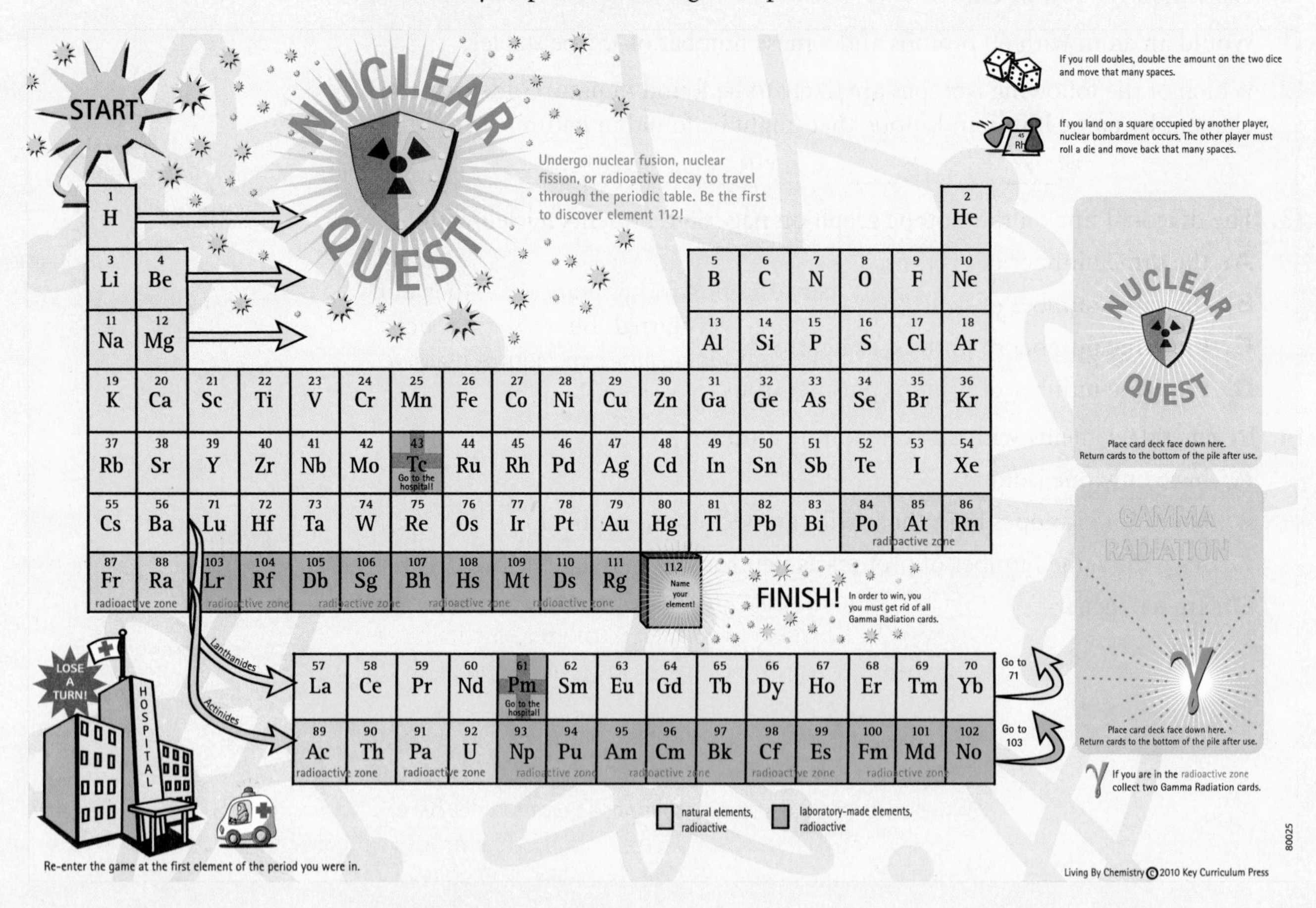

Making Sense

Write a paragraph describing what you learned about nuclear chemistry from this game.

LESSON

15 Nuclear Quest

Nuclear Reactions

Think About It

When the nucleus of an atom is changed in some way, a nuclear reaction is taking place. Nuclear reactions take place spontaneously on Earth and in the Sun. Scientists can also make nuclear reactions happen in laboratories and in nuclear power plants.

What are nuclear reactions?

To answer this question, you will explore

1. Nuclear Changes
2. Radioactive Decay
3. Fission and Fusion

Exploring the Topic

1 Nuclear Changes

Changes in the nucleus are called **nuclear reactions.** Nuclear reactions are very different from chemical reactions. Nuclear reactions involve changes to the nucleus and can change one element into another element. The nucleus of an atom is not so easy to change.

Some nuclear reactions are spontaneous—they don't require any help to get started. But changing the nucleus of an atom *on purpose* can require massive amounts of energy, so you would not be able to do this in your chemistry lab. There are several ways the nucleus of an atom can change. The nucleus can lose particles, it can be split into smaller nuclei, or it can combine with another nucleus or particle. Nuclear changes can happen only under certain circumstances. They happen spontaneously in radioactive elements, they can be made to happen with great difficulty in a nuclear reactor by scientists, and they happen all the time in the cores of stars.

In what ways can a nucleus of an atom change?

When do nuclear reactions happen?

Nuclear chemistry occurs in the fiery cores of stars.

2 Radioactive Decay

Recall that radioactive isotopes are unstable. By ejecting a particle, the nucleus of a radioactive atom can become stable. The process of ejecting or emitting pieces from the nucleus of an atom is called **radioactive decay.**

Scientists have identified the various *subatomic particles,* particles smaller than the atoms themselves, that are emitted from or shot out of the nucleus during radioactive events. These include alpha particles and beta particles.

Important to Know Isotopes can be radioactive, but it is usually incorrect to say that *elements* are radioactive. Many elements have both stable and radioactive isotopes. For example, carbon-12 and carbon-13 are stable while carbon-14 is radioactive. ◀

Alpha Decay

Alpha decay involves the ejection of an alpha particle from the nucleus of an atom. An **alpha particle** consists of two protons and two neutrons and can be represented by the Greek letter α, alpha. Because an alpha particle has two protons, it is the same as the nucleus of a helium atom.

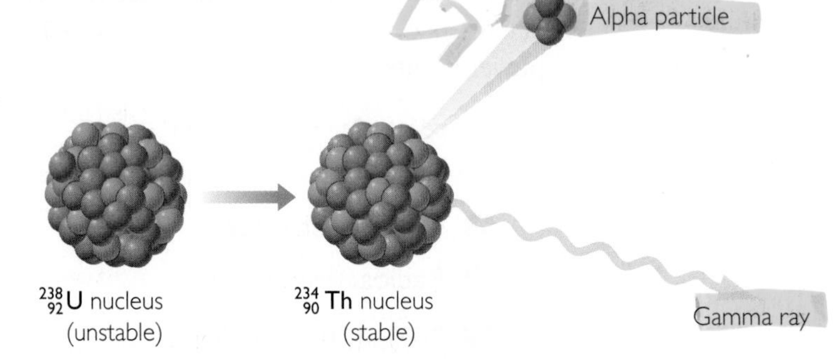

An alpha particle carries two protons away from the nucleus of an atom. This changes the identity of the element. Thus, when an atom of uranium, U, atomic number 92, undergoes alpha decay, an atom of thorium, Th, atomic number 90, is formed. This is why alpha decay causes you to move two spaces backward on the Nuclear Quest game board. Notice that a gamma ray is often released as a result of alpha decay.

Periodic Table

1 H 1.008																α	2 He 4.003
3 Li 6.941	4 Be 9.012											5 B 10.81	6 C 12.01	7 N 14.01	8 O 16.00	9 F 19.00	10 Ne 20.18
11 Na 22.99	12 Mg 24.31											13 Al 26.98	14 Si 28.09	15 P 30.97	16 S 32.07	17 Cl 35.45	18 Ar 39.95
19 K 39.10	20 Ca 40.08	21 Sc 44.96	22 Ti 47.87	23 V 50.94	24 Cr 52.00	25 Mn 54.94	26 Fe 55.85	27 Co 58.93	28 Ni 58.69	29 Cu 63.55	30 Zn 65.39	31 Ga 69.72	32 Ge 72.64	33 As 74.92	34 Se 78.96	35 Br 79.90	36 Kr 83.80
37 Rb 85.47	38 Sr 87.62	39 Y 88.91	40 Zr 91.22	41 Nb 92.91	42 Mo 95.94	43 Tc (98)	44 Ru 101.1	45 Rh 102.9	46 Pd 106.4	47 Ag 107.9	48 Cd 112.4	49 In 114.8	50 Sn 118.7	51 Sb 121.8	52 Te 127.6	53 I 126.9	54 Xe 131.3
55 Cs 132.9	56 Ba 137.3	71 Lu 175.0	72 Hf 178.5	73 Ta 180.9	74 W 183.8	75 Re 186.2	76 Os 190.2	77 Ir 192.2	78 Pt 195.1	79 Au 197.0	80 Hg 200.6	81 Tl 204.4	82 Pb 207.2	83 Bi 209.0	84 Po (209)	85 At (210)	86 Rn (222)
87 Fr (223)	88 Ra (226)	103 Lr (262)	104 Rf (267)	105 Db (268)	106 Sg (271)	107 Bh (272)	108 Hs (277)	109 Mt (276)	110 Ds (281)	111 Rg (280)							

57 La 138.9	58 Ce 140.1	59 Pr 140.9	60 Nd 144.2	61 Pm (145)	62 Sm 150.4	63 Eu 152.0	64 Gd 157.3	65 Tb 158.9	66 Dy 162.5	67 Ho 164.9	68 Er 167.3	69 Tm 168.9	70 Yb 173.0
89 Ac (227)	90 Th 232.0	91 Pa 231.0	92 U 238.0	93 Np (237)	94 Pu (244)	95 Am (243)	96 Cm (247)	97 Bk (247)	98 Cf (251)	99 Es (252)	100 Fm (257)	101 Md (258)	102 No (259)

HISTORY CONNECTION

Marie Curie was the first woman to receive the Nobel Prize, for her groundbreaking work on radioactive substances. She actually won two Nobel Prizes—one in Physics (1903) and one in Chemistry (1911). Like other pioneers, she was unaware of the dangers of radioactive samples and sometimes carried them around in her pockets, which probably cut short her life.

BIG IDEA The elemental identity of an atom is determined by the number of protons it has.

CAREER CONNECTION

X-rays are similar to gamma rays except that they are not generated by radioactive samples. Using quick exposures, doctors and dentists can image your bones. However, long-term exposure can lead to illness and even death, so when patients are having x-rays taken, they wear a lead shield to protect their organs and health professionals leave the room.

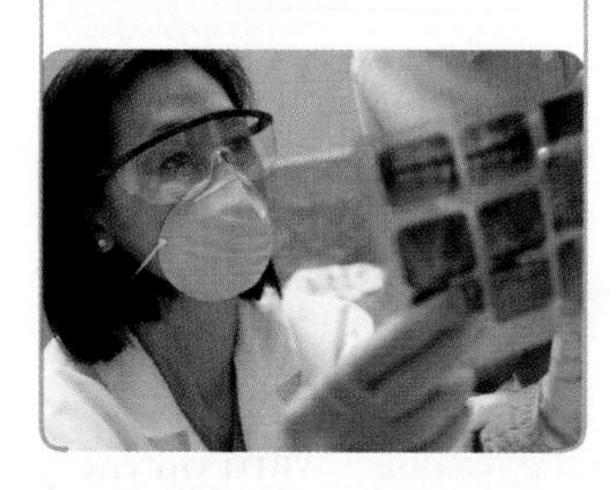

Beta Decay

Beta decay involves the ejection of an electron from the nucleus of an atom. The ejected electron is called a **beta particle.** Beta particles can be represented by the Greek letter β, beta. You may wonder how an electron got into the nucleus of an atom. In fact, electrons *do not* exist by themselves within the nucleus. However, in beta decay a neutron can split into two parts, becoming an electron and a proton. The electron is ejected and the proton stays behind in the nucleus.

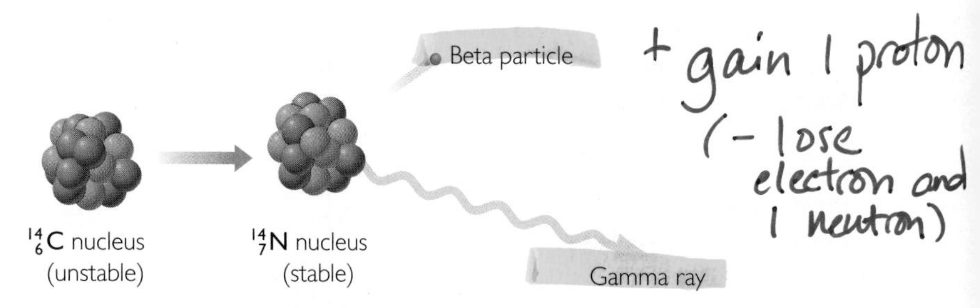

A neutron splits into a proton and an electron.

Removal of an electron from the nucleus of an atom through this unique process leaves that atom with one less neutron, one more proton, and a new identity. This is why beta decay causes you to move one space forward in the Nuclear Quest game. Beta decay can also result in the release of gamma rays.

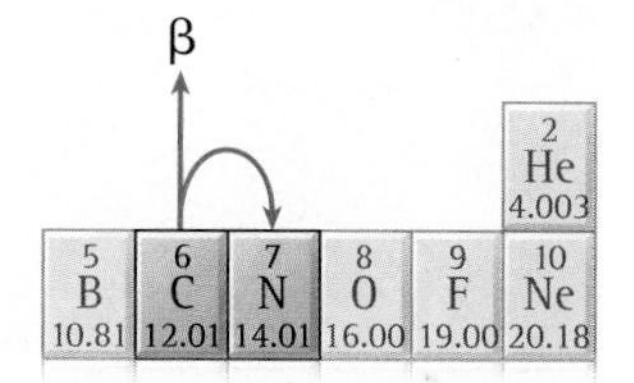

Example

Plutonium-241

The radioactive isotope plutonium-241, $^{241}_{94}Pu$, decays by emitting a beta particle. What isotope is formed?

Solution

The atomic number of plutonium is 94, so it has 94 protons. In beta decay, a neutron becomes a proton. This increases the number of protons in the nucleus by one, but decreases the number of neutrons by one. The atomic mass stays the same because protons and neutrons have the same mass. The atomic number of the new atom is 95, and the new mass is still 241 amu. Element 95 is americium, Am.

$$^{241}_{94}Pu \rightarrow {}^{241}_{95}Am$$

The new isotope is americium-241.

Half-Life

Carbon-14 has a **half-life** of 5730 years. This means that it will take 5730 years for half of a carbon-14 sample to decay to nitrogen-14. It will take another 5730 years for half of the remaining carbon-14 to decay, and so on. By measuring the amount of carbon-14 remaining compared to the amount of stable carbon, the age of the sample can be determined.

CAREER CONNECTION

Carbon-14 is a rare radioactive isotope of carbon that builds up to a steady level in living things as long as they are alive. When they die, the carbon-14 begins to decay at a predictable rate. Scientists have learned how to determine the age of a carbon fossil or artifact by measuring the amount of carbon-14 left in a sample. This process is called *carbon dating*.

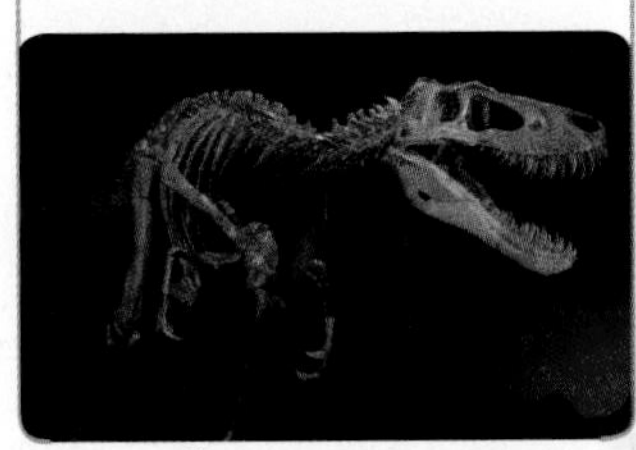

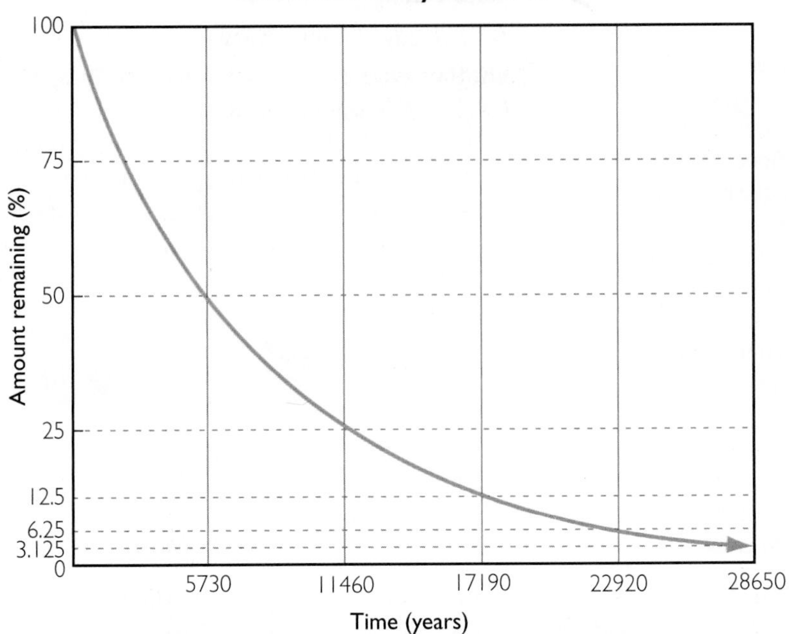

Radiation

The particles and rays that are emitted during nuclear reactions are forms of **radiation.** So alpha and beta particles are both forms of radiation. Gamma rays are often emitted during radioactive events. **Gamma rays** are a kind of radiation similar to light, microwaves, and x-rays except they are much higher in energy, so they can be very dangerous. Gamma rays are represented by the Greek letter γ, gamma. When gamma rays are emitted, the identity of the emitting atom does not change.

Naturally occurring radiation does exist, with low levels coming from sources in the Earth, as well as from the Sun and beyond. Humans have learned how to use certain types of radiation for a variety of scientific purposes. However, radiation from nuclear reactions can be harmful to your health.

HEALTH CONNECTION

Although radiation can be dangerous, it can also be used to help people. For example, some cancer patients receive radiation treatments to kill cancerous cells so that it kills them but doesn't damage the rest of the body. Radiation can also be used to kill bacteria and viruses on food, particularly meat and vegetables. This helps reduce the risk of food poisoning.

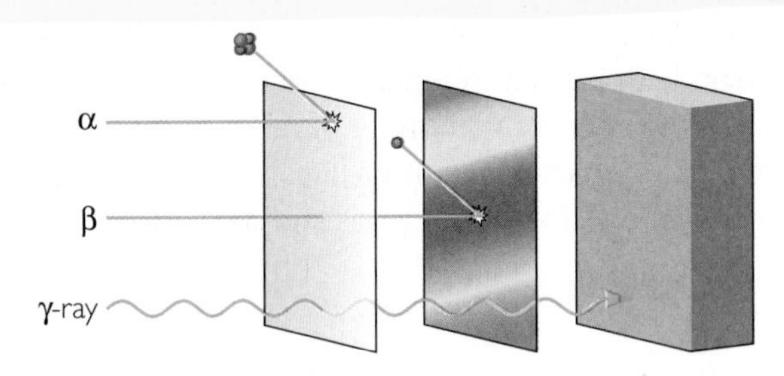

A sheet of paper, or human skin, stops alpha particles, and a sheet of aluminum foil stops beta particles. Gamma rays can penetrate several inches of lead before stopping.

❸ Fission and Fusion

The splitting apart of a nucleus is called **fission.** The result is two atoms with smaller nuclei. Fission can occur spontaneously when the nucleus of an atom is unstable. Scientists can also make

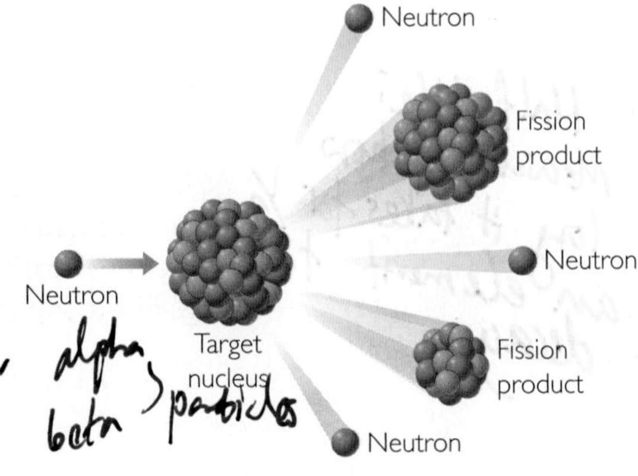

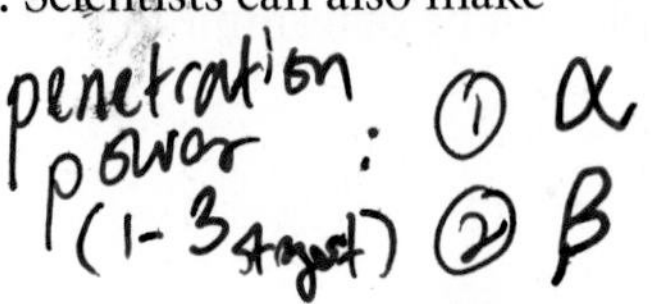

nuclear fission happen by shooting neutrons at a nucleus. Fission reactions are accompanied by radiation.

Another type of nuclear reaction is **fusion.** During fusion, nuclei join to form a larger nucleus, resulting in an atom of a different element. Temperatures of around 100,000,000 °C are required before nuclei can successfully fuse together. So the natural conditions on Earth do not support fusion.

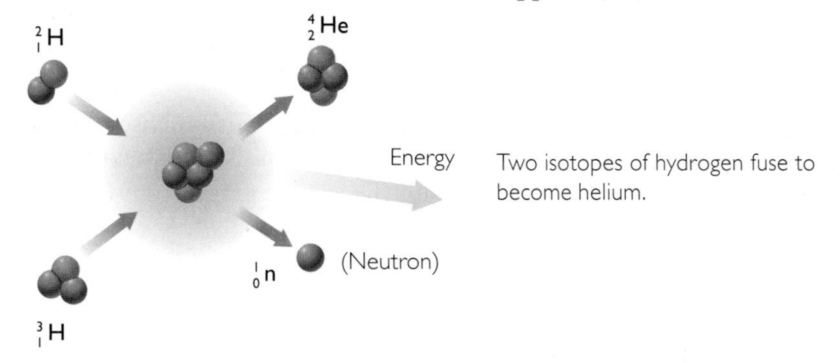

Two isotopes of hydrogen fuse to become helium.

Key Terms

nuclear reaction
radioactive decay
alpha decay
alpha particle
beta decay
beta particle
radiation
gamma ray
fission
fusion
half-life

Lesson Summary

What are nuclear reactions?

Radioactive decay, fission, and fusion are all forms of nuclear reactions. When radioactive decay is taking place, particles are emitted from the nucleus. Fission is the process by which the nucleus breaks into smaller nuclei. Fusion is the process by which nuclei combine to make a larger nucleus. Nuclear reactions can involve changes in atomic number, which mean changes in atomic identity. Radiation, such as the emission of alpha particles, beta particles, and gamma rays, usually accompanies decay and fission. All forms of radiation can cause harm to humans, but some forms are more damaging than others.

EXERCISES

Reading Questions

1. What is a nuclear reaction?
2. What is the difference between alpha decay and beta decay?
3. What type of radiation is most harmful to living things: alpha, beta, or gamma radiation? Why?

Reason and Apply

4. Explain why the mass of an atom changes when an alpha particle is emitted.
5. Explain why the mass number of an atom does not change when a beta particle is emitted.
6. An alpha particle is not a neutral atom. It has a charge of +2. Why is this the case?

7. Suppose each of these isotopes emits a beta particle. Give the isotope name and symbol for the isotope that is produced. Place a checkmark next to the symbol for isotopes that are stable. (Consult the isotope graph from Lesson 14: Isotopia.)
   a. potassium-42
   b. iodine-131
   c. iron-52
   d. sodium-24
8. Suppose each of these isotopes emits an alpha particle. Give the isotope name and symbol for the isotope that is produced. Place a checkmark next to the isotopes produced that are stable. (Consult the isotope graph from Lesson 14: Isotopia.)
   a. platinum-175
   b. gadolinium-149
   c. americium-241
   d. thorium-232
9. The following items were found at an ancient campsite: an axe with a stone blade and a wooden handle, a clay pot, ashes of a campfire, a jawbone of a deer, and an arrowhead. Which objects could be used to determine the age of the site? Explain.
10. Scientists determine that the wooden beams of a sunken ship have 67% the concentration of carbon-14 that is found in the leaves of a tree alive today. Find the age of the ship and explain how you determined it. List three possible sources of error in your determination.
11. What fraction of the original ${}^{14}_{6}C$ would be in a wooden ax handle that was 17,190 years old?

    A. $\frac{1}{4}$ B. $\frac{3}{4}$ C. $\frac{1}{8}$ D. $\frac{1}{2}$

12. Suppose ${}^{197}_{119}Pt$ undergoes beta decay. The daughter isotope then also undergoes beta decay. What is the final product?

    A. ${}^{195}_{120}Ir$ B. ${}^{197}_{118}Au$ C. ${}^{195}_{119}Hg$ D. ${}^{197}_{121}Tl$

LESSON

16 Old Gold

Formation of Elements

Think About It

The element gold is essentially a collection of identical atoms. Gold has only one stable isotope. Thus, every gold atom has 79 protons, 79 electrons, and 118 neutrons. Perhaps it is possible to make gold atoms by using nuclear processes to add or subtract protons, neutrons, and electrons. In order to understand if this is possible, it is useful to consider what processes lead to the creation of new elements.

How are new elements formed?

To answer this question, you will explore

1. Making New Elements
2. Writing Nuclear Equations
3. Nuclear Chain Reactions

Exploring the Topic

1 Making New Elements

Some nuclear reactions occur spontaneously on Earth when radioactive isotopes decay. It is much more difficult to change the nucleus of a stable atom. However, there are places with the right conditions and enough available energy to change these nuclei. Nuclear changes take place continuously in the Sun and other stars. And on Earth, scientists can carry out certain nuclear reactions in specially designed facilities like nuclear reactors or particle accelerators.

New elements can be created through nuclear fission or nuclear fusion. Whether a nucleus is split apart or joined with another one, both processes result in the formation of different elements. And both processes require an energetic push to get them started.

> **BIG IDEA** The only way to change one element into another is to change the number of protons in the nucleus.

Making Elements in the Stars

The creation of new elements through nuclear chemistry is called *nucleosynthesis.* Most nucleosynthesis takes place far from Earth, inside stars.

The Sun is a giant ball of hydrogen, H, and helium, He. The amazing light and heat energy that radiates from the Sun is the result of continuous fusion reactions. However, the Sun is not hot enough to produce elements beyond helium on the periodic table. Higher temperatures are needed for the formation of larger atoms.

Stars do not exist forever. Some stars collapse in on themselves after millions of years. Other stars explode and become what astronomers call *supernovas.*

The Owl Chemists

Where does gold comes from? Small stars like our own can produce only helium.

Larger, hotter stars can produce heavier elements up to iron.

Only a supernova explosion can produce elements heavier than iron, including gold.

ASTRONOMY CONNECTION

The Sun turns 600 million tons of hydrogen into helium every second! The energy released as a result of this nuclear fusion provides the brilliant sunlight we see every day, and the intense heat that warms our planet, even 93 million miles away. There is enough hydrogen in the Sun to keep the Sun burning for another 5 billion years.

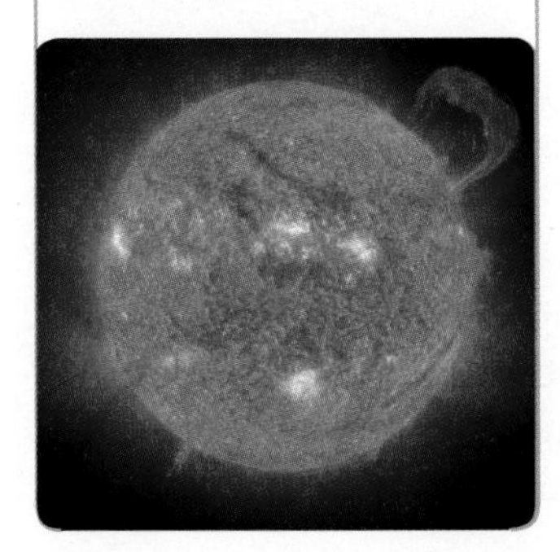

2 Writing Nuclear Equations

Nuclear reactions can be expressed as **nuclear equations.** The number of neutrons and protons in the atoms can be tracked, along with the identity of the products. Nuclear equations use isotope symbols to represent each particle. The alpha decay of an isotope of barium-140 to form xenon-136 can be written as a nuclear equation using α for the alpha particle:

$$^{140}_{56}\text{Ba} \rightarrow \alpha + ^{136}_{54}\text{Xe}$$

An alpha particle is the same as a helium nucleus, so it can also be written this way:

$$^{140}_{56}\text{Ba} \rightarrow ^{4}_{2}\text{He} + ^{136}_{54}\text{Xe}$$

In the second equation, you can actually track the protons and the mass that is removed from the barium nucleus. Notice that the equation is balanced numerically. The mass numbers on both sides are equal: $140 = 4 + 136$. The numbers of protons are also equal on both sides: $56 = 2 + 54$. In radioactive decay, the starting isotope is called the **parent isotope** and the resulting isotope is called the **daughter isotope.** In this equation, $^{140}_{56}\text{Ba}$ is the parent isotope and $^{136}_{54}\text{Xe}$ is the daughter isotope.

Beta decay can be shown in two ways as well, because a beta particle is the same as an electron, and is sometimes shown as e^-, or in isotope form as $^{0}_{-1}\text{e}$.

$$^{140}_{56}\text{Ba} \rightarrow \beta + ^{140}_{57}\text{La} \quad \text{or} \quad ^{140}_{56}\text{Ba} \rightarrow ^{0}_{-1}\text{e} + ^{140}_{57}\text{La}$$

Recall that the beta particle, β, comes from a neutron in the nucleus splitting into a proton and an electron. The new element $^{140}_{57}\text{La}$ has a higher atomic number because it has one more proton in its nucleus than barium.

Fusion equations have two starting isotopes coming together to form a new product. When a carbon-12 isotope fuses with a nitrogen-14 isotope to form an isotope of aluminum, the equation looks like this:

$$^{12}_{6}\text{C} + ^{14}_{7}\text{N} \rightarrow ^{26}_{13}\text{Al}$$

Notice once again that the equation is balanced numerically on both sides.

Example

Radium-222

Write the equation representing the alpha decay of radium-222. What is the daughter isotope?

Solution

Find radium, Ra, on the periodic table. Its atomic number is 88. Now you can write the isotope symbol for radium, $^{222}_{88}\text{Ra}$.

An alpha particle is the same as a helium nucleus. Because it is emitted, it goes on the right side of the equation. Balancing the equation gives the resulting nucleus, which has atomic number 86, which is radon, Rn.

$$^{222}_{88}\text{Ra} \rightarrow ^{4}_{2}\text{He} + ^{218}_{86}\text{Rn}$$

The daughter isotope is radon-218.

CONSUMER CONNECTION

Nuclear power became unpopular in the United States after the accident at the nuclear generating station on Three Mile Island, Pennsylvania, in 1979. In 1986, one of the nuclear reactors exploded at the Chernobyl Nuclear Power Plant in Ukraine. The radioactive fallout was 30 times that released by an atomic bomb, and a 30-kilometer zone around Chernobyl has been declared contaminated and off-limits. No new nuclear power plants have been built in the United States since the mid-1970s. However, France depends almost exclusively on nuclear power. As of 2006, there were approximately 445 nuclear power plants worldwide in 30 countries.

3 Nuclear Chain Reactions

Nuclear fission can be provoked by striking nuclei with fast-moving particles, such as neutrons. And sometimes one nuclear reaction can lead to others. A nuclear **chain reaction** can take place when the neutrons emitted strike surrounding nuclei, causing them to split apart as well.

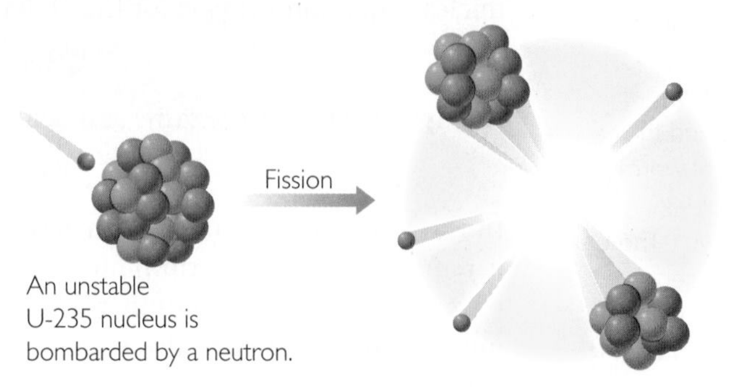

For example, natural uranium consists of three isotopes: uranium-238, uranium-235, and uranium-234. All of these uranium isotopes are mildly radioactive but are still fairly stable. Uranium isotopes can be provoked into undergoing fission reactions. Scientists do this by firing a neutron at a uranium atom. This results in two smaller isotopes being formed and three other neutrons being emitted, which in turn cause other uranium isotopes to undergo fission. The result is a chain reaction releasing enormous amounts of energy.

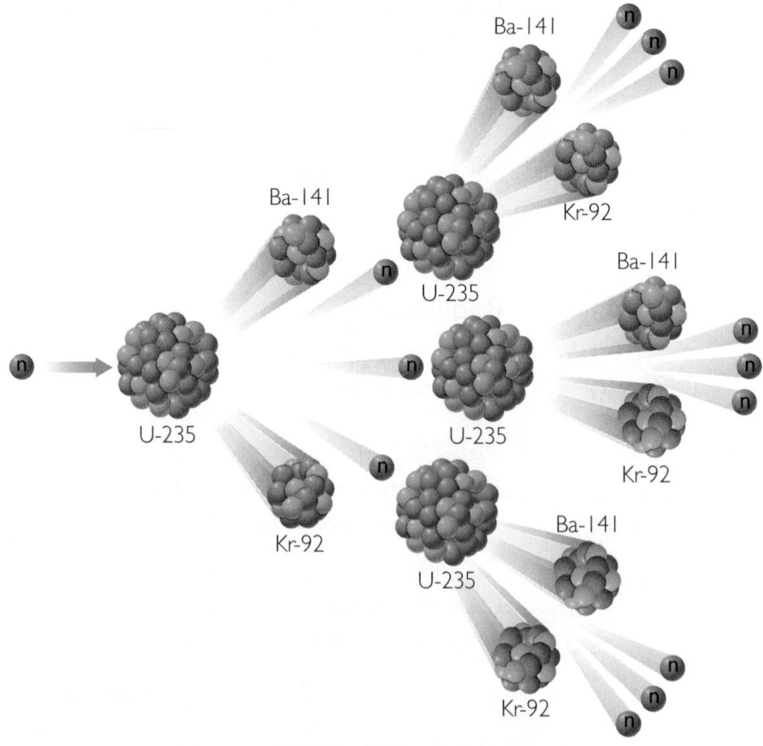

Nuclear chain reaction

Scientists have learned how to control and harness nuclear fission here on Earth. Nuclear fission is used to produce electricity, power submarines, and propel a

Topic: Nuclear Reactions
Visit: www.SciLinks.org
Web code: KEY-116

variety of large ships, such as icebreakers and aircraft carriers. Uranium-235 is the isotope most commonly used for nuclear power.

If a nuclear chain reaction is controlled and the energy is released slowly, it can be used to generate electricity. If the energy is released all at once, the result is a nuclear explosion.

Making New Gold

Most of the elements found on Earth are the products of fusion reactions in stars that exploded billions of years ago. The elements have been here for a very, very long time. There has been little change in the amounts of these elements. Small changes to the amounts of each element occur over time due to radioactive decay and fission. The gold we have on Earth came from supernova explosions that may have taken place billions of years ago. The prospect of making large quantities of new gold through nuclear chemistry appears slim.

Key Terms

nuclear equation
parent isotope
daughter isotope
chain reaction

Lesson Summary

How are new elements formed?

Elements are formed through nuclear reactions. Radioactive decay, nuclear fission, and nuclear fusion are all possible sources of new elements. The elements we have on Earth came from fusion reactions in stars and supernova explosions long ago. It is difficult to cause nuclear fusion to occur on Earth because large amounts of energy are required to fuse nuclei. Nuclear fission, on the other hand, has been harnessed as a source of power. The power that is generated by nuclear plants is a result of a controlled fission chain reaction.

EXERCISES

Reading Questions

1. Describe four processes that result in new elements being formed.
2. What is a nuclear chain reaction?

Reason and Apply

3. Write the nuclear equation for the beta decay of cerium-141.
4. Write the nuclear equation for the alpha decay of platinum-191.
5. Write a nuclear equation for the formation of iron-54 through fusion.
6. Consider the fission of uranium-235.
   a. The fission of uranium-235 begins with the addition of a neutron to the nucleus. What isotope is formed? Give the isotope symbol.
   b. After the neutron is added, the uranium atom is more unstable and undergoes fission. A possible set of products is krypton-94, barium-139, and 3 neutrons. How many protons are in these products?
   c. Were any protons lost?
7. **WEB RESEARCH** Write a paragraph explaining how a nuclear reactor works. Be sure to explain the purpose of the control rods.

SECTION

SUMMARY
A World of Particles

Key Terms

atom
atomic theory
model
half-life
electron
nucleus
proton
neutron
atomic number
isotope
mass number
average atomic mass
radioactive isotope
nuclear reaction
radioactive decay
alpha decay
alpha particle
beta decay
beta particle
radiation
gamma ray
fission
fusion
parent isotope
daughter isotope
chain reaction

Alchemy Update

Can a copper atom be transformed into a gold atom through nuclear processes?

Elements are collections of similar atoms. The number of protons in the nucleus determines the identity of an atom, for instance, all atoms of gold have 79 protons. However, atoms of an element may have different numbers of neutrons. The number of neutrons is related to the stability of the atom. Stable gold atoms have 118 neutrons.

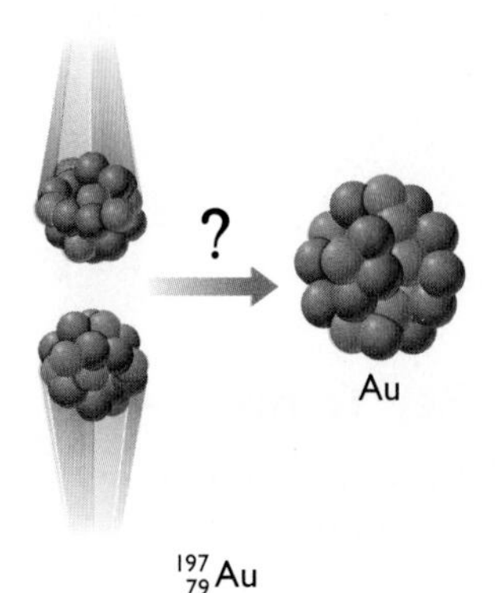

Nuclear fission, nuclear fusion, and radioactive decay can result in new elements. But nuclear fusion and nuclear fission are difficult processes to control and often involve large amounts of energy. Nuclear reactions do take place in the stars where atoms are created. So far it is not yet possible or practical to create gold atoms through nuclear reactions here on Earth.

Review Questions

1. Describe the different processes that might result in a change in atomic identity.
2. Describe how elements are formed in nature.
3. Write the nuclear equation for the beta decay of cadmium-113.
4. Write a nuclear equation for the hypothetical fusion of a copper atom with another nucleus to make gold. Can this happen? Why or why not?

WEB  RESEARCH

PROJECT *Nuclear Power*

Research nuclear power. Find out how a nuclear power plant works. Write a report including

- an explanation (with a simple drawing) of how electric power is produced from a nuclear plant.
- the major benefits and risks of nuclear power.

SECTION

IV Moving Electrons

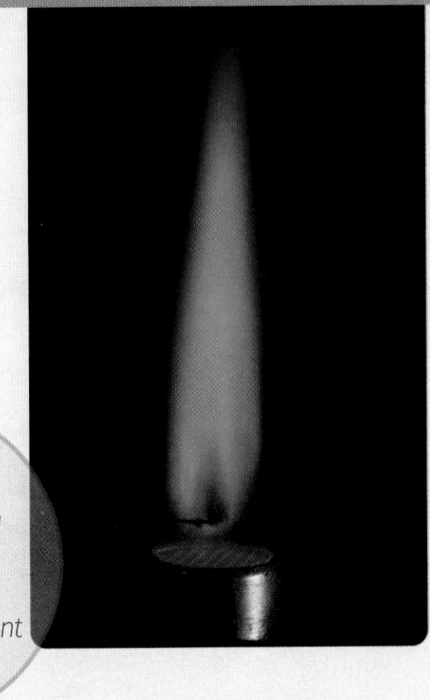

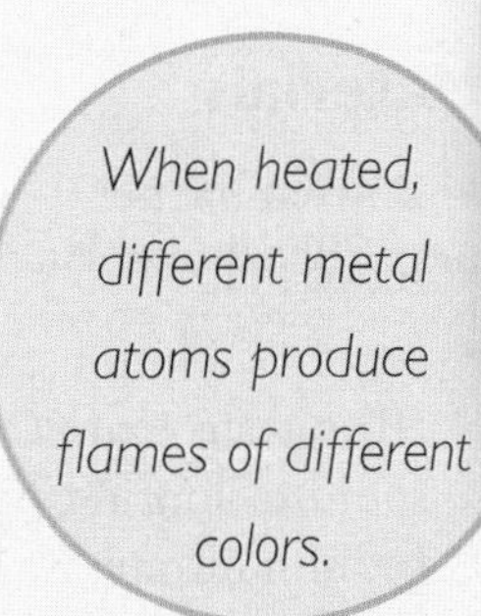

When heated, different metal atoms produce flames of different colors.

While the nucleus of an atom can be difficult to change, the electrons are a different story. Electrons can be moved around within an atom or transferred between atoms. When atoms transfer electrons, they become ions, with positive and negative charges, and form ionic compounds. Knowing the arrangement of electrons within an atom can help you to predict which ionic compounds will form.

In this section, you will study

- the systematic arrangement of electrons in an atom
- how ionic compounds are formed
- valence electrons and ionic bonding
- how the periodic table reflects electron arrangement

LAB Flame Tests

Purpose

To identify metal atoms in a variety of compounds by using a flame test and to provide evidence for the presence of certain atoms within compounds.

Materials

- Bunsen burner
- set of tongs
- copper wire
- penny
- set of 11 solutions: sodium carbonate, Na_2CO_3; potassium nitrate, KNO_3; copper (II) nitrate, $Cu(NO_3)_2$; strontium nitrate, $Sr(NO_3)_2$; potassium chloride, KCl; sodium chloride, $NaCl$; copper (II) sulfate, $CuSO_4$; strontium chloride, $SrCl_2$; sodium nitrate, $NaNO_3$; copper (II) chloride, $CuCl_2$; potassium sulfate, K_2SO_4
- 11 pieces of nichrome wire with a loop at one end

Safety Instructions

You will be working with flames and chemicals today.

Wear safety goggles.

Roll up long sleeves, tuck in loose clothing, and tie back long hair.

Know the location of the eye wash, fire blanket, and fire extinguisher.

Procedure and Observations

1. For the solutions, follow these steps and record your observations.
   - Place the end of the wire with the solution on it in the flame.
   - Record the color of the flame in your table.
   - Place the wire back in the correct solution.
2. Using tongs, place the two copper objects in the flame and observe the results.

Analysis

1. Group the substances based on the color of the flame produced. What patterns do you notice in the groupings?
2. **Making Sense** Can a flame test be used to identify a metal atom in a compound? Why or why not? What about a nonmetal atom?
3. **If You Finish Early** Copper oxide, CuO, is a black solid. It doesn't look at all like the element copper. What color flame would it produce? Draw a model of copper oxide to explain the color of the flame that is observed.

LESSON

17 Technicolor Atoms

Flame Tests

Think About It

On the Fourth of July you can watch colorful fireworks without considering the chemistry behind them. But each color that bursts forth in the sky is associated with a particular chemical compound. For example, any green sparkles you might see are probably due to a compound such as barium sulfate. But what's responsible for the color? Is it the entire barium sulfate compound, or is it one of the atoms in the compound?

What evidence is there that certain atoms are present in a compound?

To answer this question, you will explore

1. Flame Tests
2. Evidence for Atoms in Compounds
3. Excited Electrons

Exploring the Topic

1 Flame Tests

Fireworks originated in China about 2000 years ago. The legend surrounding the discovery suggests that fireworks were discovered by a Chinese alchemist who mixed charcoal, sulfur, and potassium nitrate and accidentally produced a colorful gunpowder. Since then, the noise and the bright colors associated with fireworks have been used in celebrations all over the world.

In a chemistry lab it is relatively easy to obtain the colorful flames associated with fireworks. It can be done by heating certain compounds in a flame, such as the flames in a Bunsen burner. The flame colors produced by heating four different compounds are shown here.

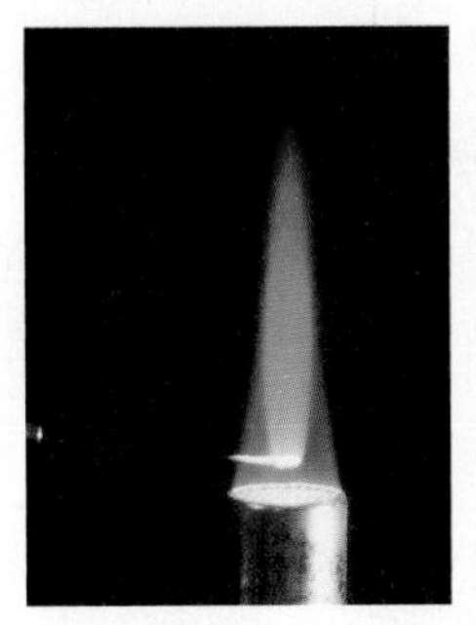

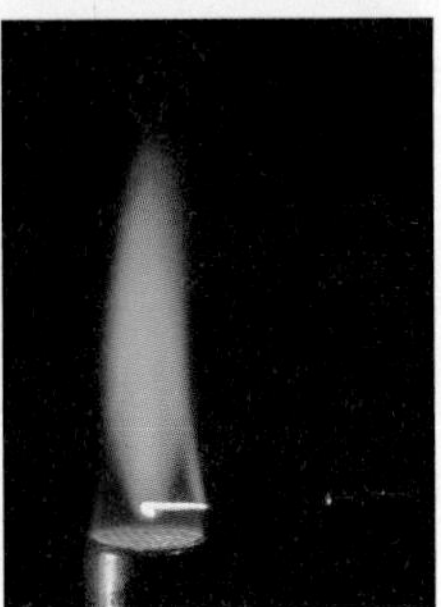

Compound Colors

Red	Orange	Yellow-orange	Green	Blue-green	Pink-lilac
lithium nitrate $LiNO_3$	calcium nitrate $Ca(NO_3)_2$	sodium nitrate $NaNO_3$	barium nitrate $Ba(NO_3)_2$	copper nitrate $Cu(NO_3)_2$	potassium nitrate KNO_3
lithium chloride LiCl	calcium chloride $CaCl_2$	sodium chloride NaCl	barium chloride $BaCl_2$	copper chloride $CuCl_2$	potassium chloride KCl
lithium sulfate Li_2SO_4	calcium sulfate $CaSO_4$	sodium sulfate Na_2SO_4	barium sulfate $BaSO_4$	copper sulfate $CuSO_4$	potassium sulfate K_2SO_4
lithium Li	calcium Ca	sodium Na	barium Ba	copper wire Cu	potassium K

HISTORY CONNECTION

The original firecracker, called *pao chuk,* was created in China around 200 B.C.E. It was a segment of green bamboo that was thrown onto a fire. Upon heating, the trapped gases inside the bamboo expanded, causing the bamboo to explode.

The colors for several more compounds are provided in the table. Take a moment to examine the data. What patterns do you notice?

Notice that each metal atom, Li, Ca, Na, and so on, is associated with a specific flame color. Lithium compounds all make a red flame, while the barium compounds all make a green flame. The nonmetal atoms in these compounds do not seem to affect the color of the compound. So $CaSO_4$ and $CaCl_2$ and $Ca(NO_3)_2$ all have the same color flame, but $CaCl_2$ and NaCl and $CuCl_2$ do not. The metal atom must somehow be responsible for the color of the flame.

❷ Evidence for Atoms in Compounds

Chemists have found these flame color patterns to be quite helpful. A **flame test** can be used to quickly confirm the presence of certain metal atoms in an unknown sample. So, a potassium compound can quickly be distinguished from a calcium compound by heating samples of the compounds.

In addition, the color patterns associated with the flame tests clearly indicate whether certain metal atoms are present within the compounds. For example, copper metal turns the flame blue-green and so does copper chloride. This is definite evidence that there are copper atoms in the copper chloride.

❸ Excited Electrons

Heating metal compounds in a Bunsen burner does not destroy the atoms in them or create new elements. This is because the temperature of a flame is not high enough to change the nuclei of metal atoms. However, the heat does have an impact on the electrons of these metal atoms.

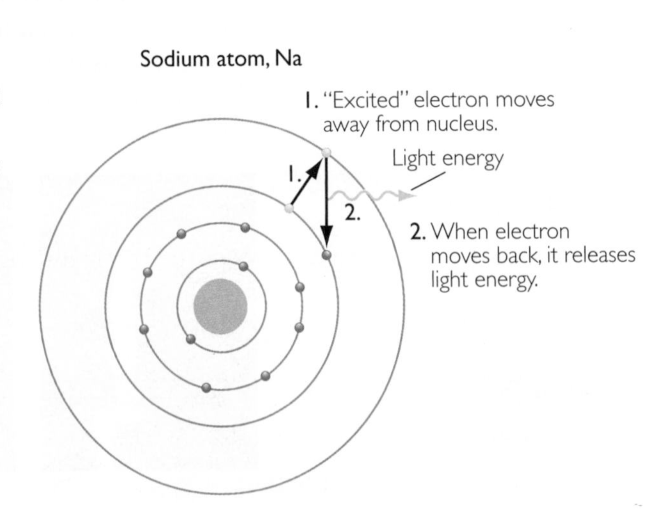

> **ASTRONOMY CONNECTION**
>
> Each element has a light "signature." Astronomers can deduce the composition of a star from the colors of light that it emits. This photo shows the light emitted by the star Arcturus, one of the brightest stars in the night sky. When the light is separated by a prism in this way, you can see that a particular shade of blue is missing because it has been absorbed by iron in the star's corona.
>
> 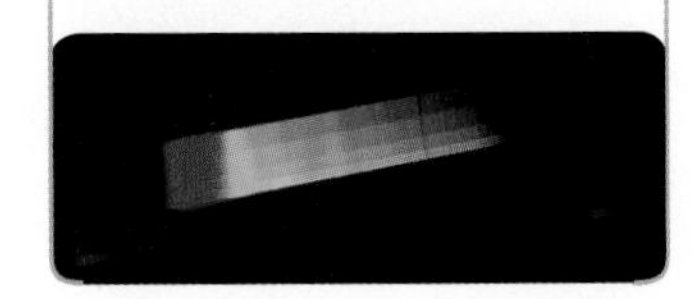

Electrons are located at different average distances from the nucleus of an atom. When heated, some electrons of metal atoms get "excited" and move to distances farther from the nucleus. This move is only temporary. When the electrons move back to their original distance from the nucleus, they release energy in the form of colored light. This is what you see during a flame test.

The fact that it is relatively easy to affect the electrons in atoms opens up new possibilities for exploration. Perhaps the solution to creating substances with the properties of gold rests in altering the electrons in some way. The nuclei of atoms cannot be changed easily, so it is time to explore ways in which the electrons in atoms can be changed. The rest of this unit explores the role of the electrons in the chemistry of atoms and compounds.

Example

Flame Colors

Which of these compounds will give similar flame colors when heated?

$NaCl$ $CaCl_2$ $SrCl_2$ $Sr(NO_3)_2$ $Cu(NO_3)_2$

Solution

The color of the flame depends on the metal. The only two compounds with the same metal are strontium chloride, $SrCl_2$, and strontium nitrate, $Sr(NO_3)_2$.

Key Term

flame test

Lesson Summary

What evidence is there that certain atoms are present in a compound?

Many metal atoms produce a characteristic colored flame when compounds containing those atoms are heated. The colors are a result of light energy produced when excited electrons return to their original distance from the nucleus in the metal atoms. Flame tests provide evidence that certain atoms are present in compounds and that they are not destroyed during chemical changes. The flame test also demonstrates that it is easier to add or remove electrons in atoms, which can be done with a small amount of energy, than to alter the protons and neutrons in the nucleus, which requires a great deal of energy.

EXERCISES

Reading Questions

1. How did the flame test provide evidence that specific atoms are present in compounds?
2. Explain what is responsible for the colors during a flame test.

Reason and Apply

3. **WEB RESEARCH** Find out why fireworks are so colorful. What substances are used to produce the colors?
4. Predict the color of the flame for the compound sodium hydroxide, NaOH. Explain your reasoning.

5. Imagine you were in charge of creating a red and purple fireworks display. Name two combinations of compounds you could use.

6. What evidence do you have from flame tests that copper is responsible for producing a flame with a blue-green color?

7. Does nitrate produce a colored flame? Explain your thinking.

8. Would it matter whether you did a flame test with sodium chloride, NaCl, in solid form or sodium chloride as an aqueous solution? Explain.

9. What flame colors would be produced by these compounds? Explain your choices.

   a. Na_2CO_3 b. $Ba(OH)_2$ c. KOH d. K_2CO_3 e. BaO

10. If two chemical samples both produce an orange flame upon testing, which statement is true?

    A. The two samples contain identical compounds.

    B. The samples both contain chlorine atoms.

    C. The samples both contain calcium atoms.

    D. The samples both contain potassium atoms.

    E. The samples contain different compounds.

11. What evidence supports the claim that chloride, Cl^-, does not cause the flame to have a color?

    A. Lithium chloride, LiCl, and sodium chloride, NaCl, have different colors.

    B. Ammonium chloride, NH_4Cl, does not cause the flame to have a color.

    C. Sodium chloride, NaCl, and sodium nitrate, $NaNO_3$, both produce flames with a yellow-orange color.

    D. All of the above.

LESSON

18 Life on the Edge

Valence and Core Electrons

Think About It

Lithium, Li, and sodium, Na, are both located in Group 1A on the periodic table because they have very similar properties. Both are soft, silvery metals that react with water. Both form compounds with chlorine in a 1:1 ratio—lithium chloride, LiCl, and sodium chloride, NaCl. What do lithium atoms have in common with sodium atoms that make them behave similarly?

Why do elements in the same group in the periodic table have similar properties?

To answer this question, you will explore

1. Electron Shells
2. Patterns in Atomic Structure
3. Valence and Core Electrons

Exploring the Topic

1 Electron Shells

Recall that atoms that are grouped together in columns on the periodic table share similar properties.

Consider lithium, Li, and sodium, Na, from Group 1A on the periodic table. You can get the following information from the periodic table squares for these two elements:

3 protons, 3 electrons
Average atomic mass = 6.941 amu

11 protons, 11 electrons
Average atomic mass = 22.99 amu

This basic information doesn't provide any evidence of similarities between atoms of lithium and atoms of sodium. However, examining the structures of these atoms in more detail reveals some patterns.

Lithium, Li

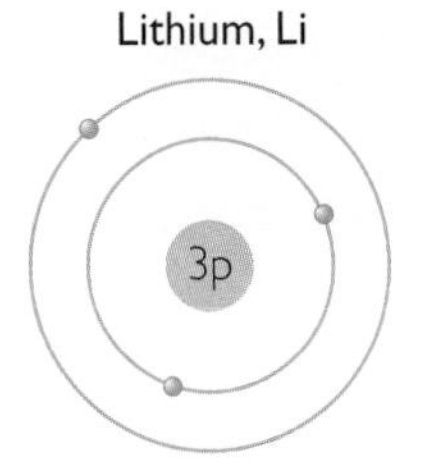

Sodium, Na

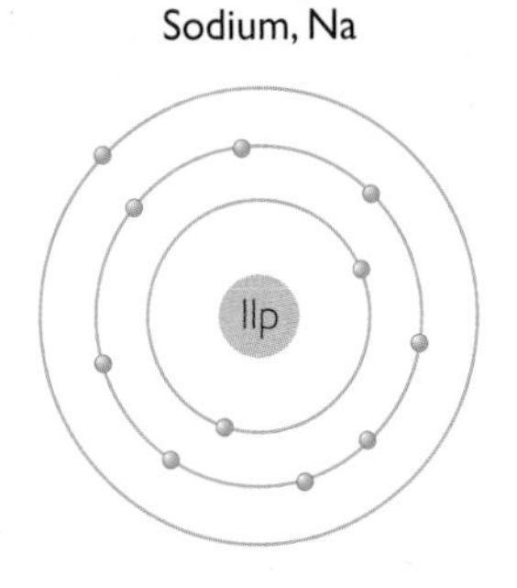

In these models, the electrons are shown orbiting in concentric circles around the nucleus. Chemists call these circles *electron shells,* and these atomic models are called *shell models.*

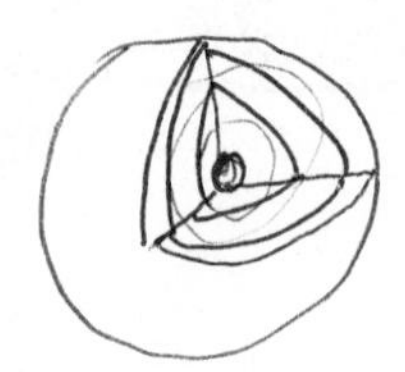

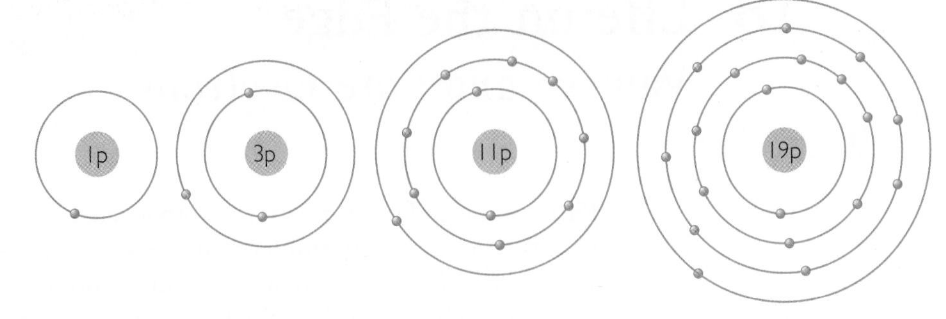

CONSUMER CONNECTION

Lithium (Li) ion batteries are rechargeable and commonly used in electronic devices.

Lithium has a total of three electrons in two shells. Sodium, on the other hand, has a total of 11 electrons in three shells. But both atoms do have one thing in common: a single electron in their outer shells. If this feature is responsible for the similar properties of lithium and sodium, you would expect the other elements in the same group to have a single electron in their outer shells.

Two more Group 1A atoms are shown in the illustration. Compare the models of hydrogen and potassium with those of lithium and sodium. The pattern holds: Each of the Group 1A elements has one electron in the outer shell of its atoms. This pattern also extends to other Group 1A elements not shown here, such as rubidium and cesium.

Important to Know Although electron configurations are easier to draw in two dimensions, keep in mind that atoms are not flat. Electron shells are really surfaces of spheres. Each model is useful in understanding different aspects of the atom. ◀

❷ Patterns in Atomic Structure

Examining the shell models of elements beyond lithium on the periodic table reveals further patterns in the outermost electrons. Shell models for the first 14 elements are shown in this illustration. A repeating pattern emerges from the diagram.

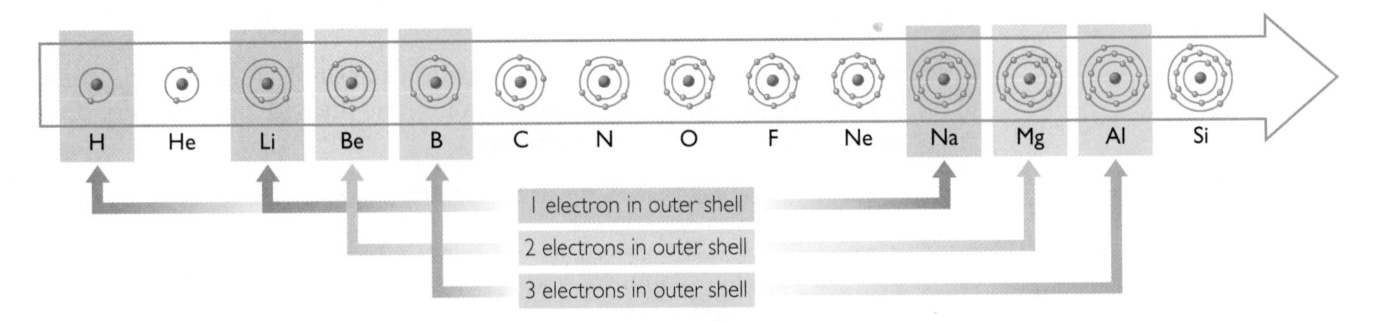

Comparing electron arrangements reveals patterns in atomic structure.

Starting with lithium, notice how the number of electrons in the outer shell repeats in a regular pattern, going from one to eight and then starting over again.

When these electron shell models are organized according to the periodic table, even more is revealed. Examine this table. The atoms in the first row, or Period 1, have only one shell. The atoms in Period 2 have two shells. The atoms in each new period of the periodic table have an additional electron shell.

	Group 1A	Group 2A	Group 3A	Group 4A	Group 5A	Group 6A	Group 7A	Group 8A
Period 1 1 shell	Hydrogen H							Helium He
Period 2 2 shells	Lithium Li	Beryllium Be	Boron B	Carbon C	Nitrogen N	Oxygen O	Fluorine F	Neon Ne
Period 3 3 shells	Sodium Na	Magnesium Mg	Aluminum Al	Silicon Si	Phosphorus P	Sulfur S	Chlorine Cl	Argon Ar
Period 4 4 shells	Potassium K	Calcium Ca	Gallium Ga	Germanium Ge	Arsenic As	Selenium Se	Bromine Br	Krypton Kr

It turns out that for main group elements, Groups 1A through 8A, everything you need to know about the placement of an atom's electrons can be decoded from the periodic table.

BIG IDEA The arrangement of atoms in the periodic table reflects the arrangement of electrons in the atom.

In the first column of the table, every element has one electron in its outermost electron shell. In the second column, each has two electrons in the outermost shell, and so on. In the eighth column, each element has eight electrons in its outermost shell. This feature turns out to be the key to many of the properties of the atoms.

❸ Valence and Core Electrons

The outermost shell that electrons occupy is very important in chemistry and is referred to as the **valence shell.** The electrons in that shell are called the **valence electrons.** The electrons that are located in all of the other inner shells are referred to as **core electrons.**

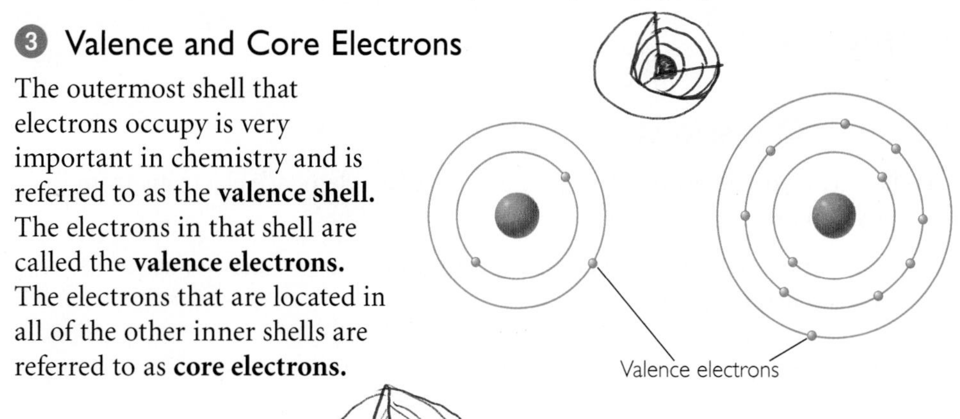

The number of electrons in the valence shell corresponds to the group number at the top of the periodic table. Thus, every Group 4A element has four electrons in its outermost shell, while every Group 8A element has eight electrons in its outermost shell.

Elements with the same number of electrons in their outermost shells have similar properties. In other words, elements with the same number of valence electrons have similar properties.

Example

Sulfur Atoms

Find the element sulfur, S, on the periodic table.

a. What group is sulfur in?

b. How many valence electrons does a sulfur atom have?

c. Explain why selenium, Se, has properties similar to those of sulfur.

Solution

Sulfur is not a metal, so it is located toward the right side of the periodic table.

a. Sulfur is in Group 6A of the periodic table.

b. The number of valence electrons is the same as the group number, so sulfur has six valence electrons.

c. Selenium, Se, is also in Group 6A. Even though it has many more total electrons than sulfur has, selenium also has six valence electrons. This is why sulfur and selenium have similar properties.

Key Terms

valence shell
valence electron
core electron

Lesson Summary

Why do elements in the same group in the periodic table have similar properties?

Elements that are grouped together in columns on the periodic table share similar properties. For main group elements, atoms in the same group have the same number of electrons in their outermost shells. The outermost shell is called the valence shell, and the electrons in that shell are called valence electrons. Electrons in the inner shells are called core electrons. The number of valence electrons in an atom of a main group element corresponds to the element's group number. Thus, the periodic table is a direct reflection of the atomic structures of the atoms.

EXERCISES

Reading Questions

1. How can you determine the arrangement of an element's electrons from its position on the periodic table?
2. If you know what group an element is in, what can you predict about its electron arrangement?

Reason and Apply

3. What do beryllium, Be, magnesium, Mg, and calcium, Ca, all have in common?
4. Draw a shell model for boron, B. Identify the core and valence electrons.
5. For main group elements, how does the number of core electrons vary across a period?
6. Consider the elements carbon, C, and silicon, Si.
   a. How many electrons does an atom of each of these elements have?
   b. Draw shell models for atoms of each of these elements.
   c. How many valence electrons do atoms of each of these elements have?
   d. How many core electrons do atoms of each of these elements have?
   e. Why are the properties of carbon and silicon similar?
7. Provide the following information for element number 17. For each answer, explain how you know.
   a. the element's name, symbol, and group number
   b. the number of protons in the nucleus
   c. the number of neutrons in the nucleus
   d. the number of electrons in a neutral atom of the element
   e. the number of electrons in a neutral atom that are valence electrons
   f. the number of electrons in a neutral atom that are core electrons
   g. the names of three other elements with the same number of valence electrons
8. Answer Exercise 7, parts a through g, for element number 50.
9. If an element is located in Group 4A of the periodic table, what can you conclude about atoms of this element?
   A. It has four valence electrons.
   B. It has four electron shells.
   C. It has four core electrons.
   D. It has properties similar to carbon.
   E. Both A and D.
10. What do the elements in Period 3 have in common?
    A. Their atoms have the same number of valence electrons.
    B. Their atoms have the same number of electron shells.
    C. Their atoms have the same properties.
    D. None of the above.

LESSON

19 Noble Gas Envy

Ions

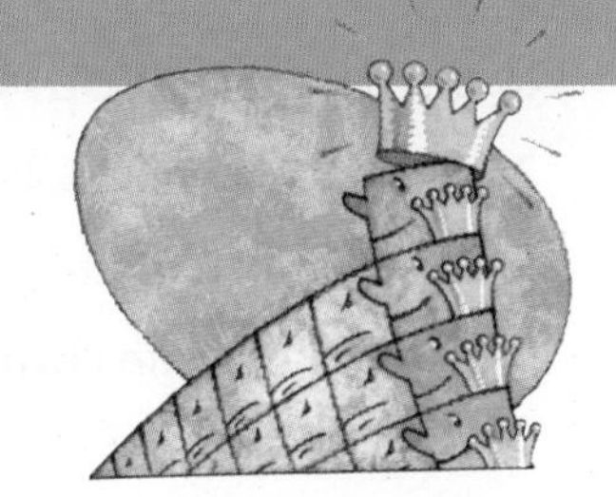

Think About It

Some atoms are more chemically stable than others. In other words, they don't readily combine with other atoms to form new compounds. The noble gases are considered the most chemically stable atoms on the periodic table. The most reactive elements are located just before and just after the noble gases, in Groups 1A and 7A. Perhaps the electron arrangements of these atoms are related to their reactivity.

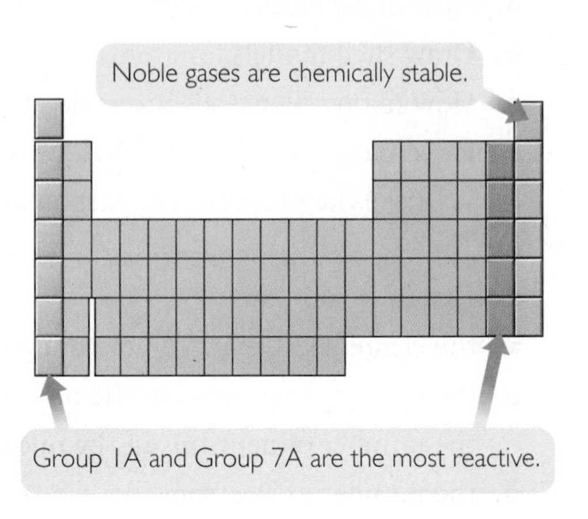

How is chemical stability related to the arrangements of electrons in atoms?

To answer this question, you will explore

1. Noble Gases
2. Noble Gas Envy
3. Patterns in Ion Charges

Exploring the Topic

1 Noble Gases

Some elements are chemically stable, or rarely react with other elements. Others are reactive and combine readily with other elements to form compounds. What makes some elements chemically stable and others reactive?

Noble gases are the most chemically stable elements. Shell models for the noble gas elements helium, He, neon, Ne, argon, Ar, and krypton, Kr, are shown below. Take a moment to compare the valence electrons of each.

INDUSTRY CONNECTION

Noble gases are so stable that they rarely form compounds. At one time it was thought that they *never* formed compounds with other atoms. But in 1962 a British scientist, Neil Bartlett, created xenon hexafluoroplatinate, a yellow solid, by accident. Chemists have not yet been able to create any compounds with either helium or neon.

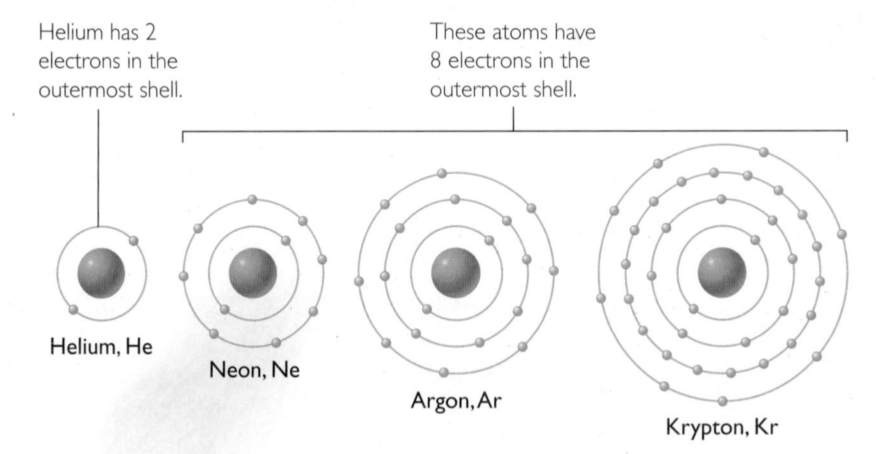

Helium has two valence electrons, which is the maximum for the first shell. The remaining noble gases each have eight valence electrons. The stability of the noble gases is due to the number of valence electrons they have.

BIG IDEA Noble gases are chemically stable. This is because they have a full outer shell.

2 Noble Gas Envy

Consider two highly reactive elements, sodium, Na, and fluorine, F. The outer shell of a fluorine atom, F, has seven electrons. This is just one short of the eight electrons that neon, Ne, has in its outer shell. The outer shell of a sodium atom, Na, has one electron. This is just one more electron than neon has in its valence shell.

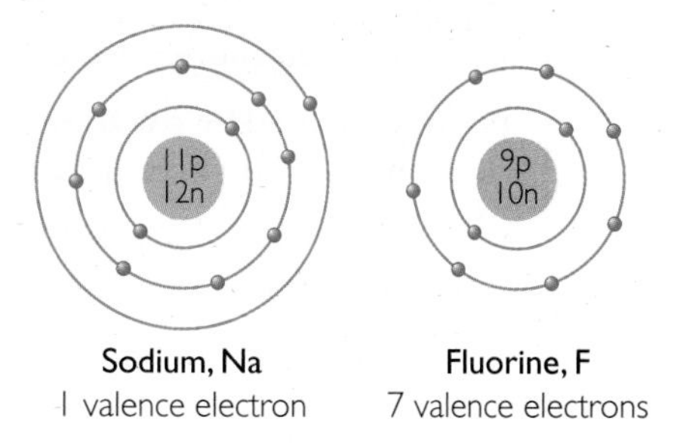

Now examine what happens to these two atoms when they combine to form a compound. Sodium, Na, gives one electron to fluorine, F.

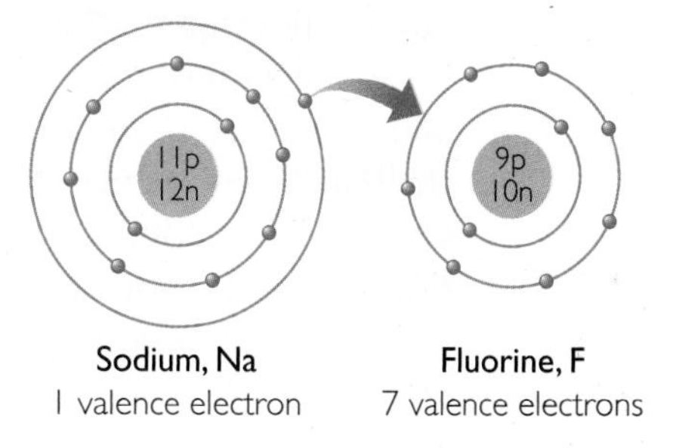

Now both atoms have an electron arrangement like that of neon, Ne.

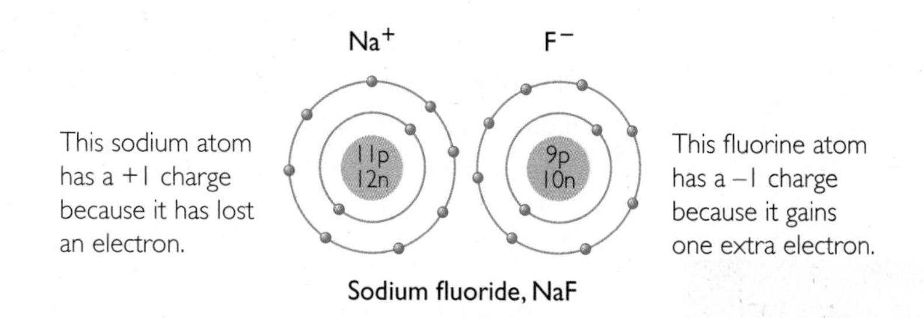

Both ions now have an electron arrangement like the noble gas neon. They form an ionic compound, NaF.

The movement of an electron from one atom to the other alters the balance of charges on both atoms. Fluorine now has more electrons than protons. Sodium has more protons than electrons. The atoms are now called **ions** because they possess a charge. The sodium ion has a charge of +1 and its symbol is written as Na^+. The fluorine ion has a charge of −1 and its symbol is written as F^-. Because

INDUSTRY CONNECTION

Sodium and chlorine are both highly reactive. In elemental form, sodium is a soft, silvery metal, and chlorine is a greenish poisonous gas. However, they are never found in elemental form in nature. Frequently, they are found as sodium chloride, which is used as table salt.

Na^+ and F^- have opposite charges, they attract one another. So the movement of an electron from one atom to the other forms a new compound, sodium fluoride, NaF. As a result of this electron transfer, sodium atoms are now bonded to fluorine atoms.

Important to Know When an atom loses or gains electrons, the result is a charge on the atom. The rest of the atom stays the same, and the identity of the element does not change. ◀

Example

Calcium Oxide, CaO

Consider the compound calcium oxide, CaO.

a. Draw shell models for calcium, Ca, and oxygen, O.

b. With arrows, show how electrons can be transferred so that each atom has the same electron arrangement as that of a noble gas.

c. What are the charges on the calcium, Ca, and oxygen, O, ions after electrons have been transferred?

Solution

a. You can draw a shell model for a neutral atom, an atom that has no charge, using the element number and the position on the periodic table. Calcium is in Period 4 and Group 2A, so it has four shells and two valence electrons. Oxygen is in Period 2 and Group 6A, so it has two electron shells and six valence electrons.

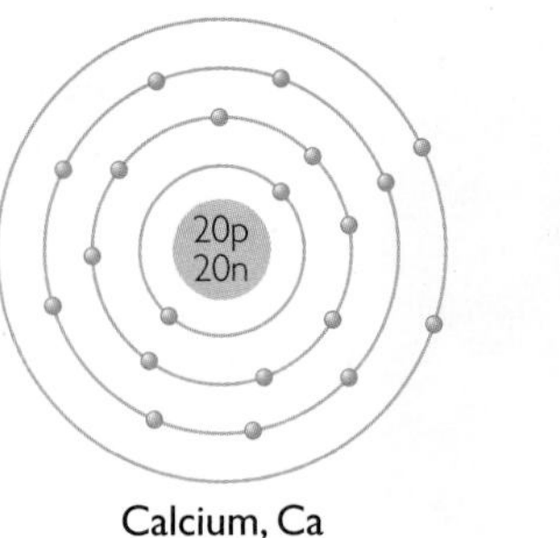

Calcium, Ca

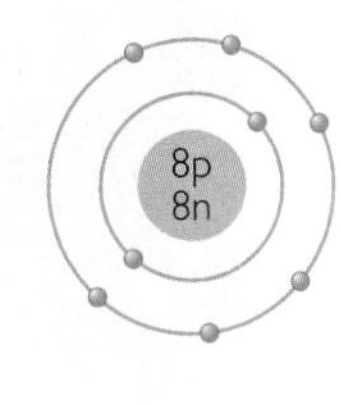

Oxygen, O

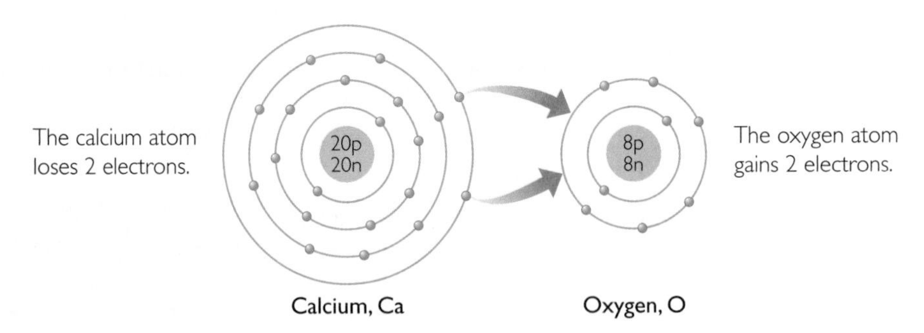

b. If calcium transfers two electrons to oxygen, they will both have noble gas electron configurations. The electron arrangement of oxygen now resembles that of neon, Ne. The electron arrangement of calcium now resembles that of argon, Ar.

c. Because calcium has lost two electrons, it now has more protons than electrons, giving it a +2 charge. The ion can be written as Ca^{2+}. Oxygen has gained 2 electrons, so its charge is −2. Its ion can be written as O^{2-}.

CONSUMER CONNECTION

Neon signs contain neon gas. When electricity is run through the gas, the electrons get excited. When they come back down to their normal energy level, they emit red light. Real neon signs glow red. Signs that glow other colors are usually made of colored glass and filled with argon.

In the calcium and oxygen example, each ion has an electron arrangement that is identical to that of a noble gas. It is reasonable to assume that there is some advantage to having this sort of electron arrangement.

Noble gases are very stable the way they are, without reacting or exchanging any electrons with other elements. Apparently, other atoms can achieve some of the stability of the noble gases by exchanging electrons and becoming ions.

❸ Patterns in Ion Charges

When atoms lose or gain electrons they become ions, with negative or positive charges. Below is a portion of the periodic table with the various charges on the ions filled in. In every case, the ion that forms has an electron arrangement identical to the noble gas it is nearest to in the table.

Take a moment to examine the table. What patterns do you notice?

Ion Charges on Main Group Elements

1A	2A	3A	4A	5A	6A	7A	8A
1+	2+	3+	4+ (or 4−)	3−	2−	1−	
1 H^+							2 He
3 Li^+	4 Be^{2+}	5 B^{3+}	6 C^{4+}	7 N^{3-}	8 O^{2-}	9 F^-	10 Ne
11 Na^+	12 Mg^{2+}	13 Al^{3+}	14 Si^{4+}	15 P^{3-}	16 S^{2-}	17 Cl^-	18 Ar
19 K^+	20 Ca^{2+}	31 Ga^{3+}	32 Ge^{4+}	33 As^{3-}	34 Se^{2-}	35 Br^-	36 Kr

Notice that the elements on the left of the table, which are mostly metals, tend to form ions with a positive charge. Ions with a positive charge are also called **cations.** The valence shells of these atoms have fewer electrons in them than those of the elements on the right. Thus, it is easier for these elements to form compounds by giving up these few electrons.

On the other hand, the elements on the right of the table, which are mostly nonmetals, tend to form ions with a negative charge. Ions with a negative charge are called **anions.**

For the first four groups of the periodic table, the ion charge is the same as the group number. For Groups 5A through 7A, the ion charge is negative and goes from −3 to −1.

Key Terms

ion
cation
anion

Lesson Summary

How is chemical stability related to the arrangements of electrons in atoms?

Noble gases have particularly stable atoms. This is attributed to their electron arrangements. Helium has two electrons, which is the maximum for the first shell. The remaining noble gases all have eight electrons in their outermost shells. Other atoms on the periodic table are more reactive than the noble gases. However, they reach greater chemical stability by gaining or losing valence electrons. The goal is to end up with an electron arrangement similar to that of a noble gas. Atoms do this by combining with other atoms to form new compounds. When atoms lose or gain electrons the atom becomes an ion, with a charge. Ions with a positive charge are called cations. Ions with a negative charge are called anions.

EXERCISES

Reading Questions

1. Explain the difference between an anion and a cation.
2. Explain what is meant by noble gas envy.

Reason and Apply

3. How many electrons, protons, and neutrons does Li^{+} have?
4. Give two similarities and two differences between Cl and Cl^{-}.
5. Give two similarities and two differences between Be and Be^{2+}.
6. Which noble gas is closest to magnesium, Mg, on the periodic table? What must happen to a magnesium atom in order for it to have an electron arrangement similar to that of a noble gas?
7. Which noble gas is closest to sulfur, S, on the periodic table? What must happen to a sulfur atom in order for it to have an electron arrangement similar to that of a noble gas?
8. List four ions that have the same number of electrons as neon, Ne.
9. List four ions that have the same number of electrons as argon, Ar.
10. What charge would an arsenic, As, ion have?
11. What is the symbol of an ion with 22 protons, 24 neutrons, and 18 electrons?
12. When chlorine gains an electron to become a chloride ion with a -1 charge, it ends up with the electron arrangement of argon. Why doesn't it become an argon atom?
13. Explain why the elements on the right side of the periodic table gain electrons instead of losing them.
14. What periodic patterns do you notice for the charges on the ions?
15. Which of these ions have the correct charge? Choose all that apply.

    A. Na^{2+} **B.** Li^{+} **C.** Al^{4+} **D.** Ca^{2+} **E.** Ga^{3+}

16. Which of these ions have the same number of electrons as S^{2-}? Choose all that apply.

    A. Cl^{-} **B.** Ca^{2+} **C.**. Na^{+} **D.** O^{2-} **E.** P^{3-}

LESSON

20 Getting Connected

Ionic Compounds

Think About It

Sodium and chlorine atoms combine to form sodium chloride, NaCl, when sodium transfers a valence electron to chlorine. There are no other compounds that form between sodium and chlorine. For example, $NaCl_4$ and Na_3Cl are not possible compounds. Why is NaCl the compound that forms rather than $NaCl_4$ or Na_3Cl?

How can valence electrons be used to predict chemical formulas?

To answer this question, you will explore

1. Ionic Compounds
2. The Rule of Zero Charge
3. More Complex Ionic Compounds

Exploring the Topic

1 Ionic Compounds

There are millions of different compounds on the planet. Each one is a result of a unique combination of atoms. Thus, each one has a unique chemical formula. However, some combinations of atoms are simply not possible. Valence electrons are the key to figuring out chemical formulas and to determining which compounds are possible.

Compounds Between Metals and Nonmetals

Atoms combine with other atoms in order to achieve the stability of the noble gases. One way to accomplish this is for atoms to transfer valence electrons to other atoms. When atoms gain or lose electrons, the resulting atoms are no longer neutral. They become ions with charges. The compounds that form in this way are called **ionic compounds.** Ionic compounds, like the noble gases, are very stable.

Ionic compounds form between metal atoms and nonmetal atoms. The metal atoms, on the left side of the periodic table, tend to give up electrons and become cations, which have positive charges. The nonmetal atoms, on the right side of the periodic table, tend to accept electrons and become anions, which have negative charges. Because the cation and anion have opposite charges, they are strongly attracted to one another.

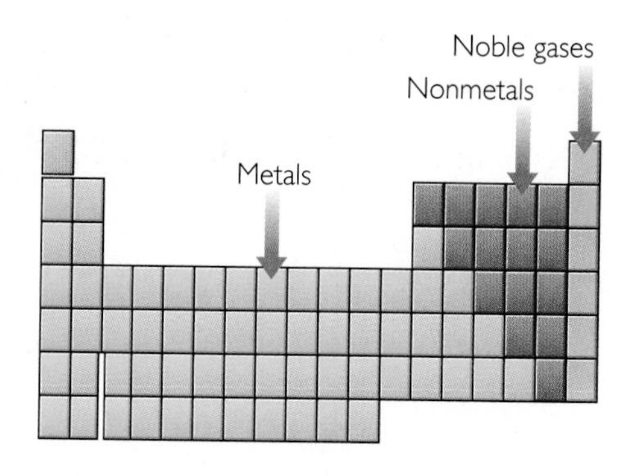

CONSUMER CONNECTION

All ionic compounds form crystals. If you look at grains of table salt with a magnifying glass, you will see that the pieces are shaped like tiny boxes.

Formulas of Simple Ionic Compounds

The tables below contain data on three simple ionic compounds. These compounds are all composed of a metal atom and a nonmetal atom in a 1:1 ratio. Pay particular attention to the total charge and the number of valence electrons on each of the atoms that combine.

Some Ionic Compounds

Compound	Metal	Cation	Number of valence electrons	Nonmetal	Anion	Number of valence electrons
NaCl	Na	Na^{+}	1	Cl	Cl^{-}	7
CaO	Ca	Ca^{2+}	2	O	O^{2-}	6
GaAs	Ga	Ga^{3+}	3	As	As^{3-}	5

Note that in each case, the charge on the cation is equal and opposite to the charge on the anion.

2 The Rule of Zero Charge

You can see there is a distinct pattern to the ions that form in ionic compounds. Every time a metal atom and a nonmetal atom bond, they form a compound with an overall zero charge. This is known as the **rule of zero charge.**

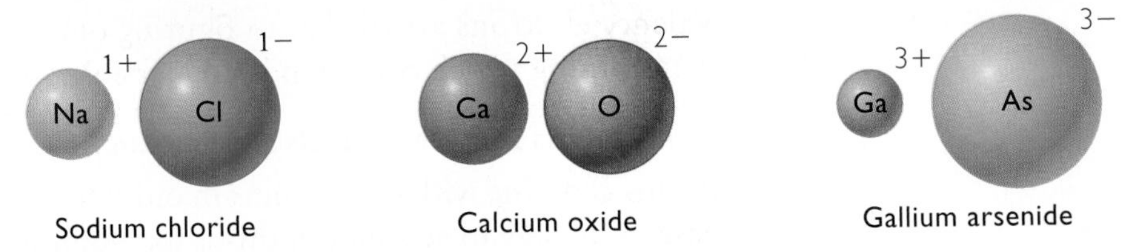

Notice how the ions in each compound add up to zero.

The charges on the metal cations and nonmetal anions add up to zero. The rule of zero charge is useful in predicting the chemical formulas for ionic compounds, which form between a metal and a nonmetal.

Predicting Ionic Bonding Using the Periodic Table

You may have noticed that the atoms that combine to form ionic compounds come from very predictable places on the periodic table. Sodium, with one valence electron, can combine with chlorine, or fluorine, or bromine, each with seven valence electrons. In fact, any Group 1A elements can combine with any Group 7A elements to form ionic compounds.

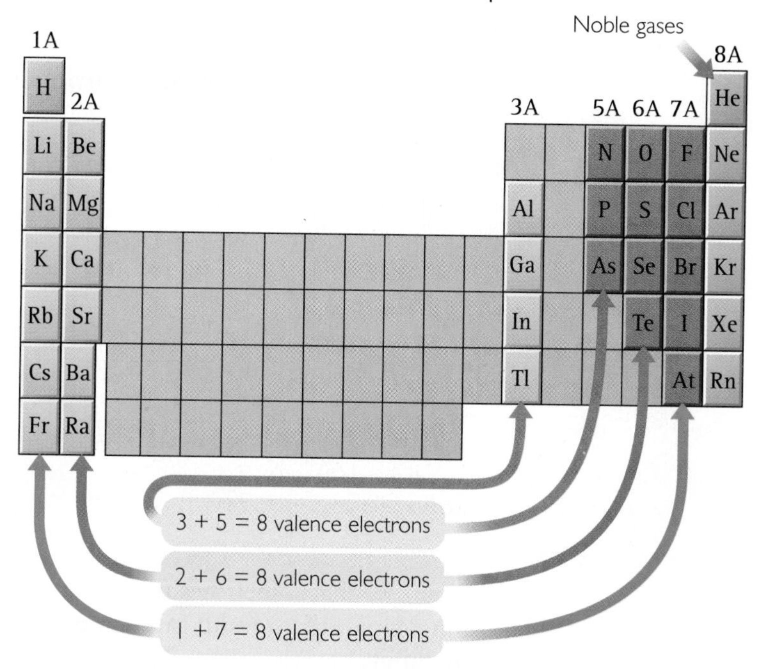

The drawing illustrates how you can use the periodic table to decide which atoms will combine in a 1:1 ratio to form an ionic compound. There are dozens of possible combinations. For example, an atom in Group 3A can form a compound with an atom in Group 5A.

Important to Know The atoms in Group 4A can either transfer or accept electrons. ◀

❸ More Complex Ionic Compounds

All of the compounds discussed in this lesson so far have a metal to nonmetal ratio of 1:1. However, it is also possible to have compounds with different ratios. For example, one magnesium atom can combine with two chlorine atoms to form magnesium chloride, $MgCl_2$. Before bonding, each magnesium atom has two valence electrons. These two electrons are transferred to two different chlorine atoms, each of which needs one more electron to have the same number of electrons as the noble gas argon, Ar.

CONSUMER CONNECTION

Energy drinks help replace necessary ions, known as electrolytes, in your body.

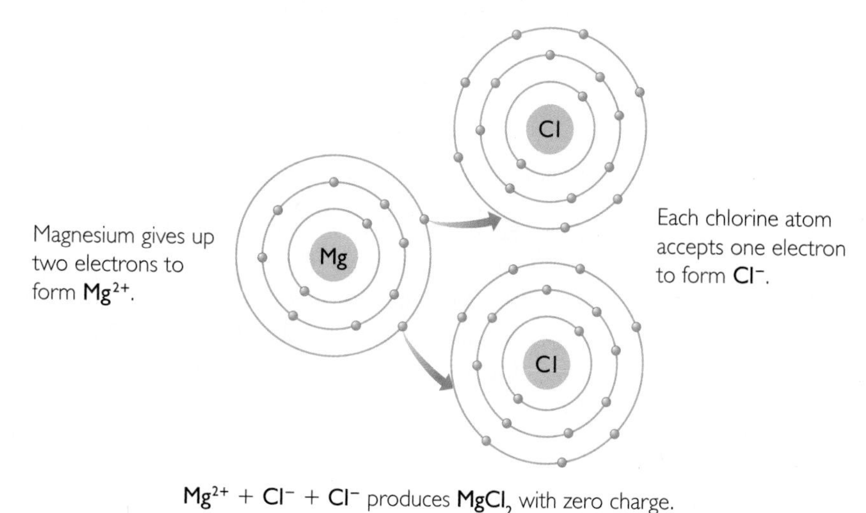

Mg^{2+} + Cl^- + Cl^- produces $MgCl_2$ with zero charge.

Notice in this table that the atoms combine in such a way that the resulting compound has a zero charge.

Some More Complex Ionic Compounds

Compound	Metal	Cations	Nonmetal	Anions
$MgCl_2$	Mg	Mg^{2+}	Cl	Cl^- Cl^-
Na_2O	Na	Na^+ Na^+	O	O^{2-}
Al_2O_3	Al	Al^{3+} Al^{3+}	O	O^{2-} O^{2-} O^{2-}

Note that the charges on cations are equal and opposite to charges on anions. For example for aluminum oxide, $+3 + 3 - 2 - 2 - 2 = 0$. The total charge on the five atoms in the compound adds up to zero.

You can see that sometimes more than one atom of a particular element must combine in order to result in a compound with a zero charge. Thus, ionic compounds in ratios of 1:2, 2:1, or even 2:3 may result.

Example

Sodium Sulfide

The chemical formula for sodium sulfide is Na_2S.

a. What is the charge on each sodium ion, Na? On each sulfur ion, S?

b. Show that the charges on the ions add up to zero.

c. Is Na_2Cl a possible ionic compound? Why or why not?

Solution

a. Sodium, Na, is in Group 1A and has one valence electron. Sodium atoms give up one electron to have a charge of +1. Sulfur, S, is in Group 6A and has six valence electrons. Sulfur atoms gain two electrons to have a charge of −2.

b. Check the charge on the compound:

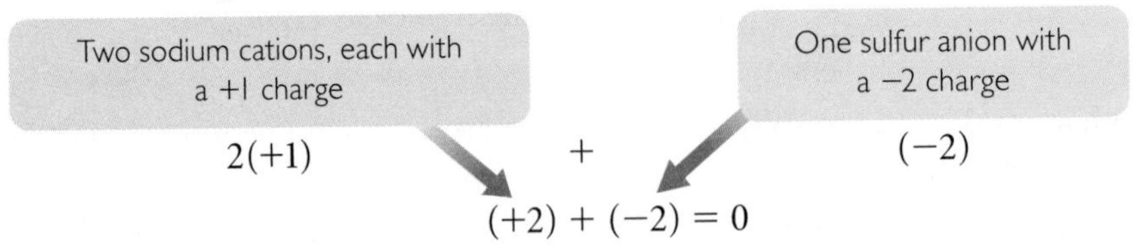

The charges on the ions add up to zero, as they should.

c. Each sodium atom has one valence electron and each chlorine atom has seven valence electrons. If two sodium atoms gave up electrons to one chlorine atom, the chlorine atom would have nine valence electrons, which is not a stable number resembling a noble gas. So Na_2Cl is not a compound that is likely to form. Also, such a compound would not have a neutral charge.

Key Terms

ionic compound
rule of zero charge

Lesson Summary

How can valence electrons be used to predict chemical formulas?

Metal atoms and nonmetal atoms combine to form ionic compounds through electron transfer. The metal atoms give up electrons. They form ions with a positive charge, called cations. The nonmetal atoms accept electrons and form ions with a negative charge, called anions. You can use the positions of the elements on the periodic table to predict the ionic compounds they can form. When an ionic compound forms, the total charge on the atoms adds up to zero. This is known as the rule of zero charge.

EXERCISES

Reading Questions

1. How can you use valence electrons to predict which ionic compounds will form?
2. How does the rule of zero charge help you predict the formula of an ionic compound?

Reason and Apply

3. Lithium nitride has the formula Li_3N.
   a. What is the charge on the lithium ion?
   b. What is the charge on the nitrogen ion?
   c. Show that the charges on the ions add up to zero.
   d. What is the total number of valence electrons in all the atoms in Li_3N?
4. Aluminum arsenide has the formula AlAs.
   a. What is the charge on the aluminum ion?
   b. What is the charge on the arsenic ion?
   c. Show that the charges on the ions add up to zero.
   d. What is the total number of valence electrons in all the atoms in AlAs?
5. For each of the compounds below, show that the charges on the ions add up to zero.
   a. KBr b. CaO c. Li_2O d. $CaCl_2$ e. $AlCl_3$
6. What are the total numbers of valence electrons in all the atoms in each of the compounds below?
   a. KBr b. CaO c. Li_2O d. $CaCl_2$ e. $AlCl_3$
7. Explain why the following compounds do not form.
   a. $NaCl_2$ b. CaCl c. AlO

ACTIVITY Salty Eights

Purpose

To make and name as many ionic compounds as possible.

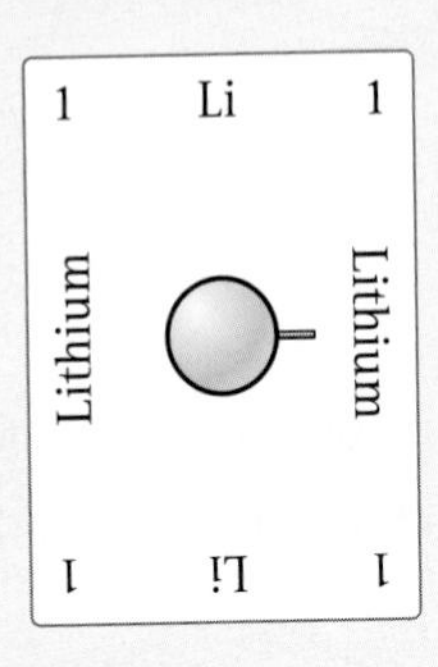

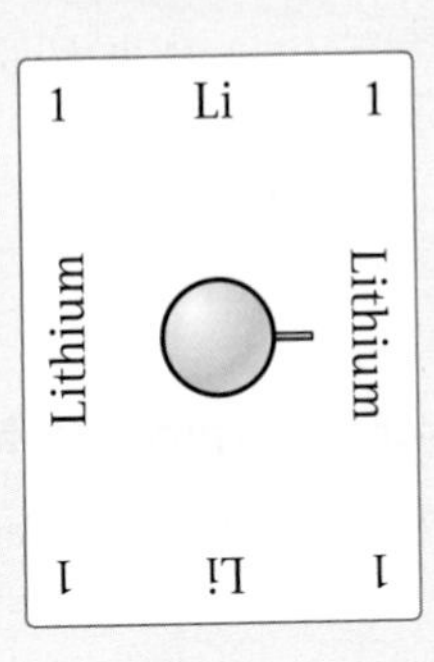

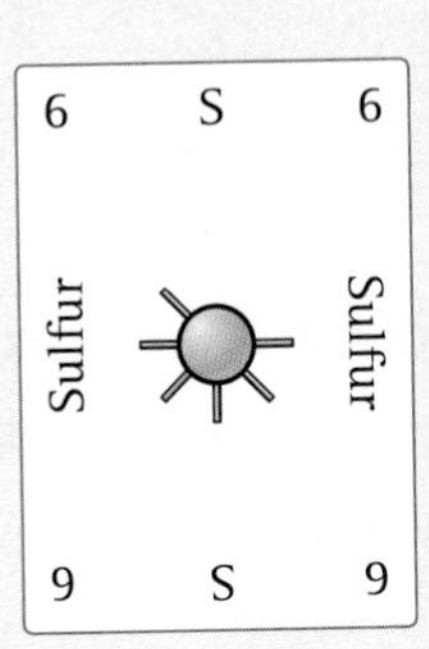

Materials

- Salty Eights cards

Instructions

1. Deal eight cards to each player and place the rest in a draw pile. On your turn, you must put down an ionic compound, a noble gas, or draw from the pile until you can. Play until one player has used all of his or her cards.
2. Keep score with a data table like this one. Point values are on the Rules card. List your compounds in the table. Write the metal first and the nonmetal second. If you use more than one atom of an element, don't forget to put the right number in the formula (Example: Li_2S – Lithium sulfide).

Cation	Anion	Compound formula	Compound name	Point value

Total:

Bonus Question Determine the compound with the highest point value you could make with these cards. Name that compound.

LESSON

21 Salty Eights

Formulas for Ionic Compounds

Think About It

There are at least 50 common metal elements. These combine with over 15 nonmetal elements to make a wide variety of ionic compounds. The task of predicting compounds might seem complex given the number of possible ways to combine a metal and a nonmetal. However, if you know how many valence electrons each atom has, you can reliably predict which ionic compounds can be made.

How can you predict chemical formulas and name ionic compounds?

To answer this question, you will explore

1. Predicting Chemical Formulas
2. Naming Ionic Compounds

Exploring the Topic

1 Predicting Chemical Formulas

An ionic compound that is made up of only one kind of metal atom and one kind of nonmetal atom is known as a *salt*. Table salt, NaCl, is the most familiar example. To predict formulas for salt compounds, follow these guidelines:

- Make sure you are combining two different elements—a metal and a nonmetal.
- No matter how many atoms you use, the total number of valence electrons must add up to 8, 16, 24, or another multiple of 8.
- Make sure that the charges on the metal cations and nonmetal anions in your ionic compound add up to zero (rule of zero charge).

This process applies only to the main group elements. Ionic compounds using transition elements are covered in Lesson 23: Alchemy of Paint.

Example 1

Practice Writing Chemical Formulas

Each of these cards represents an atom.

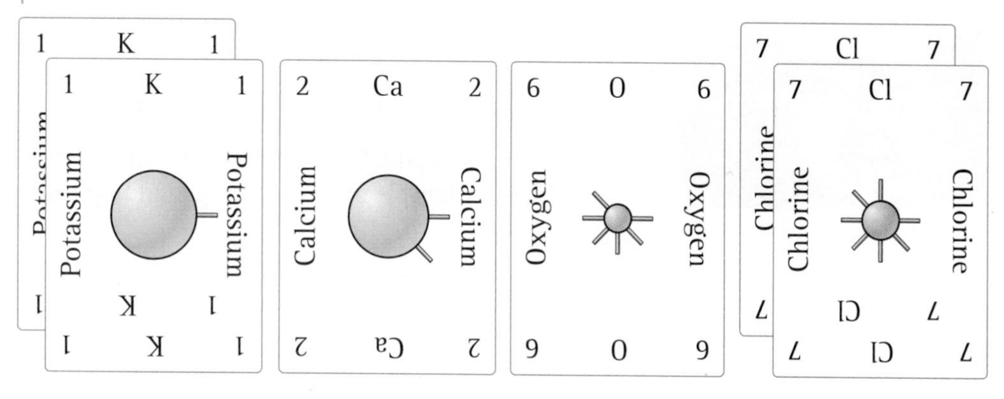

a. Which atoms are metals and which are nonmetals? How can you tell?

b. What ionic compounds can you make from two of these cards?

c. What ionic compounds could you make from three cards but only two different elements?

Solution

a. The atoms from the left side of the periodic table, with fewer valence electrons, are the metals. Thus, potassium, K, and calcium, Ca, are metal atoms. The atoms from the right side of the periodic table, which have closer to eight valence electrons, are the nonmetals. These include oxygen, O, and chlorine, Cl.

b. Look at the number of spokes on each card. The number of spokes represents the number of valence electrons. To create ionic compounds, combine a metal with a nonmetal. The number of spokes must add up to 8. Two ionic compounds are possible with these cards. These are shown here.

1 spoke + 7 spokes = 8 spokes

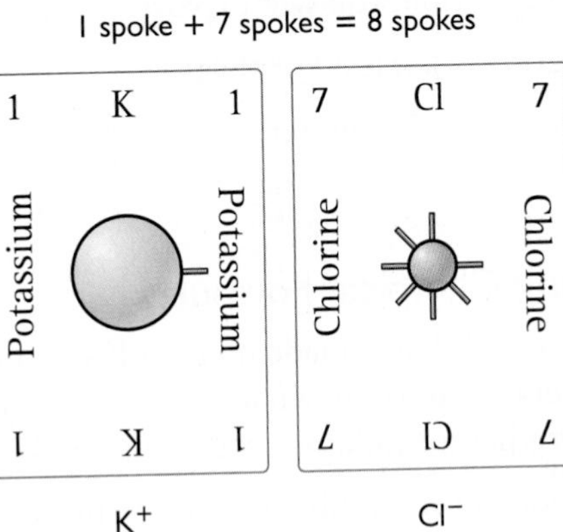

2 spokes + 6 spokes = 8 spokes

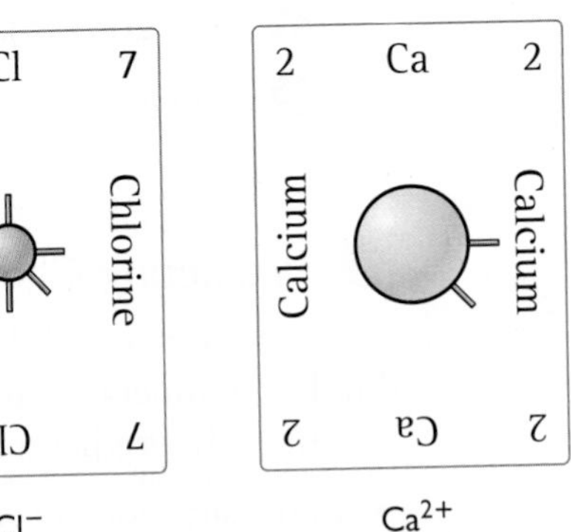

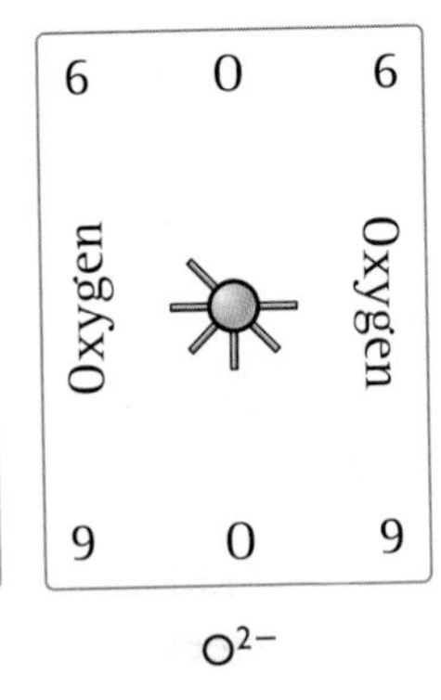

c. If two atoms with one valence electron combine with one atom with six valence electrons, the result is eight valence electrons. One possible compound is shown.

1 spoke + 1 spoke + 6 spokes = 8 spokes

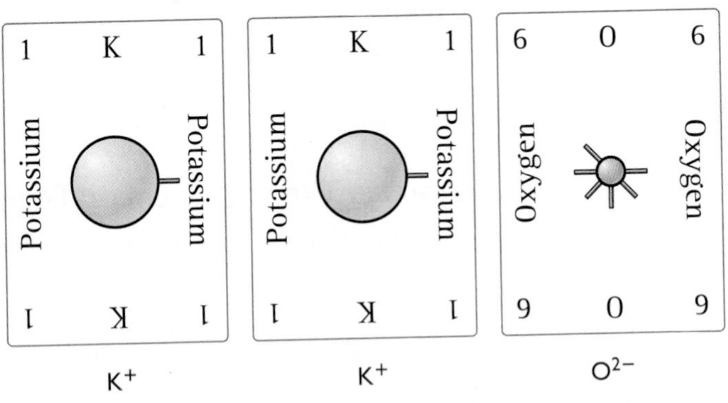

You can also make $CaCl_2$. Notice that the valence electrons add up to 16, which is a multiple of 8. The charges add up to zero.

HISTORY CONNECTION

Salt has been a valuable trade item since ancient times. For thousands of years—before refrigeration—humans depended largely on salt as a preservative for foods, especially meats and butter. The wealth of a city, country, or individual could often be linked to the amount of salt possessed or stored. Thus, people who are "worth their salt" are considered worthwhile and valuable. Salzburg, Germany, shown here, is named for its many salt mines, which contributed to its great wealth.

ENVIRONMENTAL CONNECTION

Calcium chloride and magnesium chloride are often used to remove ice on slick winter roadways in the United States. Unfortunately, this practice may have a negative effect on wildlife, causing salt poisoning in some birds, and upsetting the salt balance in lakes and rivers.

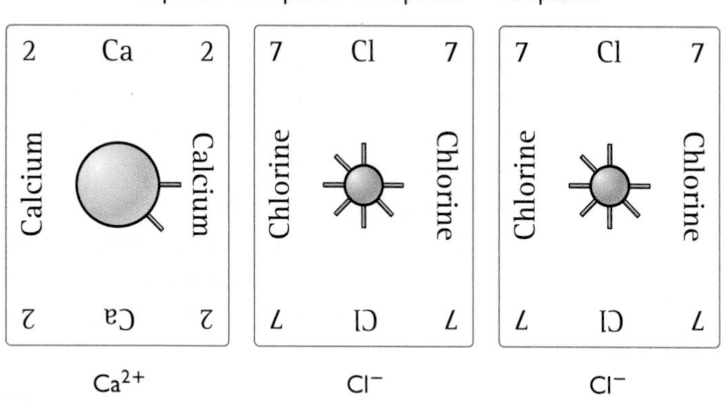

❷ Naming Ionic Compounds

When naming salt compounds, the name of the metal comes first, followed by the name of the nonmetal atom. However, the name of the nonmetal atom is altered slightly by replacing the last part of the word with the letters "-ide." Here is a list of ionic compounds with their names. Take a moment to examine them.

Compound	Name	Compound	Name
NaCl	sodium chloride	Al_2O_3	aluminum oxide
MgF_2	magnesium fluoride	GaP	gallium phosphide
Li_2S	lithium sulfide		

So in ionic compounds, sulfur becomes sulfide, nitrogen becomes nitride, and bromine becomes bromide.

Notice that the chemical name for ionic compounds does not have anything to do with the subscript numbers in the chemical formula. For example, $MgCl_2$ is simply magnesium chloride, *not* magnesium dichloride.

INDUSTRY CONNECTION

Potassium bromide is currently in use as an anti-seizure medication for dogs and cats with epilepsy.

Example 2

Naming Compounds

Write the chemical formula and name for the compound created from each pair of elements.

a. potassium and bromine
b. oxygen and calcium
c. oxygen and potassium
d. sodium and chlorine
e. sodium and oxygen

Solution

To write the formula for the compound, make sure the atoms are in the correct ratio so that the compound follows the rule of zero charge. To name the compound, write the metal atom first, followed by the nonmetal atom. Change the ending of the nonmetal atom's name to "-ide."

a. KBr, potassium bromide
b. CaO, calcium oxide
c. K_2O, potassium oxide
d. NaCl, sodium chloride
e. Na_2O, sodium oxide

Lesson Summary

How can you predict chemical formulas and name ionic compounds?

There are several important guidelines to follow in creating ionic compounds. Metal atoms are combined with nonmetal atoms. Next, the total number of valence electrons adds up to eight or a multiple of eight. Finally, the charges on the metal cations and nonmetal anions in ionic compounds add up to zero. When naming ionic compounds made from two different types of elements, the name of the metal atom comes first followed by the name of the nonmetal atom. In addition, the ending of the name of the nonmetal atom is changed to "-ide."

EXERCISES

Reading Questions

1. Explain how to use the periodic table to determine the charges on ions.
2. Explain how to use the periodic table to determine the correct formulas for ionic compounds.

Reason and Apply

3. Is each compound possible? Explain your thinking.

   a. LiCl **b.** $LiCl_2$ **c.** MgCl **d.** $MgCl_2$ **e.** $AlCl_3$

4. Give examples of six ionic compounds with a metal to nonmetal ratio of 1:1. Specify the total number of valence electrons for each compound. Name each compound.
5. Give examples of three ionic compounds with a metal to nonmetal ratio of 2:1. Specify the total number of valence electrons for each compound. Name each compound.
6. Give examples of three ionic compounds with a metal to nonmetal ratio of 1:2. Specify the total number of valence electrons for each compound. Name each compound.
7. Predict the formulas for ionic compounds between the following metal and nonmetal elements. Name each compound.

   a. Al and Br **b.** Al and S **c.** Al and As

   d. Na and S **e.** Ca and S **f.** Ga and S

8. For each compound, write the cation and anion with the appropriate charge. Then write the chemical formula for each compound.

   Example: sodium fluoride, Na^+, F^-, NaF

   a. magnesium oxide

   b. rubidium bromide

   c. strontium iodide

   d. beryllium fluoride

   e. aluminum chloride

   f. lead sulfide

LESSON

22 Isn't It Ionic?

Polyatomic Ions

Think About It

So far, we have considered ionic compounds made of only two elements, a metal and a nonmetal. However, there are some ionic compounds that consist of more than two elements. For example, sodium hydroxide, NaOH, which is commonly found in drain cleaner, consists of one metal element and *two* nonmetal elements. You might wonder about the O and the H. Are they both ions?

What is a polyatomic ion?

To answer this question, you will explore

1. Polyatomic Ions
2. Predicting Chemical Formulas for Polyatomic Ions

Exploring the Topic

1 Polyatomic Ions

Polyatomic ion	Name
OH^-	hydroxide
NO_3^-	nitrate
CO_3^{2-}	carbonate
SO_4^{2-}	sulfate
PO_4^{3-}	phosphate
BrO_3^-	bromate
NH_4^+	ammonium

Some ionic compounds contain ions that consist of two or more elements. These ions are called **polyatomic ions.** In contrast, ions that have only one element are called **monatomic ions.** *Mono* means "one" and *poly* means "many." Calcium sulfate, $CaSO_4$, is an example of a compound made up of a monatomic ion and a polyatomic ion. The calcium ion is monatomic and the sulfate ion is polyatomic.

Monatomic ion

Polyatomic ion

2^+

Ca

2^-

O S O O O

Calcium cation

Sulfate anion

$CaSO_4$

Calcium sulfate

It is important to keep in mind that polyatomic ions are a *group* of atoms that stay together. In the above compound, the entire group of atoms has a negative charge.

Names of Polyatomic Ions

Ionic compounds with polyatomic ions have specific naming rules. Each polyatomic ion has its own name, as shown in the table. When naming a compound, you simply insert the polyatomic ion name at either the beginning or ending of the chemical name. The cation is first and the anion is second. For example, a compound that ends with NO_3 is a nitrate. $NaNO_3$ is called sodium *nitrate,* and $Ca(NO_3)_2$ is called calcium *nitrate.*

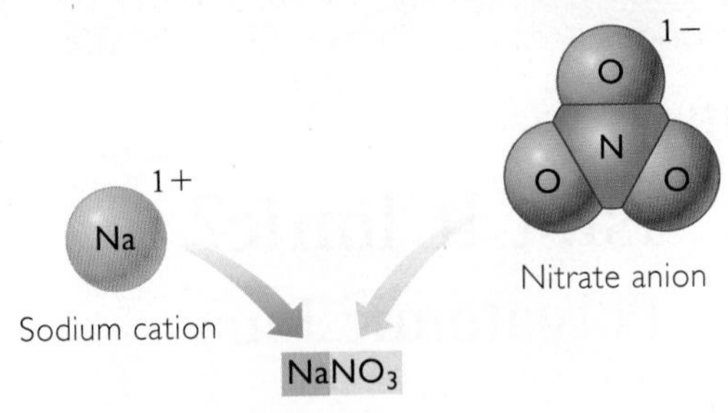

Compound	Chemical formula
calcium sulfate	$CaSO_4$
magnesium hydroxide	$Mg(OH)_2$
sodium nitrate	$NaNO_3$
calcium nitrate	$Ca(NO_3)_2$
ammonium carbonate	$(NH_4)_2CO_3$

2 Predicting Chemical Formulas for Polyatomic Ions

Several chemical formulas with polyatomic ions are listed in the table. Notice that in some cases there are parentheses around the polyatomic ion, with a subscript after the second parenthesis. Why are the formulas written this way? How do you determine the subscript?

Recall that you can use the rule of zero charge to determine the chemical formulas associated with ionic compounds. For example, a single calcium ion combines with a single sulfate ion because the charges add up to zero:

$$Ca^{2+} + SO_4^{2-} \text{ becomes } CaSO_4\text{, calcium sulfate}$$

However, like simple ionic compounds, compounds with polyatomic ions do not always combine in a 1:1 ratio. When there is more than one of the same polyatomic ion in a formula, the ion is enclosed in parentheses and a subscript number indicates how many ions are in the compound. For example, a single calcium ion combines with *two* nitrate ions:

$$Ca^{2+} + NO_3^- + NO_3^- \text{ becomes } Ca(NO_3)_2\text{, calcium nitrate}$$

Below are two more examples of ionic compounds containing polyatomic ions. Each of these compounds consists of three ions. One compound has monatomic and polyatomic ions, while the other has only polyatomic ions. The rule of zero charge applies to both.

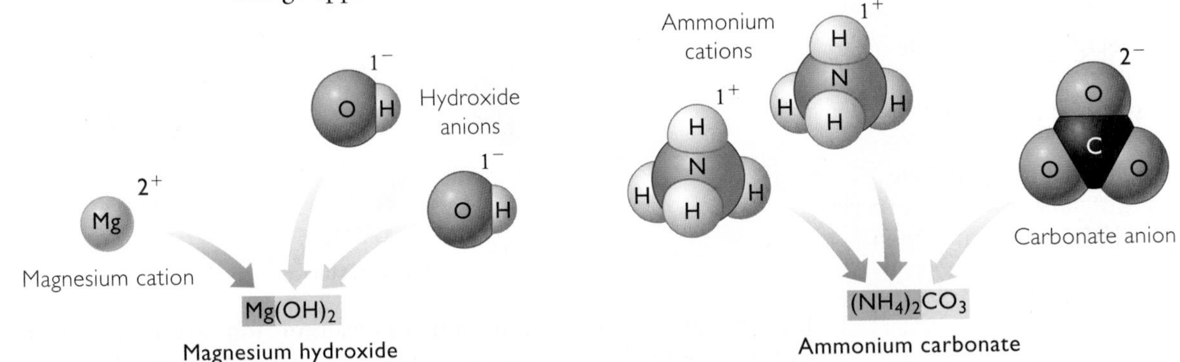

Important to Know A polyatomic ion is treated as a unit. For example, the formula for calcium nitrate is written as $Ca(NO_3)_2$ with parentheses around NO_3. Notice that it is not written as CaN_2O_6. ◀

Key Terms

polyatomic ion
monatomic ion

Lesson Summary

What is a polyatomic ion?

Some ions are composed of more than one atom. These are called polyatomic ions. The entire cluster of atoms shares the charge on a polyatomic ion. If there are two or more polyatomic ions in a chemical formula, parentheses are placed around the ion and a subscript number indicates how many ions are present. To determine the formula of a compound with polyatomic ions, you can use the rule of zero charge. Each polyatomic ion has a unique name that is used when naming the compound it is in.

CONSUMER CONNECTION

Sodium hydroxide, or lye, has many uses. It is used to straighten or curl hair but, if left in too long, it can damage hair and skin.

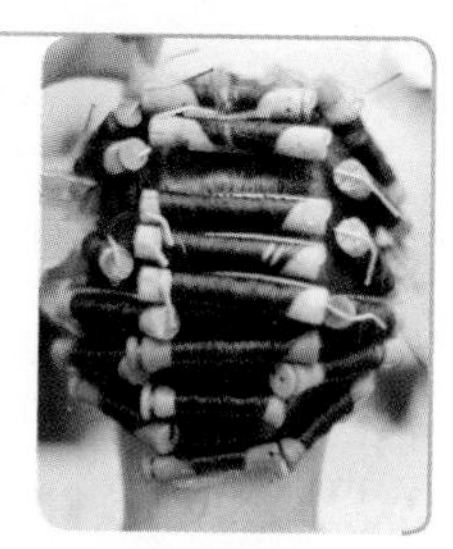

EXERCISES

Reading Questions

1. What is a polyatomic ion?
2. How can you tell from a chemical formula if there is a polyatomic ion in a compound?

Reason and Apply

3. Write the name for each ionic compound listed here.
   a. NH_4Cl
   b K_2SO_4
   c. $Al(OH)_3$
   d. $MgCO_3$
4. Write the chemical formula for each compound listed here.
   a. lithium sulfate
   b. potassium hydroxide
   c. magnesium nitrate
   d. ammonium sulfate
5. Sodium cyanide, NaCN, contains a cyanide ion. What is the charge on the cyanide ion?
6. Calcium phosphate, $Ca_3(PO_4)_2$, contains phosphate ions. What is the charge on a phosphate ion?
7. Which chemical formula does not represent a possible compound with sulfate, SO_4^{2-}? Explain your answer.
   A. Na_2SO_4 **B.** KSO_4 **C.** $Al_2(SO_4)_3$ **D.** $CaSO_4$

LESSON

23 Alchemy of Paint

Transition Metal Chemistry

Think About It

For thousands of years, human beings have expressed themselves through painting. Over time, people have discovered pigments with a wide variety of brilliant colors. The vast majority of paint pigments contain a transition metal cation.

What types of compounds are made from transition metals?

To answer this question, you will explore

1. Transition Metal Compounds
2. Charges on Transition Metal Cations

Exploring the Topic

1 Transition Metal Compounds

HISTORY CONNECTION

Ionic compounds containing transition metals were used to create ancient cave paintings. This cave painting is from Lascaux, France, and is approximately 17,000 years old.

The transition metals are named for their location in the middle of the periodic table. They consist of approximately 30 different elements. Many of the metals that we use in our daily life, such as copper, iron, nickel, silver, and gold, are located in this part of the periodic table.

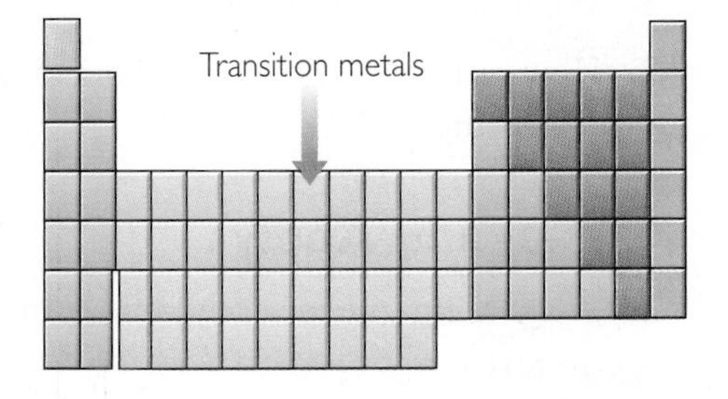

Transition metal compounds tend to be brightly colored. Hence, they are commonly found as *pigments* to color paints. For example, red ochre is a pigment made from iron (III) oxide, Fe_2O_3. It is thought to be the first pigment ever used by human beings for creating art.

The ancient alchemists worked with the transition metals a great deal, mostly because these metals were closer in their properties to gold than to other metals. As a result, alchemists occasionally discovered paint pigments while trying to create gold.

The table on the next page lists various paint pigments that you can buy at an art store. Examine the data. Pay particular attention to the different charges on the transition metal cations.

Notice that the chemical names are a bit unusual. In the middle of each chemical name for a transition metal compound is a Roman numeral: I, II, III, or IV (meaning 1, 2, 3, or 4). This Roman numeral indicates what the charge is on the transition metal cation. Thus, cobalt (II) oxide has a +2 charge on the cobalt ion, and manganese (IV) carbonate has a +4 charge on the manganese ion.

Some Pigments Containing Ionic Compounds with Transition Metals

Color	Pigment name	Chemical name	Chemical formula	Cation	Anion
blue	cobalt blue	cobalt (II) oxide	CoO	Co^{2+}	O^{2-}
earth tone	red ochre	iron (III) oxide	Fe_2O_3	Fe^{3+}	O^{2-}
dark green	viridian	chromium (III) oxide	Cr_2O_3	Cr^{3+}	O^{2-}
brown	umber	manganese (IV) dioxide	MnO_2	Mn^{4+}	O^{2-}
blue-green	malachite	copper (II) carbonate	$CuCO_3$	Cu^{2+}	CO_3^{2-}
white	titanium white	titanium (IV) dioxide	TiO_2	Ti^{4+}	O^{2-}
red	cuprite	copper (I) oxide	Cu_2O	Cu^+	O^{2-}

HISTORY CONNECTION

There is an older naming system for the transition metal ionic compounds. In this system, Fe_2O_3 and FeO were called ferric oxide and ferrous oxide. The "-ic" ending indicated the higher ion charge and the "-ous" ending referred to the lower charge ion. So Co^{2+} was cobaltous and Co^{3+} was cobaltic.

Note that copper, Cu, has a charge of +1 in copper (I) oxide, Cu_2O, and a charge of +2 in copper (II) carbonate, $CuCO_3$.

2 Charges on Transition Metal Cations

Recall that you can determine the charges on the main group metals and nonmetals from their positions on the periodic table. For example, all of the alkali metals, Group 1A, form cations with +1 charges. All of the halogens, Group 7A, form anions with −1 charges. However, you cannot simply determine the charge of a transition metal ion from its location on the periodic table. Instead, it is necessary to determine the charge of the ion from chemical formulas.

To determine the charges on transition metal cations, you use the charges on ions that you do know and apply the rule of zero charge. For example, iron combines with oxygen to form both Fe_2O_3 and FeO. Oxygen is a main group anion, so you can use the periodic table to determine that oxygen atoms form ions with a −2 charge. The rule of zero charge states that the charges on the ions in the compound should add up to zero. Working backward, you can determine the charge on the iron cations in each compound.

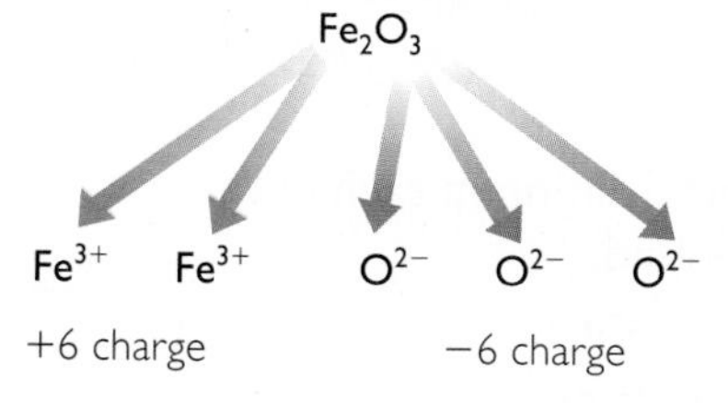

The charges on the three oxygen anions add up to −6. Each iron cation must have a charge of +3.

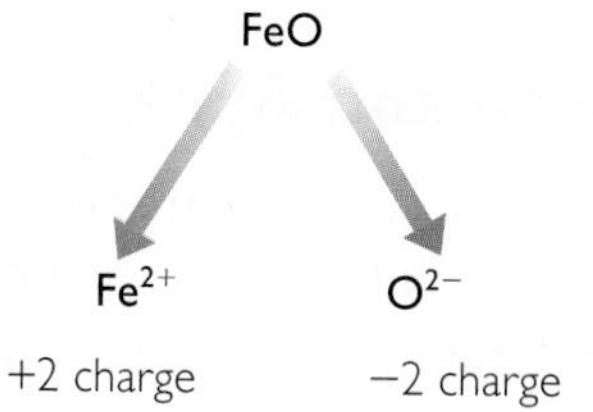

The charge on the single oxygen anion is −2. The iron cation must have a charge of +2.

These two different iron compounds are formally named iron (III) oxide and iron (II) oxide.

FINE ART CONNECTION

Some paint pigments have more complicated chemical compositions than the compounds shown in the table. For example, the pigment Egyptian blue, shown below, is calcium copper silicate, with the formula $CaCuSi_4O_{10}$. There are three different cations and one type of anion.

Important to Know Unlike the main group metals, most transition metals can form several ions with different charges. ◄

CONSUMER CONNECTION

Transition metal ions are responsible for colors in different-colored gem stones. For example, chromium ions, Cr^{3+}, are present in red ruby, green emerald, and pink topaz. Iron ions, Fe^{3+}, are present in citrine and yellow sapphire while Fe^{2+} ions make sapphires blue. The transition metals in crystals absorb certain colors of light while allowing other colors to pass through and be seen.

Example

Transition Metal Compounds

Determine the charge on the transition metal cation in each of the compounds given. Then name the compound.

a. Ag_2S **b.** $Fe(NO_3)_3$

Solution

You can determine the charge on each transition metal cation from the charges on the anions using the rule of zero charge.

a. Sulfur anions have a charge of −2. So each silver cation must have a charge of +1, Ag^+. The compound name is silver (I) sulfide.

b. The polyatomic ion is nitrate with a −1 charge. There are three nitrate ions, so the iron cation must have a charge of +3, Fe^{3+}. The compound name is iron (III) nitrate.

Lesson Summary

What types of compounds are made from transition metals?

Transition metals bond with nonmetal atoms to form ionic compounds. Unlike the main group atoms, most transition metals can have more than one ion charge. The best way to determine the charge on a transition metal cation is to work backward from the anion, whose charge is known. When naming ionic compounds, a Roman numeral is used to indicate the charge on the transition cation. Colorful paint pigments are frequently composed of transition metal compounds.

EXERCISES

Reading Questions

1. What does the Roman numeral in a chemical name indicate?

2. Explain how you determine the charge on a transition metal cation from the chemical formula.

Reason and Apply

3. Determine the charge on the transition metal cation in each of the compounds listed. Then name each compound.

a. HgS **b.** $CuCO_3$ **c.** $NiCl_2$
d. $Co(NO_3)_3$ **e.** $Cu(OH)_2$ **f.** $FeSO_4$

4. Write the cation and anion in each compound, then determine the correct chemical formula.

a. copper (II) sulfide **b.** nickel (II) nitrate **c.** iron (II) carbonate
d. cobalt (II) sulfate **e.** iron (III) carbonate **f.** chromium (VI) oxide

5. Cobalt violet is a paint pigment discovered in 1859. If the cation for this compound is Co^{2+} and the anion is PO_4^{3-}, what is the chemical formula?

LESSON

24 Shell Game

Electron Configurations

Think About It

Recall that the chemistry of the elements is closely related to the number of valence electrons in their atoms. The valence electrons are found in the outermost electron shell of an atom.

What does the periodic table indicate about the arrangements of electrons?

To answer this question, you will explore

1. Subshells in Atoms
2. Electron Configurations
3. Connecting the Periodic Table to Electron Arrangements
4. Noble Gas Shorthand

Exploring the Topic

1 Subshells in Atoms

Electrons are arranged into shells numbered $n = 1, 2, 3$, and so on. The number of electron shells in an atom is the same as the number of the period where the element is located on the periodic table. Each shell has a maximum number of electrons. For instance, the $n = 2$ shell cannot have more than 8 electrons.

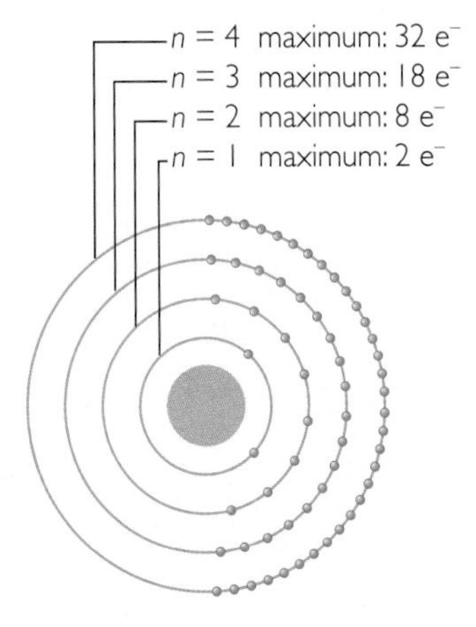

Each electron shell can have a maximum number of electrons.

Scientific evidence has led chemists to propose that electron shells are further divided into electron subshells. Imagine magnifying the basic atomic model and finding that each shell is composed of subshells. Notice that the number of subshells that a shell has is equal to n.

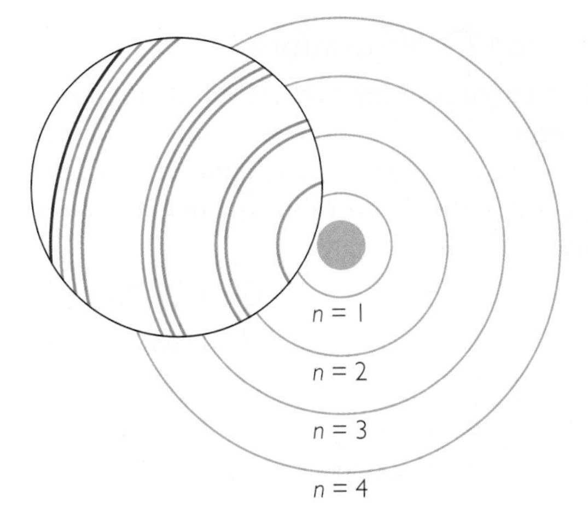

Electron shells are further divided into subshells.

The s, p, d, and f Subshells

The subshells have special names. They are called the s, p, d, and f subshells. Just like the basic shells, each subshell has a maximum capacity of electrons. The s subshells can have a maximum of 2 electrons, p subshells can have a maximum of 6 electrons, and d subshells can have a maximum of 10 electrons. Finally, f subshells can have a maximum of 14 electrons.

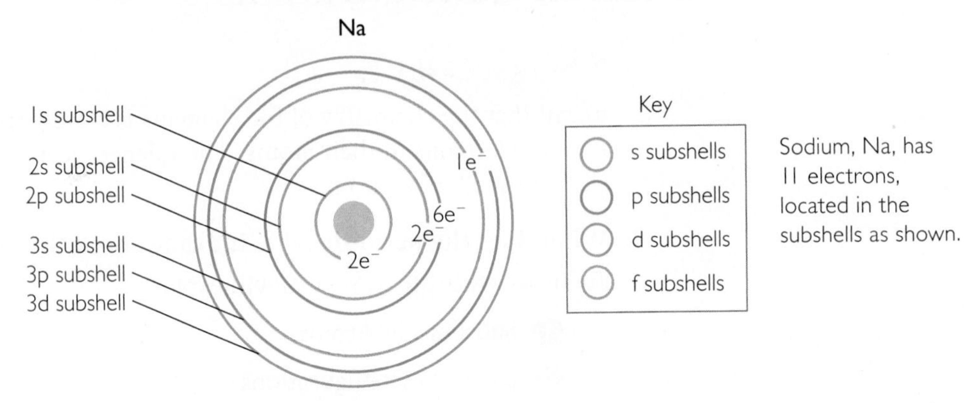

Notice that the name of each subshell is labeled using both the basic shell number and the subshell letter (1s, 2s, 2p, and so on).

Example 1

Electron Arrangements

Use the illustration of the subshells in a sodium atom above to help you answer these questions:

a. How many total electrons are there in a sodium, Na, atom? Which shells are they in?

b. How many valence electrons does sodium have? Which subshell are they in?

c. How many electrons are there in the 3s subshell of sodium? In the 3p subshell?

Solution

The atomic number of sodium is 11.

a. There are a total of 11 electrons in a neutral sodium atom. The electrons are in shells, $n = 1$, $n = 2$, $n = 3$.

b. Sodium has one valence electron, located in the 3s subshell.

c. There is one electron in the 3s subshell of sodium, and none in the 3p subshell.

2 Electron Configurations

It can be time consuming to draw subshell models of the atoms to show the arrangements of the electrons, especially for atoms with large atomic numbers. Chemists have developed a shorthand notation called an **electron configuration** to keep track of the electrons in an atom. The electron configurations for the first ten elements are shown here.

H $1s^1$ | C $1s^2 2s^2 2p^2$
He $1s^2$ | N $1s^2 2s^2 2p^3$
Li $1s^2 2s^1$ | O $1s^2 2s^2 2p^4$
Be $1s^2 2s^2$ | F $1s^2 2s^2 2p^5$
B $1s^2 2s^2 2p^1$ | Ne $1s^2 2s^2 2p^6$

Nitrogen has a total of 7 electrons.

Each subshell is written using the shell number and the subshell letter. In addition, the number of electrons in each subshell is indicated with a superscript number.

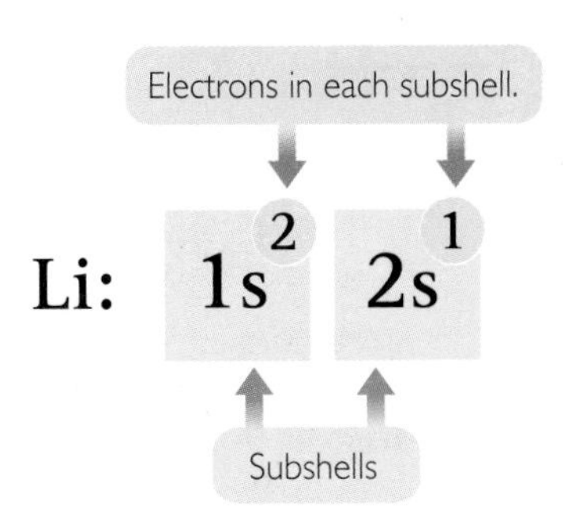

Notice that the superscript numbers add up to the total number of electrons for that atom.

The sequence in which electrons fill up the subshells is 1s, 2s, 2p, 3s, 3p. After the element argon the pattern changes slightly.

Topic: Electron Configuration

Visit: www.SciLinks.org

Web code: KEY-124

Example 2

Electron Configuration of Sulfur

Write the electron configuration for sulfur, S.

Solution

Sulfur is located in the third row in Group 6A. The atomic number of sulfur is 16, so there are 16 electrons that need to be distributed in subshells beginning with the 1s subshell.

The electron configuration of sulfur is $1s^22s^22p^63s^23p^4$.

3 Connecting the Periodic Table to Electron Arrangements

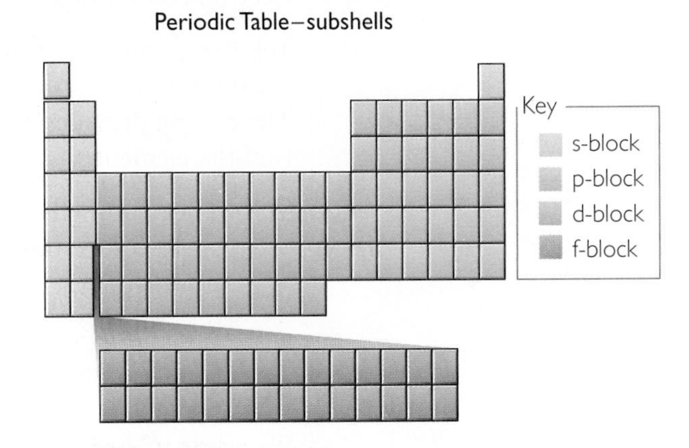

An outline of the periodic table appears here with color-coding to show the subshell for the outermost electron of each element. For example, any element located in the green area will have its outermost electron(s) in a p subshell.

As you proceed across the periodic table from one element to the next, one additional proton and one additional electron are added, along with one or more neutrons. Each additional electron goes into a specific subshell. If an atom is located in the orange areas of the table, the last electron is placed into an s subshell. If an atom is located in the blue area of the table, the last electron is placed into a d subshell. And so on.

The elements in each block have related properties. The elements in the s-block are reactive metals. The elements in the d-block tend to form colorful compounds that are used as pigments. The elements in the p-block tend to form colorless compounds.

Decoding the Table

In order to write out an electron configuration for a specific element, you can simply "read" from the periodic table, moving across from left to right and then down to the next row. For example, the sequence of subshells for argon, Ar, is 1s, 2s, 2p, 3s, 3p. The electron configuration for argon is $1s^2 2s^2 2p^6 3s^2 3p^6$.

Periodic Table—subshells

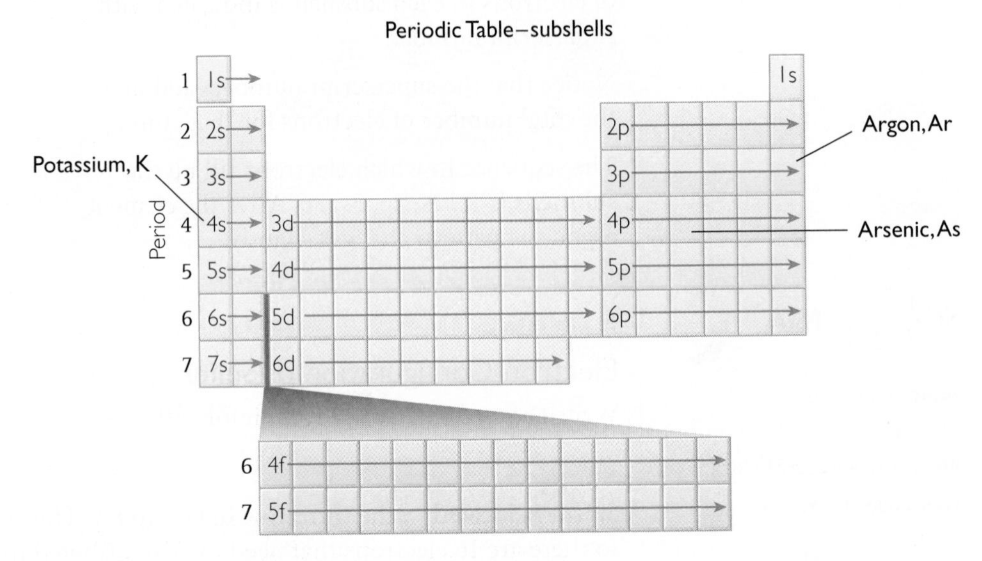

Everything runs smoothly until you reach the fourth row of the periodic table. After argon, you might expect the next electron to be in the 3d subshell. However, this does not happen. The next element is potassium, K. Like the other elements in Group 1A, potassium has one electron in the s subshell. Thus, you place an electron in 4s before 3d. The electron configuration for potassium is $1s^2 2s^2 2p^6 3s^2 3p^6 4s^1$. The electron configuration for arsenic is $1s^2 2s^2 2p^6 3s^2 3p^6 4s^2 3d^{10} 4p^3$.

You may have noticed that you only have to look at the ending of each electron configuration in order to figure out the identity of the element associated with it. The ending provides you with the exact spot on the periodic table where you can find the element.

Important to Know The s subshells fill with electrons before the d subshells from the previous shell. For example, the 4s subshell fills before the 3d subshell, the 5s subshell fills before the 4d subshell, and so on. ◀

Example 3

Electron Configuration of Cobalt

Write the electron configuration for cobalt, Co.

Solution

Locate cobalt on the periodic table. It is element number 27 and is located in the fourth period of the periodic table.

Simply trace your finger across the periodic table of subshells, writing the subshells as you go. When you get to cobalt, stop writing. Every subshell up to the 4s subshell is completely filled. In addition, cobalt has seven electrons in the 3d subshell. The answer is $1s^2 2s^2 2p^6 3s^2 3p^6 4s^2 3d^7$.

You can check that you have the correct electron configuration by adding the superscript numbers to make sure there are 27 electrons.

④ Noble Gas Shorthand

Depending on the element, the electron configuration can be lengthy to write. Plus, each element just repeats the electron configuration of the previous element, but adds one more electron.

Rather than repeat the same thing every time, chemists have devised a quicker way to write out electron configurations. They use the noble gas at the end of each period as a placeholder to symbolize all of the filled subshells before that place on the table. Using this "shorthand" method, the electron configuration of cobalt is $[Ar]4s^2 3d^7$.

Shorthand notation allows you to make some interesting comparisons. Notice that the noble gas shorthand notation emphasizes the valence electrons. Using this method, it is easy to see that each element in Group 2A has two valence electrons, both located in an s subshell.

Group 2A Elements

Element	Symbol	Electron configuration
beryllium	Be	$[He]2s^2$
magnesium	Mg	$[Ne]3s^2$
calcium	Ca	$[Ar]4s^2$
strontium	Sr	$[Kr]5s^2$
barium	Ba	$[Xe]6s^2$
radium	Ra	$[Rn]7s^2$

Example 4

Electron Configuration of Selenium

Find the element Selenium, Se, element 34, on the periodic table.

a. What is the electron configuration of selenium?

b. Write the electron configuration using noble gas shorthand.

c. In what subshells are selenium's valence electrons?

Solution

a. Selenium is in the p-block, in Period 4. The electron configuration of selenium is $1s^2 2s^2 2p^6 3s^2 3p^6 4s^2 3d^{10} 4p^4$.

b. The noble gas that comes before selenium is argon, Ar. Thus, the noble gas shorthand for this configuration is $[Ar]4s^2 3d^{10} 4p^4$.

c. Selenium's valence electrons are in subshells 4s and 4p.

Key Term

electron configuration

Lesson Summary

What does the periodic table indicate about the arrangements of electrons?

Electrons in atoms are arranged into basic shells labeled $n = 1, 2, 3$, and so on. These shells are divided into subshells. The number of subshells in each shell is equal to n.

The subshells are referred to as s, p, d, and f subshells. The s, p, d, and f subshells can have a maximum of 2, 6, 10, and 14 electrons, respectively. Chemists use electron configurations to specify the arrangements of electrons in subshells. The periodic table provides the information needed to write electron configurations.

EXERCISES

Reading Questions

1. What are electron subshells?
2. What is an electron configuration?
3. How is the arrangement of electrons in an atom related to the location of the atom on the periodic table?

Reason and Apply

4. How many subshells are in each shell: $n = 1$, $n = 2$, $n = 3$, $n = 4$?
5. What is the total number of subshells for elements in Period 5 of the periodic table?
6. Draw a subshell model for each of the following elements, putting the electrons in their appropriate places.

   a. sodium, Na **b.** neon, Ne **c.** carbon, C **d.** vanadium, V
7. What is the outermost subshell for bromine, Br?
8. Name an element with electrons in the f subshell.
9. Consider the element with the atomic number 13.
   a. What is the electron configuration for the element with atomic number 13?
   b. How many valence electrons does element 13 have? How do you know?
   c. How many core electrons does element 13 have? How did you figure that out?
10. Explain why the chemical properties of argon, krypton, and xenon are similar, even though there are 18 elements between argon and krypton, and 32 elements between krypton and xenon.
11. Write the electron configurations for each of these atoms. Then write it using the noble gas shorthand method.

    a. oxygen **b.** chlorine **c.** iron **d.** calcium
    e. magnesium **f.** silver **g.** silicon **h.** mercury
12. You should be able to figure out the identity of an atom from its electron configuration alone. Describe at least two ways you could do this.
13. Which elements are described by these electron configurations?

    a. $1s^22s^22p^63s^23p^64s^23d^4$ **b.** $1s^22s^22p^63s^23p^2$
    c. $1s^22s^22p^3$ **d.** $1s^22s^22p^63s^23p^64s^23d^{10}4p^65s^24d^{10}5p^66s^1$
    e. $1s^22s^22p^63s^23p^64s^23d^{10}4p^65s^24d^{10}5p^66s^24f^{14}5d^{10}6p^2$ **f.** $[Kr]5s^24d^9$

SECTION

IV

SUMMARY

Moving Electrons

Alchemy Update

What do the electrons of an atom have to do with its properties?

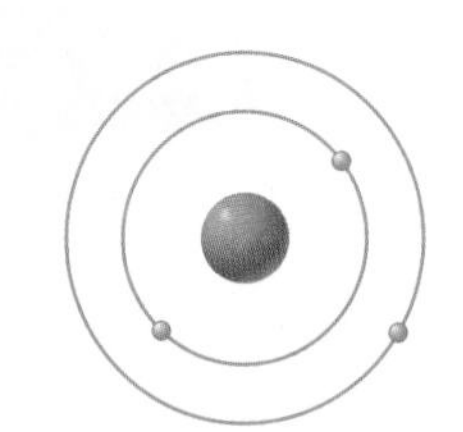

The outermost electron shell contains an atom's valence electrons. Main group elements have the same number of valence electrons as their group number. Valence electrons are responsible for many of an element's properties, so elements in a group have similar properties. While you cannot change copper atoms into gold atoms, you can combine elements into compounds with new properties.

The electrons in an atom are fairly easy to change, unlike the parts of a nucleus. Atoms become more stable when they transfer electrons to other atoms, forming ionic compounds.

Key Terms

valence shell
valence electron
core electron
ion
cation
anion
ionic compound
rule of zero charge
polyatomic ion
monatomic ion
electron configuration

Review Questions

1. Explain the importance of valence electrons.
2. What are ionic compounds, and how do they form?
3. How is the arrangement of electrons within atoms connected to the periodic table?
4. Write the electron configuration for aluminum. Then write it using noble gas shorthand.
5. For each compound listed, identify the anion and cation and the charge on each.
   a. magnesium chloride, $MgCl_2$
   b. calcium nitrite, $Ca(NO_2)_2$
6. Three polyatomic ions are listed here.

   silicate, SiO_3^{2-} chlorite, ClO_2^- bicarbonate, HCO_3^-

   a. Write the chemical formulas for sodium silicate, sodium chlorite, and sodium bicarbonate.
   b. Write the chemical formulas for calcium silicate, calcium chlorite, and calcium bicarbonate.

CONSUMER CONNECTION

Sodium bicarbonate, $NaHCO_3$, is commonly known as baking soda. When it is used in baking, it reacts with other ingredients to release carbon dioxide. For example, baking soda is what causes the batter to rise and bubble when you make pancakes.

SECTION

V

Building With Matter

Sugar dissolves in water, but glass does not. These properties give clues about how the atoms in these substances are bonded to each other.

When atoms form bonds between them, new substances with new properties are formed. There are four basic types of bonds found in the world around us. Each type of bond is a result of a different distribution of electrons within the substance. In addition, each type of bond is associated with certain properties. The chemical formula of a substance can be used to determine the type of bond in that substance.

In this section, you will study

- the basic types of bonds between atoms
- the role of electrons in bonding
- the properties and types of elements associated with each type of bond
- the formation of new substances through bonding

LAB Classifying Substances

Purpose

To investigate the properties of substances.

Materials

- bulb with wires
- 9-volt battery with snap connector
- wire with stripped ends
- 100 mL beakers
- paper clips
- salt, $NaCl$
- sand, SiO
- paraffin wax, $C_{20}H_{42}$
- calcium chloride, $CaCl_2$
- copper, Cu
- copper (II) sulfate, $CuSO_4$
- aluminum foil, Al
- sucrose, $C_{12}H_{22}O_{11}$
- distilled water, H_2O

Predictions and Data

Predict whether each substance will conduct electricity and whether it will dissolve in water. Make a table for your predictions and data.

Procedure

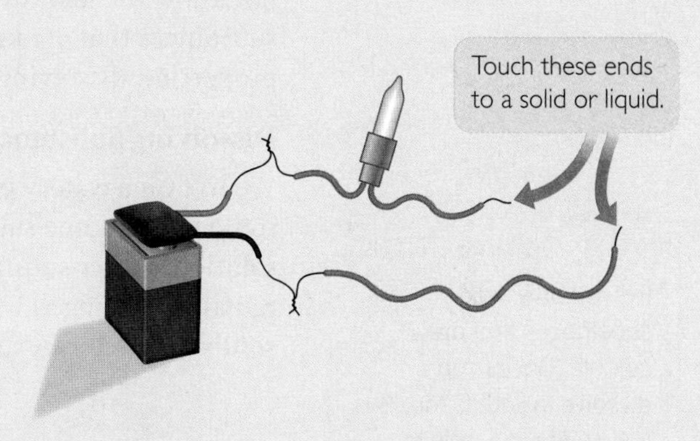

1. Assemble your conductivity tester as shown.
2. Take your conductivity tester to each station. Test the pure substance for conductivity.
3. Next, observe the substance in water. Did the substance dissolve?
4. Finally, test the substance mixed with the distilled water in the second beaker for conductivity. Record your results in a data table.

Analysis

5. Group the substances according to their properties. How do your results compare with your predicitons? What do the substances in each group have in common? Summarize your findings in a paragraph.

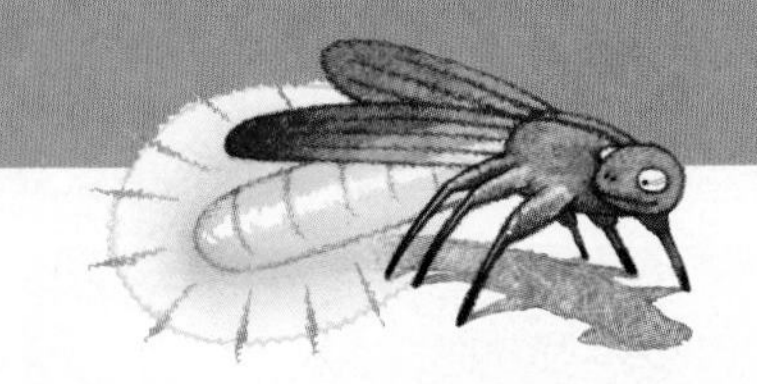

LESSON

25 You Light Up My Life

Classifying Substances

Think About It

If you accidentally drop a piece of jewelry into a tub of water, you should have no trouble getting it back again. However, if you drop a cube of sugar into the water, it will dissolve. Some substances dissolve in water and others do not. So, dissolving is one property of matter that you can use to sort substances into general groups.

How can substances be sorted into general categories?

To answer this question, you will explore

1. Dissolving and Conductivity
2. Testing and Sorting Substances

Exploring the Topic

1 Dissolving and Conductivity

There are millions of different substances in our world. Even chemists do not know the identities of all the possible compounds. However, many of the substances that are known can be sorted into general categories by looking at two properties: dissolving and conductivity.

Dissolving Substances in Water

We live on a watery planet. Most substances come into contact with water at some point. Some substances, like sugar, **dissolve** in water to make an aqueous solution. Other substances, like a gold ring, do not dissolve in water and will remain unchanged when placed in water. A substance that dissolves in water is **soluble** in water. A substance that does not dissolve is **insoluble** in water.

Many substances can be sorted into one of two categories based on whether they dissolve in water.

Conductivity

By examining a second property, it is possible to sort matter even further. The property of **conductivity** has to do with whether a substance will conduct electricity. Electrical conductivity requires the movement of ions or electrons. Copper wire is a great conductor of electricity. Your entire home is wired with hundreds of feet of

BIOLOGY CONNECTION

It is not only solid substances that are soluble. Gases can dissolve in liquids too. This is how fish are able to survive underwater. The ocean is full of dissolved oxygen that fish can extract from the water using their gills.

copper or aluminum wiring that carries electricity from the power lines outside to the electrical outlets in your home. The human body is another good conductor of electricity. However, it uses bodily fluids to conduct electricity, not wiring.

Some substances do not conduct electricity. The electrical wires in your house are covered with a coating that does not conduct electricity. That way, you can plug in a stereo or a lamp and not get an electrical shock.

The light bulb doesn't light up because there's a break in the wire.

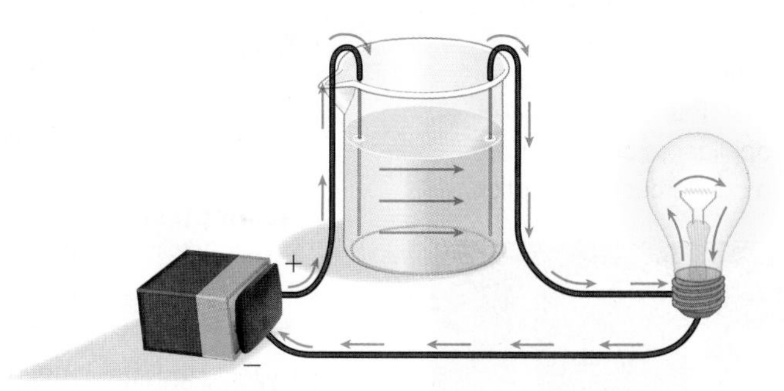

Even though there is a break in the wire, the bulb lights up because the liquid in the beaker conducts electricity.

Electrical conductivity can be tested by setting up a simple electrical circuit. Wires connect the terminals of a battery to a light bulb. When the circuit is complete, and an electrical current is flowing, the bulb will light up. If the flow of current is interrupted by the presence of a substance that does not conduct electricity, the light bulb will not light up.

❷ Testing and Sorting Substances

You can use a two-step test to sort substances by both properties. First, drop a substance into water to find out if it dissolves. Next, test it for conductivity. This allows you to sort all matter into four basic categories as shown in the illustration below.

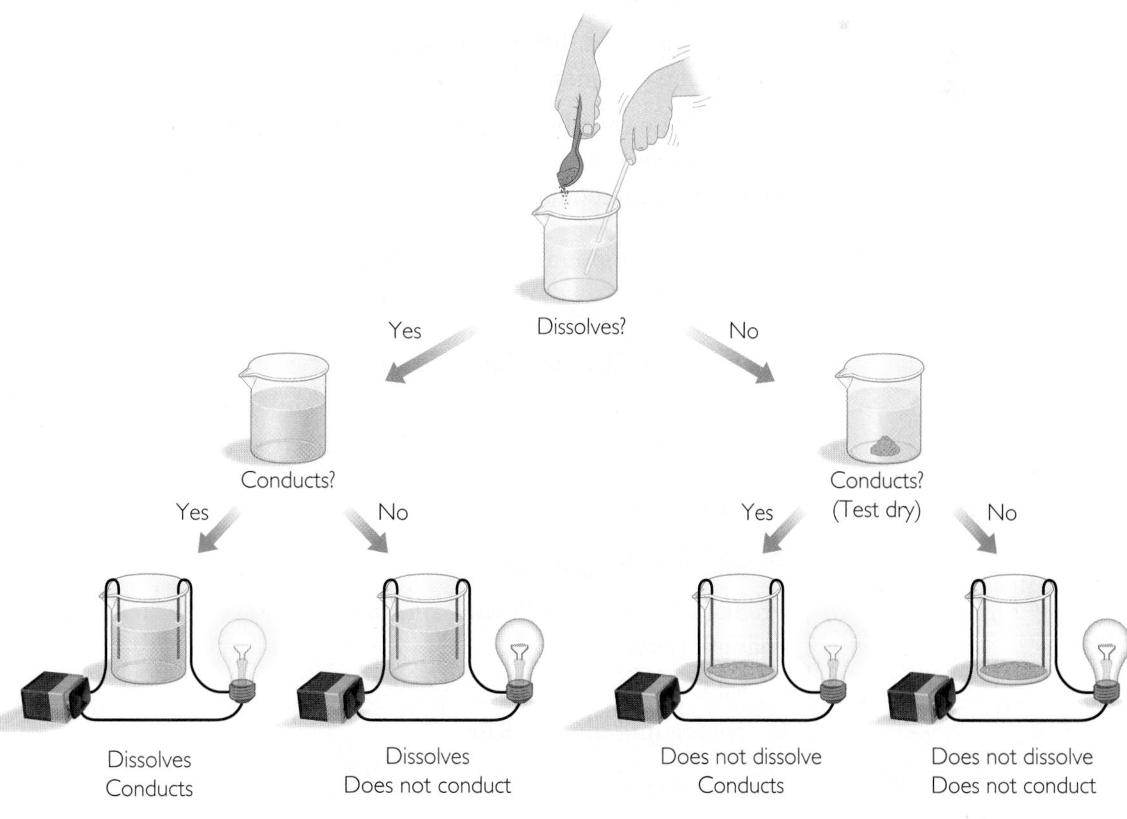

CONSUMER CONNECTION

Water that is 100% pure does not conduct electricity. However, the water that comes from your tap is loaded with dissolved ionic compounds, so it conducts electricity very well. This is why electrical appliances and water are a dangerous combination.

Examples of substances in each of the four categories are given in the table. Take a moment to look for patterns in the chemical formulas of the substances in each category. Note whether substances are made from metal atoms, nonmetal atoms, or some combination.

Types of Substances

Properties	dissolves: yes conducts: yes	dissolves: yes conducts: no	dissolves: no conducts: yes	dissolves: no conducts: no
Examples	salt, NaCl; calcium chloride, $CaCl_2$; copper sulfate, $CuSO_4$	water, H_2O; sugar, $C_{12}H_{22}O_{11}$; ethanol, C_2H_6O	gold, Au; copper, Cu; aluminum, Al	sand, SiO_2; paraffin, $C_{20}H_{42}$
Types of atoms	metal and nonmetal atoms	nonmetal atoms only	metal atoms only	nonmetal atoms only

Here are some patterns you might notice:

- The substances that are *soluble* in water and *conduct* electricity are made up of a metal element and one or more nonmetal elements. Remember, these are called ionic compounds.
- The substances that are *soluble* in water but *do not conduct* electricity are often made up of carbon, hydrogen, and oxygen atoms joined together. These are all nonmetal elements.
- The substances that are *insoluble* in water but *do conduct* electricity are made of metallic elements.
- The substances that are *insoluble* in water and *do not conduct* electricity are made up of nonmetal atoms.

Notice that only substances that contain metal atoms will conduct electricity. Also, substances that are made entirely of metal atoms will not dissolve in water. Many ionic compounds dissolve in water.

Important to Know Ionic compounds conduct electricity only when they are dissolved in water. Dry ionic solids do not conduct electricity. ◄

In Lesson 26: Electron Glue, you will discover that these two properties, solubility and conductivity, are directly related to the manner in which the individual atoms in these substances are linked.

Example

Predicting Properties

Predict whether the following substances dissolve in water and whether they conduct electricity.

a. lead, Pb

b. potassium bromide, KBr

Solution

a. Lead is a metal element. Because it is made of only metal atoms, it does not dissolve in water, but it does conduct electricity.

b. Potassium bromide consists of a metal and a nonmetal. It is an ionic compound. Therefore, it does dissolve in water, and the mixture does conduct electricity. The dry compound does not conduct electricity.

Key Terms

dissolve
soluble
insoluble
conductivity

Lesson Summary

How can substances be sorted into general categories?

Most substances on the planet can be sorted into four categories based on two properties: whether they are soluble in water and whether they conduct electricity. As you will discover in coming lessons, the properties of conductivity and solubility are directly related to the way in which atoms are connected to each other.

EXERCISES

Reading Questions

1. What does insoluble mean?
2. Describe a way to determine whether or not a substance conducts electricity.

Reason and Apply

3. What generalization can you make about a substance that is soluble in water and conducts electricity once it has dissolved? Explain your thinking.
4. Do all solid substances containing metals conduct electricity? Explain your reasoning.
5. If a substance does not conduct electricity as a solid, does this mean that it will not conduct electricity if it dissolves? Explain your reasoning.
6. Predict whether each substance listed will conduct electricity, dissolve in water, and/or conduct electricity once it has dissolved. Explain your thinking in each case.

   a. $C_3H_6O(l)$ acetone
   b. $Ti(s)$ titanium
   c. $LiNO_3(s)$ lithium nitrate
   d. $CuZn(s)$ bronze

Coal and diamond are two naturally occurring forms of carbon.

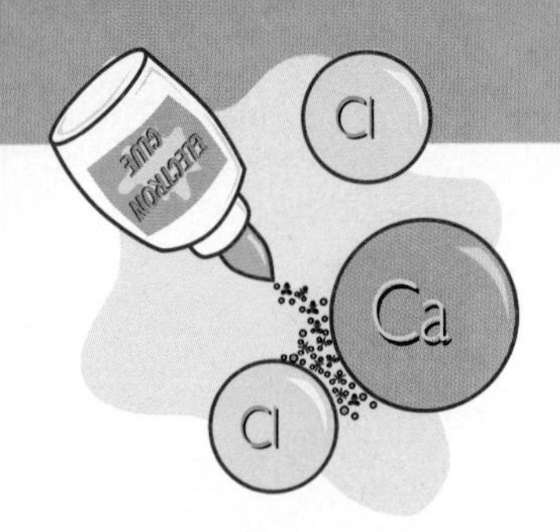

LESSON
26 Electron Glue
Bonding

Think About It

All of the objects in our everyday world, including ourselves, are made up of individual atoms. But what holds those atoms together? Why don't objects just crumble into piles of individual atoms? Something must be holding the atoms together. And why is the desk in your classroom solid, while water simply runs through your fingers? Something about the way atoms are connected must give substances the properties we observe.

How are atoms connected to one another?

To answer this question, you will explore

1. Bonds: The "Glue" Between Atoms
2. Types of Bonding
3. Relating Bonds and Properties

HISTORY CONNECTION

Stephanie Kwolek started working for DuPont in 1946. In 1965, she succeeded in creating synthetic fibers of exceptional strength due to their extremely strong bonds. This led to Kevlar, the material used in bulletproof vests.

Exploring the Topic

1 Bonds: The "Glue" Between Atoms

Chemists call the attraction that holds atoms together a **chemical bond.** As you will discover, several different types of chemical bonds exist. All bonds involve the electrons in some way. A chemical bond is essentially an attraction between the positive charges on the nucleus of one atom and the negative charges on the electrons of another atom. This attraction is so great that it keeps the atoms connected to one another.

2 Types of Bonding

Recall from Lesson 25 that most substances can be divided into four categories, based on their physical properties. These four categories can be explained by different models of bonding.

A bond is a force of attraction, so it is not possible to see the actual bonds between atoms. However, a model can help to explain how atoms are bonded in substances. Models can also help us to understand how bonding accounts for certain properties of substances that we observe.

BIG IDEA There are four main types of chemical bonds between atoms.

The types of bonding are called *ionic, molecular covalent, metallic,* and *network covalent.* Take a moment to locate the valence electrons in each model on the next page. The red spheres represent the nuclei of atoms and the core electrons, while the blue areas suggest where the valence electrons are located.

Four Types of Bonding

MODEL 1: IONIC

Properties of ionic substances:

Dissolve in water

Conduct electricity when dissolved

Tend to be brittle solids

Made of metal and nonmetal atoms combined

In **ionic bonding,** the valence electrons are *transferred* from one atom to another. Metal atoms transfer their valence electrons to nonmetal atoms.

MODEL 2: MOLECULAR COVALENT

Properties of molecular covalent substances:

Some dissolve in water, some do not

Do not conduct electricity

Some are liquids or gases

Made entirely of nonmetal atoms

In **molecular covalent bonding,** the valence electrons are shared between pairs or groups of atoms. This creates small stable units, called molecules, within the substance.

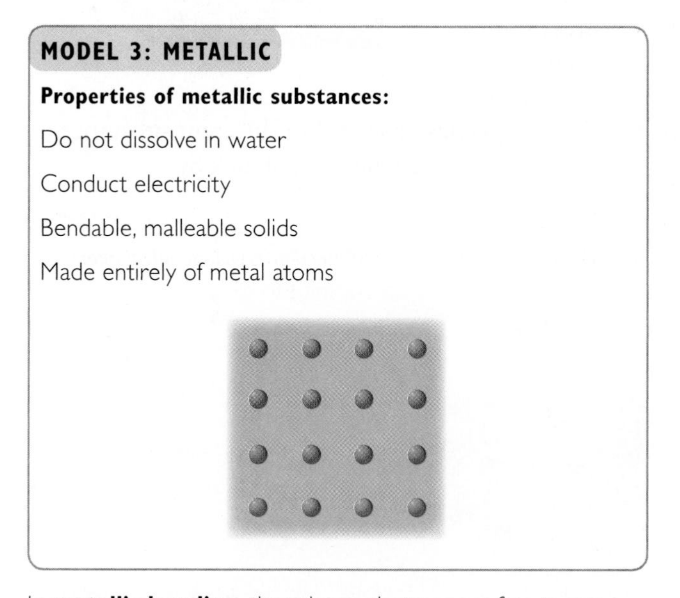

In **metallic bonding,** the valence electrons are free to move about the substance.

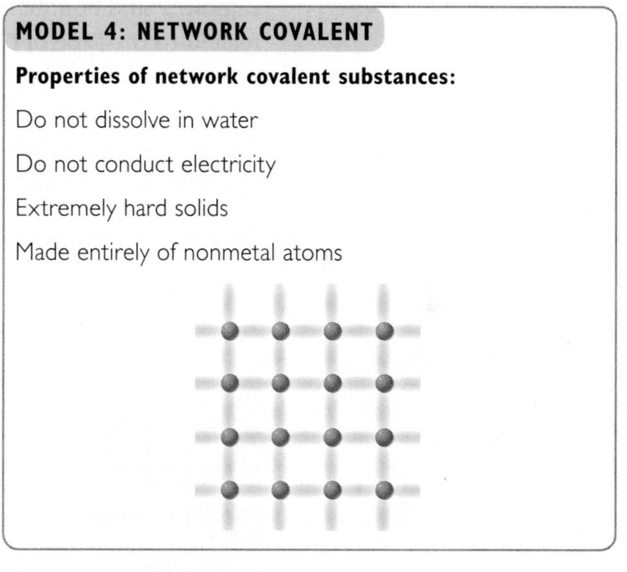

Network covalent bonding is similar to molecular covalent bonding, but the valence electrons are shared throughout the entire substance.

You have already been introduced to ionic compounds. These all have ionic bonding in which metal atoms transfer valence electrons to nonmetal atoms. The resulting oppositely charged ions are strongly attracted to each other. This attraction is what holds the ions together.

In **covalent bonding,** the nucleus of one atom is attracted to the valence electrons of another atom. Unlike ionic bonding, one atom does not transfer an electron to the other. Instead both atoms *share* the valence electrons between them.

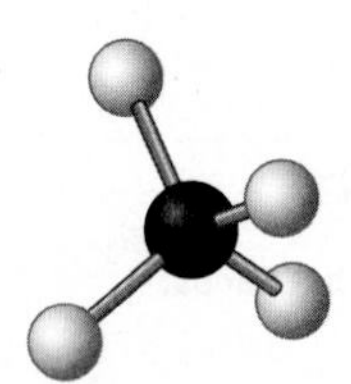

Methane, CH_4

Covalent bonding can happen in two different ways. In molecular covalent bonding, the atoms bond to form individual clusters called **molecules,** such as the methane molecule shown here.

In network covalent bonding, the valence electrons are shared between atoms but form a highly regular extended network, creating a very durable structure. Diamond consists of carbon atoms that are covalently bonded in a network.

Diamond consists of carbon atoms that are covalently bonded in a network.

In a metal, the valence electrons are distributed throughout the substance in what is sometimes called a "sea" of electrons. The valence electrons are free to move throughout the substance. The atoms are bonded by the attraction between the positively charged atoms and the negatively charged "sea" of electrons.

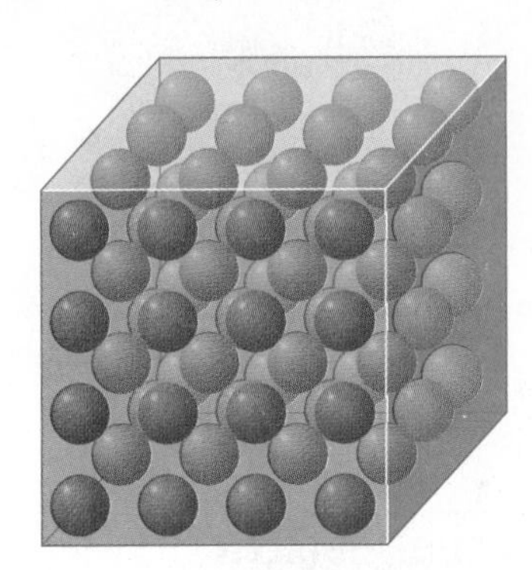

3 Relating Bonds and Properties

Some properties of substances, such as solubility and conductivity, are directly related to the type of bonds the atoms in the substances have. Therefore, it is possible to match the bonding with the physical properties observed in different substances. Examine what happens to each type of substance when it is struck by a hypothetical hammer.

Ionic substances

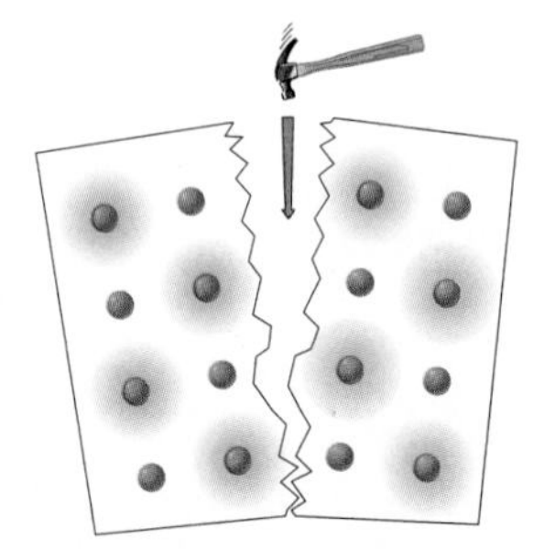

Hard but brittle, tend to fracture along planes of atoms

Network covalent substances

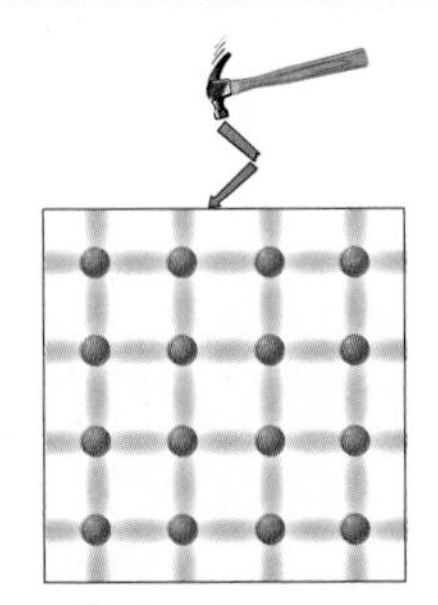

Durable, rigid, difficult to break

Metallic substances

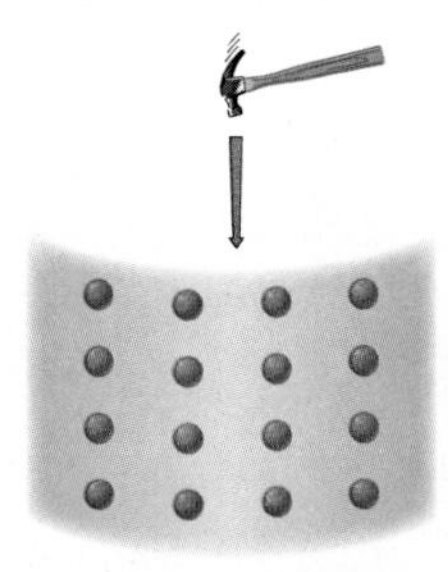

Bendable, malleable

Molecular covalent substances

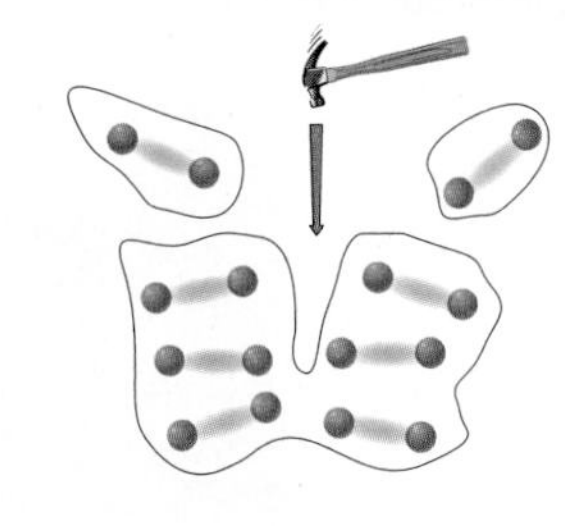

Often gases or liquids, or soft solids

Notice that the hardest substance is a solid with network covalent bonding. This is because bonding in these substances is in an organized network.

Bonding can also help to explain the properties of dissolving and conductivity. Examine the illustration on the top of the next page representing dissolving. Water is represented by the lighter blue areas. Ionic solids and molecular covalent substances dissolve in water. Metallic solids and network covalent solids do not.

INDUSTRY CONNECTION

Metals can be combined into alloys. For example, CuZn is brass, a copper-zinc alloy. Chrome-moly is an alloy of steel, chrome, and molybdenum, used to make bicycle frames that are stronger and lighter than those made of steel. Steel itself is an alloy of iron and carbon.

Ionic substances	Network covalent substances	Metallic substances	Molecular covalent substances
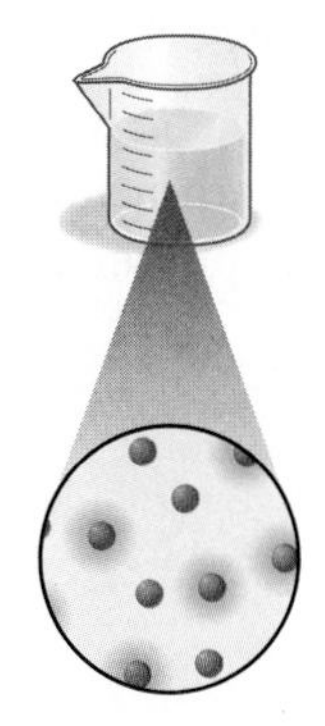	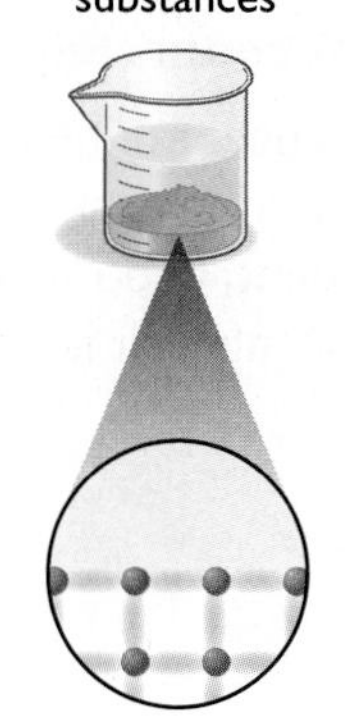	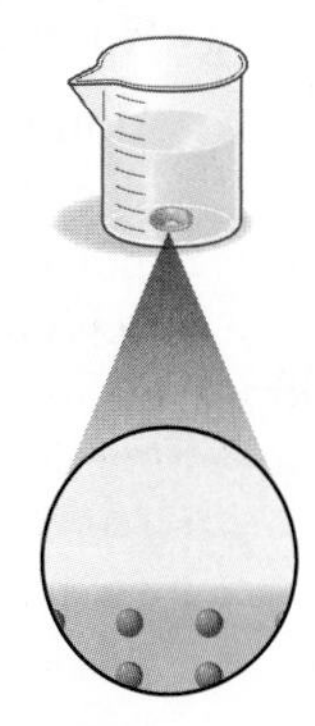	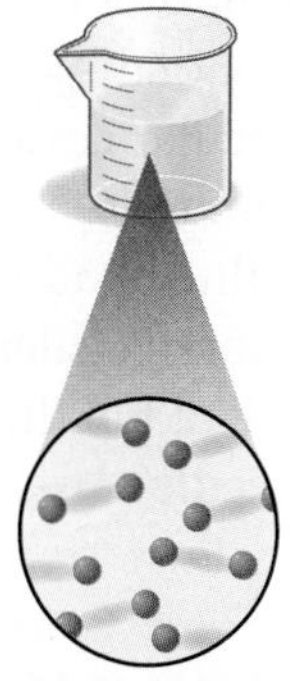
Dissolve into metal and nonmetal ions	Do not dissolve	Do not dissolve	Dissolve and molecules scatter in water

Conduction requires the movement of a charged particle, either an ion or an electron. Metals conduct electricity because the valence electrons are free to move throughout the solid. Ionic compounds that have been dissolved in water conduct electricity because the cations and anions are free to move in the solution. Network covalent solids and molecular covalent substances do not conduct electricity. The charge cannot move in these substances because the electrons are "stuck" between the atoms and are not available to move.

The periodic table is a valuable tool in figuring out bonding. You can use the table to determine if the elements in a compound are metals, nonmetals, or both.

- Ionic compounds, such as salts, are made from metal and nonmetal elements.
- Metallic compounds, such as brass, are made only of metal atoms.
- Network covalent compounds, such as diamonds, and molecular covalent compounds, such as methane, are made from nonmetals.

CONSUMER CONNECTION

Chalk, which is brittle, is an example of an ionic substance. The chemical name and formula for chalk is calcium carbonate, $CaCO_3$.

Example

Identifying the Type of Bonding

Determine the bonding in each of the following substances. What general physical properties can you expect of each substance?

a. magnesium chloride, $MgCl_2$

b. rubbing alcohol, C_3H_8O

Solution

a. Magnesium chloride, $MgCl_2$, is an ionic compound with ionic bonding because it is made of a metal and a nonmetal element. It is probably brittle, dissolves in water, and conducts electricity when dissolved.

b. Rubbing alcohol, C_3H_8O, is a molecular covalent substance with molecular covalent bonding because it is made entirely of nonmetal atoms and is a liquid.

Key Terms

chemical bond
ionic bonding
molecular covalent bonding
metallic bonding
network covalent bonding
covalent bonding
molecule

Lesson Summary

How are atoms connected to one another?

Atoms in substances are held together by chemical bonds. Chemists have identified four main types of bonding within substances: ionic, network covalent, molecular covalent, and metallic. Many properties of substances correspond to the type of bonding that is present.

EXERCISES

Reading Questions

1. Explain why substances do not simply crumble into piles of atoms.
2. Name the four types of bonding and explain them in your own words. Be specific about the location of the valence electrons.

Topic: Chemical Bonding
Visit: www.SciLinks.org
Web code: KEY-126

Reason and Apply

3. Determine the type of bonding in each substance.
   a. zinc, Zn(*s*)
   b. propane, $C_3H_8(l)$
   c. calcium carbonate, $CaCO_3(s)$
4. Based on physical properties, which of these substances is an ionic compound? Explain your reasoning.
   A. hair gel B. silver bracelet C. motor oil D. baking soda
5. You observed nitrogen dioxide, $NO_2(g)$, when you dissolved copper, Cu(*s*), in nitric acid in the Lab: The Copper Cycle. How would you classify the bonding in $NO_2(g)$? Explain.
6. Which statement is true?
   A. Aqueous solutions of calcium chloride, $CaCl_2$, conduct electricity.
   B. Glass, made of silicon dioxide, SiO_2, does not dissolve in water.
   C. Ethanol, C_2H_6O, dissolves in water but does not conduct electricity.
   D. Brass, also called copper zinc, CuZn, conducts electricity.
   E. All of the above are true.
7. Suppose you have a mixture of sodium chloride, NaCl, and carbon, C. Explain how you can use water to separate the two substances.
8. Explain why copper is used as wire, but copper chloride is not.
9. Explain why carbon is a solid and not a gas.
10. Will each of these substances dissolve in water? Explain your thinking.
   a. Ca b. $NaNO_3$ c. Si d. CH_4 e. $CuSO_4$

LAB Electroplating Metals

Safety Instructions

Wear safety goggles at all times.

The solution contains acid. Handle carefully. Rinse the nickel strips after they have been in the copper solution.

Purpose

To use electrochemistry to extract metals from ionic compounds in solution.

Materials

- copper sulfate plating solution
- 250 mL beaker
- 2 nickel strips (cut into about 1-by-3-in. strips)
- 1.5-volt D-cell battery with holder
- 2 insulated wires with alligator clips

Procedure

1. Set up the electroplating apparatus as shown. Observe what happens.

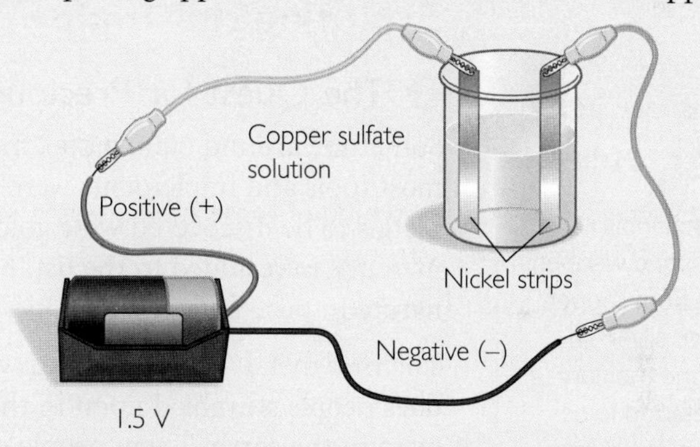

2. Switch the sides of the battery to which the two alligator clips are attached. Wait at least one minute or until you notice a change.
3. Reverse the wiring back to its original position.

Observations

1. What did you observe when you hooked up the nickel strips to the battery?
2. What happened when you reversed the flow of electricity?
3. Where does the copper come from that ends up on the nickel strip?
4. What is in the copper sulfate solution?
5. Write a short paragraph explaining your observations.
6. **Making Sense** Are copper atoms and copper ions the same element? Explain your thinking.
7. **If You Finish Early** Consider a sample of gold chloride, $AuCl_3$. Explain what procedure you might follow in order to extract solid gold from the compound.

LESSON

27 Electrons on the Move

Electroplating Metals

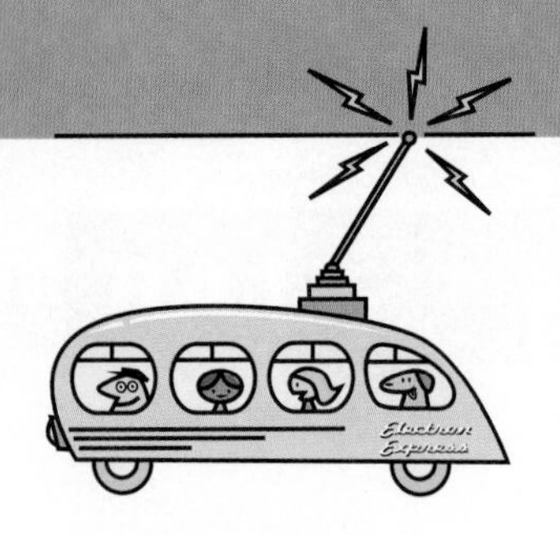

Think About It

Metals are important in our lives. They are used for everything from electrical wiring, jewelry, and soda cans, to cars, airplanes, and bridges. Most metals are dug out of the ground as ionic compounds. In ionic compounds, the metal atoms are bonded to nonmetal atoms. Through the ages, people have struggled to extract the pure metals from these compounds by figuring out ways to separate the metal atoms from the nonmetal atoms.

How can you extract an element from a compound?

To answer this question, you will explore

1. The Quest for Precious Metals
2. Electroplating Elemental Metals

Exploring the Topic

1 The Quest for Precious Metals

Sometime around 6000 B.C.E., metals came into widespread use. Before that, most tools and implements were made from earth, wood, and stone. The first metals to be discovered were gold, copper, and silver. Later, lead, tin, iron, and mercury were added to the list. Most of these metals are not readily found in nature in pure form.

The first metals that were used were the ones people stumbled upon in their pure form in the earth. Then, people discovered that, using heat they could extract metals from other compounds dug out of the earth. For example, if iron (III) oxide, Fe_2O_3, is heated in the presence of charcoal, the oxygen in the compound is removed as CO_2 gas. This leaves behind solid iron, Fe. This process requires very high temperatures.

For nearly 7000 years, there were only a handful of metals in widespread use. Other metals, such as aluminum, Al, could not be extracted easily by heating and remained unavailable.

HISTORY CONNECTION

The first metal tools, implements, and weapons were made from copper. Gold, copper, silver, tin, lead, iron, and mercury are often referred to as the Metals of Antiquity. For some 7000 years, these seven metals were the only ones in widespread use. These metals were known to the Mesopotamians, Greeks, Egyptians, and Romans.

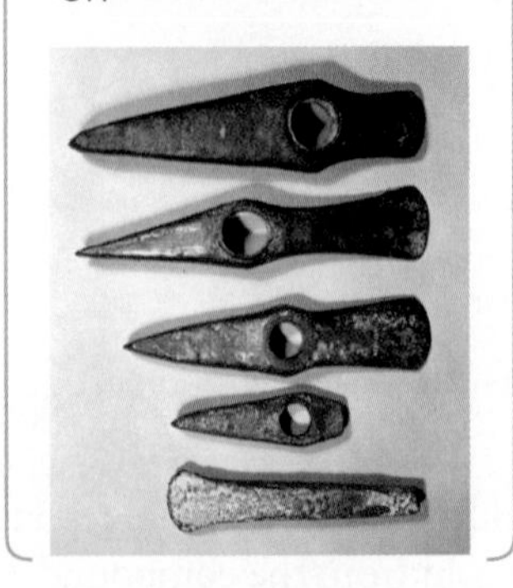

2 Electroplating Elemental Metals

Using Electricity to Extract Metals

Over time, the construction of better furnaces has resulted in the extraction of a variety of metals. More recently, it was discovered that electricity could be used to

> **CONSUMER CONNECTION**
>
> Certain kinds of batteries contain solutions of anions and cations and have positive and negative terminals, called the anode and cathode. When the battery is connected, the anions flow toward the anode and the cations toward the cathode, producing an electrical current.

extract metals from compounds. Recall that electrons are transferred from metal atoms to nonmetal atoms in ionic compounds. Scientists discovered they could essentially "give" electrons back to the metal ions, re-forming the neutral metal atoms once again. For example, copper metal can be extracted from a solution of copper sulfate by running an electrical current through the solution. A simple setup is shown here.

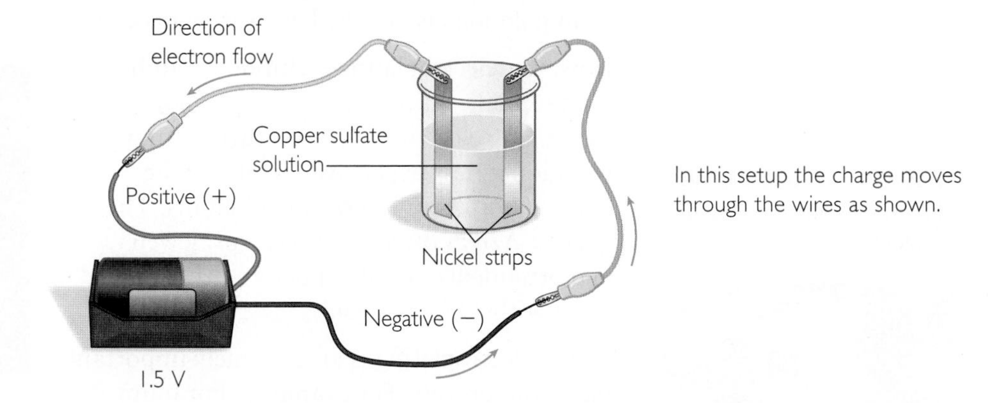

In this setup the charge moves through the wires as shown.

Once the battery is hooked up, one nickel strip has a positive charge and the other has a negative charge. In this electrical circuit, the electrons move from the negative terminal of the battery to the negatively charged nickel strip. Electrons also move from the other nickel strip toward the positive terminal of the battery. Ions moving in solution complete the circuit.

The positive copper ions, Cu^{2+}, are attracted to the nickel strip with the negative charge. Electrons are added to the copper cations in solution. Each copper ion that gains enough electrons becomes elemental copper. When the copper cations become elemental copper atoms, they come out of solution and coat the negatively charged nickel strip. This process is called **electroplating.**

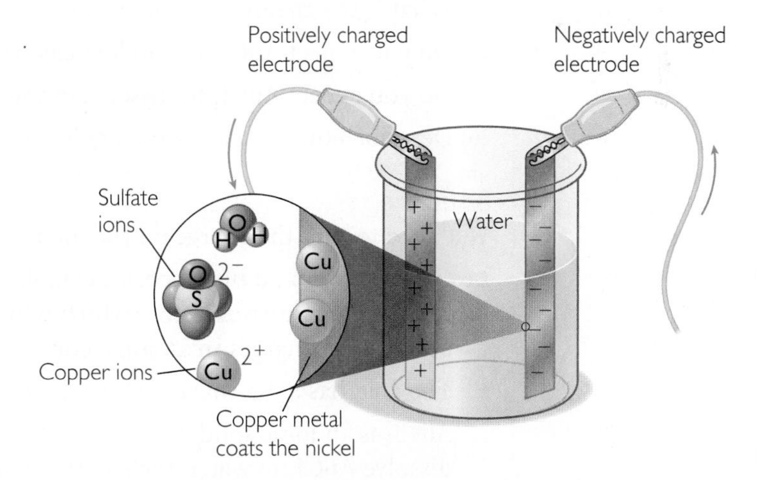

When the experiment shown is completed, the negatively charged nickel becomes coated with the pure copper metal.

You can reverse the process and produce metal cations from metal atoms. The battery has a positive and a negative terminal. By switching the wires in the test setup, you can change the direction the electrical charge moves. This causes the reaction to reverse. The copper coating comes off of the first nickel strip and reenters the solution. Copper from the solution plates onto the other electrode.

Electroplating Gold

Is it possible to design an apparatus similar to the one in the Lab: Electroplating Metals to extract gold atoms from ionic compounds containing gold ions? Yes, by all means. However, gold is one of those few elements that *is* commonly found in nature in its elemental form. It is gold *compounds* that are rarely found in nature. This is because gold is fairly stable as it is. So, harvesting gold atoms from gold ions is not the best way to make a pile of gold bars.

INDUSTRY CONNECTION

Decorative chrome plating is used on some cars and motorcycles.

However, electroplating techniques can be used to make a little bit of gold go a long way. Using electroplating or other plating techniques, less expensive metals can be covered with thin layers of gold atoms. The ancient Aztecs often made crowns, jewelry, and ornaments out of copper and then plated them with a layer of gold.

Copper bowl plated with 24 karat gold

Today, electroplating is an extremely important industrial process. For example, aluminum metal has been in widespread use only recently because chemists finally found a way to extract aluminum metal from ionic compounds using electrochemical methods. Electroplating is also used to coat electronic parts in computers.

Example

Plating Gold from $AuCl_3$

Suppose that you have a sample of $AuCl_3(s)$.

a. What is the charge on the Au ions in $AuCl_3$?
b. What is the chemical name of $AuCl_3$?
c. Do you expect $AuCl_3$ to conduct electricity as a solid? Explain your thinking.
d. Do you expect $AuCl_3$ to dissolve in water? Explain your thinking.
e. Describe how you can extract gold from $AuCl_3$.

Solution

You need to find the charge on the main group element first.

a. Cl is in Group 7B in the periodic table, so it has seven valence electrons. Thus, it needs to gain an electron, which will make it Cl^-. So the Au becomes Au^{3+} because the charges in an ionic compound must add up to zero.
b. The gold has a +3 charge. Thus, the name is gold (III) chloride.
c. $AuCl_3$ is an ionic solid. Ionic solids do not conduct electricity. However, if you dissolve $AuCl_3$ in water, the solution will conduct electricity.
d. Because $AuCl_3$ is ionic, you can assume it will dissolve in water, forming Au^{3+} and Cl^- ions.
e. Gold can be plated onto a piece of nickel by using a battery to move electrons through a solution containing $AuCl_3$.

Key Term

electroplating

Lesson Summary

How can you extract an element from a compound?

Most metal atoms are found in nature combined with other atoms, in the form of ionic compounds. Some metals can be extracted from ionic compounds through heating. Elements can also be extracted from ionic compounds by using electricity to move electrons between atoms. Using electroplating, a variety of substances that could be considered "as good as gold" can be created.

EXERCISES

Reading Questions

1. Describe three ways in which people have managed to obtain pure metals for use in making tools and other objects.
2. Explain how to plate copper metal onto a nickel strip. Write a simple procedure and draw a labeled sketch showing how to carry it out.
3. Explain how you can *remove* a metal coating from an object.

Reason and Apply

4. **Lab Report** Complete a lab report for the copper plating experiment. In your report, give the title of the experiment, purpose, procedure, observations, and conclusions.

(Title)

Purpose: (Explain what you were trying to find out.)

Procedure: (List the steps you followed.)

Observations: (Describe what you observed during the experiment.)

Conclusions: (What can you conclude about what you were trying to find out? Provide evidence for your conclusions.)

5. Explain how the copper plating experiment supports the claim that Cu^{2+} is just a Cu atom that is missing two electrons.
6. If you measure the mass of the nickel strip before and after plating copper, do you expect the mass to change? Explain your thinking.
7. Suppose that a classmate claimed that the nickel in the electroplating lab changed into copper. Provide an argument to prove that this is not true.

SECTION

V

SUMMARY

Building With Matter

Alchemy Update

How can substances with new properties be made?

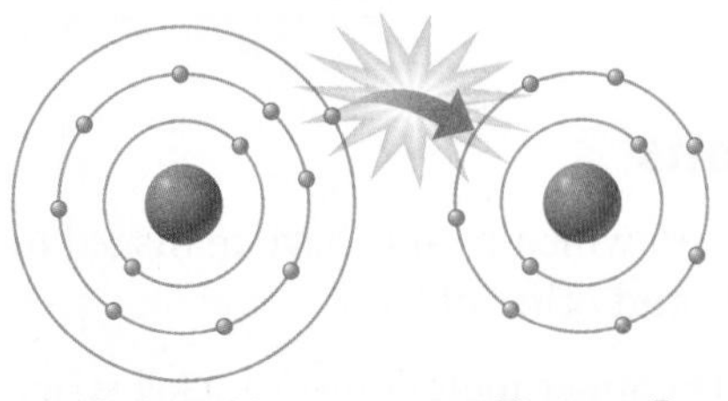

Key Terms

dissolve
soluble
insoluble
conductivity
chemical bond
ionic bonding
molecular covalent bonding
metallic bonding
network covalent bonding
covalent bonding
molecule
electroplating

A bond is an attraction between the nucleus of one atom and the electrons of another atom. When atoms bond together, new substances with new properties are made. Conversely, elements can be extracted from compounds by breaking the bond.

There are four types of bonding, depending on the location of electrons within the atoms in the substance: ionic, molecular covalent, metallic, and network covalent. Certain properties are associated with each type of bonding.

Atoms form bonds with other atoms creating a variety of substances with new properties. It is possible to make compounds that look like or behave like gold, such as iron pyrite, FeS_2, also called fool's gold. In addition, you can use chemical bonding to make substances that are even more valuable than gold.

Review Questions

1. What does the phrase "as good as gold" mean as it applies to this class?
2. Describe how you could experimentally determine the type of bonding found in a substance.
3. Predict whether isopropanol, $C_3H_8O(l)$, will conduct electricity. State your reasoning.
4. Make a table like this one. Fill it with the type of bonding that fits the category.

	Dissolves	Does not dissolve
Conducts		
Does not conduct		

UNIT

1

Alchemy Review

Matter is anything that has substance and takes up space. Mass is a measure of the substance of matter. Volume is a measure of the space that matter takes up. The density of a substance is the mass per unit of volume. Everything on the planet is composed of elements or compounds made from combinations of these elements. The elements are organized into a chart called the periodic table according to their properties and atomic structure. The periodic table holds a wealth of information about the elements and their atoms. Elements that are located in the same column of the periodic table tend to have similar behavior and reactivity. The noble gases are very stable elements with extremely low reactivity.

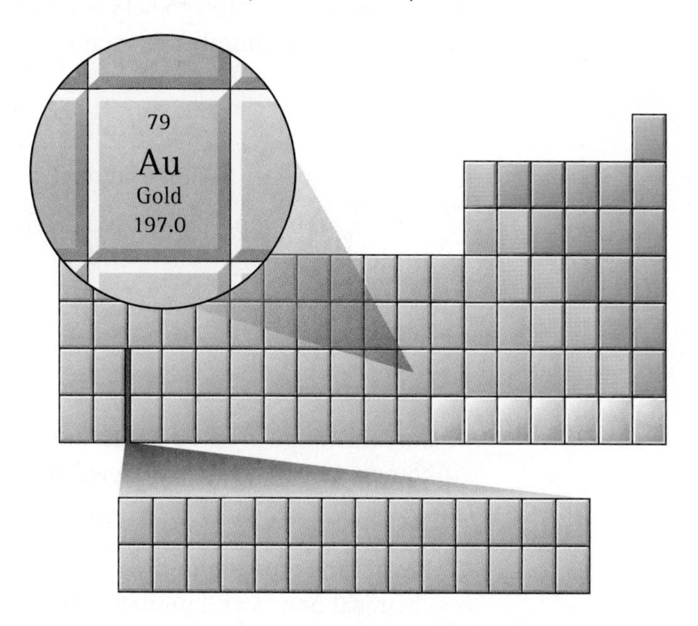

Matter is composed of individual building blocks called atoms. Atoms are much too small to be seen, so experimental evidence has led to various models of the atom. In the center of each atom is a dense nucleus consisting of protons and neutrons. Most of the mass of an atom is located in its nucleus. Most atoms have more neutrons than protons in their nuclei, which keeps the nucleus more stable. Electrons are located at specific distances from the nucleus, called shells. The electrons in the outermost shell of an atom are called its valence electrons and are key to its behavior. Atoms will sometimes lose or gain valence electrons, forming ions with positive and negative charges, and achieving the electron arrangements of the noble gases.

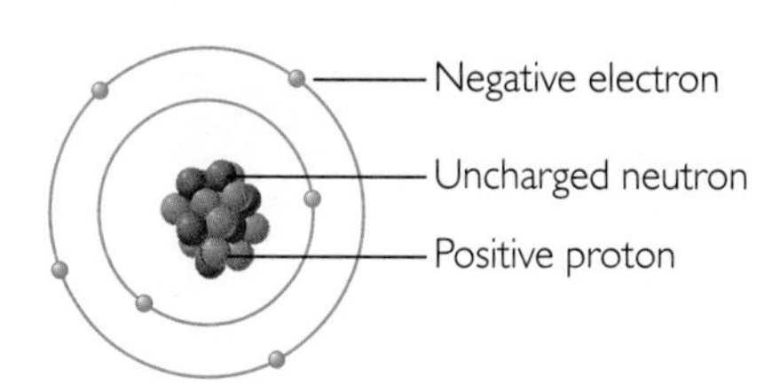

A compound is a substance that contains atoms of more than one type of element, bonded together. There are millions of compounds in existence, and many yet to be discovered. A bond is an attractive force that keeps atoms connected to one another. The attraction occurs between the positively charged nuclei and the negatively charged electrons in different atoms. Usually it is the valence electrons that are involved in bonding. There are four main types of substances on the planet: ionic compounds, molecular covalent compounds, metal substances, and network covalent substances. These four groups correspond to the four main types of bonds that are found between atoms which also determine the solubility and conductivity of the substance.

Review Exercises

1. What is the difference between an element and a compound?
2. Consider two objects that each weigh 20.0 g. One is made of lead and has a density of 11.4 g/mL. The second object is made of aluminum and has a density of 2.7 g/mL. Which object takes up more space? What volume does each occupy?
3. Use the periodic table to help you find the atomic symbol, atomic number, group number, number of protons, and number of electrons for these elements:

   a. lithium **b.** bromine **c.** zinc

   d. sulfur **e.** barium **f.** carbon
4. Describe nuclear fission and nuclear fusion.
5. What is an isotope? How can you figure out the most common isotope of an element?
6. Describe how you would use the periodic table to help you predict the type of bond between two atoms.
7. What are cations and anions?
8. Sort the following atoms into metals and nonmetals. Then name three ionic compounds that can be formed from some combination of these atoms.

   Na S Cl Sr Mg Se I Cu
9. Predict the type of bonding in each of the following substances. Provide their chemical names and predict which ones will conduct electricity.

   a. $AlCl_3$ **b.** O_2 **c.** AgOH **d.** Pt
10. Explain how you can determine the charge on a transition metal ion.
11. A substance does not dissolve in water and does not conduct electricity. What type of bond is holding it together?
12. What is required to change one element into another element?

UNIT

2 Smells

Each flower puts out a scent that attracts specific insects. Not all flowers smell pleasant. The corpse flower, which blooms once every three years, smells like decaying flesh.

Why Smells?

The sense of smell is a familiar and important part of our lives. It helps us detect pleasant smells, as well as unpleasant smells that alert us to possible dangers. Sometimes we detect things with our sense of smell that we can't see. Sometimes we detect something all the way across the room, or even in another room. But how does it work? The way atoms are connected in molecules, and the structures of the molecules, have a great deal to do with the properties of those molecules. When you understand the chemistry of molecules, you will begin to answer the question "What is the chemistry of smell?"

In this unit, you will learn

- how atoms form molecules
- to predict the smell of a compound
- to interpret molecular models
- how the nose detects different molecules
- about amino acids and proteins

SECTION

I Speaking of Molecules

Our sense of taste is closely related to our sense of smell. For this reason, many chefs have very sophisticated senses of smell.

The nose is able to identify hundreds of different smells. Most substances that smell are made of molecules. So, how does the nose differentiate one molecule from another? When you analyze the molecular composition and structure of compounds that smell, a variety of patterns emerge. Knowing the number and types of atoms in molecules, and the way atoms are connected in molecules, is vital to understanding the chemistry of smell.

In this section, you will study

- different representations for molecules
- bonding patterns in molecular covalent compounds
- how molecular structure relates to smell

LESSON

1 Sniffing Around

Molecular Formulas

Think About It

The topic of smell is both friendly and familiar. Most of us have had many experiences with smells: smells that attract, smells that disgust, smells that evoke strong memories of vacations, holidays, or other experiences. It turns out that our noses are fantastic chemical detectors. Learning more about your nose and how it works will help you learn more about chemistry.

What does chemistry have to do with smell?

To answer this question, you will explore

1. The Sense of Smell
2. Classifying Smells
3. Molecular Formula, Name, and Smell

Exploring the Topic

1 The Sense of Smell

Someone opens a perfume bottle across the room. After a few seconds you notice the sweet smell of the perfume. How did the smell get to your nose and how did your nose detect it?

The smells that come from a vial of perfume and end up in your nose are made of molecules. Molecules are groups of nonmetal atoms that are covalently bonded together. In this unit, you will examine mechanisms that your nose uses to detect and distinguish various smells.

In order to be consistent, you will always work with the consensus smell. In other words, you will work with the smell reported by the majority of individuals in the class.

2 Classifying Smells

To compare smells, we need a common language for describing them. How would you describe the smells of the items in the pictures?

BIOLOGY CONNECTION

A moth uses its antennae to detect smells, while a snake uses its tongue.

Scientists classify smells by placing similar-smelling substances in the same category. For example, apples, bananas, strawberries, and pineapples might all fit into a category called "fruity" smelling. Roses, geraniums, and lilies might all fit into a category called "floral" smelling. Additionally, fruity- and floral-smelling substances might be grouped together in a larger category called "sweet" smelling.

Not all scientists agree on the specific names for smell categories or on the number of different categories that should exist. As you begin studying smells, you will start with three categories: sweet, minty, and fishy.

3 Molecular Formula, Name, and Smell

A closer investigation reveals connections among the chemical name, the chemical formula, and the smell of a substance. For a molecular covalent compound, the chemical formula is also called the **molecular formula,** since it describes the make-up of each molecule.

Important to Know A molecular formula is just a chemical formula for a molecular covalent compound. Recall that chemical formulas can describe other compounds that are not molecules. ◀

Examine the drawings of the five vials. Each vial contains a cotton ball soaked in a different molecular compound. See if you can find any connections between the smells and their chemical names and molecular formulas.

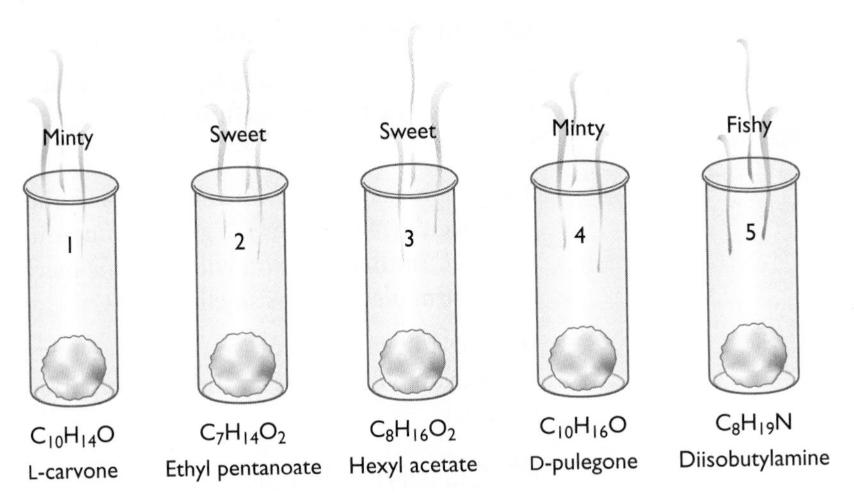

Here are some patterns that you might notice.

All the molecules

- contain carbon and hydrogen atoms
- have more hydrogen atoms than carbon atoms
- are composed of three different types of atoms

Sweet-smelling molecules

- contain two oxygen atoms
- have twice as many hydrogen atoms as they do carbon atoms
- have two-word names ending in "-ate"

HEALTH CONNECTION

Dogs are trained to sniff out everything from termites and drugs to bombs and natural gas leaks. There is some evidence that some dogs may even be able to detect when a person has developed certain cancers.

Minty-smelling molecules

- contain one oxygen atom
- contain ten carbon atoms
- have even numbers of carbon and hydrogen atoms
- have names ending in "-one"

Fishy-smelling molecules

- contain a nitrogen atom
- do not contain oxygen atoms

Perhaps *every* molecule with two oxygen atoms smells sweet. Perhaps *every* molecule that smells fishy always contains a nitrogen atom. These are reasonable hypotheses based on very little data. You can test these hypotheses by gathering more data.

Example

Predicting Smell

How would you expect a molecule with the molecular formula $C_9H_{18}O_2$ to smell? Explain.

Solution

The molecule has two oxygen atoms. Based on data given so far, this suggests that the molecule smells sweet. It would be useful to know the name to confirm this. So far, the names of sweet-smelling molecules end in "-ate."

Key Term

molecular formula

Lesson Summary

What does chemistry have to do with smell?

The smells that your nose detects are made of molecules. The molecular formulas and chemical names of molecular substances are directly related to the way these substances smell. You may be able to predict the smell of a substance by examining its chemical formula, its chemical name, or both.

EXERCISES

Reading Questions

1. How do scientists classify smells? What purpose might this serve?
2. From the evidence so far, how are molecular formulas related to smell? How are chemical names related to smell?

Reason and Apply

3. Name five substances from home that do not fit into the smell categories of sweet, minty, or fishy. What new categories would you put them in?

4. Make a list of five things from home that have no smell.
5. In a paragraph or two, write about *one* of the following. You may use a cartoon format instead, if you wish.
   a. Describe a memory that you have that relates to smell. Describe the smell itself and what it reminds you of. (Avoid using adjectives that are not informative, such as "weird.")
   b. If you were a smell superhero, how would you use your smells or smell-ability to assist the world?
   c. What would your life be like if you couldn't smell?

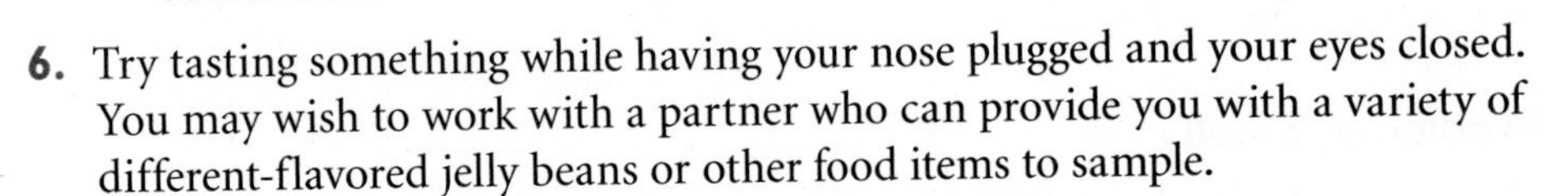

6. Try tasting something while having your nose plugged and your eyes closed. You may wish to work with a partner who can provide you with a variety of different-flavored jelly beans or other food items to sample.
   a. Describe your experience. How difficult is it to taste something when you can't smell it and you don't know what it is?
   b. How do you think taste is related to smell?
7. Predict the smells of these molecules. Explain your reasoning.
   a. methyl octenoate, $C_9H_{18}O_2$
   b. monoethylamine, C_2H_7N
   c. ethyl acetate, $C_4H_8O_2$
8. What do you think 1,5-pentanedithiol ($C_5H_{12}S_2$) would smell like? How confident of your answer are you? Explain why.
9. What makes something smell the way it does? Discuss your own theory about why different things smell different. Include questions you still have.

BIOLOGY CONNECTION

Not every human being experiences smell in exactly the same way. Some people may smell one odor completely differently from other people, or may not smell that one odor at all. These exceptions are of particular interest to smell researchers, who use this information to figure out how many distinct smells the nose can detect.

LESSON

2 Molecules in Two Dimensions

Structural Formulas

Think About It

Would you rather smell sweaty gym socks or fresh pineapple? The choice is pretty clear. But what if you were asked if you would rather smell ethyl butyrate or hexanoic acid? How would you know which compound is safe to sniff? Both molecules have the same molecular formula: $C_6H_{12}O_2$.

How can molecules with the same molecular formula be different?

To answer this question, you will explore

1. Predicting Smells
2. Structural Formula
3. Orientations of Structural Formulas

Exploring the Topic

1 Predicting Smells

In the previous lesson, you arrived at some generalizations that relate smell to molecular formula and chemical name. What happens if you use these generalizations to make predictions about some new compounds? Consider the five new compounds listed in the table.

Vial	Chemical name	Molecular formula	Predicted smell
1	pentyl acetate	$C_7H_{14}O_2$	sweet
2	pentanoic acid	$C_5H_{10}O_2$	sweet
3	propyl acetate	$C_5H_{10}O_2$	sweet
4	ethylthiol	C_2H_6S	a new smell category
5	hexylamine	$C_6H_{15}N$	fishy

All three have two oxygen atoms. Two end in "-ate" and one in "acid": sweet smelling?

Contains a sulfur atom: new smell category?

Contains a nitrogen atom: fishy?

The predictions based on molecular formula and chemical name alone are reasonably accurate. Four out of five are correct. Ethylthiol has a skunk smell, so it belongs in a new smell category. The only prediction that is incorrect is the pentanoic acid in vial 2. It has a molecular formula *identical* to the molecular formula of a sweet-smelling molecule, but it is a very putrid compound. Apparently two compounds can have identical formulas, different names, and very different smells. Perhaps this is why its name does not end in "-ate" like the other sweet-smelling compound.

As you continue to study molecules, some of the patterns discovered here will hold up, while others will be revised or abandoned. You can continue to refine your ideas about smell so that you can better understand the chemistry of smell.

BIG IDEA Science involves coming up with ideas based on observations and then refining these ideas based on further observation.

2 Structural Formula

As it turns out, two compounds can have identical molecular formulas, but smell incredibly different. This is because having the same molecular formula does not guarantee that the molecules themselves are identical.

To show how the individual atoms in a molecule are bonded to each other, chemists use what is called a structural formula. A **structural formula** is a diagram or drawing that shows how the atoms in a molecule are arranged and where they are connected. The lines between the atoms represent covalent bonds holding the atoms together. Take a moment to examine the structural formulas for ethyl butyrate and hexanoic acid below. What differences do you see?

BIOLOGY CONNECTION

The spray of a skunk is a defensive adaptation. The spray contains the compound ethylthiol. Predators will usually leave a skunk alone once they have been blasted with this stinky substance that burns the eyes and nose and is difficult to get rid of.

These compounds have the same molecular formula, but different smells.

$C_6H_{12}O_2$

```
    H   H   H   O       H   H
    |   |   |   ||      |   |
H - C - C - C - C - O - C - C - H
    |   |   |           |   |
    H   H   H           H   H
```

Ethyl butyrate, pineapple smell

```
    H   H   H   H   H   O
    |   |   |   |   |   ||
H - C - C - C - C - C - C - O - H
    |   |   |   |   |
    H   H   H   H   H
```

Hexanoic acid, sweaty gym sock smell

Notice that the oxygen atoms are located in different places within the two molecules. In one molecule the oxygen atoms are in the middle of the molecule, whereas in the other molecule the oxygen atoms are at one end.

Isomers

In some compounds, the atoms can be connected only a certain way. For instance, the formula C_2H_6 has only one structural formula.

```
    H   H
    |   |
H - C - C - H
    |   |
    H   H
```

C_2H_6

As the number and types of atoms in a molecule increase, there are more possible ways to connect the atoms. Molecules with the same molecular formula and different structural formulas are called **isomers.** The formula C_2H_6O has two isomers, shown here.

```
    H   H                   H       H
    |   |                   |       |
H - C - C - O - H       H - C - O - C - H
    |   |                   |       |
    H   H                   H       H
```

C_2H_6O

Ethyl butyrate and hexanoic acid are considered isomers of each other. The formula $C_6H_{12}O_2$ has many more isomers. In fact, there are over 25 different molecules with the molecular formula $C_6H_{12}O_2$, each with unique properties.

Drawings of structural formulas are flat and two-dimensional. However, molecules are not flat. They take up three dimensions in space.

Ethanol

```
    H   H
    |   |
H — C — C — O — H
    |   |
    H   H
```

Important to Know Structural formulas are used to represent molecules that are covalently bonded. They are not used to represent ionic compounds. ◂

3 Orientations of Structural Formulas

There are several ways to draw the *same* structural formula. Consider these two drawings.

```
    H
    |
H — C — H             H   H   H
    |                 |   |   |
H — C — H         H — C — C — C — H
    |                 |   |   |
H — C — H             H   H   H
    |
    H
```

These two molecules have the same molecular formula and each atom has the same connections. The structural formula has just been turned 90° in space. These molecules are *not* isomers; they have the same structural formula.

Here are four structural formulas. How many different compounds are represented?

```
                                                H
                                                |
    H   H   H               H   H   H           O   H   H             H       H   H
    |   |   |               |   |   |           |   |   |             |       |   |
H — C — C — C — O — H   H — C — C — C — H   H — C — C — C — H     H — C — O — C — C — H
    |   |   |               |   |   |           |   |   |             |       |   |
    H   H   H               H   O   H           H   H   H             H       H   H
                                |
                                H
```

Molecule 1 | Molecule 2 | Molecule 3 | Molecule 4

If you build three-dimensional models of these molecules, you will find that there are only three different compounds here. The first and the third structural formulas are identical. All the atoms in the first and third structures are connected in the same way.

The illustration on the next page shows how molecule 1 can be turned and rotated to match molecule 3 without taking the model apart.

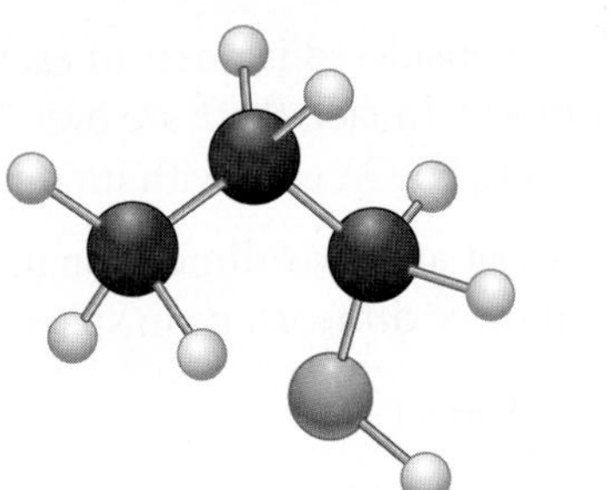

Start with molecule 1.

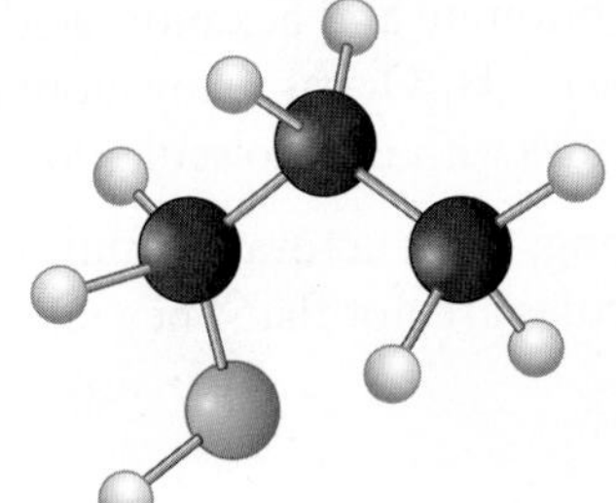

Turn it around.

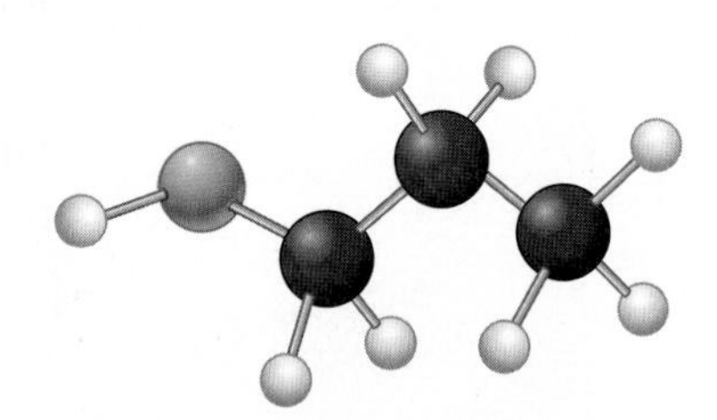

Rotate the carbon atom on the left. This is the same as molecule 3.

Structural formulas show only the sequence in which one atom is connected to the next. The way atoms are connected in a molecule is like a charm bracelet. The middle charm is always between the same two charms regardless of how you hold the bracelet. It does not matter if the oxygen atom is drawn on the left or right side of the molecule. It also does not matter if the O–H is drawn horizontally or vertically. It is only the number and type of connections that matter.

Example

Isomers of C_3H_9N

How many different isomers are represented below?

```
    H   H   H
    |   |   |
H — C — C — C — N — H
    |   |   |   |
    H   H   H   H
```

Molecule 1

```
    H   H       H
    |   |       |
H — C — C — N — C — H
    |   |   |   |
    H   H   H   H
```

Molecule 2

```
    H   H   H   H
    |   |   |   |
H — C — N — C — C — H
    |       |   |
    H       H   H
```

Molecule 3

```
        H   H
         \ /
          N
    H     |     H
    |     |     |
H — C  —  C  —  C — H
    |     |     |
    H     H     H
```

Molecule 4

Solution

There are three isomers. Molecules 2 and 3 are identical. So there are three different compounds represented here.

Key Terms

structural formula
isomer

Lesson Summary

How can molecules with the same molecular formula be different?

Compounds with identical molecular formulas may have different structures. Chemists use drawings called structural formulas to show how the atoms in molecules are connected to each other. Each line in a structural formula represents a covalent bond between atoms. When molecules have the same molecular formula but different structural formulas, they are called isomers. Isomers represent different compounds with different properties. Because smell is a property, compounds with the same molecular formula may smell different from each other because they are different substances.

EXERCISES

Reading Questions

1. What information does a structural formula provide?
2. What are isomers?

Reason and Apply

3. If you are given a structural formula of a molecule, do you also know its molecular formula? Explain your thinking.
4. The words *isotope* and *isosceles* also have the prefix "iso-." How are their meanings similar to that of the word isomer?
5. Structural formulas for six molecules are shown below. Write the molecular formula for each of the molecules.

```
a.         O   H   H
           ||  |   |
     H—O—C—C—C—H
               |   |
               H   H
```

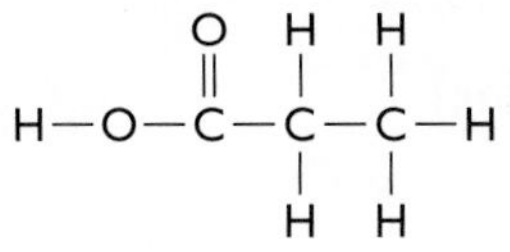

```
b.   H   O   H   H   H
     |   ||  |   |   |
   H—C—C—C—C—C—H
     |       |   |   |
     H       H   H   H
```

```
c.   H   O   H   H
     |   ||  |   |
   H—C—C—C—C—H
     |       |   |
     H       |   H
           H—C—H
             |
             H
```

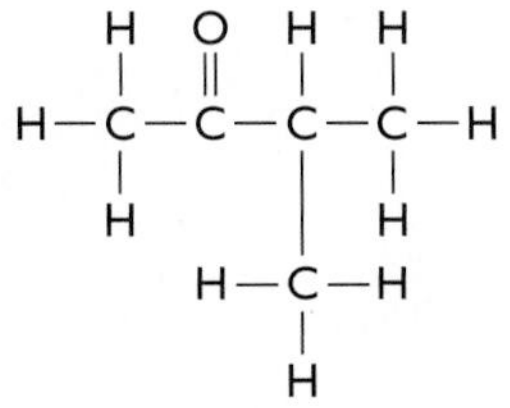

```
d.   H   H   O       H
     |   |   ||      |
   H—C—C—C—O—C—H
     |   |           |
     H   H           H
```

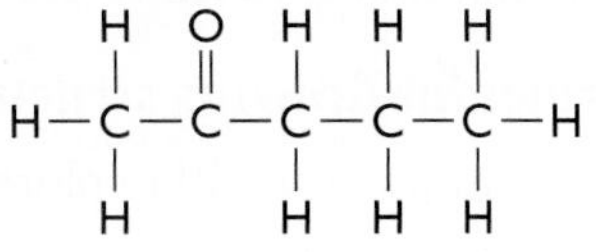

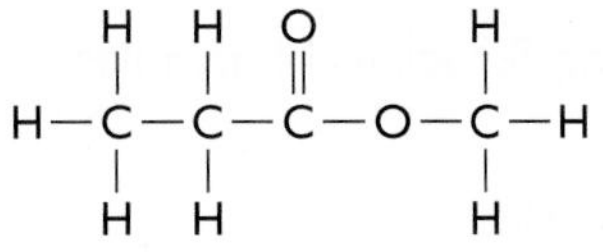

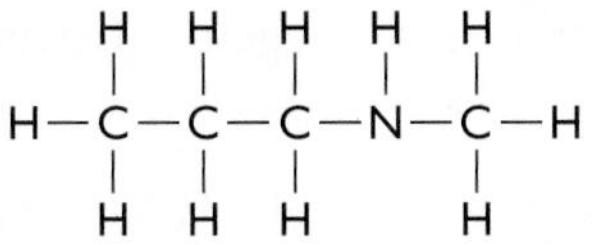

```
e.   H   H   H   H   H
     |   |   |   |   |
   H—C—C—C—N—C—H
     |   |   |       |
     H   H   H       H
```

```
f.           H
             |
   H—C—O—C—H
     ||      |
     O       H
```

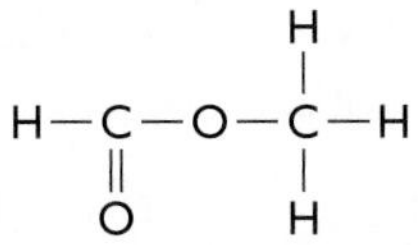

6. Two molecules have the same molecular formula yet one smells sweet and the other smells putrid. Explain how you think this might be possible.
7. For each of the molecules in Exercise 5, draw an isomer of the compound.

LESSON

3 HONC if You Like Molecules

Bonding Tendencies

Think About It

In order to understand the chemistry of smell, it is necessary to understand how molecules are put together. If you examine the structural formulas of molecules, you can see patterns in the way the atoms are connected. For example, hydrogen atoms are always arranged around the outside of the molecule. Plus, hydrogen atoms are always connected to other atoms with only one line, while carbon atoms are always connected to other atoms with more than one line.

```
  H  O     H  H
  |  ||    |  |
H-C--C--O--C--C-H
  |        |  |
  H        H  H
```

What are the rules for drawing structural formulas?

To answer this question, you will explore

1. The HONC 1234 Rule
2. Drawing Structural Formulas

Exploring the Topic

1 The HONC 1234 Rule

The structural formulas for three molecules are shown here. Take a moment to count the number of times each type of atom—hydrogen, oxygen, nitrogen, and carbon—is connected to other atoms.

```
  H H H H H O
  | | | | | ||
H-C-C-C-C-C-C-O-H
  | | | | |
  H H H H H
```

$C_6H_{12}O_2$

$C_{10}H_{18}O$

```
  H H H   H
  | | |   |
H-C-C-C-N-C-H
  | | |   |
  H H H   H
```

$C_4H_{11}N$

- Every **H**ydrogen atom has **one** line connecting it to other atoms.
- Every **O**xygen atom has **two** lines connecting it to other atoms.
- Every **N**itrogen atom has **three** lines connecting it to other atoms.
- Every **C**arbon atom has **four** lines connecting it to other atoms.

This information is sometimes referred to as the **HONC 1234 rule.** Within most molecules, hydrogen makes one bond, oxygen makes two bonds, nitrogen makes three bonds, and carbon makes four bonds. The bonds in a structural formula are represented by lines. One line connecting two atoms is called a *single bond.* A pair of lines connecting the same two atoms, as in C=O, is called a *double bond.*

The HONC 1234 rule tells you how four of the most common nonmetal atoms will bond. You will learn about the bonding of other nonmetal atoms like sulfur, S, and chlorine, Cl, in later lessons.

Example 1

HONC 1234

Are the following molecules correct according to the HONC 1234 rule? If not, what is wrong with them?

```
    H  H  H            H        H
    |  |  |            |        |
 H—C—C—O—H        H—C—H—C—H
    |  |  |            |        |
    H  H  H            H        H
   Molecule 1         Molecule 2
```

Solution

Both molecules are incorrect according to the HONC 1234 rule. In molecule 1, there is an oxygen atom with four bonds. It should have only two bonds. In molecule 2, there is a hydrogen atom with two bonds. It should have only one. The hydrogen atom cannot go in the middle of the molecule, between two carbon atoms.

ASTRONOMY CONNECTION

At room temperature, ethane, C_2H_6, is a flammable gas. On Earth, ethane is one of the components of natural gas. It has also been detected in the atmospheres of Jupiter, Saturn, Uranus, and Neptune.

2 Drawing Structural Formulas

The HONC 1234 rule is all you need in order to draw the structural formulas for thousands of different molecules correctly. Some examples are given here and on top of the next page.

Structural Formula for C_2H_6

Consider a molecule with two carbon atoms and six hydrogen atoms. Its molecular formula is C_2H_6. What structural formula would this molecule have? Start by connecting the two carbon atoms with a single bond. The "carbon backbone" is generally a good place to start when drawing molecules.

Step 1: Connect the carbon atoms:

```
C—C
```

Step 2: Add hydrogen atoms. Make sure each hydrogen atom has just one bond and each carbon atom has a total of four bonds.

```
   H  H
   |  |
H—C—C—H
   |  |
   H  H
```

The structural formula shown above is consistent with the HONC 1234 rule. This is the only structure you can make out of this molecular formula. It represents a substance called *ethane.*

Structural Formula for C_2H_6O

Consider a molecule with two carbon atoms, six hydrogen atoms, and one oxygen atom. Its molecular formula is C_2H_6O. What structural formula could this molecule have?

CONSUMER CONNECTION

Ethanol, C_2H_6O, can be made from plants such as sugarcane, corn, and switchgrass, and can be used as an alternative fuel for automobiles. It can be used alone or blended with gasoline. However, it currently takes about a liter of fuel to produce one liter of ethanol, which may offset its environmental benefit.

Start by connecting the carbon atoms. Then add the oxygen atom.

Step 1: Connect the carbon atoms: C—C

Step 2: Add the oxygen atom to the carbon chain: C—C—O

Step 3: Add hydrogen atoms last:

```
    H   H
    |   |
H—C—C—O—H
    |   |
    H   H
```

Make sure your structural formula follows the HONC 1234 rule. Notice that hydrogen atoms are generally the last atoms you add when creating a structural formula.

This molecule is called *ethanol* or *ethyl alcohol.*

A Second Structure—An Isomer

Ethanol is not the only possible structure for this particular molecular formula. The atoms in C_2H_6O can also be arranged a different way, to form a completely different molecule. This new structure is created by placing the oxygen atom between the two carbon atoms.

Step 1: Connect the carbon and oxygen atoms: C—O—C

Step 2: Add hydrogen atoms:

```
    H       H
    |       |
H—C—O—C—H
    |       |
    H       H
```

This structure *also* follows the HONC 1234 rule and is a correct structural formula for C_2H_6O. This molecule is called *dimethyl ether.* Ethanol and dimethyl ether are isomers. They have the same molecular formula but different structural formulas. Dimethyl ether has different properties from ethanol, including smell.

Not all structural formulas are simple to figure out. As the number and type of atoms increase, there are more possible ways to connect the atoms.

Example 2

Structural Formula for $C_4H_{11}N$

Draw a structural formula for the molecular formula $C_4H_{11}N$.

Solution

Start by making a chain of the carbon atoms. Place the nitrogen atom anywhere in the chain. Finish by adding hydrogen atoms so that every carbon atom has four bonds and the nitrogen atom has three bonds. Two possible solutions are shown here, but many more are possible.

```
    H   H       H   H
    |   |       |   |
H—C—C—N—C—C—H
    |   |   |   |   |
    H   H   H   H   H
```

```
        H
        |
      H\C/H
    H   |   H   H
    |   |   |   |
H—C—N—C—C—H
    |       |   |
    H       H   H
```

Key Term

HONC 1234 rule

Lesson Summary

What are the rules for drawing structural formulas?

The individual atoms in molecules are not connected randomly. Each atom in a molecule has a tendency to bond a specific number of times. The HONC 1234 rule describes the bonding patterns of hydrogen, oxygen, nitrogen, and carbon atoms. The HONC 1234 rule can be a valuable tool for creating structural formulas from molecular formulas.

EXERCISES

Reading Questions

1. What is the HONC 1234 rule?
2. Explain why one molecular formula can represent more than one structural formula.

Reason and Apply

3. Use the HONC 1234 rule to create possible structural formulas for molecules with these molecular formulas. Remember that it is easiest to start with the carbon atoms.

   a. $C_3H_8O_2$ **b.** $C_4H_{11}N$ **c.** C_4H_{10} **d.** $C_5H_{12}O$

4. Is each structural formula correct according to the HONC 1234 rule? For any molecules that don't follow the HONC 1234 rule, repair the incorrect structural formula. (*Note:* There may be more than one correct way to repair a formula.)

```
      H                H                 H   H
      |                |                 |   |
a. H—C—H       b. H—O—H        c. H—C—C—H
      |                |                 |   |
   H—C—H          H—C—H              O   H
      |                |                 |
   H—C—H          H—C—H              H
      |                |
      O                H
      |
      H

      H   H               H
      |   |               |
d. H—C—C—N—H      e. H—C—O—H
      |   |               |
      O   H            H—C—H
      |                   |
      H                   H
```

5. Think about how molecules might interact with your nose. Why do you think molecules with different structural formulas have different smells?

ACTIVITY Connect the Dots

Purpose

To investigate the role of electrons in covalent bonding.

Materials

- Lewis dot puzzle pieces

Instructions

The puzzle pieces you have been given represent Lewis dot symbols. The puzzle pieces allow you to pair up electrons and create molecules.

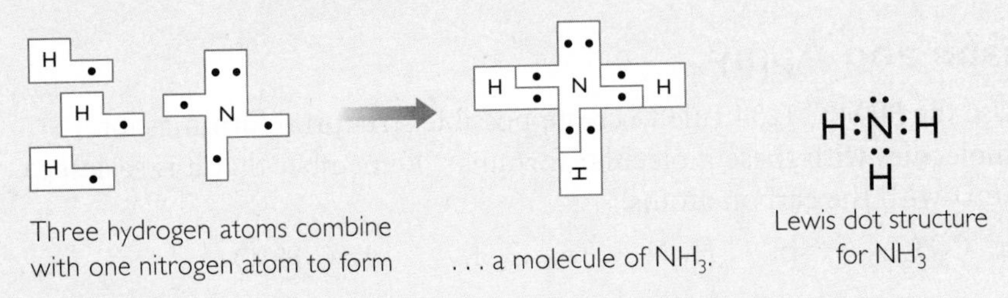

Three hydrogen atoms combine with one nitrogen atom to form . . . a molecule of NH_3.

Lewis dot structure for NH_3

Part 1: Create Molecules

1. Use the puzzle pieces to construct the molecules given below. Then draw the Lewis dot structure for each molecule, leaving off the outline of each puzzle piece.

   PH_3 HOCl F_2 CH_3Cl

2. Use the puzzle pieces to create more molecules following the directions. For each molecule, draw the Lewis dot structure and write the molecular formula.

   a. Use one S atom and as many H atoms as you need.

   b. Use one Si atom and as many F atoms as you need.

   c. Use two O atoms and as many H atoms as you need.

3. Use the puzzle pieces to construct a molecule with the molecular formula C_2H_6. Draw its Lewis dot structure and its structural formula.

4. Use the puzzle pieces to construct all the possible isomers of C_3H_8O. Draw Lewis dot structures for each isomer. Do the molecules follow the HONC 1234 rule?

5. Use the puzzle pieces to design your own molecule with at least five carbon atoms. Draw its Lewis dot structure. What is the molecular formula of your designer molecule? Does it obey the HONC 1234 rule?

Part 2: Valence Electrons

Find a puzzle piece for each type of atom. Put hydrogen and helium aside. Sort the rest of the puzzle pieces according to the periodic table.

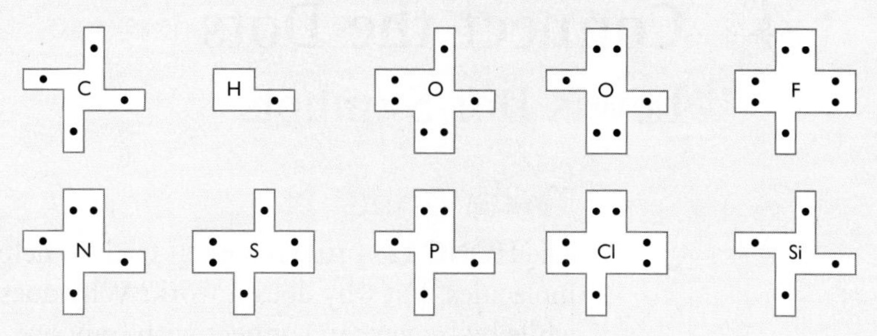

Record your sort by copying it into a grid like the one here. Include the symbols and the dots.

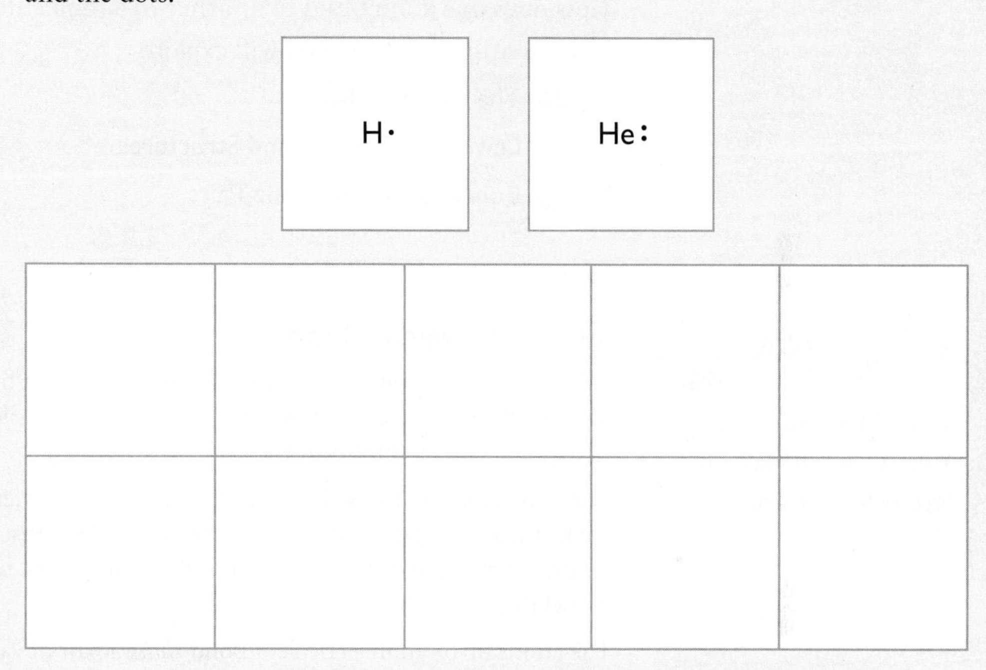

1. List two patterns that you notice.

2. Making Sense Using what you've learned, explain why the HONC 1234 rule works.

3. If You Finish Early Draw the Lewis dot structures of two different molecules with the molecular formula C_2H_7N.

LESSON

4 Connect the Dots

Lewis Dot Symbols

Think About It

The HONC 1234 rule is a great trick to help you figure out the structures of molecules. But why does it work? Why does carbon connect with four atoms, while hydrogen can connect with only one atom? To answer these questions, let's take a closer look at bonds.

How does one atom bond to another in a molecule?

To answer this question, you will explore

1. The Covalent Bond
2. Lewis Dot Symbols and Structures
3. Bonded Pairs and Lone Pairs

Exploring the Topic

SCILINKS® NSTA

Topic: Chemical Bonding

Visit: www.SciLinks.org

Web code: KEY-126

1 The Covalent Bond

All of the smelly compounds we have studied so far have been molecules. In fact, most of the substances that smell are made up of molecules, not ionic, metallic, or network covalent compounds.

The bonds that are found in molecules are called covalent bonds. Covalent bonds are formed between the atoms of nonmetallic elements. There are only about 15 nonmetallic elements in the periodic table. That is not many compared to the number of metallic elements.

The atoms involved in a covalent bond *share* a pair of valence electrons between them. The drawing below shows the difference between covalent bonds and ionic bonds.

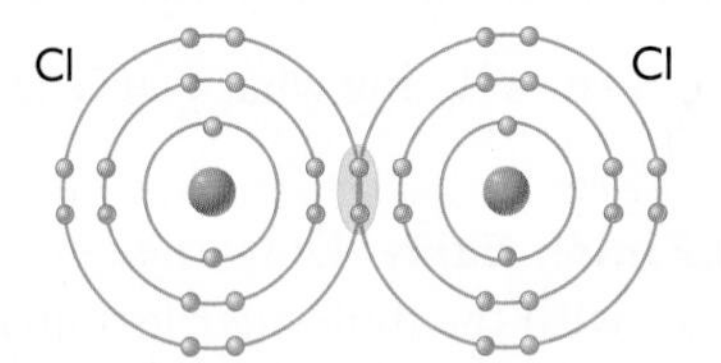

Covalent bonding: electron sharing

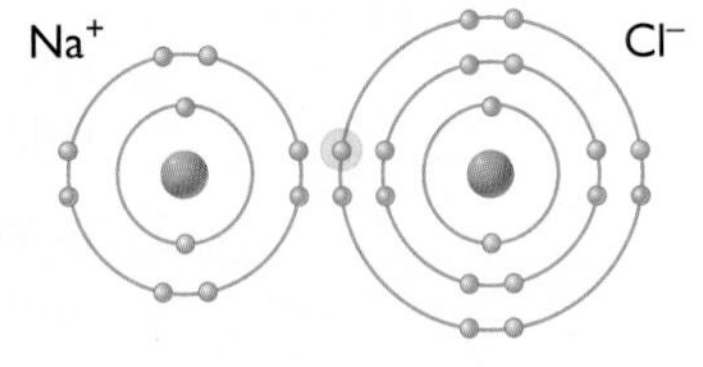

Ionic bonding: electron transfer

In covalent bonds, the nonmetal atoms are *sharing* electrons. As a result, the nonmetal atoms are tightly bound together but the atoms do not become ions with charges.

BIG IDEA A covalent bond is one in which nonmetal atoms *share* one or more pairs of electrons with one another. An ionic bond is one in which a metal atom *gives up* electrons to a nonmetal atom.

The Lewis dot symbol is named after a chemist who was a professor at the University of California at Berkeley. Gilbert Newton Lewis first introduced the idea of shared electrons and covalent bonds in 1916. He was the first scientist to use a system of dots to represent valence electrons in atoms.

❷ Lewis Dot Symbols and Structures

Structural formulas are one way to show the structure of a molecule on paper. Each line in a structural formula can be replaced with a pair of dots to represent the electrons that are being shared.

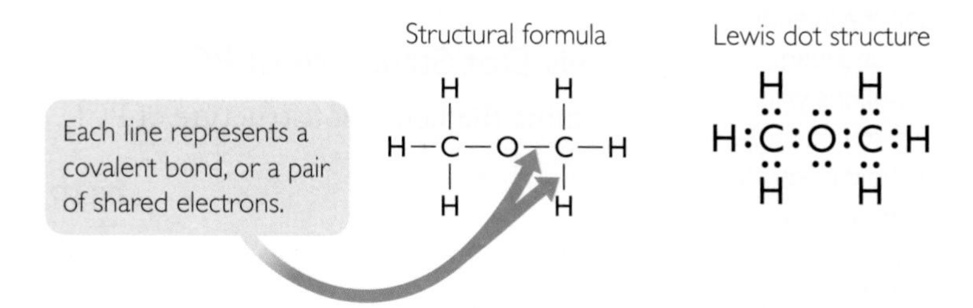

A drawing of a molecule that uses dots to represent the valence electrons is called a **Lewis dot structure.** Above is a molecule of dimethyl ether, C_2H_6O. Both the structural formula and the Lewis dot structure represent the same molecule.

Lewis dot structures keep track of every valence electron in every atom of a molecule. If you break the molecule apart into its individual atoms, you can see where each valence electron came from.

A **Lewis dot symbol** is the symbol of an element with dots to show the number of valence electrons that a single atom of that element has. Examine these Lewis dot symbols for hydrogen, oxygen, nitrogen, and carbon. Notice that some electrons are not paired up. These electrons are referred to as *unpaired electrons* and they are available for bonding. The unpaired electrons form covalent bonds with other atoms. These symbols can help to explain the HONC 1234 rule and the bonding of nonmetal atoms.

1 valence electron	6 valence electrons	5 valence electrons	4 valence electrons
H·	·Ö:	·N̈·	·Ċ·
1 electron available for sharing	2 electrons available for sharing	3 electrons available for sharing	4 electrons available for sharing

Recall that you can determine the number of valence electrons in an atom by locating its element on the periodic table. Hydrogen is in Group 1A and has one valence electron. Carbon is in Group 4A and has four valence electrons. Once you know the number of valence electrons an atom has, it is easy to draw a Lewis dot symbol for it.

❸ Bonded Pairs and Lone Pairs

Lewis dot symbols can be used to draw Lewis dot structures and structural formulas for molecules. Imagine bringing together the Lewis dot symbols for hydrogen and carbon to form methane, CH_4.

H
H:C:H
H

Because carbon bonds four times, it will bond with four hydrogen atoms. Once they are bonded, every valence electron in all of the atoms is paired up. The pairs of electrons between atoms are called **bonded pairs.** There are four bonded pairs in the methane molecule shown at right.

A pair of electrons in a molecule that are not shared between atoms is called a **lone pair** of electrons. In a molecule of ammonia, the nitrogen atom has one lone pair of electrons.

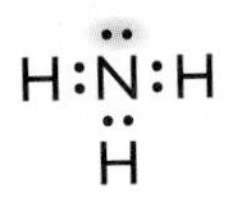

CONSUMER CONNECTION

The smell of ammonia is powerful and irritating. Ammonia is toxic and can damage the interior of your nose and lungs. However, solutions with dissolved ammonia make good household cleansers. And small amounts of ammonia are used in smelling salts to revive a person who has fainted.

Example

Lewis Dot Structure of PCl_3

Examine the Lewis dot structure of PCl_3, phosphorus trichloride.

:Cl:P:Cl:
:Cl:

a. Draw the Lewis dot symbols for phosphorus and chlorine.

b. How many bonds does a phosphorus atom make? How many does each chlorine atom make?

c. How many bonded pairs does this molecule have? How many lone pairs?

d. How many covalent bonds does this molecule have?

Solution

a. The periodic table can help you figure out how many valence electrons phosphorus and chlorine have. Phosphorus is in Group 5A and chlorine is in Group 7A, so they have five and seven valence electrons, respectively.

·P·   ·Cl:

b. Phosphorous has three unpaired electrons, so it makes three bonds. Chlorine has only one unpaired electron, so it makes one bond.

:Cl:P:Cl:
:Cl:

The three bonded pairs of electrons are circled.
All of the other pairs of electrons are lone pairs.

c. The molecule has three bonded pairs and ten lone pairs.

d. Each bonded pair represents one covalent bond. There are three covalent bonds in the molecule.

Key Terms

Lewis dot symbol
Lewis dot structure
bonded pair
lone pair

Lesson Summary

How does one atom bond to another in a molecule?

Molecules are made of nonmetal atoms that are covalently bonded. A covalent bond is a bond in which a pair of electrons is shared by two atoms. Lewis dot symbols keep track of the number of valence electrons in each atom. They can help you to predict covalent bonding between atoms. A Lewis dot structure shows how the atoms in an entire molecule are bonded together. The valence electrons that are involved in bonding are called a bonded pair. The valence electrons that are paired up in a molecule, but do not take part in bonding, are called lone pairs.

EXERCISES

Reading Questions

1. How are an ionic bond and a covalent bond different? Similar?
2. What is a Lewis dot structure? Explain how you would create one.

Reason and Apply

3. Draw the Lewis dot symbols for these elements:

   Te I K Bi In Pb

   a. Arrange them in order of their group numbers on the periodic table.

   b. Determine how many covalent bonds each element would make.

4. Germanium, antimony, selenium, and bromine each bond to a different number of hydrogen atoms.

   GeH_4 SbH_3 H_2Se HBr

   a. Draw Lewis dot symbols for Ge, Sb, Se, and Br.

   b. Draw a Lewis dot structure for each molecule.

   c. Explain the pattern in the number of hydrogen atoms.

5. Draw Lewis dot structures for the molecules listed here.

   a. $TeCl_2$ b. HI c. $AsBr_3$ d. SiF_4 e. F_2

6. How many lone pairs does each of the molecules in Exercise 5 have?
7. In your own words, explain why the HONC 1234 rule works.
8. Draw Lewis dot puzzle pieces for Si, P, S, and Cl. What rule would you make for Si, P, S, and Cl? What would be the name of this bonding rule?

LESSON 5 Eight Is Enough

Octet Rule

Think About It

When atoms bond covalently to form molecules, they share electrons to obtain an electron arrangement similar to a noble gas atom. Lewis dot structures can help you to discover how atoms share electrons to form molecules.

How do atoms bond to form molecules?

To answer this question, you will explore

1. The Octet Rule
2. Double and Triple Bonds

Exploring the Topic

1 The Octet Rule

When carbon, nitrogen, oxygen, and fluorine combine with hydrogen, they form these molecules:

```
     H
     |
  H—C—H        H—N—H        H—O—H        H—F
     |            |
     H            H
```

All of the above compounds are extraordinarily different. However, their Lewis dot structures reveal a striking similarity. Once carbon, nitrogen, oxygen, and fluorine are bonded, they are each surrounded by eight valence electrons.

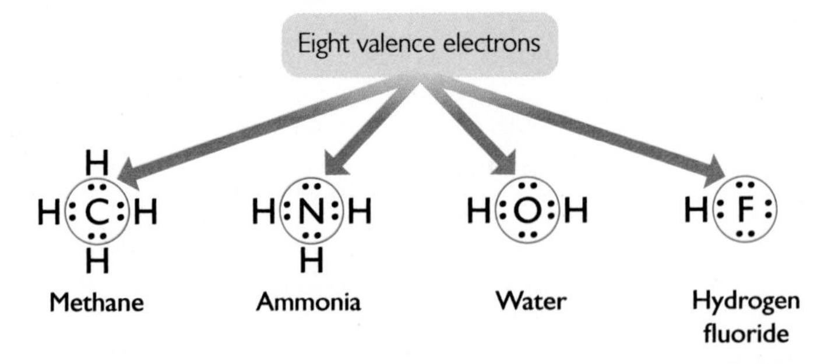

This tendency to bond until eight valence electrons surround an atom is called the **octet rule.** (The word *octet* comes from *octo,* which is Latin for "eight.") After these atoms are bonded, they all resemble a noble gas in their electron arrangements. Recall that the electron arrangements of the noble gases are very stable.

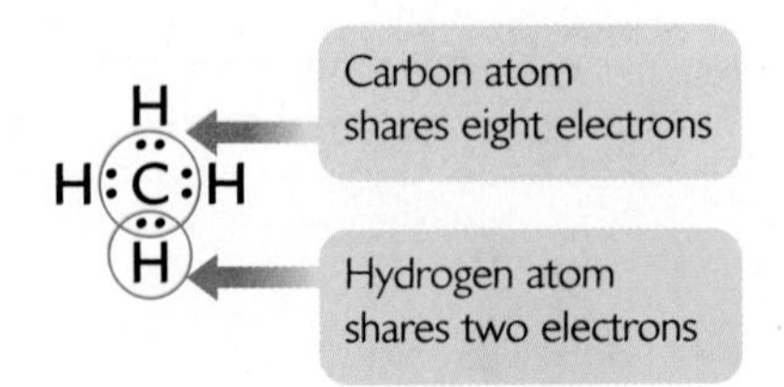

CONSUMER CONNECTION

Saturated fats are called "saturated" because the carbon atoms are bonded with as many hydrogen atoms as possible. Saturated fats are found in meat, dairy products, and tropical oils such as palm oil and coconut oil and are usually solid at room temperature. In contrast, vegetable oils are usually monounsaturated or polyunsaturated, meaning that they have one or more double bonds between carbon atoms and are therefore bonded with fewer hydrogen atoms. They are usually liquid at room temperature.

Note that hydrogen is an exception to the octet rule. Each hydrogen atom shares two electrons, not eight. A covalently bonded hydrogen atom resembles the noble gas helium, He, which has only two valence electrons.

F_2 SCl_2 CH_4O

Periodic Trends

The periodic table can be used to predict covalent bonding in molecules. In the illustration, Lewis dot symbols are shown for nonmetal elements in the second, third, and fourth rows. Notice that the elements in the same group have similar Lewis dot symbols and bond in a similar way. For example, selenium, Se, which is below oxygen, O, has two unpaired electrons and forms two bonds, just like oxygen.

4A	5A	6A	7A	8A
				He
C	N	O	F	Ne
Si	P	S	Cl	Ar
Ge	As	Se	Br	Kr

Elements in the same group have the same number of valence electrons and therefore have similar Lewis dot symbols. Helium is an exception.

2 Double and Triple Bonds

There is more than one way to satisfy the octet rule through bonding. Quite a few of the structural formulas you have examined so far have a bond with two lines. This type of bond is called a **double bond.** In a Lewis dot structure, a double bond is represented by four dots instead of the usual two. A double bond contains four electrons that are being shared between atoms. Methyl methanoate, $C_2H_4O_2$, is an example of a molecule that contains a double bond. Its structural formula and Lewis dot structure are shown here.

A double bond means four shared electrons.

The carbon and oxygen atoms with double bonds are surrounded by a total of eight valence electrons each, just like the atoms with single bonds.

Example

Carbon Dioxide

Draw the Lewis dot structure and structural formula for carbon dioxide, CO_2.

Solution

Step 1: Start with the Lewis dot symbols. Bring the atoms together.

Incorrect

Step 2: Check to see if the octet rule is satisfied. This Lewis dot structure isn't correct. Move the remaining unpaired electrons to create double bonds.

Incorrect   Correct

Step 3: Check your molecule again to see that each atom in it satisfies the octet rule and still has the right number of valence electrons.

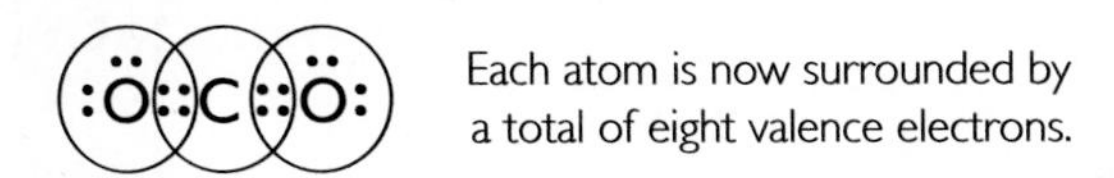

Step 4: To make the structural formula, replace each pair of dots with a line.

O=C=O

ENVIRONMENTAL CONNECTION

Some of the most common substances in the world around you are molecules with double and triple bonds. The air you breathe is composed of a mixture of nitrogen gas, oxygen gas, and a sprinkling of carbon dioxide and other gases. In fact, the air is 78% nitrogen gas.

If two atoms can share four electrons, can they share six electrons, or even more? The answer is yes.

Nitrogen gas, N_2, has a **triple bond** between the two nitrogen atoms, so each nitrogen atom has six shared valence electrons. Notice that there are also two lone pairs in this molecule. Oxygen gas, O_2, has a double bond between the oxygen atoms. The four lone pairs in the oxygen molecule have been adjusted slightly in this Lewis dot structure, to space the electrons more evenly around the atoms. Quadruple bonds are also possible, but quite rare.

N≡N	O=O	H–H
:N:::N:	:O::O:	H:H
Triple bond	Double bond	Single bond

Lesson Summary

Key Terms
octet rule
double bond
triple bond

How do atoms bond to form molecules?

When nonmetal atoms bond, they share electrons to obtain the same electron arrangement as a noble gas atom. Nonmetal atoms will share electrons with other atoms so that both atoms end up sharing a total of eight valence electrons each. This is called the octet rule. Hydrogen still fits the sharing pattern, but it ends up with only two valence electrons, like the noble gas helium. Atoms can also satisfy the octet rule by forming double and triple bonds in which they share four or six valence electrons.

EXERCISES

Reading Questions

1. Explain why nitrogen bonds with hydrogen to form NH_3, but not NH_2 or NH_4. Use Lewis dot structures to support your argument.
2. What is the octet rule, and how can you use it to create a molecular structure?

Reason and Apply

3. List three nonmetal elements that combine with only one fluorine atom to satisfy the octet rule.
4. List two nonmetal elements that combine with three hydrogen atoms to satisfy the octet rule.
5. Draw Lewis dot structures for these molecules. Notice that in part d and part f, the formulas are written in a way that emphasizes the structure of the molecule.

   a. CF_4 **b.** CH_3Cl **c.** $SiCl_2H_2$
   d. CH_3OH **e.** $HOCl$ **f.** CH_3NH_2
6. Use the octet rule to draw Lewis dot structures for all the stable molecules with the molecular formula C_3H_8O. (*Hint:* There are three total molecules.)
7. Consider the molecules C_2H_2, N_2H_2, and H_2O_2.
   a. Draw a Lewis dot structure for each one. What pattern do you notice?
   b. What can you do to check that your Lewis dot structures are correct? Name at least two ways.
8. Consider the molecules C_2H_4 and N_2H_4. Draw a Lewis dot structure for each of the molecules.
9. Which is more likely to exist in nature, a molecule of CH_3 or a molecule of CH_4? Explain your reasoning.

LESSON

6 Where's the Fun?

Functional Groups

Think About It

It makes sense that the structure of a molecule would affect its properties. The atoms in ethyl butyrate (pineapple smell) and hexanoic acid (dirty socks smell) are connected differently, so they behave differently when they enter your nose. But what about molecules that smell similar? What is it about these structures that causes them to have similar properties?

What does the structure of a molecule have to do with smell?

To answer this question, you will explore

1. Relating Smell to Molecular Structure
2. Functional Groups
3. Classifying Molecules

Exploring the Topic

1 Relating Smell to Molecular Structure

Putrid-Smelling Molecules

Consider two molecules in the same smell category. Both hexanoic acid and butyric acid smell putrid. The first smells like stinky feet while the second smells like a carton of spoiled milk. Take a moment to compare the structural formulas of hexanoic acid and butyric acid.

Hexanoic acid, $C_6H_{12}O_2$

```
    H   H   H   H   H   O
    |   |   |   |   |   ||
H — C — C — C — C — C — C — O — H
    |   |   |   |   |
    H   H   H   H   H
```

Butyric acid, $C_4H_8O_2$

```
    H   H   H   O
    |   |   |   ||
H — C — C — C — C — O — H
    |   |   |
    H   H   H
```

Notice that both molecules have two oxygen atoms bonded on the end in an identical fashion. They both contain the structural feature highlighted.

Perhaps all putrid-smelling molecules have this same structural feature. Two more examples are shown here.

Isopentanoic acid, $C_5H_{10}O_2$

```
    H   H   H   O
    |   |   |   ||
H — C — C — C — C — O — H
    |   |   |
    H   |   H
    H — C — H
        |
        H
```

Isobutyric acid, $C_4H_8O_2$

```
            O   H   H
            ||  |   |
H — O — C — C — C — H
                |   |
                |   H
            H — C — H
                |
                H
```

CONSUMER CONNECTION

The ingredients list on most of your household products reads like a chemical inventory. If you examine the labels of your shampoos, cleansers, hair gels, mouthwash, and so on, you should be able to recognize some of the chemical names.

You can see that this same structural feature is present in isopentanoic acid and isobutyric acid as well.

Sweet-Smelling Molecules

Now look at the structural formulas for two sweet-smelling molecules.

Ethyl butyrate, $C_6H_{12}O_2$

```
   H  H  H  O     H  H
   |  |  |  ||    |  |
H—C—C—C—C—O—C—C—H
   |  |  |        |  |
   H  H  H        H  H
```

Ethyl acetate, $C_4H_8O_2$

```
   H  O     H  H
   |  ||    |  |
H—C—C—O—C—C—H
   |        |  |
   H        H  H
```

Just like the putrid-smelling molecules, these sweet-smelling molecules have an identical structural feature, which is highlighted in yellow. This feature is similar to the one found in putrid-smelling molecules, but it is slightly different.

2 Functional Groups

The structural features that groups of molecules have in common are called **functional groups.** A functional group often stands out as an unusual or unique portion of a molecule. The functional group found in the putrid-smelling molecules is called a *carboxyl* functional group. The functional group found in the sweet-smelling molecules is called an *ester* functional group.

Two other functional groups are associated with fishy smells and minty smells.

Amine group, fishy

Ketone group, minty

Example 1

Functional Group and Smell

The structural formulas of two molecules are shown here. Predict how the molecules will smell. What is your reasoning?

```
   H  O     H  H  H  H
   |  ||    |  |  |  |
H—C—C—O—C—C—C—C—H
   |        |  |  |  |
   H        H  H  H  H
```

```
   O
   ||
H—C—O—H
```

Solution

This molecule contains an ester functional group. It is probably a sweet-smelling molecule.

```
 O
 ||     |
—C—O—C—
        |
```

This molecule contains a carboxyl functional group. It is probably a putrid-smelling molecule.

```
 O
 ||
—C—O—H
```

Other Functional Groups

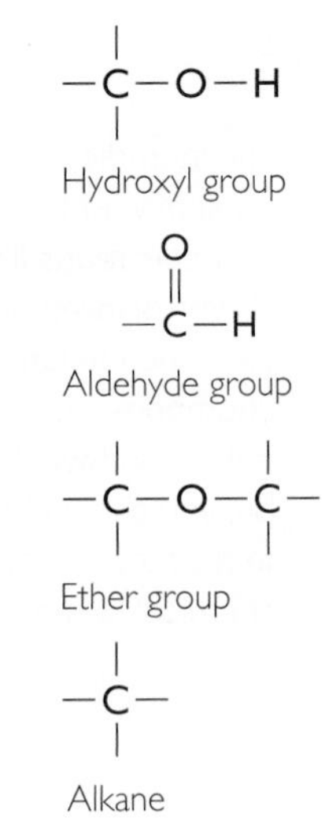

There are other functional groups besides the four discussed so far. For instance, a molecule may have a hydroxyl group (—OH). Molecules containing this feature are called *alcohols.* A molecule containing an aldehyde group is similar to a ketone, except that the carbon that is double bonded to the oxygens is between a carbon and a hydrogen.

Another type of molecule is one containing a single oxygen located between two carbon atoms. Molecules containing this feature are called *ethers.*

Some molecules have no functional groups. For example, *alkanes* consist of only carbon and hydrogen atoms connected with single bonds.

3 Classifying Molecules

The easiest way to classify molecules is by functional group. All the molecules that have ester functional groups are referred to as "esters." All the sweet-smelling compounds you have encountered so far contain an ester functional group and have two-word names that end in "-yl" and "-ate." Eth**yl** butyr**ate,** eth**yl** acet**ate,** and hex**yl** acet**ate** all smell sweet. The putrid-smelling compounds have a carboxyl group and names that end with "-ic acid." Butyr**ic acid** and hexano**ic acid** both smell putrid.

You may be able to identify the smell of a molecule by paying attention to its chemical name. Examine the compounds in the table. Each of these molecules has a unique smell even though they all have three carbon atoms and at least six hydrogen atoms. Their names are also very different.

Name	Compound type	Functional group	Smell	Common use or source
propane C_3H_8	alkane	no functional group	no smell or gasoline-like	fuel in camp stoves
propionic acid $C_3H_6O_2$	carboxylic acid	carboxyl	putrid	in sweat
ethyl formate $C_3H_6O_2$	ester	ester	sweet	in raspberries

Name	Compound type	Functional group	Smell	Common use or source
acetone C_3H_6O	ketone	ketone 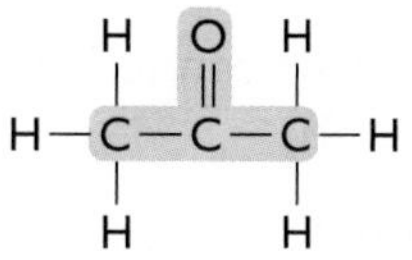	sweet	nail polish remover 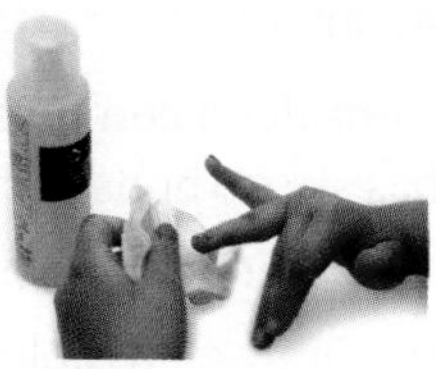
isopropanol C_3H_8O	alcohol	hydroxyl 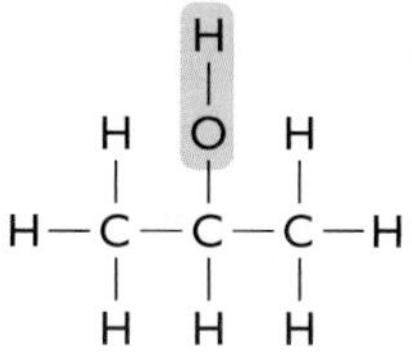	medicinal	rubbing alcohol 
timethylamine C_3H_9N	amine	amine 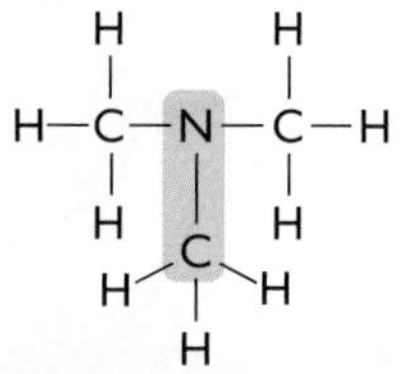	fishy	in bad breath

Compounds that have the same functional group tend to have similar properties. For example, esters smell sweet, dissolve in water, and change phase from liquid to gas fairly easily. There are exceptions. For example, larger alkanes tend to have a gasoline smell, while smaller ones like the one in the table have no smell.

Key Term

functional group

Lesson Summary

What does the structure of a molecule have to do with smell?

Based on what has been studied so far, compounds that have similar smells also appear to have similar structural features. These features are called functional groups. Functional groups have names. Compounds are frequently named according to the functional groups they contain. If you identify the functional group in a compound, you may also be able to determine how that compound will smell.

EXERCISES

Reading Questions

1. What is a functional group?
2. What information would you want to have in order to figure out the smell of a compound? Explain why.

Reason and Apply

3. Explain why C_2H_4 has fewer hydrogen atoms than C_2H_6.
4. Create a four-carbon molecule that is
   a. an alkane **b.** a carboxylic acid **c.** an alcohol **d.** an ester
5. Consider a compound called hexanol. Chemists can tell from its name that it has six carbons (hex-) and it is an alcohol (-ol).
   a. What is the molecular formula of hexanol?
   b. Draw a possible structural formula for hexanol.
   c. Is the molecular formula or the structural formula more useful in determining the smell of hexanol? Explain.
6. **Research** Examine the ingredients lists on some household products, such as shampoo, lotion, or cleanser. Write down the names of compounds and functional groups that you were able to identify.
7. If you were a chemist and you wanted to invent a new smell, what would you think about doing or creating?
8. If you were a chemist and you wanted to change the smell of a molecule, what might you try to do to that molecule?
9. Explain what you think is going on in this cartoon.

10. Structural formulas for six molecules are shown here. Identify (by name) the functional group in each molecule.

```
          O   H   H
          ||  |   |
a. H—O—C—C—C—H
              |   |
              H   H
```

```
     H   O   H   H   H
     |   ||  |   |   |
b. H—C—C—C—C—C—H
     |       |   |   |
     H       H   H   H
```

```
     H   O   H   H
     |   ||  |   |
c. H—C—C—C—C—H
     |       |   |
     H       |   H
           H—C—H
             |
             H
```

```
     H   H   O       H
     |   |   ||      |
d. H—C—C—C—O—C—H
     |   |           |
     H   H           H
```

```
     H   H   H   H   H
     |   |   |   |   |
e. H—C—C—C—N—C—H
     |   |   |       |
     H   H   H       H
```

```
                 H
                 |
f. H—C—O—C—H
     ||          |
     O           H
```

11. Write the molecular formula for each of the molecules in Exercise 10.
12. **Lab Prep** Read the Lab: Ester Synthesis on pages 173–175, paying special attention to the safety instructions. List all the steps that require caution in this lab. For each one, explain how to do the step safely and describe what you should do in case of an accident. Be prepared to pass a safety quiz before the lab.

LAB Ester Synthesis

Safety Instructions

Always wear safety goggles.

Extreme care should be taken when handling 18 M sulfuric acid.

Heating must be done slowly and carefully. When heating is finished, hot plates should be turned off immediately.

During cleanup place the products in designated waste containers.

Purpose

To create new smells and analyze the products of a chemical change.

Materials

- 50 mL beaker
- hot plate
- 3 microscale test tubes
- boiling stones
- 3 plastic pipettes
- organic acids and alcohols

Procedure

1. Fill a 50 mL beaker with about 30 mL of water. Drop in a boiling stone. Place the beaker on a hot plate and bring the water to a *gentle* boil.

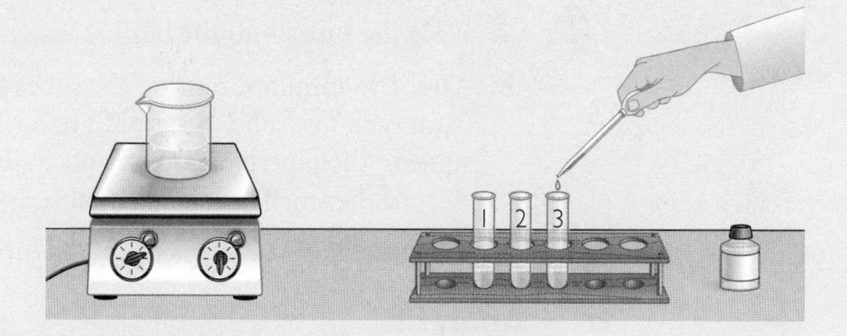

2. Label your test tubes 1, 2, and 3. Make a data table to record data.

Data Table

Test tube	Smell of carboxylic acid	Smell of alcohol	Smell of mixture before heating	Smell of mixture after heating
1				
2				
3				

3. Carefully smell the acids by wafting and record the smells. Add five drops of the appropriate carboxylic acid to each tube according to the table in Step 4.

4. Carefully smell the alcohols by wafting and record the smells. Add ten drops of the appropriate alcohol to each tube according to the table here.

Test tube	Carboxylic acid	Alcohol
1	acetic acid	isopentanol
2	acetic acid	butanol
3	butyric acid	ethanol

5. Add one drop of the concentrated sulfuric acid, H_2SO_4, to each tube. Slide a boiling stone into the mixture in each tube. Carefully smell each mixture by wafting and record the smells.
6. Cut three plastic pipettes so they are shorter than the length of a test tube. Put each one in a test tube with the stem down, so that the bulb *loosely* seals off the tube.

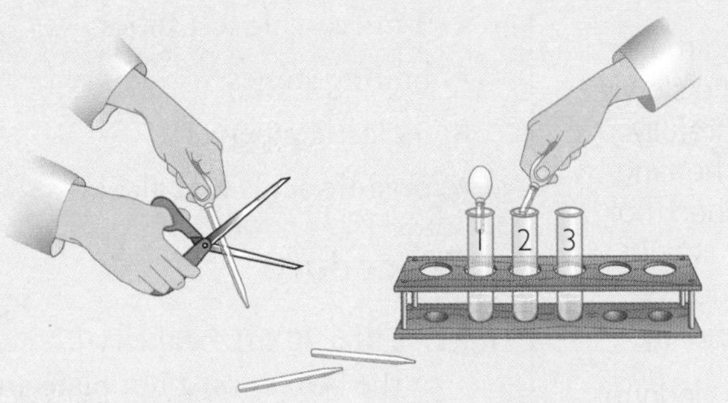

7. Place the tubes into the boiling water and heat for five minutes.
8. After five minutes, remove the tubes from the water. Remove the pipette from each test tube. It should not have any liquid in it, just vapors. Carefully squeeze the pipette near your nose so that you can waft and smell the vapors. Record the smells in the data table.

9. Clean up. Dispose of chemicals as directed by your instructor.

Analysis

1. How did the smell of the mixtures before heating compare to the smell of the mixtures after heating? What do you think happened to the molecules to change the smell?
2. Based on the smell of the mixtures after heating, what functional group must be in the final molecules that were produced? Draw it on your paper.
3. Using each pair of structural formulas, draw a new molecule that contains the functional group you stated in Question 2. These are chemical reactions, so you are allowed to break bonds and make new ones.

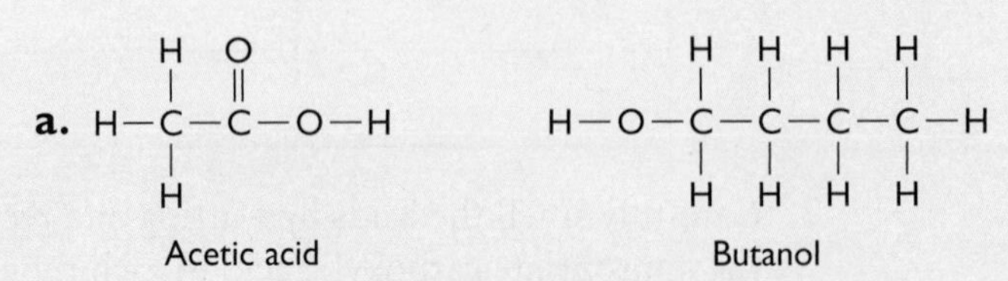

b.

```
  H O
  | ||
H-C-C-O-H
  |
  H
```

Acetic acid

```
           H
         H\|/H
           C
    H  H   |  H
    |  |   |  |
H-O-C--C---C--C-H
    |  |   |  |
    H  H   H  H
```

Isopentanol

c.

```
  H H H O
  | | | ||
H-C-C-C-C-O-H
  | | |
  H H H
```

Butyric acid

```
    H H
    | |
H-O-C-C-H
    | |
    H H
```

Ethanol

4. Are there any pieces that you did not use to make the sweet-smelling molecule in Question 3? If so, what molecule could you make with these pieces?

5. Write the chemical equation for each reaction in Question 3. Make sure that each equation is balanced (that there are the same number of carbon, hydrogen, and oxygen atoms on both sides of the arrow).

$$C_2H_4O_2 + C_4H_{10}O \xrightarrow{H_2SO_4}$$

6. Imagine that you used the following acid and alcohol in the lab to create a sweet-smelling molecule. Draw the structural formula for the sweet-smelling product. (This product is called octyl formate and smells like pineapples. Water is also produced.)

```
  O
  ||
H-C-O-H
```

Formic acid

```
    H H H H H H H H
    | | | | | | | |
H-O-C-C-C-C-C-C-C-C-H
    | | | | | | | |
    H H H H H H H H
```

Octanol

7. What are the molecular formulas for the sweet-smelling products in Question 5? Draw the structural formulas for these two molecules next to each other. Why do you think the molecules in Questions 5 and 3 smell different?

8. Making Sense In your own words, describe what you think happened on a molecular level to the two compounds that took part in these reactions. What is your evidence?

9. If You Finish Early See if you can figure out how the products of these reactions are named. What would the name of the product in Question 5 be?

LESSON

7 Create a Smell
Ester Synthesis

Think About It

Body odor, stinky shoes, moldy carpets, dog odors, musty basements, bad breath—these are just a few of the smells that we consider unpleasant. But the structures of these molecules are not so different from the structures of sweet-smelling substances. Molecules that have a carboxyl functional group are very similar to molecules that have an ester functional group. Perhaps there is a way to change one type of molecule into the other.

How can a molecule be changed into a different molecule by using chemistry?

To answer this question, you will explore

1. Transforming Smells
2. Chemical Reactions
3. Chemical Synthesis

Exploring the Topic

1 Transforming Smells

Every year, consumers spend millions of dollars to purchase products that will deal with unwanted odors. Often these products do nothing more than cover up a foul smell with a pleasing smell. But some products use chemistry to change the molecular structure of the smelly molecules.

Butyric acid, $C_4H_8O_2$, is a carboxylic acid with a putrid smell. However, the difference between this molecule and a sweet-smelling molecule is minimal. Butyric acid can be chemically changed into a sweet-smelling ester.

How can this . . .

Butyric acid

become this?

Methyl butyrate

You can see that to accomplish this task, you would have to somehow remove the hydrogen atom that is attached to an oxygen atom in the butyric acid molecule and attach a carbon atom with three hydrogen atoms in its place. This procedure is possible in the chemistry laboratory.

CONSUMER CONNECTION

The putrid-smelling compound butyric acid is formed when butter goes bad. In fact, *butyric* means "from butter." Butyric acid is also found in Parmesan cheese and vomit.

CONSUMER CONNECTION

An entire industry is devoted to the creation of fragrances and flavors for use in our foods and commercial products. Some companies even manufacture room odorizers to inject a pleasant aroma into public environments or businesses. These aromas include the smell of fresh brewed coffee, baking bread, or an evergreen forest. The idea is to create environments that evoke pleasant memories or experiences for the user, and to stimulate the appetite or the pocketbook.

❷ Chemical Reactions

In the laboratory you mixed some substances together with a catalyst and heated them, following a specific set of instructions. The result was a molecule that no longer smelled bad.

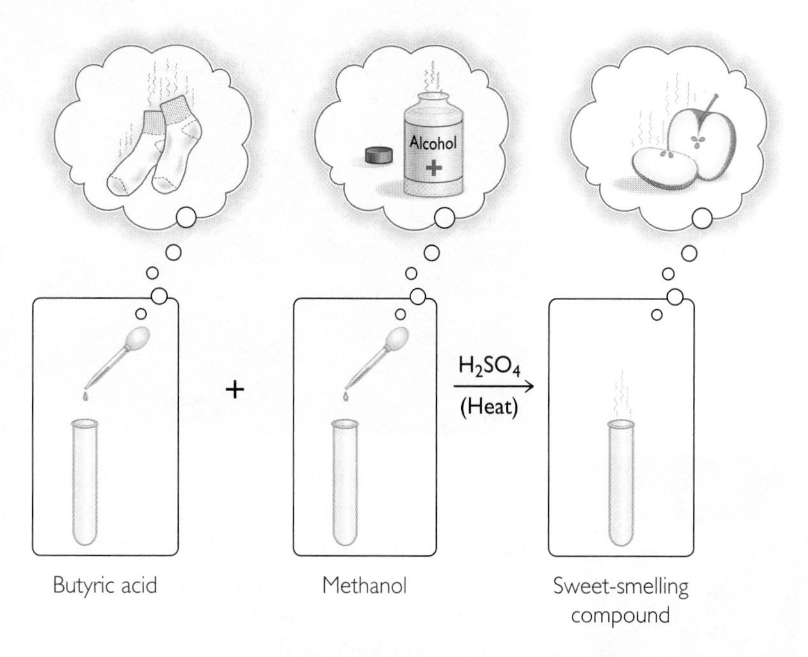

When butyric acid and methanol are heated and mixed, no change is visible to the eye. However, your nose tells you that a dramatic change in properties has occurred. The foul smell of sweaty socks disappears and the sweet smell of apples fills the room.

The change in smell is clear evidence that a chemical reaction has occurred. A **chemical reaction** is the process by which matter is changed so that new substances are formed. For new substances to form, the atoms must rearrange. Some bonds between atoms must be broken and new bonds must be formed. The result of a chemical reaction is a new compound with entirely different properties. In this case, one of the new properties is a different smell.

The detection of a sweet-smelling compound is evidence that you have made an ester. In fact, methyl butyrate smells like apples. In Lesson 8: Making Scents, you will examine this transformation in greater detail.

❸ Chemical Synthesis

Chemists often work in the laboratory with the goal of producing a specific compound. This process is called synthesis. **Synthesis** is the process of producing a chemical compound, usually by combining two simpler compounds. Many of the products that we have come to depend on are products that chemists have synthesized, like plastics, fabrics, cosmetics, deodorants, cleaning products, vitamins, and medications.

Sometimes chemists work to synthesize a molecule that is very rare or hard to find. For instance, many lifesaving, anti-cancer compounds have been discovered in plant and animal life in the rain forests of South America. It may be difficult to harvest these compounds or find enough of a substance to treat the many patients who are in need. Chemists study the structure of a compound like this and work to create it in the laboratory. Most of the medicines in use today are synthesized in laboratories.

The U.S. National Cancer Institute has identified 3000 plants containing compounds with anti-cancer properties. Most are found in the rainforest. Here, a guide points out a medicinal plant.

Sometimes the point of synthesis is to create a new molecule that is *not* found in nature. All of the plastics that we have in our lives are synthesized from small molecules that have been strung together through chemical reactions. These compounds are valued for their amazing properties. Many plastics are durable, flexible, and easily molded into a variety of useful shapes—everything from shoes to computer keyboards.

Key Term
synthesis

Lesson Summary

How can a molecule be changed into a different molecule by using chemistry?

In order to transform a putrid-smelling molecule into a sweet-smelling molecule, it is necessary to transform a carboxylic acid molecule into an ester molecule. This transformation requires a chemical reaction in which bonds between certain atoms are broken and new bonds are made. The result of any chemical reaction is a different compound or compounds than the compound you started with. And when the functional group changes, the smell changes. Chemists control chemical reactions in the laboratory to synthesize specific molecules with valuable properties.

EXERCISES

Reading Questions

1. What would you do in order to cause an alcohol and an acid to react?
2. What is chemical synthesis?

Reason and Apply

3. Why do you think the foul smell of butyric acid disappeared when you mixed it with methanol and heated the mixture?

4. The structural formulas for the molecules used in the Lab: Ester Synthesis are given in the table. Identify the functional groups.

Test tube	Organic acid	Structural formula	Alcohol	Structural formula
1	acetic acid	H O | ‖ H—C—C—O—H | H	isopentanol	H | H—C—H H H | H | | | | H—O—C—C—C—C—H | | | | H H H H
2	acetic acid	H O | ‖ H—C—C—O—H | H	butanol	H H H H | | | | H—C—C—C—C—O—H | | | | H H H H
3	butyric acid	H H H O | | | ‖ H—C—C—C—C—O—H | | | H H H	ethanol	H H | | H—C—C—O—H | | H H

5. Explain how you can convert acetic acid into an ester. Be specific about how the molecule needs to change.

6. **Lab Report** Write a lab report for the Lab: Ester Synthesis. In your report, include the title of the experiment, purpose, procedure, observations, and conclusions.

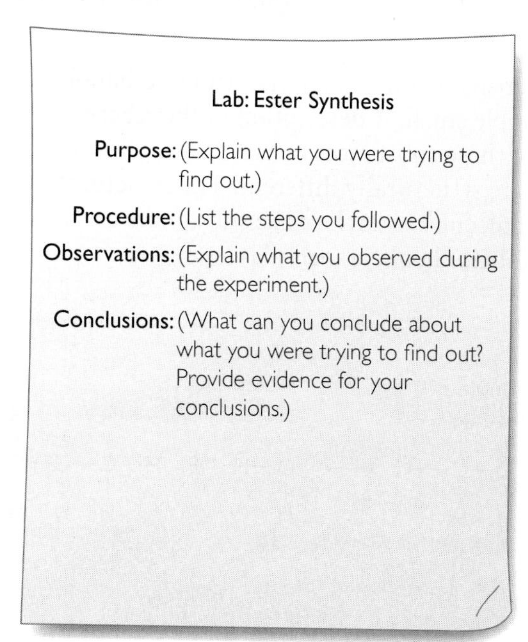

Lab: Ester Synthesis

Purpose: (Explain what you were trying to find out.)

Procedure: (List the steps you followed.)

Observations: (Explain what you observed during the experiment.)

Conclusions: (What can you conclude about what you were trying to find out? Provide evidence for your conclusions.)

LESSON

8 Making Scents

Analyzing Ester Synthesis

Think About It

When you heat a mixture of a carboxylic acid and an alcohol, a chemical reaction takes place. You can smell the outcome. A really putrid-smelling substance is transformed into one that smells sweet. An ester has been produced. But what has happened to the original substances that were mixed together?

What happened to the molecules during the creation of a new smell?

To answer this question, you will explore

1. Chemical Equations
2. The Role of the Catalyst
3. Ester Synthesis

Exploring the Topic

1 Chemical Equations

Chemists use **chemical equations** to keep track of changes in matter. These can be certain physical changes or changes due to chemical reactions. A chemical equation usually uses chemical formulas to describe what happens when substances are mixed together and new substances with new properties are formed. In class you used structural formulas to describe what happened to the molecules that were combined.

Consider what happens if you add butyric acid to ethanol and create a pineapple smell. A description of the reaction is given here. Each of these substances consists of a collection of molecules that are structurally different. The structural formulas for the molecules in each of the substances are also shown along with their chemical formulas.

```
    Butyric acid                      Ethanol            Water         Ethyl butyrate
  H   H   H   O                      H   H                           H   H   H   O       H   H
  |   |   |   ||                     |   |                           |   |   |   ||      |   |
H-C---C---C---C-O-H   +   H-O-C---C-H   ---->   O-H   +   H-C---C---C---C-O-C---C-H
  |   |   |                          |   |            |              |   |   |           |   |
  H   H   H                          H   H            H              H   H   H           H   H

            Reactants                                      Products
```

Butyric acid, $C_4H_8O_2$, is added to ethanol, C_2H_6O, to produce water, H_2O, and ethyl butyrate, $C_6H_{12}O_2$.

In a chemical reaction the substances that are mixed together are called the **reactants.** The new substances that are produced are called the **products.** The reactants in this reaction are butyric acid and ethanol. One product that is easily detected by smell is ethyl butyrate. However, this is not the only product. Although it is difficult to observe with the five senses, water is also a product of this reaction. The chemical equation helps to make this clear. The chemical equation shows that in a chemical reaction, no matter is created or destroyed.

BIG IDEA Matter cannot be created or destroyed. Matter is conserved.

The next illustration highlights the areas of the reactant molecules that change during the chemical reaction. You can see that a hydroxyl, −OH group, breaks off from the butyric acid molecule. In addition, the H−O bond in the ethanol molecule must also break. A new bond forms as the larger molecular pieces come together to form ethyl butyrate.

Finally, you can see that the H− and −OH pieces that are left over combine to form H_2O, water.

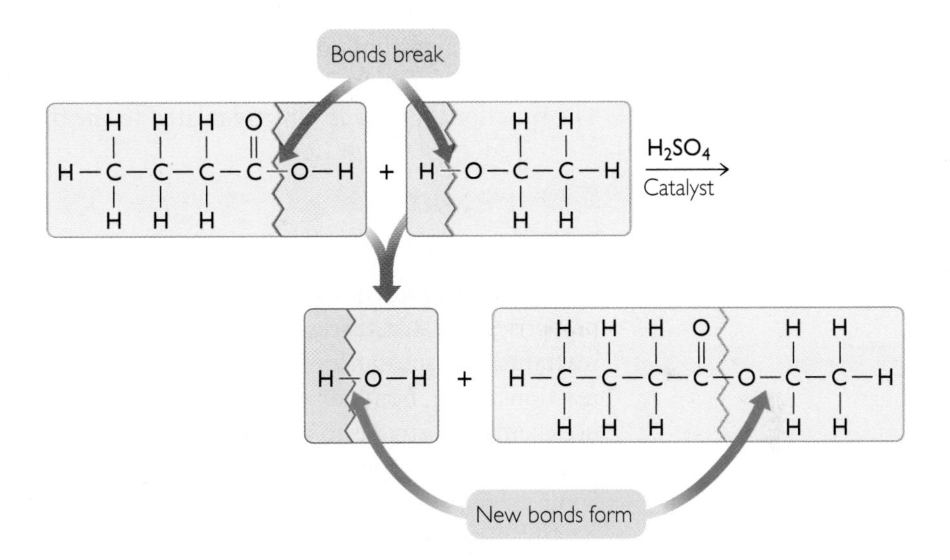

For every butyric acid molecule that reacts with an ethanol molecule, one molecule of water is produced along with one molecule of ethyl butyrate.

The complete chemical equation for this chemical reaction is

$$C_4H_8O_2 + C_2H_6O \longrightarrow H_2O + C_6H_{12}O_2$$

There are the same number of each type of atom on both sides of the equation. All of the matter is accounted for.

CONSUMER CONNECTION

Esters are found in a number of household products, from cleansers and air fresheners to hand lotions and shampoos. Their primary role is to make a product smell a certain way, although some are detergents as well.

2 The Role of the Catalyst

There's one more ingredient in the chemical reaction that we have not yet discussed. When you try to carry out this reaction in a laboratory, you will find that simply mixing butyric acid and ethanol does not produce ethyl butyrate.

CONSUMER CONNECTION

A catalyst is often used in an automobile. It is located in the catalytic converter in the muffler of the car. Its job is to help break down some of the harmful products in engine exhaust. Platinum is one of the elements used in catalytic converters.

In fact, nothing happens. In order to make the reaction happen, it is necessary to add sulfuric acid, H_2SO_4, and heat the mixture.

$$C_4H_8O_2 + C_2H_6O \xrightarrow{H_2SO_4} H_2O + C_6H_{12}O_2$$

The sulfuric acid does not get used up during the reaction. It is still present at the end. In this reaction the sulfuric acid is simply helping the reaction along. Chemists refer to a substance that assists a chemical reaction as a **catalyst.** A catalyst speeds up a reaction but the catalyst itself is not consumed by the reaction. In chemical equations a catalyst is written above the arrow.

3 Ester Synthesis

Many different acids and alcohols can be brought together to form an ester and water. The general description of this reaction is

$$\text{acid} + \text{alcohol} \longrightarrow \text{water} + \text{ester}$$

Take another look at the illustration showing the bonds breaking. It is not necessary for every single bond in the acid and alcohol molecules to break apart in order to form the ester. All of the C—H bonds remain connected during this reaction.

During chemical reactions, the major changes in the molecules often occur where the functional groups are located. The part of the molecule that is easiest to change is the functional group. This makes sense with what you already know: The functional group is directly related to the properties of a molecule.

Key Terms

chemical equation
reactant
product
catalyst

Lesson Summary

What happened to the molecules during the creation of a new smell?

When chemical reactions occur, new compounds are produced with different properties and structures. The substances that are combined are called the reactants. The substances that are produced are called the products. When chemical reactions occur, bonds are broken and new bonds are formed. Chemical equations use chemical formulas to track the changes that occur during chemical reactions. Because matter is conserved, all atoms are accounted for in a chemical equation. Some chemical reactions require a catalyst to help them get started or proceed more rapidly. A catalyst is not consumed by the chemical reaction it is assisting.

EXERCISES

Reading Questions

1. Explain why converting a carboxylic acid to an ester might be useful.
2. Describe what happens during a chemical reaction.
3. What is a catalyst?

Reason and Apply

4. Write a paragraph answering the question "How can scientists use chemistry to create compounds with specific smells?" Include evidence to support your answer.

5. Below are the structural formulas for four esters. Write the correct molecular formula for each one.

a.
```
      H   H   H   H   H       O   H   H
      |   |   |   |   |       ||  |   |
   H—C—C—C—C—C—O—C—C—C—H
      |   |   |   |   |           |   |
      H   H   H   H   H           H   H
```

b.
```
                            H
                            |
                         H—C—H
      H   O         H   H   |   H
      |   ||        |   |   |   |
   H—C—C—O—C—C—C—C—H
      |             |   |   |   |
      H             H   H   H   H
```

c.
```
      O       H   H   H   H   H   H   H   H
      ||      |   |   |   |   |   |   |   |
   H—C—O—C—C—C—C—C—C—C—C—H
              |   |   |   |   |   |   |   |
              H   H   H   H   H   H   H   H
```

d.
```
      H   O         H   H   H   H
      |   ||        |   |   |   |
   H—C—C—O—C—C—C—C—H
      |             |   |   |   |
      H             H   H   H   H
```

6. Ester molecules are named for the alcohol and acid that form them using the convention

(alcohol name)yl (acid name)ate

For example, methanol + ethanoic acid combine to make methyl ethanoate. Name the esters formed from

a. isopropanol + methanoic acid

b. caproic acid + butanol

c. salicylic acid + ethanol

CAREER CONNECTION

A flavorist's job is to create or re-create flavors. To create flavors for foods, toothpaste, chewable medications, lipgloss, and other products the flavorist will use his or her knowledge of the available chemical ingredients to create a flavor compound. The flavorist may test many different compounds until the right flavor is found.

SECTION I

SUMMARY

Speaking of Molecules

Smells Update

The investigation into the chemistry of smell has challenged you to explore how molecules are put together. So far you know that the smell of a substance is related to its molecular formula, chemical name, and any functional groups present.

```
   O
   ‖     |
 —C—O—C—
         |
```

This is based on the data investigated so far. Further investigation may lead you in new directions.

Key Terms

molecular formula
structural formula
isomer
HONC 1234 rule
Lewis dot symbol
Lewis dot structure
bonded pair
lone pair
octet rule
double bond
triple bond
functional group
synthesis
chemical equation
reactant
product
catalyst

Review Exercises

1. What functional groups are present in each of these molecules?

```
     H  H     O  H  H  H  H  H                H  H  O  H  H
     |  |     ‖  |  |  |  |  |                |  |  ‖  |  |
a. H—C—C—O—C—C—C—C—C—C—H           b. H—C—C—C—C—C—H
     |  |        |  |  |  |  |                |  |     |  |
     H  H        H  H  H  H  H                H  H     H  H

     H  H  H  H  O                            H  H  H
     |  |  |  |  ‖                            |  |  |
c. H—C—C—C—C—C—O—H                 d. H—C—C—C—N—H
     |  |  |  |                               |  |  |  |
     H  H  H  H                               H  H  H  H
```

2. Draw the Lewis dot structure and the structural formula for each of these molecules.

   a. SiF_4 b. CO_2 c. CH_4 d. SF_2
   e. C_2H_4 f. C_2H_2 g. C_2H_6

3. How many lone pairs are in each of these molecules?

   a. CO_2 b. SiF_4 c. CH_4

4. Draw at least two structural formulas for the molecular formula C_3H_6O.

5. From what you've learned so far, how is molecular structure related to smell?

PROJECT *Functional Groups*

Research a functional group. Choose a functional group and find out as much as you can about molecules that contain it. Create a poster that has these details:

- A large drawing of the functional group along with its name.
- The structural formulas for at least five molecules that possess your functional group, along with their chemical names.
- A brief description of the properties associated with this group of compounds.

SECTION

II

Building Molecules

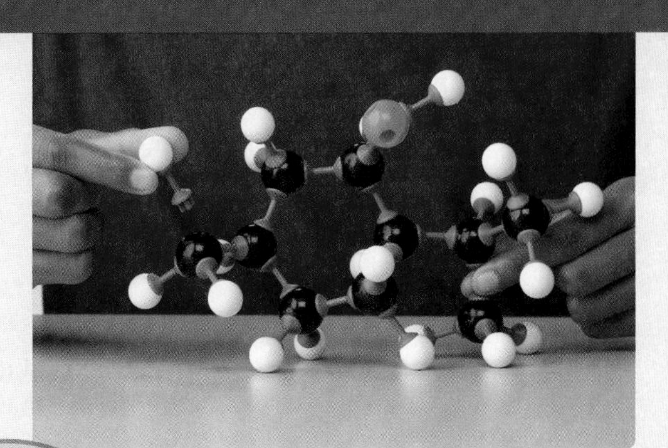

Molecular models help chemists to picture something too tiny to see.

Two compounds with the same functional group may smell different. In addition to molecular structure, the *shape* of a molecule seems to be related to smell. Molecular shape is determined by the bonds between atoms as well the locations of electrons. Because we experience different molecular shapes as different smells, it seems that the nose can somehow detect these differences.

In this section, you will study

- three-dimensional molecular models
- the role of valence electrons in determining molecular shape
- the receptor site theory and have a chance to develop your own model of how a nose works

LESSON

9 New Smells, New Ideas

Ball-and-Stick Models

Think About It

The structural formula is a valuable source of information about a molecule. You can use the structural formula to identify functional groups in a molecule. Knowing the functional group also helps you to predict molecular properties including smell. However, sometimes just knowing the functional group is not enough to predict the smell of a compound.

What three-dimensional features of a molecule are important in predicting smell?

To answer this question, you will explore

1. New Smell Molecules
2. Three-Dimensional Models

Exploring the Topic

1 New Smell Molecules

Below are three compounds, each with a distinctive smell. In spite of their wide range of smells, each of the three compounds contains the same functional group.

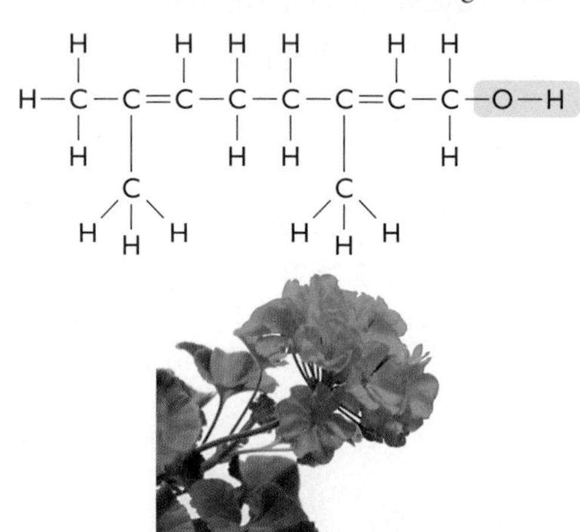

Geraniol, $C_{10}H_{18}O$

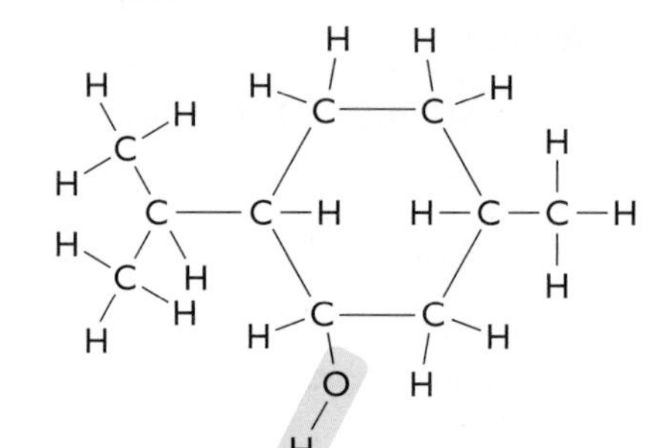

Menthol, $C_{10}H_{20}O$

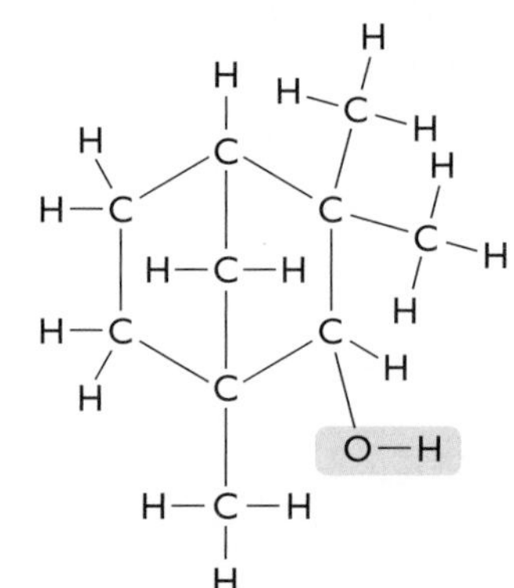

Fenchol, $C_{10}H_{18}O$

Take a moment to locate their functional groups. Each of these molecules has a hydroxyl functional group, and each name ends in "-ol." All three molecules are alcohols. So, why are the three smells so different? Something besides just functional group affects the smell of a compound. The answer lies in the three-dimensional shape of the molecules of each compound.

❷ Three-Dimensional Models

It is difficult to show a three-dimensional drawing of a molecule on a flat piece of paper. With a molecular model set, you can build a three-dimensional structure.

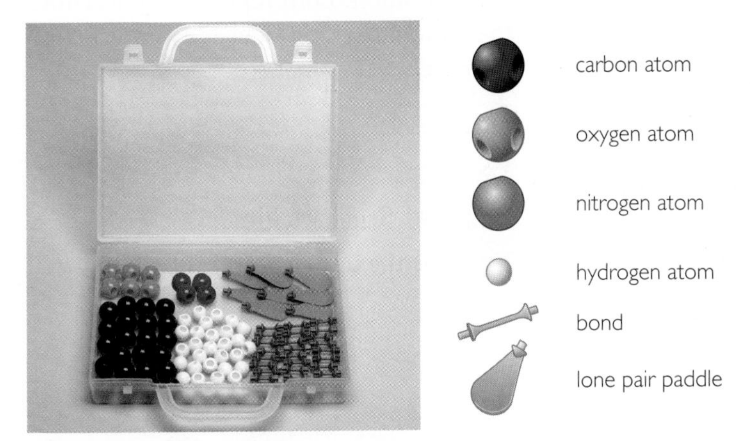

Notice that the bonds in a model kit look like little sticks. And the atoms are small spheres. These models are called **ball-and-stick models.**

A picture of a ball-and-stick model for ethyl acetate, $C_4H_8O_2$, is shown here. The carbon atoms are shown in black, the hydrogen atoms are in white, and the oxygen atoms are red.

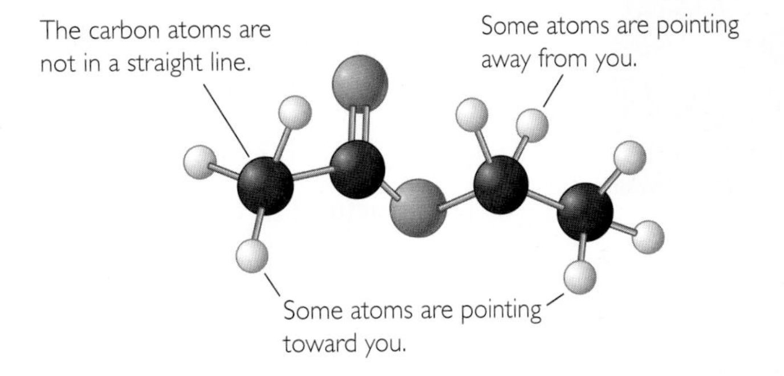

Illustrations of ball-and-stick models have some drawbacks. Some atoms may be partially visible or entirely hidden in an illustration. However, they convey more information about molecular shape than structural formulas do. Using a real model remains the best way to examine the three-dimensional shape of a molecule. The two representations for a molecule of citral, $C_{10}H_{16}O$, are shown for comparison. Look for similarities and differences between the two representations.

```
    H H H         H H H
     \|/           \|/
      C             C
H     |     H  H    |
|     |     |  |    |
H—C—C=C—C—C—C=C—H
|       |  |  |    |
H       H  H  H    C—H
                   ‖
                   O
```

Structural formula

Ball-and-stick model

Citral, $C_{10}H_{16}O$

CONSUMER CONNECTION

There are over 5,000 compounds used in the fragrance industry. How they interact with each other and with the human body must be considered when a scented product is created. Fragrance chemicals enter the human body through the skin, and by ingestion, so scents must be made safe in addition to pleasant smelling.

HISTORY CONNECTION

Abu Ali ibn Sina, also known as Avicenna, (born 980 C.E. in what is now Uzbekistan—then Persia) became a physician when he was 18. In addition to making pioneering contributions to modern medicine, Ibn Sina developed the process of extracting essential oils from plants through distillation. He worked with rose water, which became a popular fragrance and led others to experiment with other flowers.

Notice that the structural formula shows the citral molecule as flat with all of the carbon atoms in a line. The ball-and-stick representation, on the other hand, shows that the carbon atoms are not arranged in a line, but are connected in a zigzag fashion. Both representations contain much of the same information, but the ball-and-stick model adds information about the way the atoms are arranged in space.

Example

Ball-and-Stick Model

Examine the drawing of this ball-and-stick model of isopentylacetate, $C_7H_{14}O_2$.

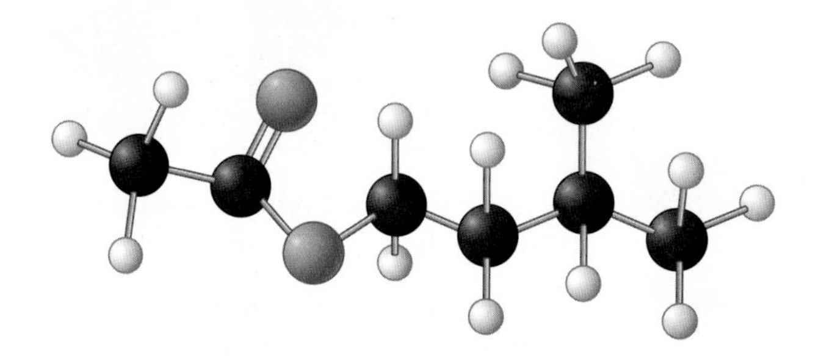

a. Draw the structural formula of this compound.

b. What functional group is in the compound?

c. Predict a smell for this compound.

Solution

a. The structural formula of this compound is

```
                          H
                        H \ | / H
                           C
      H  O        H  H     |   H
      |  ||       |  |     |   |
  H — C — C — O — C — C — C — C — H
      |           |  |     |   |
      H           H  H     H   H
```

b. This compound has an ester functional group.

c. Because it has an ester group and its name ends in "-ate," it probably smells sweet. (In fact, this molecule smells like bananas.)

Ball-and-stick models for the three compounds introduced at the beginning of this lesson are on the next page.

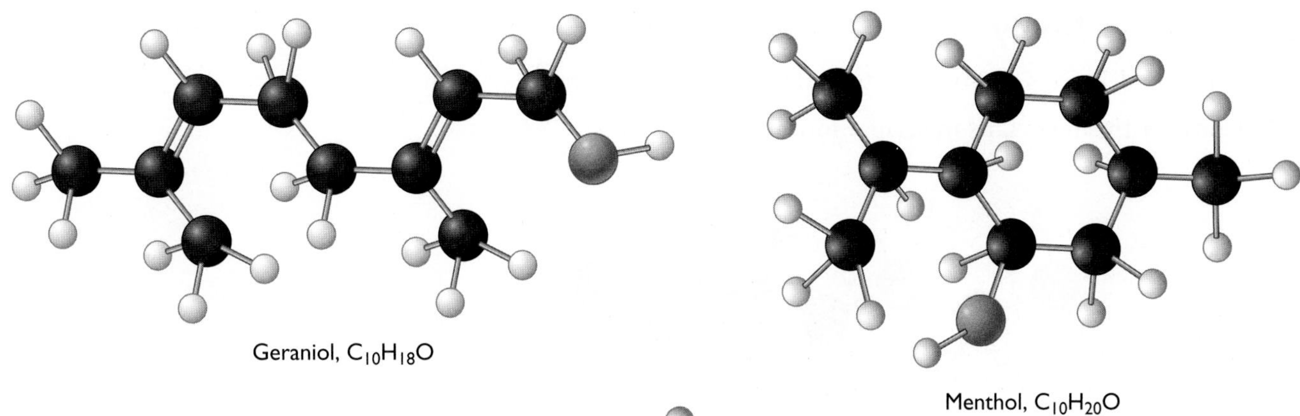

Geraniol, $C_{10}H_{18}O$

Menthol, $C_{10}H_{20}O$

Fenchol, $C_{10}H_{18}O$

The overall shapes of the molecules are quite different. The carbon atoms in geraniol are connected in a long chain, while the menthol and fenchol molecules both contain ring structures. Perhaps these overall shapes are related to the different smells of these three alcohols.

Key Term

ball-and-stick model

Lesson Summary

What three-dimensional features of a molecule are important in predicting smell?

Molecules are three-dimensional. A ball-and-stick model kit is a tool that you can use to construct models of molecules. This allows you to see how the various atoms in the molecule are arranged in space, as well as the overall shape of each molecule. Molecular shape may have something to do with smell.

EXERCISES

Reading Questions

1. What are the differences between a structural formula and a ball-and-stick model?
2. What is your theory about why the three alcohols smell different even though they have the same functional group?

Reason and Apply

3. What model pieces do you need to build a ball-and-stick model of geraniol?
4. Consider this model. Its molecular formula is $C_7H_{14}O_2$.

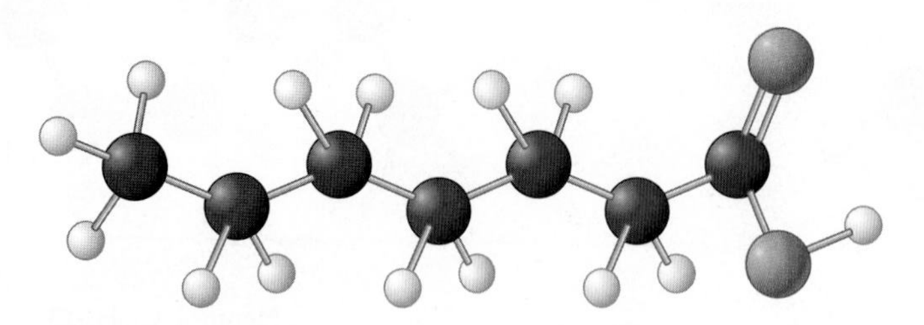

   a. Draw the Lewis dot structure.
   b. Draw the structural formula.
   c. What is the functional group in the molecule?
   d. What can you predict about the name and smell of this compound?
5. Consider this model. Its molecular formula is $C_3H_6O_2$.

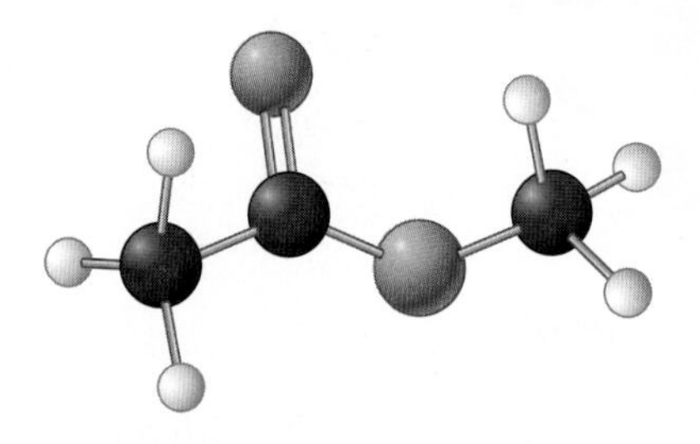

   a. Draw the Lewis dot structure.
   b. Draw the structural formula.
   c. What is the functional group in the molecule?
   d. What can you predict about the name and smell of this compound?
6. Consider this model. Its molecular formula is C_2H_7N.

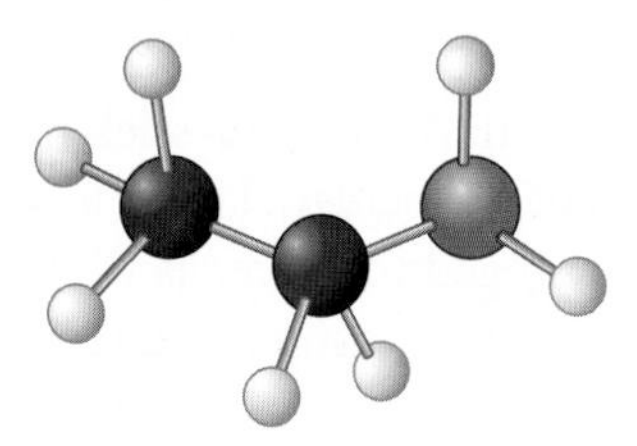

   a. Draw the Lewis dot structure.
   b. Draw the structural formula.
   c. What is the functional group in the molecule?
   d. What can you predict about the name and smell of this compound?
7. What evidence do you have that the structural formula may not always be useful in predicting smell?

ACTIVITY Two's Company

Purpose

To use three-dimensional models to visualize small molecules.

Materials

- gumdrops, marshmallows, and toothpicks
- ruler
- ball-and-stick molecular model set

Part 1: Gumdrop Molecules

1. Create a methane molecule using gumdrops, marshmallows, and toothpicks.
2. Make sure every pair of electrons in the molecule is as far away as possible from every other pair of electrons. Use a ruler to check the distances.
3. Draw Lewis dot structures for the following molecules:
   a. CH_4 **b.** NH_3 **c.** H_2O
4. How many pairs of electrons are located around the central atom of each molecule?
5. Besides the identity of the central atom, what is different about these three molecules?
6. Using gumdrops and toothpicks, create ball-and-stick models of NH_3 and H_2O.
7. Did you remember to include lone pairs? Fix your models if you need to so that lone pairs are represented. Do the lone pairs affect the shape of the molecule?
8. Compare your three gumdrop models. Describe any similarities.

Part 2: Ball-and-Stick Models

1. Use the model sets to create models of CH_4, NH_3, H_2O, and HF. Use black for carbon, white for hydrogen, red for oxygen, and blue for nitrogen.
2. Add the appropriate lone pair paddles to your models.
3. How many lone pair paddles would you need for an atom of neon? Explain your answer.
4. Draw sketches of your three-dimensional models. What is the shape of each molecule if you ignore the lone pair paddles?
5. **Making Sense** Explain how the lone pairs affect the shapes of these molecules.

LESSON 10 Two's Company
Electron Domains

Think About It

You may have noticed that there are a certain number of holes in the atoms in the molecular model kits and that the sticks can go only in certain places. So, when you connect the atoms, the molecules automatically end up with the correct three-dimensional shape. But what is it that determines the actual shape of a molecule?

How do electrons affect the shape of a molecule?

To answer this question, you will explore

1. Shapes of Molecules
2. Electron Domains

Exploring the Topic

1 Shapes of Molecules

These illustrations show models of ethanol, C_2H_6O.

Ethanol

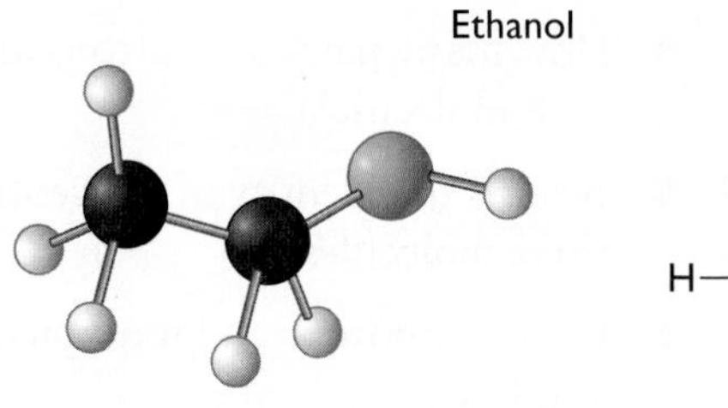

Ball-and-stick model

H H
| |
H—C—C—O—H
| |
H H

Structural formula

What accounts for the three-dimensional structure of ethanol? The answer lies in the valence electrons. They determine the ultimate shape of a molecule.

To understand molecular shape, it is useful to begin by examining a simple molecule such as methane, CH_4. While the structural formula is flat and looks like a cross, the ball-and-stick model shows that methane has a three-dimensional structure. Both models show that there are four bonds.

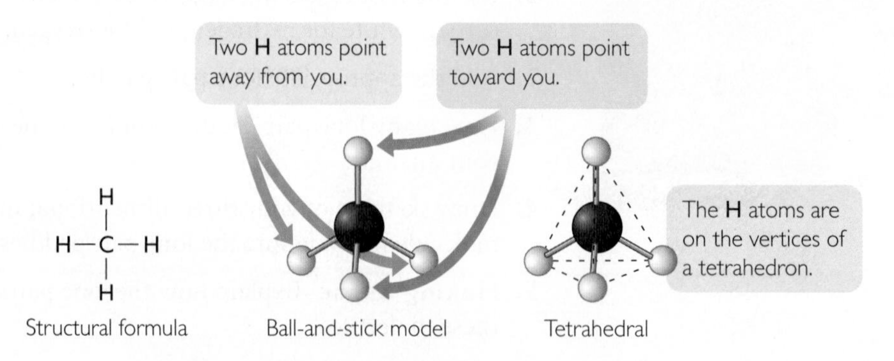

Structural formula Ball-and-stick model Tetrahedral

The word used to describe the shape of the methane molecule is **tetrahedral.** A tetrahedral molecule has a symmetrical shape, with one atom exactly in the center. The distance between any other two atoms bonded to the central atom is the same.

You can prove to yourself that a ball-and-stick model of methane is symmetrical in every direction by spinning the molecule as shown in the illustrations. The molecule looks exactly the same no matter what direction it is spun.

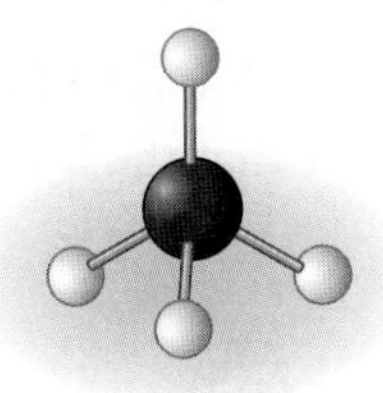

One **H** atom sticks straight up. Three **H** atoms rest on the table.

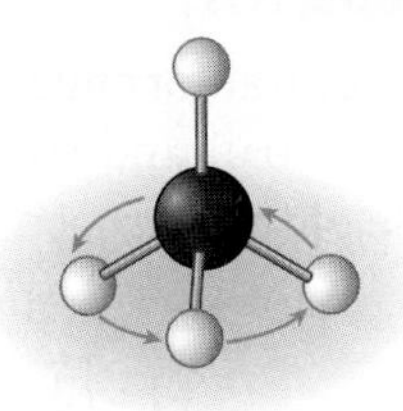

You can spin the molecule around and it looks identical.

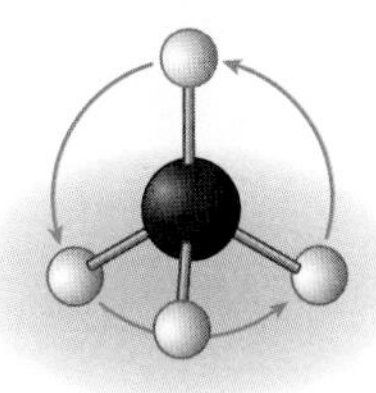

You can grab one of the **H** atoms on the table and spin the molecule around and it still looks identical.

❷ Electron Domains

Methane, CH_4

The illustration here shows a methane molecule with the shared valence electrons superimposed, or laid, over the ball-and-stick model. The bonded electron pairs occupy space between the carbon atom and each of the four hydrogen atoms. The space occupied by the electrons is called an **electron domain.** An electron domain can consist of a bonded pair of electrons, a lone pair of electrons, or multiple bonded pairs of electrons.

A methane molecule has four electron domains.

Each electron domain in the tetrahedral methane molecule is the same distance from the other three electron domains. They are as far apart from one another as possible. Because of this, the hydrogen atoms are as far apart from one another as possible. The distance between any two hydrogen atoms is the same. In addition, if you measured the angles formed between any two hydrogen atoms and the carbon atom, all the bond angles would be the same for this tetrahedral molecule. This is not accurately shown in the structural formula.

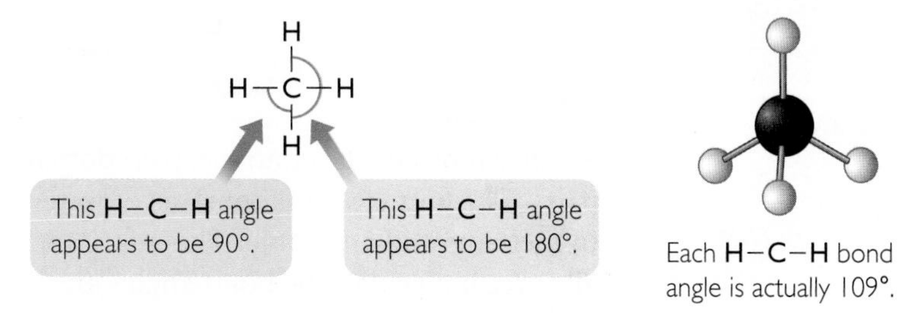

Each **H–C–H** bond angle is actually 109°.

This tendency for electron pairs to be as far apart from one another as possible is called **electron domain theory.** This theory is also referred to as valence shell electron pair repulsion theory, or *VSEPR.*

Important to Know Electrons are negatively charged, so they repel one another. However, in molecules, electrons bond in pairs. It is the electron pairs that repel one another. ◂

Ammonia, NH_3

The structural formula and ball-and-stick model for ammonia, NH_3, are shown below. The ball-and-stick model shows that the hydrogen atoms are all located on one side of the nitrogen atom.

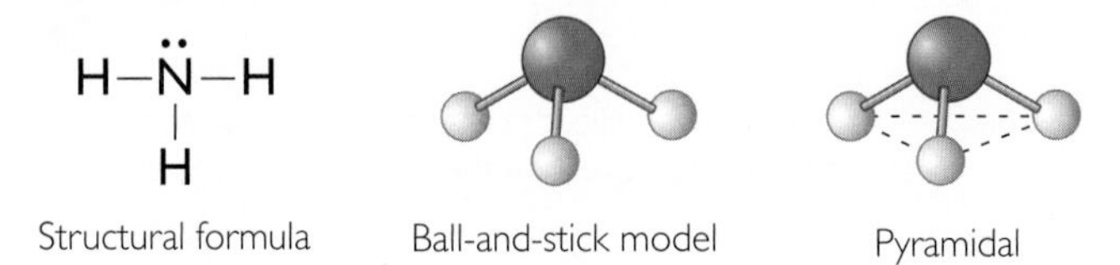

The word used to describe the shape of the ammonia molecule is **pyramidal.**

The nitrogen atom has a lone pair of electrons. Therefore, while there are only three bonds, there are four electron domains. The four electron domains are arranged in a tetrahedral shape to get as far apart as possible, similar to what happens in methane.

An ammonia molecule has four electron domains, just like methane.

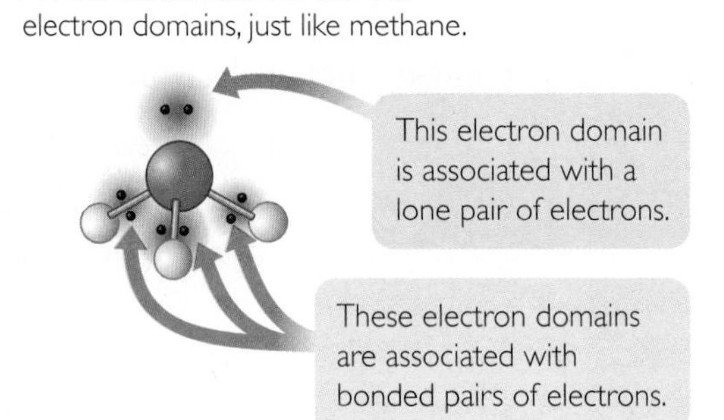

Water, H_2O

Take a look at one more molecule—water, H_2O. The word used to describe the shape of a water molecule is **bent.**

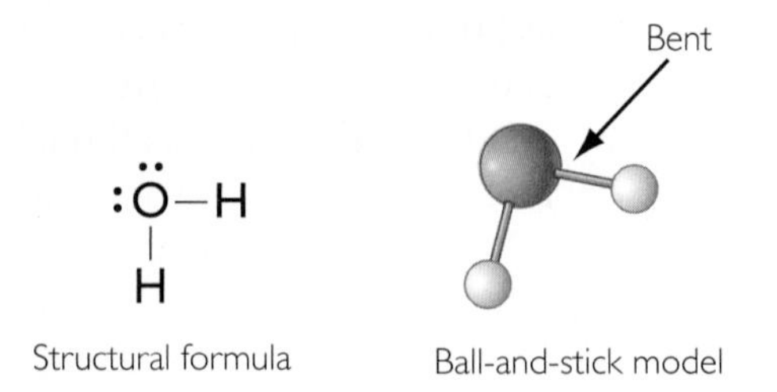

A water molecule has four electron domains, two bonding pairs and two lone pairs. The four electron domains of water are arranged in a tetrahedral shape, as in methane and ammonia. This results in a bent shape for the water molecule. In all three molecules, the bond angles are 109° or close to this.

INDUSTRY CONNECTION

Phosphine is a highly toxic, highly flammable, colorless gas used as a fumigant to kill insects that might damage stored grain.

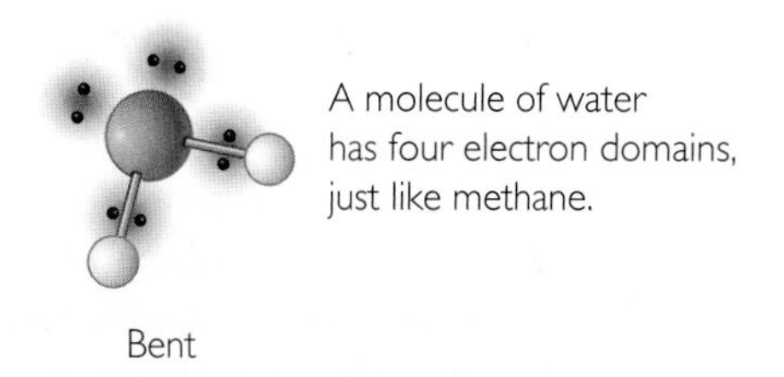

A molecule of water has four electron domains, just like methane.

Ball-and-Stick Models with Lone Pair Paddles

In a molecular model kit, the lone pairs are sometimes represented as plastic paddles. This helps you to visualize where the final shape of a molecule comes from.

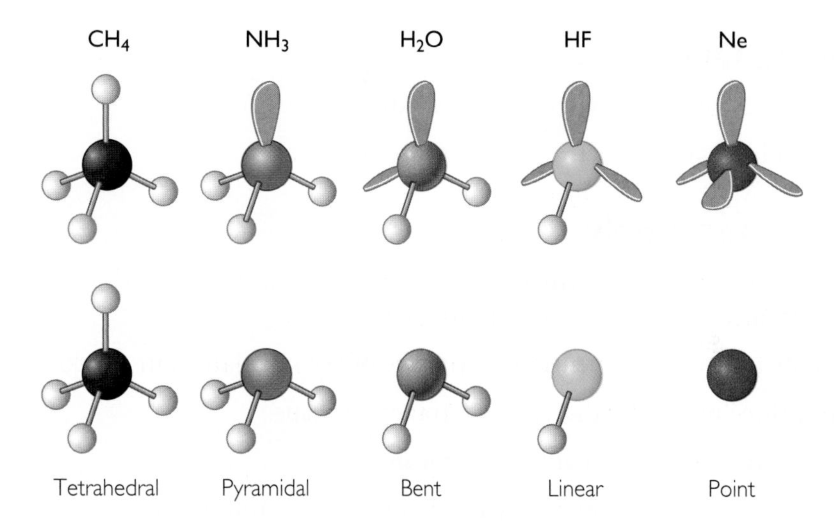

Notice that you name the shape of the molecule based only on the arrangement of atoms. The lone pairs are not considered.

Example

Phosphine, PH_3

What is the shape of phosphine, PH_3? Explain your thinking.

Solution

Begin by drawing the Lewis dot structure and a structural formula with lone pairs of phosphine. A phosphorus atom has five valence electrons and each hydrogen atom contributes one electron, for a total of eight electrons. There are three bonding pairs and one lone pair. The four atoms in phosphine are arranged in a pyramidal shape.

$$\mathrm{H:\overset{\cdot\cdot}{\underset{\cdot\cdot}{P}}:H} \quad \text{(with H below P)} \qquad \mathrm{H-\overset{\cdot\cdot}{P}-H} \quad \text{(with H bonded below P)}$$

Key Terms
tetrahedral shape
electron domain
electron domain theory
pyramidal shape
bent shape

Lesson Summary

How do electrons affect the shape of a molecule?

The three-dimensional shape of a molecule is determined by the various electron pairs in the molecule. *Both* bonded pairs of electrons *and* lone pairs of electrons affect the shape of a molecule. Each electron pair takes up space, called an electron domain. The final shape of a molecule is determined by the fact that electron domains in a molecule are located as far apart from one another as possible.

EXERCISES

Reading Questions

1. In your own words, describe a tetrahedral shape.
2. What is meant by electron domain theory?

Reason and Apply

3. If you were going to predict the three-dimensional structure of a small molecule, what would you want to know?
4. Predict the three-dimensional structure of H_2S. Explain your thinking.
5. List three molecules that have a tetrahedral shape.
6. List three molecules that have a bent shape.
7. What is the shape of arsine, AsH_3? Explain your thinking.
8. Predict the three-dimensional shape of HOCl. Explain your thinking.
9. Draw a methane, CH_4, molecule and show how it fits inside a tetrahedron. Do the same for ammonia, NH_3, and water, H_2O.

CONSUMER CONNECTION

Molecules containing *only* carbon and hydrogen atoms are generally referred to as hydrocarbons. The simplest hydrocarbon compound is methane. Notice that methane's name ends in "-ane" because it is an alkane. Many of the medium-sized alkane molecules smell like octane, a major component in gasoline.

Octane

LESSON

11 Let's Build It

Molecular Shape

Think About It

So far you have considered the shapes of small molecules containing single bonds. All of the molecules you've studied have had four electron domains. However, small molecules that have only three or even two electron domains also exist. These molecules have double and triple bonds. So, there are more molecular shapes to consider.

How can you predict the shape of a molecule?

To answer this question, you will explore

1. More Shapes
2. Larger Molecules

Exploring the Topic

1 More Shapes

Double Bonds

Formaldehyde, CH_2O, is a very simple molecule that contains one double bond. The double bond counts as only one electron domain even though it contains two pairs of bonded electrons. This means that the central atom in the formaldehyde molecule (the carbon atom) has only three electron domains surrounding it, spread out into a flat triangle. The phrase used to describe the underlying shape of the formaldehyde molecule is **trigonal planar.**

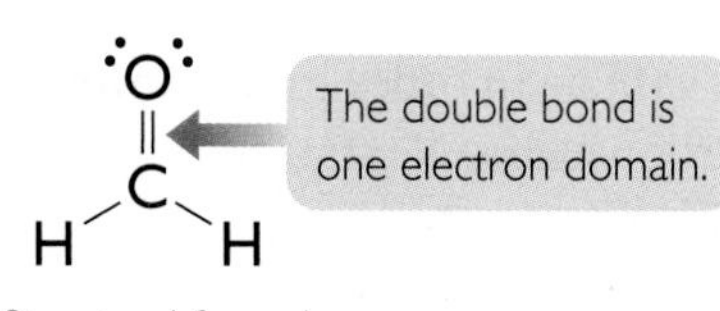

Structural formula

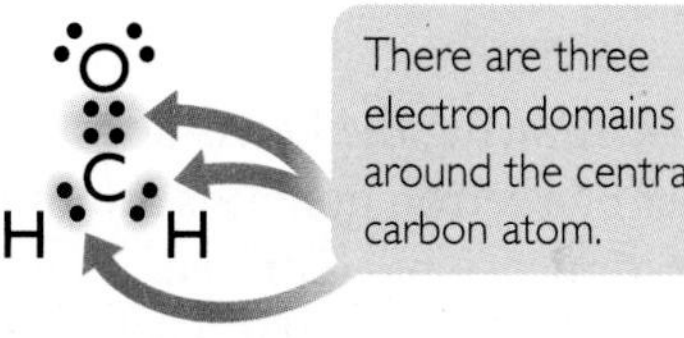

Lewis dot structure

The three electron domains spread apart into a flat triangle.

Trigonal planar

Both ammonia, NH_3, and formaldehyde, CH_2O, have four atoms but have very different shapes. The difference in shape is due to the number of electron domains around the central atom.

The three hydrogen atoms are below the nitrogen atom to make room for the lone pair on nitrogen.

Ammonia

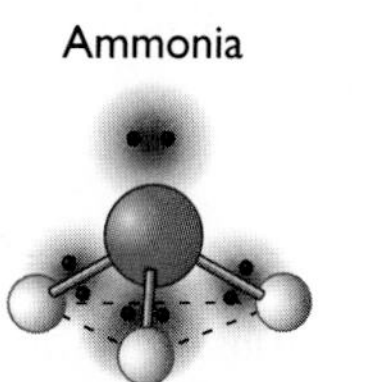

Pyramidal

Formaldehyde

Trigonal planar

There are no lone pairs, so the three electron domains push apart into a triangle.

Two Double Bonds

In a molecule with a double bond, the double bond counts as one electron domain. So a molecule like carbon dioxide, with two double bonds, has only two electron domains around the central atom and they need to be as far apart as possible. This shape is described as **linear.**

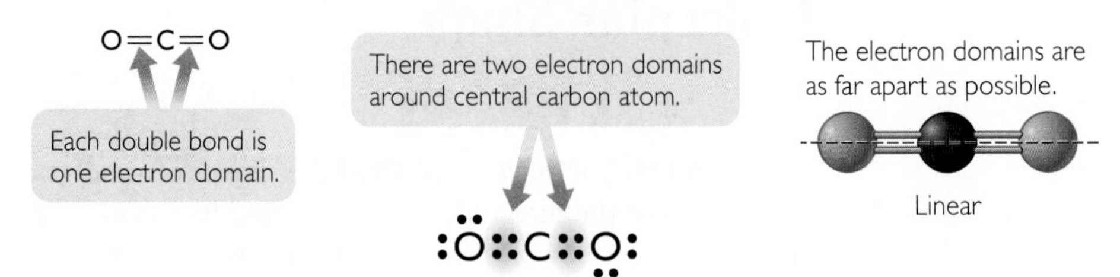

Important to Know The four electrons in a double bond remain between two atoms, so they are considered one electron domain. An electron domain can have two, four, or six electrons. ◀

Both water, H_2O, and carbon dioxide, CO_2, have three atoms. However, the shape of water is bent while carbon dioxide is linear. The difference in shape is due to the number of electron domains around the central atom.

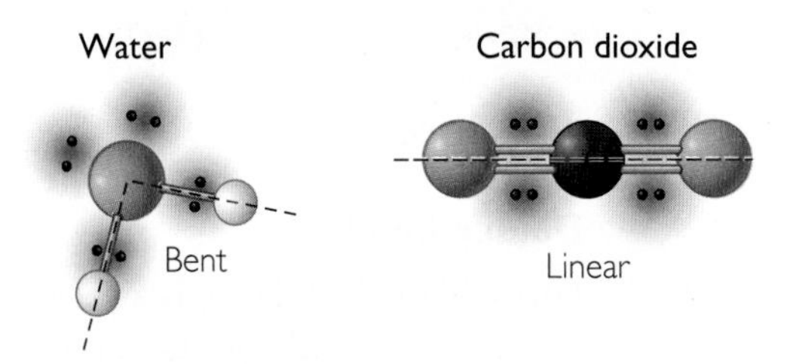

INDUSTRY CONNECTION

Ethene, C_2H_4, is also called ethylene. It is a colorless gas with a slightly sweet odor. Ethene is used in the manufacture of many common products like polyethylene plastic and polystyrene. Ethene also can be used as a plant hormone to control the ripening of fruit or the opening of flowers.

Example 1

Ethene, C_2H_4

Determine the molecular shape of ethene, C_2H_4.

Solution

Begin by drawing the Lewis dot structure for ethene in order to determine the number of electron domains around each carbon atom. Next, translate that into a three-dimensional representation of the molecule. In this case, there are no lone pairs. The molecule is flat. Each half of ethene is trigonal planar.

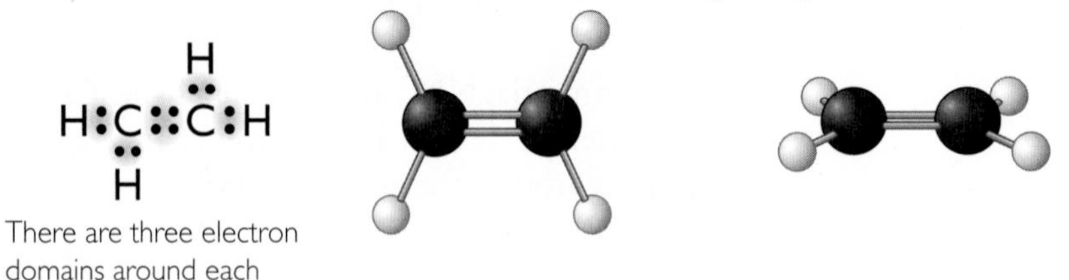

Triple Bonds

Another example of a linear molecule is hydrogen cyanide, HCN. It also has two electron domains around the central atom, like carbon dioxide. One electron domain is a single bond between carbon and hydrogen, and the other is a triple bond between carbon and nitrogen.

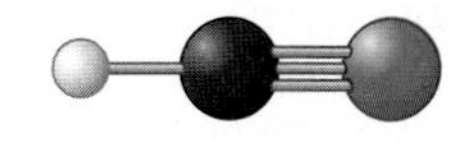

Hydrogen cyanide has one single bond and one triple bond. This results in two electron domains around the central atom.

INDUSTRY CONNECTION

Ethyne, C_2H_2, is also called acetylene. It is the highly flammable gas used by welders in oxyacetylene torches. Miners' helmets used to have carbide lamps attached to them. A carbide lamp is based on the reaction between calcium carbide, CaC_2, and water, H_2O, to produce acetylene, C_2H_2. The acetylene gas burns to produce a bright light in the dark cave.

Example 2

Ethyne, C_2H_2

Determine the molecular shape of ethyne, C_2H_2.

Solution

Begin by drawing the Lewis dot structure of ethyne in order to determine the number of electron domains around the carbon atoms. In order to satisfy the octet rule and the HONC 1234 rule, the carbon atoms must have a triple bond between them. Each carbon atom has two electron domains surrounding it. Ethyne is a linear molecule.

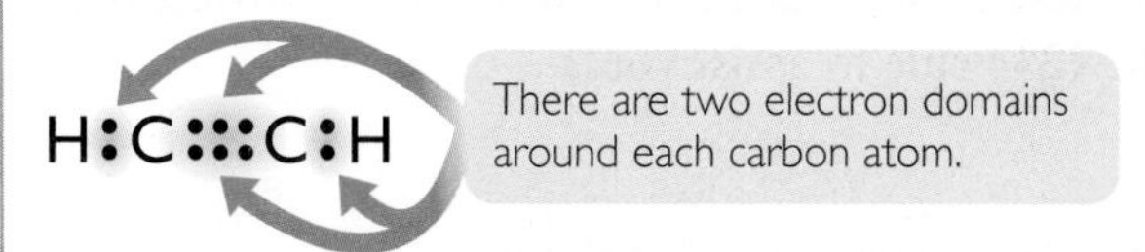

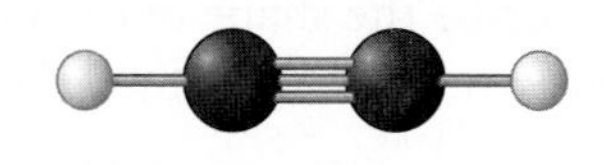

2 Larger Molecules

The shapes of all of these small molecules help to explain why chains of carbon atoms are crooked rather than straight. The illustration shows a ball-and-stick model for heptanoic acid, $C_7H_{14}O_2$, with seven carbon atoms in a chain. The arrangement of atoms around six of the carbon atoms is tetrahedral. Most of the molecule is a series of overlapping tetrahedral shapes.

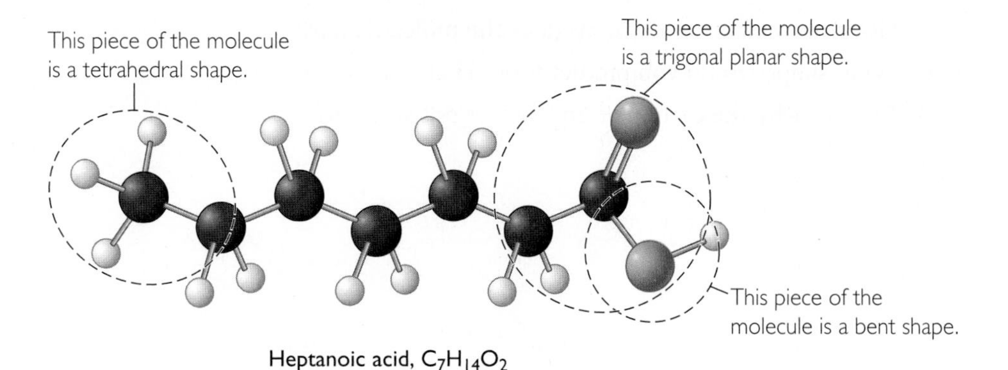

Heptanoic acid, $C_7H_{14}O_2$

The various electron domains in this molecule cause its overall shape to be crooked, or zigzag.

Key Terms

trigonal planar shape
linear shape

Lesson Summary

How can you predict the shape of a molecule?

The shape of a molecule is affected by the location and number of its electron domains. Lewis dot structures help you to determine the number of electron domains in a molecule. An atom involved in a double bond or triple bond will have fewer electron domains surrounding it. This results in molecules that have trigonal planar and linear shapes. Large molecules can have various geometric shapes within them. Areas around double bonds are flat and those around triple bonds are linear.

EXERCISES

Reading Questions

1. What shapes are possible if a molecule has three electron domains?
2. What shapes are possible for a molecule with three atoms? Explain your thinking.

Reason and Apply

3. Describe the shape of each of these molecules. (Use Lewis dot structures, the periodic table, and the HONC 1234 rule to assist you.)

   Cl_2 CO_2 H_2O
4. Which of the molecules in Exercise 3 have the same shape?
5. What shape or shapes do you predict for a molecule with two atoms? A molecule with three atoms? Four atoms? Five atoms? You can draw Lewis dot structures or electron domains to assist you in answering the questions.
6. Predict the shapes of these molecules:

   CF_4 NF_3 H_2Se H_2CS
7. Consider the butane, C_4H_{10}, molecule.
   a. Draw the Lewis dot structure for butane.
   b. How many electron domains does the molecule have?
   c. What shape would you predict for C_4H_{10}?
   d. Explain why the carbon atom chain is not straight.

LESSON

12 What Shape Is That Smell?

Space-Filling Models

Think About It

Methyl octenoate, $C_9H_{16}O_2$, is a compound that smells like violets. A ball-and-stick model of methyl octenoate shows that it is a series of overlapping tetrahedral shapes stuck together. There is a trigonal planar segment in the area of the double bonds. But how would you describe the shape of the *whole* molecule? And does the shape of the whole molecule have anything to do with its smell?

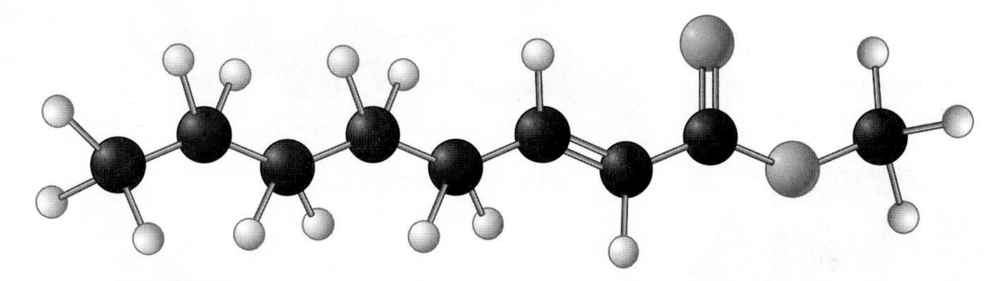

How is the shape of a molecular compound related to its smell?

To answer this question, you will examine

1. Space-Filling Models
2. Relating Shape and Smell

Exploring the Topic

SciLinks® NSTA

Topic: Molecular Modeling

Visit: www.SciLinks.org

Web code: KEY-212

1 Space-Filling Models

Most of the smell molecules you have encountered are considerably larger than three, four, or five atoms. The best way to look at the overall shape of these molecules is with a different type of model, called a **space-filling model.**

Take a moment to compare the ball-and-stick model of methyl octenoate with the space-filling model. In a space-filling model the sticks between atoms have been eliminated. There is no space between atoms. Instead, bonded atoms are shown slightly overlapping.

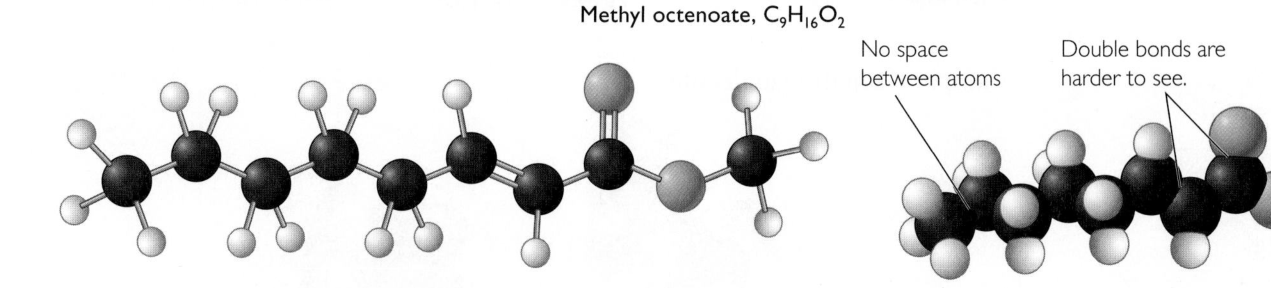

In some ways a space-filling model could be considered more accurate than a ball-and-stick model. In reality a stick has no resemblance to a bond. In a molecule, the atoms *share* electrons with neighboring atoms. This would suggest that the atoms are located very close to, or overlapping, one another. However, in an illustration of a space-filling model you can't see multiple bonds, and some atoms may be hidden behind others.

2 Relating Shape and Smell

Space-filling models for molecules that smell sweet, minty, and like camphor are shown here. Take a moment to look for patterns that may indicate a connection between the shapes of these molecules and their smells.

ENVIRONMENTAL CONNECTION

Geraniol is a sweet-smelling molecule. It is one of the molecules that make roses smell like roses.

Sweet-smelling molecules

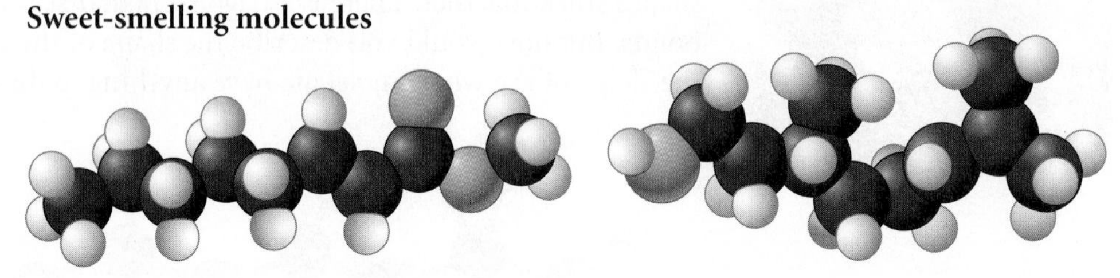

Methyl octenoate, $C_9H_{16}O_2$

Citronellol, $C_{10}H_{20}O$

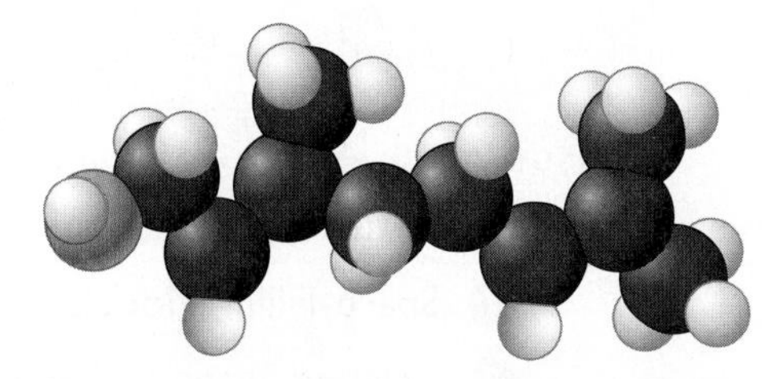

Geraniol, $C_{10}H_{18}O$

Minty-smelling molecules

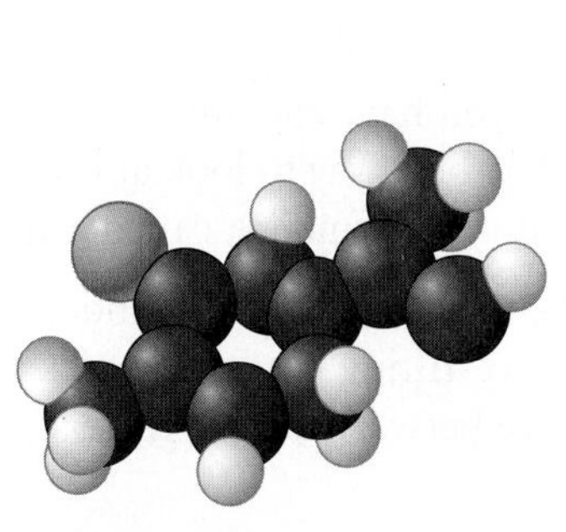

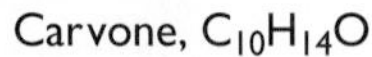

Carvone, $C_{10}H_{14}O$

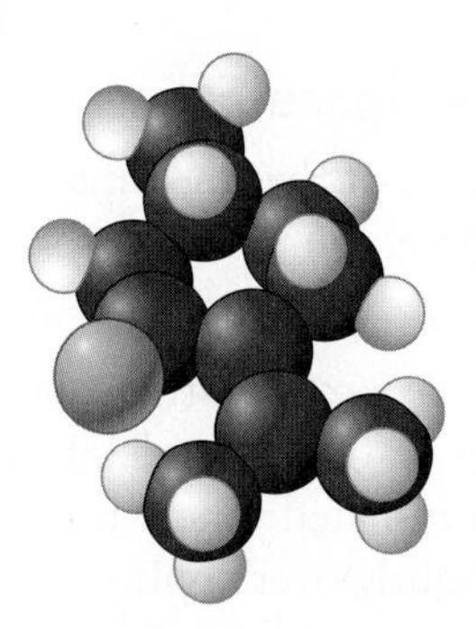

Pulegone, $C_{10}H_{16}O$

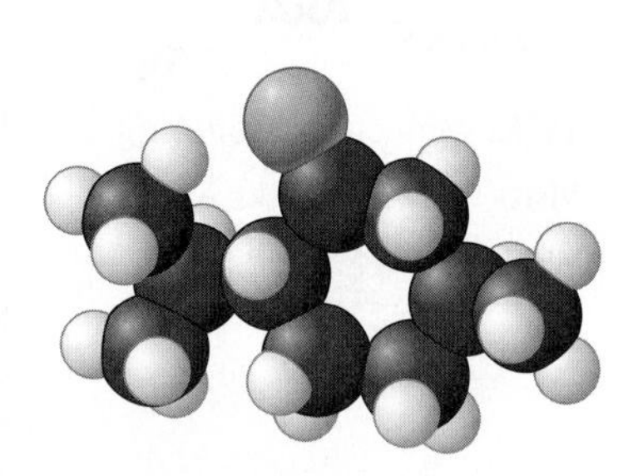

Menthone, $C_{10}H_{18}O$

Camphor-smelling molecules

Fenchol, $C_{10}H_{18}O$

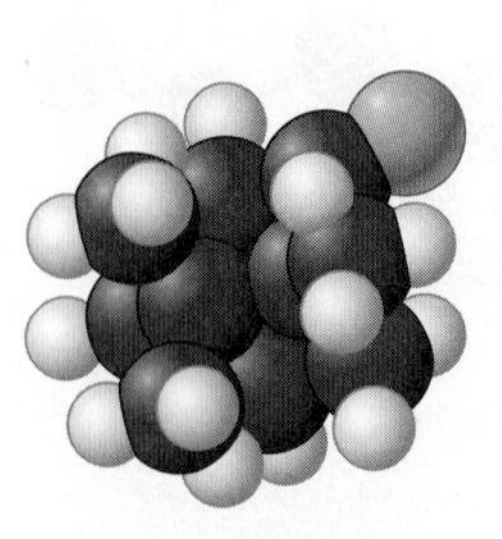

Camphor, $C_{10}H_{16}O$

The sweet-smelling molecules are all long and stringy. The minty-smelling molecules all have a six-carbon ring structure. They have a shape that resembles a frying pan. The camphor-smelling molecules are a tight cluster of atoms in the shape of a ball. So far, there appear to be three smell categories that are directly related to the molecular shape.

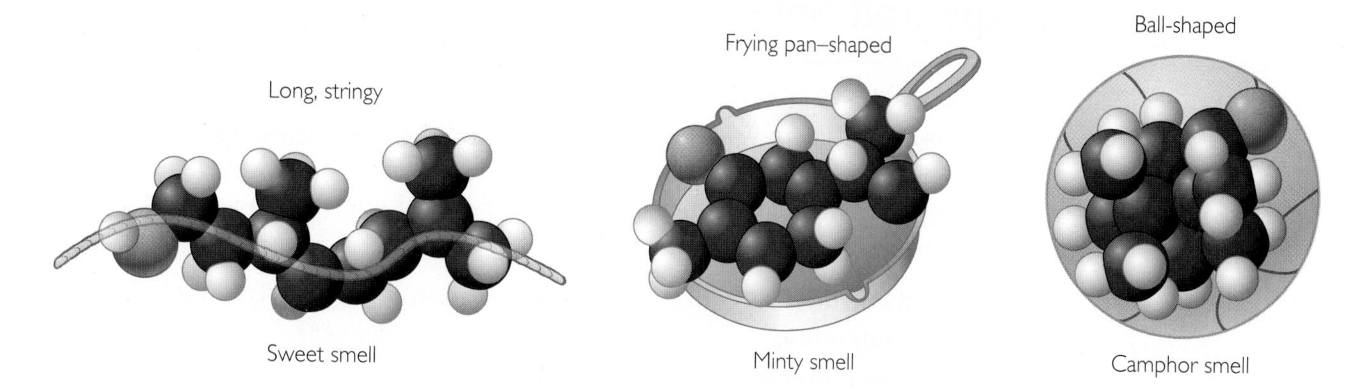

Key Term
space-filling model

Lesson Summary

How is the shape of a molecular compound related to its smell?

A space-filling model is a three-dimensional model that shows how the atoms in a molecule are arranged in space and how they fill this space. The shape of a molecule appears to be related to its smell. For example, the sweet-smelling molecules explored here are all long and stringy in shape. The minty-smelling molecules are frying pan–shaped, and the camphor-smelling molecules are ball-shaped. More data will certainly be helpful to confirm a connection between a molecule's shape and its smell.

EXERCISES

Reading Questions

1. How is a space-filling model useful?
2. How is a space-filling model different from a ball-and-stick model?

Reason and Apply

3. Draw a possible structural formula for a molecule that smells sweet and has the molecular formula $C_9H_{20}O$. What general shape does this molecule have?
4. If someone told you a molecule had a six-carbon ring, what smell would you predict for that compound?
5. Draw a possible structural formula for a molecule that smells minty and has the molecular formula $C_{10}H_{16}O$. What general shape does this molecule have?
6. If someone told you a molecule had a ball-shaped three-dimensional shape, what else would you want to know in order to determine how the molecule smells?
7. Which do you think has more influence on smell: the functional group that is present or the shape of a molecule? Explain why you think so.

LESSON

13 Sorting It Out

Shape and Smell

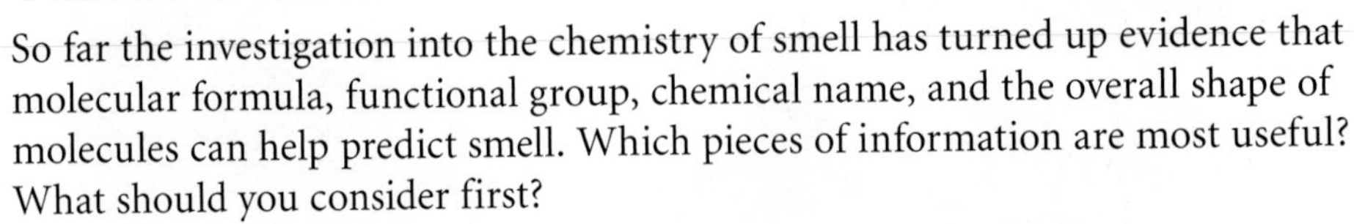

Think About It

So far the investigation into the chemistry of smell has turned up evidence that molecular formula, functional group, chemical name, and the overall shape of molecules can help predict smell. Which pieces of information are most useful? What should you consider first?

What chemical information is most useful in predicting the smell of a compound?

To answer this question, you will explore

1. A Summary of the Smell Investigation So Far
2. General Rules for Determining Smell

Exploring the Topic

1 A Summary of the Smell Investigation So Far

Clearly, the relationship between molecules and smell is quite complex. Each time you smell and examine new molecules, you find out new information and refine your hypotheses. Here is what you might conclude about smell so far.

Molecular Formula

Molecular formulas can give some indication of the way compounds smell.

Two O atoms	sweet or putrid
One N atom	fishy
Only C and H	gasoline or no smell
One O atom	sweet, minty, or camphor

Molecular formula is most useful in predicting fishy smells. It does not work so well for sweet, putrid, minty, and camphor smells.

Functional Group

Functional groups are even better than molecular formulas in narrowing down a smell. For example, compounds with a carboxyl functional group smell putrid. Compounds with an ester functional group smell sweet.

Carboxylic acids, putrid

Esters, sweet

CONSUMER CONNECTION

Citronella oil has been used for decades as an insect and animal repellent. It is found in candles, lotions, gels, and other products designed to repel mosquitoes and other pests.

Because alcohols and ketones might smell sweet, minty, or like camphor, functional group is not always correct in predicting smells.

Chemical Name

Molecules are named according to their functional groups. A carboxylic acid molecule, for instance, will have a chemical name ending in "-ic acid."

Molecular Names and Smells

Compound type	Name ending	Examples	Smell
carboxylic acid	-ic acid	butyric acid, $C_4H_8O_2$ valeric acid, $C_5H_{10}O_2$	putrid
ester	-yl -ate	ethyl acetate, $C_4H_8O_2$ propyl acetate, $C_5H_{10}O_2$	sweet
amine	-ine	hexylamine, $C_6H_{15}N$ diisobutylamine, $C_8H_{19}N$	fishy
alkane	-ane	methane, CH_4 pentane, C_5H_{12} hexane, C_6H_{14}	no smell or gasoline
alcohol	-ol	citronellol, $C_{10}H_{20}O$ menthol, $C_{10}H_{20}O$	sweet, minty, camphor
ketone	-one	menthone, $C_{10}H_{18}O$ pulegone, $C_{10}H_{16}O$	sweet, minty, camphor

Because the functional group is not useful in predicting smells for alcohols and ketones, the chemical name is also not very useful.

Molecular Shape

The overall shape of a molecule can be an important link to smell. Camphor-smelling compounds are made of ball-shaped molecules. Minty-smelling compounds are flat in overall shape. And compounds made of long and stringy molecules often smell sweet.

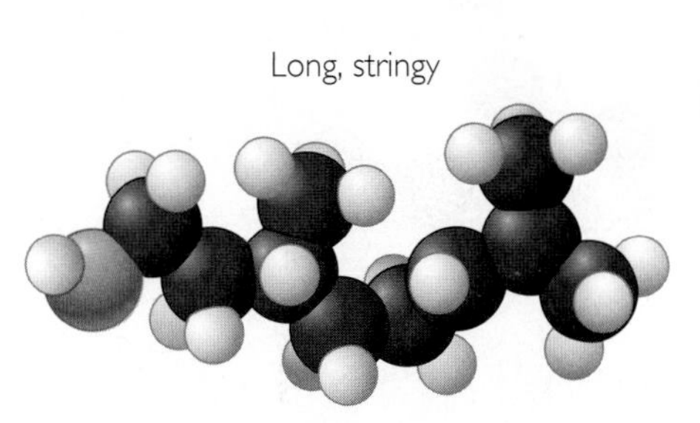

Citronellol, sweet smell

Carvone, minty smell

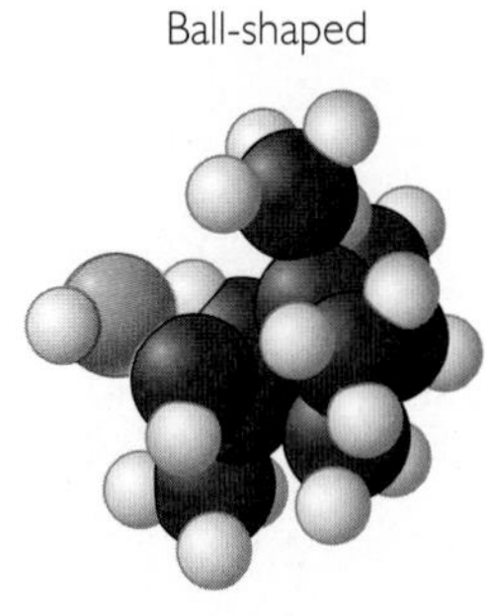

Fenchol, camphor smell

❷ General Rules for Determining Smell

Each piece of information appears to be valuable at different times, depending on the smell category. It is helpful to look for simple guidelines you can use to predict smell.

This table summarizes the chemical information of each of the five smell categories you have encountered. In each category, the most useful information for predicting that smell has been highlighted.

Summary of Chemical Information

Smell classification	Molecular formula info	Shape(s)	Functional group(s)
sweet	2 O atoms in some, 1 O atom in others	stringy	hydroxyl and ester
minty	rings, 1 O atom in all	frying pan–shaped	ketone and hydroxyl
camphor	double rings, 1 O atom in all	ball-shaped	ketone and hydroxyl
putrid	2 O atoms in all	stringy	carboxyl
fishy	1 N atom in all	stringy	amine

The only category that gives us a little trouble is the sweet-smelling category. Not every compound made of stringy molecules is sweet. However, if the compound has stringy molecules and is not a carboxylic acid or an amine, then it is sweet smelling.

Example

Predict a Smell

Examine the following molecular compound and predict its smell. What information is most useful in predicting the smell? What type of compound is this?

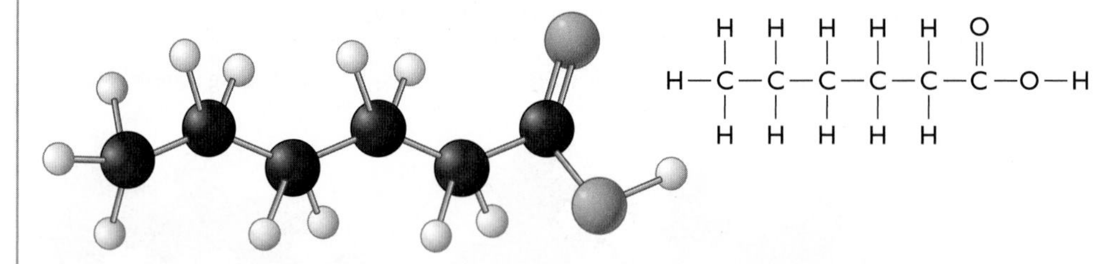

Solution

Here is what you can deduce:

- Molecular shape: Because the molecule is stringy, it probably does not have a minty or camphor smell. It may be sweet, putrid, or fishy.
- Molecular formula: There is no N atom, so it can't be fishy. It has two O atoms, so it might be sweet or putrid.

- Functional group: The carboxyl functional group indicates that the molecule smells putrid.
- The putrid smell and carboxyl group indicate that this is a carboxylic acid.

The functional group was most useful in predicting the smell of this compound.

Lesson Summary

What chemical information is most useful in predicting the smell of a compound?

By looking for patterns in the chemical information of molecules, it is possible to come up with general guidelines for certain smell categories. Some chemical information is more useful than other information depending on the smell category. Molecular shape is an accurate predictor for minty and camphor compounds. Functional group can be used to predict the smells of amines and carboxylic acids. Each piece of chemical information, including chemical formula and name, allows you to at least narrow down the possible smells of a compound.

EXERCISES

Reading Questions

1. Can the molecular formula of a compound help you to predict its smell? Explain your reasoning.
2. What one piece of chemical information would you want in order to predict the smell of a molecule? Explain your choice.

Reason and Apply

3. What is the minimum amount of information you need to know in order to determine if a compound smells sweet?
4. If someone tells you a compound smells sweet, what can you assume about its molecules? What can't you assume about its molecules?
5. If someone told you a compound was made of molecules that had a stringy three-dimensional shape, what else would you want to know in order to determine that compound's smell?

BIOLOGY CONNECTION

Scientists have found that memories evoked by specific smells seem to be stronger and more emotion-based than memories brought up by visual or auditory cues. The exact connection between smell and memory is unclear, but sensations from the smell receptors in the nose travel to areas of the brain that deal with memory.

LESSON

14 How Does the Nose Know?

Receptor Site Theory

Think About It

Imagine you are sipping a cool glass of lemonade on a hot summer day. The lemonade contains some fresh mint leaves for extra flavor. You detect both the lemon smell and the minty smell. The molecules associated with these two smells are quite different. What is happening inside your nose that allows you to detect both smells at the same time and to tell them apart? How does the nose know these molecules are different?

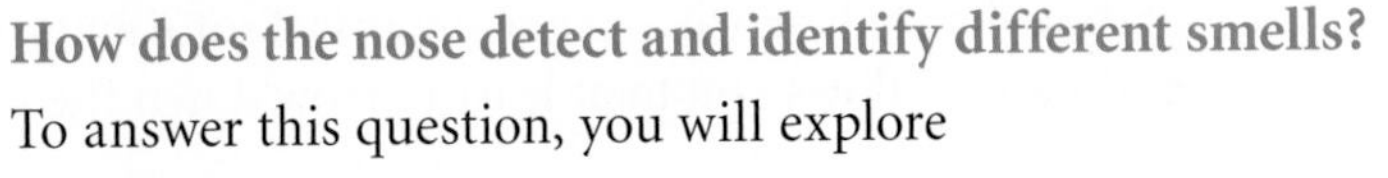

How does the nose detect and identify different smells?

To answer this question, you will explore

1. Receptor Site Theory
2. Phase Change and Molecular Stability

Exploring the Topic

1 Receptor Site Theory

The chemistry of smell is a fairly new science, and much remains to be discovered about how the sense of smell works. A number of theories exist to explain what happens in the nose when a smell is detected. While scientists are still uncertain about all of the details about how smell works, one theory seems to fit the evidence better than the others. This theory is called the **receptor site theory.** It is a widely accepted theory of how the interior of the nose detects different smells.

The receptor site theory uses the "lock and key" model. In this model, molecules are like "keys" that fit only into certain "locks." In other words, molecules have specific shapes that will fit only certain receptor sites, and the lining of the nose is covered with receptor sites.

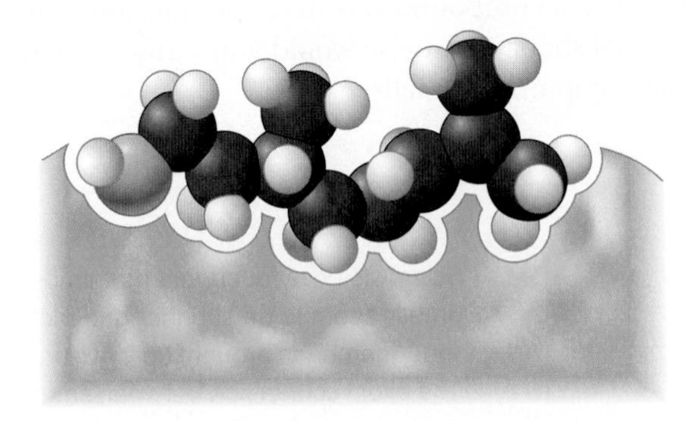

Molecules have specific shapes that fit into specific receptor sites in the nose, the way a key fits into a specific keyhole of a lock.

The lemony-smelling and minty-smelling molecules in the compounds in the glass of lemonade might fit into different receptor sites as shown in the illustrations below.

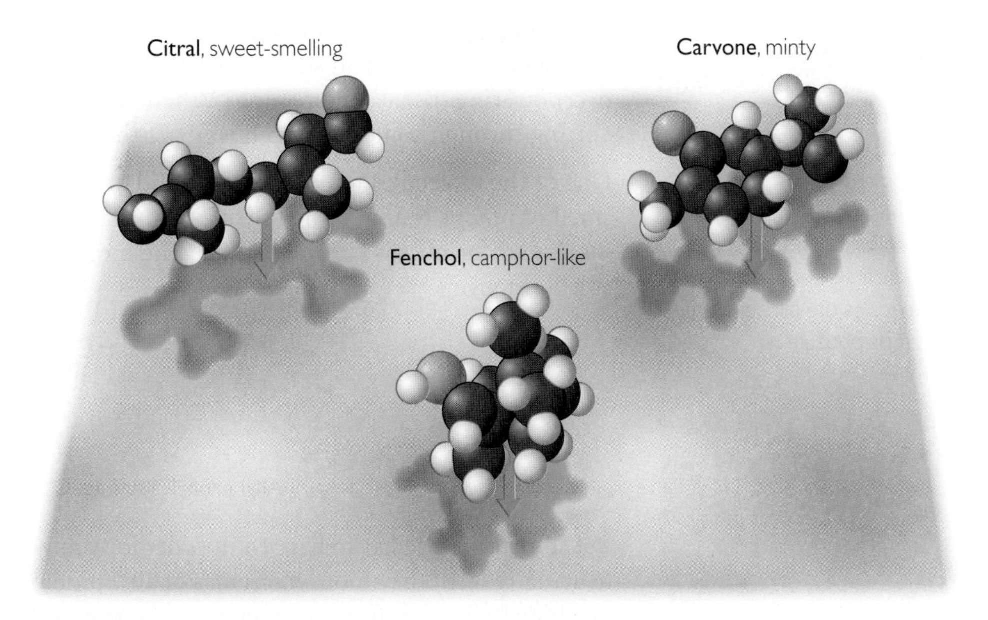

In this model, each type of molecule fits only into its own receptor sites, not into the others. Theoretically, this is how the nose distinguishes between two molecules.

Scientists are unsure how many receptor sites there are in the nose or how many different shapes the sites represent. However, they do know that receptor sites are made of large, very intricate protein molecules. These protein molecules consist of hundreds of atoms bonded together. They are much more elaborate and complex than the simple drawings of receptor sites shown in the illustrations.

The theory is that once a smell molecule has "docked" in a receptor site, nerves are stimulated and send a message to the brain. If minty receptor sites are stimulated, then the brain registers a minty smell.

It is possible that smell is much more complex than just described. It is possible that a specific smell might be the result of a combination of different molecules stimulating a combination of receptor sites.

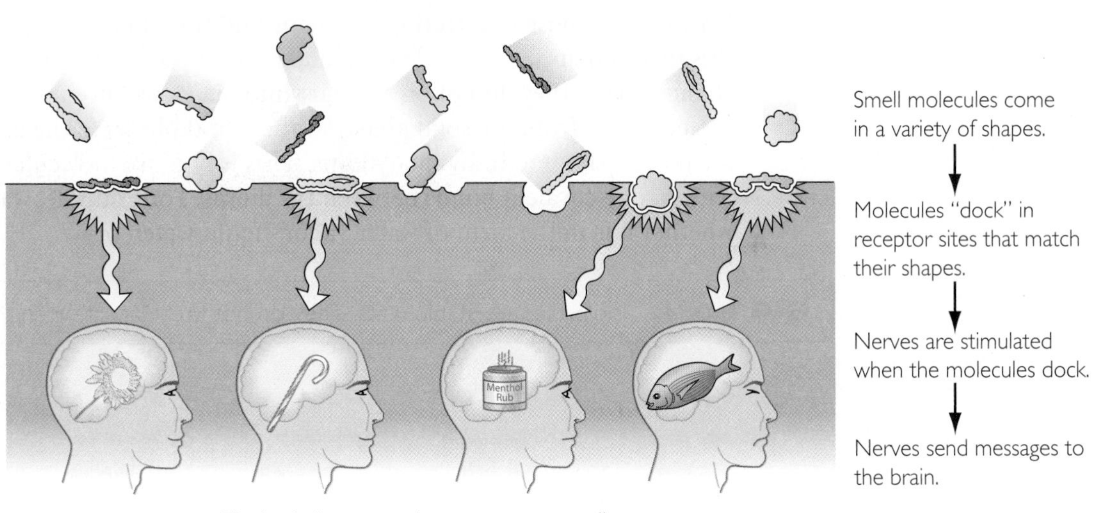

The brain interprets the messages as smells.

CONSUMER CONNECTION

A chocolate bar looks pretty solid. Still, we can smell chocolate, so we know that some of the molecules that make up chocolate must be changing phase to become a gas and entering the nose. If you melt a chocolate bar, the scent will be stronger because more molecules are escaping in the gas phase.

2 Phase Change and Molecular Stability

Smells Are Gases

When you bring an onion home from the supermarket, it doesn't give off much of a smell. But when you slice or chop an onion, the odor can become quite overpowering. What's happening? When you cut up the onion, your knife is cutting through some of the cell walls of the onion and releasing a lot of liquid.

One of the molecules that is in this liquid and is responsible for the distinctive smell of onions is allyl propyl disulfide. Its structural formula is shown here. Its molecular formula is $C_6H_{12}S_2$. This is the molecule your nose detects when you smell onions.

Allyl propyl disulfide, $C_6H_{12}S_2$

But something else has to happen in order for you to actually smell the onion. Molecules of allyl propyl disulfide have to get from the onion into your nose. The only way that can happen is if some of the molecules travel through the air to your nose. To do this, the compound changes phase to become a gas or vapor that floats through the air and enters your nostrils. There is no other possible explanation for your ability to smell an onion several feet away.

You cannot smell something unless some portion of it is in the form of a gas. This means everything you are able to smell is a substance that has become a gas.

Important to Know A substance does not have to boil in order for some of its molecules to become a gas. At the surface of molecular liquids and solids, some molecules escape and become a gas. ◂

Molecules Are Stable Units

Because your nose is detecting the shape and functional group of a molecule, this means that the molecule must enter your nose in one piece. When molecules change phase, they do not break apart into pieces or into individual atoms. Molecules go from the solid phase, to the liquid phase, to the gas phase without breaking apart into individual atoms. This is because molecules are stable units, with strong covalent bonds between the atoms. For example, water is still water whether it is in the form of water vapor, liquid water, or ice.

BIG IDEA Molecules are stable units. They do not break apart when they change phase.

Key Term

receptor site theory

Lesson Summary

How does the nose detect and identify different smells?

The receptor site theory explains how the nose detects different smells. This theory suggests that smell molecules travel to receptor sites in the lining of the nose. Once a smell molecule docks in a receptor site, it stimulates a nervous system response, sending a signal to the brain. Everything you smell is in the form of molecules in the gas phase.

EXERCISES

Reading Questions

1. Explain the receptor site theory and how it applies to smell detection.
2. What does phase change have to do with smell?

Reason and Apply

3. Here is a structural formula for an active ingredient in muscle ointment. How does this compound smell? How does your nose detect the smell?

4. A chemist creates a new molecule that has a completely different three-dimensional shape from other molecules humans have ever encountered. Would you be able smell it? Select the best answer.
   - **A.** Probably not, because synthetic compounds do not have a smell.
   - **B.** Definitely, because all esters smell sweet.
   - **C.** Probably not, because our noses would not have developed receptor sites to detect it.
   - **D.** Definitely, because parts of the molecule can break off to fit into receptor sites in our noses.
5. Some smells are similar, such as popcorn and freshly baked bread. Explain how you might account for similar smells using the receptor site theory.
6. How might the receptor site theory explain why a dog has a better sense of smell than a person?
7. Why do you think you might "get used to" a smell and hardly detect it? Use the receptor site model in your explanation.
8. Are some smells "faster" than others? Explain.

HEALTH CONNECTION

Many medications work by using a molecular lock and key mechanism. The molecules found in a medication may be designed to *stimulate* a receptor site, resulting in the desired physical response. Or a medication may have been designed to *block* receptor sites, thereby preventing a certain unwanted physical response.

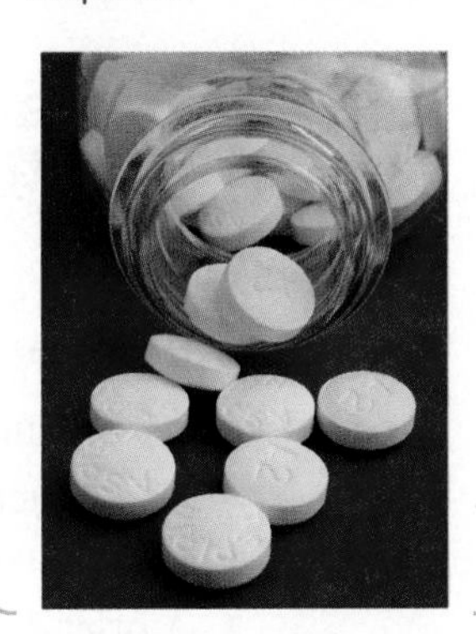

SECTION

II

SUMMARY

Building Molecules

Smells Update

The smell of a substance sometimes has more to do with molecular shape than with the functional groups that are present. In this section you learned about three-dimensional molecular structure and related it to smell. Ball-and-stick and space-filling models show how atoms are arranged within a molecule. The actual shape of a molecule is determined by the locations of electron domains.

No matter what shape a molecule has, it must fit into an appropriate receptor site in order to be smelled.

Key Terms

ball-and-stick model
tetrahedral shape
electron domain
electron domain theory
pyramidal shape
bent shape
trigonal planar shape
linear shape
space-filling model
receptor site theory

Review Exercises

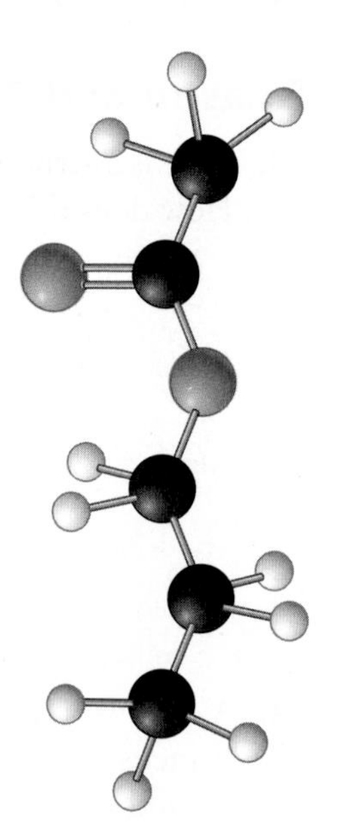

1. Examine the ball-and-stick model for propyl acetate.
   a. What is the molecular formula for this molecule?
   b. Draw the Lewis dot structure and the structural formula.
   c. What functional group is in the molecule?
   d. Predict the smell of the compound.
2. Predict the shape of each molecule. Support your answer.
   a. ammonia, NH3
   b. silicon chloride, SiCl4
   c. hydrogen sulfide, H2S
   d. hydrogen cyanide, HCN
   e. formaldehyde, CH2O
3. Explain what determines a molecule's shape.
4. Draw a possible Lewis dot structure for $C_4H_{10}O$.
5. From what you've learned so far, how is molecular shape related to smell?

PROJECT *Other Smell Classifications*

Research a smell classification not discussed in class such as musk, woody, spicy, nutty, leather, or tobacco. Create a poster that includes

- A drawing describing the smell classification and sources of the smell.
- The structural formulas and ball-and-stick models of at least five molecules in the smell classification that you investigated, along with their chemical names.
- A brief description of the properties associated with this group of molecules.

SECTION

III Molecules in Action

The water droplets on a leaf are the result of interactions between molecules in the water and the surface of the leaf.

Molecules interact with each other in a variety of ways. The process of smelling involves interactions between the molecules that enter the nose and the molecules that make up the lining of the nose. One key to molecular interactions is the fact that atoms in a molecule often do not share electrons equally. This unequal sharing creates partial positive and negative charges on different atoms in the molecule. These partial charges influence how molecules interact with each other and help to determine properties like phase, smell, and whether the substance will dissolve in another substance. Molecular size, shape, and bonding patterns all contribute to smell.

In this section, you will study

- polarity
- molecular interactions
- why some substances do not smell
- how phase, molecular size, polarity, shape, and bonding patterns relate to smell

LAB Attractions Between Molecules

Purpose

To observe the behavior of certain liquids, both near an electrical charge and as droplets.

Part 1: Testing the Liquids

With your group, test the liquid at each station with a charged wand.

1. Place the beaker directly below the burette. Open the valve slightly to create a fine flow of water from the burette into the beaker.
2. Rub a plastic wand with a cloth to create a charge on the wand. Hold the charged wand close to, but not touching, the stream of liquid. Move the wand close to and away from the stream of liquid and observe the effect.

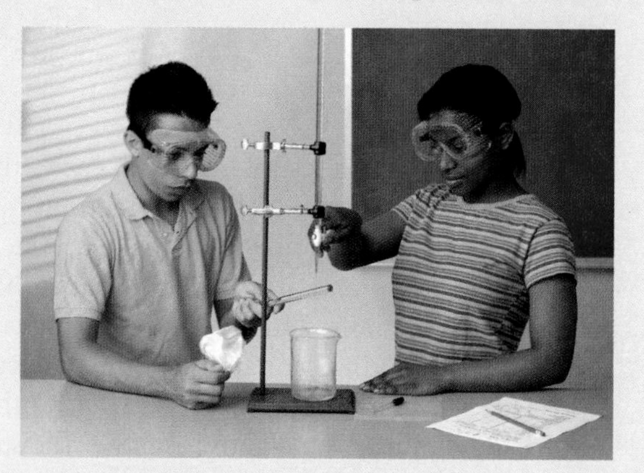

3. If a burette is close to empty, refill it carefully with the contents of the catch beaker. Make sure the valve is closed during refilling.
4. Next, place a drop of the liquid on waxed paper. Enter the results in a table similar to the one shown here.

Compound	Effect of charged wand	Behavior on waxed paper
water		
acetic acid		
isopropanol		
hexane		

Part 2: Analysis

1. Here is an artist's interpretation of what is happening between the water molecules and the charged wand. Write a paragraph describing in your own words what you think is happening in the picture.

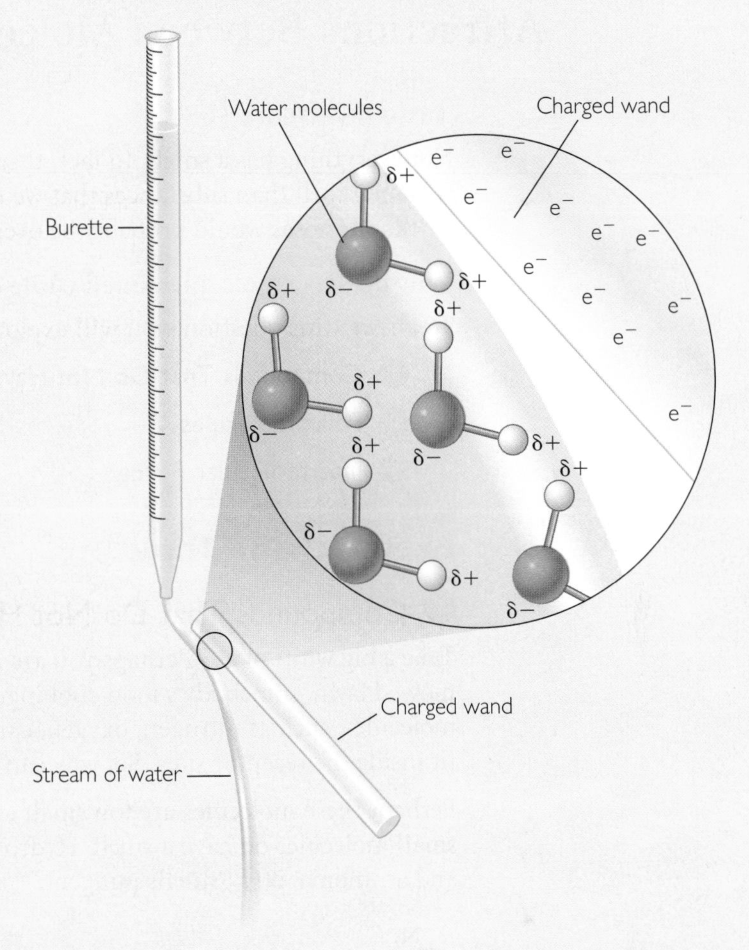

2. What evidence do you have that some of the molecules you just tested may have a charge on them?
3. How do you explain any liquids that are not attracted to the charged wand?
4. How does the behavior of the droplets relate to the charged wand experiment?
5. **Making Sense** If water molecules are carrying a partial charge on them, as shown in the following picture, how do you think a group of water molecules would behave towards each other? Draw a picture of several water molecules interacting, to illustrate your thinking. Explain your drawing.

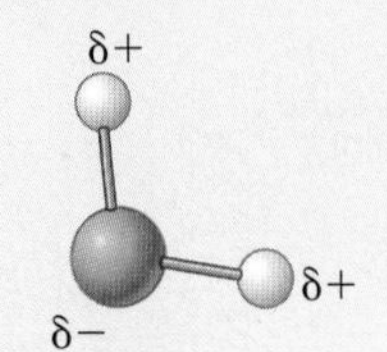

A single water molecule

LESSON

15 Attractive Molecules

Attractions Between Molecules

Think About It

Not everything has a smell. In fact, there are more substances on the planet that we *don't* smell than substances that we *do* smell. If everything around us had a smell, our noses would probably be overwhelmed most of the time.

Why do some molecules smell while others do not?

To answer this question, you will explore

1. Compounds That Do Not Have a Smell
2. Polar Molecules
3. Intermolecular Force

Exploring the Topic

1 Compounds That Do Not Have a Smell

Take a big whiff of air. Perhaps you are able to detect some odors—a freshly mowed lawn, somebody's food cooking. But the air itself does not smell. Small molecules, such as nitrogen, oxygen, carbon dioxide, and water vapor, obviously fit inside the receptor sites. So, why can't we smell them?

Perhaps these molecules are too small to stay docked in a receptor site. But some small molecules do have a smell. Hydrogen sulfide, H_2S, smells like rotting eggs and ammonia, NH_3, smells pungent.

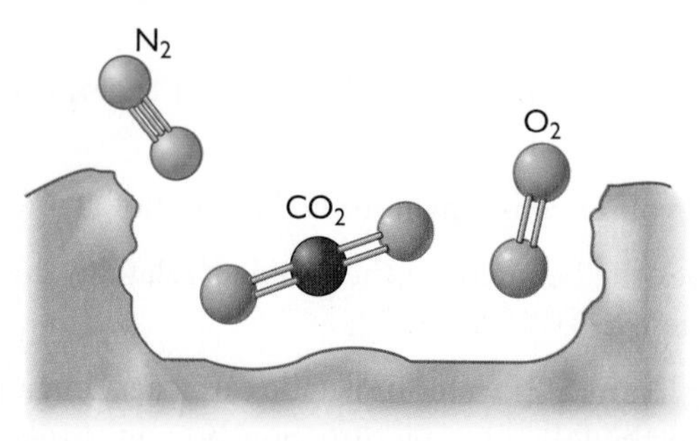

Some small molecules do not have a smell.

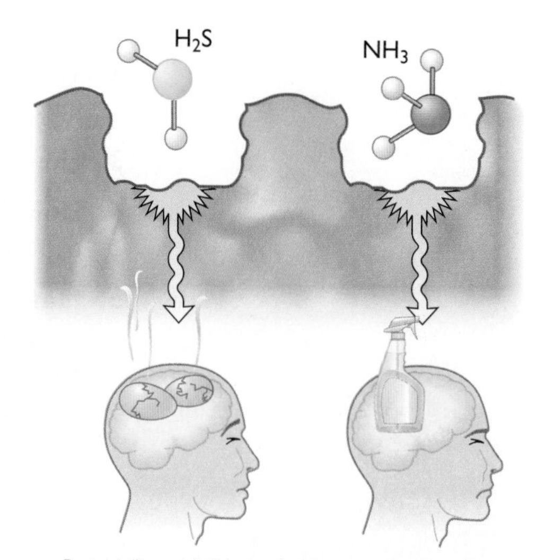

But H_2S and NH_3 both have strong smells.

So the molecular size of a compound does not explain whether we can smell it. It turns out that a molecular property called *polarity* is involved.

PHYSICS CONNECTION

You take off a wool hat, and all your hair stands on end. This happens because electrons move from your hair to the hat so that each individual hair has a positive charge. Because like charges repel one another, the hairs try to get as far apart as possible, resulting in a bad hair day. Something similar happens when a child slides down a plastic slide.

❷ Polar Molecules

When two different materials are rubbed together, some electrons can transfer from one material to the other. The result is an imbalance of positive and negative charges. One material has an excess of positive charge and the other an excess of negative charge. This is known as static electricity.

Suppose that you rub a plastic wand on a piece of cloth. The plastic wand will end up with a charge. In class you tested a number of liquids by holding a charged wand near them and observed that some of the streams bend toward the wand.

All of the liquids in the first column of the table are attracted to a charged wand. They are made up of **polar molecules.** The liquids in the second column are not attracted to a charged wand; they consist of **nonpolar molecules.**

These molecules are all **polar.** They all have a smell.

Polar molecules	Nonpolar molecules
water vinegar nail polish remover rubbing alcohol antifreeze	mineral oil hexane paint thinner motor oil

These molecules are all **nonpolar.** They do not have a smell.

Water is a polar molecule. As this illustration shows, there are partial charges at different locations on the water molecule. These **partial charges** are much smaller than the charge on an individual electron or proton. However, the charges are large enough to cause a stream of water to be attracted to a charged wand.

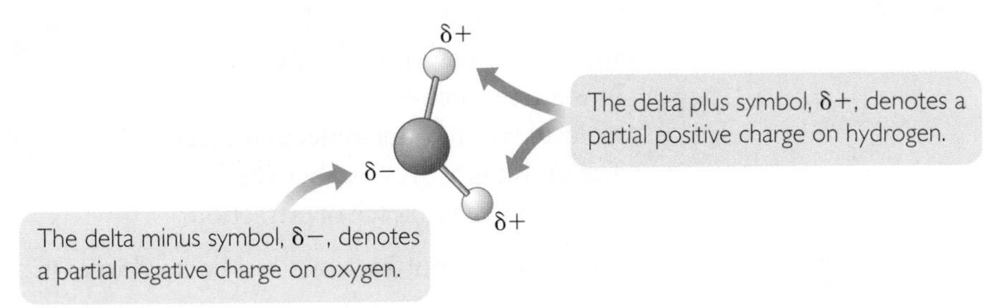

The oxygen atom has a partial negative charge, and the hydrogen atoms each have a partial positive charge.

❸ Intermolecular Force

Explaining the Charged Wand

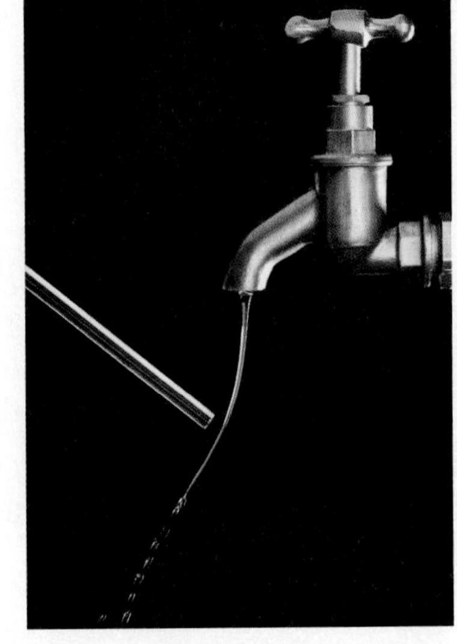

When a wand with a negative charge is placed next to a stream of water, the water molecules orient themselves in space so that their positive ends are lined up in the direction of the wand. As a result, the whole stream of water is attracted to, or pulled toward, the charged wand. The illustration on the next page shows a possible way to depict what happens to the molecules.

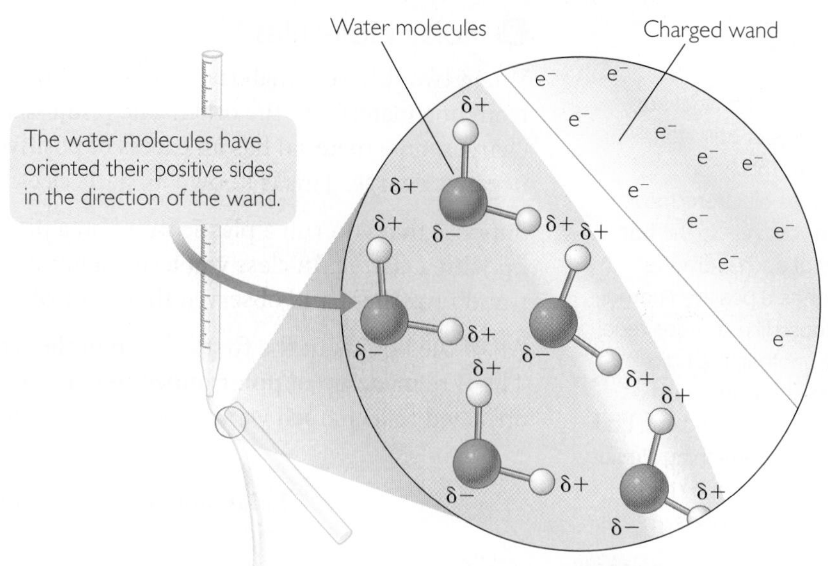

Water molecules next to a charged wand

Intermolecular Behavior

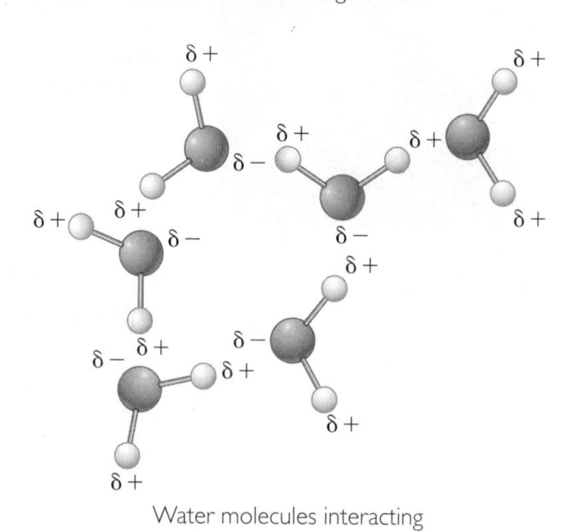

Water molecules interacting

The partial charges on polar molecules cause more than an attraction to a charged wand. They also cause attractions between molecules. As the molecules in a polar liquid tumble around they tend to align with each other because partial negative charges are attracted to partial positive charges. This attraction between individual molecules is an **intermolecular force.** The prefix *inter* means "between" and the force is a force of attraction.

Intermolecular attractions can be used to explain many observable properties.

Observation: Water beads up on waxed paper. Oil spreads out.

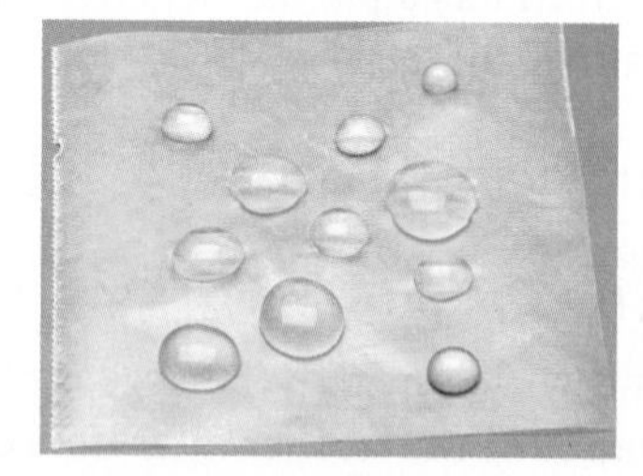

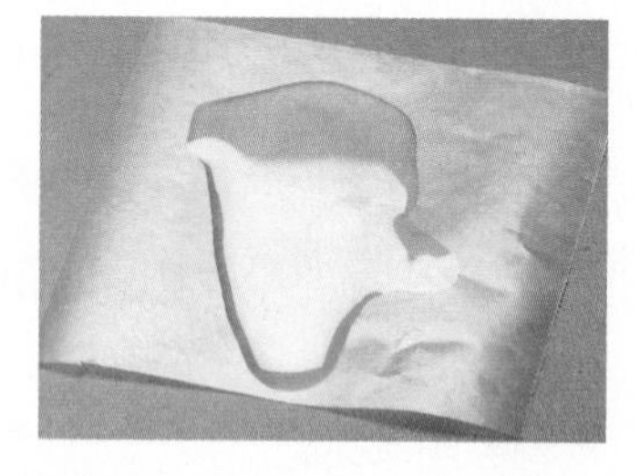

Explanation: The water molecules are polar, while the oil molecules are nonpolar. The individual water molecules are attracted to each other and "cling together." Individual oil molecules are not attracted to one another to the same extent, and they spread out.

Observation: Water is a liquid at room temperature, but methane is a gas at room temperature. They have roughly the same molecular mass.

The saying "Oil and water don't mix" is one way to remember that a nonpolar substance and a polar substance tend not to dissolve in each other.

Explanation: Water is polar and methane is nonpolar. The individual water molecules are attracted to each other and stay together as a liquid. The attractions between methane molecules are much weaker. The methane molecules spread throughout the room as a gas.

Observation: Methanol dissolves easily in water. Oil floats on top of water but does not dissolve.

Explanation: The polar methanol molecules are attracted to the polar water molecules and go into solution. The nonpolar oil molecules are not attracted to the polar water molecules.

Key Terms

polar molecule
nonpolar molecule
partial charge
intermolecular forces

Lesson Summary

Why do some molecules smell while others do not?

Molecules can be divided into two classes: polar and nonpolar. Polar molecules are attracted to a charged wand because they have partial charges distributed within the molecule. Nonpolar molecules are not attracted to a charged wand. Polarity is responsible for intermolecular attractions that affect many properties of molecules possibly including smell properties.

EXERCISES

Reading Questions

1. Explain in your own words what a polar molecule is.
2. What are intermolecular attractions?

Reason and Apply

3. **Lab Report** Write a lab report for the Lab: Attractions Between Molecules. In your conclusion, explain why some liquids were attracted to the wand and others were not as well as why some liquids beaded up on waxed paper while others did not.

(Title)

Purpose: (Explain what you were trying to find out.)

Procedure: (List the steps you followed.)

Results: (Explain what you observed during the experiment.)

Conclusions: (What can you conclude about what you were trying to find out? Provide evidence for your conclusions.)

4. Methanol, CH_3OH, is attracted to a charged wand. The hydrogen atoms have a partial positive charge, and the oxygen atom has a partial negative charge.
   a. Draw a picture showing how you predict the methanol molecules are oriented when they are attracted to a wand with negative charges.
   b. Do you expect methanol to bead up or spread out on waxed paper? Explain your thinking.
   c. Do you expect methanol to dissolve in water? Explain your thinking.
5. Hexane, C_6H_{14}, is not attracted to a charged wand, and it spreads out on waxed paper. Do you expect hexane to dissolve in water? Explain your thinking.
6. Explain why no liquids are repelled from a charged wand.

PHYSICS CONNECTION

You walk across a rug, touch a doorknob, and get a shock. This occurs because electrons move from the rug to your body. Now you have extra electrons. The doorknob is a conductor. When you touch it, electrons move to the doorknob. You sense this rapid movement of electrons as a shock.

You don't have to go to the ends of the earth to find polar molecules. They're all over the place. A polar molecule is just a molecule with a difference in electrical charge between two ends.

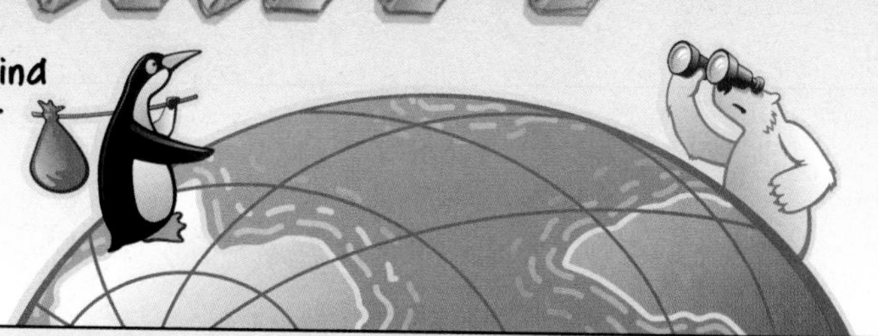

When two atoms with different electronegativity values bond, the bonding electrons spend more time around the more electronegative atom, creating a PARTIAL NEGATIVE CHARGE on that atom. The other atom then has a PARTIAL POSITIVE CHARGE, and the bond is polar.

When atoms with equal electronegativity values bond, they form nonpolar bonds. The electron-attracting strength of each atom is the same.

However, if the electronegativities of two bonded atoms are different, then their bond will be polarized—maybe a little...

...maybe a lot.

Because the elements have such varying electronegativities and can bond in many different combinations, there is really a continuum of polarity in bonding. We can break the continuum down into three categories.
NONPOLAR COVALENT
O O
O_2
H H
H_2
O
O
H
H
The clearest examples of nonpolar covalent bonds are those between identical atoms, such as in H_2, N_2, O_2, or Cl_2. Bonds between atoms with nearly the same electronegativity value, such as carbon and hydrogen, can also be considered nonpolar.

POLAR COVALENT
HF
Partial negative charge
+ H F −
Partial positive charge
H
F
In a polar covalent bond, two atoms share bonded pairs of electrons somewhat unequally. The electrons are more attracted to one atom than the other. Examples include bonds between carbon and oxygen atoms, or between hydrogen and fluorine atoms.

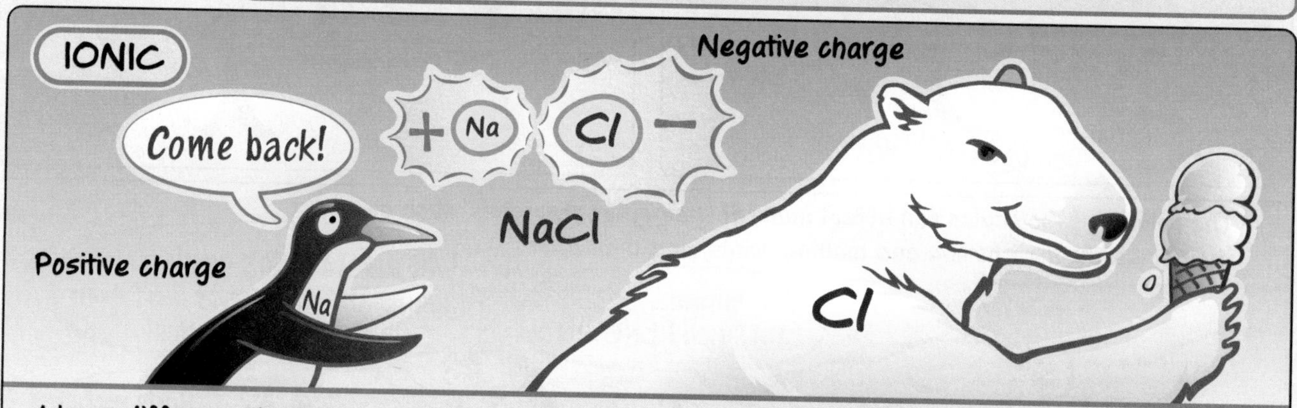
IONIC
Negative charge
Come back!
+ Na Cl −
NaCl
Positive charge
Na
Cl
A large difference in electronegativity results in the winner-take-all situation of ionic bonding. The more electronegative atom takes the bonding electrons and becomes a negative ion, while the other atom becomes a positive ion. The opposite charges on the ions attract each other.

Polar bonds between atoms create dipoles. The word dipole can refer to (1) the polarity of an individual polar bond between atoms, (2) the net polarity of an individual polar molecule that may have several polar covalent bonds within it, and (3) the polar molecule itself.

Confusing? Here are some examples:

An N_2 molecule isn't a dipole and it doesn't have any dipoles.

HCl has a dipole and it is a dipole.

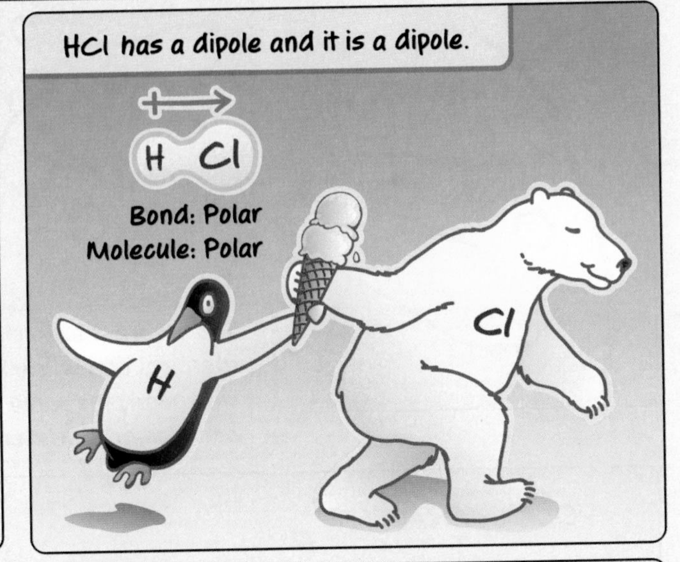

CO_2 has two dipoles but the CO_2 molecule itself is not a dipole. Its polar bonds balance each other out and make the molecule nonpolar overall.

H_2O has two dipoles. Because of its bent shape, it also has a dipole in the sense of an overall polarity.

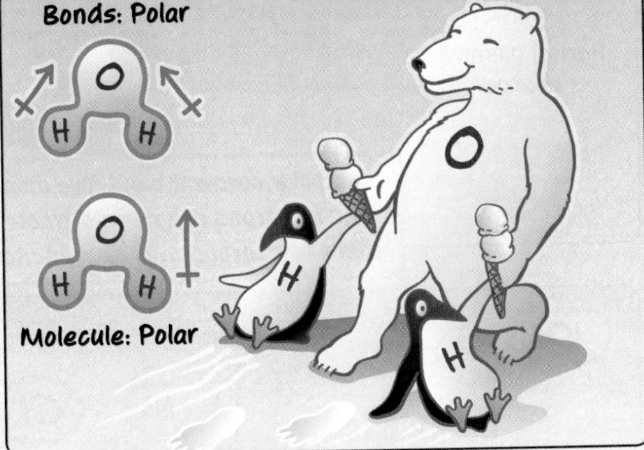

The polarity of molecules can affect many of their other properties, such as their solubility, their boiling and melting points, and their odor.

Mmmmm... you smell PENGUINY.

Why are we in this comic strip? Penguins and polar bears don't even live at the same poles!

Suits me!

LESSON

16 Polar Bears and Penguins

Electronegativity and Polarity

Think About It

Hydrogen chloride, HCl, is a colorless but very toxic gas. Its smell is described as a suffocating, acrid odor. Like most other small molecules that smell, HCl molecules are polar. But what makes an HCl molecule polar? Where do the partial charges come from on the atoms in a polar molecule?

What makes a molecule polar?

To answer this question, you will examine

1. Electronegativity
2. Nonpolar Molecules
3. Electronegativity and Bonding

Exploring the Topic

1 Electronegativity

The hydrogen atom and the chlorine atom in hydrogen chloride, HCl, form a covalent bond by sharing a pair of electrons. This cartoon represents HCl as a penguin and a polar bear. The bonded pair of electrons is represented as two scoops of ice cream. Although the penguin and the polar bear are sharing the ice cream, they are not sharing it equally.

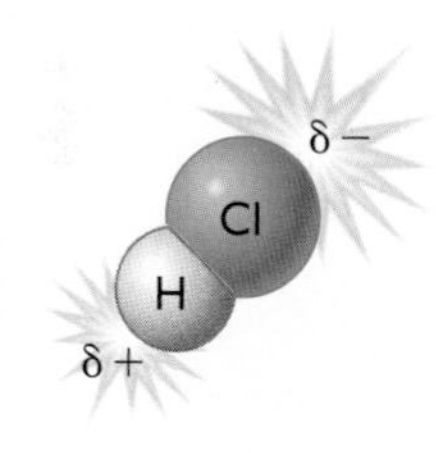

Similarly, the hydrogen atom and the chlorine atom in a hydrogen chloride molecule do not share the bonded pair of electrons equally. The chlorine atom attracts the shared electrons much more strongly than the hydrogen atom does. As a result, the shared electrons spend more time near the chlorine atom than they do near the hydrogen atom. Because of this displacement of the electrons, the hydrogen atom has a partial positive charge and the chlorine atom has a partial negative charge.

The tendency of an atom to attract shared electrons is called **electronegativity.** An atom that has a large electronegativity strongly attracts shared electrons. In this case, the chlorine atom, like the polar bear, is stronger in attracting electrons, so it is more electronegative than the hydrogen atom. The atoms that are more electronegative are the ones that end up with a partial negative charge. The atoms with less electronegativity end up with a partial positive charge. The result is a polar bond.

A polar molecule is called a **dipole,** because it has two poles: a positive end and a negative end. A dipole can be shown with an arrow starting at the positive end and pointing to the negative end of the molecule. The polar bond itself is also called a dipole.

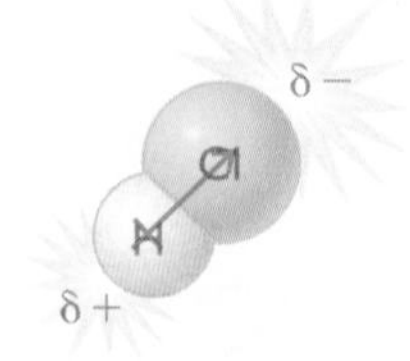

2 Nonpolar Molecules

When two atoms with identical electronegativities bond together, the attraction of the shared electrons is identical. As a result, the molecule is nonpolar. For example, H_2 and Cl_2 are both nonpolar molecules. They have no partial charges.

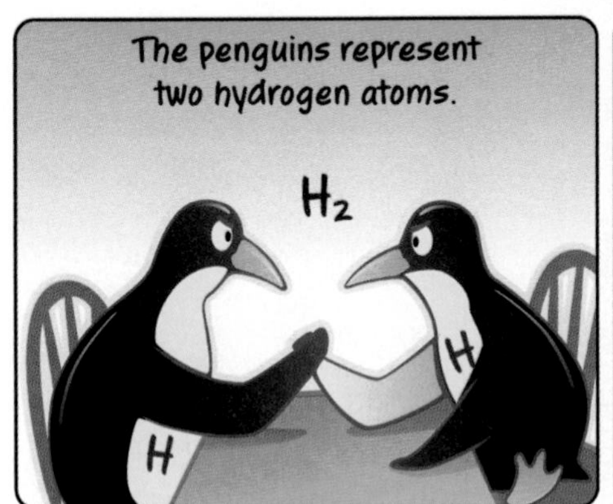

A contest between two polar bears of equal strength or two penguins of equal strength would result in a tie.

There is another way to end up with a nonpolar molecule. Examine the next illustration. Why is carbon dioxide, CO_2, a nonpolar molecule?

CO_2 is a nonpolar molecule.

The two dipoles in CO_2 balance each other, and there is no partial positive end to the molecule. So the overall molecule is nonpolar.

③ Electronegativity and Bonding

There are two types of covalent bonds: polar covalent bonds and nonpolar covalent bonds. When two atoms with different electronegativities bond together, the result is a polar covalent bond.

Different electronegativities result in polar bonds.

When the electronegativities of the atoms that bond are identical, the result is a nonpolar covalent bond.

Identical electronegativities result in nonpolar bonds.

If the electronegativities of the two atoms differ greatly, it is possible for electrons to be pulled entirely toward one of the atoms in a bond. The result is an ionic bond.

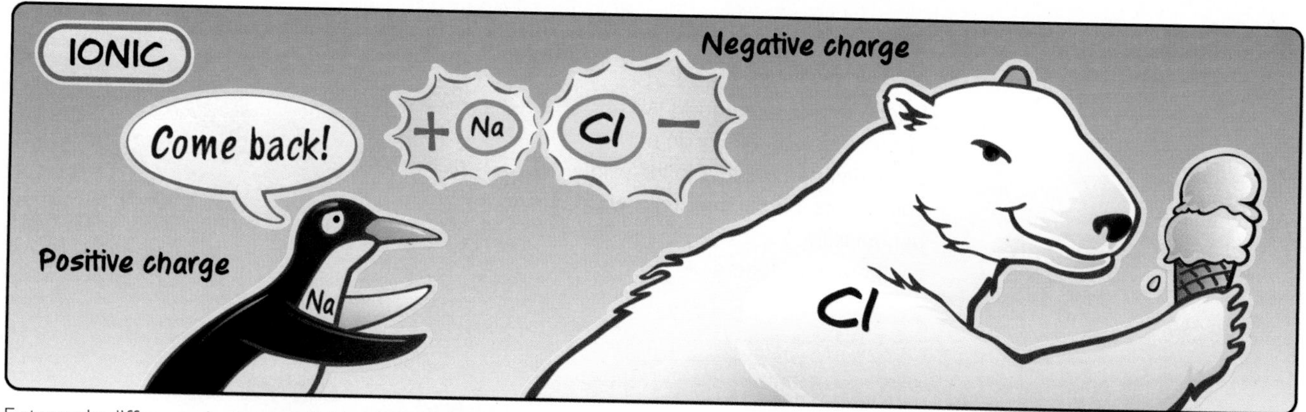

Extremely different electronegativities result in ionic bonds.

Ionic compounds represent the extreme of polar bonds, in which electrons are transferred to the more electronegative atom in the pair.

Key Terms
electronegativity
dipole

Lesson Summary

What makes a molecule polar?

Electronegativity is a measure of the ability of an atom to attract the electrons that are involved in a bond. Different atoms have different electronegativities. When two different kinds of atoms bond together, they do not attract the bonding electrons equally. The electrons are attracted to the atom with more electronegativity. The result is a polar covalent bond. When atoms with identical electronegativities bond together, the electrons are shared equally, and the result is a nonpolar covalent bond. If the two atoms involved in a bond differ greatly in electronegativity, the result is electron transfer and an ionic bond.

EXERCISES

Reading Questions

1. What is the difference between a polar covalent bond and a nonpolar covalent bond?
2. What is the difference between a polar covalent bond and an ionic bond?
3. Explain why carbon dioxide is a nonpolar molecule even though its bonds are polar.

Reason and Apply

4. Are these molecules polar covalent or nonpolar covalent?
   a. N_2 **b.** HF **c.** F_2 **d.** NO **e.** FCl
5. Using polar bears and penguins, create an illustration showing an ammonia, NH_3, molecule. (*Hint:* You may wish to start with a Lewis dot structure.)
6. Is the molecule HOCl polar or nonpolar? Use a Lewis dot structure to explain your thinking.
7. Use electronegativity to explain why some molecules are attracted to a charged wand.

NATURE CONNECTION

Polar bears and penguins live on opposite poles, so they do not encounter one another in the wild. Polar bears inhabit the Arctic, while penguins live exclusively in the southern hemisphere.

LESSON

17 Thinking (Electro)Negatively

Electronegativity Scale

Think About It

Some atoms are "greedier" than other atoms when it comes to sharing the bonding electrons between them. A bond between two atoms with very different electronegativities is more polar than a bond between atoms with similar electronegativities. How can the polarity of different bonds be compared?

How can electronegativity be used to compare bonds?

To answer this question, you will explore

1. Electronegativity Scale
2. Diatomic Molecules

Exploring the Topic

1 Electronegativity Scale

Chemists have assigned each atom a number, called an *electronegativity value.* This number corresponds to the tendency of an atom to attract bonding electrons. By using these numbers, it is possible to compare the polarity of different bonds. The table here shows the electronegativity value for an atom of each element in the periodic table.

Electronegativity scale

H 2.10																	He
Li 0.98	Be 1.57											B 2.04	C 2.55	N 3.04	O 3.44	F 3.98	Ne
Na 0.93	Mg 1.31											Al 1.61	Si 1.90	P 2.19	S 2.58	Cl 3.16	Ar
K 0.82	Ca 1.00	Sc 1.36	Ti 1.54	V 1.63	Cr 1.66	Mn 1.55	Fe 1.83	Co 1.88	Ni 1.91	Cu 1.90	Zn 1.65	Ga 1.81	Ge 2.01	As 2.18	Se 2.55	Br 2.96	Kr
Rb 0.82	Sr 0.95	Y 1.22	Zr 1.33	Nb 1.60	Mo 2.16	Tc 1.90	Ru 2.2	Rh 2.28	Pd 2.20	Ag 1.93	Cd 1.69	In 1.78	Sn 1.96	Sb 2.05	Te 2.1	I 2.66	Xe
Cs 0.79	Ba 0.89	La* 1.10	Hf 1.30	Ta 1.50	W 2.36	Re 1.90	Os 2.20	Ir 2.20	Pt 2.28	Au 2.54	Hg 2.00	Tl 1.62	Pb 2.33	Bi 2.02	Po 2.00	At 2.20	Rn
Fr 0.70	Ra 0.89	Ac* 1.10															

* Electronegativity values for the lanthanides and actinides range from about 1.10 to 1.50.

HISTORY CONNECTION

Linus Pauling, inventor of the electronegativity scale, is one of only two people (the other is Marie Curie) to receive Nobel Prizes in two different fields. In 1954, he received the Nobel Prize in Chemistry, and in 1962 he received the Nobel Peace Prize for his work in campaigning against above-ground nuclear testing.

Notice that the values generally increase from left to right and bottom to top. Noble gases are often not assigned values because they generally do not form compounds.

A scale for electronegativity was first proposed by Linus Pauling, in 1932. The scale used today by chemists ranges in value from 4.0 down to 0. The electronegativity scale allows you to compare individual atoms. For example, fluorine, with an electronegativity value of 3.98, attracts shared electrons more strongly than hydrogen, with an electronegativity value of 2.10.

2 Diatomic Molecules

The electronegativity scale also allows you to compare the polarity of bonds. In a polar covalent bond, the electrons tend to spend more time around the more electronegative atom. Several substances with only two atoms are shown below. Molecules with two atoms are called **diatomic molecules.** By looking at the numerical difference between the electronegativities of the two atoms, it is possible to compare the polarity of one bond to another. Bonds that have a greater difference in electronegativity between the two atoms are more polar.

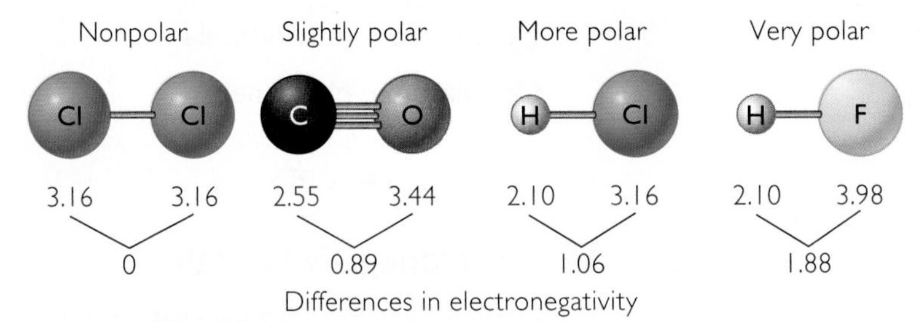

Topic: Electronegativity
Visit: www.SciLinks.org
Web code: KEY-217

When the difference in electronegativity between the two atoms is very large, the bond is no longer considered covalent. In that case, the electrons are transferred from the less electronegative atom to the more electronegative atom. A cation and an anion form that are attracted to one another in an ionic bond. As shown on the electronegativity scale, when the difference between the electronegativities of two bonded atoms is greater than about 2.1, the bond is considered ionic.

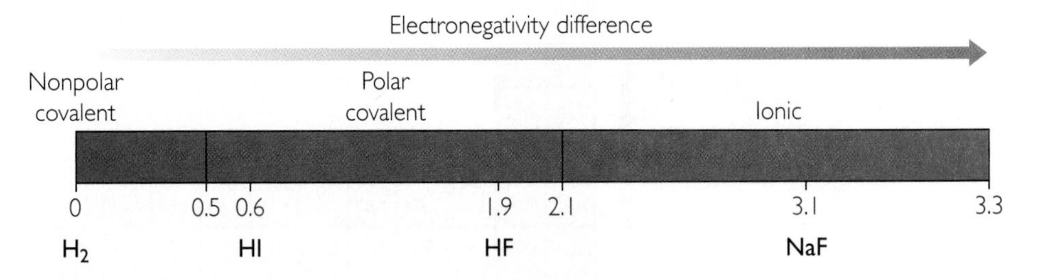

Example

Potassium and Chlorine

Predict the type of bond you would find between potassium and chlorine.

Solution

Use the electronegativity scale to find the electronegativities of potassium, K, and chlorine, Cl. It shows that the values are 0.82 for K and 3.16 for Cl. The difference is 2.34, so the bond is ionic.

Key Term

diatomic molecule

Lesson Summary

How can electronegativity be used to compare bonds?

Electronegativity values can help you to compare and classify different bonds. If there is no difference in electronegativity between the two atoms the bond is considered nonpolar covalent. The larger the electronegativity difference, the more polar the bond. The electrons in a polar bond tend to spend more time around the more electronegative atom. When the difference between the electronegativities of two bonded atoms is greater than about 2.1, the bond is considered ionic and one atom gives up an electron to the other atom, forming two ions.

EXERCISES

Reading Questions

1. Explain how electronegativity values help you to determine the polarity of a bond between two atoms.
2. How can you determine which atom in a covalent bond is partially positive?

Reason and Apply

3. Consider the following pairs of atoms. Place each set in order of increasing bond polarity. Describe the trend.

   a. Li–F Na–F K–F Rb–F Cs–F

   b. Mg–O P–S N–F K–Cl Al–N
4. Place a partial positive or a partial negative charge on each atom in the following pairs of atoms. Describe the trend.

   a. H–B b. H–C c. H–N d. H–O e. H–F
5. Is hydrogen always partially positive when bonded to another atom? Explain.
6. Name three pairs of atoms with ionic bonds. For each pair show the difference in electronegativity between the two atoms.
7. Name three pairs of atoms with polar covalent bonds. For each pair show the difference in electronegativity between the two atoms.
8. Describe or draw what happens to the electrons in a polar covalent bond, a nonpolar bond, and an ionic bond.
9. What do we mean when we say that bonding is on a continuum?

LESSON

18 I Can Relate

Polar Molecules and Smell

Think About It

Polar molecules have a potent smell. The smell of ammonia, NH_3, in some window cleaners is quite strong. In contrast, the methane gas, CH_4, in gas stoves does not have a smell even though the C—H bonds are all polar. So, why can't you smell CH_4 gas?

What does polarity have to do with smell?

To answer this question, you will examine

1. Polarity of Molecules
2. Nonpolar Molecules
3. Polarity and Smell

Exploring the Topic

1 Polarity of Molecules

If the *bonds* between atoms of a molecule are polar, what about the molecule as a whole unit? One way to tell whether a molecule is polar is to examine its overall shape. Notice that each of the three molecules shown here has an irregular shape, or some sort of asymmetry. This asymmetry makes a molecule polar.

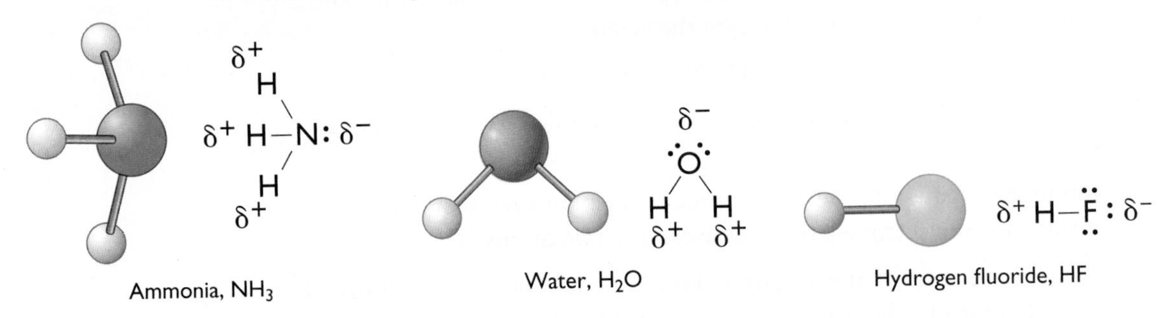

Ammonia, NH_3 Water, H_2O Hydrogen fluoride, HF

Ammonia, water, and hydrogen fluoride are all polar. In these molecules there is a partial negative charge on the nitrogen, oxygen, and fluorine atoms because these atoms are more electronegative than hydrogen atoms. The hydrogen atoms have partial positive charges. To figure out which end of a molecule has a partial negative charge and which end has a partial positive charge check the electronegativities of the atoms and the overall shape of the molecule. Consider formaldehyde, CH_2O, shown here. The numbers next to the atoms are their electronegativities.

H 2.10, C 2.55, O 3.44, H 2.10, δ+, δ−

CH_2O

Direction of dipole

ENVIRONMENTAL CONNECTION

Chlorotrifluoromethane, also known as Freon, was developed early in the 20th century and used widely as a refrigerant. However, use of Freon and other chlorofluorocarbons (CFCs) was significantly reduced in the late 1980s due to their effects on the ozone layer of the atmosphere.

The oxygen end of the formaldehyde molecule has a partial negative charge because the oxygen atom is the most electronegative atom in the formaldehyde molecule. Similarly, the hydrogen atoms have partial positive charges because they are the least electronegative atoms in the molecule.

Example 1

Phosphine, PH_3

Is phosphine, PH_3, polar? What is the direction of the dipole?

Solution

A Lewis dot structure of PH_3 shows four pairs of electrons, three bonding pairs and one lone pair. These four pairs of electrons are arranged around the P atom in a tetrahedral shape.

H
H—P:
H

Because the molecule is asymmetrical, it is polar. You can find the electronegativity values for phosphorus and hydrogen in the electronegativity scale in Lesson 7: Thinking (Electro) Negatively. The electronegativity value for P is 2.19, and for H it is 2.10. Because the electronegativity of the phosphorus atom is greater, the P atom attracts the electrons more strongly and has a partial negative charge. The hydrogen atoms have partial positive charges. The overall dipole is shown.

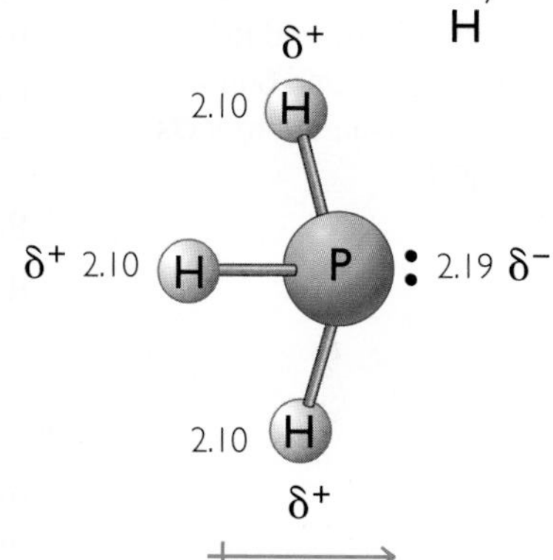

2 Nonpolar Molecules

Diatomic molecules with two identical atoms are nonpolar. The electrons between the two identical atoms are shared equally. However, these are not the only kind of nonpolar molecules. Some molecules are symmetrical. The symmetry in these molecules can make them nonpolar even though they have polar bonds within them. The polarities of the individual polar bonds balance each other out due to the shape of the molecule.

δ− δ+ δ−
O=C=O

δ+H—C—Hδ+ (with H δ+ above and below; C δ−)

δ− :F—C—F: δ− (with :F: δ− above and below; C δ+)

CO_2 CH_4 CF_4

Nonpolar molecules

CONSUMER CONNECTION

Carbon tetrachloride, CCl_4, is a tetrahedral, nonpolar molecule. But it is said to have a sweet smell. This molecule was widely used as a dry-cleaning solvent because it is very good at dissolving other nonpolar compounds, such as fats and oils. It was also widely used as a fire extinguisher and a pesticide for rodents until it was discovered to be highly toxic and banned from consumer products in 1970.

3 Polarity and Smell

Small molecules that are polar have a smell. Small molecules that are nonpolar do not have a smell. What does polarity have to do with the smell of small molecules?

Polar: Have a smell	Nonpolar: Do not have a smell	Exceptions
HF, HCl, H_2S, CH_3F, CH_2F_2, CHF_3, NH_3, PH_3, HOCl	N_2, O_2, CO_2, CH_4, CF_4	F_2, Cl_2, Br_2, CCl_4 are nonpolar but have a smell. H_2O is polar but does not have a smell.

Polar molecules have properties that distinguish them from nonpolar molecules. Polar molecules dissolve in water and are attracted to other polar molecules. One well-accepted theory is that polar molecules *dissolve* in the mucous membrane of the nose and then attach to receptor sites. The mucus is composed largely of water.

Receptor Sites in the Nose

Receptor sites in the nose contain large polar protein molecules. Scientists believe that small polar molecules are attracted to the polar receptor sites after they enter the nose. You could think of polarity as working somewhat like a magnet, with small polar molecules "sticking" to the polar part of a receptor site. Nonpolar molecules are not attracted to the same extent as polar molecules, so they may not be detected by the nose.

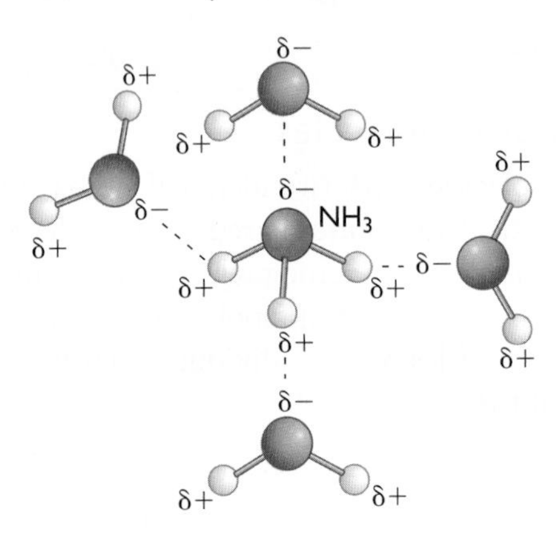

Ammonia, NH_3, dissolves in water because they are both polar liquids.

Example 2

Chlorotrifluoromethane, $CClF_3$, and Tetrafluoromethane, CF_4

Compare chlorotrifluoromethane with tetrafluoromethane. Explain why one of these molecules smells and the other does not.

Solution

Both molecules have a tetrahedral shape. In tetrafluoromethane, each C–F bond is polar. However, the four C–F bonds are distributed symmetrically so that the molecule as a whole is nonpolar. Therefore, it does not smell.

Chlorotrifluoromethane is not as symmetrical. This is because one of the F atoms is replaced by a Cl atom. The molecule is polar because F and Cl have different electronegativities. Therefore, it smells.

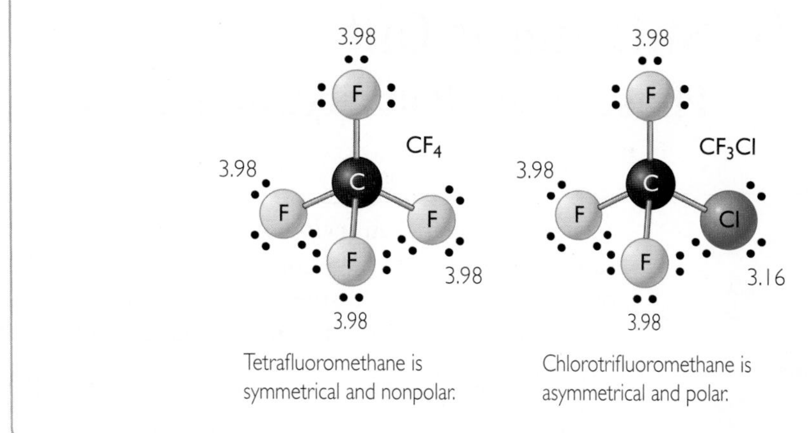

Tetrafluoromethane is symmetrical and nonpolar.

Chlorotrifluoromethane is asymmetrical and polar.

NATURE CONNECTION

Many animals, such as elephants, seem to be able to smell water at great distances.

Lesson Summary

What does polarity have to do with smell?

Small asymmetrical molecules with polar bonds are polar. Molecules that are symmetrical in every way are nonpolar. Theories suggest that the polarity of a molecule may help it to "stick" in a receptor site in the nose. Additionally, polar substances are more soluable in water. This may allow polar substances to dissolve in the mucous membrane of the nose, whereas nonpolar substances are simply exhaled undissolved and undetected.

EXERCISES

Reading Questions

1. How can you determine if a molecule is polar?
2. Describe one theory of why small nonpolar molecules do not have a smell.

Reason and Apply

3. For each of the molecules listed, draw a Lewis dot structure and indicate the shape of each molecule. Decide whether these substances smell. Explain your reasoning for each.

   a. H_2Se **b.** H_2 **c.** Ar **d.** HOF **e.** $CHClF_2$ **f.** CH_2O
4. For each of the polar molecules in Exercise 3, draw the dipole.
5. Water is an exception to our rule about small molecules. It is a polar molecule, yet you don't smell it. What do you think is going on?
6. Do you think it might be useful if you could smell the air? Explain your thinking.
7. Methane, CH_4, gas leaks can be very dangerous and can be explosive. Why do you think dimethyl sulfide, C_2H_6S, is added to natural gas that is used in homes and buildings?

LESSON

19 Sniffing It Out

Phase, Size, Polarity, and Smell

Ionic

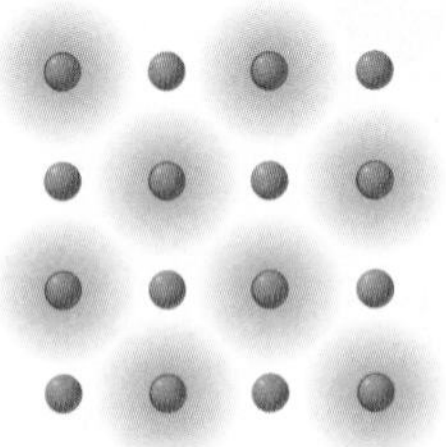

Molecular covalent

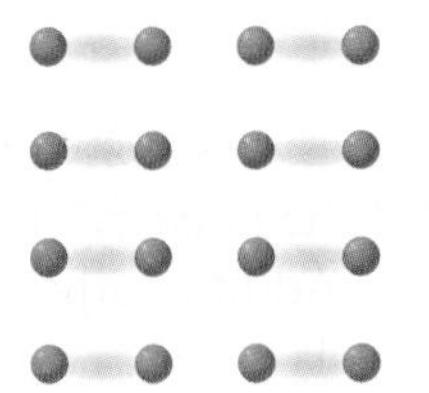

Metallic

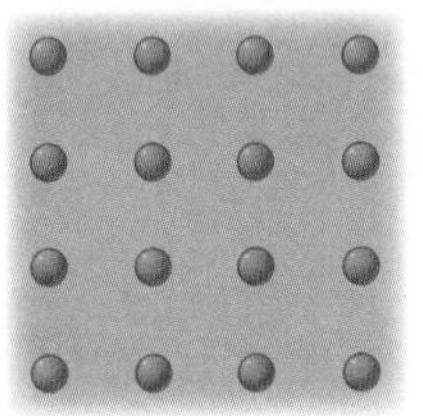

Network covalent

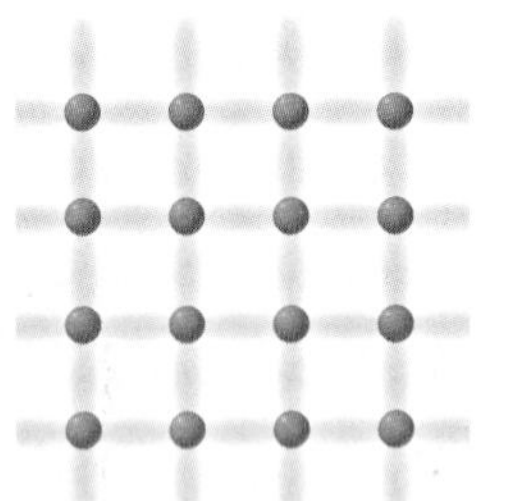

Red spheres represent nuclei and core electrons and blue areas represent bonding electrons.

Think About It

You can smell chicken frying or someone's perfume from across a room, but you can't smell a piece of paper or a binder on your desk. And cookies right out of the oven are much easier to smell than cold ones. How can you use what you have learned about molecules to understand these observations?

What generalizations can you make about smell and molecules?

To answer this question, you will explore

1. Bonding and Smell
2. Molecular Size and Smell

Exploring the Topic

1 Bonding and Smell

In Unit 1: Alchemy, you sorted matter into four classes based on whether the substance dissolved in water and whether the substance conducted electricity. This sorting led to four models of bonding: ionic, molecular covalent, metallic, and network covalent. How does the property of smell relate to bonding?

Recall that all the substances you smelled in this unit were molecules with nonmetal atoms. This evidence suggests that substances that smell fit into the molecular covalent category.

Chemical Bonding and Smell

Substance type and bonding	Smell?	Phase	Examples	
			Name	Formula
ionic metals bonded to nonmetals	**no**	solid	sodium chloride (table salt) calcium oxide (lime) calcium carbonate (chalk)	NaCl CaO $CaCO_3$
molecular covalent nonmetals bonded in molecules	**yes,** with some exceptions	gas, liquid, or solid	nitrogen ammonia menthol	N_2 NH_3 $C_{10}H_{18}O$
metallic elemental metals	**no**	solid	gold copper aluminum	Au Cu Al
network covalent nonmetals bonded in a network	**no**	solid	carbon (diamond) silicon dioxide	C SiO_2

Molecular covalent substances exist as gases, liquids, or solids at room temperature. Substances in the other three bonding categories are strictly solids (with a few exceptions, such as liquid mercury). It is reasonable to propose that substances that are ionic, metallic, or network covalent do not smell because they do not form gases easily at room temperature. This means that they cannot get into the receptor sites in our noses.

BIG IDEA Only molecular covalent substances form gases at room temperature.

Example 1

Bonding and Smell

Predict whether you would be able to smell the following substances. Explain your reasoning.

a. propanol, C_3H_8O b. iron, Fe c. copper sulfate, $CuSO_4$

Solution

a. Propanol consists of nonmetal atoms, so it is a molecular covalent substance. It probably has a smell.

b. Iron is a metallic solid. It does not have a smell.

c. Copper sulfate consists of both metal and nonmetal atoms, so it is an ionic solid. It does not have a smell.

CONSUMER CONNECTION

Propane is a colorless, odorless gas that is often used in portable stoves and travel trailers for heating and cooking purposes. Methane is the principal component of natural gas and is often used for fuel. Because propane and methane cannot be smelled, odorants are added to them to ensure that leaks can be detected.

2 Molecular Size and Smell

In the previous lesson, you learned that small molecules have a smell only if they are polar. What about medium-sized and large molecules? Take a moment to examine the data in this table.

Molecular Size, Polarity, and Smell

Molecular size and polarity	Smell?	Phase	Examples	
			Name	Formula
small nonpolar molecules	no	gas	nitrogen carbon dioxide methane	N_2 CO_2 CH_4
small polar molecules	yes	gas	hydrogen sulfide ammonia fluoromethane	H_2S NH_3 CH_3F
medium-sized polar and nonpolar molecules	yes	liquid	octane geraniol carvone pentylpropionate	C_8H_{18} $C_{10}H_{18}O$ $C_{10}H_{14}O$ $C_8H_{16}O_2$
large polar and nonpolar molecules	no	solid	1-triacontyl palmitate (beeswax) polystyrene cellulose	$C_{46}H_{92}O_2$ $C_{8000}H_{8000}$ $C_6H_{10}O_5$

Many pieces of information can be gathered from the table. First, small molecules are gases at room temperature, medium-sized molecules are liquids, and large molecules are solids. One reason for this is that large molecules tend to have more intermolecular attractions. This means that larger molecules are less likely to be gaseous and are less likely to get to your nose.

Note that a molecule is considered medium-sized if it has about 5 to 19 carbon atoms and large if it has 20 or more carbon atoms. Carbon is a unique element because it bonds easily to other carbon atoms and forms four bonds. This makes it ideal for forming long and complex molecules.

In general, all medium-sized molecules have a smell, whether they are polar or nonpolar. The medium-sized molecules form a liquid because they are attracted to one another, but they are volatile, meaning they can go easily into the gas phase. Once these molecules get inside the nose, the intermolecular attractions between the molecules and the receptor sites allow you to detect a smell.

Important to Know Some substances smell only because they are contaminated with small molecules. For example, plastic itself does not have a smell, but it is often contaminated with small molecules that are used to make the plastic. ◄

BIOLOGY CONNECTION

The grassy odor that is emitted when the lawn is mowed is cis-3-hexenal. This molecule is unstable and tends to rearrange to form trans-2-hexenal. Scientific evidence connects trans-2-hexenal with healing from stress.

Heating and Cooling

What happens when you warm or cool molecular substances? Small molecules are already gases at room temperature. But larger molecules are liquids or solids. These molecules must undergo a phase change in order to be smelled. Warming a molecular liquid or solid will generally give it a more intense aroma. This is because heating a substance increases the rate at which a substance becomes a gas. Cooling a substance has the opposite effect.

When there are more molecules in the air, you are more likely to detect them. So the temperature of a substance will usually affect the strength of its smell. For example, you are more likely to smell warm food than cold leftovers from the fridge.

Example 2

Predict the Smell

Below is the structural formula of a substance called 2-acetylthiazole. Predict whether this substance will smell at room temperature. Give evidence to support your answer.

```
    H
     \
      C—N          H
     //  \\        |
    C     C—C—C—H
   /  \  /   ||  |
  H    S     O   H
```

2-acetylthiazole

Solution

This molecule's formula is C_5H_5NSO. It has five carbon atoms. This is a medium-sized molecule, so it probably has some sort of smell. (In fact, it does. It is described as having a hazelnut or roasted popcorn smell.)

Lesson Summary

What generalizations can you make about smell and molecules?

Humans can smell some molecular substances. We cannot smell ionic, metallic, or network covalent substances. This is because the first requirement for a substance to be smelled is that it be in a gaseous form. The size of the molecule seems to be another important factor in determining smell. Most medium-sized molecules will smell. Most large-sized molecules will not smell. Much of this is due to the effect of size on phase change. For small molecules, polarity also plays a role. Small nonpolar molecules do not smell. Small polar molecules do smell.

EXERCISES

Reading Questions

1. Explain what types of substances you would expect to have a smell and why.
2. Explain what types of substances you would expect *not* to have a smell and why.

Reason and Apply

3. Predict whether you would be able to smell these substances. Explain your reasoning.

   decanol, $C_{10}H_{22}O$ lead, Pb iron oxide, Fe_2O_3 potassium chloride, KCl
4. Predict whether you would be able to smell these substances. Explain your reasoning.

   neon, Ne ethane, C_2H_6 decane, $C_{10}H_{22}$ methylamine, CH_5N
5. If a substance is capable of becoming a gas under normal conditions, should you be able to smell it? Explain.
6. Molecules that have a structure containing more than about 20 carbon atoms rarely smell. Explain why this is so.
7. Your cotton T-shirt smells when you take it out of the clothes dryer. Explain why.
8. Occasionally you might go into a restaurant or a room that smells and after a while you stop noticing the odor. Explain what you think is going on.
9. From a biological point of view, why do you think it is useful for humans to be able to smell certain substances and not others?

BIOLOGY CONNECTION

There is scientific evidence that some fish are better at smelling and tasting underwater than dogs are, out in the air. This explains why fish will often turn away from or avoid a fisherman's lure. It is said that a salmon's sense of smell can detect when a person puts their hand in the water 100 ft upstream.

SECTION

III

Key Terms

polar molecules
nonpolar molecules
partial change
intermolecular force
electronegativity
dipole
diatomic molecule

SUMMARY

Molecules in Action

Smells Update

Some substances, like hot apple cider, have a smell and other substances, like a granite tabletop, do not. The process of smelling is an interaction between molecules. In order to be smelled, a substance must become airborne and enter the nose. Once it is in the nose, it must dissolve and be detected. Receptor sites are composed of large, folded protein molecules with a polar area that links up with polar molecules.

Summary of Investigation into Smell

	Has a smell	Does not have a smell
Bonding in compound	molecular covalent	ionic, metallic, network covalent
Size of compound	medium-sized molecules and small polar molecules (see polarity)	large molecules
Polarity of compound	small polar molecules	small nonpolar molecules
Phase of compound	gas phase	no molecules in the gas phase

Review Exercises

1. Which element is more electronegative?
   a. Zn or Br b. Li or Cs c. Au or Al
2. Place the following pairs of atoms in order of increasing bond polarity. Explain your reasoning.
   H–I H–Cl H–F
3. Draw the structural formula and place a partial positive and a partial negative charge on each atom in these molecules. Which molecules are polar? Explain your reasoning.
   a. HCl b. CH_4 c. H_2O
4. At cold temperatures, hydrogen bromide, HBr, is a liquid.
   a. Draw the Lewis dot structure for HBr. Label the dipole.
   b. Would HBr be attracted to a charged wand? Would it form a round drop on waxed paper? Explain your reasoning.
   c. In HBr, what type of bond exists between H and Br?
   d. Do you expect HBr to smell? Explain.

5. Use what you know about molecules to explain why you can't smell perfume while the bottle is closed, but you can smell it once the bottle is opened.

6. Which can you smell better, cold bread or warm bread right out of the oven? Explain why.

PROJECT *Sense of Smell Study*

Plan and conduct an experiment to compare how the sense of smell differs among different groups of people.

- **Create a goal statement.** Write a sentence or two stating the goal of your study. What do you hope to discover or explore?
- **Choose subjects to study.** Pick two categories of people to study and compare. These categories should be clear and easy to determine (for example, children under the age of 12 versus adults, vegetarians versus meat-eaters, women versus men).
- **Write a proposal.** Write several sentences stating how you propose to accomplish your goal, including how you will conduct your study, how you will choose your sample of participants randomly, and how you will set up the control variables. Clear your proposal with your teacher before conducting your study.
- **Conduct your study.** Keep your data organized in a table or chart. Keep careful notes of everything that you do and what you observe.
- **Write up your results and conclusions.** Write up the results of your study and any conclusions you have come to based on your collected data. Include your data table. If possible, create a graph using your data.

SECTION

IV Molecules in the Body

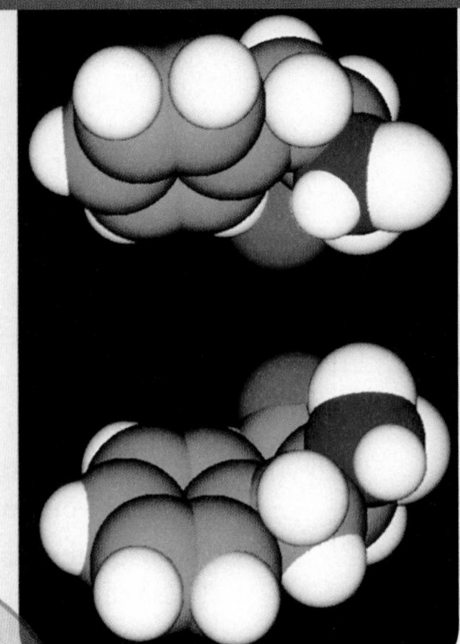

These molecules are mirror images of each other and may have very different properties.

Compounds that you smell are sensed by receptor sites in your nose. Receptor sites are made of large molecules called proteins. The protein molecules consist of amino acid molecules that are linked together to form long chains. There are a great variety of protein molecules, made of various combinations of different amino acids. Protein molecules are essential to nearly all body functions, including the sense of smell.

In this section, you will study

- mirror-image isomers
- amino acid molecules
- protein molecules

LESSON

20 Mirror, Mirror

Mirror-Image Isomers

Think About It

Limonene is a molecule that comes in two forms. The structural formulas for both forms are identical. Yet one smells like pine needles, while the other has the lemon-orange fragrance of citrus fruits. How can two molecules with the same molecular formula and the same structural formula have such distinct smells?

What are mirror-image isomers?

To answer this question, you will explore

1. Mirror Images of Molecules
2. Properties of Mirror-Image Isomers

Exploring the Topic

1 Mirror Images of Molecules

There are many things in nature that are mirror images of each other: your ears, for example, or your eyebrows. While your ears may look physically identical, your left ear would not work well on the right side of your head. Your hands and feet are other good examples. Consider your left and right hands. They are made of the same parts and have the same overall structures. But if you try to place your left hand on your right hand with your palms down, the thumbs will point in opposite directions.

Your hands are mirror images of one another. That is, when you observe your left hand in a mirror, it will look identical to your right hand and vice versa. However, you cannot *superimpose* them on each other. No matter how you position your left hand, it is never identical to your right hand.

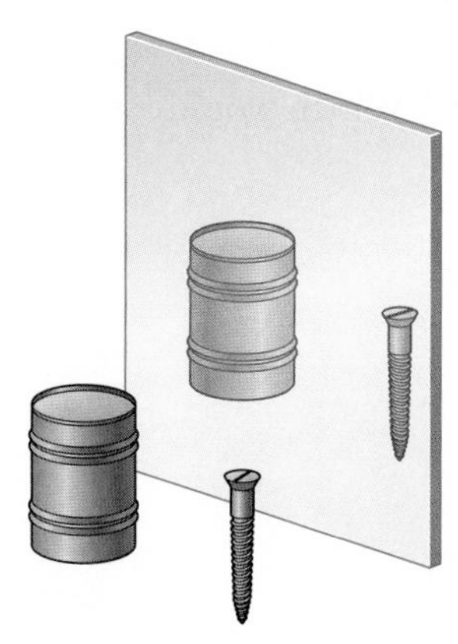

A can is identical to its mirror image, but a screw is not. (The thread spirals in the opposite direction in the image.)

Molecules with a Handedness

Just as your right hand cannot be superimposed on your left hand, some molecules have a "handedness."

> **CONSUMER CONNECTION**
>
> Naproxen is an anti-inflammatory drug used to manage pain caused by arthritis, gout, tendonitis, and several other ailments. While this drug can be helpful, its mirror-image isomer causes liver poisoning and provides no pain relief.

The two limonene molecules described earlier have identical structures that are mirror images. The mirror images are not superimposable on one another. They are actually different molecules. You could say that one molecule is left-handed and the other is right-handed. They are isomers of one another.

Determining if two molecules are mirror images or the same molecule can be visually challenging. A three-dimensional model of the molecule and an actual mirror can help.

Consider the mirror image of a simple tetrahedral molecule, tetrachloromethane, CCl_4. It is fairly easy to see that the mirror image can be superimposed on the original molecule. (In other words, you can imagine picking it up and turning or sliding it so that it matches the original.) So, tetrachloromethane does not have a mirror-image isomer. The molecule and its mirror image are identical.

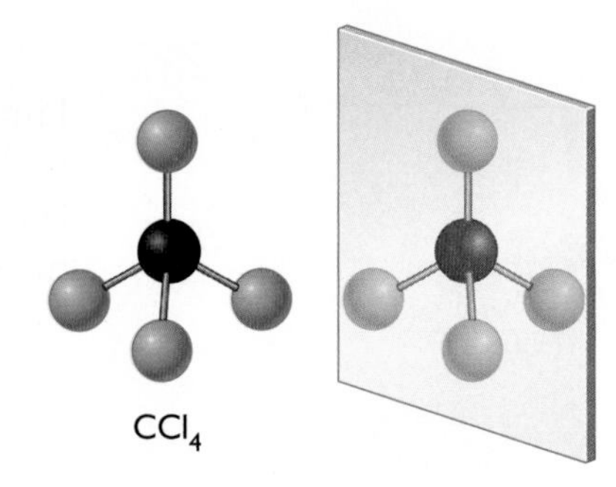

Now consider dichloromethane, CCl_2H_2, and chlorofluoromethanol, CHClFOH. Take a moment to decide if either of the mirror images can be superimposed on its original molecule.

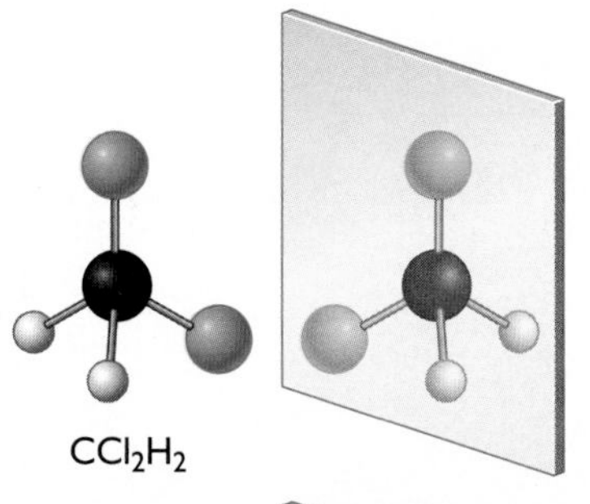

The mirror image of dichloromethane, CCl_2H_2, can be superimposed on the original molecule. However, when there are four different groups attached to a carbon atom as in the second molecule, the mirror image cannot be superimposed on the original molecule.

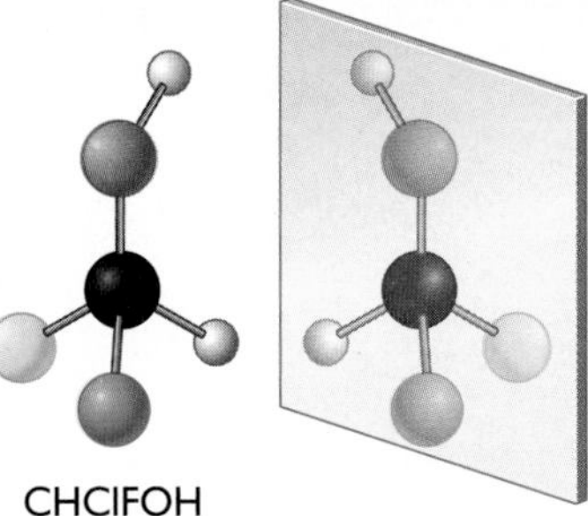

The two molecules of chlorofluoromethanol are **mirror-image isomers.** They have different properties.

Example 1

Objects in a Mirror

Which of the objects listed here look identical in a mirror? Which would look different? Explain any differences.

a. a can

b. a shoe

c. fluoromethane, CH_3F

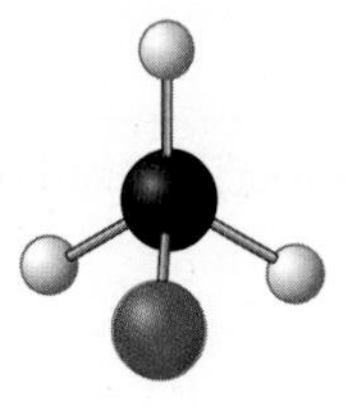

d. 2-chlorobutane, C_4H_9Cl

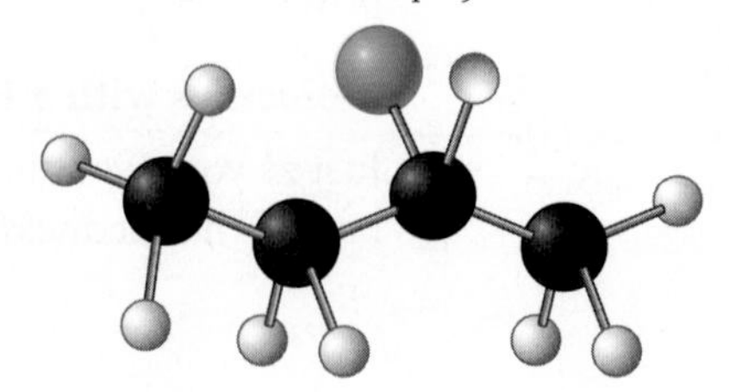

Solution

A and C look identical in a mirror. B and D do not.

In the 2-chlorobutane molecule, the second carbon atom from the left has four different groups attached to it: CH_3, H, Cl, and CH_2CH_3. Therefore, this molecule has a mirror-image isomer. When the H and Cl atoms of the two mirror-image isomers are in the same location, the CH_3 and CH_2CH_3 groups are in opposite locations.

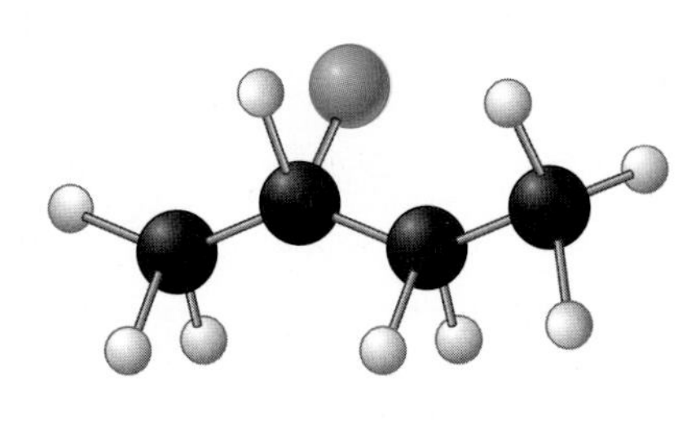

❷ Properties of Mirror-Image Isomers

Limonene

One way to figure out if a molecule has a mirror-image isomer is to check to see if any of the carbon atoms are attached to four different groups.

A structural formula for limonene, $C_{10}H_{16}$, is shown at left. The illustration at right shows that nine of the carbon atoms in a limonene molecule are not attached to four *different* groups.

The remaining carbon atom, circled in the diagram, is attached to a hydrogen atom and three carbon "groups." The two sides of the carbon ring are different from one another and considered different groups. You can conclude that limonene has a mirror-image isomer because it contains a carbon atom attached to four different groups of atoms. The two isomers are referred to as D-limonene and L-limonene.

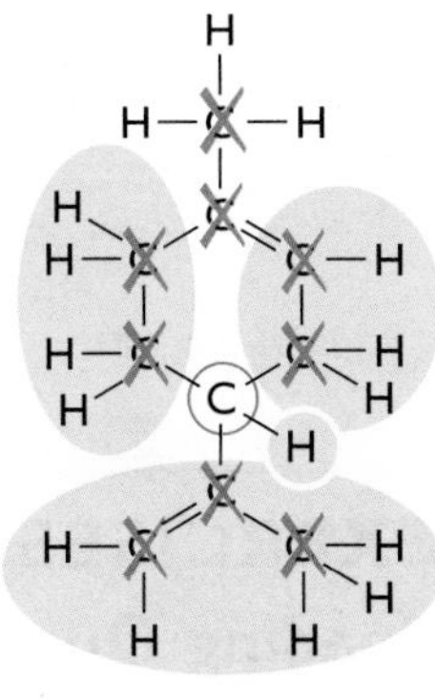

Limonene

The mirror-image isomers of limonene are shown below. The main difference between them has to do with the position of the CCH_2CH_3 group highlighted in yellow. If you flip the L-limonene molecule around and try to superimpose it onto the D-limonene molecule, the CCH_2CH_3 group will point toward you in one isomer and away from you in the other isomer. It is as if you have aligned your thumbs on your right and left hands, only to find one palm down and the other palm up.

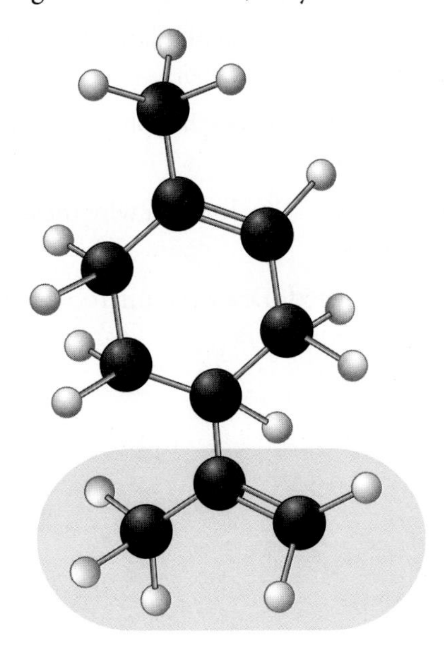

D-limonene

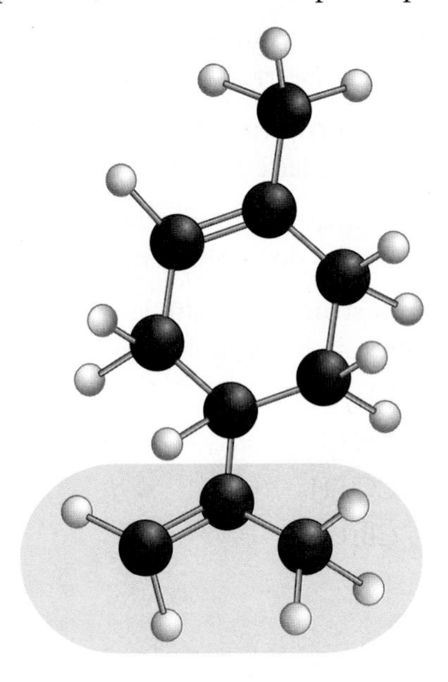

L-limonene

Different Smells

Just like your left foot does not fit into your right footprint, D-limonene does not fit into the smell receptor site for L-limonene.

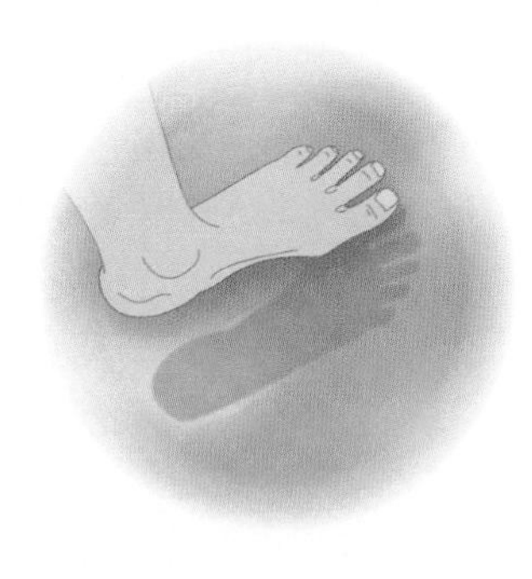

Because the mirror-image isomers do not fit into the same receptor site, you would expect them to have distinct smells. Indeed, this is the case: D-limonene smells like citrus while L-limonene smells like pine. The two molecules have the same molecular formula, the same structural formula, and the same shape, but they are mirror images of one another.

Key Term

mirror-image isomer

Lesson Summary

What are mirror-image isomers?

Nature is full of mirror images that are not identical. Some objects and some molecules can be superimposed on their mirror images. For other objects, the mirror image is not superimposable; the mirror image cannot be positioned such that it is identical to the original object or molecule. Molecules with mirror images that cannot be superimposed are called mirror-image isomers. These molecules fit into different smell receptor sites and therefore have different smells.

EXERCISES

Reading Questions

1. What do you need to do to the mirror image of the letter "D" to superimpose the mirror image on the original letter?
2. What types of molecules have mirror-image isomers?

Reason and Apply

3. List five objects that look different in a mirror and five that do not.
4. Draw mirror images of the capital letters in the English alphabet. Explain why symmetrical objects do not have mirror-image isomers.
5. Draw mirror images of the lowercase letters in the English alphabet. Which lowercase letters are mirror images of one another?
6. Draw the structural formulas for CH_2O and ClHCO. Explain why these two molecules do not have mirror-image isomers.
7. Draw structural formulas for the molecules listed. Draw the mirror image of each. Circle the molecules with mirror-image isomers.
   a. CH_4
   b. CFH_3
   c. $CClFH_2$
   d. CBrClFH

8. Examine the molecules here. Which have mirror-image isomers? Explain your reasoning.

a. butyric acid

```
    H  H  H  O
    |  |  |  ||
 H—C—C—C—C—O—H
    |  |  |
    H  H  H
```

b. ethyl formate

```
    O        H  H
    ||       |  |
 H—C—O—C—C—H
             |  |
             H  H
```

c. 2-butanol

```
          H
          |
       H  O  H  H
       |  |  |  |
    H—C—C—C—C—H
       |  |  |  |
       H  H  H  H
```

9. Two structural formulas are shown.

```
  H           H  H  H     H     H  H
  |           |  |  |     |     |  |
H-C——C=C-C-C——C——C-C-O-H
  |    |      |  |    |     |  |
  H  H-C-H    H  H  H-C-H   H  H
       |              |
       H              H
```

Citronellol

```
  H        H H H      H H
  |        | | |      | |
H-C——C=C-C-C——C=C-C-O-H
  |    |      | |   |       |
  H  H-C-H    H H H-C-H     H
       |            |
       H            H
```

Geraniol

a. Write the molecular formulas for citronellol and geraniol.

b. Molecules with the structural formula for citronellol can smell either like insect repellant or like flowers. Explain why there are two distinct smells.

c. All molecules with the structural formula for geraniol smell like roses. Explain why there is only one smell for this molecule.

HEALTH CONNECTION

These two molecules are isomers. The molecule with hydrogen atoms on opposite sides of the molecule is called a *trans* isomer. Most natural fatty acids are *cis* isomers. But, when unsaturated fatty acids are hydrogenated industrially to make products like shortening, they often change into *trans* isomers. Research indicates these trans fats are linked to poor cardiovascular health and should be avoided in our diet.

```
H         H
 \       /
  C = C
 /       \
CH3       CH3
```

Cis isomer

```
H         CH3
 \       /
  C = C
 /       \
CH3       H
```

Trans isomer

LESSON

21 Protein Origami

Amino Acids and Proteins

Think About It

Smelly molecules are small or medium-sized molecules that travel as a gas into your nose. Once in your nose, these molecules interact with receptor sites, which send a signal to your brain. Receptor sites are also built of atoms bonded together to create a "pocket" into which a specific smell molecule can fit.

What is a receptor site made of?

To answer this question, you will explore

1. Amino Acids
2. Proteins

Exploring the Topic

1 Amino Acids

Amino acids are the building materials for many of the structures in the body, especially those related to the functioning of cells. They are important molecules in all living organisms. As the name implies, amino acids have two functional groups: an amine group and a carboxyl group. The general formula for an amino acid is $C_2H_4NO_2R$, where R is simply a placeholder in the formula for a variety of possible groups of atoms. The group consists of carbon and hydrogen, and sometimes nitrogen, and/or oxygen.

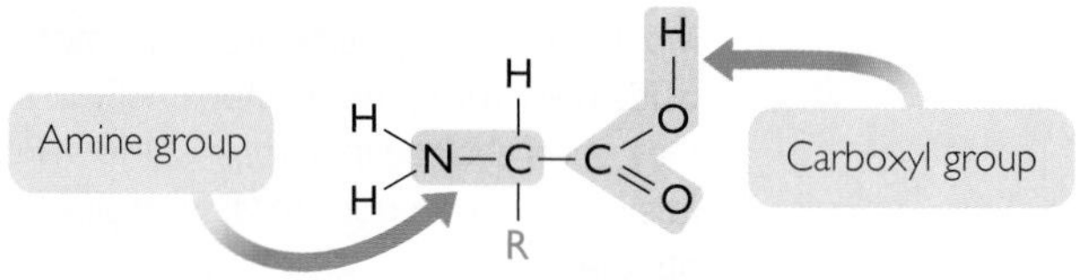

Amine group

Carboxyl group

The structure of an amino acid molecule

Here are structural formulas for three amino acids with the R groups highlighted. Notice that everything else about them is identical.

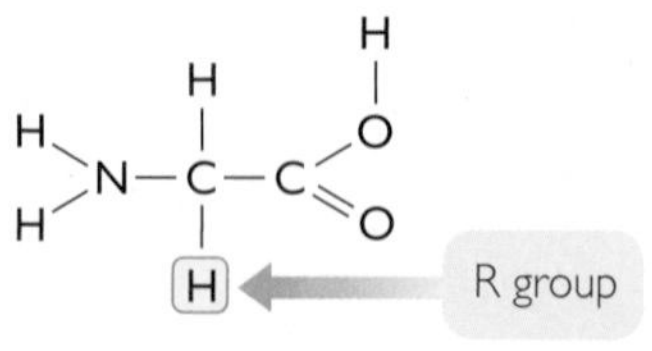

R group

Glycine

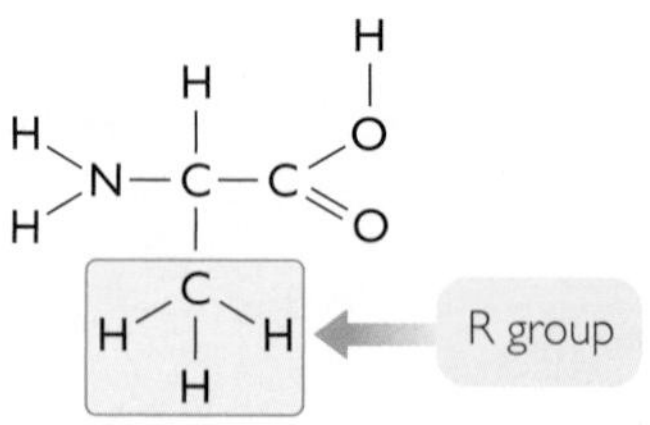

Alanine

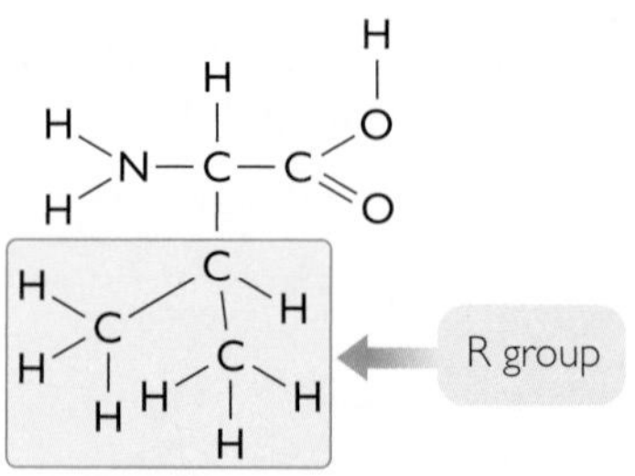

Valine

You might notice that the carbon atom in the center of the amino acid molecule has four different groups attached to it. Therefore, amino acids have mirror-image isomers and can be left-handed or right-handed molecules. There's a trick to figuring out the handedness of an amino acid. The illustration shows you how.

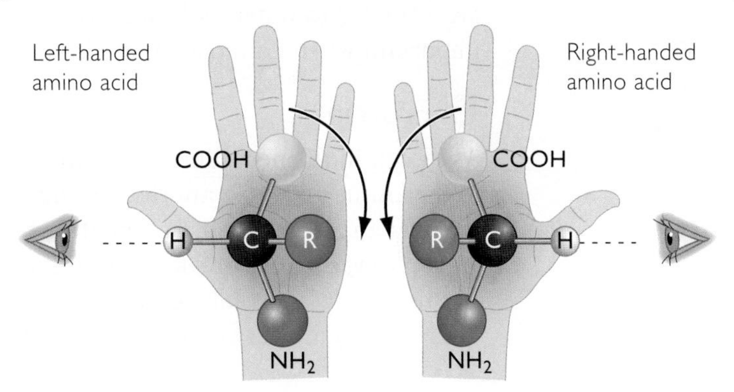

Left-handed amino acids: When you point the H atom toward your eye, the COOH group, the R group, and the NH_2 group are arranged clockwise. This is sometimes referred to as the CORN rule for **CO**OH, **R**, and **N**H_2.

BIOLOGY CONNECTION

The human body uses 20 amino acids. Of those 20 amino acids, 10 to 12 can be produced by the body and aren't necessary in our diets. The other 8 to 12 cannot be produced by the body, so we need to get them from food. Sources of these amino acids—in the form of protein—include meat, eggs, grains, legumes, and dairy products.

The human body uses only the left-handed amino acids. Both mirror images are available. How did life begin selecting only one of the two mirror images? This is a question of heated debate. Currently, no one has a good explanation.

There are 20 amino acids, with different R groups, that your body needs. Your body produces about half of these amino acids. You get the others as part of your diet. Six of the amino acids are listed in this illustration. These amino acids are placed into two categories: either hydrophobic ("water fearing") or hydrophilic ("water loving"). Take a moment to determine the difference.

Hydrophobic ("water fearing") amino acids

Glycine | Alanine | Valine

Hydrophilic ("water loving") amino acids

Aspartic acid | Asparagine | Glutamine

The R groups for the amino acids classified as hydrophobic contain only carbon and hydrogen atoms. These R groups are nonpolar. They are not attracted to water molecules. The R groups for the amino acids classified as hydrophilic contain atoms such as oxygen and nitrogen. These R groups are polar and are attracted to water molecules. As you will see, the nature of the R group is important when amino acids are linked to form proteins.

❷ Proteins

As shown in the next illustration, the carboxyl group of one amino acid can link to the amine group on another amino acid. The bond between two amino acids is called an **amide bond,** or a **peptide bond.** In the process, two hydrogen atoms and an oxygen atom break off from the amino acids and form a water molecule. In this way, it is possible to form long chains of amino acids called **proteins.** Because they are very large molecules, proteins are often called *macromolecules.*

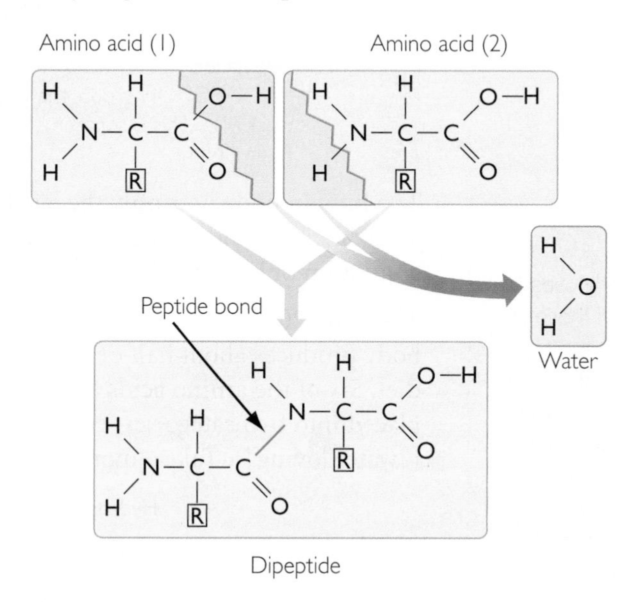

INDUSTRY CONNECTION

Polymers are long chain molecules made of smaller molecules. Proteins are polymers and so are plastics. For example, polyvinylchloride (PVC) is a polymer made of vinyl chloride molecules that covalently bond in a long chain.

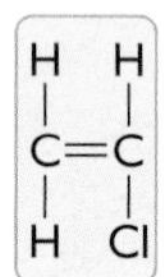

Vinyl chloride

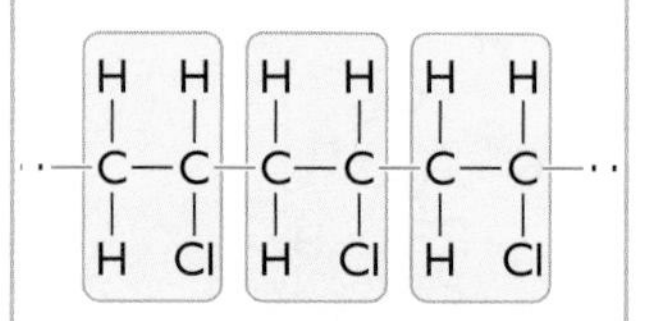

Polyvinyl chloride

PVC is used to make pipes for plumbing and many other products used in housing construction.

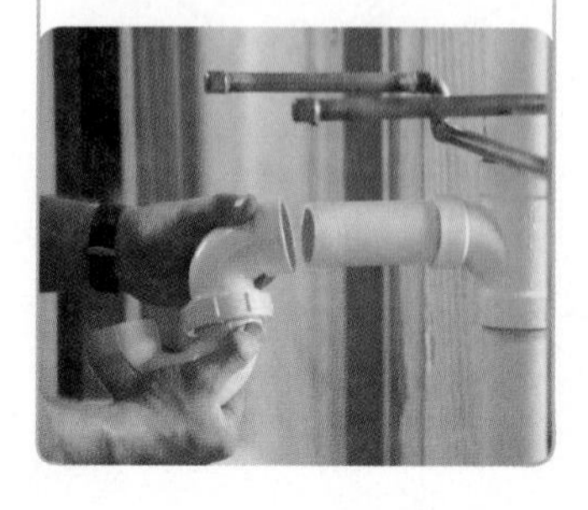

Proteins are essential in all living organisms. Protein molecules send messages throughout your body, they help the immune system fight off infections, and they are building blocks of many parts of your body, including organs, hair, and muscles. Proteins have a wide variety of structures and functions because many different sequences of amino acids are possible.

As the chain of amino acids forms, it twists and bends. In the photograph, each bead represents an amino acid. The entire chain of beads represents a protein molecule. Notice that amino acids can repeat.

The R groups on each amino acid play a major role in determining the resulting structure of the protein. The hydrophobic R groups tend to cluster in the center of the molecule. The hydrophilic R groups tend to be on the outside of the molecule, near the watery environment.

Smell receptor sites are protein molecules that have a pocket that fits a particular smell molecule. When the smell molecule attaches to the pocket, it changes the structure of the protein just a bit. This small change is enough to affect a neighboring protein, which triggers a signal to the brain.

Key Terms

amino acid
peptide bond (amide bond)
protein

Lesson Summary

What is a receptor site made of?

There are 20 different amino acid molecules that are essential building blocks of the human body. Amino acids have both a carboxyl functional group and an amine functional group. Amino acids all have mirror-image isomers, but only left-handed amino acid molecules are used by the human body. The carboxyl group on one amino acid can link with the amine group on another amino acid to form a peptide bond. In this way, a long chain of amino acids called a protein forms. A large variety of protein molecules exist because the amino acids can be linked in a variety of sequences. Some amino acids in a protein chain are polar and others are nonpolar. This affects the folding of a chain of amino acids in the protein. Smell receptor sites are composed of proteins.

EXERCISES

Reading Questions

1. What is an amino acid?
2. What is a protein?

Reason and Apply

3. Draw a diagram to show how glycine and alanine combine to form a peptide bond.
4. Imagine a protein molecule composed of 200 glycine molecules and 200 aspartic acid molecules. Describe how the protein is folded. What is on the inside and what is on the outside?
5. **WEB RESEARCH** Describe three different sources of amino acids in a person's diet.
6. Describe how forming a peptide bond is similar to forming an ester from an acid and an alcohol.

BIOLOGY CONNECTION

The average human can distinguish among approximately 10,000 different odors. The number of receptor sites in the nose is estimated to be around 1000 by some researchers. This means that some smells must be stimulated by a combination of different receptors, much like a musical chord on a piano has a distinct sound.

SECTION

IV

Key Terms

mirror-image isomer
amino acid
peptide bond (amide bond)
protein

SUMMARY

Molecules in the Body

Smells Update

Receptor sites exist throughout living systems. They serve as the mechanism for triggering many biological responses and processes, including smell. Receptor sites can be very specific, allowing only certain molecules to dock. So, the molecular structure of the receptor site is important. Some molecules have mirror-image isomers that have different properties, including smell. Amino acid molecules exist as mirror-image isomers. Amino acids combine to form a vast array of proteins, including those that make up receptor sites.

Review Exercises

1. How do you determine if two molecular models represent mirror-image isomers?
2. Draw the structural formula for these molecules. Which have mirror-image isomers? Explain your reasoning.

   a. CBrClFH **b.** CH_4 **c.** CH_2Cl_2
3. Explain how you could use your feet to show that the "handedness" of a molecule can determine if it fits into a receptor site.
4. What are amino acids and what purpose do they serve in your body?
5. Using the receptor site model, explain how a person can smell L-carvone but not its mirror-image isomer D-carvone.

PROJECT *Modeling a Receptor Site*

Build a space-filling model of a smell molecule from a compound in one of the five smell classifications you have studied. Use clay, plaster, sand, or other materials to build a receptor site for this smell molecule. Describe other molecules that might fit into the receptor site you built. Then create a space-filling model of a molecule that will not fit into the receptor site you built.

Your project should include

- the receptor site model that you built, labeled with the smell that it detects and a description of molecules that fit into it.
- a space-filling model of a molecule that fits into the receptor site and one that does not (label which is which).

UNIT

2 Smells Review

Smell chemistry is an active area of research. Scientists are still refining theories about how smell works. However, one thing is clear: Smell is all about molecules. In order to understand smell, it is important to understand what molecules are and how they are put together.

The smell of a compound is greatly dependent on its composition and structure. This is true of other molecular properties besides smell. Some molecules do not smell because they are too large or heavy to become airborne, or because they are nonpolar and do not stick in receptor sites. Other molecules have a type of smell that depends on what atoms or functional groups are present. The smell of a molecule may also depend on the shape of the entire molecule.

By exploring the topic of smell, you have studied covalent bonding and the general rules governing how atoms come together to form the amazing structures called molecules.

Review Exercises

1. What are isomers?
2. Draw two isomers with the molecular formula C_2H_4O.
3. A molecule can be described using a molecular formula, a structural formula, a ball-and-stick model, or a space-filling model. What information does each of these provide?
4. Consider the structural formula for methyl salicylate. Name two functional groups that are in this molecule.

```
       H     H
        \   /
         C=C      O       H
        /   \     ||      |
    H--C     C--C--O--C--H
        \\   //           |
         C--C             H
        /    \
       H      O
              |
              H
```

HISTORY
CONNECTION

In 2004, Richard Axel and Linda Buck shared a Nobel prize for their advances in understanding the mechanism of smell. Axel and Buck found that mammals have approximately 1000 different genes for olfactory receptors, which led to a better understanding of of smell.

5. Here is a ball-and-stick model of methyl salicylate.

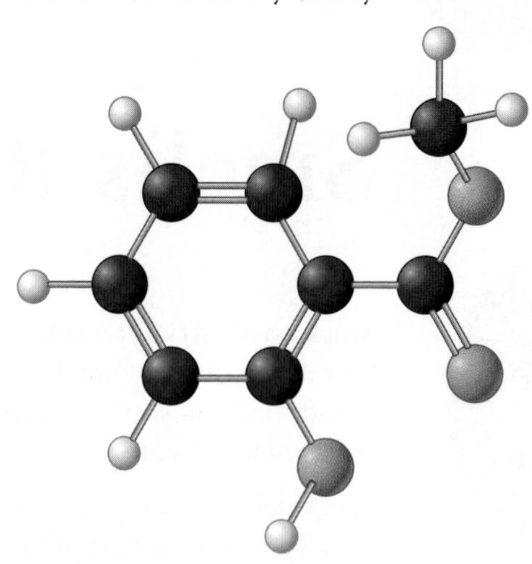

a. Does it follow the HONC 1234 rule?

b. Write the molecular formula for methyl salicylate.

c. Describe the overall shape of this molecule.

d. Find the oxygen atom that is bonded to two carbon atoms. Explain why this bond is bent.

6. A polar molecule has the molecular formula C_3H_8O.

a. Draw the structural formula.

b. Draw the Lewis dot structure and label the shared electrons and the lone pairs.

c. Identify the dipoles.

d. Name the geometric shape around each carbon atom.

e. Discuss what the name of this compound might be. Explain your reasoning.

7. What is a dipole? Explain why methane, CH_4, has four dipoles but has no overall dipole.

8. Table salt, NaCl, has no smell because

A. It is not made up of molecules.

B. It is an ionic compound.

C. No part of it enters the gas phase at room temperature.

D. All of the above.

E. None of the above.

UNIT

3 Weather

Water changes phase and is transported around the planet. It may then show up as snow or rain.

Why Weather?

Thunderstorms dump great quantities of rain, fog seeps into a bay at nightfall, warm temperatures entice us to the beach, and hurricanes devastate coastal communities. For better or worse, the weather is a part of our everyday lives. Physical change is at the core of weather. Weather occurs when matter undergoes changes in location, density, phase, temperature, volume, and pressure. When you understand the relationships between these changes in matter, you will begin to answer the question "What is the chemistry of weather?"

You will also learn

- about proportional relationships
- about temperature scales and how thermometers work
- the effects of changing temperature, pressure, and volume on matter
- about the behavior of gases
- how to read weather maps and make weather predictions

SECTION

I Physically Changing Matter

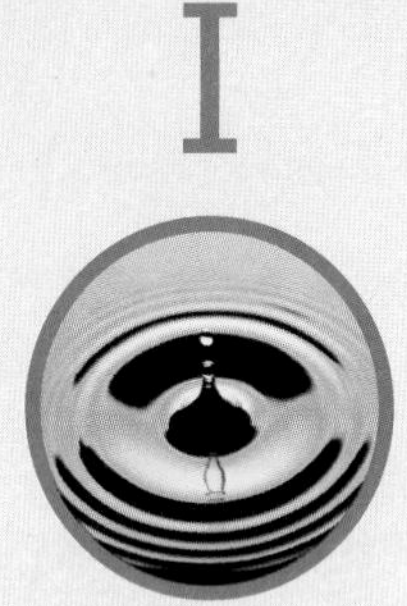

A view of Earth's atmosphere from space shows weather in action.

Weather is the result of interactions among Earth, the atmosphere, water, and the Sun. It is about the movement of matter around the planet. Water, for example, changes phase from liquid to gas and then back to liquid as it moves from oceans to the atmosphere, to clouds, and down to Earth as rain. The atmosphere is composed of different gases, so the study of the behavior of gases is important in understanding weather. The gas laws are mathematical equations that describe the relationship between changing temperature, volume, and pressure of gases. Understanding the behavior of matter in relation to temperature allows weather experts to make predictions about the weather.

In this section, you will study

- how to read basic weather maps
- proportional relationships
- density and phase changes in matter
- temperature scales and thermometers
- the relationship between volume and temperature of gases

LESSON

1 Weather or Not

Weather Science

Think About It

You have probably seen a weather forecast, either on television or in the newspapers. You may have wondered—what are highs and lows? What are fronts? What does the jet stream have to do with the weather?

What causes the weather?

To begin to answer this question, you will explore

1. Earth, Air, Water, and Sun
2. Weather Maps

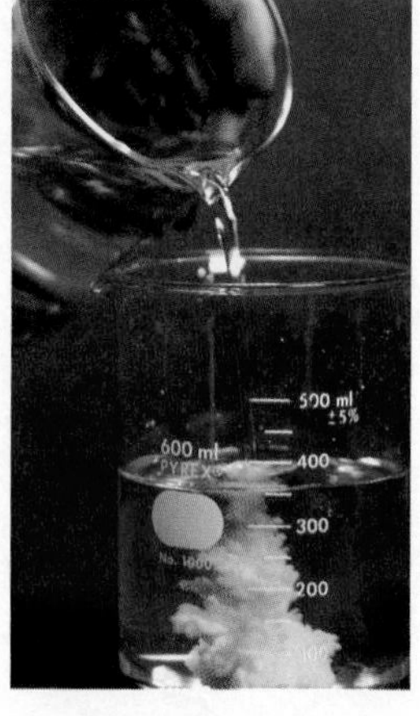

Heating and reshaping a substance are both physical changes. Mixing chemicals to form a new compound is a chemical change.

Exploring the Topic

1 Earth, Air, Water, and Sun

This lesson launches an investigation into the connection between chemistry and weather. **Weather** refers to clouds, winds, temperature, and precipitation in a region at any given time. It differs from *climate,* which describes the overall weather trends of a region over long periods of time. Weather is caused by the interaction of Earth, the atmosphere, water, and the Sun. It is about the movement of matter.

From outer space you can see clouds above the surface of the planet. A satellite picture of Earth also makes it quite clear that our planet's surface is mostly water.

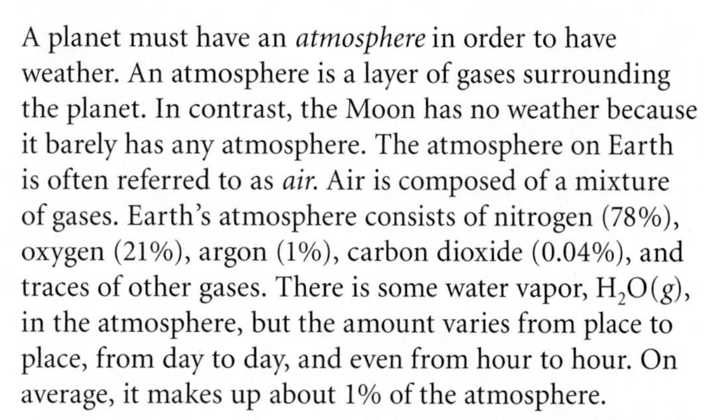

A planet must have an *atmosphere* in order to have weather. An atmosphere is a layer of gases surrounding the planet. In contrast, the Moon has no weather because it barely has any atmosphere. The atmosphere on Earth is often referred to as *air.* Air is composed of a mixture of gases. Earth's atmosphere consists of nitrogen (78%), oxygen (21%), argon (1%), carbon dioxide (0.04%), and traces of other gases. There is some water vapor, $H_2O(g)$, in the atmosphere, but the amount varies from place to place, from day to day, and even from hour to hour. On average, it makes up about 1% of the atmosphere.

The weather we experience is caused by physical changes in the atmosphere. **Physical changes** alter the form a substance takes, but do not change the identity of the substance. Examples of physical changes are changes in volume, temperature, shape, size, and pressure. These are different from chemical changes, which produce new substances. The heating of the Earth by the Sun is the root cause of most of the physical changes in Earth's atmosphere.

The **phase changes** of water between solid, liquid, and gas are types of physical changes that play an important role in weather. For example, rain is evidence that a phase change from gas to liquid has taken place.

BIG IDEA Weather is the result of physical changes to matter.

❷ Weather Maps

Meteorologists study the weather. They use maps to keep track of atmospheric conditions and to predict the weather. Weather information for a specific day is shown on the maps here. What patterns do you observe as you compare the six maps?

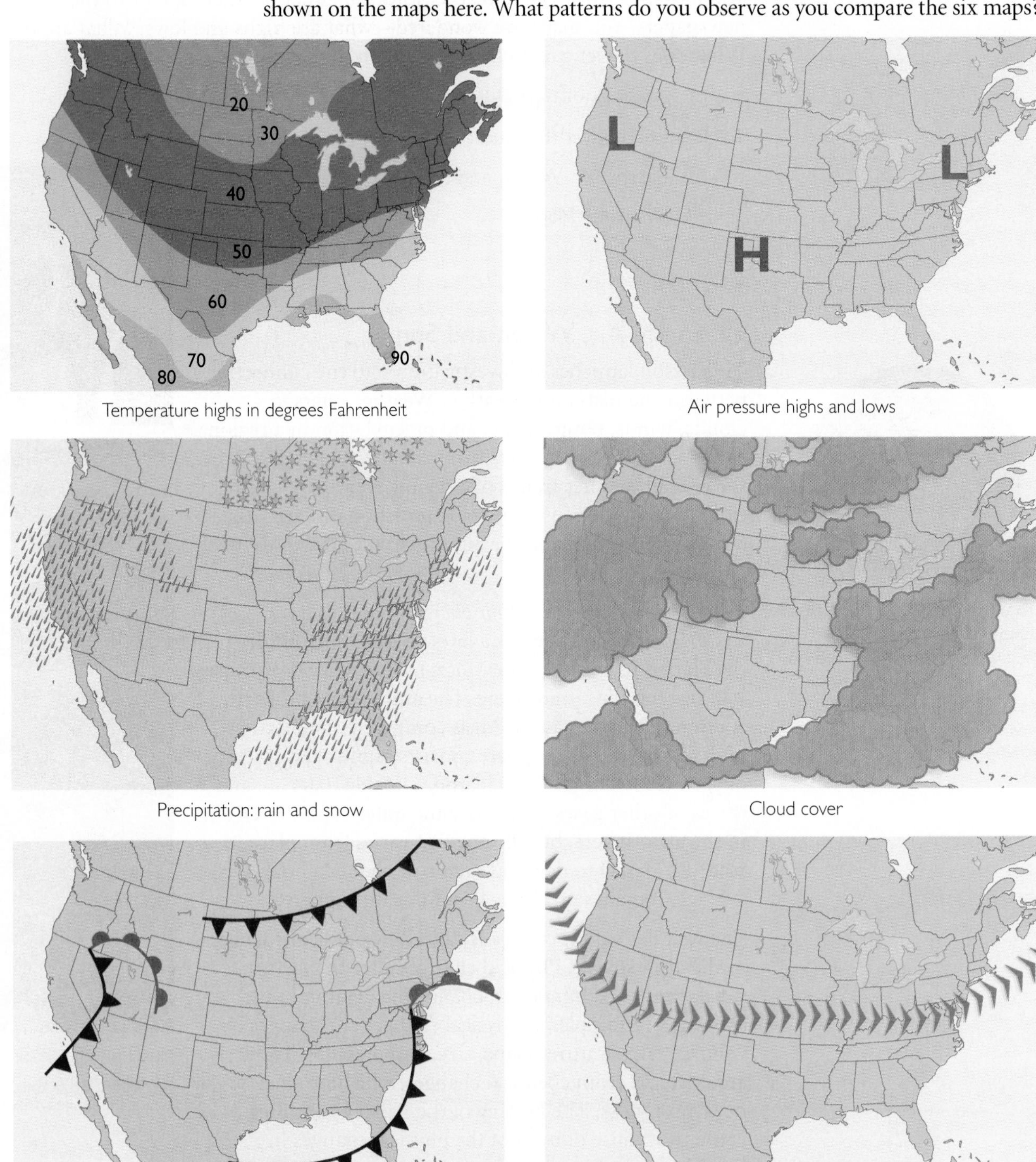

Temperature highs in degrees Fahrenheit

Air pressure highs and lows

Precipitation: rain and snow

Cloud cover

Fronts

Jet stream

GEOLOGY CONNECTION

Earth's atmosphere is divided into several distinct layers. Almost all of the weather on the planet takes place in the layer of atmosphere closest to Earth. This layer is called the *troposphere* and is about 14 kilometers high (that's about 9 miles).

In addition to familiar features such as temperature and precipitation, weather maps often show a curve called the jet stream. The jet stream shows the location of high-altitude winds. These winds are at least 57 miles per hour (up to 190 mi/h) and are in the upper atmosphere above 20,000 feet in altitude (that's four miles up). These winds have a great effect on what happens in the air below them and generally "steer" storms.

Here are some features to take note of:

- In North America, temperatures are colder in the north and warmer in the south.
- The shape of the jet stream coincides with the contour lines on the temperature map.
- If it is raining or snowing, then there must be clouds present, but clouds do not necessarily mean it will rain or snow.
- Fronts are located in areas of low pressure.
- Clouds are more likely to form near fronts, so it is more likely to rain and snow there.
- Warm fronts and cold fronts are associated with low pressure systems.
- There is no precipitation for regions with high pressure. High pressure is associated with sunny days. A region with high pressure can be either hot or cold.

SCILINKS® NSTA

Topic: Weather

Visit: www.SciLinks.org

Web code: KEY-301

Example

Weather Maps

Describe the weather in these areas using the six weather maps in this lesson. Include the temperature, air pressure, cloud cover, precipitation, and storm fronts.

a. West Coast   b. middle of the United States

Solution

a. West Coast: The temperature is about 60 °F, slightly colder to the north. The air pressure is low. There are storm fronts and it is cloudy, with rain to the south and snow to the north.

b. Middle of the United States: It is about 40 °F and sunny. The air pressure is high, there are no clouds, and there is no precipitation.

Key Terms

weather
physical change
phase change

Lesson Summary

What causes the weather?

Weather is about the movement of air and water. The gases in our atmosphere and the water on Earth's surface undergo physical changes and move around the planet. Water exists in all three phases (solid, liquid, and gas) on Earth. Water shows up as snow, ice, rain, clouds, fog, and water vapor. Meteorologists use detailed maps to monitor the physical changes taking place in our atmosphere and to keep track of the movement of moisture and air around the planet. Information on weather maps includes temperature, air pressure, cloud cover, precipitation, warm and cold fronts, and the location of the jet stream.

EXERCISES

Reading Questions

1. What substances are necessary in order for a planet to have weather?
2. Describe three variables that meteorologists study in order to make weather predictions.
3. What is a physical change?

Reason and Apply

4. Name three types of maps that meteorologists use. Describe the maps and the information that they contain.
5. Consult a newspaper, the Internet, or a weather report on television.
   a. Name three states that are currently experiencing a cold front (or warm front) moving through.
   b. Which states are experiencing high-pressure systems? What is the weather forecast in those states?
   c. If you were traveling to New York tomorrow, would you pack a raincoat? Explain.
6. Nitrogen, N_2, oxygen, O_2, carbon dioxide, CO_2, and water vapor, $H_2O(g)$, are all gases found in the atmosphere.
   a. Draw the structural formula for each and include lone pair electrons.
   b. Draw the Lewis dot structure for water.
   c. Which molecules are polar and which molecules are nonpolar?
   d. Which one of the four substances is naturally found as a liquid, a solid, and a gas on Earth?

ENVIRONMENTAL CONNECTION

On other planets, the compounds that are involved in the weather may be quite different from those found on Earth. The atmosphere of Mars is mostly carbon dioxide. The polar ice caps on Mars are made of frozen carbon dioxide. Jupiter has clouds made of ammonia. In fact, Jupiter may be one big atmosphere with no real surface.

On Venus the clouds are made of sulfuric acid, not water, and the temperature is a blistering 890 °F, or 475 °C.

LESSON

2 Raindrops Keep Falling

Measuring Liquids

Think About It

After a rainstorm, rain in the gutter or in a puddle can be several inches deep, but on the sidewalk and grass it is not that deep. The water runs off the sidewalks, runs down hills, collects in depressions, and soaks into grass. So, how do you determine how much rain fell?

How do meteorologists keep track of the amount of rainfall?

To answer this question, you will explore

1. Rain Gauges
2. Volume Versus Height
3. Proportional Relationships

Exploring the Topic

1 Rain Gauges

Suppose you have three rain gauges that are different sizes. You place them next to one another out in the rain. After a few hours these containers might look like this:

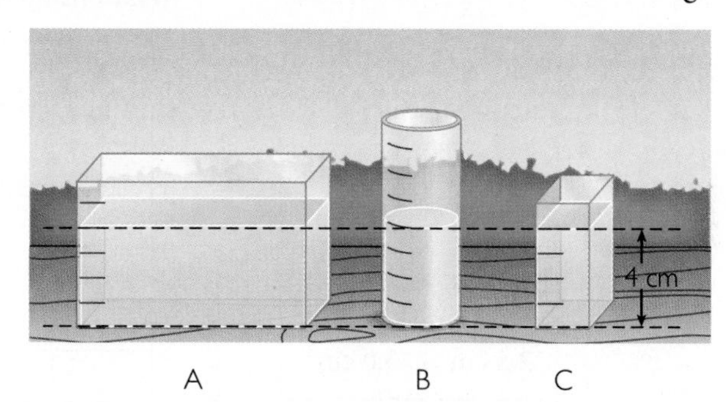

Sample data for these three containers are given in the table below.

Container	Area of the base (cm^2)	Height of rain (cm)	Volume of rain (cm^3)
A	10.0 cm^2	4.0 cm	40.0 cm^3
B	2.5 cm^2	4.0 cm	10.0 cm^3
C	2.0 cm^2	4.0 cm	8.0 cm^3

You can see that the volume of rain in each container is quite different. However, the *height* of rain is the same. This is why meteorologists report height of rainfall, in centimeters or inches, not volume of rainfall. The height of the rainfall is the same for these containers, but the volume is not.

❷ Volume Versus Height

Imagine that you place a rain gauge with a square base of 2.0 cm^2 in a steady rain for several hours. Each hour, you go outside to measure how much rain has collected in the rain gauge.

METEOROLOGY CONNECTION

Amount of rainfall is not proportional to time. This is because rainfall is not necessarily steady. It can rain a lot and then a little.

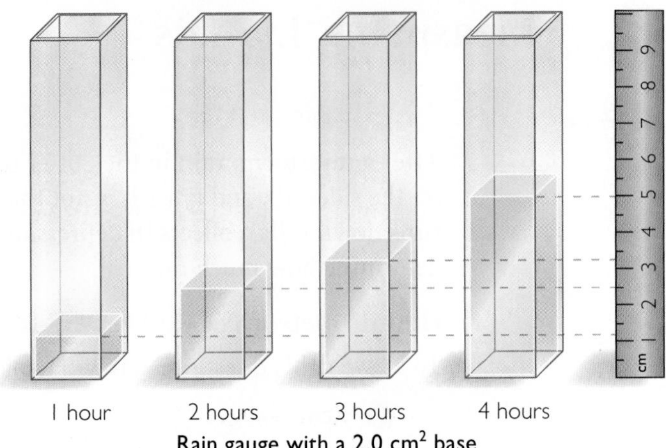

Rain gauge with a 2.0 cm^2 base

The graph shows how volume changes as height changes, or volume versus height. *Versus* means compared with and is abbreviated vs. Both the volume and the height of the rain in the gauge increase as more rain is collected. Take a moment to examine the data.

The graph shows that volume increases in a steady and predictable way in relationship to height. And once some data points are known, others can be predicted.

Height (cm)	Volume (cm^3)
1.2 cm	2.4 cm^3
2.5 cm	5.0 cm^3
3.3 cm	6.6 cm^3
5.0 cm	10.0 cm^3

Relationship of Volume and Height

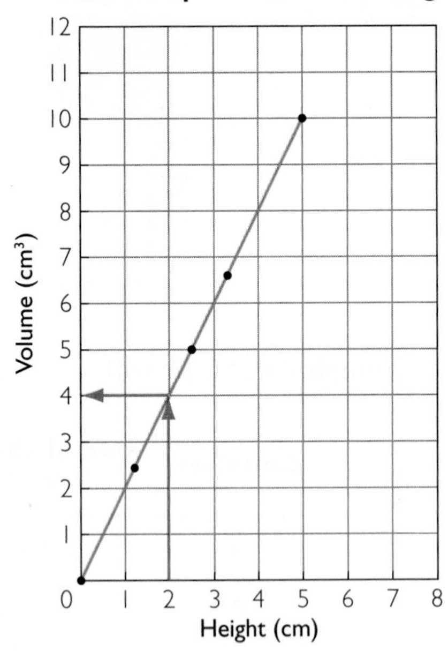

Using the Graph

You can use this graph to predict the volume of water corresponding to various heights of water in the rain gauge. Each point on the line corresponds to a different volume and height.

Suppose you want to predict the volume of water when the height is 2.0 cm. You can either use the graph to find the answer or use math to calculate it. [For a review of this math topic, see **Math *Spotlight: Graphing*** on page 625.]

Method 1: First, find the height, 2.0 cm, on the *x*-axis. Move vertically from this spot until you reach the slanted line. Now move horizontally until you reach the *y*-axis. This point on the line corresponds to a volume of 4.0 cm^3.

Method 2: Notice that for every data point in the table the volume is always 2 times the height. The volume for a height of 2.0 cm is 4.0 cm^3. Similarly, the volume for a height of 6 cm would be $6 \cdot 2 = 12$ cm^3.

Important to Know In the laboratory, volume is measured in milliliters, mL, in a graduated cylinder. Conveniently, 1 mL = 1 cm^3. ◄

AGRICULTURE CONNECTION

Rain gauges often have a wide funnel to collect water into a narrow tube. Because the collection area is large, the height corresponding to 1 inch of rain in the rain gauge is much larger than an actual inch. These rain gauges are designed so that a farmer can see these "large inches" of water from inside his or her home. There is no need to walk out to the field in the rain.

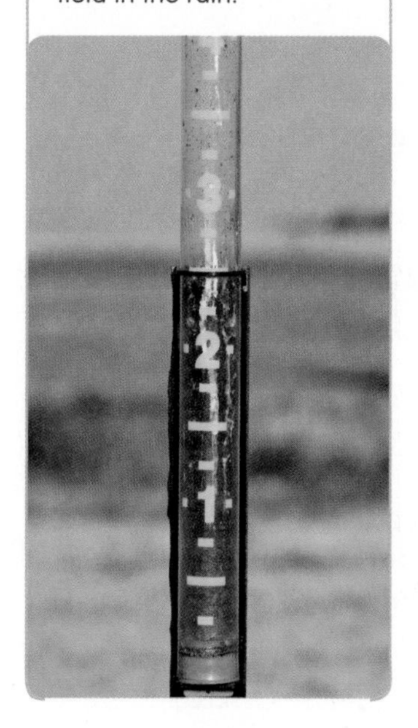

3 Proportional Relationships

The relationship between the volume of a container and its height is a **proportional** one. This means that volume and height are related to one another by a single number called the **proportionality constant.** When the height is multiplied by this number, you get the volume. In the example we just used, the height and volume of rain in the rain gauge were related by the number 2.0. In other words, for this container, the volume is always 2.0 times the height, and 2.0 is the proportionality constant.

$$\text{Volume of rain} = (\text{proportionality constant}) \cdot (\text{height of rain})$$

A proportionality constant can be represented by the letter k in a math formula.

$$\text{Volume} = k \cdot \text{height}$$

In the case of the rain gauge, the proportionality constant is also equal to the area of the base.

$$\text{Volume} = 2.0\ \text{cm}^2 \cdot \text{height}$$

Proportionality constant, k

The graph provides another way to look at proportional relationships. Whenever two variables are proportional to each other, the graph of the data points is *always* a straight line passing through the origin (0, 0). [For review of this math topic, see **MATH** ***Spotlight: Ratios and Proportions*** on page 627.]

Volume and height are proportional for containers with a uniform shape and parallel walls.

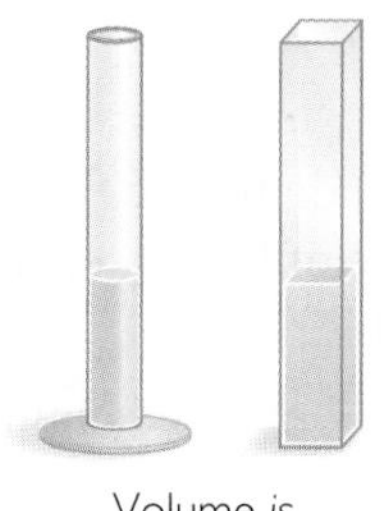

Volume *is* proportional to height.

Volume *is not* proportional to height.

Key Terms

proportional
proportionality constant

Lesson Summary

How do meteorologists keep track of the amount of rainfall?

Everywhere around the world, height of rainfall is measured. If volume were used to report rainfall, each meteorologist would report a different number, depending on the area of the base of the rain gauge used. Height and volume for a specific rain gauge are related by a proportionality constant, k.

EXERCISES

Reading Questions

1. Suppose you have a cylindrical rain gauge with a base area of 2.0 cm^2. Explain two different ways you can determine the volume when the height of water is 3.0 cm.

2. Explain in your own words why meteorologists prefer to measure rain in inches or centimeters, not in milliliters or cubic centimeters.

Reason and Apply

3. If the amount of rainfall increases, do both the volume and height of water in the rain gauge keep track of this increase? Explain your thinking.

4. If you use a beaker for a rain gauge, and the weather station uses a graduated cylinder, will both instruments give the same volume? The same height?

5. If a large washtub, a dog's water dish, and a graduated cylinder were left outside during a rainstorm, would the three containers all have the same volume of water in them after the storm? Explain why or why not.

6. Inches are used in the United States, but centimeters are used by scientists and in the rest of the world. Look at a ruler that is marked in both inches and centimeters.

a. Make a graph of centimeters to inches so that the y-axis is centimeters and the x-axis is inches.

b. Convert 12 in. to centimeters.

c. How many inches is 10 cm?

d. How many inches is 1 cm?

7. A student placed the same empty rain gauge outside before five different rainstorms. She measured the height and volume of the water in the gauge after each storm. Her results are in the table.

a. Which rainstorm dropped the most rain? Did you use height or volume to answer this question? Why?

b. What pattern do you notice?

c. Draw a graph of the data to show that the volume and height are proportional.

d. Explain why the data points do not all lie exactly on a straight line.

e. Predict the volume for a height of 6.0 cm. Show your work.

Storm number	Height (cm)	Volume (cm^3)
1	1.0 cm	2.5 cm^3
2	2.5 cm	6.3 cm^3
3	0.5 cm	1.1 cm^3
4	8.0 cm	20.0 cm^3
5	5.0 cm	12.5 cm^3

LESSON

3 Having a Meltdown

Density of Liquids and Solids

Think About It

Much of the country depends on snowfall for the year's water supply. The mountains get many feet of snow every winter, and this snow eventually melts and travels from creek to stream to river to reservoir. Scientists measure the snowpack, or the total amount of snow that has accumulated on the ground, to predict the amount of water that will be available for consumption the rest of the year. But is a foot of snow in the mountains the same as a foot of rainfall?

How much water is present in equal volumes of snow and rain?

To answer this question, you will explore

1. Density and Phase
2. Converting Snowfall to Rainfall
3. The Density of Ice

Exploring the Topic

1 Density and Phase

Just as rainfall is measured in inches, snowfall can also be measured in inches. All you need is a ruler. However, scientists who study water distribution are more interested in how much liquid water the snow represents.

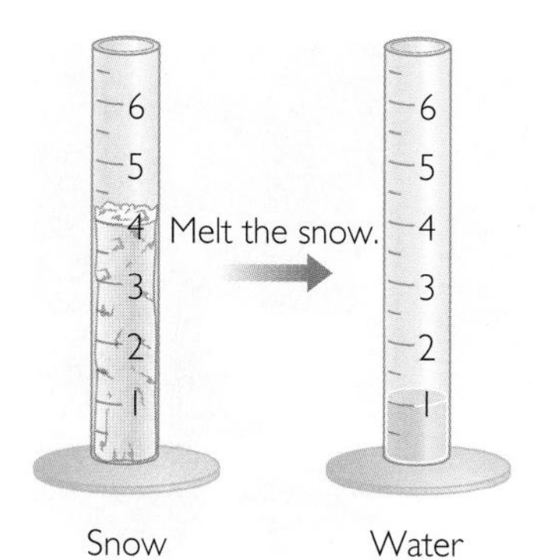

Mass is not lost, but the volume changes.

When snow melts, its volume decreases. However, its mass is still the same. This is because the same water molecules are present in the frozen snow and in the melted snow.

If the volume of the sample changes but its mass stays the same, then mathematically, the density of the sample has to change. The density of snow must be lower than the density of water.

Recall that density is the mass per unit volume.

$$D = \frac{m}{V}$$

Unlike water, snow has a wide range of densities. Snow can be fluffy or packed. So, the volume of liquid water contained in any snow sample depends on the density of the snow being considered.

Mass and Volume Are Proportional

The equation for density relates mass and volume.

As long as the density of the substance doesn't change, its mass and volume are proportional. The proportionality constant here is the density, D.

$$m = DV$$

Because mass and volume are proportional to each other for a specific substance, the data points lie on a straight line that goes through the origin. Examine the graphs of mass versus volume for water, snow, and ice. How does the slope of the line relate to the density?

Form	Density
rain	1.0 g/mL
snow	0.1–0.5 g/mL
ice	0.92 g/mL

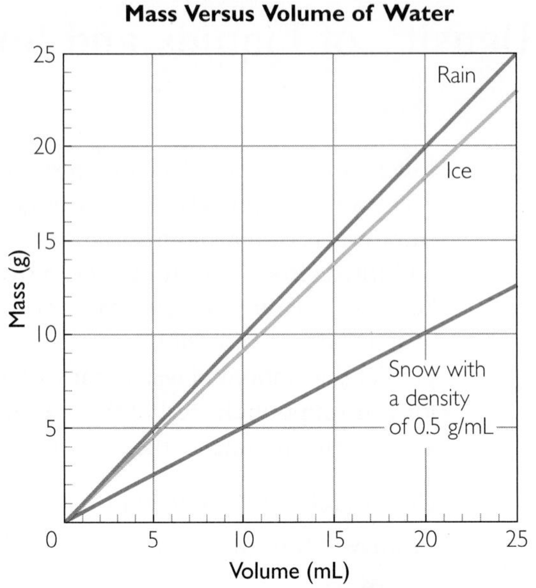

Using the Graph

The line for rain is steepest because liquid water has the greatest density. For every 1.0 mL increase in rain volume, the mass of the water increases by 1.0 g. The line for snow is the least steep because the density of snow is less than that of both rain and ice. For snow, a 1 mL increase in volume corresponds to only a 0.5 g increase in mass.

BIG IDEA Matter changes density when it changes phase.

ENVIRONMENTAL CONNECTION

The snowpack density can vary from about 0.1 g/mL to 0.5 g/mL depending on conditions. Dry, fluffy snow is less dense than wet, slushy snow, and therefore easier to shovel.

❷ Converting Snowfall to Rainfall

Calculating the Density of Snow

One way to determine the volume of water in the snowpack is to melt the snow. But melting a large amount of snow is inconvenient.

Another way to study the water volume of the snowpack is to determine the density of the snow and calculate the volume of liquid water it represents. When new snow has fallen in the mountains, hydrologists take samples of the snow with a large aluminum tube. They push the empty tube all the way down into the snow. Then they weigh the tube with the snow still inside. The density of the snow is the mass divided by the volume.

Scientists called hydrologists measure snowpack by weighing an aluminum tube filled with snow.

Example 1

Volume of Water From Snow Melt

Suppose you have a volume of 20 mL of snow with a density of 0.25 g/mL. What volume of liquid water do you get when you melt this amount of snow?

SPORTS CONNECTION

Skiers and snowboarders enjoy fresh "powder." This is the term for untouched, freshly fallen snow with a particularly low density. Powder is perfect for landing on because it forms a natural cushion. This means that it does not hurt as much when you land on it compared with snow that is compacted.

Solution

When the snow melts, its mass will not change. Use the density equation to convert between mass and volume.

Use the volume and density of snow to find its mass.
$$m_{snow} = D_{snow} V_{snow} = (0.25 \text{ g/mL}) \cdot (20 \text{ mL}) = 5 \text{ g}$$

The mass of snow is equal to the mass of water.
$$m_{snow} = m_{water} = 5 \text{ g}$$

Rearrange the density equation to find the volume of water.
$$V_{water} = \frac{m_{water}}{D_{water}} = \frac{5 \text{ g}}{1.0 \text{ g/mL}} = 5 \text{ mL}$$

When 20 mL of snow with a density of 0.25 g/mL melts, you get 5 mL of water.

You can also solve this problem using a graph. The graph here shows mass versus volume for liquid water, and snow with a density of 0.25 g/mL.

Converting Snow to Liquid Water

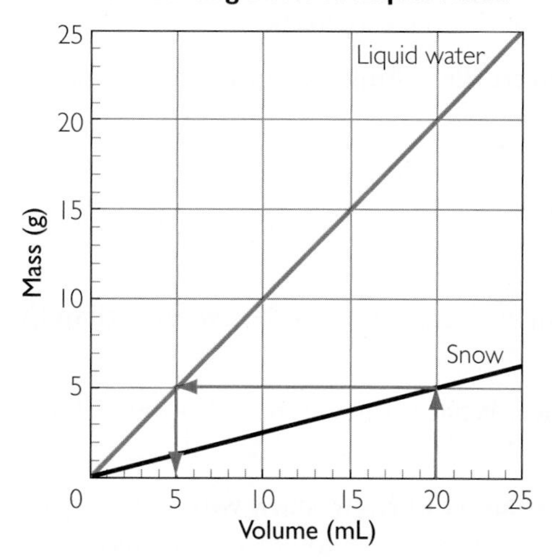

Using the Graph

The arrows on the graph show how to convert from snow volume to water volume. You begin with 20 mL of snow, which is equivalent to 5 g of snow. When snow melts, the mass remains the same. 5 g of water is equivalent to 5 mL of snow.

❸ The Density of Ice

The density of ice is 0.92 g/mL. It is less than the density of water, which is 1.0 g/mL. Because water becomes less dense when it freezes, its volume increases. This unique property of water is due to a special kind of intermolecular force called *hydrogen bonding.* In both liquid and solid water, the hydrogen atoms in one water molecule attract the highly electronegative oxygen atoms in another. When water freezes, the molecular motion slows down and the molecules lock into a hexagonal structure due to hydrogen bonding.

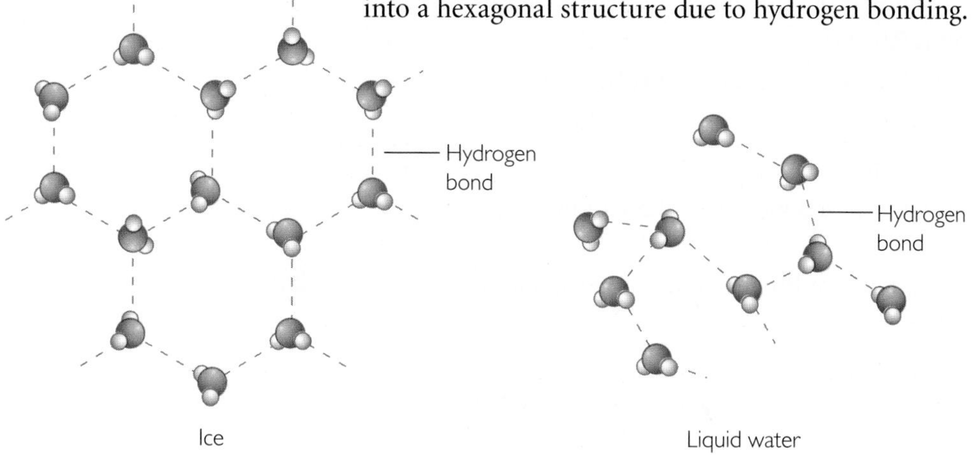

Ice

Liquid water

Topic: Density
Visit: www.SciLinks.org
Web code: KEY-105

Lesson Summary

How much water is present in equal volumes of snow and rain?

The density of snow is less than the density of liquid water. You can use mathematical equations or graphs to convert the volume of snow to the volume of liquid water. To use graphs, you need to plot two lines of mass versus volume, one for snow and another for liquid water. The slope of each line is the density of that substance. To use mathematical equations, you need to know the density of the snow and the density of liquid water.

EXERCISES

Reading Questions

1. Explain how to use mathematical equations to convert the volume of snow in a snowpack to the volume of water in the snowpack.
2. Explain how to use graphs to convert the volume of snow in a snowpack to the volume of water in the snowpack.

Reason and Apply

3. How are snow and ice different?
4. When ice melts, will the liquid water occupy more or less volume than the ice? Explain your thinking.
5. Suppose that you melt 24 mL of ice. What is the volume of liquid water that results?
6. Which has more water for equal volumes of snow, snow with a density of 0.5 g/mL or snow with a density of 0.25 g/mL? Explain your thinking.
7. Suppose you have a box with a volume of 17.5 mL.
   a. If you fill this box with ice, what mass of ice do you have? (The density of ice is 0.92 g/mL.)
   b. If you fill this box with liquid water, what mass of liquid water do you have? (The density of liquid water is 1.0 g/mL.)
8. Suppose that you have a box that is full and contains 500 grams of a substance.
   a. What is the volume of the box if the substance inside is corn oil? (The density of corn oil is 0.92 g/mL.)
   b. What is the volume of the box if the substance inside is lead? (The density of lead is 11.35 g/mL.)
9. Lead, Pb, is more dense than iron, Fe.
   a. Which occupies a larger volume, 4.3 g of Pb or 4.3 g of Fe? Explain your thinking.
   b. Which has a larger mass, 2.6 mL of Pb or 2.6 mL of Fe? Explain your thinking.

ENVIRONMENTAL CONNECTION

Most substances contract when they freeze, becoming denser. Water is unique in that it expands when it freezes, becoming less dense. This is why ice floats in water. Otherwise, icebergs would sink!

LAB Thermometers

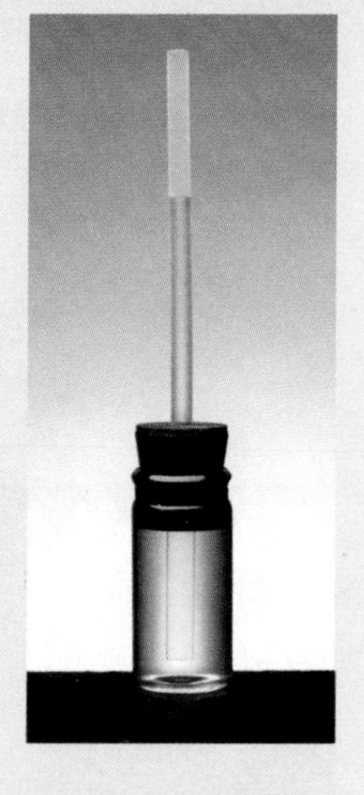

Purpose

To examine how the volume of a liquid and a gas change in response to temperature.

Part 1: Liquid Thermometer

Procedure

Build a thermometer like the one shown by following the instructions on the handout. Then mark the liquid level in the straw for these conditions.

- room temperature
- vial warmed by your hand
- ice water
- ice water with 1 tablespoon of salt per 200 mL water
- boiling water (do not allow the thermometer to touch the bottom of the beaker)

Observations and Analysis

1. What did you observe? What is happening to the liquid in the vial to make it move up and down in the straw?
2. Create a scale for the thermometer. Assign numbers for the places you marked on the straw for boiling water and ice water. What numbers did you choose and why?
3. Based on your newly created temperature scale, estimate the temperature in the room. How did you arrive at your answer?

Part 2: Gas Thermometer

Materials

- 250 mL beakers (3)
- ice
- 10 mL graduated cylinder
- food coloring
- test tube holder—wire
- hot plate

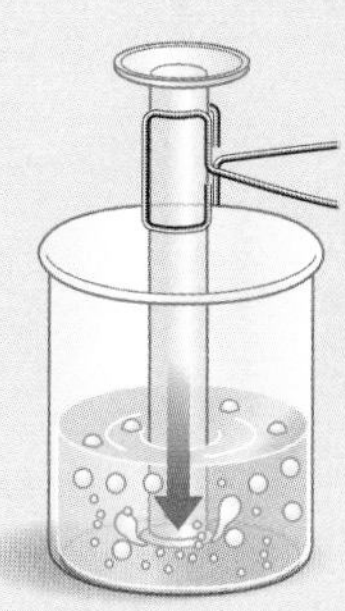

Procedure

1. Set up three beakers: one with ice water, one with room-temperature water, and one with hot water (cool it so it is no longer steaming). Add 1–2 drops of food coloring to each beaker.
2. Hold the 10 mL graduated cylinder with a test tube holder. Invert it and immerse it in each beaker for one minute.

Making Sense

Describe how a thermometer works.

LESSON

4 Hot Enough

Thermometers

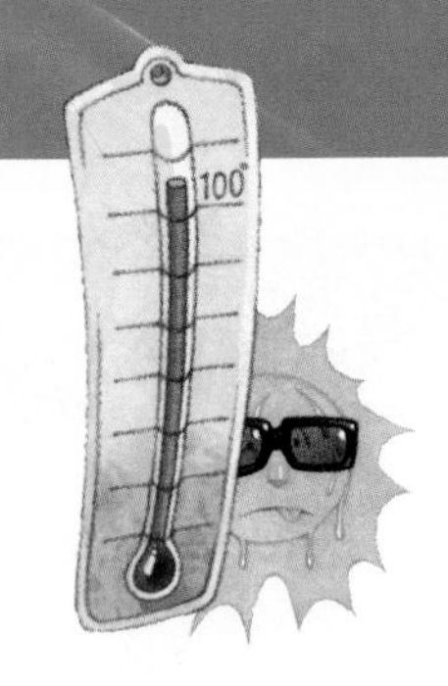

Think About It

On a single day a thermometer in Anchorage, Alaska might read −15 degrees, while a thermometer in Sydney, Australia reads 35 degrees, and a thermometer in your classroom reads 68 degrees. It is interesting to consider how a small glass tube with some red liquid inside can indicate how hot or cold it is.

How is temperature measured?

To answer this question, you will explore

1. Changes in Volume
2. Creating Temperature Scales
3. Fahrenheit Versus Celsius

Exploring the Topic

1 Changes in Volume

Matter responds to increasing temperature in a variety of ways. A wad of bubble gum gets softer, a "mood" ring changes color, and the air in a balloon expands. By systematically tracking any one of these changes, you could make a thermometer.

The liquid crystals in a mood ring change color in response to changes in temperature.

However, in order to make a *good* thermometer, it is better to track a physical change that is reproducible and easily measurable.

One way to make a thermometer is to fill a thin glass or plastic tube with a liquid. As the liquid in the thermometer is heated, its volume and height increase. As the liquid is cooled, the volume and height decrease. Because volume and height are proportional, you can detect the increase in volume as an increase in height.

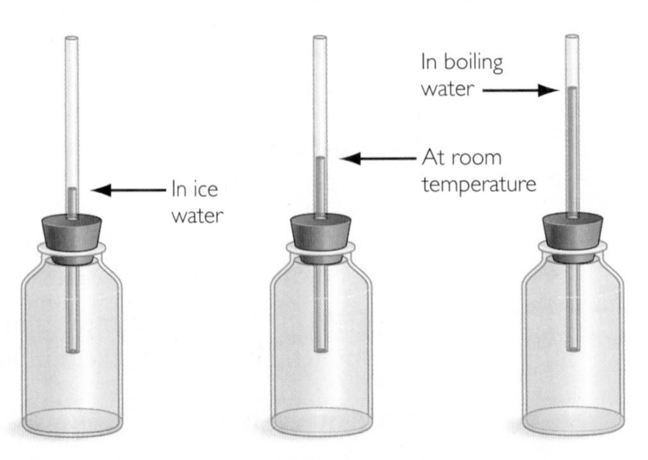

HISTORY CONNECTION

The first liquid thermometers were constructed with liquid mercury. Mercury is a good choice because it remains a liquid for a wide range of temperatures. Alcohol is another good choice. In contrast, water is not a good choice because it can freeze in the thermometer and expand, breaking the glass.

Gas Thermometers

Gases also expand and contract upon heating and cooling. A gas thermometer requires a container with a flexible volume. Suppose you have a sample of air in hot water trapped in an inverted graduated cylinder. If you heat the trapped air sample, some of the air will bubble out of the cylinder. This is evidence that the air is expanding. If you cool the trapped air sample in ice water, water will rise into the cylinder. This is evidence that the air sample contracts as it cools, leaving room for some water to move up into the cylinder. The expansion and contraction of air is important in understanding the weather.

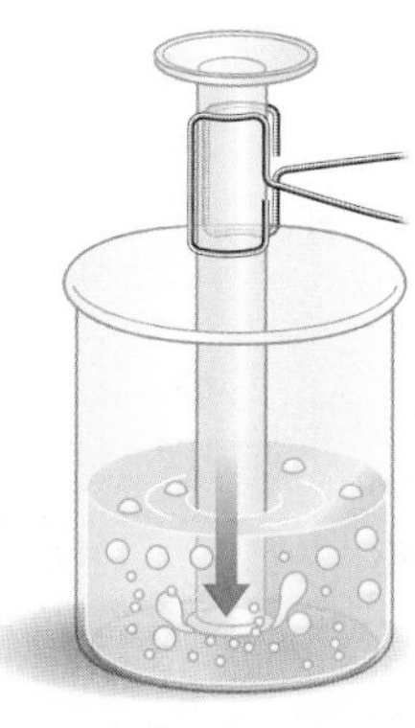
In hot water

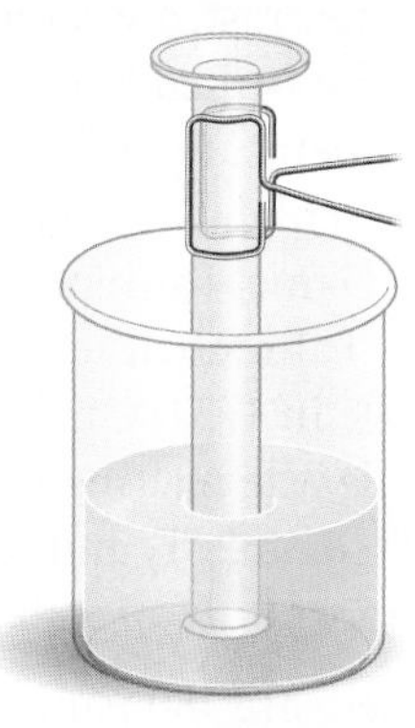
In room-temperature water

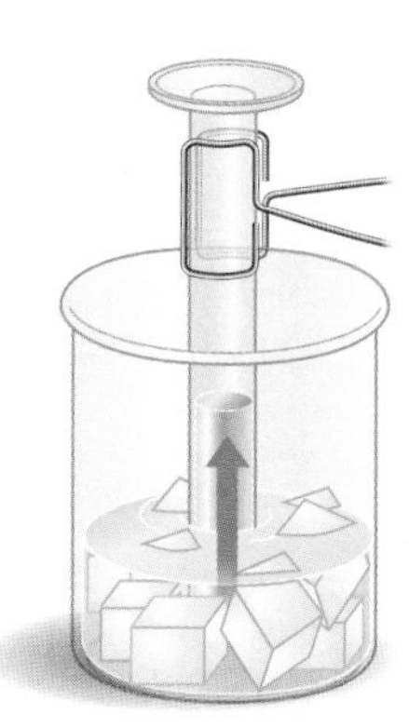
In ice water

BIG IDEA Most matter expands in volume as it is heated and contracts as it is cooled. In solids, the change is usually very small. In gases, the change is dramatic.

2 Creating Temperature Scales

To make a thermometer, you could put a colored liquid in a glass or plastic tube. Then you need to construct a scale, or number line, on the tube in order to assign numbers to different temperatures. You can construct a scale by measuring two temperatures, such as the boiling point and freezing point of water. It is typical to measure water because water is easily accessible and always boils and freezes at the same temperatures at sea level. The temperature at which a substance melts or freezes is called the **melting point,** or the **melting temperature.** The temperature at which it boils or condenses to a liquid is called the **boiling point,** or the **boiling temperature.** Each time you measure the height of the liquid at the melting point and boiling point of water, you will get the same two heights on your thermometer.

Topic: Temperature Scales
Visit: www.SciLinks.org
Web code: KEY-304

Once you have two marks, you can divide the length between the marks into a round number of intervals. Each interval is one unit of temperature. This provides numerical values for other temperatures besides the melting and boiling points of water. For instance, you can use your thermometer to measure the temperature of the room. Someone else who makes a thermometer using

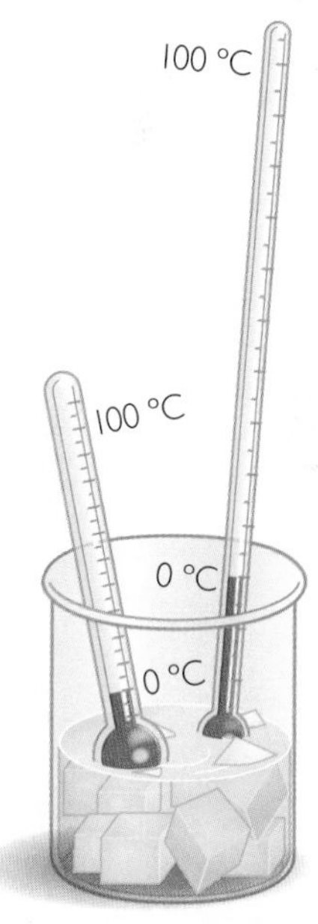

These two thermometers are different sizes but have the same temperature scale. Both show the same temperature.

COOKING CONNECTION

Water boils at a lower temperature when you are at higher altitudes. Because the water is at a lower temperature, it takes longer to cook noodles in the mountains.

the same points and intervals you used will be using the same scales and you can accurately compare temperatures.

3 Fahrenheit Versus Celsius

One commonly used temperature scale is the Celsius scale. This scale sets the melting temperature of ice at 0° and the boiling temperature of water at 100° (as measured near sea level as opposed to in the mountains). This scale uses degrees Celsius, °C, as units. The Swedish astronomer Anders Celsius created this temperature scale in 1747. Most of the world and most scientists use the Celsius temperature scale.

In contrast, a weather report in the United States gives temperature in degrees Fahrenheit, °F. German physicist G. Daniel Fahrenheit invented the Fahrenheit temperature scale in 1724. On the Fahrenheit scale, the melting temperature of ice is 32 °F, and the boiling temperature of water is 212 °F.

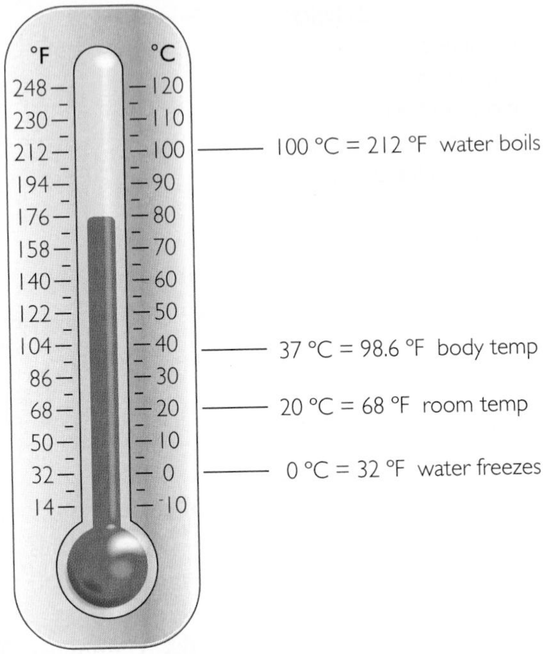

There are 180 Fahrenheit degrees between 0 °C and 100 °C. This means that a Fahrenheit degree unit is a smaller change in temperature than a Celsius degree unit.

°F: 248, 230, 212, 194, 176, 158, 140, 122, 104, 86, 68, 50, 32, 14
°C: 120, 110, 100, 90, 80, 70, 60, 50, 40, 30, 20, 10, 0, -10

100 °C = 212 °F water boils
37 °C = 98.6 °F body temp
20 °C = 68 °F room temp
0 °C = 32 °F water freezes

In this course, you will use the Celsius scale. It is useful to know how to convert between the two scales. The formula given here allows you to convert from degrees Celsius, C, to degrees Fahrenheit, F.

$$F = \frac{9}{5}C + 32$$

[For review of this math topic, see **MATH** ***Spotlight: SI Units of Measure*** on page 616.]

Example

Weather Forecast

The weather forecast in Tokyo, Japan, calls for a 60% chance of precipitation with temperatures reaching 30 °C, while in Washington, D.C., the weather forecast calls for a 70% chance of precipitation with temperatures reaching 50 °F.

a. Which city will be warmer? Justify your answer by converting the temperatures to the same scale.

b. Assuming that there is precipitation in each of these cities, will rain or snow be seen in each place? Explain your thinking.

Solution

In order to compare temperatures, you need to convert them to the same scale.

a. You can convert 30 °C to °F:

$$F = \frac{9}{5}C + 32$$

Substitute the known value. $F = \frac{9}{5}(30) + 32$

Solve for the unknown value. $F = 86$

So, 30 °C = 86 °F. You could also convert 50 °F to °C:

Tokyo is warmer because 86 °F (30 °C) is higher than 50 °F (10 °C).

b. Because ice and snow melt at 32 °F (0 °C), you would expect rain rather than snow in both cities.

Key Terms

melting point (melting temperature)

boiling point (boiling temperature)

Lesson Summary

How is temperature measured?

Matter changes volume in response to changes in temperature. Matter generally expands when heated and contracts when cooled. This property can be used to measure temperature: The volume of liquid in a thermometer changes as the temperature changes. To set a temperature scale, you need to start with at least two points. Once these two temperature points are noted on a scale, it is possible to figure out where any other temperature would be on that scale.

EXERCISES

Reading Questions

1. Explain how the height of a liquid can be used to measure temperature.
2. Describe in your own words how to construct a temperature scale.

Reason and Apply

3. What are the advantages of the Celsius temperature scale over the Fahrenheit temperature scale?
4. When the temperature is 0 °C, is it also 0 °F? Explain.
5. Convert −40 °C to °F. Show your work.
6. Which is larger, one Celsius degree or one Fahrenheit degree? Explain.
7. The doctor tells you your body temperature is 40 °C. Are you sick? Show your work and explain your answer.
8. You will be traveling to Hawaii where the forecast is for a temperature of 31 °C during the day, dropping to 28 °C overnight. Your friend recommends that you bring clothing for warm weather. Is this a good recommendation? Why or why not?
9. Create a graph comparing the Celsius and Fahrenheit scales.
   a. Plot the freezing point and the boiling point of water on your graph. Draw a straight line connecting the two points.
   b. Use the graph to determine the temperature in °F when it is 10 °C.
   c. If the weather forecast says it is 55 °F in Washington, D.C., what is the temperature in °C?

LESSON

5 Absolute Zero

Kelvin Scale

Think About It

The coldest temperature recorded on Earth was in Antarctica: a chilling −89 °C. The temperature is even lower on other planets that are farther away from the Sun. Researchers have recorded temperatures as low as −235 °C on the surface of Triton, a moon of Neptune!

How cold can substances become?

To answer this question, you will explore

1. Absolute Zero
2. Kelvin Scale
3. Molecules in Motion

Exploring the Topic

1 Absolute Zero

The volumes of most substances decrease as the temperature decreases. But there is a limit to how much a substance can shrink. It does not make sense that matter would have a negative volume. Hypothetically, the lowest temperature possible would correspond to a volume of *zero.*

Consider an example. The volume of a gas inside a flexible container is measured as the gas is cooled. Several data points are then plotted on a graph.

Volume Versus Temperature

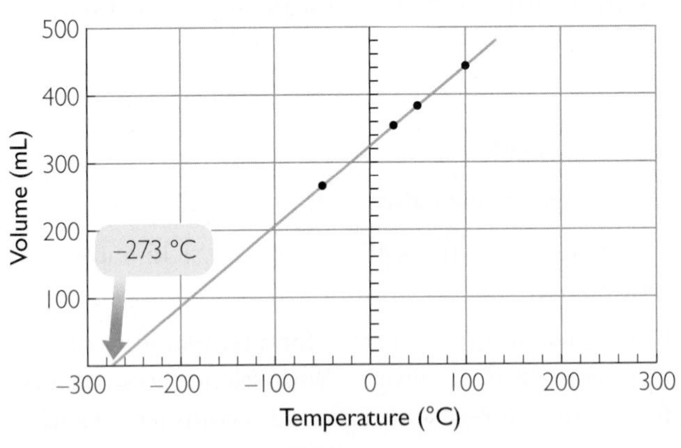

Using the Graph

This graph shows that as the temperature decreases, the volume decreases in a predictable way. The data points collected lie more or less on a straight line. If you extend this line to the *x*-axis, you can determine the theoretical temperature of the gas at zero volume.

HISTORY CONNECTION

In 2003, researchers at the Massachusetts Institute of Technology cooled a sample of sodium gas to 5000 picokelvins, or 0.0000000005 kelvin. This is the coldest temperature ever recorded in a laboratory.

More precise measurements reveal −273.15 °C as the temperature at zero volume. Scientists hypothesize that this value corresponds to the lowest temperature possible, and call it **absolute zero.** In actuality, as the temperature is lowered a gas would condense to a liquid and then to a solid well above this temperature, so zero volume is a hypothetical point. No one has ever caused a substance to reach absolute zero, but scientists have come very close—within a small fraction of a degree.

② Kelvin Scale

On the Celsius scale the temperature at which the volume of gas is theoretically zero is −273 °C. If you want to create a scale where the temperature is 0 when the volume is 0, you must add 273 to each temperature. This new temperature scale is called the **Kelvin scale.** One kelvin is the same size as one Celsius degree.

$$K = C + 273$$

$$C = K - 273$$

For example, to convert 20 °C to kelvins, add 273. This temperature corresponds to 293 K.

Note that the word degree, or the symbol for degree, °, is not used with the Kelvin scale. The unit of temperature on the Kelvin scale is the kelvin, K.

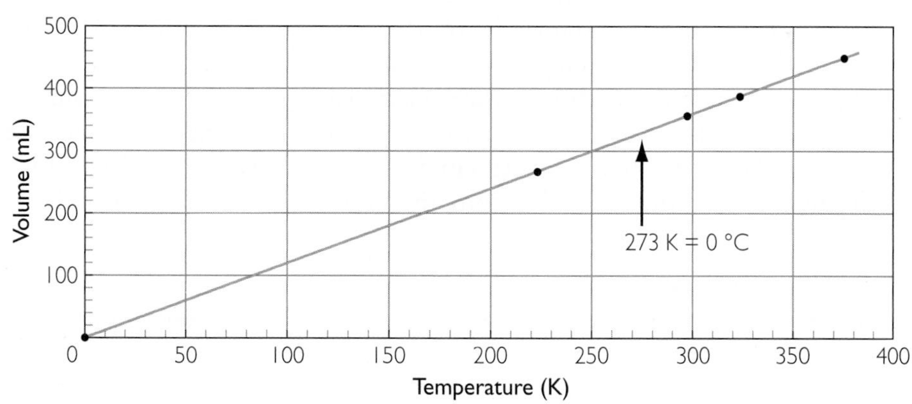

Using the Graph

The Kelvin scale sets the zero point at absolute zero. As a result, all temperatures are positive when they are expressed in kelvins. Notice that this graph is the same as the previous graph, but the *x*-axis has been shifted by 273. The value for 0 °C is 273 K.

HISTORY CONNECTION

The Kelvin temperature scale is named after William Thomson, 1st Baron Kelvin, also known as Lord Kelvin, who proposed the scale in 1848.

Room temperature is often considered to be 20 °C, or 68 °F, so 293 K is a temperature you will often find in problems. If you subtract 273 from a temperature in kelvins, you will be back at the Celsius temperature.

③ Molecules in Motion

Gases are quite different from liquids and solids. The molecules and atoms in solids and liquids are held close together by forces between molecules and atoms. Liquids lie in the bottom of any container they are in, and solids generally maintain their own shape without a container. In contrast, the molecules in gases expand to fill whatever space they are in. This indicates that there are essentially no forces between the molecules in a gas.

So, why is it that gas molecules are found throughout their container and not just at the bottom of it? And why is it that the space a gas occupies expands so dramatically as temperature increases? To explain many of the properties of gases, scientists rely on a model called the **kinetic theory of gases.** This model proposes that gas particles—atoms or molecules—are in constant motion. The word kinetic comes from a Greek word, *kinetos*, meaning "moving."

BIG IDEA On a particle level, all matter is in constant motion.

ASTRONOMY CONNECTION

The Boomerang Nebula is the coldest place known in the universe. It is 5000 light-years from Earth in the constellation Centaurus and has a temperature of 1 K.

These illustrations show the motion of three gas particles at three different times based on the kinetic theory of gases.

Three gas particles are moving in a container.

The gas particles travel in straight lines. The directions are random.

The gas particles bounce off walls and each other.

The gas particles move with different speeds.

One characteristic of gases described by the kinetic theory is that not all the gas particles are moving with the same speed. For example, the gas particle shown in blue moves more from frame to frame than the one shown in gray. However, the kinetic theory focuses on the average speed for all the particles at a given temperature. Indeed, the **temperature** of a gas can be defined as a measure of the *average* kinetic energy of the gas particles. When the temperature increases, the average speed of the gas particles increases. Scientists hypothesize that if you could cool matter to absolute zero, the atoms in the substance would stop moving.

BIG IDEA Temperature is a measure of the average speed of the atoms or molecules in a sample.

Key Terms

absolute zero
Kelvin scale
kinetic theory of gases
temperature

Lesson Summary

How cold can substances become?

The Kelvin scale assigns a value of 0 K to the hypothetical temperature of a gas with zero volume. This point is called absolute zero and is at −273 °C. Scientists consider this to be the lowest hypothetical temperature that matter can reach. They hypothesize that all motion stops at absolute zero. The kinetic theory of gases describes the motions of the gas particles. The atoms or molecules in a gas are constantly moving with an average speed that increases with increasing temperature.

EXERCISES

Reading Questions

1. What is absolute zero? Why is it considered a hypothetical temperature?
2. What advantages does the Kelvin scale have over the Celsius scale?
3. How does the kinetic theory of gases explain temperature?

Reason and Apply

4. What are the freezing and boiling temperatures of water in degrees Celsius, kelvins, and degrees Fahrenheit?

5. Which unit is the smallest: one Celsius degree, one kelvin, or one Fahrenheit degree? Explain your thinking.

6. Would you describe each of these temperatures as warm, hot, or cold?

   a. 100 K b. 60 °C c. 250 K d. 25 °C e. 300 K

   f. −100 °C g. 400 K

7. Convert each of the Kelvin temperatures in Exercise 6 to degrees Celsius, and vice versa.

8. What do you think is the highest temperature that can be reached by a substance? What is your reasoning?

9. Here are a few common temperatures on the Fahrenheit scale. Convert each of these to the Kelvin scale.

   a. 95 °F (a hot day) b. 350 °F (oven temperature) c. 5 °F (freezer)

10. The temperature on the surface of Venus is 736 K. Convert this temperature into degrees Fahrenheit and degrees Celsius. Describe what the atmosphere of Venus might be like.

11. According to the kinetic theory of gases, particles in a sample of gas

    A. Move randomly, in curved paths, at different speeds and bounce off each other and the walls of their container.

    B. Move randomly, in straight-line paths, at different speeds and bounce off each other and the walls of their container.

    C. Move randomly, in straight-line paths, at the same speed, and bounce off each other and the walls of their container.

    D. Move randomly, in curved paths, at the same speed, and bounce off each other and the walls of their container.

12. Use the kinetic theory of gases to explain why gases expand upon heating and shrink upon cooling.

HISTORY CONNECTION

The French inventor and physicist Guillaume Amontons first proposed the existence of absolute zero in 1702. Although he became deaf in his early teens, he did not allow this to stop him from having a productive scientific career. In addition to his work on temperature and pressure, Amontons developed the gas thermometer.

LESSON

6 Sorry, Charlie

Charles's Law

Think About It

The vent for a heater in your house is typically placed near the floor. This is because hot air rises and cooler air descends. If the vent were near the ceiling, the room would still be cold near the floor because the hot air would remain near the ceiling. So hot air must be less dense than cold air. For a similar mass, it occupies a larger volume than cooler air.

How can you predict the volume of a gas sample?

To answer this question, you will explore

1. Predicting Gas Volume
2. Charles's Law

Exploring the Topic

1 Predicting Gas Volume

A *piston* is a movable part that traps a sample of gas within a cylinder. It can move up and down if the volume of the gas changes. The illustration shows how the volume of nitrogen gas changes as the temperature changes. At first, the volume of the gas is 600 mL at a temperature of 300 K. When the cylinder is cooled, the gas contracts and the piston moves down. Then when the cylinder is heated, the gas expands and the piston moves up.

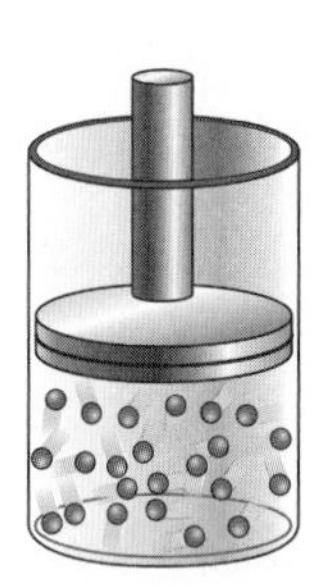

$V_1 = 600$ mL
$T_1 = 300$ K

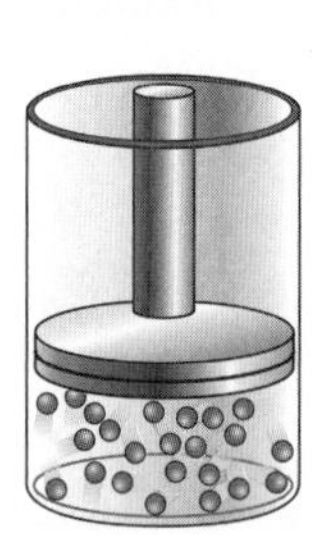

$V_2 = 400$ mL
$T_2 = 200$ K

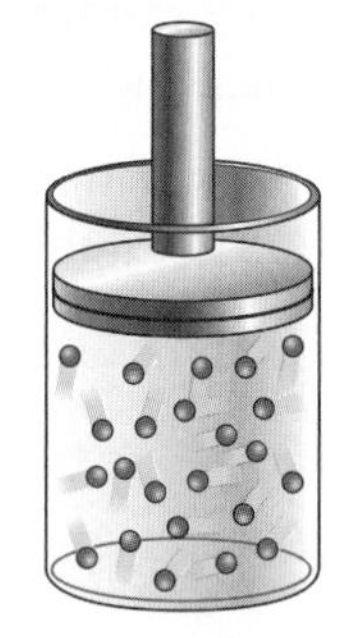

$V_3 = 800$ mL
$T_3 = 400$ K

Subscripts are used to indicate corresponding volumes and temperatures for the same gas under different conditions. For example, V_1 is the volume at the first temperature, T_1. Notice that if you determine the ratio of the volume of the gas to the temperature of the gas, the ratio is the same at all three of these temperatures.

$$\frac{V_1}{T_1} = 2.0 \text{ mL/K} \qquad \frac{V_2}{T_2} = 2.0 \text{ mL/K} \qquad \frac{V_3}{T_3} = 2.0 \text{ mL/K}$$

In fact, if nothing else changes, the ratio of volume to temperature for this gas will be 2.0 mL/K at any temperature. The relationship between the volume and

HISTORY CONNECTION

The French Montgolfier brothers invented the first hot air balloon. They demonstrated their creation in June 1783. The brothers thought that they had discovered a new gas that was lighter than air. In fact, the gas in their balloon was just air. It was less dense than the surrounding air because it was heated.

temperature for this sample of gas is proportional. So the volume and temperature for this gas are related by the proportionality constant, $k = V/T$, or 2.0 mL/K. With this proportionality constant, you can determine the volume of this gas at any temperature. [For a review of this math topic, see **MATH** ***Spotlight: Ratios and Proportions*** on page 627.]

Changing the Amount of Gas

Suppose you start with a *different* amount of gas in the same cylinder. For this new sample of gas, the proportionality constant, $k = V/T$, is 1.5 mL/K.

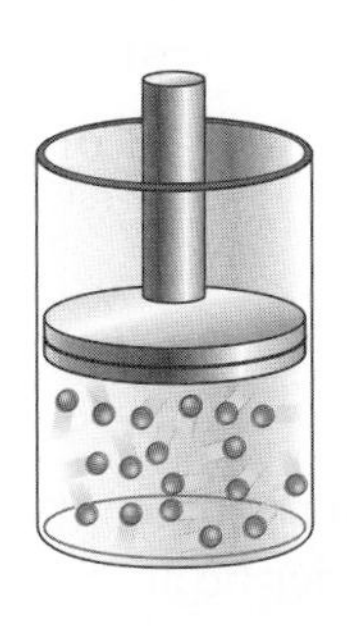

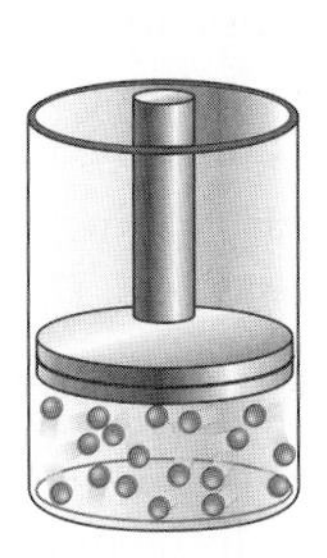

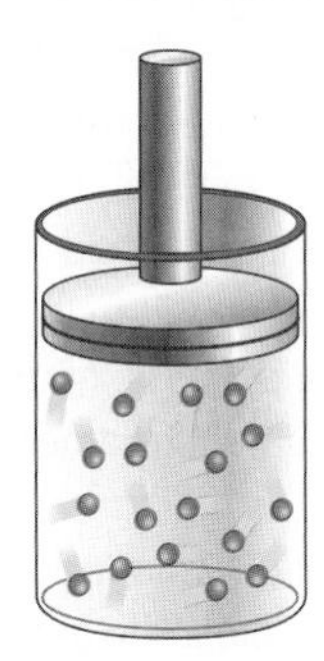

$V_1 = 450$ mL	$V_2 = 300$ mL	$V_3 = 600$ mL
$T_1 = 300$ K	$T_2 = 200$ K	$T_3 = 400$ K
$\frac{V_1}{T_1} = 1.5$ mL/K	$\frac{V_2}{T_2} = 1.5$ mL/K	$\frac{V_3}{T_3} = 1.5$ mL/K

Again, you can use $k = V/T$ for this gas sample to calculate the volume at any temperature.

Important to Know Each gas sample has a unique value for the proportionality constant, $k = V/T$, depending on the amount of gas in the sample. ◄

If you graph volume versus temperature for the two gas samples described previously, the result will be two different lines with different slopes.

Volume Versus Temperature

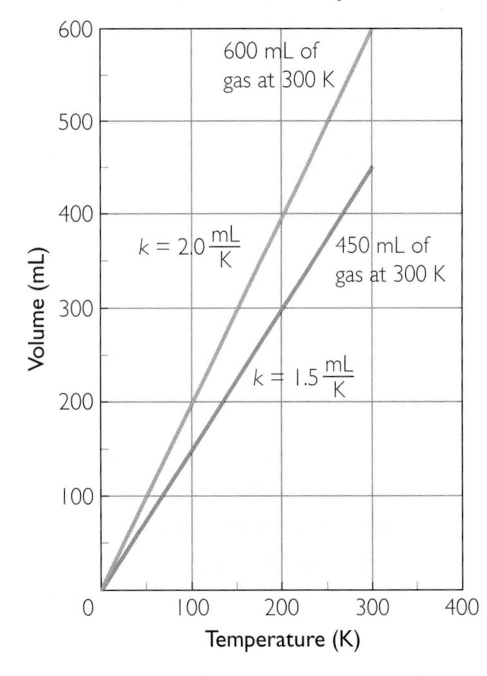

Using the Graph

Because the relationship between temperature and volume is proportional, the lines pass through the origin. The slope of each line is equal to the proportionality constant for that sample. You can use the graph to predict the volume of either gas sample at any temperature.

❷ Charles's Law

Jacques Charles was a French inventor, scientist, mathematician, and very active balloonist. Because of his interest in hot air balloons, he gave a lot of thought to the relationship between the volume and temperature of a gas. In 1787, he described the proportionality of gas volume and temperature mathematically. This description is now known as **Charles's law.**

HISTORY CONNECTION

Hydrogen gas is less dense than air. Jacques Charles took advantage of this when he invented the hydrogen balloon. In August 1783, his balloon ascended to nearly 3000 feet (914 meters). Terrified villagers destroyed the balloon when it landed outside of Paris. Charles built another hydrogen balloon, and he and a friend made the first piloted flight that December.

Charles's Law

If pressure and the number of particles of a gas stay the same, then volume is proportional to the Kelvin temperature.

$$V = kT \text{ or } k = V/T$$

Example 1

Calculate Gas Volume

Imagine that a cylinder with a piston contains 10 mL of air at 295 K. What will the volume of gas in the cylinder be at 550 K?

Solution

Calculate the volume by using Charles's law or by drawing a graph. You can predict that the volume will increase when the temperature increases. So the answer should be greater than 10 mL.

Using the formula

Find the value of k. $k = \frac{V}{T} = \frac{10 \text{ mL}}{295 \text{ K}} = 0.034 \text{ mL/K}$

Solve for the volume at 550 K. $V = kT = 0.034 \text{ mL/K} \cdot 550 \text{ K} = 18.5 \text{ mL}$

Graphical analysis

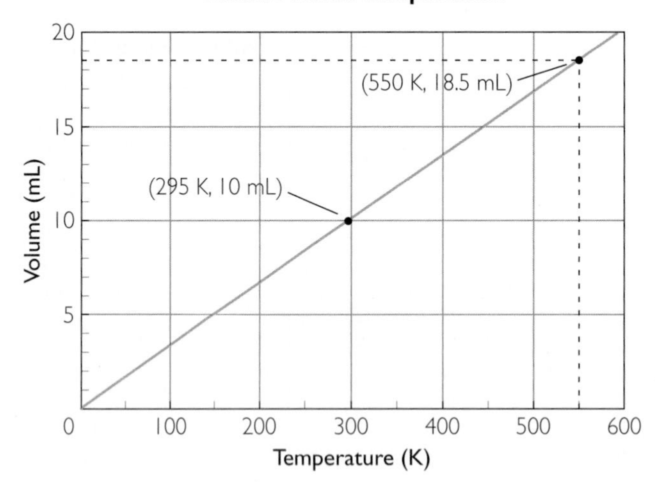

First, plot the data point for the initial conditions. Draw a line through this point and the origin, since at 0 K the y-value is theoretically 0 mL. Use the line to find the volume when the x-value is 550 K. Both methods give the same answer. At 550 K, the volume is 18.5 mL.

Topic: Gas Laws
Visit: www.SciLinks.org
Web code: KEY-306

Example 2

Calculate Gas Temperature

A balloon contains 500 mL of helium at 20 °C. You want to increase the volume to 550 mL. What temperature in °C will give you a volume of 550 mL for this sample of gas?

Solution

You can predict that the temperature will be higher for a larger volume.

Convert the temperature to kelvins. $K = C + 273$
$= 20 + 273$
$= 293\ \text{K}$

Determine the proportionality constant, k. $k = \frac{V}{T} = \frac{500\ \text{mL}}{293\ \text{K}} = 1.7\ \text{mL/K}$

Rearrange the equation in terms of T and solve for temperature at 550 mL. $T = \frac{V}{k} = \frac{550\ \text{mL}}{1.7\ \text{mL/K}} = 324\ \text{K}$

Convert back to degrees Celsius. $C = K - 273 = 324 - 273 = 51\ °\text{C}$

A temperature of 51 °C is needed to give a volume of 550 mL. The temperature is larger for a larger volume, as predicted.

[For a review of this topic, see **Math *Spotlight: Solving Equations*** on page 620.]

Key Term
Charles's law

Lesson Summary

How can you predict the volume of a gas sample?

The volume and temperature of a sample of gas are proportional to each other if nothing else changes (such as pressure or the amount of gas). This relationship between the volume and temperature of a gas is described by the equation $V = kT$ and is known as Charles's law. The proportionality constant, k, is different for each different amount of gas.

EXERCISES

Reading Questions

1. Explain how to determine the proportionality constant, k, for a sample of gas.
2. Explain how to determine the volume of a gas at a certain temperature using the proportionality constant, k.

Reason and Apply

3. A sample of gas in a cylinder has a volume of 620 mL at 293 K. If you allow the piston to move while you heat the gas to 325 K, what will the volume of the gas be? Check your answer by drawing a graph.
4. A sample of gas in a cylinder has a volume of 980 mL at a temperature of 27 °C. If you allow the piston to move while you heat the gas to 325 K, what will the volume of the gas be at 130 °C?

5. A sample of gas in a cylinder with a piston has a volume of 330 mL at 280 K. What temperature will you need to heat it to in order to change the volume to 220 mL?

6. A 2.0 L gas sample at 20 °C must be cooled to what temperature for the volume to change to 1.0 L? Show at least two different ways to solve this problem.

7. Imagine that you have a huge helium balloon for a parade. Around noon, it is 27 °C when you fill the balloon with helium gas to a volume of 25,000 L. Later in the day, the temperature drops to 15 °C.
   a. What is the proportionality constant, $k = V/T$, at the beginning of the day?
   b. Calculate the volume of the balloon when the temperature has dropped to 22 °C.
   c. What will the proportionality constant, $k = V/T$, be at the end of the day when the temperature is 15 °C? Explain your answer.

8. Would this cup make a good rain gauge? Explain your thinking.

HISTORY CONNECTION

On May 6, 1937, the airship *Hindenburg* caught fire as it landed at an airfield in New Jersey. The *Hindenburg* was kept afloat with massive chambers filled with hydrogen, which is less dense than air but is also highly flammable. Thirty-six people died when a fire on board spread rapidly, though flammable paint also may have contributed to the spread of the fire.

ACTIVITY Density, Temperature, and Fronts

Purpose

To investigate the movement of air masses and their role in determining the weather.

Part 1: Weather Maps

Reexamine the weather maps from Lesson 1: Weather or Not to answer these questions.

1. Examine the first map here. What relationships do you see among fronts, clouds, and precipitation?
2. Examine the second map. What relationship do you see between fronts and lows? Between fronts and highs?

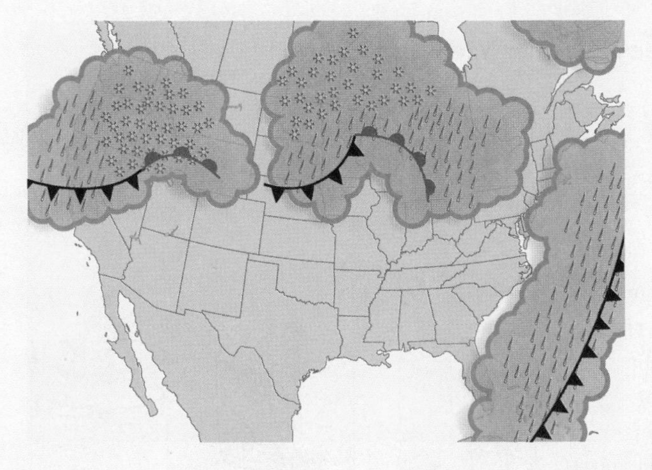

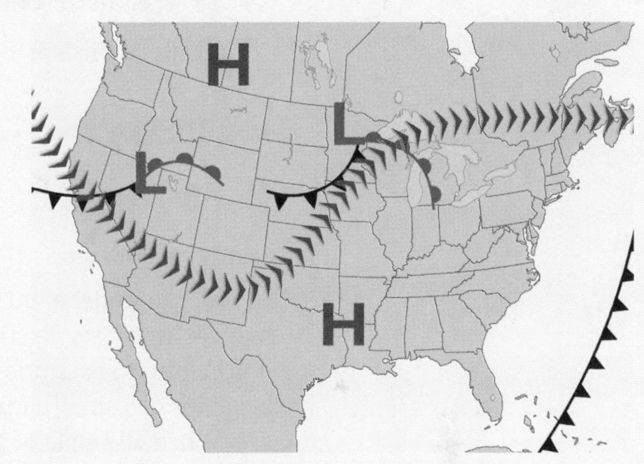

3. Describe any connections you find between fronts and the jet stream.
4. Where would you expect to see warm and cold air masses on each map?

Part 2: Warm and Cold Fronts

1. Why is a cold air mass denser than a warm air mass?
2. Explain why clouds might form when a warm air mass collides with a cold air mass.
3. **Making Sense** What does air density have to do with weather fronts?
4. **If You Finish Early** Eighty percent of the air in our atmosphere is made up of nitrogen gas, N_2, while water vapor makes up only 1% of the air. Why doesn't it rain liquid nitrogen instead of rainwater?

LESSON

7 Front and Center

Density, Temperature, and Fronts

Think About It

Meteorologists have identified large masses of air that have a consistent temperature and water content. These **air masses** form in particular locations on the Earth's surface and cover thousands of square miles. They have a great influence on the weather. Weather fronts occur at the boundaries of these air masses, where warm and cold air meet.

How do weather fronts affect the weather?

To answer this question, you will explore

1. Air Density
2. Weather Fronts
3. Relationships Between Weather Variables

Exploring the Topic

1 Air Density

Small Air Masses

In order to understand a large mass of air on the Earth's surface, it is helpful to examine the behavior of a small sample of gas. When a sample of gas gets warmer, it expands to fill a larger volume. Notice that there are fewer molecules per milliliter of volume in the cylinders as the temperature increases. The number of molecules does not change, but the spacing between those molecules does change. The density, $D = m/V$, decreases because the same mass, m, is divided by a larger volume, V, resulting in a smaller number.

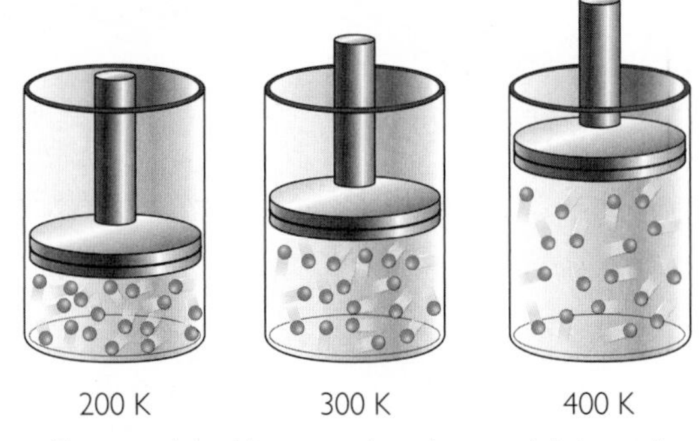

Air expands and becomes less dense as it is heated.

Large Air Masses

In Unit 1: Alchemy, you found that every substance has a specific density. For example, the density of gold is 19.3 g/mL, and the density of water is 1.0 g/mL. This is true only at a specified temperature. The density of a substance changes as the temperature changes. For liquids and solids, this change in density is very small. For gases, the change is much larger.

When air gets warmer it not only expands, it also rises, because warm air is less dense then colder air. This is what happens when a warm air mass meets a cold air mass. This fact is key to understanding weather fronts and storms.

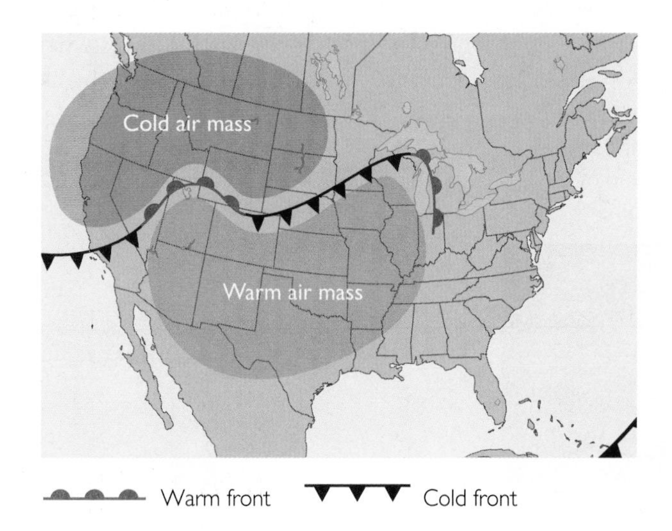

❷ Weather Fronts

Warm and Cold Air Meet

When a cold air mass meets a warm air mass, two different weather situations can arise. These two situations are shown in the illustrations and are described here.

Cold front. Cold air overtakes warm air.

Warm front. Warm air overtakes cold air.

Cold Fronts

A cold front occurs when cold air overtakes warm air. The warm air is suddenly pushed up as the advancing cold air forces its way underneath it. The warm air cools at higher altitudes, which causes water vapor in the air to condense into tiny liquid cloud droplets. The clouds associated with cold fronts can be seen directly in the area of the advancing front. They form quickly and are often thick, puffy clouds, similar to the ones seen before thunderstorms. Heavy rains are often associated with cold fronts.

Clouds at the leading edge of a cold front

Warm Fronts

A warm front occurs when warm air overtakes cold air. The warm air gradually flows up and over the cold air because the warm air is less dense. The warm air cools at higher altitudes and water vapor condenses to form clouds. The clouds associated with warm fronts usually form ahead of the place where the air masses meet.

Clouds at the leading edge of a warm front

NATURE CONNECTION

Cold fronts are responsible for squalls, tornadoes, strong winds, and other destructive weather.

You will see high, wispy clouds when a warm front is still hundreds of miles away. As the warm front gets closer, the clouds get thicker and lower, and precipitation falls. Steady, light rain over a large area is associated with warm fronts.

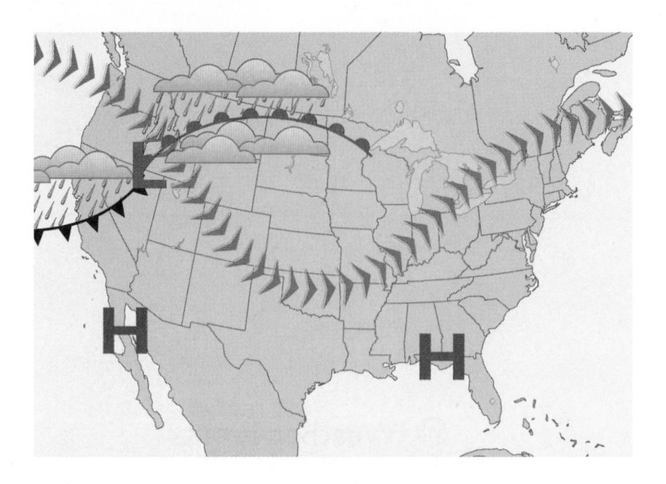

❸ Relationships Between Weather Variables

The clouds, high and low pressure systems, precipitation, and the jet stream can be added to a weather map that contains some fronts. Take a moment to draw some conclusions about how these weather variables relate to one another.

- Fronts occur at the boundaries between warm and cold air masses.
- Clouds and steady, light rain form ahead of a warm front. Clouds and heavy showers form at and behind a cold front.
- On weather maps, L's are closely associated with fronts while H's appear to be away from the fronts. H's represent areas of high pressure and are associated with clear skies. L's represent areas of low pressure and are associated with storms and cloudy skies.

When air masses form over Earth, they get their properties from the area beneath them. In other words, if an air mass forms over water, it will tend to be full of moisture; if it forms over land, it will tend to be dry.

Key Term

air mass

Lesson Summary

How do weather fronts affect the weather?

When a cold air mass and a warm air mass meet, the warm air rises and layers on top of the cold air. This is because warm air is less dense than cold air. A cold front occurs when a cold air mass overtakes a warm air mass, and a warm front occurs when a warm air mass overtakes a cold air mass. Clouds and precipitation occur with both warm and cold fronts. Warm fronts produce steady, light rain, and cold fronts produce sudden, heavy showers.

ENVIRONMENTAL CONNECTION

Air masses are not always moving. Sometimes they stay in place for a long time. If the hot, dry continental tropical air mass stays over the Great Plains, it may not rain for months, and a serious drought can occur.

EXERCISES

Reading Questions

1. Explain why hot air rises.
2. Explain the difference between a warm front and a cold front.

Reason and Apply

3. Suppose you have two gas samples in flexible containers with the same outside pressure and the *same amount of gas* in each. Sample A is at a temperature of 25 °C and Sample B is at 5 °C. Which of these statements is true?

   A. Sample A occupies a larger volume and has a smaller density.

   B. Sample A has a greater density and a smaller volume.

   C. Sample B has molecules moving with the greater average speed.

4. A cold front is approaching your hometown and is due to arrive tomorrow. What kind of weather would you expect to observe?
5. A warm front is approaching your hometown and is due to arrive tomorrow. What kind of weather would you expect to observe?
6. The continental polar air mass overtakes the maritime tropical air mass.

   a. What kind of front develops?

   b. What happens to the air masses when they meet?

   c. What sort of weather would you expect and where?

7. **WEB RESEARCH** Look in the newspaper or on the Internet. Find a recent weather map with at least one warm front and one cold front.

   a. Find a warm front on the map. What weather is predicted for tomorrow in the region of the warm front?

   b. What weather is predicted in the direction in which the warm front is moving?

   c. Find a cold front on the map. Describe the weather forecast given for the region near the cold front.

   d. What weather is predicted for tomorrow in the region close to the cold front?

   e. What weather is predicted in the direction in which the cold front is moving?

SECTION

I

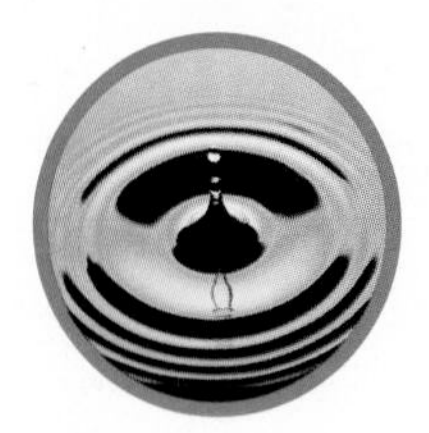

SUMMARY

Physically Changing Matter

Weather Update

The water and gases that make up Earth's atmosphere change density and temperature as they move about the planet. Physical changes of matter cause the weather we experience. Most matter expands when heated and contracts when cooled. When heated enough, matter changes phase, from solid to liquid to gas. The density of gases is dramatically lower than the density of either liquids or solids. Changes in temperature and density play a role in the weather.

Key Terms

weather
physical change
phase change
proportional
proportionality constant
melting point (melting temperature)
boiling point (boiling temperature)
absolute zero
Kelvin scale
kinetic theory of gases
temperature
Charles's law
air mass

Review Exercises

1. What is 50 °F in kelvins?
2. Use the kinetic theory of gases to explain temperature.
3. A gas sample in a flexible container has a volume of 650 mL at a temperature of 27 °C. If the pressure stays the same when the sample is heated to 80 °C, what will be the new volume?
4. You fill a balloon with helium to a volume of 3 L in the morning, when the temperature is 23 °C. At the end of the day, the temperature drops to 10 °C.
   a. What is the proportionality constant, $k = V/T$, at the start of the day?
   b. What is the proportionality constant at the end of the day? Explain your thinking.
   c. What will the volume of the balloon be when the temperature is 10 °C?
5. a. What is the volume of 5.2 g of solid CO_2 if the density of solid CO_2 is 1.56 g?
   b. What will the volume of the 5.2 $g(s)$ be if all of the solid changes to a gas? The density of $CO_2(g)$ is 0.0019 g/mL. Explain your reasoning.

PROJECT *Different Thermometers*

Different thermometers use different properties of matter to measure temperature. Research how one of these thermometers works. Write a description and include a diagram.

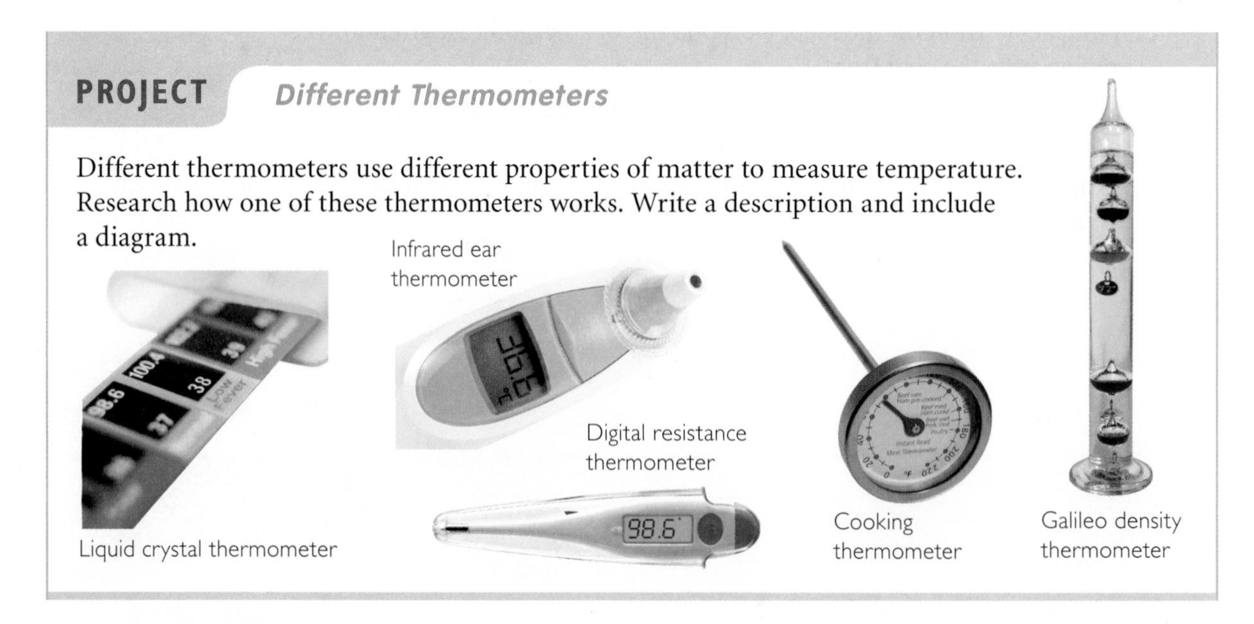

SECTION

II Pressing Matter

The hot air inside a balloon has extremely low density, and it causes the balloon to rise. Density differences cause warm air masses to rise in the atmosphere in a similar way.

The atoms and molecules in solids, liquids, and gases are always in motion. However, gas molecules are free to move to fill the entire space of their container. Gas molecules collide with one another and with anything else they come in contact with, causing pressure. When samples of gases are placed in containers, it is possible to measure the relationships between gas pressure, temperature, and volume. The air pressure of the atmosphere has a great deal of influence on the weather.

In this section, you will study

- the density of gases
- the behavior of gas particles
- the relationships between gas pressure, temperature, and volume
- high- and low-pressure weather systems

LESSON

8 It's Sublime

Gas Density

Think About It

Our planet is unique in the solar system for having so much liquid water on its surface. The weather and the life on our planet depend on the movement of moisture around the planet. The phase changes of water are mostly responsible for this movement. When water changes phase from liquid to gas, the airborne water molecules spread out in the atmosphere and the density changes dramatically.

How do the densities of a solid and a gas compare?

To answer this question, you will explore

1. Density of Gases
2. Molecular View
3. Comparing Density and Phase of Water

Exploring the Topic

1 Density of Gases

An airplane glides easily through the air at around 500 miles per hour. The density of air must be very low for objects to pass so easily through it. Determining the density of a gas is not as straightforward as determining the densities of liquids and solids. This is because it is not easy to measure the mass of a gas. It is hard to put a gas on a balance, and many gases, like helium, rise up in the air around them. One approach is to measure the mass of a substance while it is a liquid or a solid, and then turn it into a gas. Provided you don't let any escape, the substance will have the same mass as a gas as it did as a solid. You can then measure the volume of the gas to determine its density.

Sublimation of Dry Ice

$CO_2(s)$ is commonly referred to as "dry ice" because it does not go through a liquid phase when it changes to a gas.

One substance that you can use to study phase changes is carbon dioxide, CO_2. Carbon dioxide turns *directly* from a solid to a gas at the extremely low temperature of −78 °C (−108 °F). This means the temperature in your classroom is warm enough to turn solid carbon dioxide into a gas.

When a solid changes directly into a gas without forming a liquid, the phase change is called **sublimation**. In class, you gathered data on the density of $CO_2(g)$ by finding the mass of a chunk of $CO_2(s)$ and then placing it into a plastic bag and allowing it to sublime. The volume of the gas is huge compared to the volume of the same mass of solid.

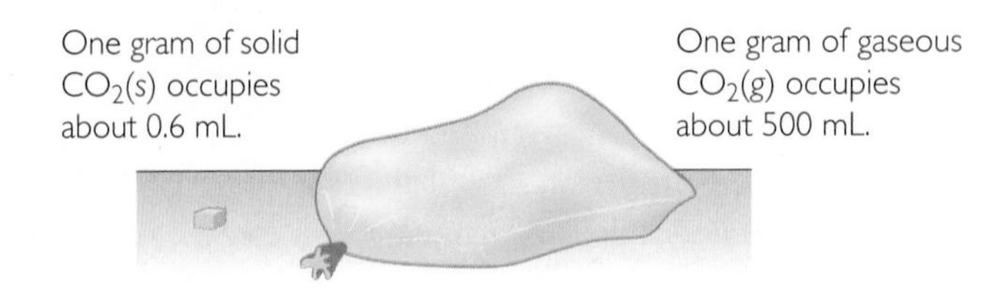

The density of the carbon dioxide changes dramatically as the phase changes. $CO_2(s)$ has a density of 1.56 g/mL. In contrast, $CO_2(g)$ has a very small density of 0.0019 g/mL. The volume of the gas is nearly 800 times as large as the volume of the solid!

BIG IDEA The densities of gases are extremely low compared to the densities of liquids and solids.

2 Molecular View

Why did the bag puff up so much as the $CO_2(s)$ sublimed? Because you cannot see gases, you must depend on hypotheses about how the molecules in a gas behave. Consider four possible hypotheses and the corresponding models for the sublimation of solid carbon dioxide.

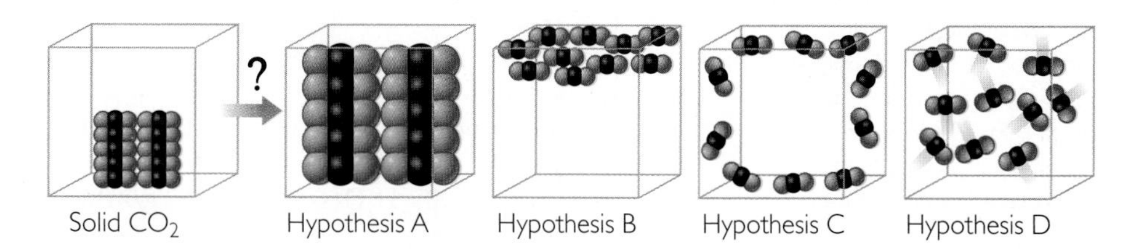

Solid CO_2 Hypothesis A Hypothesis B Hypothesis C Hypothesis D

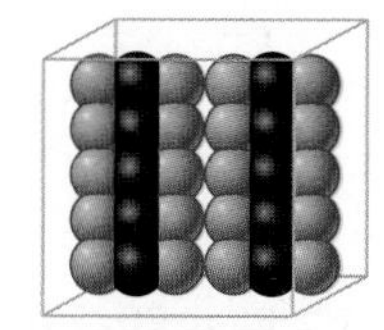

Hypothesis A

Hypothesis A: The CO_2 molecules become larger in the gas phase compared with the solid phase.
Evidence against Hypothesis A: You can move freely through a gas, so the molecules must not take up much space. The mass stays the same, so it is difficult to imagine how the molecules would get bigger. The evidence doesn't support Hypothesis A.

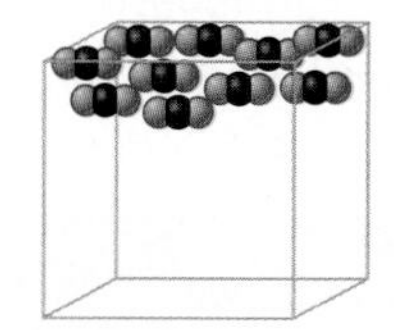

Hypothesis B

Hypothesis B: The CO_2 molecules rise to the top of the container.
Evidence against Hypothesis B: The bag filled with $CO_2(g)$ was pushed out in all directions. This indicates that the molecules are spread throughout, not just at the top. For example, the air in a room is everywhere, not just near the ceiling. The evidence doesn't support Hypothesis B.

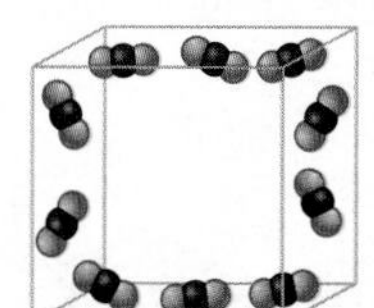

Hypothesis C

Hypothesis C: The molecules go to the outer edges of the bag, pushing the sides out.
Evidence against Hypothesis C: If gases went to the outer edges of their containers, you would walk into a room and find all the air on the walls and none in the middle. Since you can breathe everywhere in the room, this indicates that gas molecules are spread throughout. The evidence doesn't support Hypothesis C.

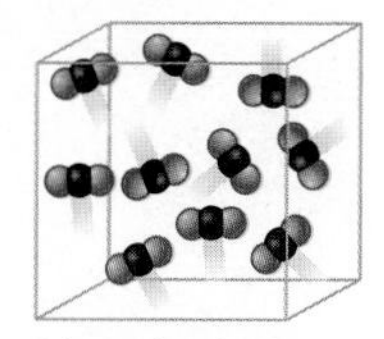

Hypothesis D

Hypothesis D: The CO_2 molecules spread apart from one another and bounce around inside the bag.
Evidence supporting Hypothesis D: This model is consistent with the large volume and the low density of the gas. This model is also consistent with the bag staying puffed out in all directions and offering resistance when you push on it. This evidence supports Hypothesis D.

Model D can still be improved upon. It shows ten molecules in a small cubic space. In actuality the gas molecules are much tinier, much farther apart, much more numerous, and moving very fast. When a gas is trapped in a bag, the moving molecules bang into the bag and push it out. If you were to remove the bag, the molecules would quickly spread throughout the room. This is what happens to water when it **evaporates,** or changes into a vapor, and spreads out in the atmosphere.

BIG IDEA A sample of gas expands to fill whatever space it is in.

Clouds could be considered a mixture of water vapor and liquid water. It takes 15 million cloud droplets to form one drop of rain. The extremely low density of clouds explains why airplanes can move easily through them.

CONSUMER CONNECTION

Sublimation is the secret to freeze-dried food such as the food that astronauts eat in space. After the food is frozen, you can get the water to sublime by removing the air around it with a vacuum pump. The result is a dried-out food with a curious texture.

3 Comparing Density and Phase of Water

The table shows the density for some different forms of water. Water has three phases: solid, liquid, and gas. But these phases can take a few different forms. For example, snow and ice both represent solid forms of water.

Different Forms of Water

Substance	Chemical formula	Density (g/mL)	Volume of 1000 g (mL)
rain	$H_2O(l)$	1.0 g/mL*	1,000 mL
water vapor	$H_2O(g)$	0.0008 g/mL*	1,250,000 mL
ice	$H_2O(s)$	0.92 g/mL†	1,087 mL
snow	$H_2O(s)$	0.1–0.5 g/mL	2,000–10,000 mL
clouds	$H_2O(l)$	0.001–0.002 g/mL	500,000–1,000,000 mL

*The density is for 25 °C. †The density is for 0 °C.

Notice what happens to the *volume* of 1000 grams of water as it goes from liquid to water vapor. The volume increases more than a thousand-fold when evaporation occurs.

The density landscape illustration shows the relationship between the location of matter and its density.

Substances that are less dense float in substances that are denser. For example, ice floats in liquid water. When the temperature of a mass of air increases, the air expands, resulting in a lower density. This change in density causes the warmer air to rise above cooler air. These changes are related to the movement of matter on the planet and are essential to causing different weather.

Key Terms

sublimation
evaporation

Lesson Summary

How do the densities of a solid and a gas compare?

Gases form when solids sublime and liquids evaporate. During both sublimation and evaporation, molecules spread far apart from each other. This increases the volume of the substance by about 1000 times. It also decreases the density of the substance dramatically. Solids are generally denser than liquids. Gases are much less dense than both solids and liquids. They expand to fill whatever container they are in.

EXERCISES

Reading Questions

1. How does the density of a gas compare with the density of a solid?
2. Draw a molecular view of carbon dioxide gas.

Reason and Apply

3. In your own words, define sublimation and evaporation.
4. Describe three ways to show that gases exist.
5. When water freezes, the water molecules move apart very slightly. What evidence can you provide to support this claim?
6. Draw a molecular view for water vapor, liquid water, and ice.
7. What is the volume of 6.4 g of $CO_2(s)$? (density = 1.56 g/mL)
8. What is the mass of 3.5 L of $CO_2(g)$? (density = 0.0019 g/mL)
9. How many grams of $CO_2(s)$ would you need in order to fill a 6.5 L bag with $CO_2(g)$?
10. A person has 15 g of dry ice and wants to completely fill a bag that has a volume of 8 L with carbon dioxide gas. Is this enough dry ice to fill the bag with $CO_2(g)$? Explain.

ASTRONOMY CONNECTION

The lowest temperature ever recorded on Earth was −129 °F in 1983 in Antarctica. At this temperature, $CO_2(g)$ condenses to form solid dry ice. The southern polar ice cap on Mars contains dry ice because the temperature there is below the sublimation temperature of CO_2.

LESSON

9 Air Force

Air Pressure

Think About It

There is a layer of air surrounding Earth nearly 80 miles thick called the atmosphere. It is made up predominantly of colorless, odorless gases, which you can move through. The air molecules are tiny and they are spaced far apart, so most of the time it feels as if nothing is there. However, the weight of all that air above you exerts a great deal of pressure on you.

What evidence do we have that gases exert pressure?

To answer this question, you will explore

1. Evidence of Air Pressure
2. Explanation for Air Pressure
3. The Atmosphere

Exploring the Topic

1 Evidence of Air Pressure

When you fill the tires on your car with air, the rubber is stretched quite tightly. The pressure exerted by the air in the tires is large enough to push the entire car up off the ground. Indeed, the air pressure is holding up more than 2000 pounds. How can air be so powerful?

ASTRONOMY CONNECTION

Atmospheric pressure varies from planet to planet. On Venus, for instance, the atmosphere is much denser than on Earth. The atmospheric pressure is 90 atm at the surface of Venus. This would crush a human being. On Mars, however, the atmospheric pressure at the surface is just 0.007 atm. On either Venus or Mars, you would need a very good pressure suit to keep your body at the pressure it is used to.

The ability of automobile tires to remain inflated while holding up so much weight provides evidence that air is pushing on the inside of the tire in all directions. This "pushing" property of a gas is called **pressure.** Pressure is defined as a force over a specific surface area. A gas exerts pressure on all surfaces it comes in contact with.

There are many ways to demonstrate that gases exert pressure on the things around them. Here are two examples.

Submerged Paper

When a plastic cup is turned upside-down and submerged in water, the paper inside the cup stays dry. There is air trapped in the cup that pushes out in all directions. The air pushes on the water with enough pressure to keep the water out of the cup.

Balloon in a Bottle

Suppose you try to blow up a balloon inside a bottle. Even with a great deal of effort, the balloon will inflate only a tiny bit. The balloon does not inflate because there is air inside the bottle. Even though it looks like

PHYSICS CONNECTION

A gas can even push against another gas. In fact, gas with a high pressure pushes the space shuttle into the air.

"nothing is there," the bottle is full of air, and when you try to inflate the balloon, you are pushing on this air, which takes a lot of work!

❷ Explanation for Air Pressure

Collisions of gas molecules with surrounding objects are what we experience as air pressure. There are huge numbers of molecules in an automobile tire, many more than are shown in the illustration. The tiny push from each one adds up to a lot, enough to hold a car up.

The pressure of the air inside your tires is expressed in pounds per square inch, or lb/in^2. This unit is sometimes written as *psi,* for pounds per square inch. That is, each square inch of surface area on the tire experiences a certain number of pounds of force from the air inside. A tire on a racing bicycle might be inflated to a pressure of 100 lb/in^2.

A tire filled with air is firm because the air molecules are colliding with the inside walls of the tire, thereby pushing the walls out.

Example

Air Pressure in Car Tires

Suppose a car weighs 2000 pounds. Each of the four tires touches the road over an area that is about a 4-by-4-inch square. What air pressure do you need in each tire to hold up the car?

Solution

Each of the four tires needs to push up ¼ of the 2000 pounds, or 500 pounds. A 4-by-4-inch square has an area of 16 in^2, so each tire is in contact with a 16 in^2 area of the ground. Pressure is force per unit area.

$$\begin{aligned} \text{Pressure} &= \text{force/area} \\ &= 500 \text{ lb}/16 \text{ in}^2 \\ &= 31 \text{ lb/in}^2 \end{aligned}$$

The pressure in each tire needs to be at least 31 lb/in^2.

❸ The Atmosphere

Earth's atmosphere is a mixture of gases, mostly nitrogen, with some oxygen, carbon dioxide, water vapor, and argon. This mixture of gases is what we call air.

The density of the Earth's atmosphere changes as you travel up in altitude—the air becomes thinner at higher elevations. This concept will be covered more extensively in later lessons.

BIG IDEA The atmosphere is a mixture of gases, including gaseous water.

Although we barely notice it, the air around us exerts pressure on us all the time. **Atmospheric pressure** is air pressure that is always present on Earth as a result of air molecules colliding with objects on the planet. The pressure due to the atmosphere is the equivalent of a 14.7-pound weight pushing on every square inch of surface. We say that the pressure due to the atmosphere is 14.7 pounds per square inch, or 14.7 lb/in^2. This pressure is measured at sea level.

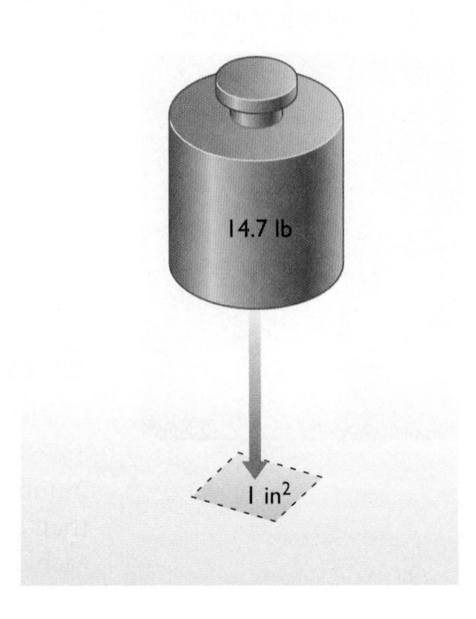

Scientists also use a unit called an **atmosphere,** or **atm,** to measure air pressure.

CONSUMER CONNECTION

Suction cups are misnamed. There is nothing pulling inside the suction cup. The atmosphere is pushing on the outside, and there are fewer molecules on the inside to push back.

Atmospheric Pressure

At sea level and 25 °C, there is 1 atm of pressure.

$$1 \text{ atm} = 14.7 \text{ lb/in}^2$$

The pressure of the atmosphere played a role in all of the demonstrations you completed in class. The air pressure mat is particularly good at demonstrating how much pressure the air around us exerts on objects. If the mat has a surface area of 100 square inches and standard air pressure from the atmosphere is 14.7 lb/in^2, the total pressure on the mat from the air is 1,470 pounds! No wonder no one can lift it.

Key Terms

pressure
atmospheric pressure
atmosphere (atm)

Lesson Summary

What evidence do we have that gases exert pressure?

The Earth is surrounded by an atmosphere of gases. We call this mixture of gases air. Although we cannot see it, the air around us exerts pressure. Gas molecules are constantly moving and colliding with anything they come in contact with. Air pressure is defined as the force over a surface area. This force is caused by the collisions of the gas molecules. Gas pressure is measured in pounds per square inch, lb/in^2, or in atmospheres, atm. At sea level and 25 °C, the atmospheric pressure is 14.7 lb/in^2, or 1 atm.

EXERCISES

Reading Questions

1. Give three pieces of evidence that air exerts pressure.
2. Explain what causes air pressure.

Reason and Apply

3. High up in space there are no molecules to collide with the outside walls of a balloon. Therefore, the air pressure inside the balloon is much greater than the air pressure outside of the balloon. Describe what would happen to an inflated balloon if it were suddenly put in space.
4. Calculate the weight in pounds of the atmosphere pushing down on the back of your hand.
5. When you fly in an airplane, it is common for you to feel painful pressure in your ears. Explain what might be happening in terms of air pressure.
6. What is the minimum pressure you need in bike tires to hold up a person who weighs 150 lb? Assume that the tires touch the ground in a square that is 1.5 in. by 1.5 in.
7. The air pressure from the atmosphere measures 0.5 atm at an altitude of 18,000 ft. How much pressure is this in pounds per square inch? Calculate the weight in pounds on the back of your hand at this altitude.
8. **WEB RESEARCH** Go to a library or do a Web search to find out how gases, such as helium or oxygen, are transported for industrial use.
9. If a car runs over the end of your foot, it doesn't break your toes. Explain why.

HISTORY CONNECTION

Atmospheric pressure was demonstrated in 1654 when Otto von Guericke, of Magdeburg, Germany, used a vacuum pump to remove almost all of the air from the space between two copper hemispheres that were 1.6 ft in diameter. The air pressure holding the two halves together was so strong that two teams of horses could not pull them apart. However, when air was let back in, the hemispheres fell apart easily.

LAB Boyle's Law

Safety Instructions

The cap on the tip of the syringe should always be pointed down, away from eyes. Wear safety goggles.

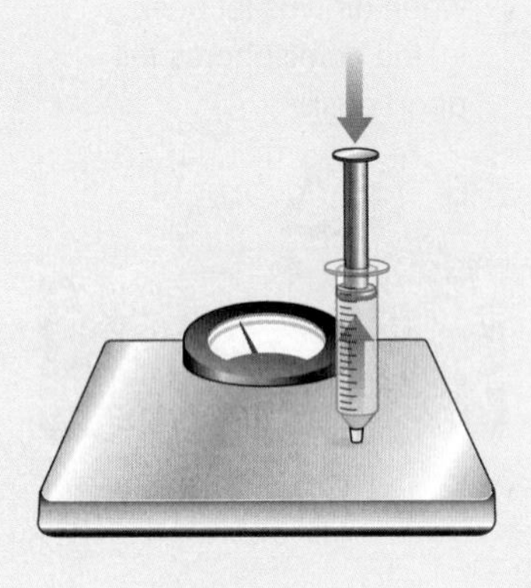

Purpose

To observe and quantify the relationship between gas pressure and gas volume.

Materials

- 50 mL plastic syringe with cap screwed on tight
- bathroom scale

Procedure

1. Start with the syringe at 50 mL. Make sure the cap is on tight.
2. Hold the syringe with the tip on top of a bathroom scale.
3. Repeat these steps for at least five different volumes.
   - One person depresses the plunger by a few milliliters.
   - A second person reads the exact volume.
   - A third person reads the number of pounds that is exerted on the bathroom scale.
   - Everyone records the volume and weight data in a table like this one.

Data

Trial	Volume (mL)	Weight that you apply (lb)	Pressure that you apply (lb/in^2)	Atmospheric pressure (lb/in^2)	Total pressure (lb/in^2)
1				14.7 lb/in^2	
2				14.7 lb/in^2	

Analysis

1. Estimate the cross-sectional area inside of the syringe in square inches. Determine the pressure applied in pounds per square inch. Calculate the rest of the values needed to complete your table.
2. Graph the total pressure versus volume. What happens to the pressure of the gas as the volume of the gas decreases?
3. **Making Sense** Using today's observations, explain how the pressure and volume of a gas change in relation to each other.

LESSON

10 Feeling Under Pressure

Boyle's Law

Think About It

A flat basketball does not bounce very well. However, you can push air into the ball with a pump. The air you trap inside pushes on the skin of the ball with a lot of pressure. The amount of air you push into the ball determines how firm and bouncy the ball is. What would happen if you pushed the same amount of air into a smaller ball?

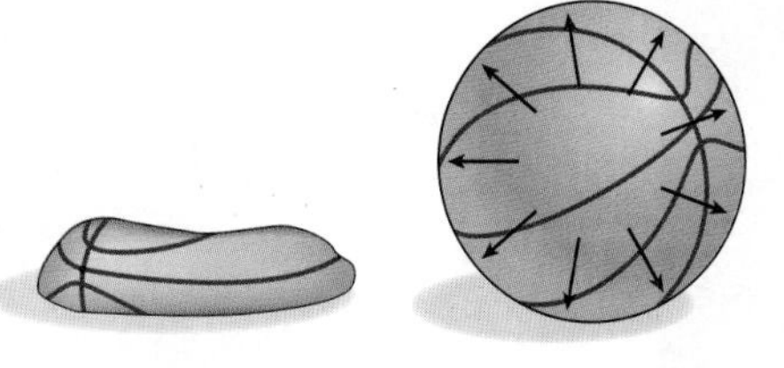

How does gas volume affect gas pressure?

To answer this question, you will explore

1. The Syringe and Scale
2. Graphing Pressure–Volume Data
3. Boyle's Law

Exploring the Topic

1 The Syringe and Scale

In class you examined how the pressure of a gas trapped inside a container is related to the volume the gas occupies. A syringe makes a good container for such an investigation, because it can be sealed off and the volume of air trapped inside the syringe can be measured.

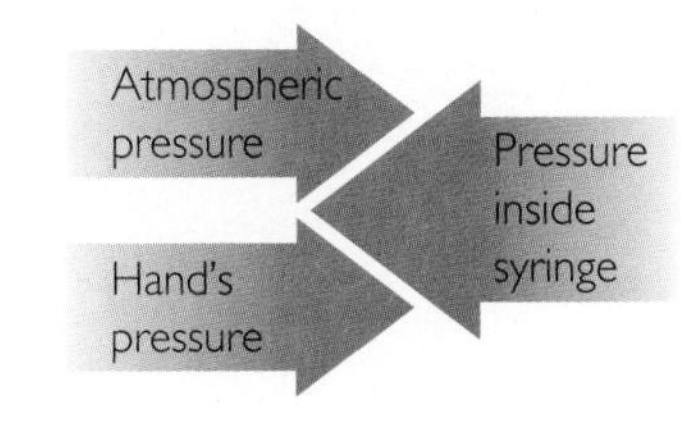

If you push down on the plunger of the syringe, you can feel the pressure of the gas inside the syringe. Air molecules take up space and as you squeeze them into a smaller space, they push back more and more. This is because the same number of molecules occupies a smaller volume, resulting in more collisions between the molecules and the container. More collisions mean greater gas pressure.

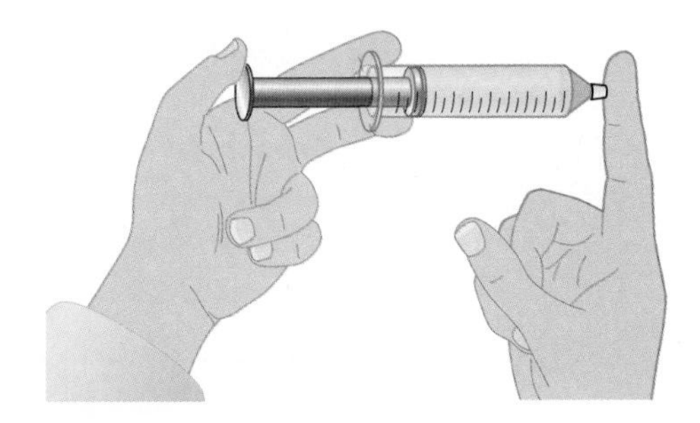

Topic: Gas Pressure
Visit: www.SciLinks.org
Web code: KEY-310

SPORTS CONNECTION

The basketball used by professional players is inflated to a pressure sufficient to make it rebound to a height of 5.4 feet when it is dropped from a height of 6 feet. This test is conducted in a basketball factory, and then that air pressure is stamped on the outside of the ball.

Gathering Data

You can learn more about the relationship between the volume of a gas and its pressure by using a capped syringe and a bathroom scale. As you push down on the plunger of the syringe, the gas pressure inside the syringe increases, and the weight on the scale increases. You can record this quanitative, or numerical, data. To calculate the pressure of the gas, divide the measured weight on the scale by the surface area of the plunger in pounds per square inch, or lb/in^2. Weight is a measure of force.

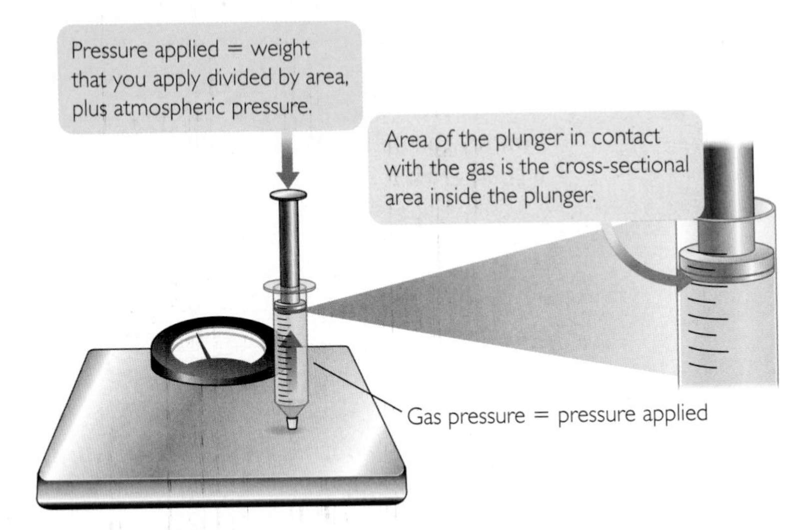

Note that the pressure you apply plus atmospheric pressure is equal to the gas pressure inside the syringe. When you push down, the gas pushes back with the same pressure that is being applied to it.

Important to Know You need to add 14.7 lb/in^2 to the pressure that you apply because the atmosphere is also pushing on the syringe. ◄

Data for this experiment are given in the table. Notice that as the volume decreases, the pressure of the gas inside the syringe increases.

Pressure and Volume Data

Trial	Volume (mL)	Total pressure (lb/in^2)
1	100 mL	15 lb/in^2
2	80 mL	19 lb/in^2
3	60 mL	25 lb/in^2
4	40 mL	38 lb/in^2
5	30 mL	50 lb/in^2
6	20 mL	75 lb/in^2

BIG IDEA If you squeeze a sample of gas into a smaller container, its pressure will increase.

BIOLOGY CONNECTION

During takeoff and landing on an airplane, or at other times when your altitude changes, you may feel the need to "pop" your ears. Your middle ear contains a pocket of air. A tube that leads from the back of the nose to the ear supplies this air. When you swallow, air goes through the tube and into the pocket, so that the pressure inside your ear is the same as the air pressure. You experience the change in pressure for the gas inside your ear to match a change in air pressure as a "pop."

2 Graphing Pressure–Volume Data

Many of the relationships you have explored so far, like mass versus volume and volume versus temperature of a gas, are proportional relationships. The relationship between pressure and volume is not directly proportional. A graph of the data for this experiment results in a curved line that does not go through the origin.

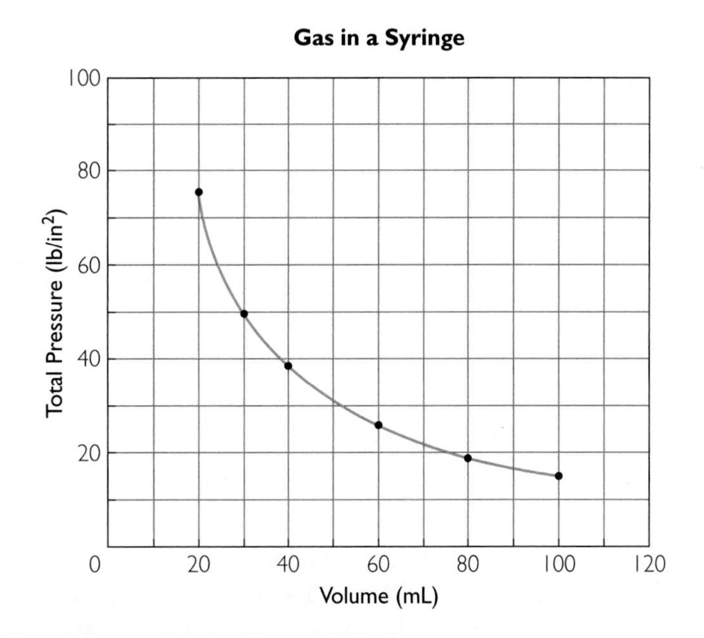

Gas pressure and gas volume are **inversely proportional** to one another. In other words, when one variable increases, the other decreases, and vice versa.

When the volume of the gas in the syringe is very small, the gas pressure is quite high. When the volume of the gas is large, the gas pressure is quite low. However, as long as there are gas molecules in a container, neither the gas pressure nor the gas volume can ever reach zero.

INDUSTRY CONNECTION

Gases are often stored in tanks at high pressure or even at low temperatures so that they're liquid, because otherwise the containers needed for them would be huge. These tanks are usually made of thick metal so that they don't change volume.

3 Boyle's Law

In 1662, a British scientist named Robert Boyle discovered that gas pressure and volume are inversely proportional to each other. This relationship is known as **Boyle's law.**

Boyle's Law

The pressure P of a given amount of gas is inversely proportional to its volume V, if the temperature and amount of gas are not changed. The relationship is $PV = k$ or $P = k(1/V)$, where k is the proportionality constant.

Examine the Pressure and Volume Data table. If you multiply pressure by volume, the product will always be around 1500. (It will vary slightly due to rounding and slight errors in measurement.) Boyle's law expresses this relationship as $PV = k$. In this experiment the proportionality constant, k, is 1500 mL · lb/in^2. [For a review of this math topic, see **MATH** ***Spotlight: Ratio and Proportions*** on page 627.]

Boyle's law can be used to solve problems involving gas pressure and gas volume. Once you have one set of pressure and volume measurements for a gas, you can determine the proportionality constant. With the proportionality constant, you can determine the pressure of the gas at any volume.

Example

Gas in a Syringe

Determine the pressure of the gas inside this syringe if the volume were reduced to 10 mL.

Solution

Start with Boyle's law.	$P = k\left(\frac{1}{V}\right)$
Substitute values for k and V.	$= 1500\ \text{mL} \cdot \text{lb/in}^2\left(\frac{1}{10\ \text{mL}}\right)$
Solve.	$= 150\ \text{lb/in}^2$

You can check this answer: $PV = 150\ \text{lb/in}^2 \cdot 10\ \text{mL} = 1500\ \text{mL} \cdot \text{lb/in}^2$.

Key Terms

inversely proportional
Boyle's law

Lesson Summary

How does gas volume affect gas pressure?

If you squeeze a gas into a smaller volume, it will exert more pressure on its container because a smaller space means more collisions between the molecules and the container. If you let a gas spread out into a huge volume, the gas pressure will get very low because there will be fewer collisions between the molecules and the container. Gas pressure and gas volume are inversely proportional. This means that as one variable increases, the other decreases, and vice versa. This relationship is described by Boyle's law: $P = k(1/V)$.

EXERCISES

Reading Questions

1. Explain how you can use a scale to measure the pressure of the gas in a syringe.
2. Describe how pressure varies as the volume of a trapped gas changes.

Reason and Apply

3. **Lab Report** Write a lab report for the Lab: Boyle's Law. In the results section, provide two graphs of the data. Plot pressure versus volume and plot pressure versus the inverse of the volume, $1/V$. Also discuss how accurate your results are and what factors might contribute to experimental error in this lab. In the conclusion section, explain the relationship between pressure and volume. Explain how your data relate to Boyle's law.
4. You are using a bicycle pump to fill a bicycle tire with air. It gets harder and harder to push the plunger on the pump the more air is in the tire. Explain what is going on.

Title

Purpose: (Explain what you were trying to find out.)
Procedure: (List the steps you followed.)
Results: (Explain what you observed during the experiment.)
Conclusions: (What can you conclude about what you were trying to find out? Provide evidence for your conclusions.)

5. How is the relationship between gas pressure and gas volume different from the relationship between gas volume and gas temperature?

6. Imagine you blow up a balloon before going scuba diving. You put on your gear and descend to 30 ft with the balloon. The total pressure from the ocean at that depth is 2 atm. Assume temperature is constant.

   a. What happens to the volume of the balloon?

   b. What happens to the pressure of the air on the inside of the balloon?

7. This table shows experimental data for gas volume and pressure.

Trial	Volume (mL)	Pressure applied (lb/in^2)
1	60 mL	10 lb/in^2
2	40 mL	15 lb/in^2
3	30 mL	20 lb/in^2
4	15 mL	40 lb/in^2
5	10 mL	60 lb/in^2

   a. Show that $P = k(1/V)$ for each trial. Remember to add 14.7 lb/in^2 to the values in the last column, to account for atmospheric pressure.

   b. Show that $PV = k$ for each trial.

   c. Create a graph of P versus V. What does the graph tell you about the relationship between gas pressure and volume?

   d. Create a graph of P versus $1/V$. What does the graph tell you about the relationship between gas pressure and inverse volume?

   e. Use the graphs from parts c and d to estimate the volume when the pressure is 30 lb/in^2.

   f. Use the graphs from parts c and d to estimate the pressure when the volume is 50 mL.

LESSON

11 Egg in a Bottle

Gay-Lussac's Law

Think About It

Gases are trapped in balloons, soda cans, pickle jars, bicycle tires, and in your lungs if you hold your breath. Some of these containers are flexible and can vary in size, while others are rigid. A gas sample can be described by pressure, volume, and temperature. The gas might be heated, cooled, compressed, taken underwater, left outside overnight, or taken up in an airplane.

How does gas pressure change in flexible and rigid containers?

To answer this question, you will explore

1. Gay-Lussac's Law
2. Gas Law Problems

Exploring the Topic

1 Gay-Lussac's Law

Gases are sometimes stored in rigid gas containers. For example, underwater divers wear a metal gas tank strapped to their backs. In order to supply air to the diver for a sufficient amount of time, a lot of air is pushed or compressed into the tank. The result is that the gas inside the tank is at a high pressure.

With rigid containers the pressure of the air inside the container changes if the temperature changes. Indeed, there is a risk of an explosion if you heat the gas inside a pressurized tank.

Pressure Versus Temperature

Consider a tank containing helium. If this tank is placed in the sun and the temperature increases, the pressure of the gas inside the tank also increases.

Helium Tank Data

Temperature (K)	Volume (L)	Pressure inside tank (atm)	P/T (atm/K)
293 K	55 L	130 atm	0.44
313 K	55 L	138 atm	0.44
283 K	55 L	125 atm	0.44
240 K	55 L	106 atm	0.44

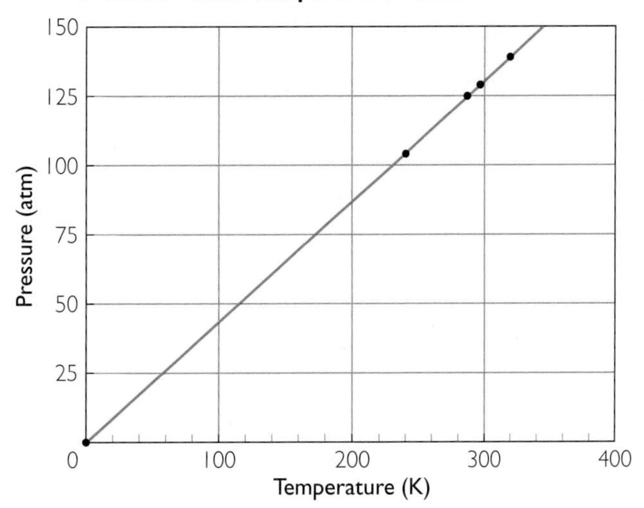

Using the Graph

Because the pressure and the Kelvin temperature are proportional, the graph is a straight line through the origin. If you used the Celsius temperature, the graph would still be a straight line, but it would not go through the origin.

BIG IDEA If you heat a sample of gas in a closed container, its pressure will increase.

If you graph the helium tank data, the points lie on a straight line. The straight line through the data points and the point (0, 0) indicate that there is a proportional relationship between pressure and temperature. Joseph Louis Gay-Lussac, a French scientist, described this proportional relationship in the early 1800s.

HISTORY CONNECTION

Guillaume Amontons (1663–1705) investigated the relationship between pressure and temperature in gases before Gay-Lussac did, though his thermometers were not accurate or precise enough for him to find the exact relationship. He also speculated about the concept of absolute zero, later formalized by Lord Kelvin.

Gay-Lussac's Law

The pressure P and the Kelvin temperature T of a gas are proportional when the volume and the amount of gas do not change. The relationship is $P = kT$. The value of the proportionality constant, k, depends on the specific gas sample.

Example 1

Propane Gas Tank

A 5.0 L propane gas tank on a camping stove contains propane gas at a pressure of 2.0 atm when the temperature is 14 °C (about 57 °F) in the morning. During the day, the tank warms up to a temperature of 34 °C (about 93 °F). What is the pressure of the propane gas inside the tank when the gas temperature is 34 °C?

Solution

The volume of the tank does not change. The temperature of the gas increases causing the pressure of the gas to increase.

You can use the subscript 1 for the first set of conditions, and subscript 2 for the second set of conditions. These are sometimes called initial conditions and final conditions. You need to find P_2, the final pressure.

Initial Conditions		Final Conditions
$P_1 = 2.0$ atm		$P_2 = ?$
$T_1 = 14$ °C	→	$T_2 = 34$ °C
$V_1 = 5.0$L		$V_2 = 5.0$L

SPORTS CONNECTION

The tire pressure in race cars is extremely important. Too much gas pressure and the tires don't have enough contact with the track. Too little gas pressure and the car doesn't handle well. But the tires heat up considerably during a race, increasing the pressure. The average temperature of a tire being removed during a pit stop is about 100 °C.

The variables that are changing are P and T. Use Gay-Lussac's law: $P = kT$. Recall that the temperature must be in kelvins.

Convert the temperature to the Kelvin scale.

$$K = 273 + C$$
$$T_1 = 273 + 14 = 287 \text{ K (in the morning)}$$
$$T_2 = 273 + 34 = 307 \text{ K (during the day)}$$

Determine the value of k using P_1 and T_1.

$$k = \frac{P_1}{T_1} = \frac{2.0 \text{ atm}}{287 \text{ K}} = 0.0070 \text{ atm/K}$$

Use k to solve for P_2.

$$P_2 = k\,T_2$$
$$= 0.0070 \text{ atm/K} \cdot 307 \text{ K}$$
$$= 2.15 \text{ atm}$$

At 34 °C, the pressure of the propane gas in the tank has increased to 2.15 atm.

2 Gas Law Problems

Gas pressure, gas volume, and gas temperature can change depending on the conditions. But no matter what, the gases must obey the gas laws!

There are two general types of containers: rigid and flexible. A rigid container has a volume that doesn't change. A flexible container can vary in size.

Rigid containers

A can and a glass bottle are rigid containers.

Flexible containers

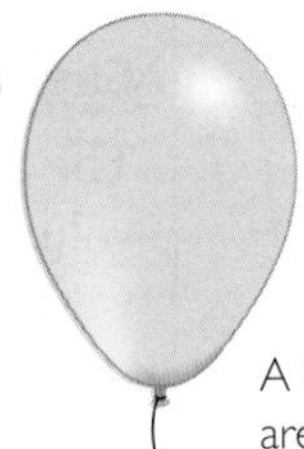

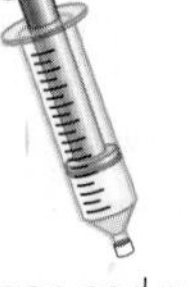

A balloon and syringe are flexible containers.

The type of container has an effect on what happens to the gas inside when the conditions change. For example, Boyle's law and Charles's law apply when a gas sample is in a flexible container, which can change volume. Gay-Lussac's law applies for gas samples in rigid containers.

So far you've been introduced to three of the gas laws. These gas laws apply when any two of the variables, P, V, or T, are changed and everything else stays the same.

Important to Know All temperatures must be converted to the Kelvin scale when completing gas law problems. ◀

Charles's law	$V = kT$	P and amount of gas do not change.	$k = \frac{V}{T}$
Gay-Lussac's law	$P = kT$	V and amount of gas do not change.	$k = \frac{P}{T}$
Boyle's law	$P = k\left(\frac{1}{V}\right)$	T and amount of gas do not change.	$k = PV$

SPORTS CONNECTION

Because tire pressure is so important, many race cars use pure nitrogen gas in their tires. This keeps the moisture in the tires to a minimum. The race car drivers do not want the major fluctuations in tire pressure that are the result of phase changes.

Note that the proportionality constant, k, is a generic symbol and is different for each gas sample and for each gas law.

Example 2

Balloon With Air

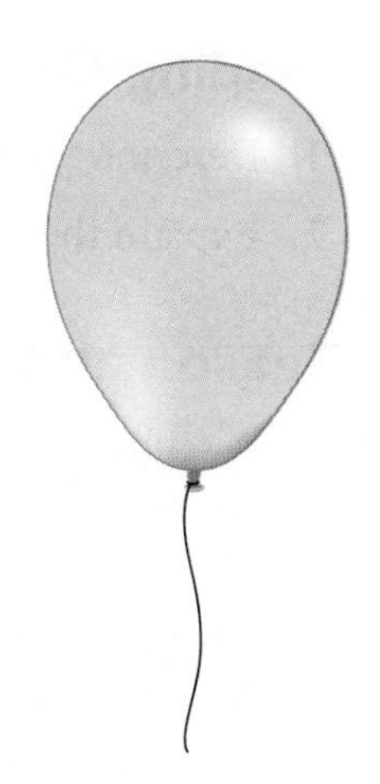

Imagine that you fill a balloon with air to a volume of 1.5 L. The air is at a temperature of 20 °C. You place the balloon in a refrigerator for half an hour until the air in the balloon is at a temperature of 10 °C. What is the new volume of the balloon?

Solution

First figure out which gas law to use. The external pressure on the balloon remains the same because the air pressure in the refrigerator is the same as the air pressure outside. Use Charles's law. You can predict that V will decrease because T decreases.

Initial Conditions		Final Conditions
$P_1 = 1.0$ atm		$P_2 = 1.0$ atm
$T_1 = 20$ °C	→	$T_2 = 10$ °C
$V_1 = 1.5$ L		$V_2 = ?$

Convert temperature to the Kelvin scale.

$$K = 273 + C$$

$$T_1 = 273 + 20 = 293 \text{ K}$$

$$T_2 = 273 + 10 = 283 \text{ K}$$

Determine the value of k.

$$k = \frac{V_1}{T_1} = \frac{1.5 \text{ L}}{293 \text{ K}} = 0.0050 \text{ L/K}$$

Use k to determine V_2.

$$V_2 = k\,T_2 = 0.0050 \text{ L/K} \cdot 283 \text{ K} = 1.4 \text{ L}$$

The balloon has shrunk to a volume of 1.4 L.

Key Term

Gay-Lussac's law

Lesson Summary

How does gas pressure change in flexible and rigid containers?

Gases can be trapped in an assortment of containers and subjected to various conditions of pressure, temperature, and volume. The gas laws allow you to calculate new values for gas temperature, gas pressure, and gas volume when two of these three variables change while the other remains the same. Charles's law and Gay-Lussac's law are proportional relationships: $V = kT$ and $P = kT$. Boyle's law is different because P and V are inversely proportional: $P = k(1/V)$. The proportionality constant, k, is the key to solving gas law problems.

EXERCISES

Reading Questions

1. Describe three ways in which you can change a gas sample.
2. Explain the difference between flexible and rigid containers.

Reason and Apply

3. A scuba-diving tank holds 18 L of air at a pressure of 40 atm. If the temperature does not change, what volume would this same air occupy if it were allowed to expand until it reached a pressure of 1.0 atm?
4. Imagine you fill a balloon with air to a volume of 240 mL. Initially the air temperature is 25 °C and the air pressure is 1.0 atm. You carry the balloon with you up a mountain where the air pressure is 0.75 atm and the temperature is 25 °C.
   a. When the balloon is carried up the mountain, what changes? What stays the same?
   b. The air pressure on the outside of the balloon has decreased. Can the air pressure on the inside decrease so that the pressures are equal? Why or why not?
   c. What happens to the volume occupied by the air inside the balloon? Explain your thinking.
   d. Solve for the new volume of the balloon.
5. The air inside a 180 mL glass bottle is at 1.0 atm and 25 °C when you close it. You carry the glass bottle with you up a mountain where the air pressure is 0.75 atm and the temperature is 5 °C.
   a. The air pressure on the outside of the glass bottle has decreased. What happens to the volume of air inside of the bottle? Explain your thinking.
   b. Do you expect the temperature of the air on the inside of the bottle at the top of the mountain to cool to 5 °C? Explain your thinking.
   c. What happens to the pressure inside the glass bottle? Explain your thinking.
   d. Solve for the new pressure of the gas.

COOKING CONNECTION

A pressure cooker can cook food in a fraction of the time it takes with a regular pot. Because the lid fits tightly, a higher pressure is achieved, resulting in a higher temperature.

LESSON

12 Be the Molecule

Molecular View of Pressure

Think About It

Imagine you have three balloons, one filled with sand, a second filled with water, and a third filled with air. The air balloon has the least mass but it has the largest volume. It is also the most uniform and spherical, indicating that the gas molecules are pushing out in all directions.

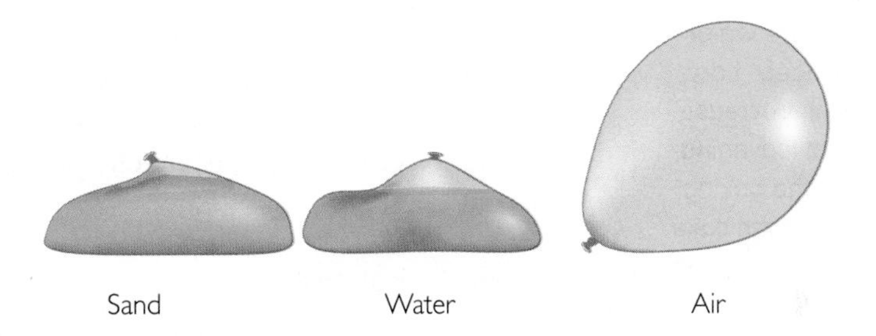

How do molecules cause gas pressure?

To answer this question, you will explore

1. Kinetic View of Pressure
2. Changing Gas Pressure

Exploring the Topic

Topic: Gas Pressure
Visit: www.SciLinks.org
Web code: KEY-310

1 Kinetic View of Pressure

The atoms that make up sand are tightly packed together. Forces between the water molecules keep them close to each other but not quite as close as the molecules in solids. Inside the air balloon, however, there is a lot of movement and activity. The gas molecules are moving rapidly and bouncing off each other and the walls of the balloon.

BIG IDEA Gas pressure is caused by the collisions of molecules or atoms.

Collisions Cause Pressure

According to the kinetic theory of gases, gas molecules are in constant motion. The gas molecules in a sample move at many hundreds of miles per hour in straight line paths. Even though they are very tiny, the molecules inside the container hit each other and the walls of their container. There are so many molecules hitting the walls that the collisions add up to a measurable pressure.

Note that gas pressure is caused by molecules hitting the walls of the container only. When gas molecules collide with each other, they may change the speed or direction, but this does not affect the overall gas pressure.

CONSUMER CONNECTION

Have you ever wondered why boxes of cake mix sometimes include high-altitude directions? At higher altitudes, water boils at a lower temperature, and the air bubbles from the leavening expand more, causing cakes to expand too much while they're cooking. To adjust for high altitudes, bakers usually add more water (so the cake doesn't dry out too much), decrease the amount of leavening, and increase the baking temperature (so the cake won't have as much time to rise before it sets).

The illustration shows a gas trapped in a container with a piston that can move up and down. The temperature and pressure of the gas inside the container are indicated on a thermometer and pressure gauge. Notice that the gas inside the container exerts enough pressure on the piston to hold it up. The piston does not fall downward, even though the gas molecules are very far apart.

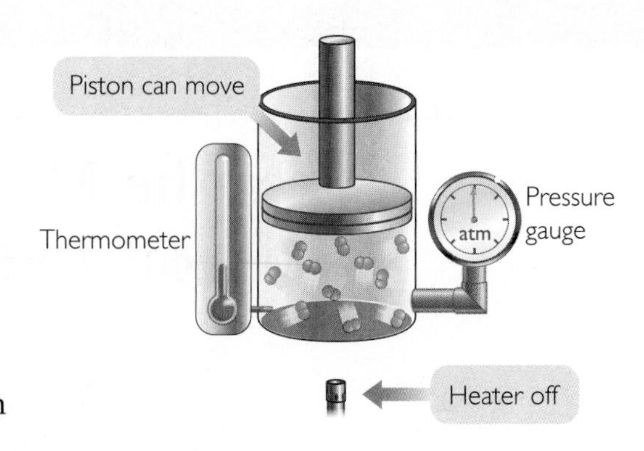

❷ Changing Gas Pressure

Look again at the illustration. Suppose you keep the temperature the same but decrease the volume of the cylinder. What happens to the gas molecules inside the container? There is less space for the same number of molecules. The pressure increases. If you keep the volume the same but heat the gas inside the container, the gas molecules move faster. The pressure increases again. The drawings here illustrate these two changes.

CHANGING VOLUME

Action: The piston is pushed in to decrease the volume. The temperature is held constant by allowing heat to exchange with the surroundings.

Observation: The gas pressure increases.

Explanation: When the gas molecules are forced into a smaller space, they hit the walls of the container more often. The gas pressure is larger. This change is associated with Boyle's law.

CHANGING TEMPERATURE

Action: The gas is heated. The volume is held constant by clamping the piston so that it cannot move.

Observation: The gas pressure increases.

Explanation: At higher temperatures, the gas molecules move faster. They hit the walls of the container harder and more often, which increases the gas pressure. This change is associated with Gay-Lussac's law.

Lesson Summary

How do molecules cause gas pressure?

The collisions of gas molecules with the walls of a container cause gas pressure. If the volume decreases while the temperature stays the same, the molecules collide with the walls more frequently. This causes an increase in gas pressure. If the volume stays the same and the temperature increases, the molecules move faster. They collide with the walls more frequently and they hit the walls harder. This causes an increase in gas pressure.

EXERCISES

Reading Questions

1. Use the kinetic theory of gases to explain why increasing the gas volume decreases the gas pressure.
2. Use the kinetic theory of gases to explain why decreasing the gas temperature decreases the gas pressure.

Reason and Apply

3. Suppose that the pressure of a gas in a cylinder has increased. What might have changed to cause this? Explain your thinking.
4. Why is it dangerous to heat a gas in a sealed container?
5. Can you decrease the volume of a gas to zero? Why or why not?
6. When you fly in a commercial airplane you often feel the change in air pressure in your ears. It feels painful. Use the kinetic theory of gases to explain what you think is happening.
7. At sea level, the pressure of trapped air inside your body is 1 atm. It is equal to the pressure of air outside your body at sea level. Imagine you do a deep sea dive. You descend slowly to a depth where the pressure outside your body is 3.5 atm.
   a. How does the volume of your lungs compare at 3.5 atm with the volume at sea level?
   b. Why is it dangerous to hold your breath and ascend quickly?
8. As a helium balloon floats up into the sky, the pressure of the atmosphere decreases as altitude increases. What do you expect will happen to the volume of the balloon? Explain your thinking.
9. A gas is trapped in a cylinder with a piston as shown. The gas pressure is 1.0 atm, the volume is 500 mL, and the temperature is 300 K.
   a. Draw the cylinder after the volume has been changed to 1000 mL. The gas temperature is 300 K.
   b. Does the pressure inside the cylinder increase or decrease? Explain your thinking.
   c. Which gas law applies?
10. **WEB RESEARCH** Look up information on steam engines in a book or on the Web. Explain how a steam engine works.

BIOLOGY CONNECTION

The diaphragm is a muscle that separates the chest from the abdomen. When the diaphragm moves up, air flows out of your nose and mouth. When the diaphragm moves down, the volume of the chest cavity increases, allowing air to flow in.

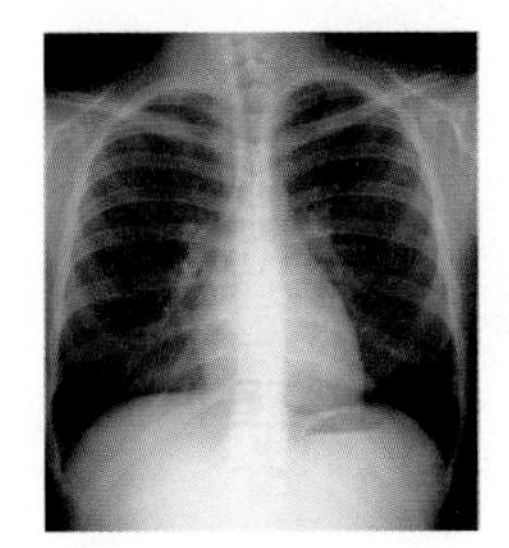

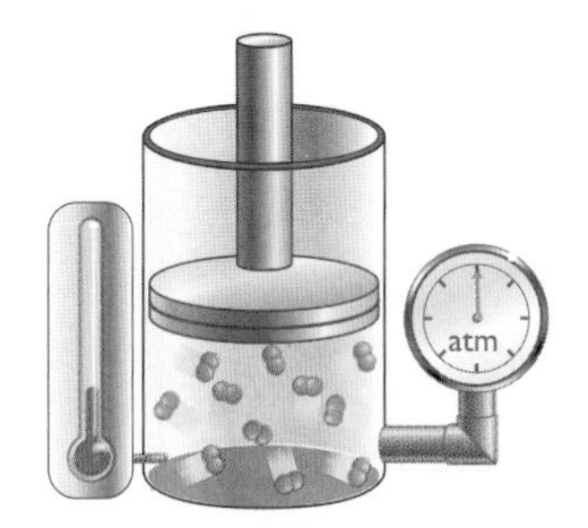

LESSON

13 What Goes Up

Combined Gas Law

Think About It

In some circumstances more than two gas variables change at once. For example, if you released a balloon at sea level and it rose into the atmosphere, you would be dealing with three changing gas variables: pressure, temperature, and volume. What effect does this have on the balloon?

What are the relationships among pressure, volume, and temperature for a sample of gas?

To answer this question, you will explore

1. Changing Pressure, Volume, and Temperature
2. Combined Gas Law

Exploring the Topic

1 Changing Pressure, Volume, and Temperature

Variations in Pressure and Temperature with Altitude

Both the temperature and the pressure of the atmosphere decrease with altitude. These two conditions (temperature and pressure) will naturally vary as a balloon rises and will affect the overall volume of the balloon. The sample of gas inside the balloon responds to the conditions outside the balloon.

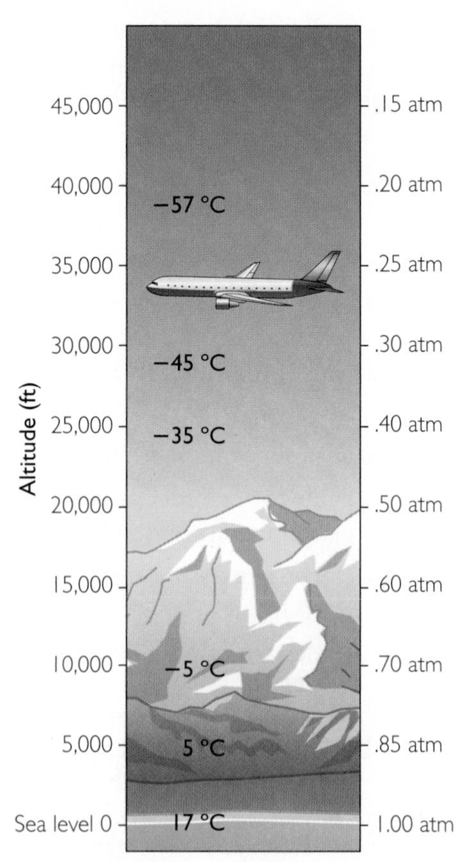

The illustration indicates the approximate average pressure and temperature of the atmosphere at various altitudes. At an altitude of 40,000 ft, the air pressure is only 0.2 atm and the air temperature is around a chilly −57 °C (−70 °F). Chances are a party balloon will burst before reaching that altitude.

Weather Balloons

Weather balloons are large balloons filled with helium or hydrogen gas that carry instruments aloft to study the atmosphere at high altitudes. As the weather balloon rises, the instruments measure atmospheric conditions like the temperature, relative humidity, and

METEOROLOGY CONNECTION

A weather balloon has three parts: the elastic balloon, the instrument package, called a *radiosonde,* and the parachute. Weather balloons are released twice a day from sites all around the world. There are about 900 release sites worldwide; 95 of these sites are in the United States. Data from weather balloons is fed into National Weather Service supercomputers across the country. This information is used to help predict weather around the country.

pressure. This information is relayed back to a weather station by a transmitter. Wind speed and wind direction can be calculated from the collected data.

In order to follow the changes to the gas inside a weather balloon, you must first know the starting conditions of the balloon. Imagine a clear cool morning, where the air temperature is 17 °C and the atmospheric pressure is 1 atm. A team of meteorologists inflates a weather balloon with helium to a volume of 8000 L. If released, this weather balloon will rise no matter what the outside conditions. This is because the balloon is full of helium and helium is less dense than air.

Conflicting Outcomes

Consider the changing conditions on the volume of the rising weather balloon. Charles's law indicates that the volume of the balloon will shrink with decreasing temperature. However, Boyle's law indicates that as the air pressure outside the balloon decreases, the balloon will expand in volume.

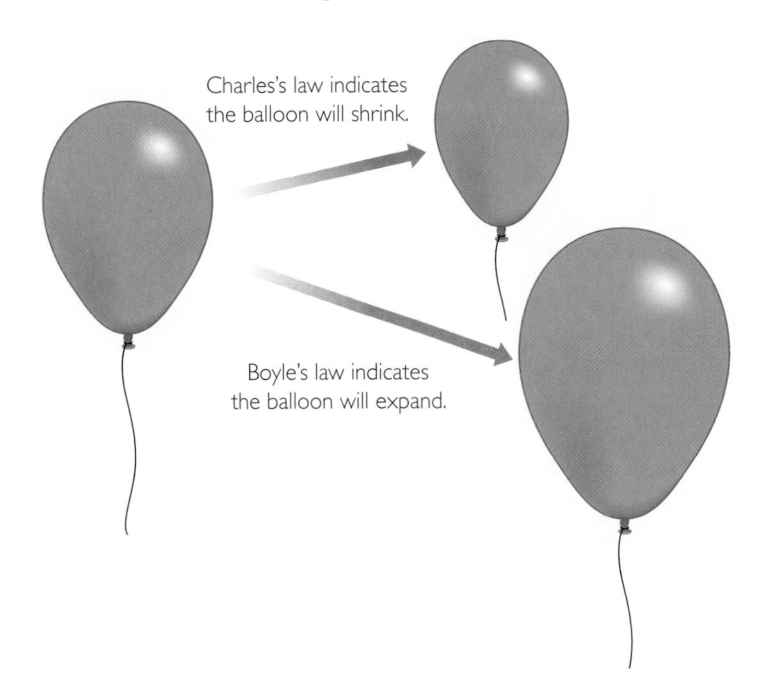

There are two conflicting outcomes. Will the balloon expand or shrink?

2 Combined Gas Law

For a sample of gas, the relationship among gas pressure, temperature, and volume can be expressed in a mathematical equation called the **combined gas law.**

Combined Gas Law

If you know the temperature, T, pressure, P, and volume, V, of a gas, you can determine k. Then you can use k to determine the volume of the gas for other pressures and temperatures.

$$k = \frac{PV}{T}$$

Example

Weather Balloon at 25,000 Ft

A weather balloon is filled at sea level with 8000 L of helium. The pressure at sea level is 1 atm and the temperature is 17 °C. Calculate the volume of the balloon at 25,000 ft when the atmospheric pressure has decreased to 0.4 atm.

Solution

Because pressure, volume, and temperature are all changing, you need to use the combined gas law. The first step in using the combined gas law is to make note of all the values for *P*, *V*, and *T*. Notice that we already know five out of the six values. Remember to convert °C to K. Then it is a matter of solving for V_2.

$P_1 = 1.0$ atm	$P_2 = 0.40$ atm
$V_1 = 8000$ L	$V_2 = ?$
$T_1 = 290$ K	$T_2 = 238$ K

Use the combined gas law.

$$k = \frac{P_1V_1}{T_1} = \frac{P_2V_2}{T_2}$$

Determine the proportionality constant, *k*.

$$k = \frac{P_1V_1}{T_1}$$

$$= \frac{1.0 \text{ atm} \cdot 8000 \text{ L}}{290 \text{ K}}$$

$$= 27.6 \frac{\text{atm} \cdot \text{L}}{\text{K}}$$

Use *k* to solve for V_2.

$$V_2 = \frac{kT_2}{P_2}$$

$$V_2 = \frac{\left(27.6 \frac{\text{atm} \cdot \text{L}}{\text{K}}\right)(238 \text{ K})}{0.40 \text{ atm}}$$

$$V_2 = 16{,}422 \text{ L}$$

The balloon has expanded to about 16,000 L at this altitude.

[For a review of this math topic, see **MATH** ***Spotlight: Accuracy, Precision, and Significant Digits*** on page 617.]

Pressure Versus Temperature

The example shows that decreasing air pressure has a greater effect on a weather balloon than decreasing temperature does. The volume of the balloon continues to expand as the balloon goes up in altitude, in spite of the decreasing temperature. The balloon eventually ruptures. This is why there is a parachute on every weather balloon, so that the weather instruments can be carried safely to the ground.

Key Term

combined gas law

Lesson Summary

What are the relationships among pressure, volume, and temperature for a sample of gas?

If there is a situation involving a gas in which volume, temperature, and pressure are all changing, then you can use the combined gas law to determine the effects of changing two variables on the third variable. The value of the proportionality constant, k, is equal to PV/T. Therefore, $P_1V_1/T_1 = P_2V_2/T_2$. Remember that the combined gas law works only in situations where the amount of gas involved remains the same.

EXERCISES

Reading Questions

1. What is the combined gas law? When do you apply it?
2. Does the temperature of the atmosphere have any effect on the volume of a weather balloon as it rises? Explain.

Reason and Apply

3. A gas collected in a flexible container at a pressure of 0.97 atm has a volume of 0.500 L. The pressure is changed to 1.0 atm. The amount of gas and the temperature of the gas do not change.
   a. Will the volume of the gas increase or decrease? Explain.
   b. Which equation should you use to calculate the new volume of the gas?
   c. Calculate the volume of the gas at a pressure of 1.0 atm.
4. The pressure in an automobile tire was 1.0 atm at 21 °C. After driving for an hour, the tire heated up to 55 °C. Assume that the tire remained at constant volume and no gas escaped from the tire.
   a. Did the pressure inside the tire increase or decrease? Explain.
   b. Which equation should you use to calculate the new pressure of the tire?
   c. Calculate the pressure of the tire at 55 °C.

Topic: Gas Laws
Visit: www.SciLinks.org
Web code: KEY-306

5. The helium inside a balloon has a volume of 1.5 L, a pressure of 1.0 atm, and a temperature of 25 °C. The balloon floats up into the sky where the pressure is 0.95 atm and the temperature is 20 °C.
   a. Will the volume of the balloon increase or decrease? Explain.
   b. Which equation should you use to calculate the new volume of the balloon?
   c. Calculate the volume of the balloon at 20 °C and 0.95 atm.
6. Someone leaves a steel tank of nitrogen gas in the sun. The tank was under a pressure of 150 atm at 27 °C to begin with. After several hours the internal temperature rises to 55 °C. What is the pressure in the tank now?
7. An example of a trapped gas is a car airbag. The airbag is designed to inflate to 65 L if you have an accident in a location where the temperature is 25 °C and the air pressure is 1.0 atm.
   a. Determine the size of the airbag if you have an accident in the mountains where the temperature is −5 °C and the air pressure is 0.8 atm.
   b. Suppose you have an accident in a location in which the airbag inflates to 60 L. What has to be true about the temperature and air pressure in this location?

LAB High and Low Air Pressure

Purpose

To find the connection between air pressure and the weather forecast.

Materials

- 2 L plastic bottle, or vacuum chamber and vacuum pump
- warm tap water
- long safety matches

Safety Instructions

Safety goggles must be worn at all times.

Procedure

1. Put a small amount (about 15 mL) of warm water into the plastic bottle.
2. Light a match. Blow it out and then hold it inside the bottle to collect some smoke. Quickly remove the match and put the cap tightly on the bottle. Shake the chamber to add moisture to the air inside.
3. Squeeze the bottle and release. Pump a few times while observing the air inside the bottle.
4. Repeat the experiment, but this time use 10 mL of cold water (above ~80 °C). Next, repeat the experiment with a dry bottle. Do not add water, just create smoke, close the bottle, and squeeze and release.

Observations

1. What did you observe inside the bottle when you squeezed and released the bottle?
2. What happens to the pressure, volume, and temperature of the air inside the bottle when you squeeze it? When you release it?
3. What gas law was operating during this experiment? Explain your choice.
4. How does this gas law explain what's going on in the bottle?
5. What did you observe when you used a dry bottle?

LESSON 14 Cloud in a Bottle

High and Low Air Pressure

Think About It

Weather maps typically contain information about high air pressure and low air pressure. You often hear meteorologists refer to high- and low-pressure systems. But how does this information help you figure out whether it will be rainy or sunny?

How are areas of high and low air pressure related to the weather?

To answer this question, you will explore

1. High and Low Air Pressure
2. Cloud Formation

Exploring the Topic

1 High and Low Air Pressure

Meteorologists measure air pressure at different locations on the Earth's surface. These readings of air pressure are summarized on an air pressure map similar to this one.

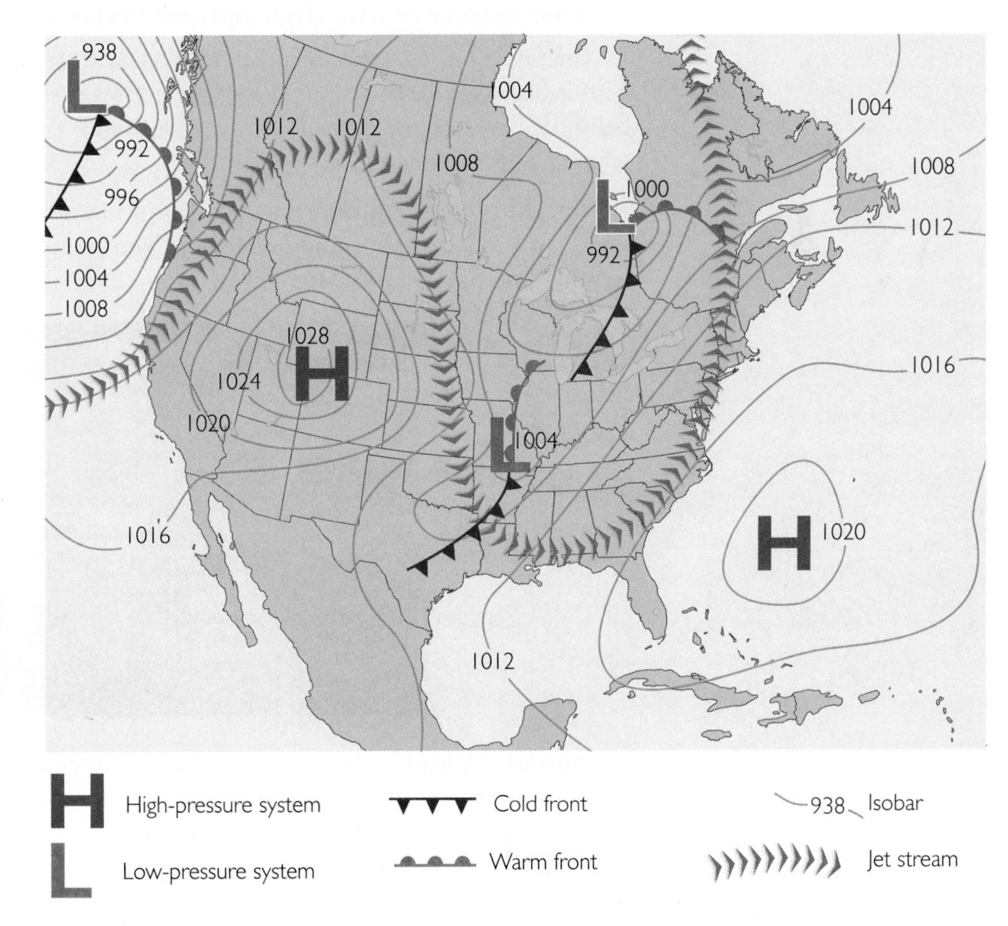

The lines labeled with numbers on the map are called *isobars*. They mark areas with the same air pressure. The numbers correspond to air pressure measurements in millibars. One atmosphere is equal to 1013.25 millibars. Notice that each new line represents an air pressure difference of 4 millibars.

Low-pressure areas are associated with lower numbers, high-pressure areas with higher numbers. However, high and low air pressure are measured relative to the surrounding air. In other words, a high-pressure air system is not associated with a certain number, but simply shows an area that stands out as having higher air pressure than the areas around it. If you examine the map, you can also see that the isobars encircle areas with high- and low-pressure systems.

The map also shows the jet stream as a series of blue arrows traveling from west to east across the continent. Warm and cold fronts are also included. Take a moment to look for patterns.

- The jet stream weaves between the H's and L's instead of going through the middle of them.
- The jet stream appears to follow the general curvature of the isobar lines. For this particular map, this means the jet stream will guide western storms up into Canada.
- The location of the weather fronts is also associated with air pressure. Notice that the weather fronts appear to be centered on the low-pressure areas.
- There are no weather fronts near the high-pressure areas, so you wouldn't expect to see any storms in these areas.

Weather Associated with High and Low Pressure

In general, low-pressure areas are associated with fronts, clouds, and precipitation. This is because when warm and cold air masses meet, the warm air rises, leaving behind an area of low pressure. Areas of lower air pressure coincide with areas where fronts are located.

In fact, the centers of all storms are areas of low air pressure. When forecasters say a low-pressure area is moving toward your region, this usually means cloudy weather and precipitation are on the way.

In contrast, high-pressure areas are places where skies are clear and pleasant. When forecasters say that a high-pressure system has moved into your area, it means you can expect clear skies and sunny days. High-pressure centers mean that denser air must be sinking, which inhibits precipitation and cloud formation.

② Cloud Formation

When you look up at the sky and see big, puffy, white clouds, it's hard to believe that they are just a bunch of condensed water droplets suspended in the air. Because they are a mixture of water droplets and air, our eyes see them as white or sometimes as gray. But what causes clouds to form in the first place?

Because there is no volume or container associated with Earth's atmosphere, it makes sense to explain weather by focusing on pressure and temperature changes in air masses.

Warm air rises because it is less dense than cold air. As it rises, its temperature and pressure decrease. As the water vapor cools, it condenses into droplets, forming clouds.

Enough water vapor must be present in the air in order for clouds to form. This is why more clouds form over rainforests than over deserts. On warm days, large puffy clouds are observed at lower altitudes. On cold days, if there are clouds at all, they tend to be wispy and high in the sky.

Lesson Summary

How are areas of high and low air pressure related to the weather?

Air pressure varies across the surface of Earth. On air pressure maps. areas of high and low air pressure are indicated with H's and L's. Areas of high pressure indicate clear skies and sunny days. Areas of low pressure are associated with storms. In order for clouds to form, there must be moisture in the air. If this moisture is in a warm air mass, it travels upward, cools off, and undergoes a phase change, forming clouds composed of tiny water droplets. Clouds form under conditions of low temperature and low pressure.

EARTH SCIENCE CONNECTION

Above the equator, winds around high-pressure areas move in a clockwise direction. The winds around low-pressure areas move in a counterclockwise direction. This is one reason why the jet stream loops up and over highs and down and under lows as it travels across the United States. Below the equator, winds around high- and low-pressure systems move in the opposite directions.

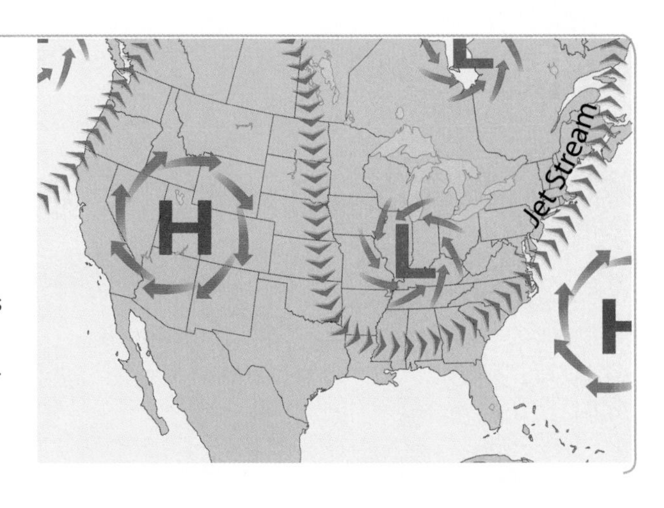

EXERCISES

Reading Questions

1. What weather is associated with high air pressure? Low air pressure?
2. Describe how a cloud forms. Start with water in a lake.

Reason and Apply

3. One atmosphere of pressure, 1 atm, is equivalent to 1,013.25 millibars. Convert the pressure from millibars to atmospheres for the two highs and the two lows on the air pressure map on page 317. What do you notice about these values?

4. **WEB RESEARCH** Find out about different types of clouds and the conditions under which they form. Try to explain how the type of cloud relates to the conditions.

SECTION

II

SUMMARY

Pressing Matter

Weather Update

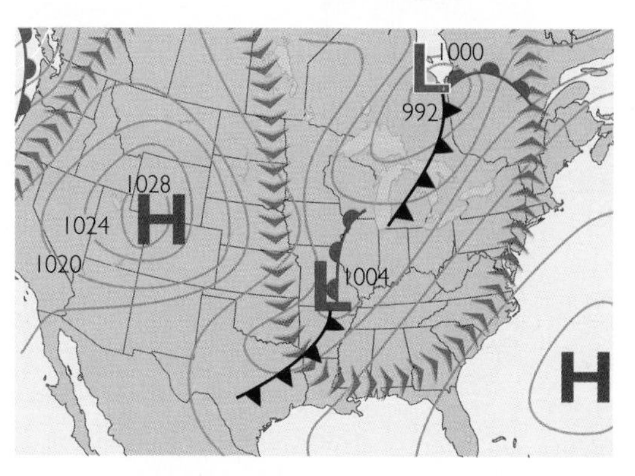

Air pressure at the Earth's surface is constantly shifting and moves from place to place and day to day. These changes and variations in atmospheric pressure greatly influence the weather. Gas molecules are free to move to fill up the space they occupy. They collide with each other and whatever objects they come in contact with, resulting in pressure. When a gas is heated, the molecules move faster and collide more frequently. So gas pressure is directly related to gas temperature.

Studying gas behavior in small containers helps us to explain the behavior of gases in the atmosphere. Three gas laws relate gas pressure to temperature and volume—Charles's law, Gay-Lussac's law, and Boyle's law. The combined gas law relates pressure, volume, and temperature.

Key Terms

sublimation
evaporation
pressure
atmospheric pressure
atmosphere (atm)
inversely proportional
Boyle's law
Gay-Lussac's law
combined gas law

Review Exercises

1. Use the kinetic theory of gases to explain how temperature affects the pressure of a gas.
2. Explain why a weather balloon pops when it reaches a sufficiently high altitude.
3. A cylinder with a movable piston contains 26.5 L of air at a pressure of 1.5 atm. If the temperature is kept the same but the volume of the container is expanded to 50.0 L, what is the new pressure of the gas?
4. A gas sample in a rigid container has a pressure of 3.0 atm at 350 K. If the temperature is raised to 450 K, what will the new gas pressure be?
5. A sample of gas in a flexible container has a volume of 7.5 L, a pressure of 2.5 atm, and a temperature of 293 K. What will the new pressure be if the volume expands to 10.0 L and the temperature is raised to 315 K?

PROJECT *High and Low Pressure*

In a newspaper or on the Web, find a weather map of the United States.

1. Find the areas of high and low pressure.
2. Explain where the air is moving up and where the air is moving down.
3. Where do you expect to find stormy weather? Explain your thinking.

SECTION

III Concentrating Matter

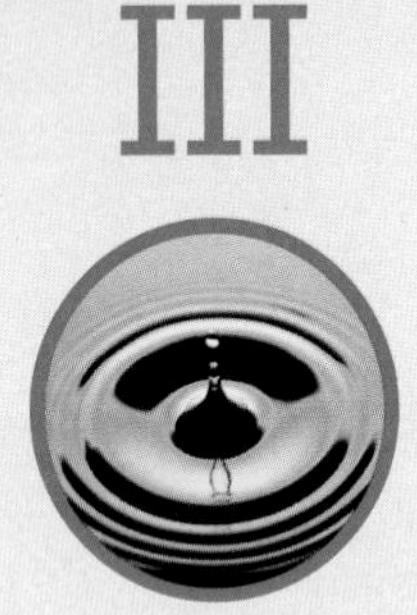

Lightning is an extreme weather phenomenon that typically occurs during thunderstorms.

Gas molecules are too small and numerous to count. However, the mathematical relationships among gas temperature, pressure, volume, and number of molecules allow scientists to determine the exact number of molecules in a gas sample. Knowing the number of molecules of water in a particular volume of air helps scientists to predict when precipitation will occur. Air pressure and air temperature greatly affect the amount of water vapor in the air. When moist air cools, the water vapor in the air changes from a gas to a liquid. When extreme conditions of temperature and pressure occur, hurricanes can form.

In this section, you will study

- variations in the atmosphere with changing altitude
- how to determine the numbers of gas molecules in a sample
- the relationships among pressure, temperature, volume, and number of molecules in a gas
- humidity
- conditions that lead to hurricane formation

LESSON

15 *n* Is for Number

Pressure and Number Density

Think About It

Airplanes fly at about 35,000 ft (~6.6 mi) above Earth. At this altitude the air pressure is less than 0.25 atm and the air temperature is less than −45 °C. It is very cold, and the air is very thin up there! You might wonder how many oxygen molecules are available at this air pressure to breathe.

How is the number of gas molecules in a sample related to pressure?

To begin to answer this question, you will explore

1. Pressure and Number Density
2. Measuring Air Pressure

Exploring the Topic

1 Pressure and Number Density

The illustration provides information on the *troposphere,* or lower atmosphere, where clouds form and weather takes place. The containers with spheres show how the density of air molecules decreases with altitude.

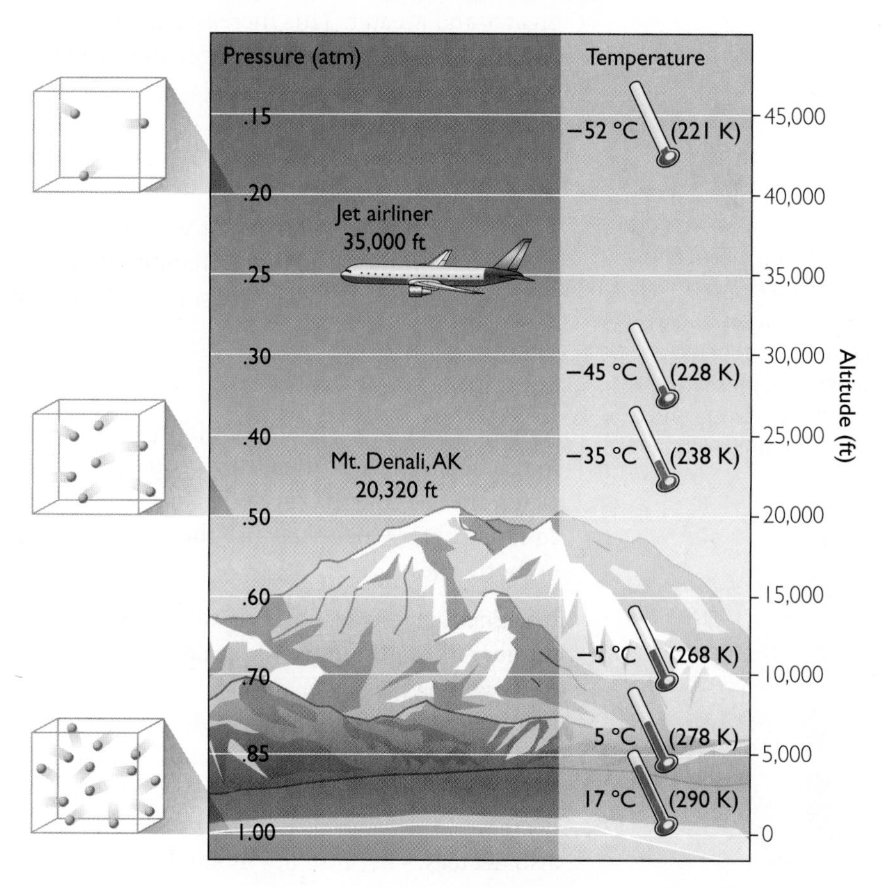

EARTH SCIENCE CONNECTION

The stratosphere is the layer in the atmosphere above the troposphere. It extends from about 10 miles to 30 miles above Earth's surface. Very few airplanes venture into the stratosphere because there are not enough gas molecules to create the lift needed to support the plane.

Chemists use the variable n to refer to the number of molecules, atoms, or particles they are considering. The number of air molecules per unit of volume is referred to as the **number density,** n/V.

The number density of air molecules is relatively low at high altitudes. This is why it gets more difficult to breathe at higher altitudes. There are fewer air molecules per volume of air.

If there are fewer gas molecules in a certain volume, then the air pressure will be lower. Air pressure is proportional to number density. This relationship can be expressed by the equation

$$P = k\left(\frac{n}{V}\right)$$

The decrease in air pressure with increasing altitude is due mainly to a decrease in the number density of gas molecules.

BIG IDEA The more gas molecules you have in a space, the greater the gas pressure will be.

Example

Bicycle Tire

Use the kinetic theory of gases to explain why the air pressure increases as you pump up a bicycle tire.

Solution

As you pump up a tire, the number of gas molecules per unit of volume becomes greater and greater. This increases the number of collisions between molecules and the walls of the tire. The pressure inside the tire increases until the tire is almost rigid.

HISTORY CONNECTION

The "weather glass" is an old-fashioned instrument for predicting the weather. The liquid is higher in the spout than in the bottle. This indicates that the atmospheric pressure is lower than average and a storm is approaching.

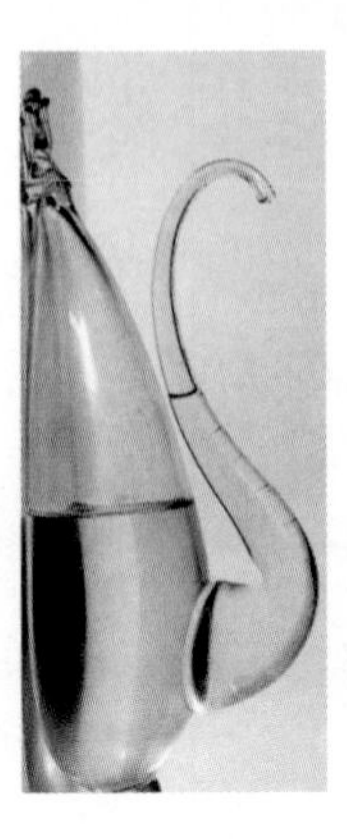

2 Measuring Air Pressure

There are a number of different ways to measure air pressure. Commonly, either a manometer or a barometer is used. Both instruments measure the height of a liquid in a tube.

Manometer

A manometer consists of a liquid in a U-tube as shown in the illustration. If the U-tube is partially filled with water, the water levels on both sides of the U will be at equal heights. This is because the air pressure is the same on both sides.

Imagine that you seal off one side of the U-tube with a stopper so that the number of molecules in the sealed-off area cannot change. The heights of the water will be the same, at least at first. However, if the outside air pressure increases, it will push on the water in the open end causing the water to move up on the side with the stopper, compressing the air on that side. The air pressure of the trapped gas increases because n/V increases. If the outside air pressure decreases, the water in the open end will rise. The air pressure of the trapped gas also decreases because n/V is smaller.

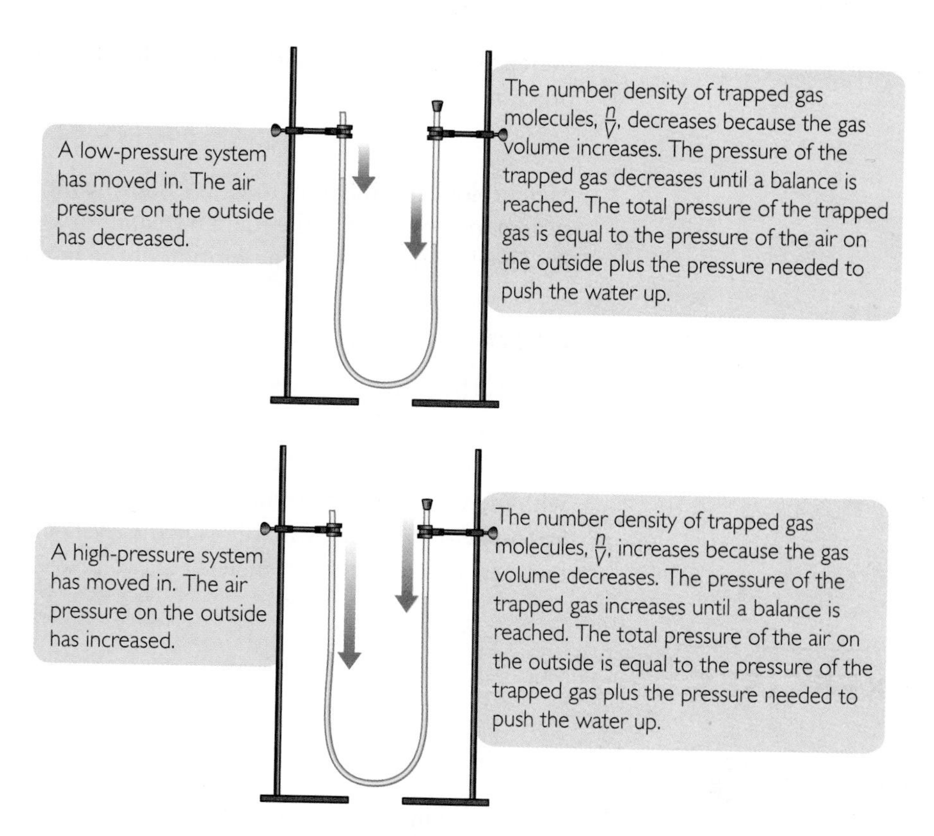

In a U-tube manometer, the water level on the closed side *rises* when the air pressure in the atmosphere increases. The water level on the closed side *falls* when the air pressure in the atmosphere decreases.

HISTORY CONNECTION

Evangelista Torricelli invented the barometer. Pressure is still sometimes reported in torrs. A torr is equivalent to a millimeter of mercury, or mm Hg.
1 atm = 760 mm Hg = 760 torr

Barometer

A barometer consists of a tube filled with liquid, which is inverted in a dish of the same liquid. The surface of the pool of liquid is exposed to the atmosphere. There is no gas pressure inside the tube.

The barometer shown in the illustration measures the difference in pressure on two surfaces of a liquid. The pressure on the liquid inside the tube is close to zero, so it is the mass of liquid in the tube that balances the air pressure on the outside.

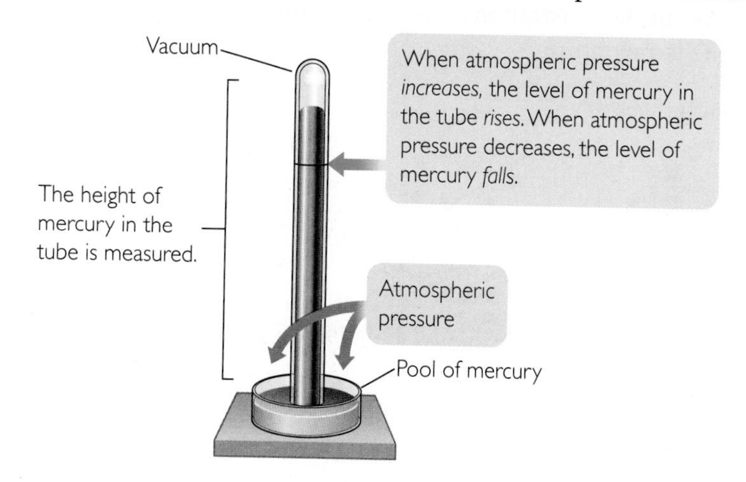

Important to Know The liquid does not spill out of the inverted tube because the force of the air pushing on the surface of the liquid is greater than the weight of the liquid pushing down in the inverted tube. ◄

Mercury is often the liquid of choice for barometers due to its high density. Air pressure is sometimes expressed in inches of mercury or millimeters of mercury.

$$1 \text{ atm} = 29.9 \text{ in Hg} = 760 \text{ mmHg}$$

Meteorologists will sometimes say that the mercury is *rising* when a high-pressure system arrives, or that the mercury is *falling* when a low-pressure system moves in. They are referring to the motion of the liquid mercury in the inverted tube.

Key Term

number density

Lesson Summary

How is the number of gas molecules in a sample related to pressure?

Number density is a measure of the number of molecules, atoms, or particles per unit of volume, n/V. The number density of air molecules in the atmosphere decreases with increasing altitude resulting in lower gas pressure. It is possible to monitor atmospheric pressure by monitoring the level of a liquid in a tube that is sealed at one end. Manometers and barometers are two instruments that measure gas pressure.

EXERCISES

Reading Questions

1. How can you account for the decrease in air pressure with increasing altitude?
2. Use the kinetic theory of gases to explain the relationship between number density and gas pressure.
3. Explain the difference between a manometer and a barometer.

Reason and Apply

4. What effect does changing the number of gas particles in a container have on the gas pressure in the container? Use the kinetic theory of gases in your explanation.
5. Name three ways that you can increase the pressure of a tire.
6. The boxes in the illustrations show tiny samples of air. Assume they are at the same temperature.

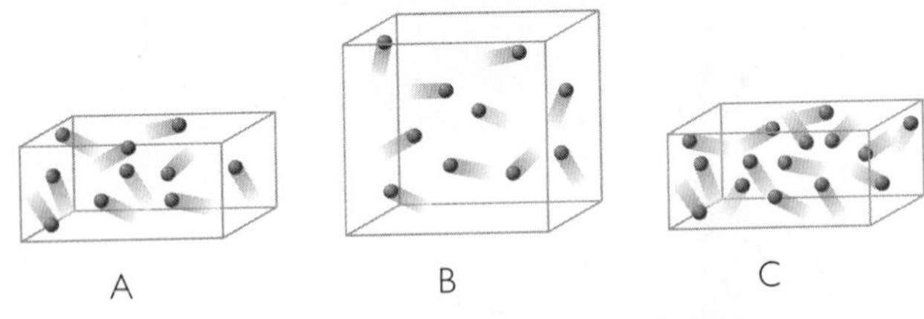

   a. List the samples in order of increasing gas pressure. Explain your reasoning.
   b. Sketch a volume of air that has a pressure in between the pressures in boxes A and B.
7. Explain how the mercury level in a barometer changes when a high-pressure system moves into a region.
8. Explain why the liquid in the barometer does not spill out.

CONSUMER CONNECTION

An aneroid barometer contains a small flexible metal box called a cell. Small changes in air pressure cause the cell to expand or contract. These changes are amplified by mechanical levers and translated to a needle on a dial.

LESSON

16 STP

The Mole and Avogadro's Law

Think About It

Suppose you have two balloons, one filled with helium and the other with carbon dioxide. The pressure, temperature, and volume of the two gases are identical. However, the masses of the balloons are different. While the helium balloon floats, the carbon dioxide balloon sinks. How can you figure out the number of gas particles in each balloon?

How do chemists keep track of the number of gas particles?

To begin to answer this question, you will explore

1. Counting Gas Particles
2. Avogadro's Hypothesis

Exploring the Topic

1 Counting Gas Particles

If you take two air samples of identical volume in the same room, you might predict that both samples will contain the same number of air molecules. A smaller air sample from this room will contain fewer air molecules than a larger air sample.

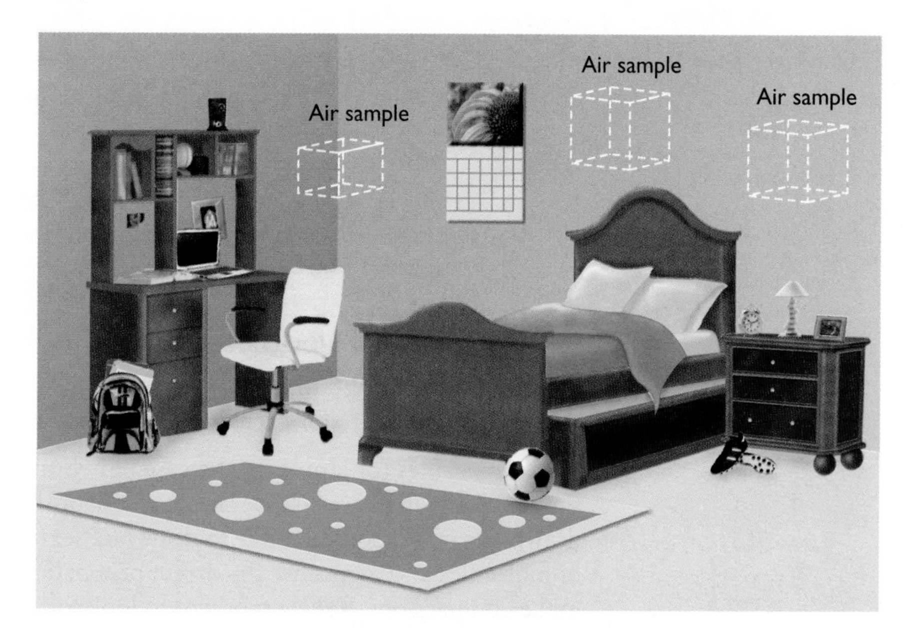

Because the air molecules are spread uniformly over the space the gas occupies, the number of air molecules, n, is proportional to the volume, V, the gas occupies.

SPORTS CONNECTION

A hot air balloon is not a closed container. The flame from a propane burner heats the air inside the balloon. As the gas is heated, it expands and some moles of gas molecules escape through the opening at the base. This decreases the number density, n/V, and the mass density, m/V, of the gas inside the balloon. The lower mass density of the captured gas causes the balloon to rise.

A Mole of Gas Particles

Gas particles are far too numerous to count individually. So, chemists have defined a counting unit for small particles called a **mole.** The abbreviation for mole is mol. A mole represents a very large number of gas particles. A mole is equal to 602,000,000,000,000,000,000,000 molecules. This is 602 sextillion! This number is also called **Avogadro's number,** after Amadeo Avogadro, a 19th century Italian scientist.

The number of gas particles in a sample is referred to in terms of moles of gas. Keep in mind that when referring to 1 mole, 2 moles, or half a mole of gas particles, these quantities all refer to very large numbers. For example, 2 moles of gas particles is 1,204,000,000,000,000,000,000,000 particles (1,204 sextillion). Half a mole of gas particles is 301,000,000,000,000,000,000,000 (301 sextillion).

Standard Temperature and Pressure, STP

Chemists have figured out how to calculate the number of gas particles in a sample of gas if they know the temperature, volume, and pressure of the sample. However, when comparing gases, it is convenient to define just one set of conditions. Chemists have chosen a gas pressure of 1 atm and a gas temperature of 273 K as the conditions to compare gases. These conditions are called **standard temperature and pressure,** or **STP.** At STP, 1 mole of gas molecules occupies a volume of 22.4 L.

Example 1

Two Moles of Air Molecules

Suppose you want to collect 2.0 mol of air molecules at STP. What size box do you need?

Solution

Because 1 mol of gas molecules has a volume of 22.4 L at STP, 2 mol has a volume of 44.8 L. So, you need a box with a volume of 44.8 L.

BIG IDEA The mole is a counting unit. One mole of gas particles at standard temperature and pressure occupies a volume of 22.4 liters.

2 Avogadro's Hypothesis

Several cylinders contain samples of different gases. The cylinders are of varying sizes at STP. The data for each cylinder is shown in the table on the next page.

A few molecules (or atoms, in the case of helium and neon) are drawn in the table to help you visualize the gas. Obviously, only a few gas particles are drawn because it is impossible to draw 602 sextillion. However, the numbers of molecules and atoms are in correct proportion to each other so that you can make comparisons. Take a moment to look for patterns.

Notice that the mass of the gas sample depends on the identity of the gas. A hydrogen molecule has a mass of 2.0 amu, while each neon atom has a mass of 20 amu. So, 0.25 mole of neon gas has ten times the mass of 0.25 mole of hydrogen gas.

Various Gases at STP

Sample	1	2	3	4	5
Gas	H_2	Ne	CO_2	O_2	He
n (moles)	0.25	0.25	0.50	0.50	1.0
Volume	5.6 L	5.6 L	11.2 L	11.2 L	22.4 L
Mass	0.50 g	5.0 g	22.0 g	16.0 g	4.0 g

The volume of each cylinder is determined by the number of gas particles. For example, a 22.4-liter sample at STP always has twice as many gas particles as an 11.2-liter sample, regardless of the identity of the gas.

Avogadro's law is extremely useful. If two gases have the same pressure, volume, and temperature, then they have the same number of gas molecules *independent of the identity of the gases.* This generalization is known as **Avogadro's law.** Amadeo Avogadro first proposed this hypothesis in 1811. The reverse is also true: if two gas samples with the same pressure and temperature have the same number of molecules, then they occupy the same volume.

BIG IDEA If two gas samples have the same pressure, temperature, and volume, they contain the same number of particles, even if you are comparing two different types of gases.

Example 2

Helium and Carbon Dioxide Balloons

Suppose you have two balloons, one filled with helium and the other with carbon dioxide. The pressure, temperature, and volume of the two gases are identical.

a. Why is the mass of the carbon dioxide balloon greater?

b. What do you know about the number of atoms in the balloons?

Solution

a. Even though the number of CO_2 molecules is identical to the number of He atoms, the mass of the carbon dioxide balloon is greater because individual molecules of carbon dioxide, CO_2, have a greater mass than atoms of helium, He.

b. There are more atoms in the CO_2 balloon. The number of molecules of CO_2 is equal to the number of atoms of He according to Avogadro's law. Because each molecule of CO_2 consists of three atoms, there are three times as many atoms in the CO_2 balloon than in the He balloon.

Key Terms

mole
Avogadro's number
standard temperature and pressure (STP)
Avogadro's law

Lesson Summary

How do chemists keep track of the number of gas molecules?

Gas molecules move randomly and are distributed uniformly in the space they occupy. At standard temperature and pressure, STP, the gas pressure is 1.0 atm and the temperature is 273 K. At STP in a volume of 22.4 L, there is 1 mole of gas molecules. One mole represents 602,000,000,000,000,000,000,000 molecules. This is 602 sextillion, also called Avogadro's number. Avogadro's law states that equal volumes of any gas at the same pressure and temperature have the same number of gas particles, independent of the identity of the gas.

EXERCISES

Reading Questions

1. Explain why chemists invented the unit called a mole.
2. Explain Avogadro's law.

Reason and Apply

3. Suppose you have 22.4 L of the following gases at STP: neon, Ne, argon, Ar, and xenon, Xe.
   a. How many atoms are there in each gas sample?
   b. What is the number density of atoms, n/V, for each sample?
   c. Which sample has the largest mass? Explain your reasoning.
   d. Which sample has the largest mass density, m/V?
4. Suppose you have 22.4 L of the following gases at STP: hydrogen, H_2, nitrogen, N_2, and carbon dioxide, CO_2.
   a. How many molecules are there in each gas sample?
   b. What is the number density of molecules, n/V, for each sample?
   c. Which sample has the largest mass? Explain your reasoning.
   d. Which sample has the largest number of atoms? Explain your reasoning.
   e. Which sample has the largest mass density, m/V?
5. Which has more atoms, 8.0 g of helium, He, or 40.0 g of argon, Ar? Explain your reasoning.
6. Which has more particles, a balloon filled with 10 L of oxygen, O_2, gas or a balloon filled with 15 L of hydrogen, H_2, gas? Explain your reasoning. Assume STP.
7. At 25 °C, which balloon has the greater volume, an oxygen, O_2, balloon at 1.2 atm with a mass of 16.0 g, or a helium, He, balloon at 1.2 atm with a mass of 2.0 g?

LESSON

17 Take a Breath

Ideal Gas Law

Think About It

Climbing Mount Everest is extremely difficult without bringing along a tank of oxygen to help you breathe. As you climb higher and higher, the air pressure gets lower and lower. At 29,000 ft, on top of Mount Everest, the air pressure is only 0.33 atm. You might wonder how many oxygen molecules are available at this pressure.

How can you calculate the number of moles of a gas if you know *P*, *V*, and *T*?

To begin to answer this question, you will explore

1. Pressure, Volume, Temperature, and Number
2. Ideal Gas Law

Exploring the Topic

1 Pressure, Volume, Temperature, and Number

Four variables describe gas samples: pressure, volume, temperature, and number of moles of atoms or molecules. Any two variables can be related to each other mathematically if the other two are kept the same. You might begin to wonder if you can relate all four variables in a single equation. The combined gas law relates three of the variables: pressure, volume and temperature.

So far you've learned that at STP, there is 1.0 mole of gas molecules in a volume of 22.4 liters for *any* gas sample.

Increasing the Pressure

Suppose you want to adjust this sample of gas so that it has a pressure of 2.0 atm. In what ways can you accomplish this?

You can change the pressure to 2.0 atm in different ways.

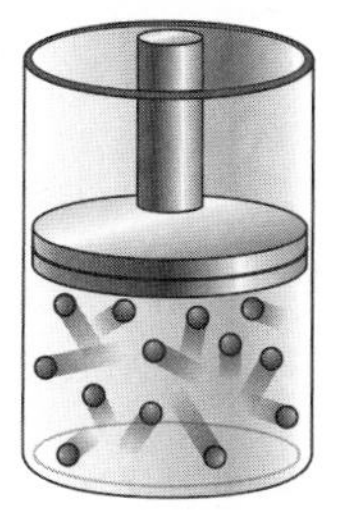

Original gas sample

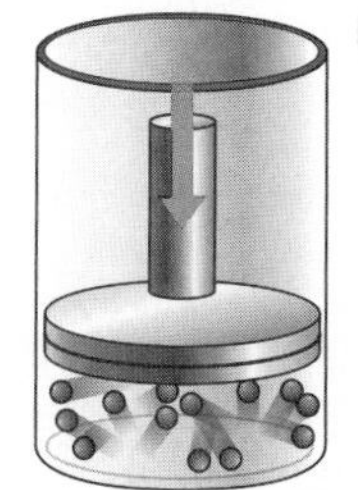

Decrease *V*. Squeeze the gas at 273 K into a volume of 11.2 L.

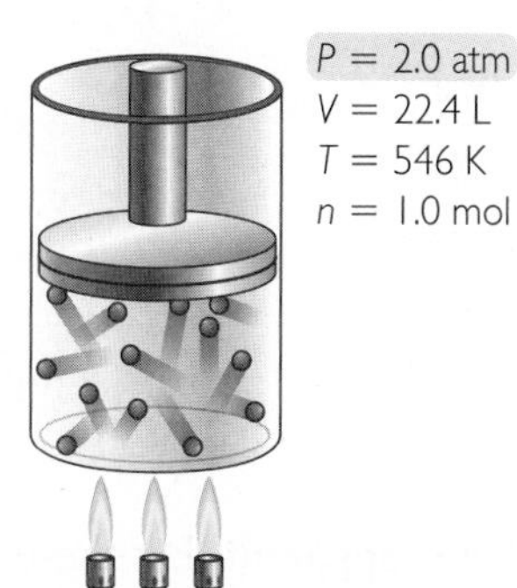

Increase *T* without changing *V*. Heat the gas in 22.4 L to a temperature of 546 K.

Increase *n* without changing *V*. Add 1.0 mol of gas molecules to the container at 273 K at a volume of 22.4 L.

So while the pressure is the same in each case you can see how the other variables have changed.

Target pressure	Volume	Temperature	Moles
2.0 atm	11.2 L	273 K	1.0 mol
2.0 atm	22.4 L	546 K	1.0 mol
2.0 atm	22.4 L	273 K	2.0 mol

Increasing the Volume

Suppose you want to adjust the original sample of gas so that it has a volume of 44.8 L. How can you accomplish this?

You can produce a sample of gas with a volume of 44.8 L in several ways.

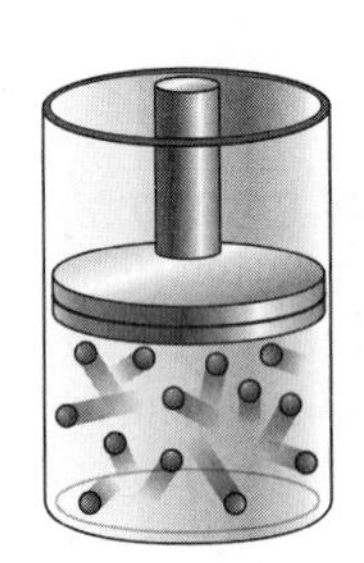

Original gas sample

P = 0.5 atm
V = 44.8 L
T = 273 K
n = 1.0 mol

Increase *V*. Expanding 1.0 mol of gas molecules at 273 K until the pressure is 0.5 atm.

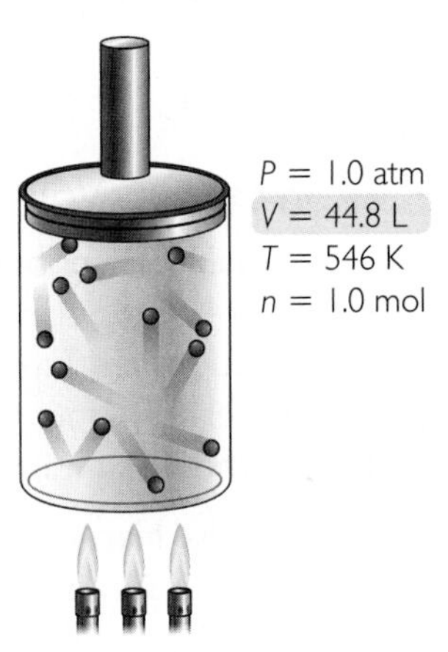

Increase *T*. Heating 1.0 mol of gas molecules at 1.0 atm to a temperature of 546 K.

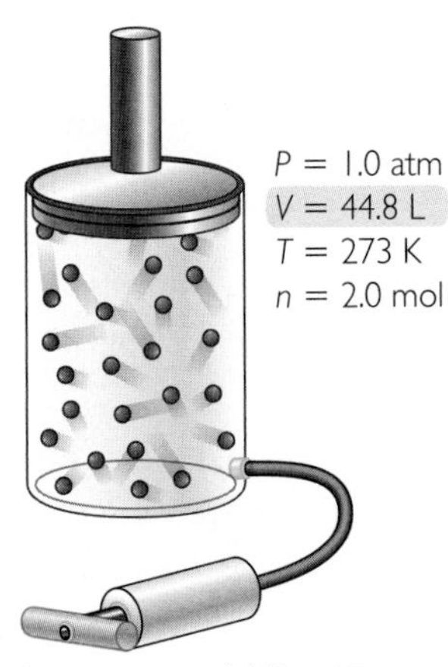

Increase *n*. Adding 1.0 mol of gas molecules to the container at STP, to arrive at 2.0 total moles.

Pressure	Target volume	Temperature	Moles
0.5 atm	44.8 L	273 K	1.0 mol
1.0 atm	44.8 L	546 K	1.0 mol
1.0 atm	44.8 L	273 K	2.0 mol

It is apparent that these variables are mathematically related. In fact, if you combine the equation for the combined gas law with the relationship described by Avogadro's hypothesis, the result is a mathematical relationship known as the ideal gas law.

❷ Ideal Gas Law

The **ideal gas law** relates the pressure, volume, temperature, and number of moles for a gas sample.

> **Ideal Gas Law**
>
> The product of the pressure, P, and volume, V, of a gas is proportional to the product of the number of moles, n, and the temperature, T.
>
> $$PV = nRT$$
>
> The two products are related by the constant, R.
>
> $$R = 0.082 \frac{\text{L} \cdot \text{atm}}{\text{mol} \cdot \text{K}}$$

The **universal gas constant,** R, is the same for all gases.

Because R has units of L · atm/mol · K, you can use this value of R only if volume is measured in *liters*, pressure in *atmospheres*, temperature in *kelvins*, and number of gas molecules in *moles*. When these units are used, R is always equal to 0.082 L · atm/mol · K.

If you know three of the four variables for any gas sample, you can use the ideal gas law to calculate the fourth variable. This is most useful in determining the moles of gas molecules in a gas sample because the pressure, volume, and temperature can be easily measured.

SCILINKS® NSTA

Topic: Gas Laws

Visit: www.SciLinks.org

Web code: KEY-306

Example

Moles of Air on Mount Everest

How many moles of air are in a 0.5 L breath on top of Mount Everest? The pressure is 0.33 atm and the temperature is 254 K.

Solution

The ideal gas law relates all these quantities. Insert the values for P, V, T, and R into the equation and solve for n.

$$PV = nRT$$

$$(0.33 \text{ atm})\ (0.5 \text{ L}) = n\left(0.082 \frac{\text{L} \cdot \text{atm}}{\text{mol} \cdot \text{K}}\right)(254 \text{ K})$$

$$n = \frac{(0.33 \text{ atm})(0.5 \text{ L})}{\left(0.082 \frac{\text{L} \cdot \text{atm}}{\text{mol} \cdot \text{K}}\right)(254 \text{ K})}$$

$$n = 0.008 \text{ mol of air}$$

So there is only 0.008 mol of air molecules in a 0.5 L breath of air atop Mount Everest.

Key Terms

ideal gas law
universal gas constant, R

Lesson Summary

How can you calculate the number of moles of a gas if you know P, V, and T?

Any sample of gas can be described by four variables: pressure, volume, temperature, and moles. The ideal gas law relates these four variables to each other mathematically: $PV = nRT$, where R is the universal gas constant. The value of R is 0.082 L · atm/mol · K for all gas samples if units of atmospheres, liters, moles, and kelvins are used. If you know three of the four variables for any gas sample, you can use the ideal gas law to calculate the fourth variable.

EXERCISES

Reading Questions

1. What is the ideal gas law?
2. Describe when you might want to use the ideal gas law.

Homework

3. How many moles of hydrogen, H_2, gas are contained in a volume of 2 L at 280 K and 1.5 atm?
4. What volume would 1.5 mol of nitrogen, N_2, gas occupy at standard temperature and pressure?
5. Find the pressure of 3.40 mol of gas if the gas temperature is 40.0 °C and the gas volume is 22.4 L.
6. How many moles of helium, He, gas are contained in a 10,000 L weather balloon at 1 atm and 10 °C?
7. Suppose you have 1.0 mol of gas molecules in 22.4 L at STP. Describe three ways you can get a gas pressure of 0.50 atm.
8. Will the pressure of helium, He, gas be the same as the pressure of oxygen, O_2, if you have 1 mol of each gas, each at a volume of 22.4 L and each at 273 K? Explain your thinking.

HEALTH CONNECTION

Altitude sickness occurs at high altitudes when people cannot get enough oxygen from the air. It causes headaches, dizziness, lethargy, nausea, and lack of appetite. It can be quite dangerous for mountain climbers.

LESSON

18 Feeling Humid

Humidity, Condensation

Think About It

Exercising can make you work up a sweat. Sweating helps to regulate your body temperature. When a breeze blows across your sweaty forehead, you may notice your skin feels cooler. However, on a humid day, with a lot of moisture in the air, it seems as if no amount of sweating helps you to cool down.

What is humidity and how is it measured?

To answer this question, you will explore

1. Evaporation and Condensation
2. Humidity
3. Relative Humidity

Exploring the Topic

1 Evaporation and Condensation

After a summer rainstorm, puddles of water on the ground often disappear quickly. The rainwater is evaporating. Recall that evaporation is a phase change from a liquid to a gas.

Evaporation is the reverse process of condensation, when water vapor becomes a liquid. These two processes, evaporation and condensation, are both occurring wherever water is present. In fact, there is a competition between the two processes that can result in net evaporation or net condensation.

The rates of both evaporation and condensation depend mainly on temperature and the amount of water vapor already in the air.

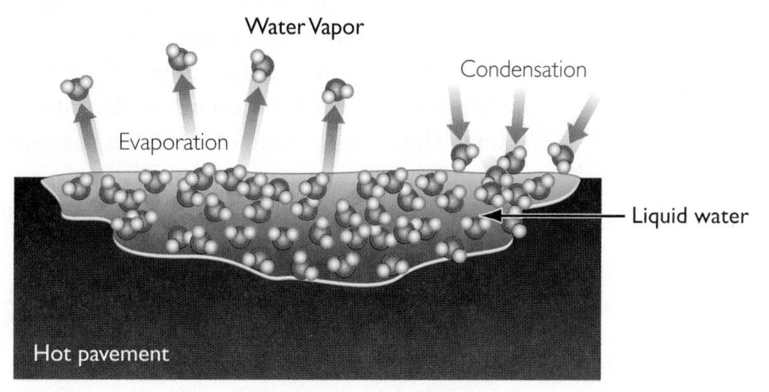

Cloud formation, rain and snowfall, fog, frost, and the appearance of dew are all events associated with the condensation of water vapor out of the atmosphere.

Important to Know The water in rain puddles does not have to boil to go into the gas phase. Evaporation takes place on the surface of a liquid all the time. Try leaving a glass of water on a table. The water will gradually disappear, leaving the glass empty. ◄

BIOLOGY CONNECTION

Only mammals sweat. Humans and other primates have sweat glands all over their bodies. Dogs, cats, and mice have sweat glands only on their feet. Rabbits have sweat glands around their mouths. Dogs pant when they are hot, and the evaporation cools them down. Cats lick their paws and chests. The water evaporates from their fur, cooling them.

2 Humidity

Water vapor is present in the air around us all the time. There is more or less water vapor in the air depending on the weather conditions. Warmer air masses can have more water vapor than colder air masses.

Humidity is the number density of water vapor in the air in moles per unit of volume, or n/V. Sometimes it is expressed as grams per unit of volume (such as cm^3 or liters). For a given temperature, there is a limit to the number density of water molecules that can be in the air. At a certain point, the air becomes saturated, and no further net evaporation takes place. The graph shows the maximum number density for air temperatures between −1 °C and 40 °C. You can see that the maximum humidity depends on temperature.

On a cool day, say around 10 °C (50 °F) the maximum number density of water molecules in the air is approximately 0.5 mole per 1000 liters of air. This amount represents the maximum humidity for this temperature; the actual humidity may be at or below this value.

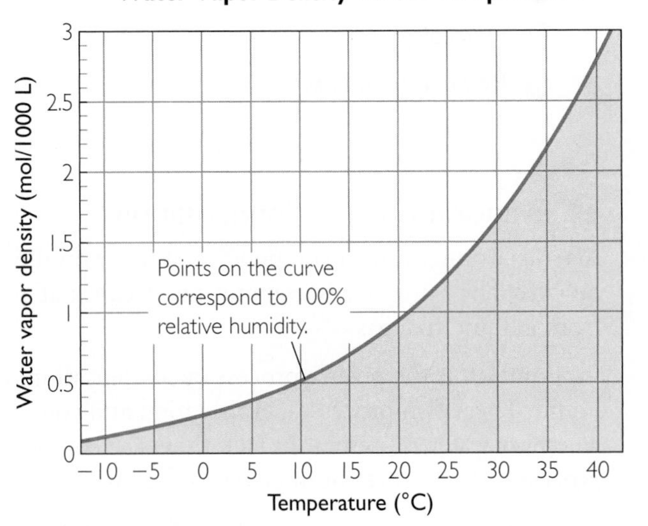

Using the Graph

This graph represents an inequality. The actual value for humidity can be any point on or below the curve. The graph shows that the total amount of water vapor that can be in the air increases with increasing temperature.

METEOROLOGY CONNECTION

You might have noticed a tendency for fog to form on cool nights after a warm, humid day. This is because a lot of water has evaporated during the day. When the temperature decreases, the amount of water vapor in the air exceeds the maximum allowable for the cooler temperature and fog forms.

If the temperature of the air drops or the humidity rises, precipitation can occur. If these two values, T and n/V, result in a data point above the curve, there will be precipitation or condensation of some sort. The condensation of water on the outside of a glass of ice water is a sign that the air next to the glass has cooled enough for the water vapor to change phase to a liquid.

Like all gases, water vapor exerts a pressure. This pressure is part of the total pressure exerted by all the gases in the atmosphere. Each gas exerts a **partial pressure**, and these partial pressures add up to the atmospheric pressure.

When air comes in contact with a cold surface, its temperature drops, causing some of the water vapor in the air to condense on the surface.

❸ Relative Humidity

When meteorologists consider humidity, they focus on the relative humidity. **Relative humidity** is the percent of the maximum humidity for a specified temperature. When the air contains the maximum amount of water vapor for the temperature of the air, it is at 100% relative humidity.

The terms *humidity* and *relative humidity* are often used interchangeably in weather reports. Meteorologists might say the humidity today is at 35% when they are really talking about the relative humidity.

METEOROLOGY CONNECTION

Weather forecasters use a sling psychrometer to determine the relative humidity. It consists of two thermometers, one dry and one wrapped in a wet cloth. The thermometers are spun through the air and then read.

During spinning, water evaporates from the cloth, cooling the wet-bulb thermometer. The amount of cooling that occurs is directly related to how much moisture is in the air. When there is more moisture in the air, less evaporation and less cooling occur.

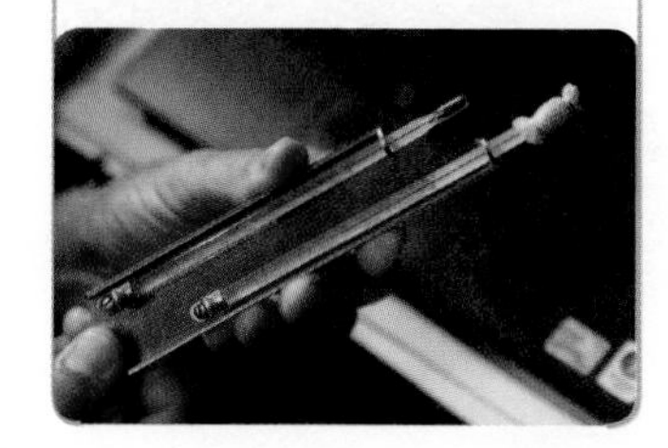

Example

Relative Humidity

Imagine that on a certain day the water vapor density is 1.6 mol/1000 L and the temperature is 30 °C.

a. What is the relative humidity?

b. At night, the temperature drops to 20 °C. Do you expect there will be precipitation?

Solution

a. You can read the maximum water vapor density on the graph Water Vapor Density vs. Temperature at 30 °C. It is 1.7 moles/1000 L. This corresponds to 100% humidity. The relative humidity is the measured water vapor density divided by the maximum possible water vapor density.

$$\frac{1.6}{1.7} = 0.94 = 94\%$$

b. The point (20 °C, 1.6 mol/1000 L) on the graph is above the 100% humidity curve, so you can expect precipitation.

Matter that is in contact with an evaporating liquid will cool off. Sweating helps cool you because the evaporation of water on your skin transfers heat away from the body. When there is more water vapor in the air it becomes harder for sweat to evaporate. This is why you often feel much hotter in humid air than in dry air.

Key Terms

humidity
partial pressure
relative humidity

Lesson Summary

What is humidity and how is it measured?

Humidity refers to the amount of water vapor in the air. It is a measure of the number density of water molecules in the air. Air temperature affects how much water vapor can be in the air. Warmer air contains more water vapor than colder air. There is an upper limit to the amount of water vapor that can be in the air at any given temperature. This upper limit is called maximum humidity, or 100% relative humidity. At 100% relative humidity, no more water can evaporate into the air because the air is saturated. You might expect rain or fog when the relative humidity is close to 100%.

EXERCISES

Reading Questions

1. What does humidity measure?
2. Explain what is meant by relative humidity.

Reason and Apply

3. What is the maximum vapor density possible at 30 °C?
4. Is it possible for the water vapor density in the air to reach 10 moles per 1000 L at a temperature of 40 °C? Explain why or why not.
5. Use the graph Water Vapor Density Versus Temperature to predict the water vapor density at 100% humidity for 35 °C.
6. What is the relative humidity if there is 0.5 mole of water vapor per 1000 L of air at 40 °C?
7. When the humidity is 25% and the temperature is 30 °C, what is the water vapor density?
8. On one day, 100% humidity corresponds to 1.9 moles of water vapor per 1000 L of air. On another day, 100% humidity corresponds to only 1.0 mole of water vapor per 1000 L of air. How can two different water vapor densities both be at 100% humidity?
9. Explain why it is easy to get dehydrated when you are exercising at a temperature of 0 °C and a relative humidity of 20%.
10. Do you feel cooler at 100% humidity at 30 °C or at 50% humidity at 30 °C? Explain your thinking.
11. Suppose the humidity is 65% during the day when the temperature is 30 °C.
    a. If the temperature drops to 25 °C at night, do you expect fog? Explain.
    b. If the temperature drops to 20 °C at night, do you expect fog? Explain.
12. On a cold winter day the relative humidity is 50% outdoors but only 5% indoors. Which answer best explains what is going on?
    A. It must be raining outside.
    B. It must be snowing outside.
    C. The heater is on inside, so the same humidity is a lower relative humidity.
    D. The heater is on inside, and it is evaporating some of the humidity in the air.

GEOLOGY CONNECTION

It is estimated that Earth has about 326 million cubic miles of water. This includes all of the water in the oceans, underground, and locked up as ice. About 3100 mi^3 of this water is in the air—mostly as water vapor, but also as clouds or precipitation—at any one time, the U.S. Geological Survey estimates.

LESSON

19 Hurricane!

Extreme Physical Change

Think About It

Water and air provide our planet with its weather and support plant and animal life. Weather itself is a result of a dynamic interplay of physical change between water, Earth's atmosphere, and energy from the Sun. There are times when these physical changes create conditions that can be catastrophic. A hurricane is one example of an extreme weather phenomenon that can be quite destructive.

What are hurricanes and what causes them?

To answer this question, you will explore

1. The Properties of Hurricanes
2. Hurricane Formation
3. Global Warming

Exploring the Topic

1 The Properties of Hurricanes

A hurricane is an enormous tropical rainstorm with powerful winds that forms over the ocean. It has a distinctive shape to it, resembling a giant rotating pinwheel. The wind speeds in a hurricane are over 75 mi/hr and may even exceed 150 mi/hr. The storm may bring torrential rains, destructive winds, and flooding to areas of land that it passes over. The sheer size of these storms is shown in this photo of a hurricane in the Gulf of Mexico.

HISTORY CONNECTION

The word *hurricane* comes from the name of a Mayan god of winds and storms called Hurakan.

These giant storms may be from 125 to 1000 mi across and 15 mi high. Additionally, they are characterized by very low air pressure at their centers. In general, the lower the air pressure, the larger the storm. Consider the data in the table for three hurricanes in the same year.

Hurricane name	Year	Wind speed (mi/h)	Pressure (atm)	Category
Hurricane Dennis	2005	150	0.918	4
Hurricane Irene	2005	98	0.962	2
Hurricane Rita	2005	173	0.885	5

HISTORY CONNECTION

Before World War II, large storms were usually referred to by their location, which meant assigning a longitude and latitude number to each storm. However, it was easy to confuse one storm with another using this method. During World War II, military weather forecasters began to give names to hurricanes.

Hurricanes are categorized according to their strength, as indicated in the table. They are accompanied by a surge in coastal waters called a *storm surge*. A storm surge can reach 20 ft or more in height, causing major coastal flooding.

❷ Hurricane Formation

Hurricanes form only over very warm ocean waters of at least 80 °F. In addition, they require a great deal of moisture in the air. For these two reasons, hurricanes originate south of the United States, in the tropical ocean waters near the equator.

You have probably heard people refer to "hurricane season." Most hurricanes form during summer and fall when the waters are warmest in tropical zones. For the United States, hurricane season stretches from June through November.

A storm must grow through several stages before it is considered a hurricane. A future hurricane actually starts out as a tropical depression, which is a clearly defined low-pressure system with winds below 38 mi/h. Some tropical depressions continue to build and become tropical storms. In tropical storms, the air circulates around a low-pressure system with winds between 38 and 74 mi/h. Finally, a tropical storm that builds to wind speeds beyond 75 mi/h is considered a hurricane.

These satellite images show the stages of hurricane formation.

It can take several days for a thunderstorm to develop into a hurricane. It begins with moist warm air from the ocean surface rising rapidly. The water vapor in this warm air condenses and forms storm clouds. The heat released by this condensation warms the cooler air above it, causing the air to rise even more. This process continues, pulling more and more warm air up into the developing storm. A spiral wind pattern begins to develop, and the hurricane takes on its characteristic shape. In the very center of the storm, an "eye" develops and cool air descends, creating a calm storm-free area. Near the ocean surface, spiral bands of rain stretch out for miles.

METEOROLOGY CONNECTION

The *eye of the storm* refers to a small area of calm in the midst of chaos. The eye of a hurricane may be between 12 and 60 mi in diameter. When the eye of the storm passes over an area, people will experience calm, clear conditions, as if there is no hurricane at all. The eye is surrounded by the eyewall, which is the most violent part of the storm.

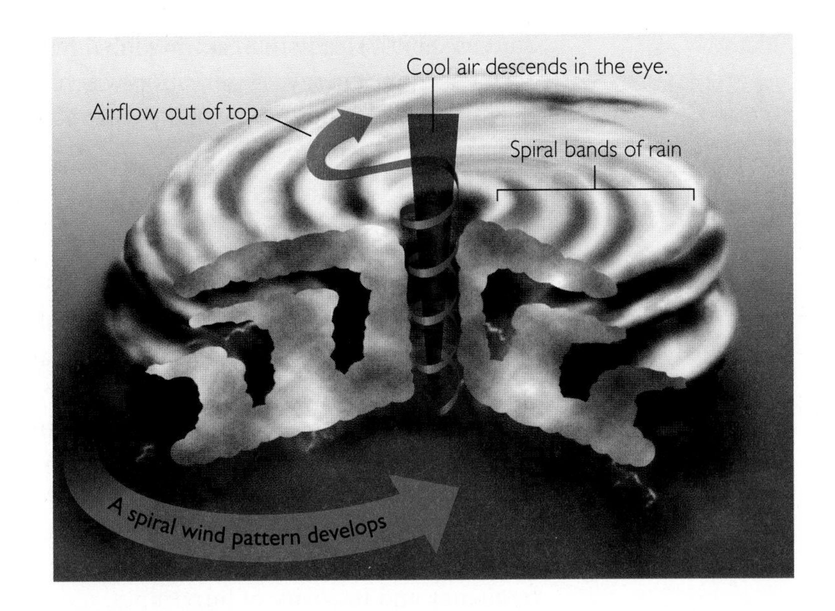

❸ Global Warming

Our planet is currently in a warming phase. The hurricane season of 2005 was an unusually destructive one in the United States. The number and severity of storms caused scientists to speculate that there is a connection between the increasing number and severity of hurricanes and the warming of the entire planet. In other words, Earth may be undergoing long-term climate change.

Experts think that the warming of the planet is worsened by our practice of using petroleum products for our energy needs. Burning gasoline, coal, and natural gas releases enormous amounts of carbon dioxide into our atmosphere. This increased carbon dioxide traps the heat of the Sun in Earth's atmosphere, causing a "greenhouse effect."

The 20th century's 10 warmest years all occurred in the last 15 years of the century. Meteorologists theorize that increases in ocean temperatures of only a few degrees may have dramatic effects on the weather of our planet, including the possibility of increased frequency and intensity of hurricanes. One degree may not sound like much, but it is sufficient to raise ocean levels by a few inches and to cause more rainfall and stronger storms.

Dramatic effects of climate change are shown in these two photos of Glacier Bay National Park, taken 60 years apart. The effects of global warming can also be seen in the loss of habitat for polar animals.

Studies show a steady increase in global temperature over the past 120 years. Scientists hope energy conservation will help to reverse or slow down this warming trend. Automobile makers are actively involved in creating vehicles that run on alternative energy sources.

Lesson Summary

What are hurricanes and what causes them?

A hurricane is a large and powerful tropical storm with intense spiraling winds. Hurricanes form over the warm tropical waters near the equator. At the very center of a hurricane is an area of extremely low air pressure. The heat released when moist warm air rises and condenses into storm clouds powers hurricane formation. Also, the low pressure caused by condensation draws more air into the storm from areas of relatively higher pressure. This cycle of evaporation and condensation of very large amounts of water is a key feature of hurricanes. Experts speculate that climate changes due to global warming are increasing the frequency and intensity of hurricanes.

EXERCISES

Reading Questions

1. What conditions are necessary for hurricane formation?
2. Write a creative paragraph describing what life on the planet may be like if global warming continues at its present pace.

Reason and Apply

3. **WEB RESEARCH** Find data showing that ocean temperature is related to the wind speed of a hurricane.
4. **WEB RESEARCH** Discuss three pieces of evidence that global warming is occurring.

AGRICULTURE CONNECTION

A greenhouse is a structure made of glass or clear plastic designed to keep plants warm in the winter. Greenhouses trap radiation from the sun and are usually several degrees warmer than the outside air.

SECTION

SUMMARY

Concentrating Matter

Weather Update

Air pressure is directly related to the number density, or number of particles per unit of volume, of the gas molecules in a sample of air. Avogadro's law states that two gas samples have the same number of molecules if temperature volume and pressure are the same. Avogadro's law and the combined gas law together form the ideal gas law. The ideal gas law relates pressure, volume, temperature, and the number of moles by a proportionality constant. This constant, R, is called the universal gas constant.

Extreme weather can occur if conditions support rapid evaporation of water molecules followed by rapid condensation of these molecules.

The amount of water vapor in the air plays a big role in determining the weather. Water vapor density, or humidity, is dependent on air temperature and air pressure. When air is at 100% humidity, chances are it is raining, snowing, or densely foggy because the air is saturated with water molecules.

Key Terms

number density
mole
Avogadro's number
standard temperature and pressure (STP)
Avogadro's law
ideal gas law
universal gas constant, R
humidity
partial pressure
relative humidity

Review Exercises

1. Explain how you can determine the number of molecules in a breath of air.
2. What type of weather do you predict if the relative humidity is 100%?
3. Suppose you have 22.4 L of helium, He, gas and 22.4 L of neon, Ne, gas at STP. Which of these statements is *false*?
   A. Both samples contain the same number of atoms.
   B. Both samples have the same number density, n/V.
   C. The two samples have different masses.
   D. The two samples have the same mass density, D.
4. How many moles of nitrogen gas, $N_2(g)$, are contained in 2 liters at 350 K and 1.5 atm?

5. Suppose the water vapor density is 1.5 moles/1000 L at a temperature of 35 °C, and the maximum water vapor density at 35 °C is 2.2 moles/1000 L. What is the relative humidity?

PROJECT *Global Climate Change*

Earth's overall climate is currently in a warming phase. Research some of the causes and effects of global warming on Earth.

- What evidence do scientists have that average global temperatures are increasing?
- What are some possible causes of global warming?
- How might global warming affect the severity and frequency of storms in a region?
- What are some possible effects of global warming on different living things and their environments?

UNIT

3

Weather Review

The atmosphere surrounding our planet is a mixture of gases, including water vapor. The weather that we experience every day is almost entirely caused by physical changes to these gases. Energy from the Sun causes water molecules to change phase and evaporate into the air. This water vapor moves around, sometimes changing phase again and forming clouds, rain, ice, snow, dew, fog, and so on.

By studying small samples of gas in containers, we can learn more about how gases in the atmosphere behave. The kinetic theory of gases explains the behavior of individual gas particles under varying conditions. For example, when a gas in a rigid container is heated, its pressure increases because the molecules move faster, increasing the force and number of collisions with whatever they contact. When a sample of gas is squeezed into a smaller volume, it also exerts more pressure. This is because the same number of rapidly moving molecules must occupy a smaller space. The pressure, volume, and temperature of a gas sample are related by the combined gas law.

Meteorologists track the variables that affect the gases in the atmosphere. For example, they record water vapor density, or humidity, and areas of differing air pressure. They study air masses and fronts where warm and cold air masses collide. They combine all the information and use it to make predictions about the weather.

Review Exercises

1. Suppose you have a 500 mL sample of liquid water and a 500 mL sample of ice. The ice has a density of 0.92 g/mL. Which has a greater mass? Explain your reasoning.
2. Convert 25 °F to kelvins. Show your work.
3. Use the kinetic theory of gases to explain why a gas expands when heated and contracts when cooled.
4. Suppose you fill a balloon with 600 mL of air when the temperature is 27 °C. Later in the day the temperature drops to 21 °C. Determine the new volume of the balloon.
5. A cylinder with a movable piston contains 2.5 L of gas at a pressure of 2 atm. If the volume is decreased to 2.1 L and the temperature stays the same, what is the new pressure of the gas?

6. Which graph represents Boyle's law? Explain your thinking.

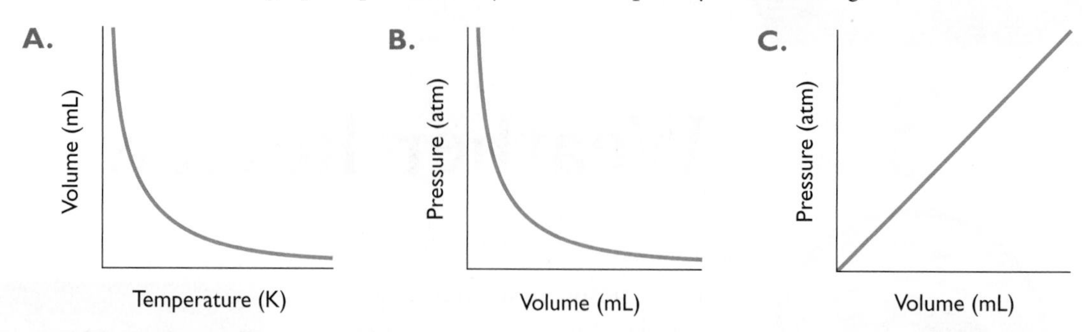

7. A propane gas tank contains propane gas at a pressure of 2.00 atm and a temperature of 30 °C. If the temperature is increased to 37 °C, what is the new pressure of the gas?

8. Suppose you fill a balloon with 75 L of air at 28 °C and a pressure of 1 atm. You then take the balloon to the top of a mountain where the pressure is 0.8 atm and the temperature is 18 °C. What is the new volume of the balloon?

9. Suppose you have 22.4 L of nitrogen gas, $N_2(g)$, at STP.
   a. What are the temperature and pressure of this sample?
   b. How many molecules of N_2 are in the sample?
   c. How many nitrogen atoms are in the sample?
   d. What is the number density, n/V, of nitrogen molecules in the sample?

10. Suppose you have a 750 mL rigid container of hydrogen gas, $H_2(g)$, at 250 K and a 750 mL rigid container of nitrogen gas, $N_2(g)$, at 250 K. A pressure gauge on each container shows that they both have the same internal pressure.
    a. What do you know about the number of individual molecules of gas in each container? Explain your reasoning.
    b. Which container would have the greater mass? Explain your reasoning.

11. Suppose you have 77.0 L of hydrogen, H_2, gas at a pressure of 3.5 atm and a temperature of 2 °C. How many moles of hydrogen are in this sample?

12. Would you feel cooler if the temperature were 28 °C at 100% relative humidity or 28 °C at 45% relative humidity? Explain your reasoning.

13. Name the gas laws. Write the formula associated with each one, and list the factors that need to stay constant in order for each law to apply.

UNIT

4 Toxins

Some animals produce toxins for self-defense, or to paralyze their prey. However, many toxins also have medicinal uses. For example, certain compounds found in snake venom can help treat a person having a heart attack.

Why Toxins?

Chemical reactions help our bodies to process food and create new tissues. However, some chemical reactions have toxic and harmful outcomes. The toxicity of a substance is highly dependent on the dose. Sometimes a small amount of a compound, such as a vitamin, can be therapeutic, but a large amount can damage your health. This unit investigates chemical changes by exploring how toxic substances are measured and tracked through their transformations.

In this unit, you will learn

- how toxins are defined
- how chemists determine toxicity
- the mechanisms by which toxic substances act in our bodies and what this has to do with chemical reactions

SECTION

I Toxic Changes

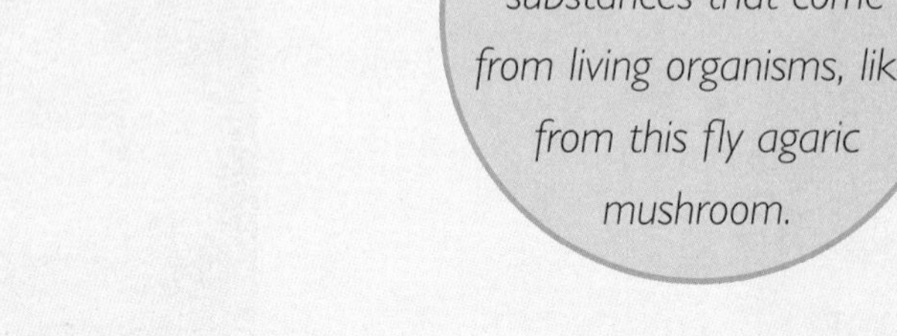

Biologists define toxins as harmful substances that come from living organisms, like from this fly agaric mushroom.

Chemical changes are necessary for life. However, some substances interact with the body in ways that have unhealthy effects. In order to monitor and understand toxic interactions, it is necessary to use chemical equations to track changes in matter.

In this section, you will study

- how to write, balance, and interpret chemical equations
- definitions of chemical and physical change
- how to classify chemical reactions
- what happens to mass when a chemical change occurs

LESSON

1 Toxic Reactions

Chemical Equations

Think About It

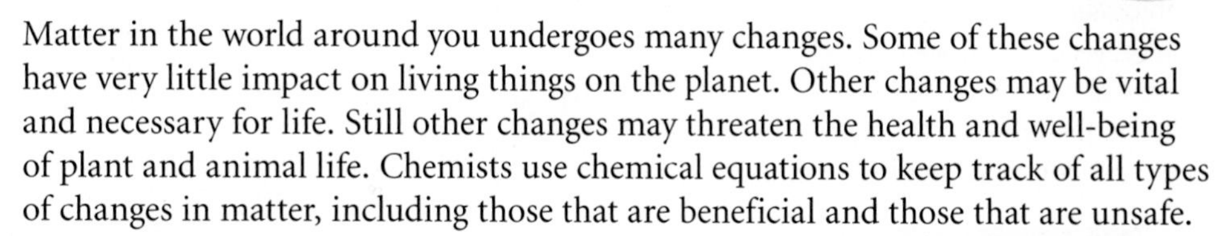

Matter in the world around you undergoes many changes. Some of these changes have very little impact on living things on the planet. Other changes may be vital and necessary for life. Still other changes may threaten the health and well-being of plant and animal life. Chemists use chemical equations to keep track of all types of changes in matter, including those that are beneficial and those that are unsafe.

How do chemists keep track of changes in matter?

To answer this question, you will explore

1. Chemical Equations
2. Toxic Substances and Their Effects

Exploring the Topic

1 Chemical Equations

A chemical equation is a chemical "sentence" that describes change, using numbers, symbols, and chemical formulas. Chemical equations describe what happens when a single substance is changed, or when two or more substances are combined and a change occurs. Once you understand how to decode chemical equations, you will be able to use them to predict what you might observe when substances are mixed.

Interpreting a Chemical Equation

In some reactions that take place in your body, the element chromium is safe and even necessary. In other reactions it is toxic. Consider what happens if you ingest chromium metal and it reacts with the hydrochloric acid in your stomach. A chemical equation can help you decode this reaction.

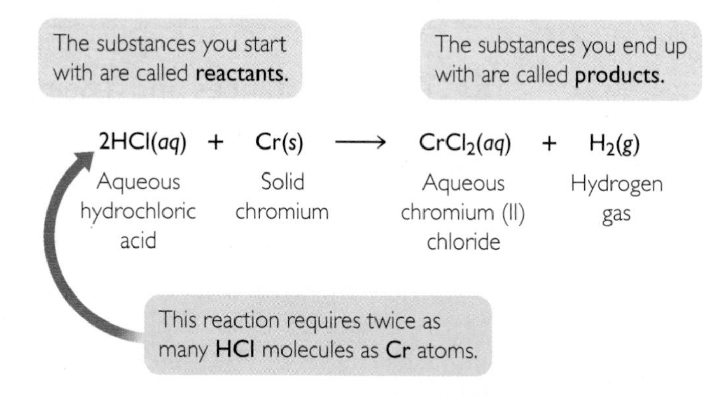

Chemical equation: $2HCl(aq) + Cr(s) \longrightarrow CrCl_2(aq) + H_2(g)$

Interpretation: Hydrochloric acid reacts with solid chromium to produce a solution of chromium (II) chloride and bubbles of hydrogen gas.

ENVIRONMENTAL CONNECTION

In larger amounts, some chromium compounds cause cancer. Since the 1980s, Erin Brockovich has brought several successful lawsuits in California to stop unsafe levels of chromium compounds in drinking water. A legal clerk at the time of her first legal case, she is now a consultant and speaker.

A chemical equation can help you anticipate what you will observe when the reactants are combined. Examine the same equation, but focus on what you expect to observe.

$$2HCl(aq) + Cr(s) \longrightarrow CrCl_2(aq) + H_2(g)$$

According to the equation, when hydrochloric acid and chromium react, the solid chromium will disappear. You would expect to see the formation of a new aqueous solution as well as some evidence that a gas was produced.

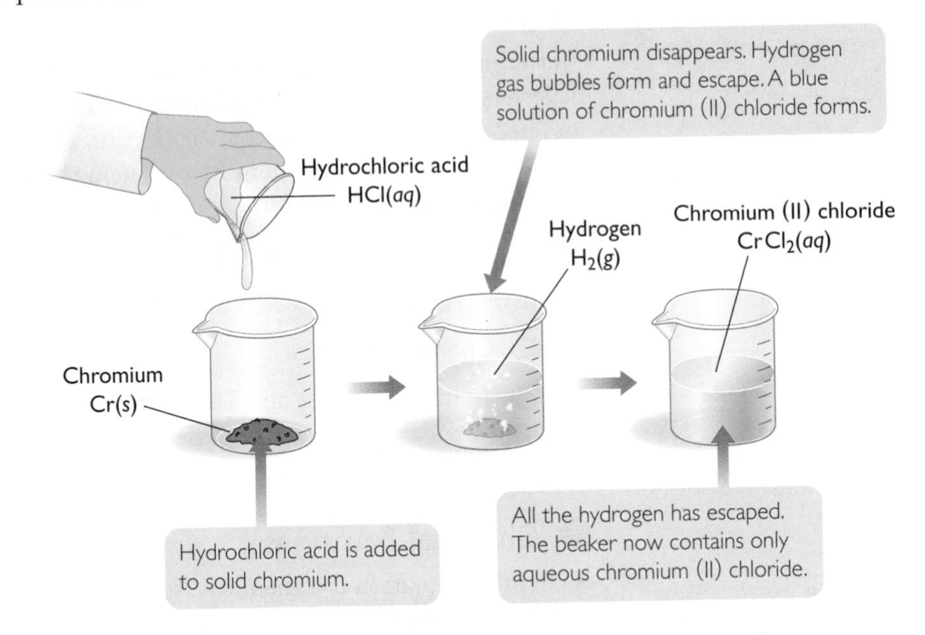

Sometimes, the changes that take place when substances are mixed are not visible to the eye. For example, death from poisoning occurs within minutes of swallowing a solution of sodium cyanide, NaCN. One successful treatment for this type of poisoning is injection of an antidote, or a remedy that counteracts the poison. The antidote for sodium cyanide is a solution containing sodium thiosulfate, $Na_2S_2O_3$. If the reaction between sodium cyanide and sodium thiosulfate were observed in a beaker, you would not be able to tell that a reaction had occurred because both reactants and both products are clear, colorless liquids.

BIG IDEA Chemical equations keep track of changes in matter.

2 Toxic Substances and Their Effects

Toxic substances enter the body in limited ways. The most common methods are through ingestion (eating or swallowing), inhalation (breathing something in), or contact with the skin. Once toxic substances enter the body, they react in a variety of ways.

Toxic substances can react with water in mucous membranes, with oxygen carried through the blood, or with stomach acid. Some toxic substances have an immediate negative effect on the well-being of a person or living thing. Other toxic substances may stay a long time in the body, becoming part of the body's chemistry, perhaps damaging the body many years later.

Toxic substances may be molecular, ionic, or metallic substances. Many small molecular substances react with the body and create acid products, which can damage and irritate tissue or upset the acidity of the blood. Toxic metals often react to form ionic compounds, which move throughout the body and compete with "good" metals that are useful to the body. Some ionic compounds form solids that clog the body's filtering systems. Of course, these are just a few of the possible types of toxic reactions that occur. Chemical equations are the primary tool you will use to track these changes.

Lesson Summary

How do chemists keep track of changes in matter?

A chemical equation tracks changes in matter. The left side of the equation contains the chemical formulas for the reactants, or the substances that are being combined. The right side of the equation contains the chemical formulas for the products, or the substances that are produced. The equation also shows the phase of each reactant and product. Decoding an equation allows you to predict what substances may be made when the reactants are combined. Chemical equations often provide more information than what you can observe with your senses.

EXERCISES

Reading Questions

1. What is the difference between a reactant and a product?
2. Are chemicals and chemical reactions important for life? Why or why not?
3. Describe in your own words what a toxic substance is.

Reason and Apply

4. Both bleach and ammonia are used for cleaning. However, it is very dangerous to mix bleach with ammonia because they react to produce sodium hydroxide and the toxic gas chloramine.

$$NaOCl(aq) + NH_3(aq) \longrightarrow NaOH(aq) + NH_2Cl(g)$$

   a. Write an interpretation of the chemical equation.
   b. What do you expect to observe?

5. Poisoning with mercury chloride can be reversed by chelation therapy. The chelating agent called EDTA, $C_{10}H_{16}N_2O_4$, is injected into the bloodstream. EDTA forms a water-soluble compound with mercury ions, allowing removal from the body through the kidneys.

$$HgCl_2(s) + C_{10}H_{16}N_2O_4(aq) \longrightarrow HgC_{10}H_{12}N_2O_4(aq) + 4HCl(aq)$$

   a. Write an interpretation of the chemical equation.
   b. What do you expect to observe?

6. Describe at least three types of effects that a toxic substance can have on the body.

LESSON

2 Making Predictions

Observing Change

Think About It

Table salt, or sodium chloride, NaCl, can enhance the flavor of soup. Even though the salt dissolves in the soup, you can still taste it. This change is described by a chemical equation:

$$NaCl(s) \longrightarrow NaCl(aq)$$

It is also possible to change the salt in far more dramatic ways. For example, when an electric current is passed through a sodium chloride solution, a toxic, green gas bubbles out of the solution. The change is described by this chemical equation:

$$2NaCl(aq) + 2H_2O(l) \longrightarrow 2NaOH(aq) + Cl_2(g) + H_2(g)$$

How can you predict what you will observe based on a chemical equation?

To answer this question, you will explore

1. Predicting Change From Chemical Equations
2. Information in Chemical Equations

Exploring the Topic

1 Predicting Change From Chemical Equations

While some chemical changes are difficult to see, many changes in matter are accompanied by observable evidence. For example, when a change takes place, you may hear fizzing, a pop, or an explosion. You might see changes in color or physical form. You may smell gases that escape, or you may feel heat. Perhaps there is even a fire.

But what if you don't have a laboratory and a lot of chemicals around? How can you predict what you might observe if you are provided only with a chemical equation?

Sugar Dissolving

Consider these two chemical equations. What type of change does each describe?

$$C_{12}H_{22}O_{11}(s) \longrightarrow C_{12}H_{22}O_{11}(aq)$$

$$C_{12}H_{22}O_{11}(aq) \longrightarrow C_{12}H_{22}O_{11}(s)$$

The only difference between these two equations is the position of the (aq) and (s) symbols. Notice that the chemical formula does not change from one side of the equation to the other. The sugar molecule changes form, but not identity. The first equation describes dissolving, in which a solid is mixed with water. The second equation describes removing water from an aqueous substance, leaving a solid.

Although the sugar changes appearance when it dissolves, the sugar itself has not changed into a different compound. The chemical equation indicates how the molecules have, or have not, changed.

Sugar Melting

Consider these two equations that involve changes to sugar.

$C_{12}H_{22}O_{11}(s) \longrightarrow C_{12}H_{22}O_{11}(l)$

$C_{12}H_{22}O_{11}(l) \longrightarrow C_{12}H_{22}O_{11}(s)$

Again the formulas are identical on both sides of the arrow, except for the symbols in parentheses, (l) and (s). These equations describe a phase change from a solid to a liquid and back to a solid. By heating sugar carefully, you can melt it to a clear liquid. When liquid sugar is cooled, it becomes a solid again.

The same compound shows up on both sides of the equation. Even though the tiny grains of sugar have been transformed into a single, large, chunk of solid sugar, the atoms within the sugar molecules have not rearranged into new compounds. Sugar is still sugar.

BIG IDEA When a substance changes phase or dissolves, its chemical formula does not change.

Sugar Decomposing

Consider another transformation of sugar.

$$C_{12}H_{22}O_{11}(s) \longrightarrow 12C(s) + 11H_2O(g)$$

Notice that the two sides of the equation look very different. There is no sugar on the right side. The sugar has been decomposed. It has been converted to solid carbon and water vapor.

If you heat sugar sufficiently (beyond melting), it will turn black, and it will release smoke. It will not taste sweet anymore, and there is no way to turn it back into sugar. These property changes, and the release of smoke, are signs that a more dramatic rearrangement of atoms has occurred.

$C_{12}H_{22}O_{11}(s) \longrightarrow 12C(s) + 11H_2O(g)$

$12C(s) + 11H_2O(l) \longrightarrow$ no change

This particular rearrangement of atoms is not easy to reverse. In other words, you cannot combine carbon with water and expect to form sugar.

Sugar Reacting

Consider one more transformation of sugar, as described by this equation:

$$C_{12}H_{22}O_{11}(s) + 8KClO_3(s) \longrightarrow 12CO_2(g) + 11H_2O(g) + 8KCl(s)$$

Interpreting the chemical formulas tells you that sugar and potassium chlorate are added together. The result is the formation of two gases, carbon dioxide and water vapor, as well as the formation of some solid potassium chloride.

Sugar and potassium chlorate react explosively.

The sugar and potassium chlorate are converted to carbon dioxide gas, water vapor, and solid potassium chloride. This change happens very quickly and releases energy in the form of light and heat.

Example

Changing Salt

Predict what you would observe for the changes described by these equations:

a. $NaCl(s) \longrightarrow NaCl(aq)$ b. $NaCl(s) \longrightarrow NaCl(l)$

c. $2NaCl(l) \longrightarrow 2Na(l) + Cl_2(g)$

Solution

a. The solid sodium chloride dissolves in water.

b. The solid sodium chloride melts to become liquid sodium chloride.

c. The liquid sodium chloride decomposes into liquid sodium and chlorine gas.

2 Information in Chemical Equations

Just as a chemical equation can help you predict what you might observe, it can also provide more information than you can get by observation alone. Consider the reaction described at the beginning of this lesson. The change that happens upon passing an electric current through aqueous sodium chloride is described by this equation:

$$2NaCl(aq) + 2H_2O(l) \longrightarrow 2NaOH(aq) + Cl_2(g) + H_2(g)$$

If you were to observe this reaction without knowing the equation, you would not be able to identify the compounds before and after the reaction. Many solutions are clear and colorless, and you might not realize that two gases are produced. A chemical equation identifies the reactants and products in a reaction. This chemical equation indicates that aqueous sodium chloride combines with water to form aqueous sodium hydroxide, chlorine gas, and hydrogen gas. An equation also identifies the form or phase a substance is in. This equation indicates that a liquid of some sort will be observed before and after the reaction, and gas bubbles will be seen as the reaction takes place.

Some information is not contained in a chemical equation. The equation above, for example, does not indicate that chlorine gas is green and toxic. The reaction of sugar with potassium chlorate, mentioned previously, is quite spectacular. The equation does not indicate that the reaction is explosive or that a purple flame is produced. An equation does not tell you whether a procedure will result in hot or cold sensations. It won't tell you how fast a change will occur or if a color change will be observed. It does not tell you if there will be an explosion, a loud noise, or an overflowing beaker. Only direct observation can provide you with all of the information about what you will see, hear, and feel.

When sugar molecules are allowed to crystallize, they form a hard candy, like lollipops. When lemon juice or fructose is added to the sugar, the molecules do not lock into place as in crystalline candy. Instead, they form soft candy like taffy or caramels, called amorphous candy.

Lesson Summary

How can you predict what you will observe based on a chemical equation?

Chemical equations contain valuable information about what is happening during a change in matter. They track the identities of the substances involved in the change, using chemical formulas. They also identify the phase of each substance. By carefully examining a chemical equation, you can predict what you might observe if the procedure were completed. A chemical equation cannot tell you everything that you will observe. It does not inform you about the speed of a reaction, if a color change will occur, or if energy is transferred.

EXERCISES

Reading Questions

1. In words, describe the difference between sugar melting and sugar dissolving in water. The formula for sugar is $C_{12}H_{22}O_{11}$.
2. Use chemical equations to describe the difference between sugar melting and sugar decomposing. The formula for sugar is $C_{12}H_{22}O_{11}$.

Reason and Apply

3. Describe what you think you would observe for these chemical equations.
   a. $Mg(s) + 2HCl(aq) \longrightarrow H_2(g) + MgCl_2(aq)$
   b. $2H_2O_2(aq) \longrightarrow 2H_2O(l) + O_2(g)$
   c. $2NaCl(aq) + Pb(NO_3)_2(aq) \longrightarrow 2NaNO_3(aq) + PbCl_2(s)$
4. Write a chemical equation for these reaction descriptions.
   a. Solid sodium chloride dissolves in water.
   b. Solid magnesium sulfide is heated to produce solid magnesium and sulfur gas.
   c. Solid titanium is heated in oxygen gas to produce titanium dioxide.
5. These are two reactions that can occur between bleach and ammonia in solution. What would you expect to observe in each case?

   $NaOCl(aq) + NH_4OH(aq) \longrightarrow NH_2Cl(g) + NaOH(aq) + H_2O(l)$

   $NaOCl(aq) + 2NH_4OH(aq) \longrightarrow N_2H_4(g) + NaCl(aq) + 3H_2O(l)$

6. **WEB RESEARCH** Look up chloroamine, NH_2Cl, and hydrazine, N_2H_4, on the Internet. Describe the toxicity of each compound.

CONSUMER CONNECTION

Many cleaning products contain either bleach or ammonia. You might think that combining the two would lead to a super cleanser. This is not so. When combined, bleach and ammonia form chloramine, a highly dangerous and toxic, strong-smelling gas. So, you should NEVER mix the two.

LESSON

3 Spare Change

Physical Versus Chemical Change

Think About It

You breathe in oxygen and it dissolves in your blood, where it binds to a molecule called hemoglobin. The hemoglobin transports the oxygen to where it is needed for reactions with carbohydrates. In the end, the oxygen and carbohydrates react to produce carbon dioxide and water. Some of these changes are physical changes and some are chemical changes.

How are changes in matter classified?

To answer this question, you will explore

1. Defining Physical and Chemical Change
2. Dissolving: A Special Case

Exploring the Topic

1 Defining Physical and Chemical Change

Examine the chemical equations in the table.

Physical change	Chemical change
$H_2O(l) \longrightarrow H_2O(s)$	$2Na(s) + Cl_2(g) \longrightarrow 2NaCl(s)$
$Br_2(l) \longrightarrow Br_2(g)$	$CH_4(g) + 2O_2 \longrightarrow CO_2(g) + 2H_2O(l)$
$I_2(s) \longrightarrow I_2(g)$	$2KClO_3(s) \longrightarrow 2KCl(s) + 3O_2(g)$

Physical Change

The changes in the left column are all **physical changes.** In each case the substance on the left side of the equation changes phase (solid, liquid, or gas), but does not change its identity. Also, there is only *one* substance involved in each of these equations, and that one substance is present on both sides of the arrow.

$H_2O(l) \longrightarrow H_2O(s)$

Liquid water freezes to become ice.

$Br_2(l) \longrightarrow Br_2(g)$

Liquid bromine evaporates to bromine gas.

$I_2(s) \longrightarrow I_2(g)$

Solid iodine sublimes to iodine gas.

When something changes in a physical way, the chemical formula does not change because you end up with the same substance you started with. Changing pressure or temperature, and grinding a substance into a powder are other examples of physical change.

Chemical Change

The three equations in the right column of the table on the previous page are chemical changes. In each case, new substances are created. The chemical formulas on the left side of each equation are different from the formulas on the right side. New substances are formed during chemical changes, so we can expect the products to have properties significantly different from the reactants.

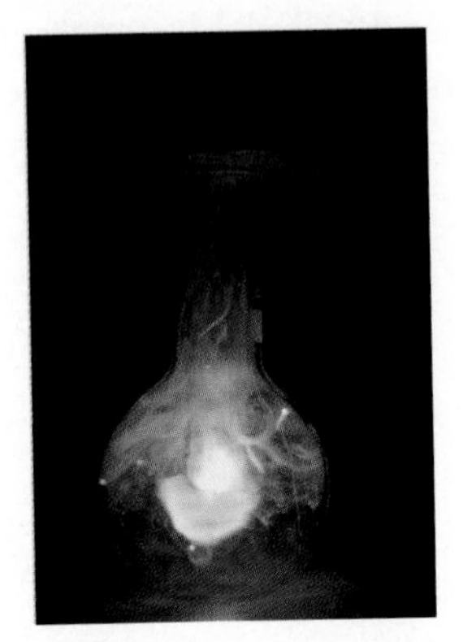

$2Na(s) + Cl_2(g) \longrightarrow 2NaCl(s)$

Solid sodium reacts with chlorine gas to produce solid sodium chloride.

$CH_4(g) + 2O_2(g) \longrightarrow CO_2(g) + 2H_2O(l)$

Methane gas and oxygen gas burn to produce carbon dioxide gas and water.

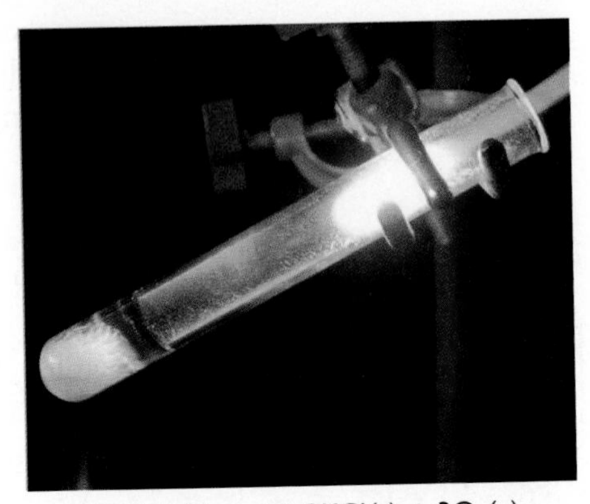

$2KClO_3(s) \longrightarrow 2KCl(s) + 3O_2(g)$

Solid potassium chlorate is heated to form solid potassium chloride and oxygen gas.

Heat, light, smoke, smells, bubbles, and changes in color often accompany chemical changes. When chemical change takes place, chemists say that a chemical reaction has occurred. In fact, the terms chemical change and chemical reaction mean the same thing.

2 Dissolving: A Special Case

Chemical equations can also represent the process of dissolving. Here are three equations that represent dissolving.

Liquid ethanol dissolves in water: $C_2H_6O(l) \longrightarrow C_2H_6O(aq)$

Gaseous hydrochloric acid dissolves in water: $HCl(g) \longrightarrow HCl(aq)$

Solid copper chloride dissolves in water: $CuCl_2(s) \longrightarrow CuCl_2(aq)$

These equations indicate that the identity of each dissolved substance has not changed. Each equation describes a physical change. The symbol (*aq*), for aqueous, on the product side of the equation means that each substance has dissolved in water.

Here is a different way to write the second and third equations listed above.

$$HCl(g) \longrightarrow H^+(aq) + Cl^-(aq)$$

$$CuCl_2(s) \longrightarrow Cu^{2+}(aq) + 2Cl^-(aq)$$

When hydrogen chloride gas and solid copper chloride dissolve in water, they break apart into ions. The second type of equation shows what happens during this change. These equations showing a substance breaking apart to form ions are necessary for the dissolving of ionic compounds and acids. Note that molecular substances like ethanol, C_2H_6O, do not break apart when they dissolve. So there is only one way to write the equation.

Ionic compounds dissolved in water have different properties than ionic solids. For example, ionic solutions conduct electricity but the solid forms do not. This makes dissolving seem like a chemical change. However, the identity of the substances has not changed. You can recover solid compounds from solutions by evaporating the water they are dissolved in. So, you can categorize dissolving as either physical or chemical change.

Important to Know You cannot always tell by observing a change whether it is a physical change or a chemical change. A chemical equation can provide more information than observation alone. ◀

SCILINKS® NSTA

Topic: Chemical and Physical Changes

Visit: www.SciLinks.org

Web code: KEY-403

Example

Magnesium Chloride and Water

Write two equations that show what happens when magnesium chloride, $MgCl_2(s)$, dissolves.

Solution

$$MgCl_2(s) \longrightarrow MgCl_2(aq)$$
$$MgCl_2(s) \longrightarrow Mg^{2+}(aq) + 2Cl^-(aq)$$

Key Term

physical change

Lesson Summary

How are changes in matter classified?

In general, changes in matter can be classified as either physical or chemical. Physical changes, including phase changes, involve a change in form without changing the identity of the substance. There is no change in chemical formula. Chemical changes, also called chemical reactions, involve the formation of new substances with new properties. The chemical formulas of these new products are different from the chemical formulas of the reactants. Dissolving can fall into both categories of change, though it is usually considered a physical change.

EXERCISES

Reading Questions

1. What is the difference between a physical change and a chemical change?
2. Explain why dissolving can be described as either a physical or chemical change.

Reason and Apply

3. Give five examples of physical changes. Explain why each example is a physical change.

4. Copy the equations 1–7 onto your paper. Match the chemical equations with their descriptions.

1. $2CH_4O(l) + O_2(g) \longrightarrow 2CH_2O(l) + 2H_2O(l)$
2. $NH_2Cl(g) \longrightarrow NH_2Cl(aq)$
3. $2C_8H_{14}(l) + 23O_2(g) \longrightarrow 16CO_2(g) + 14H_2O(l)$
4. $H_2SO_4(aq) + CaCO_3(s) \longrightarrow CaSO_4(aq) + CO_2(g) + H_2O(l)$
5. $Hg(l) \longrightarrow Hg(g)$
6. $C_{16}H_{30}O_2(s) + H_2 \longrightarrow C_{16}H_{32}O_2(s)$
7. $CaO(s) + H_2O(l) \longrightarrow Ca(OH)_2(s)$

A. The sulfuric acid dissolved in raindrops of acid rain reacts with the calcium carbonate in seashells and marble structures. This reaction produces calcium sulfate dissolved in water and carbon dioxide gas.

B. To balance the effect of acid rain, solid calcium oxide has been added to many lakes. It reacts with water to form solid calcium hydroxide, a strong base that is only slightly soluble in water.

C. Chloramine is added to our water supply in very small amounts to kill bacteria.

D. When octane and oxygen gas are burned in our cars, carbon dioxide and water come out in the exhaust.

E. Methanol, if ingested, reacts with oxygen to form formaldehyde, which is toxic. Water is also formed in this reaction.

F. Liquid mercury evaporates to produce mercury vapor.

G. Saturated fatty acids, like palmitic acid, tend to form long solids and clog people's arteries. These saturated fatty acids can be made from unsaturated fatty acid by adding hydrogen gas.

5. For equations 1–7 from Exercise 4, label each one as a physical or chemical change.

6. Classify the following two changes as physical or chemical. Explain your reasoning.
   a. $CaCO_3(s) + H_2SO_4(aq) \longrightarrow CaSO_4(aq) + CO_2(g) + H_2O(l)$
   b. $NaCl(s) \longrightarrow NaCl(l)$

ART CONNECTION

Many ancient buildings and monuments such as the Great Sphinx in Egypt, shown here, are being destroyed by acid rain because they are made of materials like limestone that dissolve in sulfuric acid.

LESSON

4 Some Things Never Change

Conservation of Mass

Think About It

During chemical and physical changes, the atoms in a substance rearrange. Sometimes a solid forms or disappears. Sometimes the substances change appearance or substances with new properties appear. But is matter really "appearing" or "disappearing"?

How does mass change during a chemical or physical change?

To answer this question, you will examine

1. Tracking Mass
2. Conservation of Mass

Exploring the Topic

1 Tracking Mass

In Unit 1: Alchemy, you learned that matter is conserved during chemical reactions. If matter is not created or destroyed during physical and chemical changes, then there should be no measurable difference in mass before and after a change. Three chemical procedures are described here. The mass before and after each change is shown on the balances.

Physical Change—Dissolving a Solid

Consider what happens when solid sodium carbonate is added to water.

$$Na_2CO_3(s) \longrightarrow Na_2CO_3(aq)$$

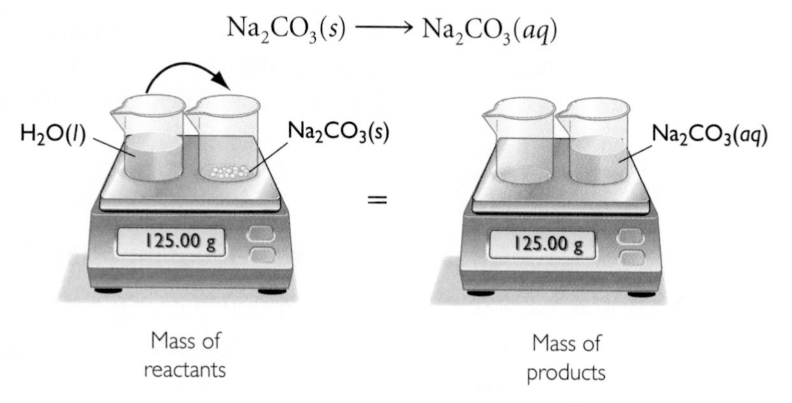

The solid seems to disappear into the liquid. However, notice that the total mass doesn't change; it is the same before and after the substances are mixed. The ions in the white solid have spread evenly throughout the water.

This change can also be represented by the chemical equation:

$$Na_2CO_3(s) \longrightarrow 2Na^+(aq) + CO_3^{2-}(aq)$$

Even though this compound breaks apart into ions, the mass is the same before and after the change.

EARTH SCIENCE CONNECTION

There are small changes in the total number of atoms on the planet. Sometimes meteorites collide with Earth, adding new matter, or meteors burn up in the atmosphere, settling as dust. Also, some elements undergo radioactive decay. However, in chemical and physical changes, the total number of atoms does not change. This photo shows a "shooting star," or meteor, and a crater from a large meteorite.

ENVIRONMENTAL CONNECTION

In 2002, roughly half the United States population had access to curbside recycling programs. Today, about 42% of all paper, 40% of all plastic soft drink bottles, 55% of all aluminum cans, 57% of all steel packaging, and 52% of all major appliances are now recycled. The photo shows a paper recycling plant.

Chemical Change—Producing a Solid

Consider a chemical change in which a new solid forms. An aqueous solution of sodium chloride is added to an aqueous solution of silver nitrate producing aqueous sodium nitrate and solid silver chloride.

$$NaCl(aq) + AgNO_3(aq) \longrightarrow NaNO_3(aq) + AgCl(s)$$
$$58\text{ g} + 170\text{ g} = 85\text{ g} + 143\text{ g}$$

Even though a solid is produced, the mass of the products is still equal to the mass of the reactants. No mass was gained or lost.

Chemical Change—Producing a Gas

Now consider a chemical change that produces a gas as a product. An aqueous solution of hydrochloric acid is added to solid magnesium metal producing an aqueous solution of magnesium chloride and bubbles of hydrogen gas.

$$2HCl(aq) + Mg(s) \longrightarrow H_2(g) + MgCl_2(aq)$$

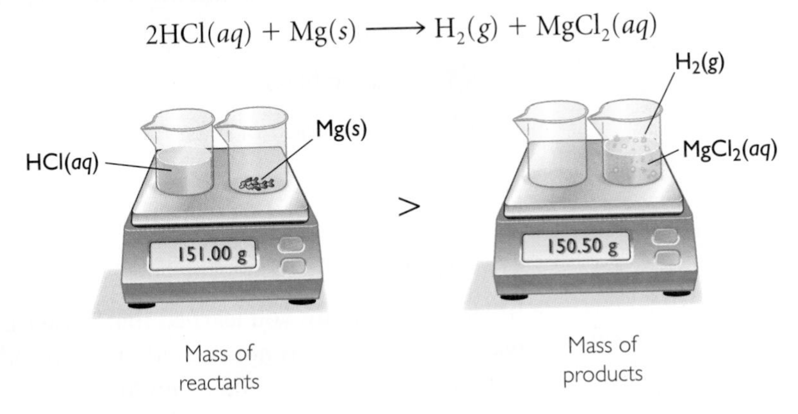

According to the balance, some mass appears to have been lost. However, notice that some gas escaped from the beaker. If you were able to trap the gas that escaped and measure its mass, it would account for the difference in masses seen here.

Important to Know It is difficult to measure the weight of a sample of gas, so it is tempting to conclude that gases do not have mass. However, as you saw in Unit 3: Weather, gases are made of molecules and they do have mass. ◄

Example

Chemical Change—Decomposing Calcium Carbonate

Examine this description of a chemical change along with its chemical equation.

Verbal description: Solid calcium carbonate is heated to produce solid calcium oxide and carbon dioxide gas.

Chemical equation: $CaCO_3(s) \longrightarrow CaO(s) + CO_2(g)$

$$50\text{ g} = 28\text{ g} + ?$$

mass of reactants = mass of products

What would you expect the mass of the carbon dioxide gas to be? Explain your reasoning.

Solution

The mass of the products on the right side must equal the mass of the reactants on the left side. So, the mass of the carbon dioxide produced is 22 grams.

❷ Conservation of Mass

The fact that the mass of the reactants is equal to the mass of the products follows the law of conservation of mass. Mass is conserved during chemical and physical changes because no atoms are created or destroyed.

Getting Rid of Garbage

People generate a lot of waste: cups, cans, bottles, candy wrappers, paper, and diapers. Most of this waste just keeps piling up in landfills. Some of the waste is quite toxic. In the United States alone, about 132 million tons of garbage are put into landfills each year. Will it be there forever?

Atoms are not created or destroyed; they are simply rearranged to form new substances. Therefore, everything on the planet is made up of a limited number of atoms. Even your body does not have the same atoms in it that it did several years ago. You are continually generating new cells of every kind and getting rid of the old ones.

The law of conservation of mass is important in considering waste disposal. You cannot get rid of the atoms in the waste that you throw away. Smelly, toxic garbage accumulates in large quantities in waste disposal sites. A small portion of it may biodegrade—that is, get broken down into harmless products by microorganisms—but the process can take decades. The waste will be there unless scientists find ways to convert more of our waste into useful products. Given this fact, it makes sense to reuse and recycle as much as possible.

ENVIRONMENTAL CONNECTION

Compounds are sometimes recycled in surprising ways. For example, PETE plastic bottles can be recycled to make t-shirts, suits and fleece!

Lesson Summary

How does mass change during a chemical or physical change?

When chemical and physical changes take place, the atoms in substances rearrange. A rearrangement may be a physical change, such as the mixing of two substances, or a chemical change, the formation of new compounds. However, the atoms involved in these rearrangements cannot be created or destroyed. A chemical equation tracks the atoms involved in chemical and physical changes. All of the atoms are accounted for, so the mass of the products is identical to the mass of the reactants. This is known as the law of conservation of mass.

EXERCISES

Reading Questions

1. Explain the law of conservation of mass.
2. Explain how the law of conservation of mass applies to garbage.

Reason and Apply

3. Below is a chemical equation along with a verbal description of the reaction.

   Verbal description: Gaseous sulfur trioxide is added to liquid water to produce aqueous sulfuric acid.

   Chemical equation: $SO_3(g) + H_2O(l) \longrightarrow H_2SO_4(aq)$

   a. Was matter lost or gained during this reaction? Explain how you could prove this by taking measurements.

   b. The left side of the equation shows one atom of S, two atoms of H, and four atoms of O. How many atoms of each element are on the right side of the equation? How does this provide evidence for the law of conservation of mass?

4. When an ice cube melts, which of these quantities will change?

   A. The number of atoms it contains

   B. Its mass

   C. Its volume

   D. All of the above

   E. None of the above

5. Explain what happens to the number of atoms, the mass, and the weight of the water in a glass when it evaporates.
6. Write a paragraph to convince a friend that mass is conserved when chemical and physical changes take place. Give evidence.
7. What would you have to do to prove that matter is conserved when a piece of paper is burned?

ENVIRONMENTAL CONNECTION

Bacteria can break down organic waste such as food scraps and dead leaves. But if garbage is piled deep in a landfill, the bacteria can't access the air and water they need to break things down. Some trash companies collect food scraps and garden waste separately, so they can be composted.

LESSON

5 Atom Inventory

Balancing Chemical Equations

Think About It

The law of conservation of mass states that mass is not lost or gained in a chemical reaction. When you write a chemical equation to describe change, it is important that the equation follow this law. So any equation describing a chemical change must account for every atom involved.

How do you balance atoms in a chemical equation?

To answer this question, you will explore

1. Balancing Chemical Equations
2. Coefficients Are Counting Units

INDUSTRY CONNECTION

Ammonia can be made by first converting methane into another carbon compound and hydrogen gas. Then an iron compound is used as a catalyst to react hydrogen with nitrogen. Ammonia produced through this process is used to fertilize approximately one-third of the agricultural crops in the world.

Exploring the Topic

1 Balancing Chemical Equations

Imagine you're in business to make ammonia, NH_3, for farmers to fertilize their crops. It would be helpful to know how much nitrogen, $N_2(g)$, and hydrogen, $H_2(g)$, to combine so that nothing is wasted. How do you make sure you have the correct amount of nitrogen and hydrogen so that you have enough of each with no extra?

A chemical equation represents the exact ratio in which reactants combine to form products. A chemical equation that accounts for all the atoms involved and shows them combining in the correct ratio is called a *balanced* chemical equation. Learning how to balance chemical equations is a necessary part of working with chemical reactions.

Formation of Ammonia

The unbalanced equation below describes how nitrogen gas and hydrogen gas react to make ammonia gas.

To balance this equation, first take an inventory of the atoms on each side.

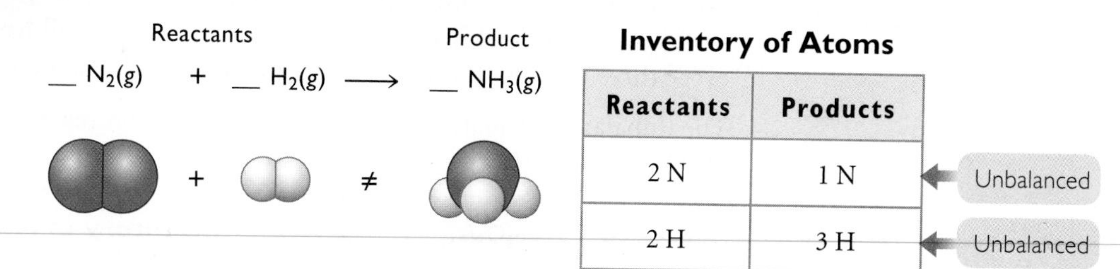

Inventory of Atoms

Reactants	Products	
2 N	1 N	Unbalanced
2 H	3 H	Unbalanced

Next, balance the atoms on each side by adding units of N_2, H_2, or NH_3. The product side needs one more N atom. The only way to accomplish this is to add a whole ammonia molecule, NH_3, to the product side.

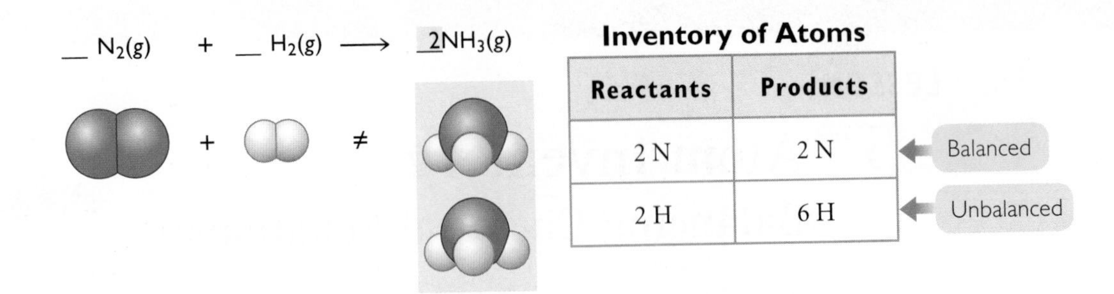

Inventory of Atoms

Reactants	Products
2 N	2 N
2 H	6 H

Take a new inventory. The N atoms are now balanced. Now the reactant side needs four more H atoms. The only way to accomplish this is to add two more molecules of H_2 to the reactant side.

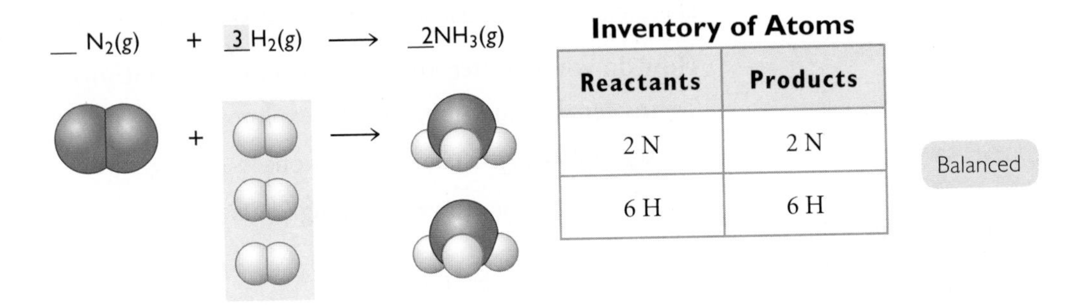

Inventory of Atoms

Reactants	Products
2 N	2 N
6 H	6 H

The equation is now balanced. The balanced equation shows that mass is conserved. There are the same numbers of N and H atoms on each side of the arrow.

In the balanced equation above, the 3 in front of the H_2 and the 2 in front of the NH_3 are called **coefficients.** Coefficients indicate how many units of each substance take part in the reaction. If there is no number in front of a chemical formula in an equation, the coefficient is understood to be a 1.

Important to Know To balance a chemical equation, you can change only the coefficients, or the numbers in front of each chemical formula. You cannot change the chemical formulas. ◀

Formation of Rust

Iron reacts with oxygen in the air to form iron (III) oxide, or rust. Iron (III) oxide is *not* a molecule. It is a highly organized collection of Fe and O atoms that are bonded to each other ionically. It is an ionic solid. There are a huge number of atoms in any piece of iron (III) oxide. But each piece will have two Fe atoms for every three O atoms. So Fe_2O_3 is the **formula unit** of iron oxide.

The unbalanced equation below describes how iron reacts with the oxygen in the air to form iron (III) oxide.

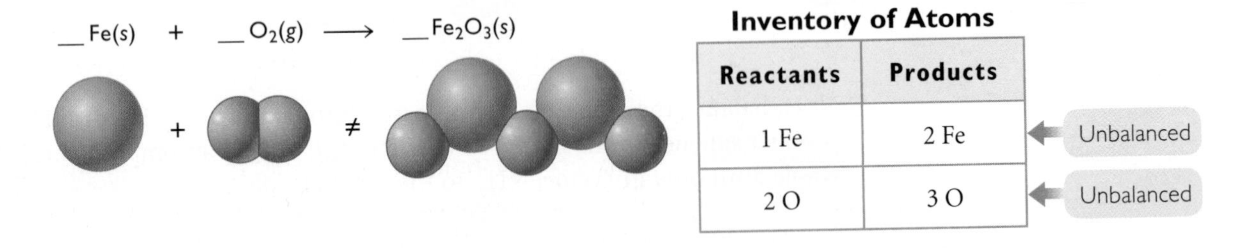

Inventory of Atoms

Reactants	Products
1 Fe	2 Fe
2 O	3 O

Balance the atoms on each side by adding molecules of O_2, atoms of Fe, or formula units of Fe_2O_3. If you add an Fe atom to the reactant side, the Fe atoms are balanced. However, the O atoms are still not equal.

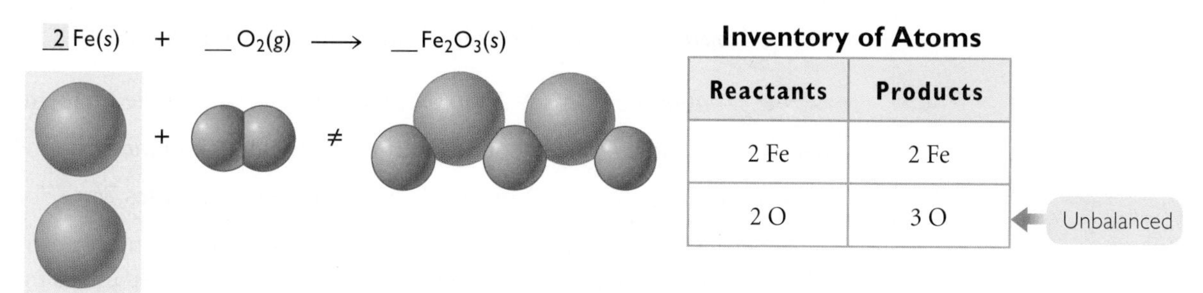

Reactants	Products
2 Fe	2 Fe
2 O	3 O

If you add two O_2 molecules to the reactant side and one Fe_2O_3 formula unit to the product side, the oxygen atoms are balanced.

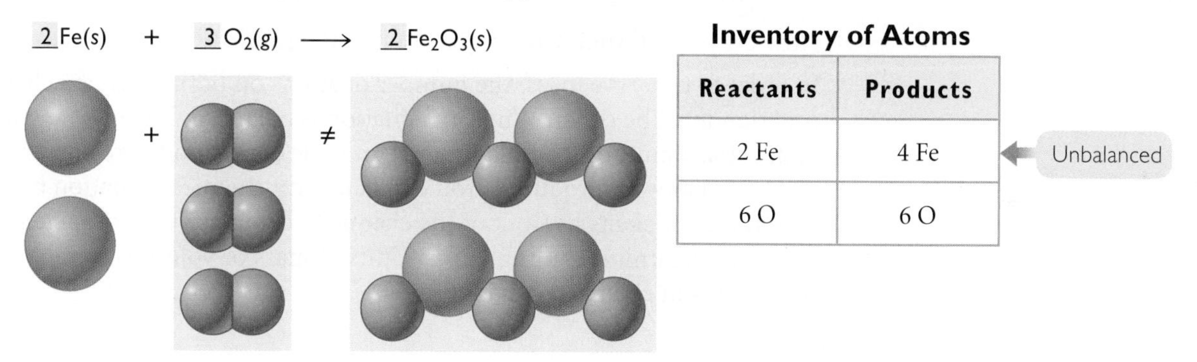

Reactants	Products
2 Fe	4 Fe
6 O	6 O

But now the iron atoms are unbalanced again. Add two more iron atoms on the reactant side, increasing the total to four. The balanced chemical reaction is shown here.

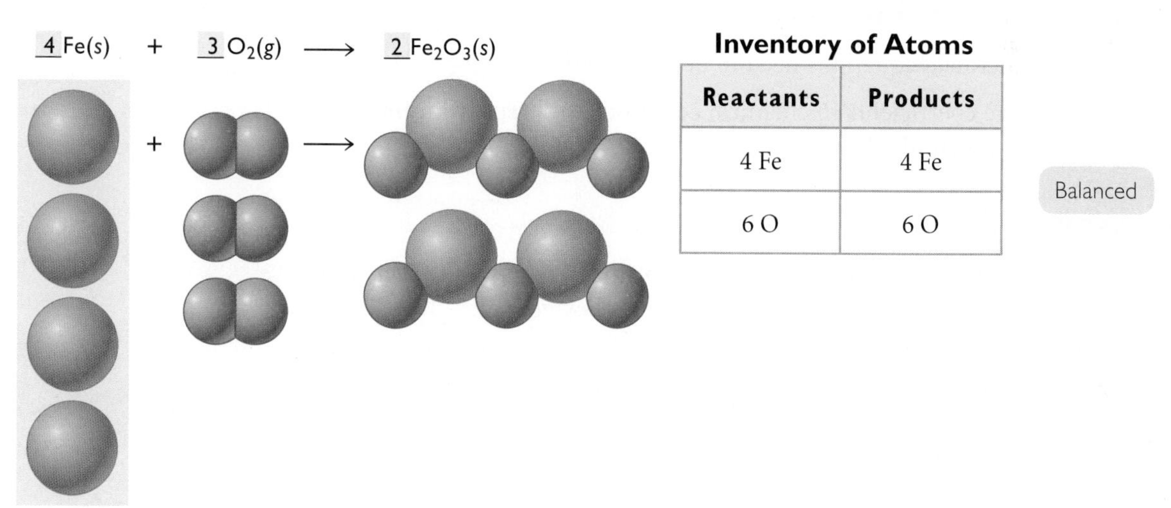

Reactants	Products
4 Fe	4 Fe
6 O	6 O

It took a few steps, but the equation is finally balanced. Four iron atoms combine with three oxygen molecules to form two formula units of iron oxide.

2 Coefficients Are Counting Units

Once you have a balanced equation, multiplying all the coefficients by any counting unit will also give a balanced equation. You can multiply the coefficients by a dozen, or a thousand, or a million.

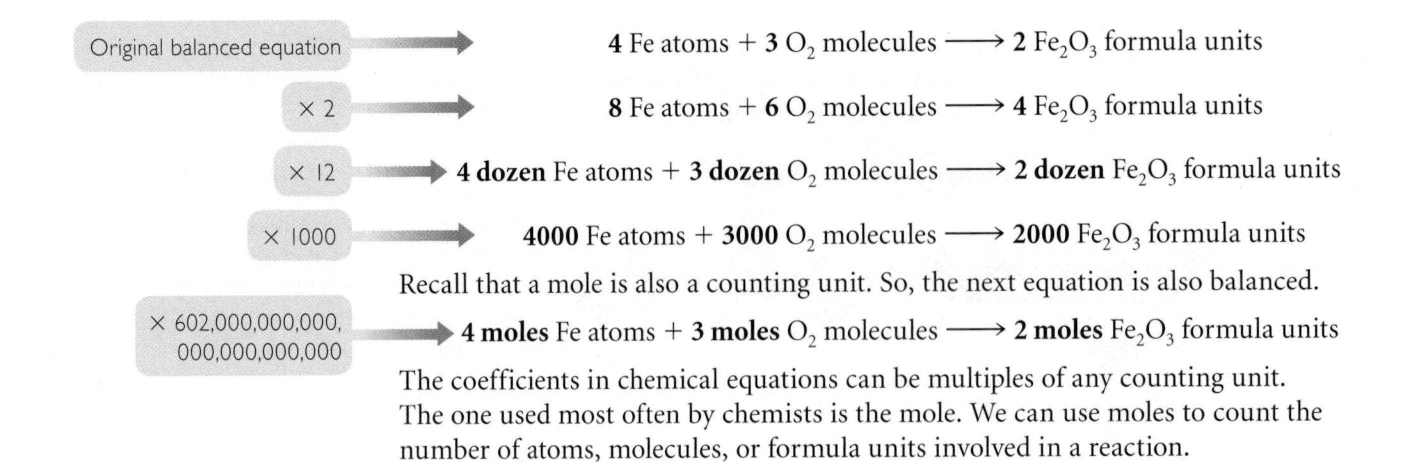

Recall that a mole is also a counting unit. So, the next equation is also balanced.

The coefficients in chemical equations can be multiples of any counting unit. The one used most often by chemists is the mole. We can use moles to count the number of atoms, molecules, or formula units involved in a reaction.

Key Terms
coefficient
formula unit

Lesson Summary

How do you balance atoms in a chemical equation?

In order to conserve mass, the number of atoms on both sides of a chemical equation must be equal. When an equation is balanced, it shows how many molecules, atoms, or formula units of an ionic compound take part in a reaction and how many are produced. You can balance a chemical equation by working with the coefficients in front of the chemical formulas. The coefficients can be any counting unit including moles. Units of mass or volume *cannot* be used as coefficients in chemical equations.

EXERCISES

Reading Questions

1. Why do chemical equations need to be balanced?

2. How are the coefficients in chemical equations different from the subscripts in chemical formulas?

Reason and Apply

3. Your recipe for banana bread calls for two ripe bananas. However, you have six ripe bananas that you want to use before they go bad.

a. How can you make banana bread with all six bananas?

b. How is this related to balancing a chemical equation?

4. Copy and balance these chemical equations.

a. $K(s) + I_2(s) \longrightarrow KI(s)$

b. $Mg(s) + Br_2(l) \longrightarrow MgBr_2(s)$

c. $KBr(aq) + AgNO_3(aq) \longrightarrow KNO_3(aq) + AgBr(s)$

d. $KClO_3(s) \longrightarrow KCl(s) + O_2(g)$

e. $C_2H_6(g) + O_2(g) \longrightarrow CO_2(g) + H_2O(l)$

f. $Al(s) + O_2(g) \longrightarrow Al_2O_3(s)$

g. $P_4(s) + H_2(g) \longrightarrow PH_3(g)$

LESSON

6 What's Your Reaction?

Types of Reactions

Think About It

Toxins work by reacting with chemicals in the body. These toxic reactions can remove compounds that are important to body function, or create new compounds that interfere with normal body processes. Classifying toxic reactions according to the different ways they react can help us to understand how toxic substances work in the body. It can also help us to come up with suitable approaches for dealing with them.

How do atoms rearrange to form new products?

To answer this question, you will explore

1. Combination and Decomposition Reactions
2. Exchange Reactions

Exploring the Topic

1 Combination and Decomposition Reactions

Imagine you are an ocean researcher and you spend your days underwater in a submarine. As you breathe you use up oxygen gas, $O_2(g)$, and release carbon dioxide gas, $CO_2(g)$. Eventually, the high level of carbon dioxide and low level of oxygen in your submarine will be dangerous to your health. How do you remove carbon dioxide and produce oxygen gas to sustain life in a submarine?

Combination Reactions

Combination reactions are utilized in air scrubbers on submarines to remove $CO_2(g)$ from the air. In a **combination reaction,** two reactants combine to form a single product. In Reactions 1 and 2, carbon dioxide combines with another compound to form a new product.

Topic: Chemical Equations
Visit: www.SciLinks.org
Web code: KEY-406

Reaction 1: Carbon dioxide gas reacts with solid sodium oxide to produce sodium carbonate (washing soda):

$$CO_2(g) + Na_2O(s) \longrightarrow Na_2CO_3(s)$$

Reaction 2: Carbon dioxide gas reacts with aqueous sodium hydroxide to produce sodium bicarbonate (baking soda):

$$CO_2(g) + NaOH(aq) \longrightarrow NaHCO_3(aq)$$

A combination reaction is sometimes represented by the general equation

$$A + B \longrightarrow C$$

Decomposition Reactions

In Reactions 3 and 4, compounds containing oxygen atoms decompose to produce $O_2(g)$ and an ionic solid. These decomposition reactions are two ways to produce oxygen. In a **decomposition reaction,** one reactant decomposes, or breaks down, to produce two or more products.

Reaction 3: Solid potassium chlorate decomposes to produce solid potassium chloride and oxygen gas:

$$2KClO_3(s) \longrightarrow 2KCl(s) + 3O_2(g)$$

Reaction 4: Solid barium peroxide decomposes to produce solid barium oxide and oxygen gas:

$$2BaO_2(s) \longrightarrow 2BaO(s) + O_2(g)$$

A decomposition reaction is sometimes represented by the general equation

$$A \longrightarrow B + C$$

On the international space station, oxygen is produced by a decomposition reaction that uses electrical energy to split water into hydrogen gas and oxygen gas:

$$2H_2O(l) \longrightarrow 2H_2(g) + O_2(g)$$

2 Exchange Reactions

Calcium is a necessary part of our diet, especially as we age and our bones become more fragile. An example of a calcium supplement is calcium carbonate, $CaCO_3(s)$. However, before the calcium in a vitamin tablet can become bone, it must be made more transportable in the bloodstream. It begins its chemical journey in the stomach.

Examine the reactions involving calcium below. What patterns do you notice in the ways the atoms in the reactants are rearranged to produce the products?

Reaction 5: Solid calcium reacts with hydrochloric acid in the stomach to produce aqueous calcium chloride and hydrogen gas:

$$Ca(s) + 2HCl(aq) \longrightarrow CaCl_2(aq) + H_2(g)$$

Reaction 6: Solid calcium carbonate reacts with hydrochloric acid (stomach acid) to produce aqueous calcium chloride and carbonic acid:

$$CaCO_3(s) + 2HCl(aq) \longrightarrow CaCl_2(aq) + H_2CO_3(aq)$$

In Reaction 5, calcium atoms replace the hydrogen atoms in the HCl molecules.

$$Ca(s) + 2HCl(aq) \longrightarrow CaCl_2(aq) + H_2(g)$$

Ca replaces H.

In Reaction 6, the polyatomic ion, CO_3^{2-}, switches places with the chlorine ion, Cl^-. The second reaction is a common way in which calcium is made available to your body through chemical changes that take place in your stomach.

$$CaCO_3(s) + 2HCl(aq) \longrightarrow CaCl_2(aq) + H_2CO_3(aq)$$

CO_3^{2-} ions change place with Cl^- ions.

In both cases the two reactants exchange atoms. The first reaction is called a **single exchange reaction** or a single replacement reaction because atoms of an element are exchanged with atoms of another element in a compound. A single exchange reaction is sometimes represented by the general equation

$$A + BC \longrightarrow AC + B$$

The second reaction is called a **double exchange reaction** or a double replacement reaction because two compounds exchange atoms with each other. A double exchange reaction is sometimes represented by the general equation

$$AB + CD \longrightarrow AD + CB$$

ENGINEERING CONNECTION

Inside submarines, air quality is a major concern. Not only must carbon dioxide be scrubbed out of the air but oxygen must be replenished and water vapor removed. Otherwise, the inside of the submarine becomes too damp from all the exhaled water vapor. For this reason, the air in a submarine is also pumped through dehumidifiers.

Example

Classify Chemical Reactions

Classify each reaction as a combination reaction, a decomposition reaction, a single exchange reaction, or a double exchange reaction.

a. $H_2(g) + Cl_2(g) \longrightarrow 2HCl(g)$

b. $CaCO_3(s) \longrightarrow CaO(s) + CO_2(g)$

c. $Sn(s) + O_2(g) \longrightarrow SnO_2(s)$

d. $CaI_2(s) + Cl_2(g) \longrightarrow CaCl_2(s) + I_2(g)$

e. $AgNO_3(aq) + NaOH(aq) \longrightarrow AgOH(s) + NaNO_3(aq)$

Solution

a. Two reactants combine to produce one product. This is a combination reaction.

b. One reactant decomposes to produce two products. This is a decomposition reaction.

c. Two reactants combine to produce one product. This is a combination reaction.

d. Chlorine, Cl, replaces iodine, I, in CaI_2. This is a single exchange reaction.

e. Silver, Ag, and sodium, Na, exchange places. This is a double exchange reaction.

Toxins sometimes act in the body in an exchange reaction by replacing an atom or atoms from a compound that exists in the body. This can have the effect of destroying necessary compounds or creating a new compound with harmful properties, or both. One way to remove toxic substances from the environment is by using combination and decomposition reactions.

Key Terms

combination reaction
decomposition reaction
single exchange reaction (single replacement)
double exchange reaction (double replacement)

Lesson Summary

How do atoms rearrange to form new products?

Chemical reactions can be classified into general categories based on how atoms in the reactants rearrange to form the products. Four general types of chemical change include combination reactions, decomposition reactions, single exchange reactions, and double exchange reactions. Toxins sometimes harm the body through chemical reactions. However, chemical reactions can also be used to remove toxins from the body or the environment.

EXERCISES

Reading Questions

1. How are combination reactions and decomposition reactions related?
2. What is the difference between a single exchange reaction and a double exchange reaction?

Reason and Apply

3. Classify these reactions as combination, decomposition, single exchange, or double exchange. Copy the equations and fill in any missing products and write a balanced equation for each reaction.
   a. $NaOH(aq) + HNO_3(aq) \longrightarrow NaNO_3(aq) +$ ____________ (l)
   b. $C_2H_4(g) + Cl_2(g) \longrightarrow$ ____________ (g)
   c. $Cl_2(g) + MgBr_2(s) \longrightarrow Br_2(s) +$ ____________ (s)
4. List four molecules and four ionic compounds in the reactions from Exercise 3.
5. Sulfur trioxide gas combines with liquid water to produce aqueous sulfuric acid:

   ____________ $(g) + H_2O(l) \longrightarrow H_2SO_4(aq)$

   Predict the missing reactant, balance the equation, and explain how these toxic substances could enter the body.
6. Solid lithium reacts with aqueous hydrochloric acid to produce hydrogen gas and aqueous lithium chloride:

   $Li(s) + HCl(aq) \longrightarrow H_2(g) +$ ____________

   Predict the missing product, balance the equation, and explain how these toxic substances could enter the body.
7. Aqueous silver nitrate and aqueous sodium hydroxide are mixed to produce aqueous sodium nitrate and solid silver hydroxide:

   $AgNO_3(aq) + NaOH(aq) \longrightarrow NaNO_3(aq) +$ ____________

   Predict the missing product, balance the equation, and explain how the reaction could help remove toxic substances from water.

SPACE SCIENCE CONNECTION

The idea of colonizing another planet, such as Mars, has sparked imaginations for a long time. Any sort of settlement on a space station or another planet would make use of technology that creates a breathable atmosphere.

SECTION

I

Key Terms

physical change
coefficient
formula unit
combination reaction
decomposition reaction
single exchange reaction (single replacement)
double exchange reaction (double replacement)

SUMMARY

Toxic Changes

Toxins Update

A toxic substance causes an undesirable chemical reaction, producing a harmful or unhealthy change in a living system. Chemical equations keep track of these changes and allow you to predict what you will observe when compounds combine. Becoming familiar with chemical equations is the first step in understanding these chemical changes.

Review Exercises

1. Why is it difficult to identify a physical or chemical change through observations alone?

2. Based on the chemical equation for a reaction, can you tell if any of the substances are toxic?

3. Consider the equation for the formation of a kidney stone.

$$Na_3PO_4(aq) + 3CaCl_2(aq) \longrightarrow Ca_3(PO_4)_2(s) + 6NaCl(aq)$$

a. Is each reactant bonded ionically or covalently? How do you know?

b. Is this is a combination reaction, decomposition reaction, single exchange reaction, or double exchange reaction?

c. Is this a chemical change or a physical change?

d. How does a balanced reaction show that matter is conserved?

e. What is the chemical name of the solid that makes up a kidney stone?

4. Copy and balance these chemical equations.

a. $N_2(g) + H_2(g) \longrightarrow N_2H_4(g)$

b. $KNO_3(s) + K(s) \longrightarrow K_2O(s) + N_2(g)$

c. $H_2SO_4(aq) + NaCN(aq) \longrightarrow HCN(g) + Na_2SO_4(aq)$

d. $H_3PO_4(aq) + Ca(OH)_2(aq) \longrightarrow Ca_3(PO_4)_2(aq) + H_2O(l)$

e. $C_3H_8(g) + O_2(g) \longrightarrow CO_2(g) + H_2O(l)$

f. $H_2S(g) + O_2(g) \longrightarrow SO_2(g) + H_2O(l)$

g. $H_2(g) + O_2(g) \longrightarrow H_2O(l)$

WEB  RESEARCH

PROJECT ***Toxins in the Environment***

Research a potentially toxic substance. (Your teacher may assign you one.) Find out where in your environment you might find this substance and describe its effects on the body. Prepare a short report.

SECTION

II

Measuring Toxins

Many compounds, such as the medications shown here, keep us healthy and even cure diseases. But these same substances can be quite harmful if taken in the wrong doses.

Any substance, even water, can be toxic if too much of it is consumed. And many substances that are considered quite harmful are actually therapeutic when administered in the correct amounts. It is the dose that is key to determining the safety of a substance. This section concentrates on how amounts of matter are measured and counted.

In this section, you will study

- how lethal doses of toxic substances are measured
- ways to count small objects by weighing them
- converting between counting units (moles) and mass units
- how Avogadro's number is used

LESSON

7 Lethal Dose

Toxicity

Think About It

Poisonous snakebites are medical emergencies. The venom from the bite is very toxic and can be deadly, especially for small children or small animals such as rabbits and mice. However, tiny amounts of snake venom have been used therapeutically as a medicine to control high blood pressure. How much snake venom is toxic and how much is therapeutic?

How much is too much of a substance?

To answer this question, you will explore

1. How Toxicity Is Measured
2. Relationship to Body Weight

Exploring the Topic

1 How Toxicity Is Measured

You have probably been warned about the dangers of dozens of different substances, from pesticides to poisonous wild mushrooms. You know better than to drink your shampoo or rub gasoline on your skin. But how is toxicity determined in the first place? And how is it that some things that are good for you, like vitamin tablets, can be toxic or even lethal in higher doses?

BIOLOGY CONNECTION

One of the chemicals in snake venom lowers the blood pressure of its prey to make it unconscious. In tiny doses the same chemical can be used to treat high blood pressure.

Measuring Toxicity

For our safety, scientists try to determine the **toxicity** of the products we buy and the substances we come into contact with. Sometimes they find the toxicity accidentally, when a person is exposed to a harmful substance. Other times it takes a great deal of evidence before it is determined that a substance is toxic. This was the case with lead-based paints.

It is clearly not a good idea to determine how toxic something is by purposefully exposing humans. Yet it is vital to know how much of a particular substance is dangerous. So, toxicities of most substances are measured by exposing laboratory animals to toxic substances. Typically, rats or mice (sometimes rabbits, dogs, or even monkeys) are used. The harm that might be done to the animals must be carefully weighed against potential harm to humans if the toxicities of substances, such as medicines, are unknown.

One way to measure toxicity is to measure the *lethal dose* of a substance, the amount of substance that can cause death. The LD_{50} of a substance is the amount of a substance that causes the death of half, or 50%, of the animals exposed to it.

The table provides the LD_{50}'s for several common substances in grams, milligrams, or micrograms (mcg) of substance per kilogram of body mass. A microgram is one millionth of a gram.

Lethal Doses

Common name	Chemical name and formula	Lethal dose (LD_{50}) per kg body mass	Toxic response
aspirin	acetylsalicylic acid, $C_9H_8O_4$	200 mg/kg (rat, oral)	gastric distress, confusion, psychosis, stupor, ringing in ears, drowsiness, hyperventilation
table salt	sodium chloride, NaCl	3 g/kg (rat, oral) 12,357 mg/kg (human, oral)	eye irritation, elevated blood pressure
castor beans	ricin—very large protein molecule, molecular mass 63,000 amu	30 mg/kg (human, oral) 3.0 mcg/kg (human, intravenous)	vomiting, diarrhea, internal bleeding, kidney and liver failure; death within minutes if injected
arsenic	arsenic (III) oxide As_2O_3	15 mg/kg (rat, oral)	*acute:* irritates eyes, skin, respiratory tract; nausea *chronic:* convulsions, tissue lesions, hemorrhage, kidney impairment
sugar	glucose, $C_6H_{12}O_6$	30 g/kg (rat, oral)	depressed activity, gastric disturbances; if diabetic, heart disease, blindness, nerve damage, kidney damage
snake venom	alpha-bungarotoxin, $C_{338}H_{529}N_{97}O_{105}S_{11}$	25.0 mcg/kg (rat, intramuscular)	paralysis, suffocation, loss of consciousness, seizures, hemorrhaging into tissues
coffee beans	caffeine, $C_8H_{10}N_4O_2$	140 mg/kg (dog, oral) 192 mg/kg (rat, oral)	acute kidney failure, nausea, psychosis, hemorrhage, increased pulse, convulsions

Castor oil is made from castor beans, which are the seeds of the castor plant.

Notice the wide range of doses—from grams, to milligrams (thousandths of a gram), to micrograms (millionths of a gram). A small LD_{50} indicates that it does not take very much of the substance to produce ill effects. Substances with small LD_{50}'s should definitely be avoided. Snake venom is one of the most dangerous toxins in the table, with an LD_{50} of 25.0 mcg/kg, while sugar, with an LD_{50} of 30 g/kg is one of the safer substances.

Dosage Determines Toxicity

You may be surprised to see lethal doses for substances like sugar and aspirin. These substances are usually considered beneficial. However, everything is toxic in a large enough dose, even life-sustaining substances like water and oxygen. And scientists have discovered that many highly toxic substances also have therapeutic value. Botulinum toxin (Botox) can help patients suffering from cerebral palsy. A very large protein isolated from the venom of the blue scorpion has been discovered to inhibit certain cancer tumors.

2 Relationship to Body Weight

You may be wondering how testing a substance on very small animals, like rats, will determine whether it is toxic to a larger mammal, such as a human. To take size into account, the LD_{50} is reported in milligrams of substance per kilogram of body weight, mg/kg. That way the lethal dose can be estimated for a mammal of any size.

MEDICINE CONNECTION

Many toxins serve as medicines when used in moderation. Atropine, for example, extracted from the deadly nightshade plant, is used to treat cardiac arrest and poisoning by organophosphate insecticides and nerve gases. However, atropine is potentially addictive and an overdose can be fatal.

Testing lab animals gives us only an approximate range for safety. Our bodies differ in many ways from the bodies of rats or rabbits. And it may seem particularly unkind to subject small animals to testing. However, the alternative would be to risk human life with untested medications. Imagine if the safety of products on store shelves had never been confirmed in any way.

Example

Caffeine

The lethal dose for caffeine is approximately 150–200 mg/kg of body mass. How much caffeine would be lethal for a 120 lb person? How many cups of coffee is this?

Structural formula for caffeine

1 cup strong coffee

Contains ~150 mg caffeine

Solution

Lethal dose is reported in mg/kg, so first convert pounds to kilograms. There are 2.2 pounds in a kilogram.

Convert body weight from pounds to kilograms. $120 \text{ lb} \cdot \frac{1 \text{ kg}}{2.2 \text{ lb}} = 54.4 \text{ kg}$

Multiply the person's weight by the lethal dose. $54.4 \text{ kg} \cdot \frac{150 \text{ mg}}{1 \text{ kg}} = 8160 \text{ mg caffeine}$

Convert number of milligrams to number of cups. $8160 \text{ mg} \cdot \frac{1 \text{ cup}}{150 \text{ mg}} = 54.4 \text{ cups}$

So, it would be difficult to ingest a lethal dose of caffeine simply by drinking coffee. (It would take about 54 cups for the 120 lb person!)

[For a review of this math topic, see **MATH** ***Spotlight: Dimensional Analysis*** on page 632.]

Key Term

toxicity

Lesson Summary

How much is too much of a substance?

Every substance on the planet is a potential toxin. By the same token, many substances that are considered toxic have health benefits when taken in the correct dose. Thus, toxicity is determined by dose. Scientists use a variety of methods to measure toxicity. One measurement is called the LD_{50}. This is the amount of a substance that kills 50% of animals exposed to it. It is commonly expressed in milligrams of substance per kilogram of body weight. A low LD_{50} means a more toxic substance.

EXERCISES

Reading Questions

1. How can scientists determine the toxicity of a substance?
2. How is toxicity related to body weight?

Reason and Apply

3. Find at least five products at home with labels that warn of toxicity. For each product, give the name of the product, and answer these questions:
   a. How does each label advise you to avoid harmful exposure to the product?
   b. What does each label tell you to do if a dangerous exposure does occur?
   c. Using all of this information, what can you hypothesize about the chemical properties of each product?
   d. Look at the recommended treatment for dangerous exposure to the product. What chemical and physical processes do you think might be involved in the treatment?
4. Ethanol is grain alcohol. The LD_{50} for ethanol is 7060 mg/kg (rat, oral).
   a. How many milligrams of ethanol would be lethal to a 132 lb adult?
   b. How many glasses containing 13,000 mg of ethanol would be lethal to a 22 lb child?
5. The LD_{50} for vitamin A is 1510 mg/kg (rat, oral).
   a. How many mg of vitamin A would be lethal to a 132 lb adult?
   b. How many vitamin tablets containing 0.40 mg of vitamin A would be lethal to an adult?
   c. **WEB RESEARCH** What are the benefits of vitamin A?

Topic: Toxicology
Visit: www.SciLinks.org
Web code: KEY-407

HEALTH CONNECTION

Fresh fruits and vegetables contain many vitamins that are essential in helping our bodies function properly. Leafy greens are rich in folic acid, a vitamin that is essential in producing and maintaining new cells. Tomatoes and bell peppers have high levels of vitamin C as well as the vision-aiding vitamin A.

ACTIVITY Counting by Weighing

Purpose

To count large numbers of small objects by weighing.

Procedure

1. Obtain a sandwich bag from your teacher. Your bag may contain items such as rice, beans, or paper clips.
2. Your challenge is to determine the number of objects in your bag without opening it.
3. Brainstorm how you will solve the challenge with the members of your group and decide what tools you might use.

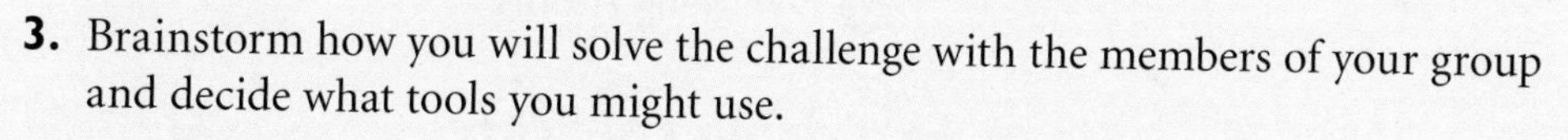

Questions

1. Describe the method you are going to use to determine the number of objects you have in your bag. Be specific.
2. Find out what other substances are in the sandwich bags of the other groups. Make a list of all the substances.
3. Which of the substances' counts do you think will be easiest to determine? Explain your reasoning.
4. **Making Sense** What do you think this activity has to do with keeping track of chemical compounds?
5. **If You Finish Early** Are 100 molecules of snake venom equivalent in toxicity to 100 atoms of arsenic? Why or why not?

LESSON

8 Make It Count

Counting by Weighing

Think About It

When a toxin enters your body, it is important for a doctor to know exactly how much toxin there is. One molecule of snake venom has over 100 times as much mass as one atom of arsenic. So, 1 milligram of snake venom contains many fewer particles than 1 milligram of arsenic. To track the toxin you need an inventory, or a count, of the atoms or molecules. However, atoms are so small that you cannot see them, let alone count them. Luckily, there are other ways of figuring out the number of atoms in a sample besides counting them one by one.

How can mass be used to count large numbers of small objects?

To answer this question, you will explore

1. Using Mass to Count
2. Weighing Atoms

Exploring the Topic

1 Using Mass to Count

Let's explore how to count objects by weighing them. The masses of a number of different small objects are provided in the table.

Substance	Mass of 1000 pieces	Mass of 10 pieces	Mass of 1 piece
rice grains	22 g	0.22 g	0.022 g
lentils	56 g	0.56 g	0.056 g
rubber bands	260.5 g	2.60 g	0.260 g
paper clips	500 g	5.00 g	0.500 g
pennies	2500 g	25.00 g	2.500 g

It is difficult to accurately measure the mass of an object if the object has a mass that is below the detection limit of the balance. Many electronic balances have 0.001 g as a lower weight limit. So, weighing heavier objects tends to yield more accurate results.

One way around this difficulty is to count out a collection of identical (or nearly identical) objects and weigh the collection. You can then find the mass of a single object by dividing the mass of the collection by the number of objects. This gives you an average mass of one object that is more accurate than if you had weighed the object individually.

Once you know the average mass of a single object, you can use it to determine exactly how many objects you have in a large sample. Consider a sample of pushpins.

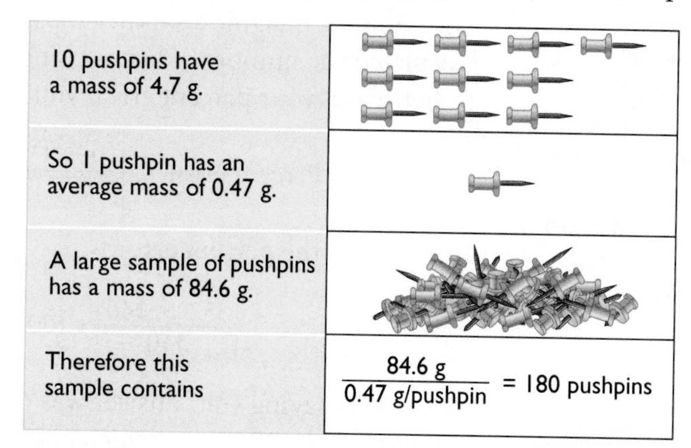

In addition to helping you overcome the detection limit of the balance, there is a second reason that it is important to find the *average* mass of the objects in a sample: There is a slight variation in the masses of the objects. For example, not all grains of rice are identical in size. Some are slightly larger and others are slightly smaller (like isotopes of atoms). So, an average mass is a more accurate predictor of the mass of a *typical* object.

Example

A 5 lb Bag of Rice

How many grains of rice are there in a 5 lb bag of rice? The average mass of a rice grain is 0.0221 g, and the bag contains exactly 5.00 lb. (1 lb = 454 g)

Solution

First, determine how many grains of rice are in the bag.

Multiply by the conversion factor.

$$\text{mass of rice in grams} = 5.00 \text{ lb} \cdot \frac{454 \text{ g}}{1 \text{ lb}}$$

$$= 2270 \text{ g rice}$$

This is the mass of all the grains of rice. Next, determine how many grains of rice are in the bag.

$$\text{Total mass of rice} = \text{average mass of 1 grain} \cdot \text{number of grains}$$

Divide the total mass of the rice by the average mass of 1 grain.

$$\text{number of grains} = \frac{\text{total mass of rice}}{\text{average mass of 1 grain}}$$

Calculate the answer.

$$= \frac{2270 \text{ g}}{0.0221 \text{ g}} = 103{,}181 \text{ grains}$$

So, rounded to three significant digits, there are about 103,000 grains.

[For a review of these math topics, see **MATH *Spotlight: Dimensional Analysis*** on page 632 and **MATH *Spotlight: Accuracy, Precision, and Significant Digits*** on page 617.]

BIOLOGY CONNECTION

Some snakebites can be harmless, others lethal. For example, the bites of garter snakes, found throughout North America, are generally harmless. However, it would take just 1/14,000 ounce of venom from an Australian brown snake, shown here, to kill a person.

Percent Error

Chemists use **percent error** to express how accurate their measurements are. For example, imagine a sandwich bag contains exactly 340 beans. Suppose you calculated the number of beans using mass and came up with 352 beans. In order to figure out your percent error, you would use this formula:

$$\text{Percent error} = \left|\frac{(\text{observed value} - \text{actual value})}{\text{actual value}}\right| \bullet 100\%$$

In this case, the percent error is

$$\left|\frac{352 - 340}{340}\right| \bullet 100\% = \left|\frac{12}{340}\right| \bullet 100\% = 3.5\%$$

Notice that saying your answer was off by 3.5% is more meaningful than saying you were off by 12 beans, which could be a lot or a little, depending on how many beans you were counting. The smaller the percent error, the more accurate your answer.

2 Weighing Atoms

You can apply the same method you used to count tiny objects to counting atoms. In order to determine the number of atoms in a 1.0 g sample, you will need to know the average mass of the atoms in the sample. The average mass of the atoms of each element is given on the periodic table in atomic mass units, or amu.

Atomic mass units are special units used for atoms because the mass of an atom in grams is so tiny. Each atomic mass unit is equal to 0.00000000000000000000000166 g. There are 23 zeros in front of the 166, making this number inconvenient to use.

Consider hydrogen atoms. Hydrogen has an average atomic mass of about 1.0 amu, or 0.00000000000000000000000166 g. How many hydrogen atoms are in a 1.0 g sample?

$$\text{number of H atoms} = \frac{\text{total mass of the sample of H atoms}}{\text{average mass of 1 H atom}}$$

$$= \frac{1.0\text{ g}}{0.00000000000000000000000166\text{ g}}$$

$$= 602{,}000{,}000{,}000{,}000{,}000{,}000{,}000\text{ H atoms}$$

So one gram of hydrogen atoms contains 602 sextillion, or 1 mole, of hydrogen atoms.

Key Term

percent error

Lesson Summary

How can mass be used to count large numbers of small objects?

It is difficult to count very small objects like atoms and molecules. One method used to get a count of small objects is to weigh a large number of the objects and divide by the average weight of one object. The average atomic mass on the periodic table is equivalent to the mass of one atom of each element. This number is an average because some atoms are isotopes with different masses. The mass of a single atom is expressed in atomic mass units rather than grams. To find the number of atoms in a sample of an element, you can divide the mass of the sample by the average mass of one atom.

EXERCISES

Reading Questions

1. Explain how you can use mass to count large numbers of objects.
2. What does the percent error tell you?

Reason and Apply

3. Recall the method you used to count the objects in the sandwich bag you received in class and the results you obtained.
   a. Explain what you were trying to find out.
   b. Explain the method you used.
   c. Show your calculations and results.
   d. Calculate the percent error. Explain how you might modify your procedure to reduce the percent error.
4. Suppose that you have 50 grams of rice and 50 grams of beans. Which sample has more pieces? Explain your thinking.
5. Suppose that you have 740 tiny plastic beads and 740 marbles. Which sample has more mass? Explain your reasoning.
6. One bean weighs 0.074 g. How many beans are in a 50-pound bag?
7. Suppose you want to fill a 500 g bag with twice as many red jelly beans as yellow. What mass of red jelly beans do you need? A jelly bean weighs 0.65 g.
8. What is the mass in grams of one copper atom?
9. What is the mass in grams of one gold atom?
10. Suppose you have 50 grams of copper and 50 grams of gold. Which sample has more atoms? Explain your thinking.
11. Suppose you calculated the number of beans using the mass. Which of these experimental values has a smaller percent error?
    A. A calculated value of 1342 when the actual value is 1327.
    B. A calculated value of 1327 when the actual value is 1342.
    C. They have the same percent error.
    D. There is not enough information to answer the question.

LESSON

9 Billions and Billions

Avogadro's Number

Think About It

Mercury is a toxic substance that accumulates in the body and damages the central nervous system and other organs. Which do you think would be worse for you, 1 mole of mercury or 10 grams of mercury? To answer this question, it is necessary to understand the relationship between these two measures.

What is the relationship between mass and moles?

To answer this question, you will explore

1. Moles and Grams
2. Scientific Notation

Exploring the Topic

1 Moles and Grams

In the world around you, a variety of different measures are used to specify amounts. For example, at the store you might buy 10 apples or a 5 lb bag of apples. One measure is a counting unit and the other is a unit of weight. To compare the prices, you would probably want to know how many apples are in the 5 lb bag.

One mole of O_2, Cu, NaCl, and H_2O

Likewise, chemists use both number and mass to specify the amount of a substance. For example, they might know the effect of either 1 mol of mercury atoms or of a 10 g sample of mercury. To compare these two quantities, they would need to know either how many moles of atoms are in each gram of mercury or the mass in grams of one mole of mercury.

The mole, abbreviated mol, is a counting unit used to count a large number of atoms. A mole of atoms is equal to 602 sextillion atoms. This number is also referred to as Avogadro's number.

The table here compares the average mass of one atom with the mass of 1 mol of atoms for five elements. Examine the table.

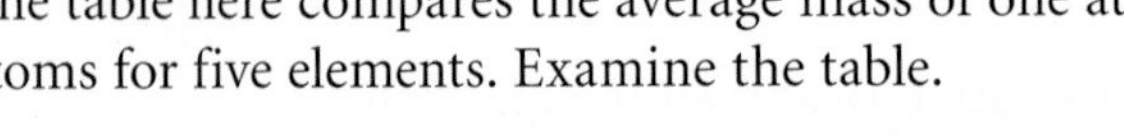

Comparing Atoms and Moles

Element	Average mass of one atom (amu)	Mass of one mole of atoms (g)
hydrogen, H	1.00 amu	1.00 g
carbon, C	12.01 amu	12.01 g
iron, Fe	55.85 amu	55.85 g
arsenic, As	74.92 amu	74.92 g
mercury, Hg	200.59 amu	200.59 g

Notice that the numerical value of the average atomic mass on the periodic table for an element is identical to the mass of one mole of atoms of that element. In other words, both values are represented by the same number. What is different about the two measurements is the units. The average mass of *one atom* of arsenic is 74.92 amu, and the mass of *one mole* of arsenic atoms is 74.92 grams. How convenient!

BIG IDEA The mass of one mole of atoms of an element in grams is numerically the same as the average atomic mass for that element.

Molar Mass

The mass of one mole of atoms of an element is called the **molar mass.** You can find the molar mass of any element by looking up the average atomic mass on the periodic table. For example, the average atomic mass of mercury is 200.6 amu. The mass of 1 mol of mercury atoms is 200.6 g.

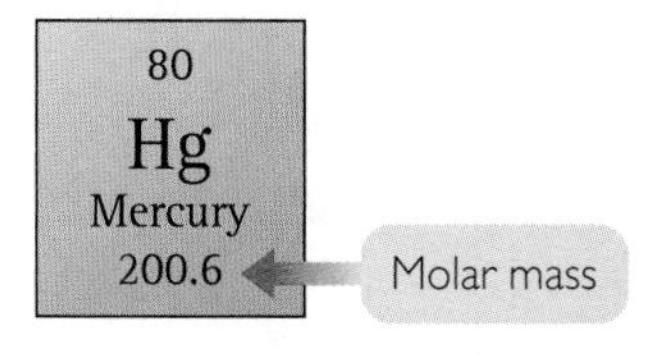

Example 1

Copper Versus Arsenic

Which has more mass, 1 mol of copper atoms or 1 mol of arsenic atoms?

Solution

Consult the periodic table to find the molar mass of each substance. The molar mass of copper is 63.55 grams per mole, and the molar mass of arsenic is 74.92 grams per mole. A mole of arsenic atoms has more mass.

Example 2

Amount of Mercury, Hg

Which is more toxic, 1 mol or 10 g of mercury, Hg?

Solution

By looking on the periodic table, you can determine that the mass in grams of 1 mol of mercury is 200.6 g. So, 1 mol of Hg is more toxic than 10 g of Hg because 1 mol represents a much larger mass.

ASTRONOMY CONNECTION

Each galaxy in the universe is estimated to have 400 billion stars in it. There are about 130 billion galaxies in the known universe. This is about 5.2×10^{22} total stars in the universe, or 0.08 mol.

2 Scientific Notation

Consider the table on the next page. It shows the mass, in grams, of different numbers of atoms of hydrogen, arsenic, and mercury. Notice that the number of atoms and the mass are shown both in longhand and in **scientific notation.** Scientific notation is a shorthand method that uses an exponent to keep track of where the decimal point should be. Small numbers, between 0 and 1, have negative exponents. Numbers greater than 1 have positive exponents. The exponent indicates where the decimal point belongs in the longhand number.

Number of Atoms and Mass of Mercury Atoms

Number of atoms	Number in scientific notation	Mass (g)	Mass in scientific notation (g)
1	1×10^0	0.00000000000000000000033 g	3.3×10^{-22} g
1000	1×10^3	0.00000000000000000033 g	3.3×10^{-19} g
1,000,000,000	1×10^9	0.00000000000033 g	3.0×10^{-13} g
602,000,000,000,000,000,000,000	6.02×10^{23} (1 mol)	200.59 g	2.0059×10^2 g

MATHEMATICS CONNECTION

A googol is 10^{100}, or 1 followed by 100 zeros. A googolplex is the largest named number. It is 10^{googol}, or 1 followed by a googol zeros.

In scientific notation, ten thousand is written 1.0×10^4. One ten-thousandth is written 1.0×10^{-4}.

$$10{,}000 = 1.0 \times 10^4$$

$$1/10{,}000 = \frac{1}{1.0 \times 10^4} = 1.0 \times 10^{-4} = 0.0001$$

[For a review of this topic, see **MATH** ***Spotlight: Scientific Notation*** on page 631.] The mass of 1000 atoms of arsenic is only 0.00000000000000000012 g or 1.2×10^{-19} g. This many arsenic atoms wouldn't even be visible to the naked eye. On the other hand, 74.92 g of arsenic contains 1 mol, or 6.02×10^{23} atoms.

When you are dealing with atoms, numbers tend to be either very tiny or very large. So it is more convenient to use scientific notation.

Key Terms

molar mass
scientific notation

Lesson Summary

What is the relationship between mass and moles?

Chemists keep track of the number of molecules and atoms they are working with by using a unit called the mole. They also use scientific notation to express numbers that are very small or very large. One mole is equal to 602 sextillion. Using scientific notation, this number is written as 602,000,000,000,000,000,000,000, or 6.02×10^{23}. The mass (in grams) of one mole of atoms of an element can be found on the periodic table. This is called the molar mass and has the same numerical value as the average atomic mass.

EXERCISES

Reading Questions

1. How do you find the average atomic mass of atoms of an element? What unit of measure is this given in?
2. What is meant by the term molar mass? What unit of measure is this given in?
3. Why do chemists convert between moles and grams?

Reason and Apply

4. Give the molar mass for the elements listed.
   a. nitrogen, N
   b. neon, Ne
   c. chlorine, Cl
   d. copper, Cu
5. Which has more mass?
   a. 1 mol of hydrogen, H, or 1 mol of carbon, C
   b. 1 mol of aluminum, Al, or 1 mol of iron, Fe
   c. 1 mol of copper, Cu, or 1 mol of gold, Au
   d. 5 mol of carbon, C, or 1 mol of gold, Au
6. Which contains more atoms?
   a. 12 g of hydrogen, H, or 12 g of carbon, C
   b. 27 g of aluminum, Al, or 27 g of iron, Fe
   c. 40 g of calcium, Ca, or 40 g of sodium, Na
   d. 40 g of calcium, Ca, or 60 g of zinc, Zn
   e. 10 g lithium, Li, or 100 g of lead, Pb
7. Which is more toxic? Explain your reasoning.
   a. 1 mol of beryllium, Be, or 10 g of beryllium
   b. 2 mol of arsenic, As, or 75 g of arsenic
   c. 2 mol of lead, Pb, or 500 g of lead
8. Copy and complete the table.

Amount	Number of moles	Number of atoms	Number of atoms in scientific notation
12 g carbon, C	1 mol	602,000,000,000,000,000,000,000 atoms	6.02×10^{23} atoms
24 g carbon, C			
40 g calcium, Ca			
20 g calcium, Ca			

EARTH SCIENCE CONNECTION

It is estimated that there are 2.0×10^{21} grains of sand on the entire Earth. This is only 0.003 mol of sand grains! In contrast, there is approximately 0.0002 mol of water molecules, or 1.2×10^{21} molecules, in a single drop of water.

LESSON

10 What's in a Mole?

Molar Mass

Think About It

Lead is a highly toxic substance that can be accidentally ingested by humans and animals. Lead atoms interfere with normal processes in the body causing disturbances in the nervous system. Suppose you have 100 g of lead (II) carbonate, $PbCO_3$, and 100 g of lead (II) chloride, $PbCl_2$. It is the lead atoms specifically that are toxic. If you are exposed to equal masses of lead carbonate and lead chloride, which substance exposes you to more lead atoms, and is potentially more toxic? To determine this, you need to determine the mass of 1 mol of each.

How can you convert between mass and moles?

To answer this question, you will explore

1. Molar Mass of Compounds
2. Comparing a Mole's Worth

Exploring the Topic

1 Molar Mass of Compounds

Counting With Moles

Any object can be counted with units such as a dozen or a million or a mole. Consider a dozen cheese sandwiches. Each sandwich has two slices of bread and one slice of cheese. Therefore, a dozen sandwiches would have a total of two dozen slices of bread and one dozen slices of cheese.

In a similar way, you can count atoms of lead in a lead compound. Because a compound consists of bonded atoms, you can simply add the molar masses of each atom in a molecule or formula unit of the compound to obtain the molar mass of the substance. For example, each formula unit of lead (II) chloride, $PbCl_2$, contains one atom of lead and two atoms of chlorine.

Molar mass of lead (II) chloride, $PbCl_2$

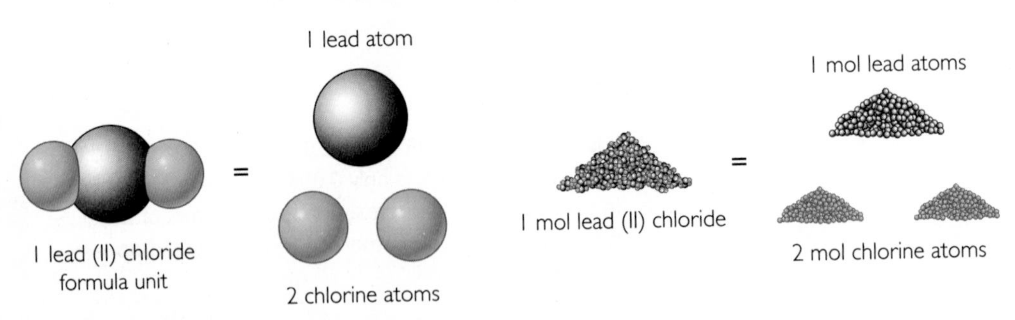

The molar masses of lead and chlorine are 270.21 g/mol and 35.45 g/mol, respectively.

HEALTH CONNECTION

Lead (II) carbonate is one of the white pigments in lead paint. Lead paint dries quickly and forms a durable shiny coating. It is still used to draw lines on pavement, but it is no longer used to paint houses or furniture due to the high toxicity of lead.

Adding the molar mass of lead and twice the molar mass of chlorine gives the molar mass of lead chloride, $PbCl_2$, which is 341.11 grams per mole, or 341.11 g/mol.

Example 1

The Mass of One Mole

Suppose you have 1 mol of lead, Pb, 1 mol of lead (II) chloride, $PbCl_2$, and 1 mol of lead (II) carbonate, $PbCO_3$. What is the mass of each sample?

Solution

You can find the molar mass of lead, Pb, on the periodic table, 207.2 g/mol.

A mole of $PbCl_2$ has 1 mol of lead atoms and 2 mol of chlorine atoms. You find the molar mass of the compound by adding these molar masses together.

$$\begin{aligned}\text{molar mass of } PbCl_2 &= 207.2 + 2(35.45)\\ &= 278.1 \text{ g/mol}\end{aligned}$$

So the mass of 1 mol of lead (II) chloride is 276.1 g. The molar mass of $PbCO_3$ can be found similarly.

$$\begin{aligned}\text{molar mass of } PbCO_3 &= 207.2 + 12.01 + 3(16.00)\\ &= 267.2 \text{ g/mol}\end{aligned}$$

So the mass of 1 mol of lead (II) carbonate is 267.2 g.

[For a review of this math topic, see **MATH** ***Spotlight: Solving Equations*** on page 620.]

2 Comparing a Mole's Worth

A mole of atoms or molecules is usually an amount that you can hold in your hand if the substance is liquid or solid. A table with a "mole's worth" of a few substances is given below.

Molar Masses

Chemical formula	Molar mass (g/mol)	Moles of what?	Equivalent to
$O_2(g)$	31.98 g/mol	oxygen molecules	22.4 L oxygen gas
$Al(s)$	26.98 g/mol	Al atoms	2 aluminum cans
$H_2O(l)$	18.01 g/mol	H_2O molecules	18 mL water
$He(g)$	4.00 g/mol	He atoms	22.4 L helium gas
$NaCl(s)$	58.44 g/mol	sodium chloride units	1/4 cup salt
$C_{12}H_{22}O_{11}(s)$	342.23 g/mol	sugar molecules	0.75 lb sugar

Note that the molar mass of oxygen gas, $O_2(g)$, is double the molar mass of oxygen found on the periodic table because oxygen gas is diatomic. Also note that the volume of a mole of O_2 gas is the same as the volume of a mole of any other gas at standard temperature and pressure, or STP.

Example 2

Toxicity of Lead Compounds

Which is potentially the most toxic: 1 g lead, Pb, 1 g lead (II) chloride, $PbCl_2$, or 1 g lead (II) carbonate, $PbCO_3$?

Solution

The molar mass of each compound was calculated in Example 1.

$$\text{molar mass of Pb} = 207.2 \text{ g/mol}$$
$$\text{molar mass of PbCl}_2 = 278.1 \text{ g/mol}$$
$$\text{molar mass of PbCO}_3 = 267.21 \text{ g/mol}$$

Because Pb has the lowest molar mass, 1 g lead, Pb, will have the largest number of moles of lead, so it will potentially be the most toxic, followed by lead (II) chloride, $PbCl_2$.

Lesson Summary

How can you convert between mass and moles?

Chemists compare moles of substances rather than masses of substances because moles are a way of counting atoms, molecules, or units in a compound. The molar mass of a substance is the mass, in grams, of one mole of the substance. The molar mass of a compound is the sum of the molar masses of the atoms in the compound.

EXERCISES

Reading Questions

1. Explain how to determine the molar mass of sodium chloride, NaCl.
2. Describe the approximate size of 1 mol of a solid, a liquid, and a gas. Give a specific example of each.

MATHEMATICS CONNECTION

In daily life you would never use the mole as a counting unit. It is just too big. For example, there is nowhere near a mole of pennies in the world, sand grains on Earth, or stars in the known universe. In other words, there are more atoms of carbon in a pencil than there are grains of sand on Earth or stars in the known universe!

Reason and Apply

3. Copy this table and use a periodic table to complete the second column

Chemical formula	Molar mass (g/mol)	Moles of what?
Ne(*g*)		1 mol Ne atoms
Ca(*s*)		1 mol Ca atoms
$CO_2(g)$		1 mol carbon dioxide molecules
$CaCO_3(s)$		1 mol calcium carbonate units
$CH_4O(l)$		1 mol methanol molecules
$C_2H_6O(l)$		1 mol ethanol molecules
$Fe_2O_3(s)$		1 mol iron oxide units

4. Which has more moles of molecules, 1.0 g methanol, CH_4O, or 1.0 g ethanol, C_2H_6O?
5. Which has more moles of metal atoms?
   a. 10.0 g calcium, Ca, or 10.0 g calcium chloride, $CaCl_2$
   b. 5.0 g sodium chloride, NaCl, or 5.0 g sodium fluoride, NaF
   c. 2.0 g iron oxide, FeO, or 2.0 g iron sulfide, FeS
6. How many grams of carbon molecules are in 1 mol of each substance?
   a. methane, CH_4
   b. methanol, CH_4O
   c. ethanol, C_2H_6O
7. What is the mass of 5 mol of iron (III) oxide, Fe_2O_3?
8. Which of these has the most chromium?
   A. 1.0 g chromium (II) chloride, $CrCl_2$
   B. 1.0 g chromium (III) chloride, $CrCl_3$
   C. 1.0 g chromium (IV) oxide, CrO_2

LESSON

11 Mountains Into Molehills

Mass-Mole Conversions

Think About It

When you get a headache, you might reach for one of the many pain relievers on the market. A bottle of aspirin might direct you to take one 325 mg tablet. A bottle of acetaminophen might say to take one 500 mg tablet. Is one of these medications stronger than the other? One way to find out is to compare the number of moles of pain reliever in each dose.

How are moles and mass related?

To answer this question, you will explore

1. Relating Mass and Moles
2. Converting Between Mass and Moles

Exploring the Topic

1 Relating Mass and Moles

You can find the mass of a substance by using a balance. However, you cannot measure the number of moles in a sample directly. You must convert the mass of a substance to moles to figure out how many atoms or molecules you have. If you want to compare amounts of two substances accurately, it is important to be able to convert between mass and moles.

Typically, you measure the mass of a substance in grams. Medications, like aspirin, are often measured in milligrams. Because there are 1000 milligrams in 1 gram, converting from milligrams to grams is simply a matter of dividing by 1000.

Consider three compounds commonly used in over-the-counter pain relievers.

Pain reliever	Molecular formula	Molar mass	Adult dose	Moles in a standard dose
ibuprofen	$C_{13}H_{18}O_2$	206.3 g/mol	400 mg	0.0019 mol
acetaminophen	$C_8H_9NO_2$	151.2 g/mol	500 mg	0.0033 mol
acetylsalicylic acid (aspirin)	$C_9H_8O_4$	180.2 g/mol	325 mg	0.0018 mol

Notice that each tablet has a different adult dose. It would appear that acetylsalicylic acid (aspirin), with a smaller, 325 mg dose, is stronger than either acetaminophen, 500 mg, or ibuprofen, 400 mg. But ibuprofen is a heavier molecule than acetylsalicylic acid. One gram of aspirin will not represent the same number of molecules as one gram of ibuprofen. In order to make an accurate comparison, you must convert the milligrams of each compound to moles.

Examine the table. Which tablet contains the most moles of pain reliever? A 500 mg acetaminophen tablet contains almost twice as many moles of pain reliever as a 325 mg tablet of acetylsalicylic acid. So aspirin does appear to be more powerful than acetaminophen. Acetylsalicylic acid has the lowest dose and uses the fewest molecules of pain reliever to get rid of your headache. Ibuprofen is almost identical in potency to acetylsalicylic acid. According to the data, acetaminophen is not as powerful as the other two compounds.

2 Converting Between Mass and Moles

The relationship between the mass in grams of a substance and moles is proportional. The proportionality constant is the molar mass, in g/mol. The molar mass will be different for each substance.

Mole to Mass Conversions

The molar mass is the mass per mole of a substance. You can use it to convert between the mass, m, and the number of moles, n.

Example 1

Converting From Moles to Mass

Imagine that a pharmaceutical company has 100 mol of acetaminophen available to be made into 500 mg tablets. The molar mass of acetaminophen is 151.2 g/mol. How many 500 mg tablets can be made?

Solution

First, figure out how many grams of acetaminophen there are in 100 mol.

Start with the mole to mass formula.	$m = \text{molar mass} \cdot n$
Substitute values and solve for mass.	$= (151.2 \text{ g/mol})\ 100 \text{ mol}$
	$= 15{,}120 \text{ g}$

Because each 500 mg tablet is 0.5 g, each gram of the compound will make two tablets.

$$(15{,}120 \text{ g})\ (2 \text{ tablets/g}) = 30{,}240 \text{ tablets}$$

So 30,240 tablets can be made from 100 mol of acetaminophen.

Topic: The Mole
Visit: www.SciLinks.org
Web code: KEY-411

Example 2

Converting From Mass to Moles

Restaurants usually have small packets of sugar to sweeten coffee or tea. Each sugar packet contains approximately 1.0 g of sucrose, $C_{11}H_{22}O_{11}$. How many moles of sucrose does this represent?

Solution

First, use the periodic table to determine the molar mass of sucrose. The molar mass of sucrose is 342.3 g/mol rounded to one decimal place. To convert from mass to moles, divide the mass in grams by the molar mass.

Rearrange the mole to mass formula to solve for n.

$$n = \frac{m}{\text{molar mass}}$$

Substitute values and solve for moles of sucrose.

$$= \frac{1.0\ \text{g}}{342.3\ \text{g/mol}} = 0.003\ \text{mol}$$

So there is 0.003 mol of sucrose in one packet.

Lesson Summary

How are moles and mass related?

Molar mass values allow you to convert between mass in grams of a substance and moles. To convert mass to moles, divide the mass of a sample in grams by its molar mass. To convert moles to mass, multiply the number of moles of the substance by its molar mass.

EXERCISES

Reading Questions

1. How can you convert between moles of a substance and grams of a substance?
2. Why might a 200 mg tablet of aspirin not have the same effect as a 200 mg tablet of ibuprofen?

Reason and Apply

3. There are 8.0 mol of H atoms in 2.0 mol of CH_4O molecules. How many moles of H atoms are there in 2.0 mol of C_2H_6O molecules?
4. List these compounds in order of increasing moles of molecules: 2.0 g CH_4O, 2.0 g H_2O, 2.0 g C_8H_{18}. Show your work.
5. Which has more moles of oxygen atoms, 153 g of BaO or 169 g of BaO_2? Show your work.
6. List these compounds in order of increasing mass in grams: 2.0 mol $SiCl_4$, 2.0 mol PbO, 2.0 mol Fe_2O_3. Show your calculations.
7. Suppose you run a company that buys copper compounds and then recycles the copper for resale. Your company wants to get the most pure copper for the lowest cost. Three different suppliers want to sell you 1 mol $CuO(s)$, 1 mol $CuCO_3(s)$, and 1 mol $Cu_2O(s)$ for the same price.
   a. Which compound has the greatest total mass? Show your work.
   b. Which compound has the greatest mass of Cu? Show your work.
   c. Assuming it costs the same to extract the copper from each compound, which compound represents the best deal for your company? Explain.
8. Suppose Container A contains 1 mol $C(s)$ and 1 mol $O_2(g)$, and Container B contains 1 mol $CO_2(g)$. Both containers are closed and both are the same size. Compare the containers in each of these ways.
   a. number of atoms
   b. number of gas molecules
   c. mass
   d. gas pressure

INDUSTRY CONNECTION

Copper is the third most used metal after iron and aluminum. Copper usage is rapidly expanding as more products are developed that contain electronic components. Modern cars contain 50 lb to 80 lb of copper and a jet airplane contains about 9000 lb of copper.

LESSON

12 How Sweet It Is

Comparing Amounts

Think About It

The safety of artificial sweeteners has been the subject of debate for years. More recently, some schools have banned the sale of both diet and regular soft drinks. Regular soft drinks are sweetened with fructose, while diet soft drinks are usually sweetened with aspartame. In order to make a healthy decision, you need to consider toxicities of fructose and aspartame and the amount of each substance in a can of beverage.

How can you use moles to compare toxicity?

To answer this question, you will explore

1. Comparing Moles of Sweeteners
2. Toxicity of Sweeteners

Exploring the Topic

CONSUMER CONNECTION

The compounds aspartame, saccharin, and sucralose are the most common artifical sweeteners in use today. Aspartame is the sweetener most commonly used in diet soft drinks. Saccharin is often used as a sweetener in toothpastes.

1 Comparing Moles of Sweeteners

It is hard for many of us to imagine a day without sugar. From cereal to flavored drinks and candy, our intake of sugar adds up. So, when the first artificial sweeteners came onto the market in the 1950s, it seemed a good idea to use this low-calorie substitute for a favorite ingredient.

Two sweeteners, fructose, $C_6H_{12}O_6$, and aspartame, $C_{14}H_{18}O_5N_2$, are shown. Notice that they are both medium-sized molecules made up of carbon, hydrogen, and oxygen atoms. The artificial sweetener, aspartame, also has two nitrogen atoms in its molecules.

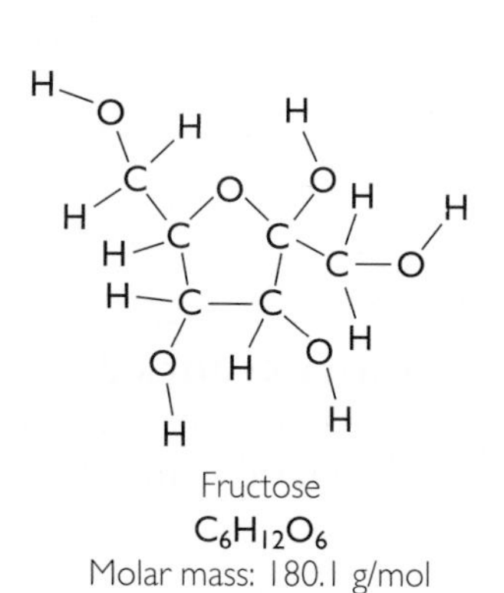

Fructose
$C_6H_{12}O_6$
Molar mass: 180.1 g/mol

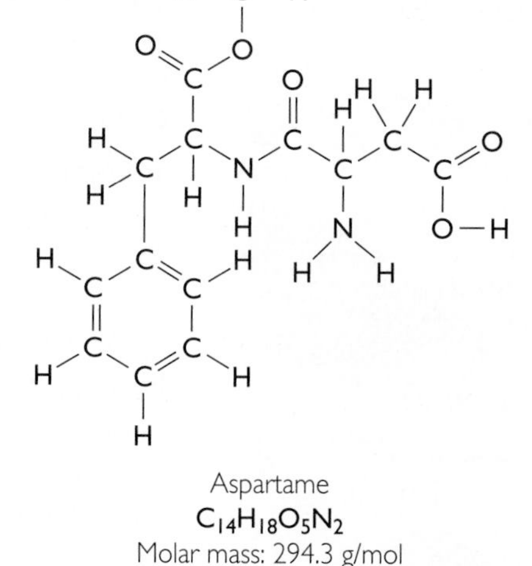

Aspartame
$C_{14}H_{18}O_5N_2$
Molar mass: 294.3 g/mol

The molar masses of fructose and aspartame represent the mass per mole of molecules. Recall that you can obtain the molar mass of a molecule by adding the molar masses of each atom in the molecule.

40 g Fructose   0.225 g Aspartame

The molar mass of fructose is less than that of aspartame. A 1 g sample of fructose will contain more molecules than a 1 g sample of aspartame because each fructose molecule has a smaller mass than an aspartame molecule.

A can of regular soft drink has about 40.0 g of fructose as a sweetener. A can of diet soft drink has about 0.225 g of aspartame as a sweetener. You can determine the number of molecules of sweetener in each can using the molar mass of each substance.

Fructose = (regular soft drink)

$$n = \frac{40 \text{ g}}{180.1 \text{ g/mol}} = 0.22 \text{ mol fructose}$$

Aspartame = (diet soft drink)

$$n = \frac{0.225 \text{ g}}{294.3 \text{ g/mol}} = 0.00076 \text{ mol aspartame}$$

A diet soft drink needs only 0.00076 mol of aspartame molecules to replace the sweetness of 0.22 mol of fructose molecules in a regular soft drink. So you need significantly less aspartame to obtain the same sweetness provided by fructose.

2 Toxicity of Sweeteners

Artificial sweeteners allow people with diabetes, and others who must avoid sugar, to enjoy foods that would otherwise be dangerous for them to eat. Artificial sweeteners do not cause cavities, and because they are not digested, artificial sweeteners do not provide extra calories to the body. However, the safety of artificial sweeteners continues to be a subject of debate.

In order to make an informed decision about whether you want to consume artificial sweeteners, it is useful to compare the toxicities of fructose and aspartame. Take a moment to review this information about each sweetener.

Fructose (sugar)	Aspartame (artificial sweetener)
$C_6H_{12}O_6$	$C_{14}H_{18}O_5N_2$
LD_{50} = 28.5 g/kg = 0.158 mol/kg	LD_{50} = 10 g/kg = 0.034 mol/kg

The LD_{50} of fructose in grams is about three times larger than the LD_{50} for aspartame (30 g per kg vs. 10 g per kg). The LD_{50} in moles is almost five times larger for fructose than for aspartame (0.158 mol per kg vs. 0.034 mol per kg). So, the sweetner aspartame is more toxic than fructose if you consume similar amounts of each.

HEALTH CONNECTION

When aspartame is broken down in the body, it produces methanol and two amino acids: aspartic acid and phenylalanine. People with a rare genetic disorder called phenylketonuria can't metabolize phenylalanine and experience toxic effects. For this reason, a warning label about phenylalanine must be placed on all products containing aspartame.

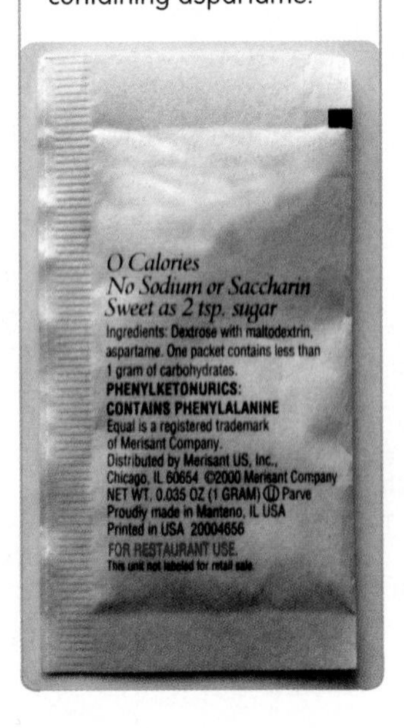

Example

How Many Cans?

How many cans of a regular soft drink does a 64 kg (141 lb) person need to drink in one sitting to exceed the lethal dose of fructose? How many cans of diet soft drink does the same person need to drink in a short time to exceed the lethal dose of aspartame?

Solution

First, determine how many grams of each sweetener are lethal for a 64 kg person. Then, take this number of grams and divide by the grams of each sweetener in a can of soft drink to get the number of cans of soft drink that would be lethal.

	Regular soft drink	**Diet soft drink**
Determine the lethal dose for each sweetener using LD_{50} values.	$64\ \text{kg}\left(25.8\ \frac{\text{g}}{\text{kg}}\right) = 1651\ \text{g}$	$(64\ \text{kg})\left(10\ \frac{\text{g}}{\text{kg}}\right) = 640\ \text{g}$
Divide by the mass of sweetener in one can to get number of cans.	$(1651\ \text{g})\left(\frac{1\ \text{can}}{40\ \text{g}}\right) = 41.3\ \text{cans}$	$640\ \text{g}\left(\frac{1\ \text{can}}{0.225\ \text{g}}\right) = 2884\ \text{cans}$

It would take about 41 cans of regular soft drink to reach the lethal dose of fructose and 2884 cans of diet soft drink to reach the lethal dose of aspartame.

It would be difficult to drink enough regular soda or diet soda to reach the lethal dose in a short period of time. So it is highly unlikely that exposure to either fructose or aspartame would be lethal in the *short term*. However, it is important to note that the LD_{50} is not a good measure of health effects from *long-term* exposure. It is probably best to limit intake of both sugar and aspartame.

Lesson Summary

How can you use moles to compare toxicity?

Chemists often find it useful to compare moles of substances rather than masses of substances. Moles allow you to compare numbers of molecules or formula units in your sample. Substances with large molar masses contain fewer molecules per gram than substances with smaller molar masses. The health effect of exposure to a toxic substance depends on both the LD_{50} and the amount of the substance.

EXERCISES

Reading Questions

1. What evidence shows that aspartame is sweeter than fructose?
2. What evidence shows that it would be difficult to exceed the lethal dose of aspartame?

Reason and Apply

3. There are 25 mg of caffeine, $C_8H_{10}N_4O_2$, in a can of regular soft drink. The LD_{50} for caffeine, $C_8H_{10}N_4O_2$, is 140 mg/kg.
   a. How many cans of regular soft drink can a 65 kg person drink in a short period of time before exceeding the lethal dose?
   b. What is the toxicity of caffeine in moles per kilogram?
4. The LD_{50} for saccharin, $C_7H_5NO_3S$, is 14.2 g/kg. If you have 1 mol of aspartame, and 1 mol of saccharin, which would be more toxic? Show your work.
5. **WEB RESEARCH** Write an argument for or against the use of artificial sweeteners. Be sure to provide evidence to support your argument. Cite at least two references that you use.

HEALTH CONNECTION

In 1958, Congress passed the Delaney Clause of the Food, Drug, and Cosmetic Act. The clause prohibited the use of all food additives found to be carcinogenic. As a result, several artificial sweeteners have been banned by the Food and Drug Administration over the years.

SECTION

II

Key Terms

toxicity
percent error
molar mass
scientific notation

SUMMARY

Measuring Toxins

Toxins Update

In order to keep the public safe, health experts measure the toxicity of substances that people might be exposed to in the course of their daily lives. The most common measurement of toxicity is called lethal dose, or LD_{50}, which is expressed in milligrams per kilogram of body weight. The higher the LD_{50} for a substance, the safer the compound is.

Chemical equations track compounds in counting units called moles. In order to determine if a certain amount of matter amounts to a lethal dose, it is necessary to convert between mass and moles using molar mass.

Review Exercises

1. Mass and moles are proportionally related. Explain what this means.
2. Why is it necessary to take a person's weight into account when considering the correct dose for a medication?
3. Why is scientific notation a useful tool for chemists?
4. Consider vitamin D-3, $C_{27}H_{44}O$, cholecalciferol. Vitamin D-3 helps the body extract calcium and phosphorous from food, plays a role in mental health, and may prevent some cancers. Most of the vitamin D you get is produced in your skin when you are exposed to the sun.

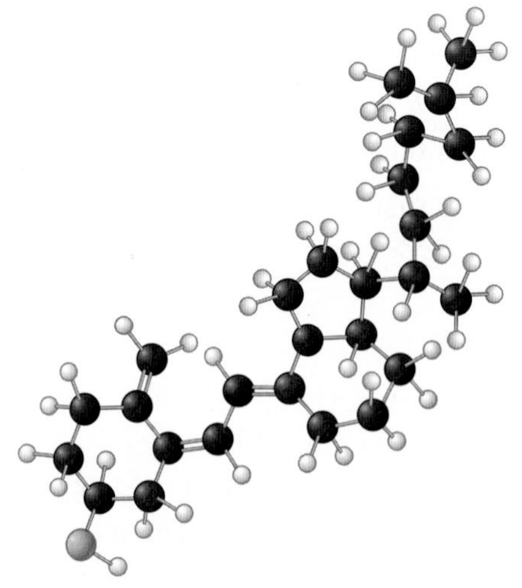

Vitamin D-3

Lethal dose: LD_{50} = 42.0 mg/kg (rat, oral)

Recommended Dietary Allowance: RDA = 0.010 mg/day

a. Determine the number of moles and molecules of vitamin D-3 in the Recommended Dietary Allowance.

b. Determine the lethal dose for a 140 lb person.

c. Determine the number of moles of vitamin D-3 in a lethal dose for a 140 lb person.

d. Determine the number of molecules of vitamin D-3 in a lethal dose for a 140 lb person.

e. Would 1.0×10^{20} molecules of vitamin D-3 represent a lethal dose for a 140 lb person? Explain your reasoning.

f. Would 400 tablets (0.010 mg each) of vitamin D-3 be a lethal dose for a 140 lb person? Explain your reasoning.

5. List these compounds in order of increasing moles of metal in each compound: 5.0 g NaCl, 5.0 g AgCl, 5.0 g LiCl. Show your work.

6. Copy and balance these chemical equations.

a. $TiCl_4(s) + Mg(s) \longrightarrow Ti(s) + MgCl_2(s)$

b. $Mg(s) + H_2O(l) \longrightarrow MgO(s) + H_2(g)$

c. $Fe(s) + CuSO_4(s) \longrightarrow FeSO_4(s) + Cu(s)$

d. $Li(s) + H_2O(l) \longrightarrow LiOH(aq) + H_2(g)$

WEB  RESEARCH

PROJECT *Lethal Dose of a Toxic Substance*

Research a toxic substance (your teacher may assign you one). Find some information about your substance. Your project should include

- the LD_{50} for the substance.
- the lethal dose for a 140 lb person expressed in grams and also in moles.
- two sources for the information you obtained. One source should be a Materials Safety Data Sheet (MSDS).

SECTION

III

Toxins in Solution

Water is one of the most valuable resources on the planet. It must be carefully monitored to make sure it is safe for public use.

Many compounds dissolve in water and can become part of our drinking water supply. An aqueous solution is one that contains water, with other ingredients uniformly mixed in. Understanding solutions allows you to monitor and track these substances.

In this section, you will study

- how to track dissolved substances in solutions
- a particle view of solutions
- how to create solutions

LAB Solution Concentration

Purpose

To explore solutions and the concentration of dissolved solids in solution.

Materials

- sugar, $C_{12}H_{22}O_{11}$, ~60g
- salt, NaCl, ~60g
- water, 200 mL
- 250 mL beakers (2)
- balance
- stirring rod
- 2 spatulas or plastic spoons
- 2 green or red gummy bears

Part 1: Dissolving Solids

Procedure

1. Make predictions. How many grams of sugar do you think you can dissolve into 100 mL of water? (1 tsp = 4 g) How many grams of salt? (1 tsp = 6 g)
2. Determine how many grams of sugar will dissolve in 100 mL of water at room temperature. Keep adding sugar and stirring until you notice solid sugar in the beaker that will not dissolve.
3. Repeat this process using table salt, NaCl, instead of sugar.

Calculations

1. Compare the number of *grams* of dissolved solid per liter of solution.
2. Compare the number of *moles* of solid per liter of solution.
3. Why does the sugar solution have the greater mass of dissolved solid per liter? Why does the salt solution have the greater number of moles of solid per liter?

Part 2: Gummy Bears

Five gummy bears were placed overnight in aqueous sugar solutions as shown.

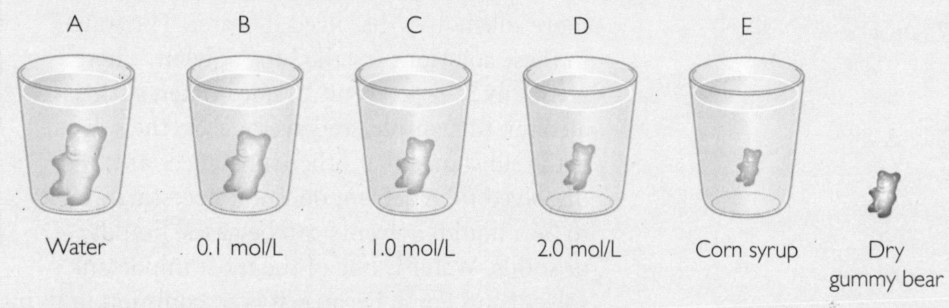

1. Describe what happens to the bears in the solutions. Which solution had the greatest effect on the gummy bears? Why do you think this is so?
2. What evidence do you have that it is water and not sugar that is moving into and out of the bears?
3. Place a gummy bear in each saturated solution you created. Predict what they will look like after 24 hours. Explain your reasoning.

LESSON

13 Bearly Alive

Solution Concentration

Think About It

When a toxic substance is in solid form, keeping track of toxic amounts is fairly simple. But what happens if the toxic substance is dissolved in water? For example, the water that comes out of your tap is probably a solution containing small amounts of dissolved chlorine. Small amounts of chlorine benefit the water supply by killing bacteria, but large amounts would be toxic to people who drink the water. It is important to have methods for tracking substances dissolved in water.

How can you keep track of compounds when they are in solution?

To answer this question, you will explore

1. Solutions
2. Solution Concentration

Exploring the Topic

1 Solutions

A **solution** is a mixture of two or more substances that is uniform throughout. This means it is uniform at the molecular, ionic, or atomic level. The ocean, a soft drink, liquid detergent, window cleaner, and even blood plasma are all examples of solutions. A glass of powdered drink has many dissolved substances in it, but they are mixed so well that the individual substances in the solution are no longer distinguishable.

Each solution listed above consists of one or more substances dissolved in water. The water in these solutions is called the **solvent.** There are many other solvents besides water, such as alcohol, turpentine, and acetone. So the solvent is the substance that other substances are dissolved *in.* A solvent does not necessarily have to be a liquid. Solvents can be gases, liquids, or solids. Water is one of the most important solvents on Earth because it is so common in living systems.

A substance that is dissolved in a solvent is called the **solute.** Solutes can be gases, liquids, or solids. In a sugar solution, the sugar is the solute and water is the solvent. In a carbonated soft drink, carbon dioxide gas is one of the solutes dissolved in the soft drink.

A solution is considered a **homogeneous mixture** because of its uniformity. Mixtures that are not uniform throughout are called **heterogeneous.** Salad dressing or mixed nuts are examples of heterogeneous mixtures.

A tossed salad is a heterogeneous mixture.

2 Solution Concentration

The **concentration** of a solution refers to the amount of a substance dissolved in a specified amount of solution. The concentration of a solution depends on both the quantity of solute and the quantity of solvent. There are several ways of measuring these quantities. A common way to measure concentration is to determine the number of moles of solute per total volume of solution in liters. This is called the **molarity** of a solution and is expressed in moles per liter, mol/L. Notice that solution concentration is the same as number density, n/V.

ENVIRONMENTAL CONNECTION

Scientists also measure the concentration of substances in solution using parts per million (ppm) or parts per billion (ppb). When water is tested for contamination, the results are often reported in ppm. A level of contamination of 1 ppm means that there is 1 mg of dissolved contaminant per liter of solution.

$$\text{Molarity} = \frac{\text{moles of solute}}{\text{liters of solution}}$$

Consider the two solutions in the illustration. Both contain a blue dye. Dyes are generally molecular solids that make a solution appear to be a certain color. The concentration of blue dye is greater in the first solution even though the volume of solution is smaller. You can tell this because the solution on the left is darker than the solution on the right.

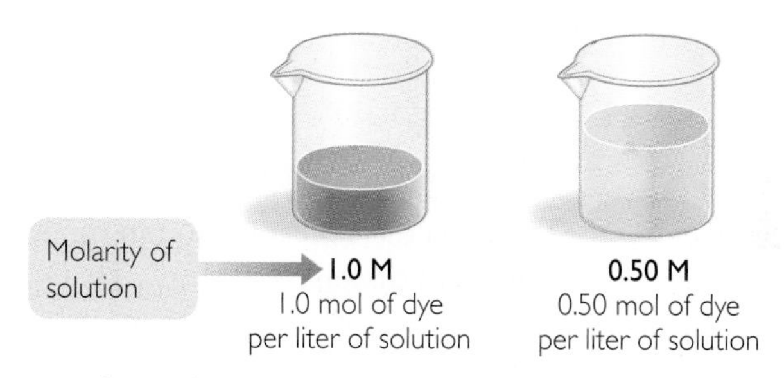

Dyes are often molecular solids. This is a sample of indigo-colored dye.

The M is an abbreviation for **molar,** which means moles of solute per liter of solution. The dye solution on the left is 1.0 molar, 1.0 M, while the one on the right is 0.50 molar, or 0.50 M.

If you want to make a solution more concentrated, you need to add solute. If you want to make a solution more dilute, or *less* concentrated, you need to add water.

Example 1

Molarity

Determine the molarity of each.

a. 0.10 mol of NaCl in 1.0 L of solution

b. 0.50 mol of NaCl in 1.0 L of solution

c. 0.10 mol of NaCl in 0.50 L of solution

d. 0.10 mol of NaCl in 250 mL of solution

CONSUMER CONNECTION

Water-soluble dyes are used in preparing a lot of our foods. One form of blue dye that goes into our food is indigo, $C_{16}H_{10}N_2O_2$. Indigo is used to dye denim cloth for blue jeans. When this dye is put in food products, it is referred to as FD and C Blue No. 2.

Solution

Molarity is the number of moles of solute divided by the liters of solution.

a. $\frac{0.10 \text{ mol}}{1.0 \text{ L}} = 0.10 \text{ mol/L} = 0.10 \text{ M}$

b. $\frac{0.50 \text{ mol}}{1.0 \text{ L}} = 0.50 \text{ mol/L} = 0.50 \text{ M}$

c. $\frac{0.10 \text{ mol}}{0.50 \text{ L}} = 0.20 \text{ mol/L} = 0.20 \text{ M}$

d. $\frac{0.10 \text{ mol}}{250 \text{ mL}} = \frac{0.10 \text{ mol}}{0.25 \text{ L}} = 0.40 \text{ mol/L} = 0.40 \text{ M}$

Calculating Molarity From Mass in Grams

In the classroom you were challenged to create a saturated sugar solution and a saturated salt solution. A **saturated solution** is one that contains the maximum amount of dissolved solute possible in a certain volume of liquid. To make a saturated solution, you must add solid and stir until no more solid dissolves. You can calculate how many grams of each solid you added if you find the mass of the solutions before and after you added the solute. The molar mass of a substance allows you to convert the mass of solute in grams to moles of solute. You can then determine the molarity of each solution.

BIG IDEA Molarity keeps track of the number of dissolved particles in a liquid.

Example 2

Finding Molarity

What is the concentration in moles per liter of a 250 mL solution that contains 16.0 g of dissolved sugar, $C_{12}H_{22}O_{11}$?

Solution

First, convert the mass of sugar to moles by dividing by the molar mass of sugar, $C_{12}H_{22}O_{11}$. Next, divide the number of moles by the volume to determine molarity.

Convert mass to moles. $\frac{16.0 \text{ g}}{342.3 \text{ g/mol}} = 0.047 \text{ mol}$

Calculate the molarity. $\frac{0.047 \text{ mol}}{0.250 \text{ L}} = 0.188 \text{ M}$

Lesson Summary

How can you keep track of compounds when they are in solution?

A solution is a mixture of two or more substances that is uniform throughout. In order to track substances that are in solution, chemists measure the number of particles per liter of liquid. This is called the concentration, or the molarity of the solution. Molarity is a measure of number density and is expressed in units of moles per liter of solution.

Key Terms

solution
solvent
solute
homogeneous mixture
heterogeneous mixture
concentration
molarity
molar
saturated solution

EXERCISES

Reading Questions

1. What does it mean that a solution is "uniform throughout"?
2. What is the difference between the concentration of a solution and the volume of a solution?

Reason and Apply

3. Find at least three solutions at home. Identify the solute and the solvent for each.
4. Find at least three mixtures at home that are not solutions. Identify the substances in each mixture.
5. Place these salt solutions, NaCl(*aq*), in order of increasing molarity.
   a. 4.0 mol per 8.0 L
   b. 6.0 mol per 6.0 L
   c. 1.0 mol per 10 L
6. Determine the molarity for each of these salt solutions, NaCl(*aq*). Then list the solutions in order of increasing molarity.
   a. 29.2 g per 0.50 L
   b. 5.8 g per 50 mL
   c. 2.9 g in 10.2 mL
7. Determine the molarity for each of these solutions. Then list the solutions in order of increasing molarity.
   a. 25 g $C_6H_{12}O_6$ per 0.25 L
   b. 50 g $C_6H_{12}O_6$ per 0.25 L
   c. 50 g $C_{12}H_{22}O_{11}$ per 0.25 L
8. Which of the substances listed here represent a heterogeneous mixture? Choose all that apply.
   A. the air in your classroom
   B. beef stew
   C. mouthwash
   D. chocolate pudding
9. How can you increase the molarity of a solution?
   A. Add solute.
   B. Add solvent.
   C. Pour out some of the solution.
   D. All of the above.
10. When expressing molarity, M stands for
   A. atoms per liter of water.
   B. mole per milliliter of solvent.
   C. mole per liter of solution.
   D. atoms per milliliter of solute.
   E. grams per liter of solution.

LESSON

14 Drop In

Molecular Views

Think About It

A tiny drop of a concentrated solution of hydrochloric acid, HCl(*aq*), will cause a severe burn if it comes in contact with your skin. In contrast, a larger volume of a more dilute HCl(*aq*) solution will cause only a minor skin irritation. How can a single drop of a solution be so powerful?

How can you use molarity to determine the moles of solute?

To answer this question, you will explore

1. Molecular Views of Solutions
2. Molarity Calculations

Exploring the Topic

1 Molecular Views of Solutions

When a toxin is dissolved in water, the amount of toxin you are exposed to depends on both the concentration of the solution and the volume of the solution. In order to better understand the relationship between solution concentration and solution volume, it is useful to explore solutions with the same volume, but different concentrations, as well as solutions with the same concentration, but different volumes.

Same Volume, Different Concentrations

This illustration shows four solutions of red dye, as well as particle views of each. Each red dot represents a red dye molecule dissolved in water. Take a moment to compare the solutions and molecular views.

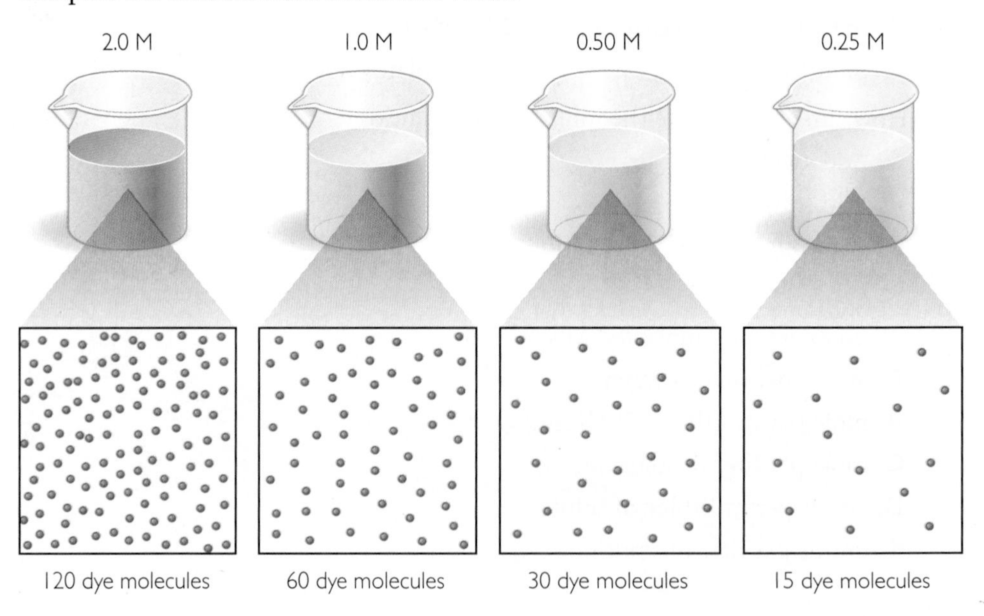

SPORTS CONNECTION

Chlorine, found in hypochlorous acid, HOCl, is toxic to humans in high concentration. However, a tiny amount can be added to a swimming pool to make the water safe. The low concentration is deadly to bacteria and other microorganisms, but not to humans—though it can irritate the eyes and skin.

Notice that each time the concentration is halved, there are half as many particles per unit of volume. So, as the solution concentration decreases, equal volumes of solution will contain fewer molecules.

The color of the red dye solution provides evidence to support this conclusion. When there are more dye molecules per unit of volume, the red color of the solution is more intense.

Same Concentration, Different Volumes

The molecular views of the two solutions in the next illustration show what happens when you pour a portion of a solution into another container. Take a moment to compare the solutions and molecular views.

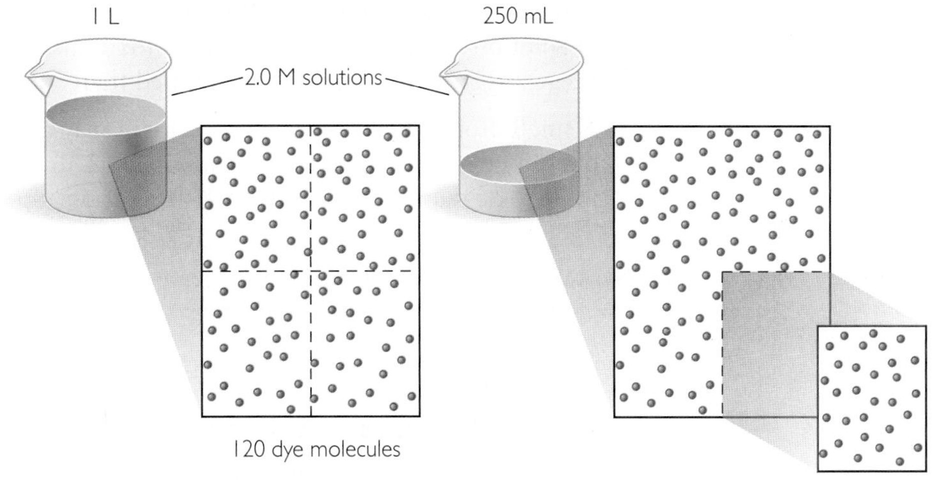

There are fewer total dye molecules when you pour a portion of the solution into the smaller container. However, the concentration of the solution has not changed. If you remove 250 mL from a 1 L solution, you have one quarter as many dye molecules. But because the concentration has not changed, the color of the red dye solution does not change.

2 Molarity Calculations

Consider the solutions below. Suppose the first two contain a concentrated solution of a toxin. The third and fourth contain a more dilute, or watered down, solution of the same toxin. Take a moment to try to put the solutions in order of most total moles of toxin to least total moles of toxin.

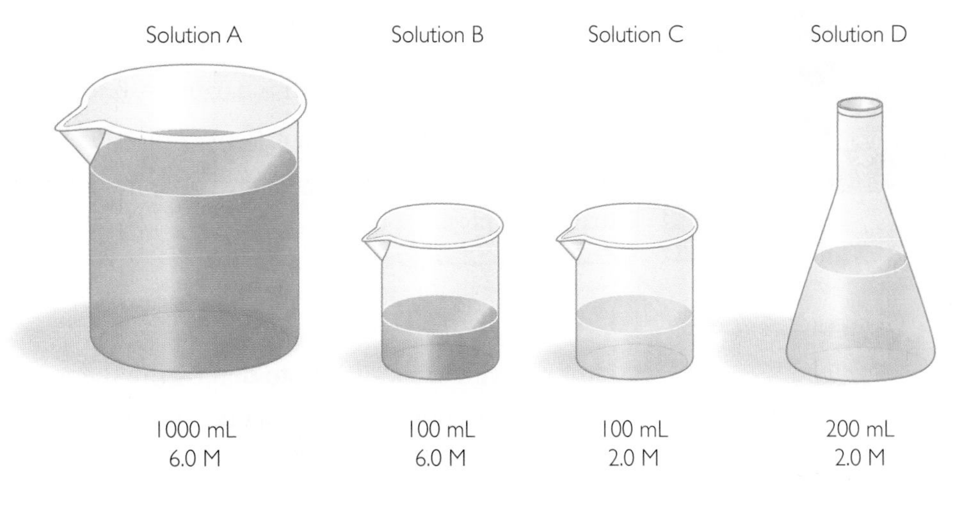

The relationship between moles of solute and volume of solution in liters is proportional. Recall that the molarity of a solution in moles/liter is the proportionality constant, k, for that solution. So, if you multiply the molarity times the volume in liters, you can determine the number of moles of solute.

$$\text{moles of solute} = k \cdot \text{volume of solution in liters}$$

$= 6.0\ \text{mol/L} \cdot 1.0\ \text{L} = \mathbf{6\ mol}$	Solution A
$= 6.0\ \text{mol/L} \cdot 0.10\ \text{L} = \mathbf{0.60\ mol}$	Solution B
$= 2.0\ \text{mol/L} \cdot 0.10\ \text{L} = \mathbf{0.40\ mol}$	Solution C
$= 2.0\ \text{mol/L} \cdot 0.20\ \text{L} = \mathbf{0.20\ mol}$	Solution D

You can see that even though the volume of Solution B is smaller than the volume of Solution D, there are more moles of toxin dissolved in Solution B. This is because Solution B is more concentrated than Solution D.

A small drop of highly concentrated toxin may be worse for you than a large quantity of dilute toxin. For example, a 2 mL drop of 12 M HCl solution is much more toxic than 50 mL of 0.10 M HCl solution.

BIG IDEA The concentration of a solution does not change with the size of the sample.

Example

Comparing Moles of Solute

Two solutions of a toxic substance are shown in the illustration.

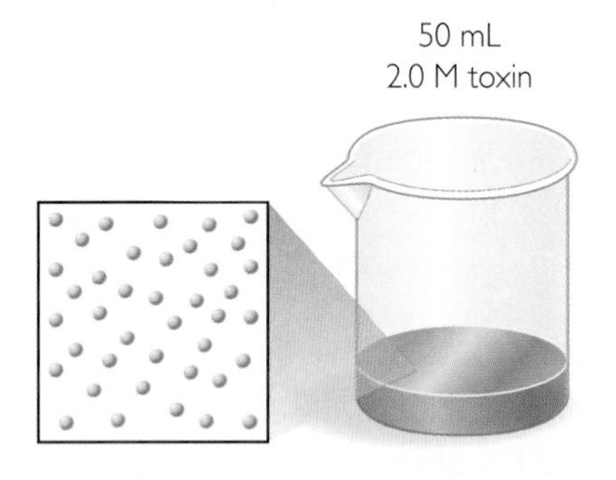

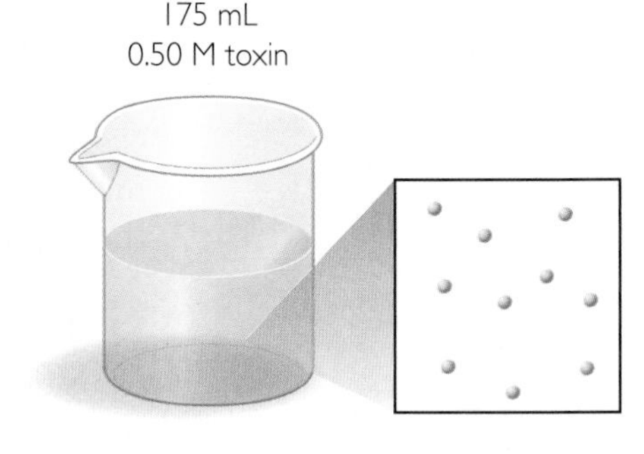

Which beaker contains the larger dose of the toxic substance?

Solution

Use the concentration of each solution to determine the moles of solute.

2.0 M solution: $2.0\ \text{mol/L} \cdot 0.050\ \text{L} = 0.10\ \text{mol}$

0.50 M solution: $0.50\ \text{mol/L} \cdot 0.175\ \text{L} = 0.0875\ \text{mol}$

So, the beaker on the left, the more concentrated solution, contains more moles of toxic substance.

Fructose is a natural sugar found in fruits and vegetables. However, when it is highly concentrated, as it is in high-fructose corn syrup, it is difficult for your body to process and too much of it can cause health problems.

Lesson Summary

How can you use molarity to determine the moles of solute?

The concentration of a solution does not change with the size of the sample. This is because concentration is a measure of number density, or moles of particles per unit of volume, a property that does not change with sample size. However, a large sample of a solution will contain more total particles than a small sample of the same solution. The relationship between moles of particles in a solution and liters of solution is a proportional one. It is described by the formula moles of solute $= k \cdot$ volume of solution, where k is the molarity of the solution. This equation can be used to calculate the exact number of moles in a sample of known concentration and volume.

EXERCISES

Reading Questions

1. How can two solutions with different volumes have the same concentration?
2. How can you figure out how many moles of solute you have in a solution with a specific concentration?

Reason and Apply

3. Draw molecular views for blue dye solutions that are 0.50 M, 0.25 M, and 0.10 M.
4. What portion of 1.0 L of 0.50 M blue dye solution has the same number of moles as 1.0 L of 0.25 M blue dye solution?
5. Glucose and sucrose are two different types of sugar. Consider these aqueous solutions:

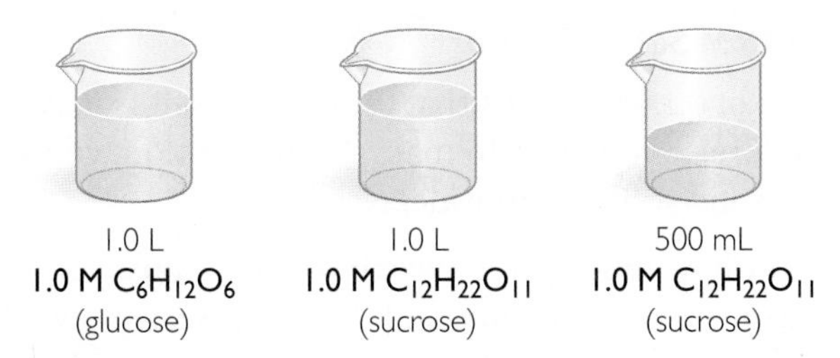

   a. Which solution has the most molecules? Explain.
   b. Which solution has the greatest concentration? Explain.
   c. Which solution has the most mass? Explain.

6. How many moles of sodium cations, Na^+, are dissolved in a 1.0 L sample of each solution listed below?
   a. 0.10 M NaCl **b.** 3.0 M Na_2SO_4 **c.** 0.30 M Na_3PO_4
7. Draw molecular views for 1.0 M NaCl, 2.0 M NaCl, and 1.0 M Na_2S. Use different symbols for each type of ion. Circle the solution(s) with the least total number of ions.
8. Determine the number of moles of solute in each of these aqueous solutions.
   a. 50 L of 0.10 M NaCl **b.** 0.25 L of 3.0 M $C_6H_{12}O_6$
   c. 35 mL of 12.0 M HCl **d.** 300 mL of 0.025 M NaOH
9. How many liters of each solution do you need to get 3.0 mol HCl?
   a. 12.0 M HCl **b.** 2 M HCl
   c. 0.5 M HCl **d.** 0.010 M HCl

LESSON

15 Holey Moley

Preparing Solutions

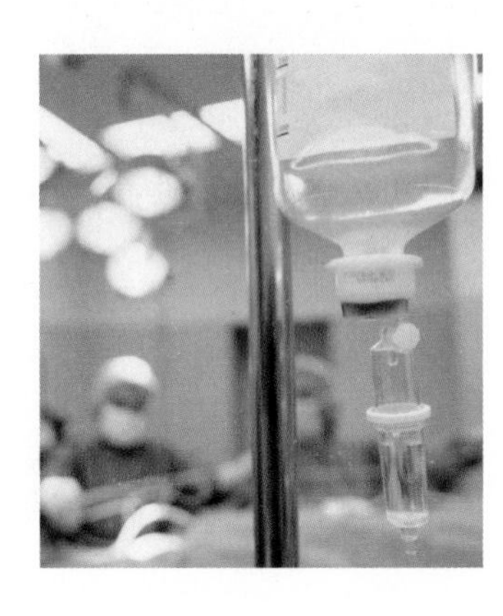

Human blood plasma contains a certain molarity of dissolved salt, NaCl. If you are taken to the hospital in an ambulance and given an intravenous, or IV, solution, it is critical that the salt concentration matches that of your blood, about 0.15 M NaCl. Otherwise, the IV could be toxic to you.

How can you create a solution with a specific molarity?

To answer this question, you will explore

1. Creating Solutions
2. Relating Mass, Moles, and Volume

1 Creating Solutions

Suppose a lab technician wants to make a 0.15 M sodium chloride solution. How should this solution be prepared?

Preparing a 0.15 M Salt Solution

A concentration of 0.15 M means that 0.15 mol of salt is dissolved in 1 L of solution. Therefore, if you want to make 1 L of solution, you will need to measure out 0.15 mol of solid sodium chloride, NaCl.

In order to know how much salt to weigh out, you must convert moles to mass using the molar mass of the compound. In this case, simply add the molar masses for Na and Cl from the periodic table. The molar mass of NaCl is 58.4 g/mol to one decimal place. Now multiply the moles of NaCl you want by the molar mass to get the number of grams of NaCl you need for your solution.

$$\text{Mass} = \text{molar mass} \cdot n = (58.4 \text{ g/mol})(0.15 \text{ mol}) = 8.76 \text{ g NaCl}$$

If you dissolve 8.76 g NaCl in water so that the total volume of solution is 1 L, you will have a solution that is 0.15 M, or 0.15 mol/L NaCl.

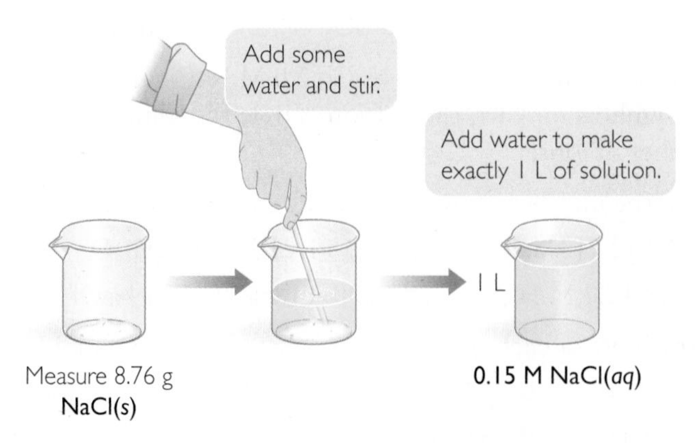

Important to Know When you dissolve one substance in another, they both take up space. So, to make 1 L of an aqueous solution, you add less than 1 L of water. ◄

LANGUAGE CONNECTION

Dextrose means "right-handed sugar." This molecule is the right-handed mirror-image isomer of glucose.

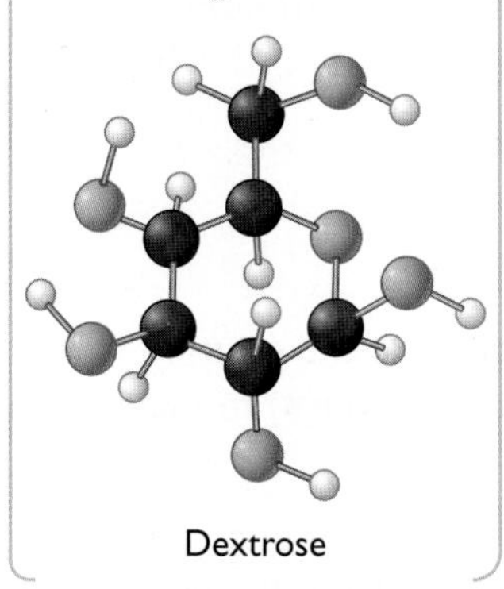

Dextrose

Example 1

Dextrose Solution

Sometimes doctors administer an IV solution containing a solution of dextrose to a patient who has low sugar or high sodium in the bloodstream. Dextrose is a type of sugar. The structural formula for dextrose is shown in the illustration.

The intravenous solution contains 150 g of dextrose in 3 L of solution. What is the molarity of this solution?

Solution

You need to convert grams of dextrose to moles of dextrose in order to determine the molarity. You know there are 150 g of dextrose in 3 L, or 50 g/L.

First, determine the molar mass of dextrose. As you can see from the ball-and-stick model, there are 6 C atoms, 12 H atoms, and 6 O atoms, so the formula is $C_6H_{12}O_6$.

Find the molar mass of dextrose, $C_6H_{12}O_6$. 180.2 g/mol

Divide grams of solute by the molar mass to obtain moles. $n = \frac{50\text{ g}}{180.2\text{ g/mol}} = 0.28\text{ mol}$

There is 0.28 mol of dextrose in 1 L of solution, so the molarity is 0.28 M.

❷ Relating Mass, Moles, and Volume

Just as it is important to convert mass to moles, it is also important to be able to convert moles to mass. The table below shows data for four samples of sucrose, $C_{12}H_{22}O_{11}$, each with a different volume and molarity. How many grams of sucrose are dissolved in each sample?

To find the moles of solute in each sample, multiply the values in the second and third columns. To find the mass of solute in each sample, multiply the fourth and fifth columns.

Sample	Molarity (moles/liter)	Volume (liters)	Number (moles of solute)	Molar mass of solute (grams/mole)	Mass of solute (grams)
1	1.0 M	1.0 L	1.0 mol	342.3 g/mol	342.3 g
2	0.10 M	1.0 L	0.10 mol	342.3 g/mol	34.23 g
3	1 M	0.1 L	0.10 mol	342.3 g/mol	34.23 g
4	0.10 M	0.1 L	0.010 mol	342.3 g/mol	03.423 g

Notice that there is more than one way to make a 0.10 M solution, for example by dissolving 0.10 mol to make 1.0 L or by dissolving 0.010 mol to make 100 mL (0.10 L).

Example 2

Guidelines for Safe Drinking Water

Arsenic, As, is an element that is found naturally in the soil. As a result, tiny levels of arsenic are often found in some water supplies. Like all substances, the toxic effect of arsenic is dependent on the dosage you receive. For this reason, the Environmental Protection Agency, or EPA, sets limits on the amounts of dissolved substances that can be present in our drinking water. The EPA has set the upper limit for arsenic in drinking water at 0.00010 g/L.

Is it safe to drink water that has a concentration of 0.000020 M As?

Solution

The EPA limit is given in grams/liter, and the concentration of the solution is in moles/liter. So you need to convert moles to grams (or vice versa) to make a comparison.

The proportionality constant between grams of arsenic and moles of arsenic is the molar mass, which you can find on the periodic table. To find the mass of arsenic, multiply the molar mass by the number of moles.

Look up the molar mass of As.	74.9 g/mol
Mass = molar mass $\cdot$ n	(74.9 g/mol)(0.000020 mol) = 0.0015 g As

The 0.000020 M arsenic solution has 0.0015 g of As per liter of solution. This amount is 15 times the concentration allowed by the EPA. So, the solution is not safe for drinking.

Topic: Solutions
Visit: www.SciLinks.org
Web code: KEY-415

Lesson Summary

How can you create a solution with a specific molarity?

Molar mass values on the periodic table allow you to convert between grams and moles. The amount in grams of solid needed to create solutions of specific molarities can be calculated using the molar mass of the solid. Likewise, the mass of solid dissolved in a solution can be determined if the concentration and volume of the solution are known.

EXERCISES

Reading Questions

1. Explain how you would prepare a solution of sucrose with a molarity of 0.25.
2. How are mass of solute, moles of solute, and volume of solution related?

Reason and Apply

3. How many grams of solute do you need to make 1 L of each of the solutions listed?

   a. 0.50 M NaCl b. 2.0 M $CaCl_2$ c. 1.5 M NaOH

4. Copy this table and complete it for solutions of glucose, $C_6H_{12}O_6$. Remember to convert milliliters to liters.

Molarity (moles/liter)	Volume (liters)	Number (moles of solute)	Molar mass of solute (grams/mole)	Mass of solute (grams)
	2.0 L			180 g
0.40 M		0.10 mol		
0.10 M		0.03 mol		
	100 mL			0.180 g

5. What volume of each of these solutions would contain 58.44 g of NaCl?

   a. 0.10 M NaCl b. 3.0 M NaCl c. 5.5 M NaCl

6. How many grams of fructose, $C_6H_{12}O_6$, are in 1 L of soft drink if the molarity of fructose in the soft drink is 0.75 M?

7. Which is more concentrated: a 1.0 L solution with 20 g of sucrose, $C_{12}H_{22}O_{11}$, or a 1.0 L solution with 20 g of glucose, $C_6H_{12}O_6$?

8. The Environmental Protection Agency has set the upper limit for fluorine, F, in drinking water at 0.0040 g/L of water. Is it safe to drink a glass of water in which the concentration of fluorine is 0.00010 M?

CONSUMER CONNECTION

Water that has a high mineral content is called hard water, while water with a low mineral content is called soft water. Soap and shampoo do not lather up well in very hard water, and don't rinse out easily in very soft water. Minerals commonly found in tap water are calcium and magnesium carbonates, and they sometimes build up as deposits in pipes and on faucets.

LESSON

16 Is It Toxic?

Mystery Solutions

Think About It

Selling bottled water is a big business. Some companies claim to get their water from mountain streams, some say their water is quite pure, and yet others claim their water contains beneficial minerals. How could you determine how pure a water sample is?

Can molarity calculations be used to identify a toxic substance?

To answer this question, you will explore

1. Accounting for the Mass of a Solution
2. Drinking Water Standards

Exploring the Topic

1 Accounting for the Mass of a Solution

Suppose that you want to know how pure a sample of water is. Examine the illustration and consider how you might use mass to make a determination of purity.

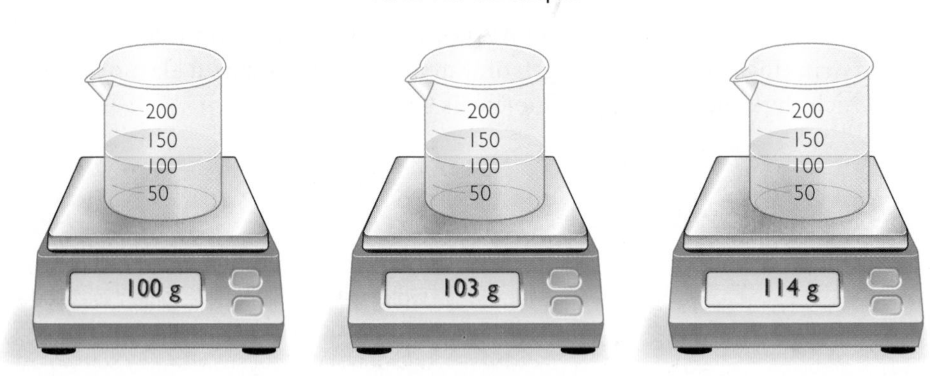

What masses do you expect if the samples are all pure water? Recall that mass per volume is equal to density. The density of pure water is 1 g/mL. If a 100 mL sample is pure water, then you expect a mass of 100 g.

The sample in the first beaker may be pure water. The other two samples must contain solutes that add more mass to the same volume, making them denser (and therefore heavier) than pure water.

The Density of a Solution

Suppose you have three samples of salt water with the concentrations shown in the illustration at the top of the next page. Take a moment to consider how the densities of the three solutions compare. Why does the ball float higher in the solution on the right?

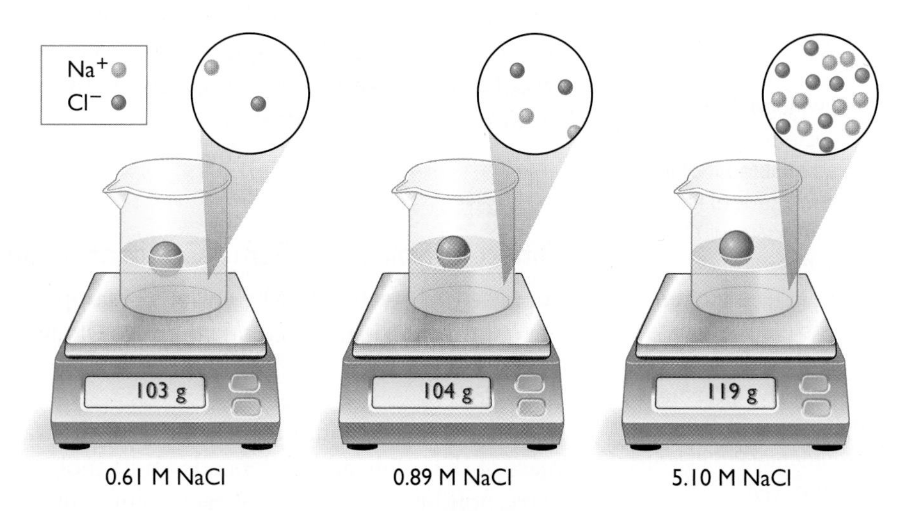

As the density of the solution increases, an object floats higher. That is why when you swim in very salty water, you float with more of your body out of the water. As the concentration increases, there are more units of NaCl per volume of solution. The higher the salt concentration, the denser the solution.

It's easy to float in the Dead Sea, the saltiest sea in the world.

Same Mass, Different Solutes

Imagine you add 10 g of three different solids to three different beakers and add enough water to make 100 mL of solution. The solutions all have approximately the same mass—the mass of about 100 mL of water plus 10 g of solute. But do they have the same concentration?

Compound	Molar mass
$CuSO_4$	249.7 g/mol
NaOH	40.0 g/mol
KCl	74.6 g/mol

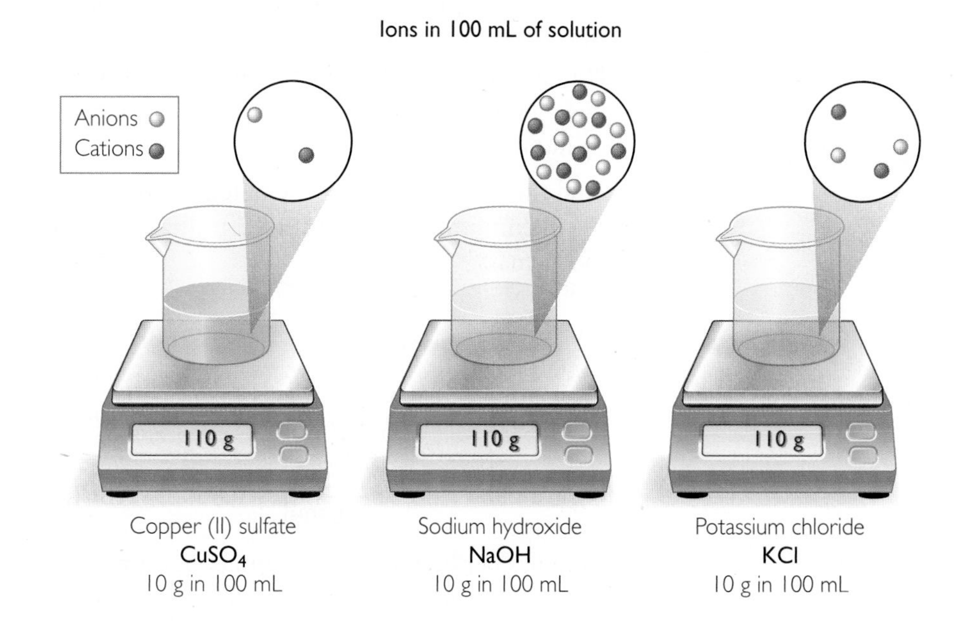

Copper (II) sulfate, $CuSO_4$ has the largest molar mass. That means there are fewer formula units of $CuSO_4$ in 10 grams compared with NaOH. While you've added the same mass to each sample, you have not added the same number of moles. Dissolving 10 grams of the compound with the higher molar mass will actually result in a lower concentration.

ENVIRONMENTAL CONNECTION

The Cuyahoga River in northeastern Ohio used to be so polluted that it caught fire several times between 1936 and 1969. In 1969, an article about the fires in *Time* magazine led to the Clean Water Act, the Great Lakes Water Quality Agreement, and the creation of the Environmental Protection Agency, or EPA.

Example

Different Solutions, Same Concentration

Suppose you create 0.10 M solutions of copper (II) sulfate, sodium hydroxide, and potassium chloride. For equal volumes, which solution will have the most mass?

Solution

The mass of each solution is the mass of the water plus the mass of solute. The mass of the solute added depends on its molar mass. To determine the mass corresponding to 0.10 mol of each solute, multiply the number of moles by the molar mass of each compound.

$$\text{Mass of } CuSO_4 = (249.7 \text{ g/mol})\ (0.10 \text{ mol}) = 25.0 \text{ g } CuSO_4$$
$$\text{Mass of NaOH} = (40.0 \text{ g/mol})\ (0.10 \text{ mol}) = 4.0 \text{ g NaOH}$$
$$\text{Mass of KCl} = (74.6 \text{ g/mol})\ (0.10 \text{ mol}) = 7.5 \text{ g KCl}$$

So, for equal volumes of solution, the $CuSO_4(aq)$ will have the most mass.

2 Drinking Water Standards

Environmental Protection Agency standards protect public health by limiting contaminants. The maximum contaminant levels for fluorine, iron, and barium in drinking water are given in mg/L as shown in the table.

Element	Maximum contaminant level
fluorine, F	4.0 mg/L
iron, Fe	0.30 mg/L
barium, Ba	2.0 mg/L

Based on the numbers in the table, you might conclude that iron, Fe, is the most toxic because the allowable level is the smallest by mass. However, these contaminants interact with your body atom by atom. So you need to compare the contaminants by number of atoms.

Find the concentration of the potential toxins in moles per liter. To do this convert the maximum contaminant level milligrams to grams, and then divide by the molar mass.

fluorine, F $\qquad \dfrac{0.0040 \text{ g/L}}{19.0 \text{ g/mol}} = 0.00021 \text{ mol/L} = 0.00021 \text{ M} = 2.1 \times 10^{-4} \text{ M}$

iron, Fe $\qquad \dfrac{0.00030 \text{ g/L}}{55.8 \text{ g/mol}} = 0.000054 \text{ mol/L} = 0.000054 \text{ M} = 5.4 \times 10^{-5} \text{ M}$

barium, Ba $\qquad \dfrac{0.0020 \text{ g L}}{137.3 \text{ g/mol}} = 0.000014 \text{ mol/L} = 0.000014 \text{ M} = 1.4 \times 10^{-5} \text{ M}$

In moles per liter, the maximum contaminant level for barium, Ba, is the smallest numerical value. This means less of it is tolerated in our drinking water, making barium the most toxic of the three elements.

Lesson Summary

Can molarity calculations be used to identify a toxic substance?

When a substance is dissolved in water, it adds to the mass of the solution. One mole of one substance may have a very different molar mass than one mole of another substance. So different solutions of the same concentration can be differentiated by weighing them. Comparing contaminants by number of atoms may give a different picture of toxicity than comparing them by mass.

EXERCISES

Reading Questions

1. Explain how you might use mass to determine if a sample of water is contaminated.
2. Explain why the density of a solution increases as the concentration increases.
3. Explain why 0.10 M $CuCl_2$ has a greater density, m/V, than 0.10 M KCl.

Reason and Apply

4. Find the molarity of each solution. Which solution has the highest molarity? The highest density? Give your reasoning.
   a. 1.0 g KCl in 1.0 L water **b.** 5.0 g KCl in 0.5 L water **c.** 10.0 g KCl in 0.3 L water
5. Find the molarity of each solution. Which solution has the highest molarity? The highest density? Give your reasoning.
   a. 10 g $NaNO_3$ in 1.0 L water **b.** 10 g $Cd(NO_3)_2$ in 1.0 L water **c.** 10 g $Pb(NO_3)_2$ in 1.0 L water
6. Find the mass of solute in each solution. Which solution has the highest density? Is it possible to calculate the density of each solution? Why or why not?
   a. 1.0 L of 0.10 M $NaNO_3$
   b. 1.0 L of M 0.10 M $Cd(NO_3)_2$
   c. 1.0 L of 0.10 M $Pb(NO_3)_2$
7. Which would weigh the most? Explain your thinking.
   A. 1.0 L of 1.0 M NaBr **B.** 500 mL of 1.0 M KCl **C.** 1.0 L of 0.5 M NaOH
8. **WEB RESEARCH** Look up the MSDS or Materials Safety Data Sheets, for sodium chloride, NaCl, and sodium sulfate, Na_2SO_4.
   a. Describe the toxic effects of each salt.
   b. What are the LD_{50}'s (mouse, oral)?
   c. Compare the toxicities based on weight in grams.
   d. Compare the toxicities based on moles.
   e. Compare the toxicities based on volume of 0.10 M solution.

SECTION

Key Terms

solution
solvent
solute
homogeneous mixture
heterogeneous mixture
concentration
molarity
molar
saturated solution

SUMMARY

Toxins in Solution

Toxins Update

It is easy for toxins to dissolve in and enter our water supply. In fact, it would be challenging to find or create a sample of "pure water"—one that had nothing else in it but H_2O molecules. To track the concentrations of different substances in solutions, scientists use moles of solute per liter of solvent. This is also called the molarity, *M*, of the solution. If you know the molarity of a solution and its volume, you can determine the exact number of grams, moles, or even molecules or ions of a substance in your water sample. Sometimes, toxic substances in solution are tracked in parts per million, or ppm. A solution with a toxic concentration of 1 ppm would have 1 mL of toxic substance per 1000 L of solution.

Review Exercises

1. Why is it useful to know exactly how many moles or grams of contaminant are in a water sample?
2. Lead is one of the few substances with a zero tolerance level in our drinking water, meaning that there should be no lead at all in our water. Why do you think this is?
3. Consider a 10 mL water sample from a nearby creek that has a 0.002 M lead (II) nitrate, $Pb(NO_3)_2$, concentration.
   a. How many lead ions are in this sample?
   b. How many grams of lead nitrate are in this sample?
   c. How many grams of lead ions are in this sample?
4. Suppose you measure 100 mL each of 1.0 M NaOH, 1.0 M KCl, and 0.5 M $PbCl_2$ solutions.
   a. Which has the greatest concentration?
   b. Which has the most mass?
   c. Which has the most moles?
   d. Which has the most ions?
5. a. Suppose you need to make a 0.5 M solution of sodium hydroxide, NaOH. Explain step by step how you would go about making this solution.
   b. Do you have to use 1 L of water when you make this solution? Explain your reasoning.

6. A mass of 47 g of sulfuric acid, H_2SO_4, is dissolved in water to prepare a 0.50 M solution. What is the volume of the solution?

7. Determine the molarity of the solutions created by following each set of directions.

   a. Dissolve 7.30 g NaCl to make 500 mL of solution.

   b. Dissolve 43.84 g NaCl to make 1 L of solution.

   c. Dissolve 25.00 g NaCl to make 1 L of solution.

   d. Dissolve 25.00 g NaCl to make 2 L of solution.

8. How many grams of magnesium sulfate, $MgSO_4$, do you need to make 100 mL of a 0.1 M solution?

WEB  RESEARCH

PROJECT *Types of Bonding*

Research a toxic substance (your teacher may assign you one). Your report should cover these topics.

- Is the bonding in your substance ionic, covalent or metallic?
- If your substance is ionic, determine the cation, and the anion. Write and label the ions and their charges. Include a particle view drawing of your substance.
- If your substance is covalent, determine if it is polar or nonpolar. Explain your reasoning. Include the structural formula and show the dipole(s).
- If your substance is metallic, find out its density. Include a particle view drawing of your substance.
- Does your substance conduct electricity? Explain your reasoning.

SECTION

IV

Acidic Toxins

A vinegaroon is similar to a scorpion, but instead of stinging, it squirts acetic acid, or vinegar, on its prey.

Some solutions can be classified as either *acids* or *bases.* They have a unique set of properties and are an extremely important part of solution chemistry. Toxic levels of acids and bases can burn the skin and damage living tissue. However, without acids and bases the body would not function properly. This section introduces you to the properties of acids and bases and shows how chemists monitor them.

In this section, you will study

- how the acidity of solutions is tracked
- theories of acids and bases
- differences between acids and bases
- the effect of dilution and neutralization on acid-base solutions
- a lab procedure for determining acid concentration

LAB Acids and Bases

Purpose

To explore the properties of acids and bases.

Safety Instructions

 Acids and bases are corrosive. Do not get solutions on skin. In case of a spill, rinse with large amounts of water. Wear safety goggles.

Materials

- 2 well plates
- cabbage juice
- 0.1 M NaCl (table salt) solution
- $C_3H_8O(aq)$ (rubbing alcohol)
- 0.1 M HCl (hydrochloric acid)
- $C_2H_4O_2(aq)$ (vinegar)
- $NH_4OH(aq)$ (window cleaner)
- small pieces of $CaCO_3(s)$ (antacid tablet)
- 2 wash bottles
- universal indicator
- 0.1 M NaOH (drain cleaner)
- $Na_2CO_3(aq)$ (washing soda)
- $C_6H_8O_7(aq)$ (lemon juice)
- distilled water
- waste container

Part 1: Testing Solutions

Procedure

1. Place three or four drops of each solution in an empty well of a well plate.
2. Add one or two drops of cabbage juice indicator to each solution. Record the color you observe in a data table like the one shown.
3. Now test three or four drops of the same nine solutions with universal indicator. Record the color you observe and the number associated with the color of the universal indicator.

Classification Data Table

Solution	Formula	Cabbage juice color	Universal indicator color	Indicator number
vinegar	$C_2H_4O_2(aq)$			
rubbing alcohol	$C_3H_8O(aq)$			

Analysis

1. Group the substances based on their responses to the cabbage juice.
2. Place the nine substances on a number line like this one, based on the number associated with the color of the universal indicator.

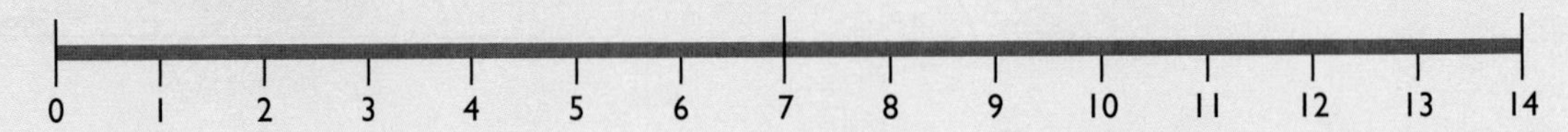

3. Does the number line match the groupings you came up with in Question 1? Would you change your groups in any way?

Part 2: Treating Indigestion

Calcium carbonate, $CaCO_3$, is a compound used to treat acid indigestion. It is found in over-the-counter antacid tablets.

Procedure

1. Place a small piece of calcium carbonate into nine wells of the well plate.
2. Add ten drops of each solution to each well. Record your observations.

Analysis

1. What generalizations can you make about these solutions based on how they responded to calcium carbonate and how they responded to cabbage juice?
2. **Making Sense** List at least four characteristics of the substances you placed on the left side of the number line. List four characteristics of the substances you placed on the right side of the number line.

LESSON

17 Heartburn

Acids and Bases

Think About It

Solutions have a wide variety of properties. Acetic acid, or vinegar, is used to flavor salad dressing, while citric acid gives a sour taste to lemon juice and orange juice. Ammonium hydroxide is used to clean windows, and sodium hydroxide is used to open clogged drains. All of these solutions are either acids or bases and can be classified into these categories based on their general properties.

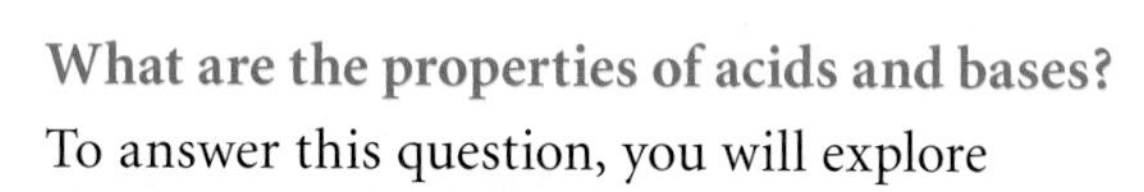

What are the properties of acids and bases?

To answer this question, you will explore

1. Acids and Bases
2. Indicators
3. The pH Scale

Exploring the Topic

1 Acids and Bases

General Properties

Acids and bases are special categories of solutions. They are extremely useful to us in our everyday lives precisely because of their unique properties. The term *acid* comes from the Latin word *acidus*, which means sour. Many of the sour tastes in our food come from the acids found in those foods. For much of history, the term *alkaline* was used instead of *base*. Bases are found in many household cleaners, from soaps to drain openers to oven cleaners. Bases have a bitter taste, as you may have noticed if you have ever accidentally tasted soap. Bases usually feel slippery to the touch. The slipperiness of bases arises from the fact that they are reacting with the fats in your skin and turning them into soap.

CONSUMER CONNECTION

Carbonic acid, H_2CO_3, and phosphoric acid, H_3PO_4, are two acids found in carbonated soft drinks. The phosphoric acid acts as a flavoring and a preservative, while the carbonic acid is a byproduct of carbonation.

In general, acids and bases are toxic, especially large quantities of concentrated solutions. It is important that acids and bases do not splash on your skin or in your eyes. Acids and bases are both corrosive and can cause a *chemical burn*. A chemical burn is one in which living tissue is damaged.

2 Indicators

Many acidic and basic solutions are colorless and odorless, which can make them difficult to detect by their appearance alone. Because these solutions can be toxic, it is useful to be able to monitor them. Luckily, there are molecular substances called **indicators** that change color when they come into contact with acids and bases. If you add a drop or two of an indicator to an unknown solution, you can tell if you have an acid or a base by the color that results.

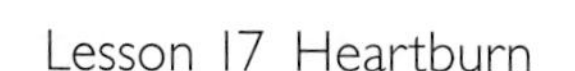

Cabbage juice is a natural acid-base indicator obtained by boiling or grinding up red cabbage. The cabbage juice changes color as it is added to various solutions, as shown in the illustration. Take a moment to look for patterns.

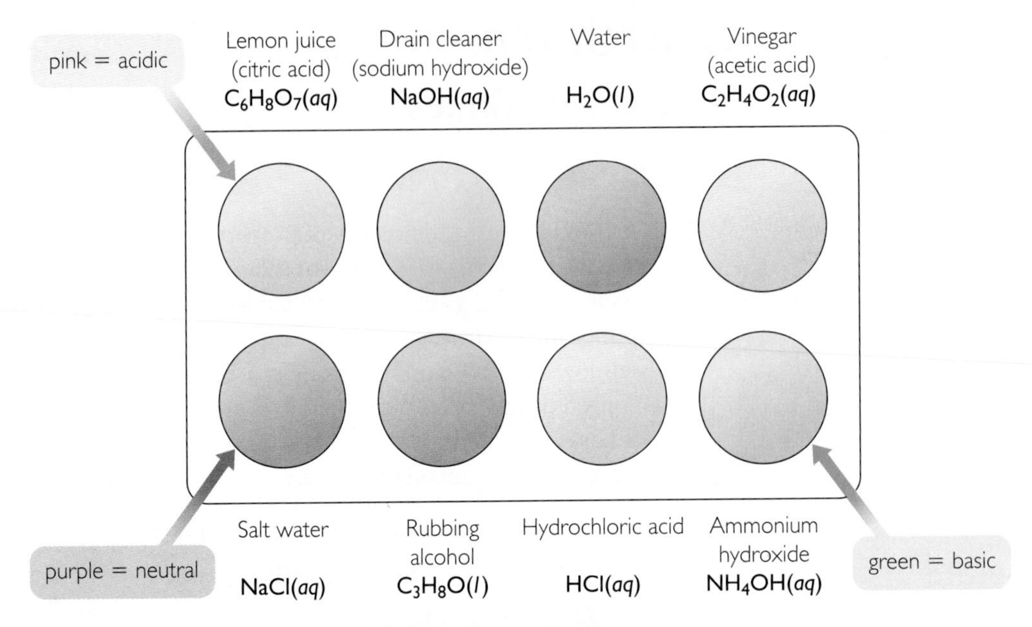

Cabbage juice turns pink when it is in contact with acids, such as citric acid, acetic acid, and hydrochloric acid. It turns green or yellow in contact with bases, such as sodium hydroxide and ammonium hydroxide. And it remains purple or blue in contact with *neutral* substances such as pure water, salt water, and alcohol. Neutral solutions are neither acidic nor basic. So cabbage juice can be used to indicate if a solution is acidic or basic or neither.

3 The pH Scale

There are dozens of different types of acid-base indicators, each with a unique color scheme. The colors associated with "universal indicator" are shown in the illustration. Notice that there is a number associated with each color. This number is referred to as the *pH number,* or just as the *pH.*

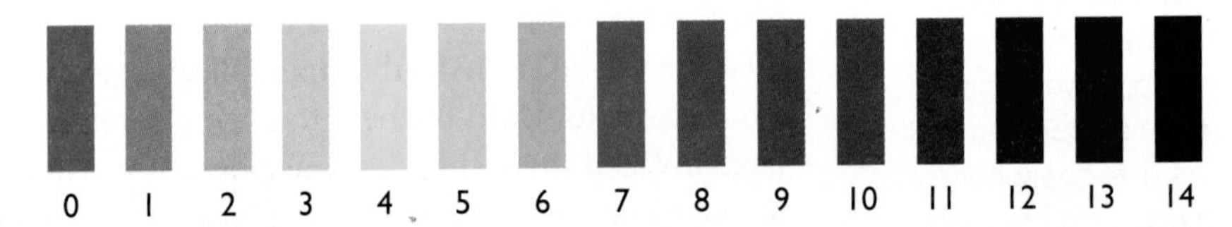

Another way to measure the pH for a solution is with paper coated with indicator, referred to as pH paper.

One common type of pH paper is called litmus paper. Litmus paper turns red in acidic solutions and blue in basic solutions.

Litmus Paper

The **pH scale** is a number line that assigns number values from 0 to 14 to acids and bases.

Substances with a pH below 7 at 25 °C are **acids.** Substances with a pH above 7 at 25 °C are **bases.** Substances with a pH at or near 7 at 25 °C are considered neutral. Stomach acid is extremely acidic, with a pH around 1. Lemon juice is

HEALTH CONNECTION

Too much acid in the stomach may result in indigestion, heartburn, or ulcers. Antacids are bases that work in the stomach to neutralize the excess acid there. Other remedies work on receptor sites in your body, preventing the secretion of excess acid in the first place.

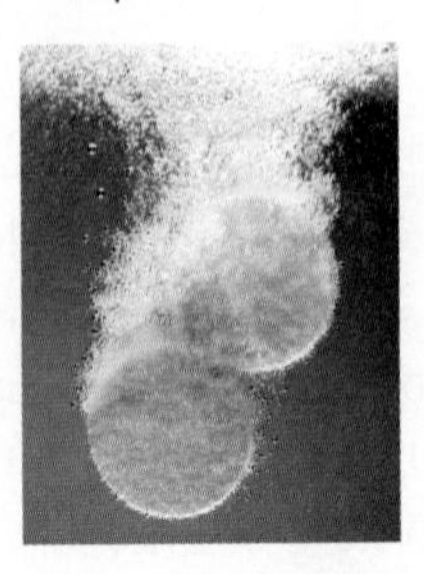

The pH scale

pH position	Substance
Acids (Increasing $[H^+]$)	Stomach acid, Lemon juice, Vinegar
7 Neutral	Salt solution, water, rubbing alcohol
Bases (Increasing $[OH^-]$)	Washing soda, Ammonium hydroxide, Drain cleaner

0 — 7 — 14

the next most acidic substance shown here, with a pH near 2.4. The most basic substance shown here is drain cleaner. The substances on either end of the pH scale are potentially more dangerous and more toxic than substances found in the middle of the scale.

Key Terms

indicator
pH scale
acid
base

Lesson Summary

What are the properties of acids and bases?

Acids and bases are special categories of solutions. The sour tastes in our food come from the acids found in those foods. Bases cause the slippery feel of soaps and detergents. Chemical compounds called indicators help chemists to identify the presence of acids and bases in solution. Indicators change color in response to acids and bases. The pH scale is a number scale that assigns values to acids and bases, between 0 and 14. Substances with a pH below 7 at 25 °C are acids. Substances with a pH above 7 at 25 °C are bases. Substances with a pH at or near 7 at 25 °C are neutral.

EXERCISES

Reading Questions

1. What are some of the observable properties of acids and bases?
2. What is the pH scale?
3. What does it mean to say that a substance has a neutral pH?

Reason and Apply

4. **Lab Report** Write a lab report for the Lab: Acids and Bases. In your report, give the title of the experiment, purpose, procedure, observations, analysis, and conclusion.

(Title)

Purpose: (Explain what you were trying to find out.)
Procedure: (List the steps you followed.)
Observations: (Describe your observations.)
Analysis: (Explain what you observed during the experiment.)
Conclusion: (What can you conclude about what you were trying to find out? Provide evidence for your conclusions.)

5. Classify each of the following solutions at 25 °C as acidic or basic based on the information provided.
   a. lemon juice tastes sour
   b. a solution of washing soda turns cabbage juice green
   c. a dilute solution of potassium hydroxide feels slippery
   d. a sugar solution has a pH of 7
   e. drain cleaner has a pH of 12

6. Examine the list of acids below. List three patterns that you notice in the names and chemical formulas.

hydrochloric acid, HCl	nitric acid, HNO_3
carbonic acid, H_2CO_3	acetic acid, $C_2H_4O_2$
sulfuric acid, H_2SO_4	citric acid, $C_6H_8O_7$

7. Examine the list of bases below. List three patterns that you notice in the names and chemical formulas.

sodium hydroxide, $NaOH$	magnesium hydroxide, $Mg(OH)_2$
potassium hydroxide, KOH	barium hydroxide, $Ba(OH)_2$
lithium hydroxide, $LiOH$	ammonium hydroxide, NH_4OH

8. **WEB RESEARCH** Find three acids and three bases used in your home.
   a. What are these acids and bases used for?
   b. Look up the chemical name and chemical formula of each acid and base.

BIOLOGY CONNECTION

The human stomach secretes hydrochloric acid, HCl, which begins the digestion process for large protein molecules and also provides a barrier against bacteria, fungi, and other micro-organisms naturally found in food and water. Stomach acid is also necessary for the proper absorption of certain minerals such as iron, zinc, and certain B-vitamins.

LESSON

18 Pass the Proton

Acid-Base Theories

Think About It

People have known for thousands of years that vinegar, lemon juice, green apples, and many other foods taste sour. However, it is only recently that scientists have presented theories to explain what all these substances have in common. In what ways are these sour-tasting substances similar, and why are they defined as acids?

How are acids and bases defined?

To answer this question, you will explore

1. Chemical Makeup of Acids and Bases
2. Acid and Base Theories
3. Strong and Weak Acids and Bases

Exploring the Topic

1 Chemical Makeup of Acids and Bases

A variety of substances dissolve in water to produce solutions that are acidic. Because the behavior of all acid solutions is similar, you might expect some similarity in the chemical makeup of these substances.

Take a moment to look for patterns in the table of common acids.

Topic: Acids and Bases
Visit: www.SciLinks.org
Web code: KEY-418

Common Acids

Name	Found in	Chemical formula	Ions in solution
hydrochloric acid	stomach acid	HCl	$H^+ \; Cl^-$
nitric acid	acid rain	HNO_3	$H^+ \; NO_3^-$
sulfuric acid	acid rain	H_2SO_4	$H^+ \; H^+ \; SO_4^{2-}$
phosphoric acid	cola	H_3PO_4	$H^+ \; H^+ \; H^+ \; PO_4^{3-}$
acetic acid	vinegar	CH_3COOH	$H^+ \; CH_3COO^-$

Here are some patterns you might notice.

- The acids are made of main group nonmetal atoms such as carbon, oxygen, fluorine, and chlorine bonded covalently.
- The only element that they all have in common is hydrogen.
- They all break apart, or **dissociate,** in solution to form at least one hydrogen cation, H^+, and an anion.

CONSUMER CONNECTION

Lye, or sodium hydroxide, is a corrosive substance found in many household items including oven cleaner and drain opener. Lye is also used for curing foods such as green olives and century eggs, and for glazing pretzels. It is also used in soap making.

You might also notice that some of the molecular formulas are written in a way that gives some information about the molecular structure. For example, acetic acid, $C_2H_4O_2$, is written as CH_3COOH. A variety of other substances dissolve in water to produce solutions that are basic. If acid solutions have H^+ ions in common, what is similar in the chemical makeup of basic solutions?

Take a moment to look for patterns in the names and chemical formulas of the common bases listed in this table.

Common Bases

Name	Found in	Chemical formula	Ions in solution
sodium hydroxide	drain cleaner	$NaOH$	$Na^+ (OH)^-$
magnesium hydroxide	antacid tablets	$Mg(OH)_2$	$Mg^{2+} (OH)^- (OH)^-$
ammonia	window cleaner	NH_3	$NH_4^+ (OH)^-$
sodium carbonate	washing soda	Na_2CO_3	$Na^+ Na^+ (HCO_3)^- (OH)^-$

Here are some things you might notice.

- Many of the bases contain a metal atom and a hydroxide ion, OH^-.
- The metal atoms are in the first two columns of the periodic table.
- Some compounds, such as ammonia and sodium carbonate, produce OH^- when dissolved in water, even though there is no OH^- in the chemical formula.
- Except for the two hydroxides, their names do not reveal that they are bases.

2 Acid and Base Theories

Arrhenius Theory of Acids and Bases

One of the early theories of acids and bases was first proposed in 1884 by a Swedish chemist named Svante Arrhenius. He defined an acid as a substance that adds hydrogen ions, H^+, to an aqueous solution. He defined a base as a substance that adds OH^- to an aqueous solution.

Brønsted-Lowry Theory of Acids and Bases

The Arrhenius definitions of acids and bases do not explain how substances without hydroxide ions, OH^-, in their formula, like ammonia, NH_3, can be bases. In the early 1920s, a Danish chemist, Johannes Brønsted, and an English chemist, Thomas Lowry, proposed a slightly different definition for acids and bases to account for this chemical behavior. They defined an acid as a substance that can donate a proton to another substance. They defined a base as a substance that can accept a proton from another substance. A proton in this case is actually a hydrogen ion, H^+. Because a hydrogen atom typically has no neutrons, if you remove the electron from a hydrogen atom to form an ion, all that's left is a proton.

When ammonia, NH_3, is added to water, it removes a hydrogen ion, H^+, from some of the water molecules, leaving behind some hydroxide ions, OH^-. An NH_3

molecule accepts a hydrogen ion and becomes NH_4^+. This makes NH_3 a Brønsted-Lowry base.

$$NH_3(g) + H_2O(l) \longrightarrow NH_4^+(aq) + OH^-(aq)$$

So dissolved ammonia really consists of ammonium and hydroxide ions. For this reason, it is often referred to as ammonium hydroxide solution. Many substances that do not contain OH^- can act as bases.

3 Strong and Weak Acids and Bases

These two illustrations show particle views of a 0.010 M hydrochloric acid, HCl, solution and a 0.010 M formic acid, HCOOH, solution. The water molecules are not shown. Take a moment to examine them.

Particle Views of a Strong and Weak Acid

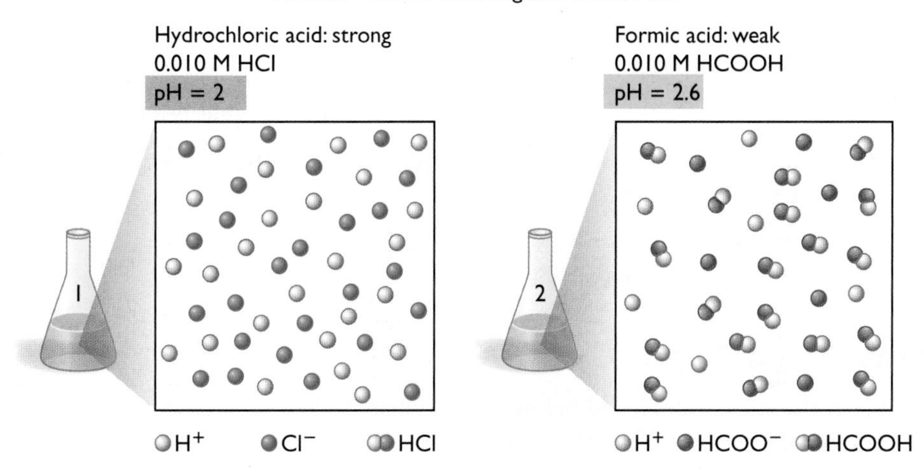

Notice that there are no molecules of HCl in the solution on the left. The HCl has dissociated completely into H^+ and Cl^- ions. However, the solution on the right contains formic acid molecules. Only some of the HCOOH molecules have dissociated into ions. This means that the concentration of H^+ ions is smaller in the formic acid solution than in the hydrochloric acid solution, even though the solutions have the same molarities.

Acids that dissociate completely in solution are called **strong acids.** Strong acids include hydrochloric acid, HCl, nitric acid, HNO_3, sulfuric acid, H_2SO_4, and hydrobromic acid, HBr. Strong acids are good conductors of electricity.

Acids that dissociate only partially in solution are called **weak acids.** These solutions are only moderate conductors of electricity. Some common weak acids are formic acid, HCOOH, acetic acid (vinegar), CH_3COOH, citric acid, $C_3H_5O(COOH)_3$, and phosphoric acid, H_3PO_4. Weak acids tend to be less corrosive because they do not dissociate completely into ions.

Bases can also be classified as strong or weak. A **strong base** dissociates completely into ions in solution and **weak bases** dissociate only partially. Some examples of strong bases include sodium hydroxide, NaOH, and barium hydroxide, $Ba(OH)_2$. Examples of weak bases include ammonia, NH_3, and aniline, $C_6H_5NH_2$.

BIOLOGY CONNECTION

Formic acid, HCOOH, is the simplest carboxylic acid. It is most famous for its presence in bee stings and ant bites. In fact, the word formic comes from *formica,* the Latin word for ant. It is also used as a preservative for livestock feed and is sometimes added to poultry feed to kill salmonella bacteria.

BIG IDEA Strong acids and strong bases dissociate completely into ions. Weak acids and weak bases dissociate only partially into ions.

Key Terms

dissociate
strong acid
weak acid
strong base
weak base

Lesson Summary

How are acids and bases defined?

The Arrhenius definition of an acid is a compound that dissociates in solution to form a hydrogen ion, H^+, and an anion. The Arrhenius definition of a base is a substance that dissociates in solution to form hydroxide ion, OH^-, and a cation. However, not all bases contain OH^-. Brønsted and Lowry defined acids as proton donors and bases as proton acceptors. The word "proton" in this case refers to an H^+ ion. A strong acid or base dissociates completely into ions in solution. A weak acid or base dissociates only partially into ions in solution.

EXERCISES

Reading Questions

1. According to the Arrhenius theory, what is an acid and what is a base?
2. How is the Brønsted-Lowry theory of acids similar to the Arrhenius theory, and how is it different?
3. What is the difference between a strong acid and a weak acid? Draw a picture showing the particle view of each.

Reason and Apply

4. Label the substances listed below as acids or bases. In each case, list the ions you would expect to find in solution.
   a. hydroiodic acid, HI
   b. formic acid, HCOOH
   c. rubidium hydroxide, RbOH
   d. hypochlorous acid, HOCl
   e. selenous acid, H_2SeO_4
   f. phosphine, PH_3
   g. perchloric acid, $HClO_4$
   h. calcium hydroxide, $Ca(OH)_2$
5. Consider a solution of hydrobromic acid, HBr.
   a. If you draw a particle view of this acid with 10 H^+ ions, how many Br^- ions would you need? Explain your thinking.
   b. Sketch a particle view of an HBr solution with 10 H^+ ions.
6. Consider a solution of magnesium hydroxide, $Mg(OH)_2$.
   a. If you draw a particle view of this base with 10 Mg^{2+} ions, how many OH^- ions would you need? Explain your thinking.
   b. Sketch a particle view of a $Mg(OH)_2$ solution with 10 Mg^{2+} ions.
7. Explain why aqueous washing soda, Na_2CO_3, is a basic solution.
8. A solution of 0.10 M hydrochloric acid, HCl, is a better conductor of electricity than 0.10 M acetic acid, CH_3COOH. Sketch the ions and molecules in both solutions to explain this observation.

LESSON

19 pHooey!

$[H^+]$ and pH

Think About It

During delivery, a baby can sometimes go into distress. So doctors monitor the vital signs to make sure the baby is okay. One test is called a fetal blood test. This test involves taking the pH of the baby's blood. If the baby's blood has a pH of 7.2 or lower, the doctors know that the baby is not getting enough oxygen and is in danger. A pH of around 7.3 indicates that the baby is safe.

How is pH related to the acid or base concentration of a solution?

To answer this question, you will explore

1. Acid Concentration
2. Relationship Between $[H^+]$ and pH
3. pH of Water and Basic Solutions

Exploring the Topic

1 Acid Concentration

If you are testing the acidity of a solution by determining pH, you are determining the hydrogen ion concentration in moles per liter. When a solution has a greater hydrogen ion concentration than water does, it is an acid. Chemists use a standard notation with brackets to symbolize concentration in moles per liter. For example, $[H^+]$ symbolizes the concentration of hydrogen ions, H^+, in moles per liter.

Particle Views of Acidic Solutions

To get an idea of how pH numbers relate to hydrogen ion concentration, it is useful to compare particle views of solutions. The illustration shows the ions in two hydrochloric acid solutions, one that is 0.010 M and the other 0.002 M. How is the hydrogen ion concentration, $[H^+]$, related to the pH?

How Does H^+ Concentration Relate to Molarity?

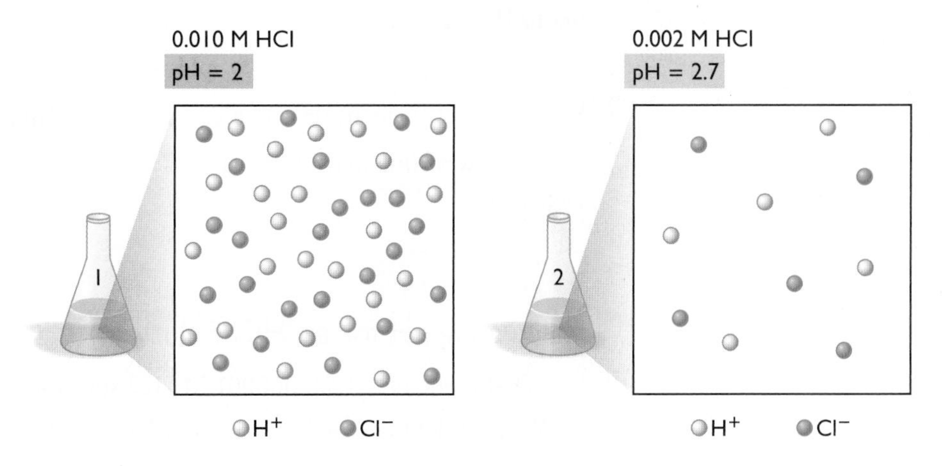

In the previous illustration, solution 1 has five times as many H^+ cations and five times as many Cl^- anions as solution 2. However, the difference in the pH is only 0.7. Small changes in pH signal large differences in hydrogen ion concentrations.

BIG IDEA The greater the concentration of H^+ ions in a solution, the lower its pH, and the more acidic it is.

2 Relationship Between [H^+] and pH

The hydrogen ion concentration of a solution is directly related to the pH of that solution. The table shows the H^+ concentration and the pH number for a few different concentrations of hydrochloric acid.

The Relationship Between [H^+] and pH Number

Molarity	[H^+]	pH
1.0 M HCl	1.0×10^0 M	0
0.1 M HCl	1.0×10^{-1} M	1
0.01 M HCl	1.0×10^{-2} M	2
0.001 M HCl	1.0×10^{-3} M	3
0.0001 M HCl	1.0×10^{-4} M	4
0.00001 M HCl	1.0×10^{-5} M	5

Notice that each time the H^+ concentration changes by a factor of 10, the pH number changes by 1 unit. You can also see that the pH number is equal to the exponent, without the negative sign. For example, 1.0×10^{-4} M HCl has a pH of 4. This is true whenever the H^+ concentration is expressed in scientific notation and the coefficient is 1.0.

Each unit of change in pH represents a tenfold difference in H^+ concentration. This is called a *logarithmic* relationship. The mathematical formula connecting pH to [H^+] is

$$pH = -\log [H^+]$$

You can use this formula and a calculator to convert between pH and [H^+].

[For review of this math topic, see **MATH *Spotlight: Logarithms*** on page 634.]

Example 1

Calculating pH for an HCl Solution

a. What is the pH of a 0.000001 M HCl solution?

b. What is the pH of a 4.2×10^{-3} M HCl solution?

Solution

a. Each HCl formula unit dissociates into an H^+ and a Cl^- ion. So the concentration of H^+ ions is 0.000001 M, which can also be written 1.0×10^{-6} M. Since the coefficient is 1.0, the pH is the exponent without the minus sign. The pH is 6.

b. Use the relationship $pH = -\log [H^+]$.

$$pH = -\log [4.2 \times 10^{-3}] = 2.4$$

The pH of the 4.2×10^{-3} molar HCl solution is between 2 and 3. This makes sense because the H^+ concentration is between 1.0×10^{-2} M and 1.0×10^{-3} M.

3 pH of Water and Basic Solutions

Basic solutions also have pH numbers associated with them. This is because they also have an H^+ concentration and pH value.

The number line below shows the relationship between pH and H^+ concentration for acidic, basic, and neutral solutions at 25 °C. The H^+ concentration in basic solutions is quite small. For example, a solution with a pH of 14 has a hydrogen ion concentration of 1.0×10^{-14} M, or 0.00000000000001 M.

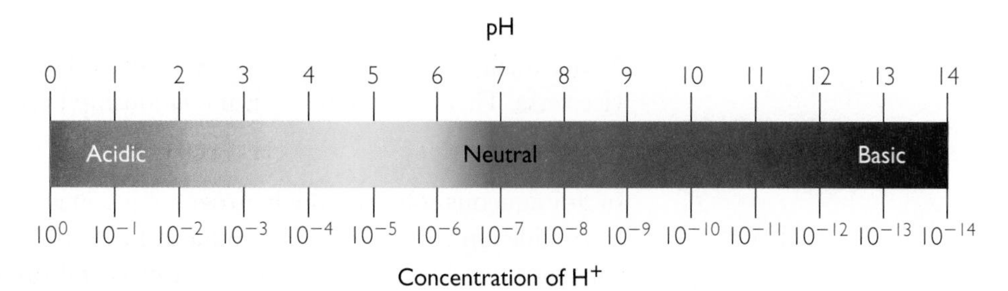

Dissociation of Water

It may seem odd that a basic solution has hydrogen ions in it. In order to understand this, it is necessary to consider water on a molecular level.

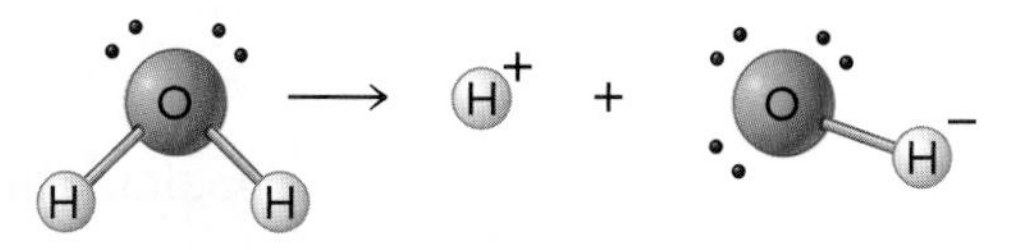

Notice that the hydrogen ion concentration in a neutral solution, such as water, is listed as 1.0×10^{-7} M H^+ at 25 °C. Water molecules dissociate slightly, forming H^+ and OH^- ions. The dissociation is so slight, there are only 2 H^+ ions and 2 OH^- ions for every one billion water molecules. This translates to an H^+ concentration of 1×10^{-7} M, or a pH of 7. Water is considered neutral because it contains equal amounts of H^+ and OH^-.

When a substance dissolves in water and the H^+ concentration, $[H^+]$, remains equal to the OH^- concentration, $[OH^-]$, we say that substance is neutral. For example, NaCl dissolves in water to form Na^+ cations and Cl^- anions. This does not change the balance of H^+ and OH^- ions already present in the water. NaCl(*aq*) is neutral.

When an acid such as HCl dissolves in water, it breaks apart into H^+ and Cl^- ions, increasing the ratio of H^+ ions to OH^- ions. Likewise, when a substance like NaOH dissolves in water, it breaks apart into Na^+ and OH^- ions. This also alters the balance of H^+ and OH^- ions in the water. In the case of NaOH(*aq*), the hydrogen concentration, $[H^+]$, is less than the hydroxide concentration, $[OH^-]$, so NaOH is considered a base.

HEALTH CONNECTION

The pH of healthy blood is always a little on the basic side and should be around 7.4. Individuals with a terminal illness, such as cancer, often have an acidic blood pH.

The table shows the concentration of H^+ and OH^- ions in a variety of sodium hydroxide solutions.

The Relationship Between $[OH^-]$ and pH Number

Molarity	$[OH^-]$	$[H^+]$	pH
water	1.0×10^{-7} M	1.0×10^{-7} M	7
0.00001 M NaOH	1.0×10^{-5} M	1.0×10^{-9} M	9
0.0001 M NaOH	1.0×10^{-4} M	1.0×10^{-10} M	10
0.001 M NaOH	1.0×10^{-3} M	1.0×10^{-11} M	11
0.01 M NaOH	1.0×10^{-2} M	1.0×10^{-12} M	12
0.1 M NaOH	1.0×10^{-1} M	1.0×10^{-13} M	13
1.0 M NaOH	1.0×10^{0} M	1.0×10^{-14} M	14

Notice that the H^+ concentration decreases as the OH^- concentration increases and vice versa. There is a mathematical relationship between $[H^+]$ and $[OH^-]$:

$$[H^+][OH^-] = 1 \times 10^{-14}$$

In *any* aqueous solution, the hydrogen ion concentration multiplied by the hydroxide ion concentration is equal to 1.0×10^{-14}. For example, when the OH^- concentration is 1×10^{-3} M, the H^+ concentration is 1×10^{-11} M.

$$(1 \times 10^{-11})(1 \times 10^{-3}) = 1 \times 10^{-14}$$

You can also see that adding the values of the exponents always results in a total of −14.

Example 2

Calculating pH for a Sodium Hydroxide, NaOH, Solution

What is the pH of a 0.000001 M NaOH solution?

Solution

A solution with a molarity of 0.000001 M NaOH has an OH^- concentration of 1.0×10^{-6} moles per liter. The H^+ concentration must be 1.0×10^{-8} M because the exponents must sum to −14. The pH is just the exponent of the H^+ concentration without the negative sign. So the pH is 8.

Lesson Summary

How is pH related to the acid or base concentration of a solution?

The pH number and the hydrogen ion concentration of a solution are related logarithmically: $pH = -\log[H^+]$. This means that when the hydrogen ion concentration changes by a factor of 10, the pH changes by 1. Water and basic solutions also have pH numbers because they contain small concentrations of

H^+ ions. Water dissociates slightly into H^+ and OH^- ions. In pure water, the H^+ concentration is 1×10^{-7} M and the pH is 7. The hydroxide ion concentration and the hydrogen ion concentration in a solution are related mathematically: $[H^+][OH^-] = 1.0 \times 10^{-14}$.

EXERCISES

Reading Questions

1. What is the relationship between pH and H^+ concentration?
2. How is the H^+ concentration related to the OH^- concentration in a solution?

Reason and Apply

3. What pH would you expect for solutions with the concentrations given here?
   a. $[H^+] = 1.0 \times 10^{-4}$ M
   b. $[H^+] = 1.0 \times 10^{-12}$ M
   c. $[OH^-] = 1.0 \times 10^{-8}$ M
4. Determine the pH for solutions with the H^+ concentrations given here.
   a. $[H^+] = 0.0014$
   b. $[H^+] = 6.0 \times 10^{-8}$ M
   c. $[H^+] = 4.2 \times 10^{-11}$ M
   d. $[H^+] = 1.5 \times 10^{-1}$ M
5. Which of the solutions in Exercise 5 are acids? Which are bases?
6. What H^+ concentration would you expect for the solutions below?
   a. pH = 9
   b. pH = 7.3
   c. pH = 2.9
   d. pH = 10.2
7. What is the pH of a 2.5 M HCl solution?
8. What is the pH of a 0.256 M NaOH solution?
9. **WEB RESEARCH** The Richter scale is a logarithmic scale for measuring the magnitude of earthquakes. How much more intense is a Richter 7.0 earthquake than a Richter 5.0 earthquake?

LESSON

20 Watered Down

Dilution

Think About It

In Unit 1: Alchemy, a "golden" penny was made with the help of a strong base called sodium hydroxide. In Unit 2: Smells, you added concentrated sulfuric acid to a mixture of chemicals in order to make a sweet-smelling ester. Whenever you work with very concentrated acids and bases, you are instructed to rinse with large amounts of water if you get any on your skin or clothing. How does this help?

How does dilution affect acids and bases?

To answer this question, you will explore

1. Diluting Acid Solutions
2. Approaching pH 7

Exploring the Topic

1 Diluting Acid Solutions

If you spill concentrated acid on your skin, it will begin to sting and burn right away. If you spill it on your clothing, you will eventually see holes in the cloth. The best thing you can do for this situation is to put a lot of water on the part of your skin or clothing where the acid spilled.

The process of adding more and more water to a solution is called **dilution.** You could say that the solution is becoming more dilute, or "watered down." This changes the number density of the solute particles in solution.

The effect of diluting a concentrated solution can be demonstrated with a glass of grape juice. The color of the grape juice becomes paler as more water is added. And the more dilute the grape juice becomes, the weaker it tastes.

Acid solutions are not necessarily colorful like grape juice, but the concept of dilution is the same: the solution becomes weaker. When you dilute an acid, you can track what is happening by testing the pH.

The illustration at the top of the next page shows a particle view of what happens when you dilute an acid solution. A few drops of red dye have been added so that you can visually keep track of the different solutions. The solution is diluted with water in the beaker. The starting solution in the test tube has a pH of 2. The particle-view squares show how the H^+ ions and Cl^- ions end up farther apart as more water is added.

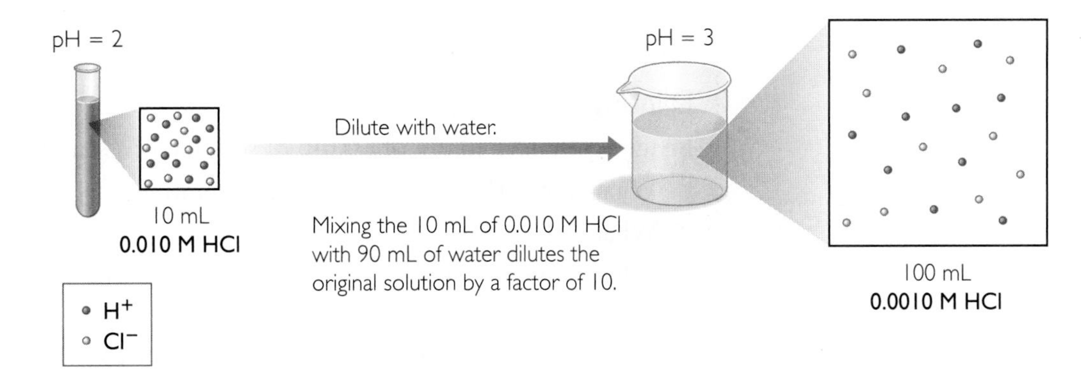

Dilution of a strong acid or base can make it safer. This is why you can swim in a swimming pool, even though the water contains both acids and bases.

Example 1

Diluting HCl

Imagine you have a 1.0 mL water sample containing 0.10 M hydrochloric acid, HCl. You want to get it to a safe pH of 6. How much water do you need to add?

Solution

The pH of 0.10 M HCl is 1. The H^+ concentration needs to change from 1.0×10^{-1} M to 1.0×10^{-6}, a factor of 10^5, or 100,000. So you need to add 99,999 mL of water to 1 mL 0.10 M HCl to get it to a safe pH of 6.

RECREATION CONNECTION

Swimming pools require a slightly basic pH, between 7.2 and 7.8. If the pH is too low, eye irritation occurs and metal pumps may corrode. If the pH is too high, the water may become cloudy and the chlorine may not work to combat algae. Sodium carbonate is added to pools to keep the pH in the right range. Indicators are used to test the pH.

❷ Approaching pH 7

Suppose you continue to dilute the 0.0010 M HCl solution in the large beaker in the previous illustration. What is the lowest H^+ concentration you can obtain? What is the lowest pH you can reach?

If you take 1 mL of 0.00010 M HCl and add 9 mL of water, the new solution will have a concentration of 0.000010 M HCl and a pH of 5. If you take 1 mL of 0.000010 M HCl and add 9 mL of water, the new solution will have a concentration of 0.0000010 M HCl and a pH of 6. You can continue this process, diluting the acid until the pH becomes 7. At this point, the liquid is no longer acidic, it is neutral.

As you dilute an acid solution, its H^+ concentration gets closer and closer to the H^+ concentration of water, 1×10^{-7} moles per liter. However, the pH does not increase beyond 7, no matter how much you dilute the solution. In other words, the solution does not become basic. You cannot start with an acidic solution and dilute it to make a basic solution.

Likewise, you cannot dilute a basic solution to make an acid solution. When you dilute a basic solution, the OH^- concentration decreases, the H^+ concentration increases, and the pH decreases toward 7.

BIG IDEA An acid can never be made into a base by diluting with water. A base can never be made into an acid by diluting with water.

Example 2
Diluting NaOH

Imagine you have 10 mL of a 0.010 M solution of sodium hydroxide, NaOH.

a. What is the H^+ concentration of this solution?

b. What is the pH?

c. What is the NaOH concentration if you dilute the 0.010 M solution by adding 90 mL of water?

d. Does the pH increase or decrease after diluting the 0.010 M NaOH solution?

e. What is the pH of the solution after adding 90 mL of water?

Solution

a. The OH^- concentration is 0.010 mole per liter, which is equivalent to 1×10^{-2} M. Because $[H^+][OH^-] = 1 \times 10^{-14}$, the H^+ concentration is 1×10^{-12} M. The exponents sum to -14.

b. $pH = -\log [H^+] = -\log 10^{-12} = 12$

c. If you add 90 mL water to 10 mL of 0.010 M NaOH, then you have 100 mL. The volume has increased tenfold. So the concentration after diluting is 0.0010 M NaOH.

d. As the OH^- concentration decreases, the H^+ concentration increases. So the pH decreases when you dilute 0.010 M NaOH to 0.0010 M.

e. The OH^- concentration is now 1.0×10^{-3} M. Because $[H^+][OH^-] = 1.0 \times 10^{-14}$, $[H^+] = 1 \times 10^{-11}$ M.

$$pH = -\log [1 \times 10^{-11}] = 11$$

So the pH has decreased, from 12 to 11, as you might expect. Diluting with water makes a solution more neutral.

Key Term
dilution

Lesson Summary

How does dilution affect acids and bases?

When water is added to an acidic solution, the result is a "watering down" of the solution. The H^+ concentration of the solution decreases, and the pH increases toward 7. When water is added to a concentrated basic solution, the OH^- concentration decreases. This increases the H^+ concentration and the pH decreases toward 7. Acids cannot be made into bases and bases cannot be made into acids by adding water.

EXERCISES

Reading Questions

1. Describe what happens to the pH of a solution when you add water to an acidic solution.
2. Explain why you cannot turn an acid into a base by diluting with water.

Reason and Apply

3. Copy this table and fill in the missing information.

Solution	$[H^+]$ in scientific notation	pH number	$[OH^-]$ in scientific notation
1.0 M HCl	1×10^{0}	0	1×10^{-14}
0.10 M HCl	1×10^{-1}	1	1×10^{-13}
0.010 M HCl			
0.0010 M HCl			
0.00000010 M HCl			
0.000000010 M HCl	1×10^{-7} M		
0.0000000010 M HCl	1×10^{-7} M		
0.0010 M NaCl			
0.00000010 M NaOH			1×10^{-7} M
0.0000010 M NaOH	1×10^{-8} M		1×10^{-6} M
0.000010 M NaOH			

4. How much water do you need to add to 10 mL of a solution of HCl with a pH of 3 to change the pH to 6?

5. Explain why the pH of a 1×10^{-9} M HCl solution is 7 even though the exponent is -9.

6. Explain why the pH of a NaCl solution does not change if you dilute the solution.

7. Imagine you have 0.75 L of a 0.10 M HCl solution.
   a. How many moles of H^+ are in the 0.75 L?
   b. If you add 0.35 L of water, what is the new concentration of the solution?
   c. What is the pH after adding 0.35 L?

8. Imagine you have 15 mL of a 0.025 M HCl solution.
   a. How many moles of H^+ are in the 15 mL?
   b. If you add 25 mL of water, what is the new concentration of the solution?
   c. What is the final pH after adding 25 mL?

LAB Neutralization Reactions

Safety Instructions

Safety goggles must be worn at all times.

Acids and bases are corrosive. Do not get any on skin or near eyes.

In case of a spill, rinse with large amounts of water.

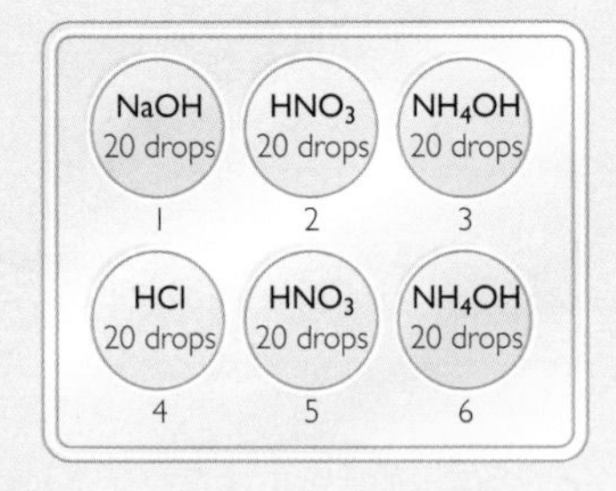

Purpose

To examine reactions between acids and bases.

Materials

- well plate
- set of five labeled dropper bottles: 0.10 M HCl (hydrochloric acid), 0.10 M HNO_3 (nitric acid), 0.10 M NaOH (sodium hydroxide), 0.10 M NH_4OH (ammonium hydroxide), and bromothymol blue indicator

Procedure

1. Add 20 drops of each solution to the well plate as specified in the illustration.
2. Add 1 drop of bromothymol blue indicator to each well plate. Bromothymol blue is yellow in acid, blue in base, and green in neutral solution.
3. Test with HCl: Try to turn the solutions in wells 1, 2, and 3 green using drops of 0.10 M HCl. Record your observations in a data table.

Solution in the well plate	Indicator color	Acid, base, or neutral?	Drops of 0.10 M HCl to turn solution green	Indicator color after mixing	Does a reaction occur?
20 drops 0.10 M HCl					

4. Test with NaOH: Try to turn the solution in wells 4, 5, and 6 green using drops of 0.10 M NaOH.

Analysis

1. What did you observe when you mixed an acid with a base?
2. Acids react with bases to form an ionic salt and water. Copy this chemical equation and label the acid, base, and ionic salt.

$$HCl(aq) + NaOH(aq) \longrightarrow NaCl + H_2O$$

3. Complete and balance the equation for the reaction in each well. If no reaction occurred, simply write "no reaction" on the products side of the equation.
4. What combinations of reactants do not yield new products?
5. Name the three salts produced in the reactions.
6. **Making Sense** List three things you learned as a result of performing this laboratory procedure.

LESSON

21 Neutral Territory

Neutralization Reactions

Think About It

There are many advertisements on television about products that are designed to treat "acid indigestion." Some of these substances claim to neutralize extra acid that may have built up in your stomach.

What happens when acids and bases are mixed?

To answer this question, you will explore

1. Neutralization Reactions
2. Predicting the Products of Neutralization Reactions

Exploring the Topic

1 Neutralization Reactions

Pure water is neutral because it has equal numbers of H^+ and OH^- ions.

Water Dissociates into H^+ and OH^- Ions

$$H_2O(l) \longrightarrow H^+(aq) + OH^-(aq)$$

In pure water at 25 °C, $[H^+]$ and $[OH^-]$ are both 1.0×10^{-7} mol/L. But if an acid or a base is added to water, it upsets the balance of ions in solution. Acidic solutions have an excess of H^+ and basic solutions have an excess of OH^-.

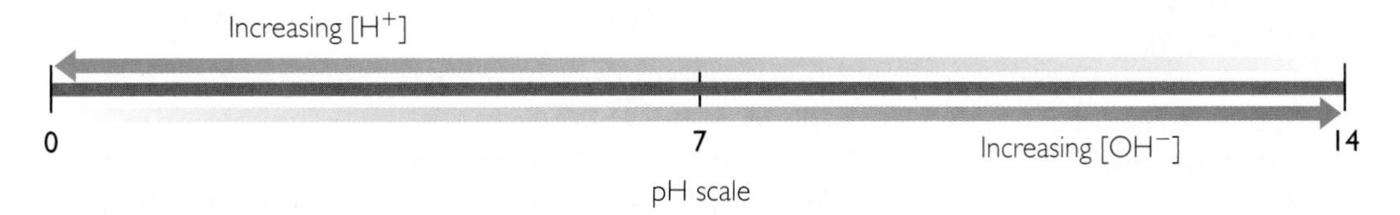

One way to reduce the toxicity of an acidic or basic solution is by dilution. If you add enough water, you can reduce the concentration of the acid or base. Extremely dilute acids and bases have a pH of 7 at 25 °C because this is the pH of pure water.

However, there is another way you can reduce the toxicity of an acidic or a basic solution. In order to counteract the effects of an acid in solution, you can add a base. Likewise, in order to counteract the effects of a base in solution, you can add an acid.

The acid and base react in solution and neutralize each other. The chemical equation for the reaction between an acid and base is shown below.

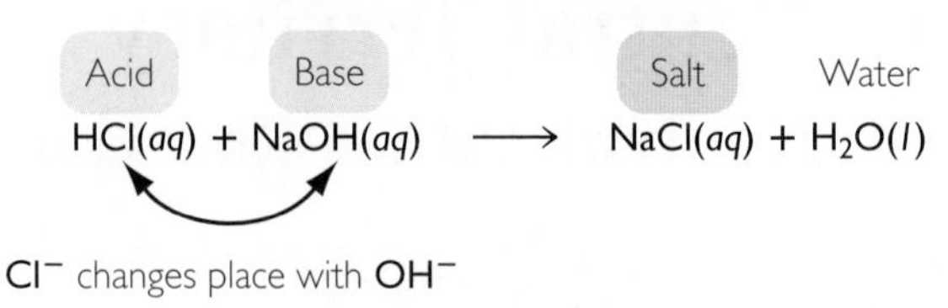

Notice that this reaction is a double exchange reaction. The H^+ and Na^+ cations exchange anions. The result is the production of a salt, NaCl, and water. The excess H^+ from the acid combines with OH^- from the added base to form H_2O. The reaction is referred to as a **neutralization reaction.**

BIG IDEA Acids and bases neutralize each other, producing a salt and water.

So, one way to make an acidic solution safe is by adding a base. Likewise, you can make a basic solution safe by adding an acid.

Not all neutralization reactions produce neutral solutions. For example, if you use an acidic solution that is very concentrated and a basic solution that is not very concentrated, there will be leftover H^+ ions after mixing. There will not be enough OH^- ions to neutralize all of the H^+ ions. But some neutralization will have taken place. As a result of mixing, the solution will be closer to neutral than either of the starting solutions.

2 Predicting the Products of Neutralization Reactions

Several neutralization reactions are shown. Notice the patterns in the products of the reactions.

$$HCl(aq) + KOH(aq) \longrightarrow KCl(aq) + H_2O(l)$$
$$HNO_3(aq) + NaOH(aq) \longrightarrow NaNO_3(aq) + H_2O(l)$$
$$H_2SO_4(aq) + 2NaOH(aq) \longrightarrow Na_2SO_4(aq) + 2H_2O(l)$$
$$2HCl + Mg(OH)_2(aq) \longrightarrow MgCl_2(aq) + 2H_2O(l)$$
$$H_2SO_4(aq) + Ca(OH)_2(aq) \longrightarrow CaSO_4(aq) + 2H_2O(l)$$

HEALTH CONNECTION

Some antacids are made of calcium carbonate. Calcium carbonate is the main component of eggshells and seashells and can also be used as an inexpensive calcium supplement.

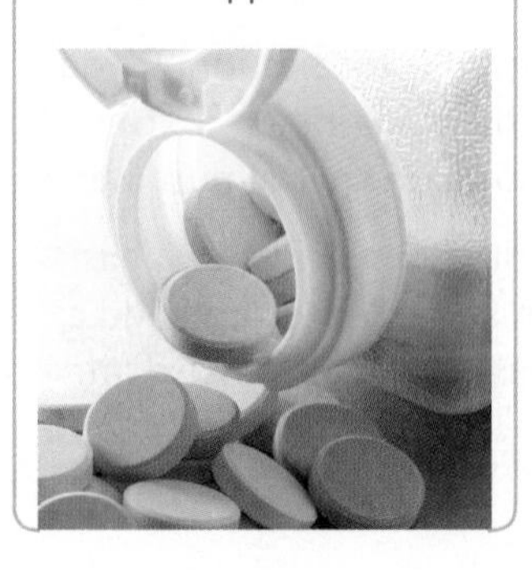

Each reaction results in the production of an ionic compound (a salt) plus water. The salt consists of the cation of the base and the anion of the acid. For example, the cation in potassium hydroxide, KOH, is potassium, K^+. The anion in nitric acid, HNO, is nitrate, NO_3^-. The K^+ and NO_3^- produce the ionic compound, potassium nitrate, KNO_3, which is a salt.

Some acids transfer more than one H^+ ion. For example, each mole of sulfuric acid, H_2SO_4, transfers two moles of H^+ ions. This is because there are two hydrogen ions in solution for every one sulfate anion, SO_4^{2-}. Likewise, each mole of calcium hydroxide, $Ca(OH)_2$, transfers two moles of OH^- ions. There are two hydroxide ions in each formula unit of calcium hydroxide to balance the charge on the calcium cations, Ca^{2+}.

Example

Producing Potassium Sulfate, K_2SO_4

Which of the acids and bases in this table could you mix in order to make potassium sulfate, K_2SO_4? Write a balanced chemical equation for the reaction.

Acids	Bases
HNO_3	KOH
H_2SO_4	NaOH
HBr	$Mg(OH)_2$

Solution

One way to use a neutralization reaction to make potassium sulfate is

$$H_2SO_4(aq) + 2KOH(aq) \longrightarrow K_2SO_4(aq) + 2H_2O(l)$$

Notice that you need 2 mol of KOH in order to balance the equation properly.

Relieving Stomach Acid

A common over-the-counter remedy for excess stomach acid uses a neutralization reaction. Milk of magnesia is a white mixture containing magnesium hydroxide, $Mg(OH)_2$. The mixture looks like milk because magnesium hydroxide, a white solid, does not dissolve completely in water. Instead, the white solid is suspended in the liquid.

Magnesium hydroxide is a base. It neutralizes stomach acid, HCl, to produce a salt and water as shown by this reaction.

$$2HCl(aq) + Mg(OH)_2 \longrightarrow MgCl_2(aq) + 2H_2O(l)$$

$$\text{stomach acid} + \text{milk of magnesia} \longrightarrow \text{magnesium chloride} + \text{water}$$

Key Term

neutralization reaction

Lesson Summary

What happens when acids and bases are mixed?

A neutralization reaction is a reaction in which an acid and a base react in aqueous solution to produce an ionic compound (a salt) and water. The pH approaches 7 because the H^+ from the acid and the OH^- from the base combine to form H_2O. It is relatively easy to predict the products of a neutralization reaction. The salt that forms is made from the cation of the base and the anion of the acid.

EXERCISES

Reading Questions

1. Describe two ways to make a strong acidic solution safer.
2. What is a neutralization reaction?

Reason and Apply

3. **Lab Report** Complete a laboratory report for the Lab: Neutralization Reactions. In your report, give the title of the experiment, purpose, procedure, observations, results, and conclusion. In your conclusion, explain how you can use chemical equations to predict the products of the reactions you carried out.

(Title)

Purpose: (Explain what you were trying to find out.)

Procedure: (List the steps you followed.)

Observations: (Describe your observations.)

Conclusion: (What can you conclude about what you were trying to find out?)

4. Predict the products for the reactions given below. Be sure to balance each equation.
   a. $HF(aq) + NaOH(aq) \longrightarrow$
   b. $HCl(aq) + Mg(OH)_2(aq) \longrightarrow$
   c. $HF(aq) + NH_4OH(aq) \longrightarrow$
   d. $CH_3COOH(aq) + NaOH(aq) \longrightarrow$
   e. $HNO_3(aq) + Mg(OH)_2(aq) \longrightarrow$
   f. $HNO_3(aq) + NH_4OH(aq) \longrightarrow$
   g. $CH_3COOH(aq) + Mg(OH)_2(aq) \longrightarrow$
   h. $CH_3COOH(aq) + NH_4OH(aq) \longrightarrow$
   i. $H_2SO_4(aq) + HNO_3(aq) \longrightarrow$
   j. $H_2SO_4(aq) + Mg(OH)_2(aq) \longrightarrow$
5. Suppose you mix 1 mol of sulfuric acid, H_2SO_4, with 1 mol of sodium hydroxide, NaOH. Why does the pH of the solution remain below 7?
6. Suppose you mix 1 mol of sodium hydroxide, NaOH, with 1 mol of magnesium hydroxide, $Mg(OH)_2$. Why does the pH of the solution remain above 7?
7. Which of these substances might be useful in neutralizing a lake damaged by acid rain?
   A. H_2SO_4
   B. CH_3COOH
   C. $CaCl_2$
   D. $Ca(OH)_2$
8. Which combination of reactants would result in a neutralization reaction with sodium nitrate, $NaNO_3$, as one of the products?
   A. $Mg(NO_3)_2 + NaOH$
   B. $HNO_3 + NaOH$
   C. $CH_3OH + NaOH$
   D. $HNO_3 + NaCl$

LESSON

22 Drip Drop

Titration

Think About It

When sulfur dioxide, SO_2, and nitrogen oxide, NO, two components of air pollution, come in contact with water in the atmosphere they are converted to sulfuric acid, H_2SO_4, and nitric acid, HNO_3. The acids then fall as acidic rain or snow. Acid precipitation is highly destructive, sometimes causing the complete elimination of fish and insect species from lakes or streams. Acid rain also makes the soil unsuitable for plant life, killing off whole sections of forest. Water scientists, called hydrologists, regularly study the acidity of lakes and streams to monitor the extent of the acid rain problem.

How can you use a neutralization reaction to figure out acid or base concentration?

To answer this question, you will explore

1. Titrations
2. Particle Views of Titrations
3. Titration Calculations

Exploring the Topic

1 Titrations

A Titration Setup

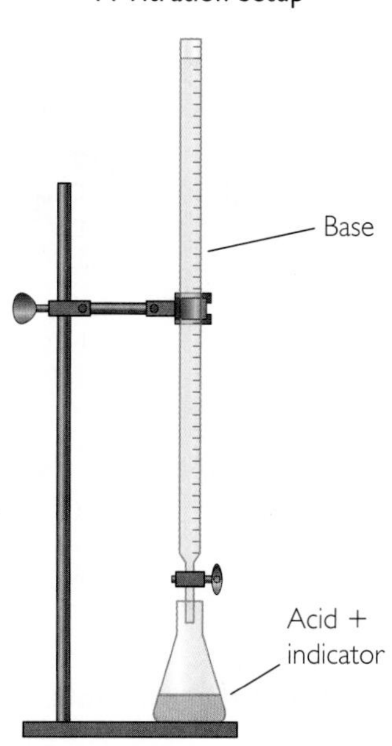

To track the spread of acid rain, scientists take water samples from lakes and test them for H^+ concentration. One way to determine the concentration of a strong acid in a water sample is to use a method called a **titration.** A titration is a neutralization reaction that is monitored with an acid-base indicator. For example, a strong base with a known concentration is added to a strong acid sample with an unknown concentration. The base is added until the indicator changes color.

The indicator provides a visual signal that the solution has reached the **equivalence point,** the point at which the moles of base added have neutralized the moles of acid. By keeping track of the exact volume of base that is added to a known volume of acid, you can figure out the unknown concentration of acid in the sample.

The illustration shows a titration setup. The long thin tube is a *burette.* A valve on the burette allows you to regulate how much solution goes into the beaker. During a titration, solution is added from the burette until the indicator changes color. You determine the volume of solution added by noting the change in the height of the solution in the burette.

Phenolphthalein is an indicator that changes from colorless to bright pink in a base.

ENVIRONMENTAL CONNECTION

Acid rain creates unwanted chemical changes in soil, causing damage to trees and other plants. Around the world, acid rain and acid snow have removed calcium and magnesium from the soil and have increased levels of aluminum. When aluminum is absorbed into plants, it displaces calcium and other nutrients that the plants need.

❷ Particle Views of Titrations

This illustration shows what happens as equal volumes of 1.0 M HCl, hydrochloric acid, and 1.0 M NaOH, sodium hydroxide, are mixed together.

$$HCl(aq) + NaOH(aq) \longrightarrow NaCl(aq) + H_2O(l)$$

● H^+ ○ Na^+
□ Cl^- ■ OH^-

Mixture of the 2 drops

1 drop 1.0 M HCl + 1 drop 1.0 M NaOH → 2 drops 0.5 M NaCl

Neutralization of 1.0 M NaOH by 1.0 M HCl

When you add one drop of 1.0 M HCl to one drop of 1.0 M NaOH, H^+ ions from the acid and OH^- ions from the base combine to form H_2O. The Na^+ ions and Cl^- ions from the original drops are dissolved in the water. Notice that the concentration of Na^+ ions and Cl^- ions has decreased because each ion has spread out over twice the volume, equal to two drops of solution.

The next illustration shows what happens when equal volumes of 1.0 M HCl and 2.0 M NaOH are mixed together.

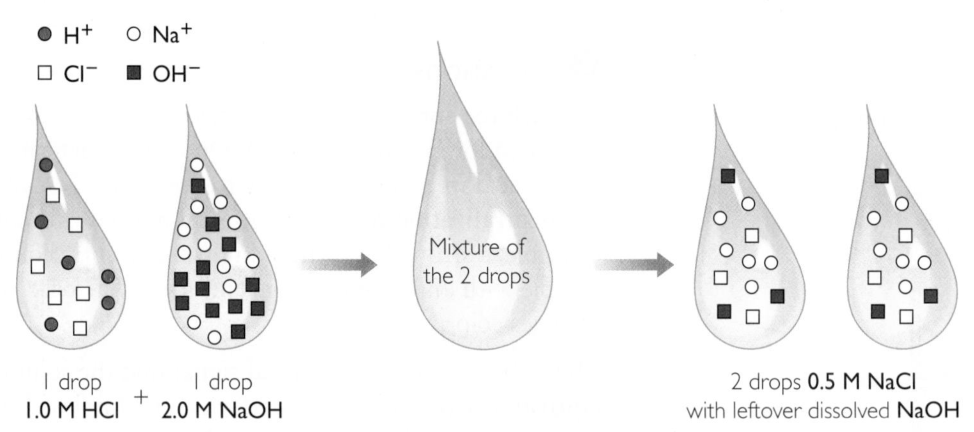

Neutralization of 1.0 M HCl by 2.0 M NaOH

In this example, the base is more concentrated than the acid so the two-drop mixture will have leftover dissolved NaOH in addition to dissolved 0.5 M NaCl.

Because the 2.0 M NaOH is twice as concentrated as the 1.0 M HCl, you only need half as much of it to neutralize the HCl.

Moles of H^+ in 1.0 M HCl (10 mL)	Moles of OH^- in 2.0 M NaOH (5 mL)
moles H^+ = molarity • volume = 1.0 M • 0.010 L = 0.005 mol	moles OH^- = molarity • volume = 2.0 M • 0.005 L = 0.005 mol

Bromothymol blue indicator is yellow in acidic solution, green in neutral solution, and blue in basic solution.

3 Titration Calculations

Suppose you take a 100 mL water sample from a lake contaminated with sulfuric acid, H_2SO_4. You want to determine the concentration of the sulfuric acid in the lake. So you titrate the sample with 2.0 molar sodium hydroxide, 2.0 M NaOH, until the solution is neutral. At the equivalence point, you know that the moles of base you added neutralized moles of acid that were in the unknown solution.

You need an indicator that changes color when the solution is neutral, so you add a drop of bromothymol blue to the lake water sample. You add NaOH slowly until the indicator turns green. Because you know the volume and molarity of the NaOH added, you can calculate the moles of acid per liter of solution in the lake water sample.

Example

Titration of H_2SO_4

A 100 mL sample of aqueous sulfuric acid, H_2SO_4, is titrated with 2.0 M NaOH. After 50 mL of NaOH are added, the indicator changes color at pH 7. What was the starting concentration of the H_2SO_4?

Solution

Begin by writing a balanced chemical equation.

$$H_2SO_4(aq) + 2NaOH(aq) \longrightarrow Na_2SO_4(aq) + 2H_2O(l)$$

$$\text{sulfuric acid} + \text{sodium hydroxide} \longrightarrow \text{sodium sulfate} + \text{water}$$

Find the total number of moles of NaOH that were used to neutralize the H_2SO_4.

$$\begin{aligned}\text{moles NaOH} &= (\text{molarity of NaOH})(\text{volume of NaOH added}) \\ &= (2.0\ \text{mol/L})(0.050\ \text{L}) = 0.10\ \text{mol NaOH}\end{aligned}$$

Since 2 mol of NaOH are needed to neutralize every 1 mol of H_2SO_4, the moles of H_2SO_4 in the sample must be half the number of moles of the added NaOH.

$$\text{moles of } H_2SO_4 = \frac{1}{2}(\text{moles of NaOH}) = \frac{1}{2}(0.10\ \text{mol}) = 0.05\ \text{mol}$$

Finally, find the molarity of the sulfuric acid solution.

$$\begin{aligned}\text{molarity of sulfuric acid} &= \frac{n}{V} \\ &= 0.050\ \text{mol}/0.100\ \text{L} \\ &= 0.50\ \text{M } H_2SO_4\end{aligned}$$

Key Terms

titration
equivalence point

Lesson Summary

How can you use a neutralization reaction to figure out acid or base concentration?

A titration is a chemical procedure carried out between an acid and a base in order to determine the concentration of either the acid or the base. A titration is a neutralization reaction that is monitored with an indicator. The volume of acid and the volume of base used in the procedure are carefully recorded. If the molarity of either the acid or the base is known, the molarity of the other can be determined.

EXERCISES

Reading Questions

1. Describe how you might use a titration to figure out the concentration of potassium hydroxide in a water sample.
2. What is the role of an indicator in titration?

Reason and Apply

3. How many mL of 0.1 M NaOH would be required to neutralize 2.0 L of 0.050 M HCl?
4. The table represents a series of titrations using 0.10 M NaOH to determine the concentrations of five different samples of HCl. Copy the table and fill in the missing values.

Initial volume of HCl	Volume of 0.10 M NaOH added	Total moles of NaOH	Total moles of HCl	Initial HCl concentration
1.0 L	1.0 L	0.10 mol	0.10 mol	0.10 M
100 mL	200 mL	0.020 mol	0.020 mol	
50 mL	200 mL			
50 mL		0.0025 mol	0.0025 mol	0.050 M
100 mL	73 mL			

5. A student mixes 100 mL of 0.20 M HCl with different volumes of 0.50 M NaOH. Are the final solutions acidic, basic, or neutral? Explain your thinking.
   a. 100 mL of 0.20 M HCl + 20 mL of 0.50 M NaOH
   b. 100 mL of 0.20 M HCl + 40 mL of 0.50 M NaOH
   c. 100 mL of 0.20 M HCl + 60 mL of 0.50 M NaOH
6. Imagine you use 0.95 M NaOH to titrate several water samples. The volume of base needed to neutralize a specified amount of acid is given. Determine the acid concentration and the pH for each.
   a. 25 mL acid, 46 mL NaOH
   b. 10 mL acid, 17 mL NaOH
   c. 25 mL acid, 12 mL NaOH

SECTION

IV

SUMMARY

Acidic Toxins

Toxins Update

Acids and bases are present in living systems and are a valuable part of the chemistry of life. They are toxic under certain conditions, such as when they are too concentrated, or when they upset the pH balance that must be maintained for proper health. The pH number is a measure of the H^+ concentration of a solution, and the lower the pH, the more acidic a solution is.

There are two main approaches to dealing with toxic acids and bases. One is dilution. This moves the pH toward neutral. Another approach is neutralization. Acids and bases neutralize one another.

Key Terms

indicator
pH scale
acid
base
strong acid
weak acid
strong base
weak base
dilution
neutralization reaction
titration
equivalence point

Review Exercises

1. What are the main differences between the Arrhenius and Brønsted-Lowry definitions of acids and bases?
2. Name three substances that you might use to neutralize a hydrochloric acid solution. Write the balanced chemical equation for each reaction.
3. Explain how a titration procedure works to help you identify the H^+ concentration of a sample solution.
4. Lemon juice has a pH around 2. This is quite acidic and can damage the tissue of the eye.
   a. What is the concentration of hydrogen ions, $[H^+]$, in lemon juice?
   b. What is the concentration of hydroxide ions, $[OH^-]$?
5. Sodium bicarbonate, $NaHCO_3$, is a base.
   a. Explain how sodium bicarbonate can be classified as a base, even though it has no OH^- ion.
   b. Explain why you had sodium bicarbonate on hand during the Lab: The Copper Cycle from Unit 1: Alchemy.

RESEARCH

PROJECT *Dissolving Toxins*

Choose a toxic substance (your teacher may assign you one). Find some information about your toxin. Your report should cover these topics.

- Does your substance dissolve in water? If it does not dissolve would it sink or float?
- Draw a particle view drawing of your substance in water, even if it does not dissolve.
- Is your substance acidic, basic, or neutral? If your substance cannot be described in this way, explain why.

SECTION

V Toxic Cleanup

Chemical reactions are often used to remove toxic substances from a solution. Here, a solid forms when an ionic solution is added.

One way to remove specific ions from a solution is to use chemical reactions to transform these ions into solid products. Solid compounds can be easily filtered out of a solution. The human body uses this same approach with many toxic substances, by forming solids that are then passed out of the body as waste. Mass-mole conversions allow you to determine how to create the maximum amount of product from a chemical reaction.

In this section, you will study

- solids in solutions
- soluble and insoluble ionic compounds
- how to determine the product yield of a reaction
- how to use the "mole tunnel" to perform stoichiometric calculations

LESSON

23 Solid Evidence

Precipitation Reactions

Think About It

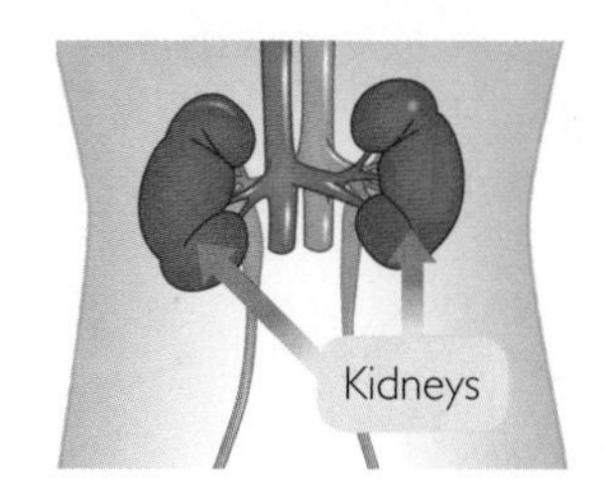

Sometimes reactions between dissolved substances result in the formation of a solid. In your kidneys, for example, dissolved substances can react to form solid calcium oxalate. If there is enough of this solid, a painful blockage called a kidney stone can form.

Which substances precipitate from aqueous solutions?

To answer this question, you will explore

1. Precipitation of Ionic Solids
2. Solubility
3. Toxicity of Precipitates

Exploring the Topic

1 Precipitation of Ionic Solids

Sometimes certain anions and cations combine and come out of solution as a solid, or a **precipitate.** This kind of reaction is called a **precipitation reaction.**

The formation of kidney stones is an example of a precipitation reaction. If the concentrations of calcium ions, Ca^{2+}, and polyatomic oxalate ions, $C_2O_4^{2-}$, in urine become too great, calcium oxalate, CaC_2O_4, precipitates as kidney stones.

$$CaCl_2(aq) + Na_2C_2O_4(aq) \longrightarrow 2NaCl(aq) + CaC_2O_4(s)$$

Dissolved calcium chloride	Dissolved sodium oxalate	Dissolved sodium chloride	Solid calcium oxalate

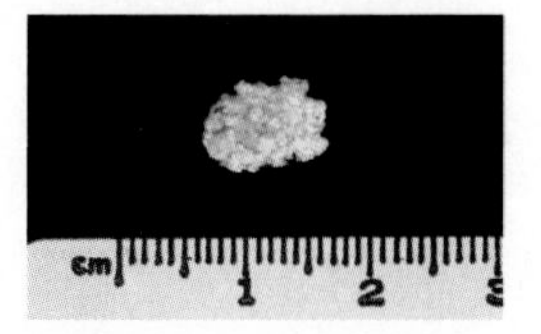

Kidney stone crystal

The equation for the formation of kidney stones shows that it is a double exchange reaction. Many other similar chemical combinations result in precipitation reactions.

Example 1

Precipitation of Lead Iodide

An aqueous solution of lead nitrate, $Pb(NO_3)_2(aq)$, is mixed with an aqueous solution of potassium iodide, $KI(aq)$. The result is a bright yellow solid, lead iodide, $PbI_2(s)$, in a clear solution. Write a balanced chemical equation for this precipitation reaction.

WEATHER CONNECTION

Precipitation can refer to either water coming out of the atmosphere as rain or snow, or compounds coming out of an aqueous solution as solids.

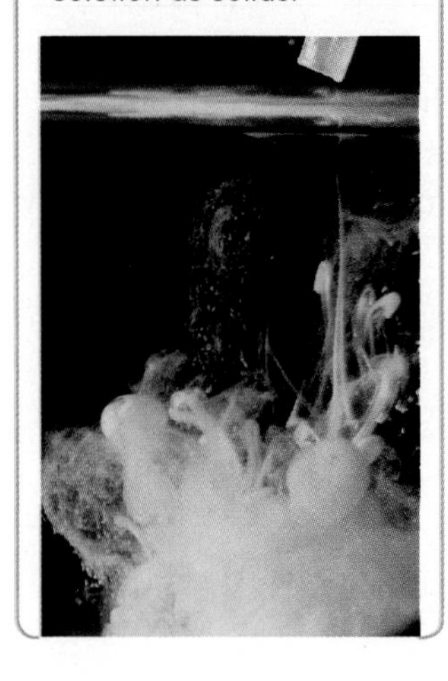

Solution

This is a double exchange reaction. The lead and potassium cations exchange anions with one another. Begin by writing an equation for what you know.

$$\underset{\text{lead nitrate}}{Pb(NO_3)_2(aq)} + \underset{\text{potassium iodide}}{KI(aq)} \longrightarrow \underset{\text{lead iodide}}{PbI_2(s)} + \underset{\text{clear solution}}{?}$$

The potassium ions, K^+, and the nitrate ions, NO_3^-, combine in a 1:1 ratio to form KNO_3.

$$\underset{\text{lead nitrate}}{Pb(NO_3)_2(aq)} + \underset{\text{potassium iodide}}{KI(aq)} \longrightarrow \underset{\text{lead iodide}}{PbI_2(s)} + \underset{\text{potassium nitrate}}{\mathbf{KNO_3(aq)}}$$

Balance the equation.

$$\underset{\text{lead nitrate}}{Pb(NO_3)_2(aq)} + \underset{\text{potassium iodide}}{\mathbf{2}KI(aq)} \longrightarrow \underset{\text{lead iodide}}{PbI_2(s)} + \underset{\text{potassium nitrate}}{\mathbf{2}KNO_3(aq)}$$

2 Solubility

Ionic substances vary significantly as to how much they will dissolve in water. For some ionic substances, large quantities dissolve in water. These substances have a high solubility. For other ionic substances, only very small quantities dissolve. These substances have a low solubility. Substances with a low solubility tend to form precipitates in aqueous solutions.

BIG IDEA Some ionic solids are more soluble than others. When a compound reaches the limits of its solubility, undissolved solid is visible.

You can use a solubility table such as the one shown here to determine the solubility of various ionic compounds. To use the table, combine a cation from the rows on the left with an anion from the columns on the right to determine if the compound formed is very soluble (S), insoluble, or not very soluble (N).

Solubility Trends

		Anions						
		NO_3^-	Cl^-	OH^-	SO_4^{2-}	CO_3^{2-}	$C_2O_4^{2-}$	PO_4^{3-}
Cations	Most alkali metals, such as Li^+, Na^+, K^+, NH_4^+	S	S	S	S	S	S	S
	Most alkaline earth metals, such as Mg^{2+}, Ca^{2+}, Sr^{2+}	S	S	N	S	N	N	N
	Some Period 4 transition metals, such as Fe^{3+}, Co^{3+}, Ni^{2+}, Cu^{2+}, Zn^{2+}	S	S	N	S	N	N	N
	Other transition metals, such as Ag^+, Pb^{2+}, Hg^{2+}	S	N	N	N	N	N	N

EARTH SCIENCE CONNECTION

Stalactites and stalagmites are caused by precipitation of solids from water that drips from cave walls. The word *stalagmite* comes from the Greek word for "drip." These beautiful cave structures are often formed from calcium carbonate.

Recall that in Unit 1: Alchemy, you characterized ionic solids as soluble in water. This is true for many ionic solids. But as you can see from the solubility table, some ionic solids are not very soluble.

Example 2

Predicting Solid Products

Suppose that you combine aqueous calcium nitrate, $Ca(NO_3)_2$, with aqueous sodium phosphate, Na_3PO_4, and there is a double exchange reaction. Do you expect a precipitate to form?

Solution

First, write a balanced chemical equation for this reaction. You know that the two aqueous cations exchange anions:

$$3Ca(NO_3)_2(aq) + 2Na_3PO_4(aq) \longrightarrow Ca_3(PO_4)_2(?) + 6NaNO_3(?)$$

Next, use the solubility table to determine if there is a precipitate. According to the table, $NaNO_3$ is soluble in water, so it remains dissolved. $Ca_3(PO_4)_2$ is insoluble, so it forms a precipitate.

$$3Ca(NO_3)_2(aq) + 2Na_3PO_4(aq) \longrightarrow Ca_3(PO_4)_2(\mathbf{s}) + 6NaNO_3(\mathbf{aq})$$

Reactions in aqueous solutions usually involve ions in solution. To describe what is happening, you can write a **complete ionic equation,** which shows all of the dissolved ions involved in the reaction. Examine the complete ionic equation for the reaction in Example 2.

$$3Ca^{2+}(aq) + 6NO_3^{-}(aq) + 6Na^{+} + 2PO_4^{3-}(aq) \longrightarrow Ca_3(PO_4)_2(s) + 6Na^{+} + 6NO_3^{-}$$

An ion that appears on both sides of a complete ionic equation and does not directly participate in the reaction is called a **spectator ion.** For example, Na^+ and NO_3^- are spectator ions in this reaction. To write a more efficient equation for this reaction, you can cancel the spectator ions on each side of the equation:

$$3Ca^{2+}(aq) + \cancel{6NO_3^{-}}(aq) + \cancel{6Na^{+}} + 2PO_4^{3-}(aq) \longrightarrow Ca_3(PO_4)_2(s) + \cancel{6Na^{+}} + \cancel{6NO_3^{-}}$$

The result is a **net ionic equation** that describes the reaction in terms of only the ions that are involved in the reaction:

$$3Ca^{2+}(aq) + 2PO_4^{3-}(aq) \longrightarrow Ca_3(PO_4)_2(s)$$

CAREER CONNECTION

Doctors use a type of treatment called *chelation therapy* to treat patients who have had long-term exposure to metals like arsenic, mercury, or lead, usually in the workplace. In chelation therapy a compound is introduced into the bloodstream of the patient. This compound bonds with heavy metals, forming water-soluble products that are then passed out of the body.

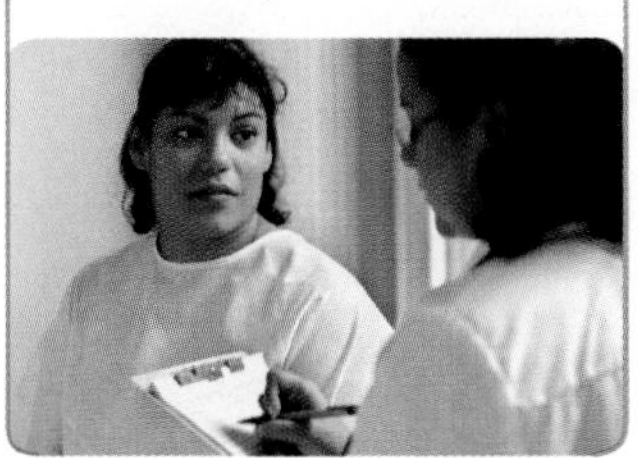

3 Toxicity of Precipitates

There are positive and negative aspects to the solubility of toxic substances. On the one hand, if a substance is not very soluble, it might not react with anything, and it might pass through the body relatively unnoticed. For example, some metals that can be ingested go through our bodies without causing harm. (Imagine a child swallowing a nickel.) However, many metal compounds are soluble, and once our bodies absorb them, they can do long-term damage.

Key Terms

precipitate
precipitation reaction
complete ionic equation
spectator ion
net ionic equation

Lesson Summary

Which substances precipitate from aqueous solutions?

A precipitate is a solid that forms when a reaction in an aqueous solution forms a compound that is not very soluble. Most precipitation reactions are double exchange reactions involving ionic compounds. The cations exchange anions in solution, and one of the substances precipitates as a solid. Precipitation depends on the solubility of a compound. Some ionic compounds are not very soluble, so they precipitate from aqueous solutions. Precipitates in your body include bones, teeth, and kidney stones.

EXERCISES

Reading Questions

1. Describe what a precipitate is and how it forms.

2. What does solubility have to do with precipitation from solution?

3. Explain what a spectator ion is.

Reason and Apply

4. Circle the ionic solids that are soluble in water.

$LiNO_3$ KCl $MgCl_2$ $Ca(OH)_2$ $RbOH$

$CaCO_3$ Li_2CO_3 $PbCl_2$ $AgCl$

5. Name solutions you can combine that will precipitate each of the compounds listed.

a. $CaCO_3$ **b.** $FePO_4$ **c.** $HgCl_2$ **d.** $AgCl$

6. Write balanced chemical equations for these reactions. Be sure to indicate which substances are aqueous and which are solid.

a. $NaCl(aq) + Hg(NO_3)_2(aq)$ **b.** $NaOH(aq) + CaCl_2(aq)$

c. $K_2CO_3(aq) + LiNO_3(aq)$ **d.** $Zn(NO_3)_2(aq) + Na_3PO_4(aq)$

e. $NaCl(aq) + Ca(NO_3)_2(aq)$

7. Refer to your answers for Exercise 6. For each equation that represents a precipitation reaction, write both complete and net ionic equations.

LAB Mole Ratios

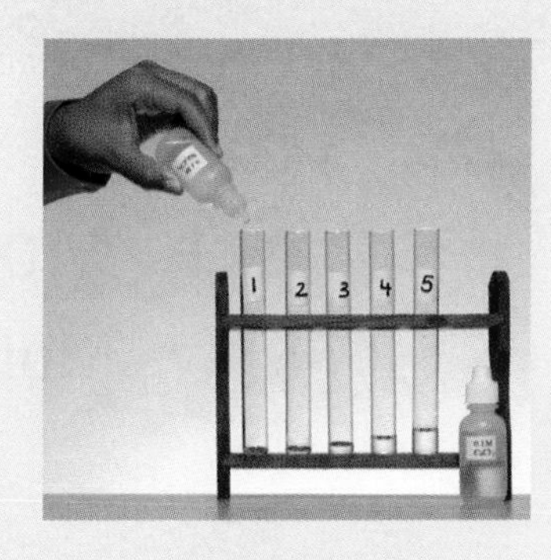

Safety Instructions

 Safety goggles must be worn at all times.

Purpose

To examine how the ratio of reactants affects the amount of products.

Materials

- 12 × 75 mm test tubes (10)
- test tube rack
- marker (to label test tubes)
- 2 dropper bottles with 0.10 M solutions of calcium chloride, $CaCl_2$, and sodium carbonate, Na_2CO_3

Procedure

1. Number five small test tubes from 1 to 5.
2. Add drops of 0.10 M $CaCl_2(aq)$ to the test tubes as indicated in the table below.
3. Swirl each test tube gently in order to mix the reactants. Allow the solids to settle for about ten minutes while you complete the table and continue the analysis.

	Tube 1	Tube 2	Tube 3	Tube 4	Tube 5
Drops of 0.10 M $CaCl_2$	4	6	12	18	20
Drops of 0.10 M Na_2CO_3	20	18	12	6	4
Ratio of $CaCl_2$ to Na_2CO_3	1:5				

Observations and Analysis

1. Write the balanced chemical equation for the exchange reaction of aqueous calcium chloride with aqueous sodium carbonate. Predict the ratio of moles of $CaCl_2(aq)$ to $Na_2CO_3(aq)$ that will produce the maximum amount of $CaCO_3(s)$.
2. After the solids have settled determine which tube has the most $CaCO_3(s)$. Identify the ratio in the table that resulted in the largest volume of solid.
3. **Making Sense** How do the balanced chemical equations compare to your observations?
4. **If You Finish Early** Suppose you mix calcium chloride, $CaCl_2(aq)$, with sodium oleate, $NaC_{18}H_{33}O_2(aq)$. What products do you expect? What ratio of drops would give you the largest amount of precipitate? (*Note:* One product is calcium oleate, commonly referred to as soap scum.)

LESSON

24 Mole to Mole

Mole Ratios

Think About It

Water treatment plants in our communities process millions of gallons of water a day. These plants remove toxic substances from the water through a series of chemical and physical treatments. For example, impure water often contains dissolved metal cations. These can be removed from water through precipitation of insoluble metal salts. Once the metals are in solid form, they can be safely separated by filtration. Of course, for maximum effectiveness the treatment plant should remove as much of each metal from the water as possible.

How can you convert all the reactants to products?

To answer this question, you will explore

1. Moles of Product
2. Excess and Limiting Reactant

Exploring the Topic

1 Moles of Product

The list of metals that can show up in our water supplies is quite long, including lead, silver, mercury, chromium, copper, zinc, cadmium, and tin. According to the solubility table in the previous lesson, many metals form hydroxides and carbonates that are not soluble. So a good way to remove the more toxic metal cations from our water supplies is by addition of hydroxides or alkali metal carbonates to precipitate hazardous metal cations.

Imagine a water sample that contains unwanted copper ions. The goal is to figure out how to remove as many of those ions per liter of solution as possible through precipitation. Sodium hydroxide, NaOH, is chosen as a reactant because it is soluble and contains a hydroxide ion. When hydroxide ions enter the solution, solid copper (II) hydroxide, $Cu(OH)_2$, will precipitate. The question remains: How much NaOH should be added, and in what concentration?

The next illustration shows what happens when a solution of 1.0 M $CuSO_4$ is mixed with a solution of 1.0 M NaOH in different ratios. Take a moment to examine the outcomes.

INDUSTRY CONNECTION

Once a hydroxide ion has been used to remove the metal ions from a water sample, an acid needs to be added to neutralize the remaining hydroxide ions.

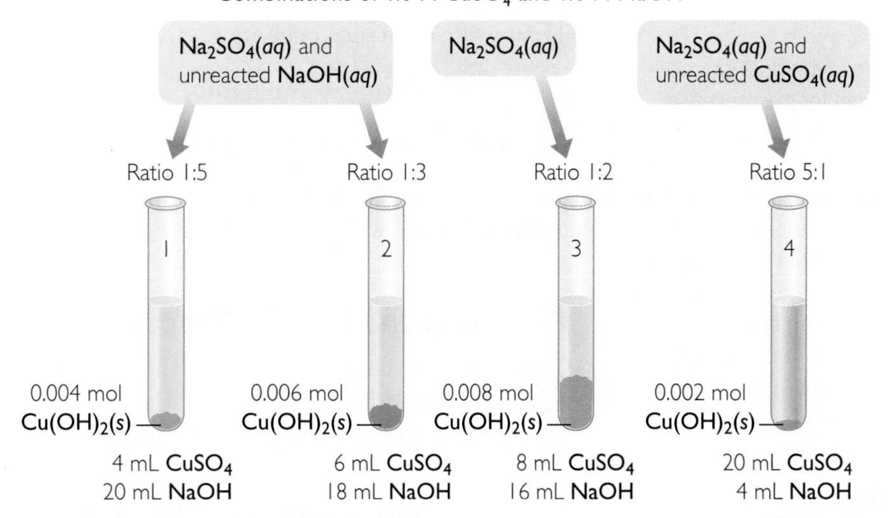

In each case, a total of 24 mL of solution were combined. However, the amounts of 1.0 M $CuSO_4$ and 1.0 M NaOH that are combined vary from test tube to test tube. The combination that produces the maximum amount of solid precipitate is in test tube 3, where 8 mL of 1.0 M $CuSO_4$ combined with 16 mL of 1.0 M NaOH. This results in 0.008 mole of solid copper hydroxide.

After the reaction, all four test tubes contain aqueous sodium sulfate, $Na_2SO_4(aq)$, and a blue solid, copper (II) hydroxide, $Cu(OH)(s)$. Therefore, sodium sulfate and copper (II) hydroxide must be products of this reaction. The solutions in three of the test tubes are almost colorless, while the solution in test tube 4 is pale blue. This means there is unreacted $CuSO_4(aq)$ left over in the last test tube.

Chemical equation: $CuSO_4(aq) + 2NaOH(aq) \longrightarrow Cu(OH)_2(s) + Na_2SO_4(aq)$

Mole Ratio

The coefficients in a chemical equation indicate the proportions in which substances react, in moles or other counting units. The coefficients in this particular equation indicate that for every 1 mol of $CuSO_4$, 2 mol of NaOH are required in order to make the maximum amount of products. The test tube containing the most amount of solid (test tube 3) corresponds to this 1:2 ratio of reactants. This ratio is often referred to as the **mole ratio.** So, using the ratio given by the balanced chemical equation produces the maximum amount of product.

BIG IDEA In order to get the maximum amount of product from a reaction, reactants must be mixed in the correct proportions.

Example 1

Precipitation of Silver Hydroxide

An aqueous solution contains 0.020 mol of silver nitrate, $AgNO_3(aq)$. Predict the number of moles of $NaOH(aq)$ you will need to precipitate all of the silver, Ag, as silver hydroxide, AgOH.

BIOLOGY CONNECTION

One of the best ways to test for the presence of heavy metals in a person's body is through hair samples. Human hair is a permanent record of some of the substances that have passed through someone's body. A three-inch strand of hair will give a six-month history of what's going on chemically in the body.

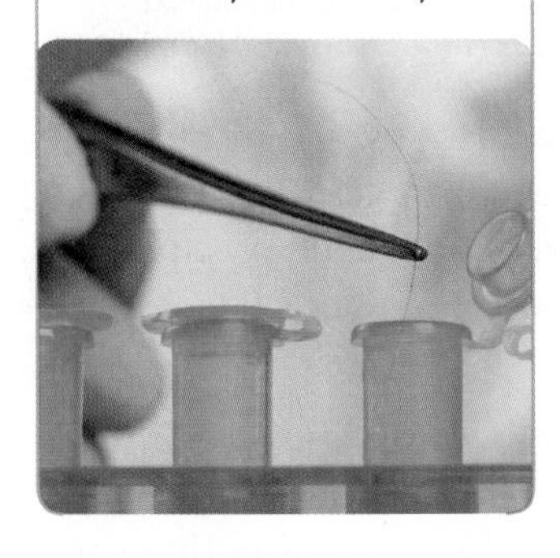

Solution

First, write a balanced chemical equation for the reaction.

$$AgNO_3(aq) + NaOH(aq) \longrightarrow AgOH(s) + NaNO_3(aq)$$

The mole ratio of $AgNO_3$ to NaOH is 1:1.

The ratio given by the balanced equation is 1:1, and you have 0.020 mol of $AgNO_3$. So you will need 0.020 mol of NaOH to precipitate all of the silver.

2 Excess and Limiting Reactant

Look again at the four test tubes. Using the molarity and volume data for each reactant, you can determine the number of moles of each reactant. The table shows the moles of each reactant and the ratios in which they are combined.

Test tube	Moles of $CuSO_4$	Moles of NaOH	Ratio	Left over	Runs out
1	0.004 mol	0.020 mol	1:5	NaOH	$CuSO_4$
2	0.006 mol	0.018 mol	1:3	NaOH	$CuSO_4$
3	0.008 mol	0.016 mol	**1:2**	neither	neither
4	0.020 mol	0.004 mol	5:1	$CuSO_4$	NaOH

Notice that in test tubes 1, 2, and 4, one of the reactants runs out before the other. Only in test tube 3 have both reactants been used up completely. Test tubes 1 and 2 no longer have copper ions in solution, but the solution is now toxic due to the excess NaOH. Test tube 4 still has copper ions in solution.

If you mix the reactants in a ratio other than the mole ratio, a reaction will still occur, but one of the reactants will run out and the other one will have some left over. The reactant that runs out is called the **limiting reactant,** or **limiting reagent.** This is because it limits how much product you can make.

To remove all of the copper ions from the solution, it is best to mix the reactants in the mole ratio determined by the balanced chemical equation.

If you want to remove copper ions (or any other metal ions) from water in a water treatment plant, it would be best to add sodium hydroxide in the mole ratio specified by the chemical equation. In order to do this, you will first have to determine the concentration of the copper ions in your water in moles per liter so you can match that concentration with the appropriate concentration of reactant.

Example 2

Precipitation of Calcium Phosphate

Aqueous calcium chloride, $CaCl_2(aq)$, reacts with aqueous sodium phosphate, $Na_3PO_4(aq)$, forming a precipitate of calcium phosphate, $Ca_3(PO_4)_2(s)$.

a. If 12 mol of $CaCl_2(aq)$ react with 8 mol of $Na_3PO_4(aq)$, will there be any reactant left over? If so, which reactant?

BIOLOGY CONNECTION

Calcium phosphate is similar to the solid that forms bones and teeth. The calcium cations and phosphate anions dissolved in your blood, deposit as solids to form your skeleton.

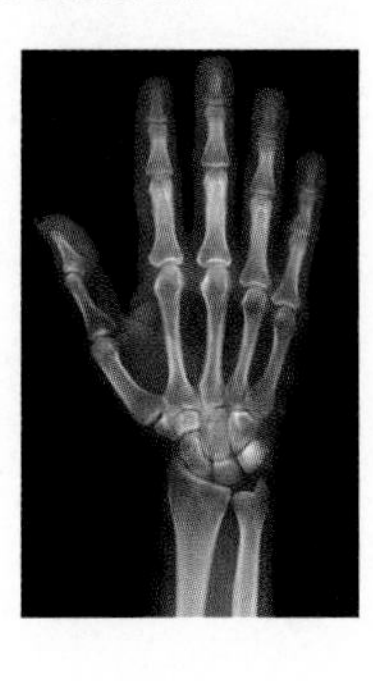

b. If 12 mol of $CaCl_2(aq)$ react with 16 mol of $Na_3PO_4(aq)$, will there be any reactant left over? If so, which reactant?

c. If 12 mol of $CaCl_2(aq)$ react with 4 mol of $Na_3PO_4(aq)$, will there be any reactant left over? If so, which reactant?

Solution

First, write a balanced chemical equation for the reaction. Because atoms are conserved, $NaCl(aq)$ must be the other product.

$$3CaCl_2(aq) + 2Na_3PO_4(aq) \longrightarrow Ca_3(PO_4)_2(s) + 6NaCl(aq)$$

The mole ratio for $CaCl_2$ and Na_3PO_4 is 3:2. For any other ratio, there will be left over reactant.

a. 12 mol $CaCl_2$ to 8 mol Na_3PO_4 is a ratio of 3:2. So all the reactants will be converted to products.

b. 12 mol $CaCl_2$ to 16 mol Na_3PO_4 is a ratio of 3:4. There will be some Na_3PO_4 left over.

c. 12 mol $CaCl_2$ to 4 mol Na_3PO_4 is a ratio of 4:1. There will be some $CaCl_2$ left over.

Key Terms

mole ratio
limiting reactant (limiting reagent)

Lesson Summary

How can you convert all the reactants to products?

Coefficients in chemical equations represent the proportions in which moles of reactants combine and products form. The ratio of two coefficients is also called the mole ratio. In order to produce the maximum amount of product, moles of reactants should be combined in the exact proportions specified by the coefficients. Mass and volume amounts cannot be substituted for coefficients. If there is not enough of a reactant, resulting in other reactants being left over, that reactant is called a limiting reactant.

EXERCISES

Reading Questions

1. Why are coefficients important in chemical equations?
2. Explain how to create the maximum amount of product from a reaction.

Reason and Apply

3. For each of the reactions listed, determine the mole ratio of reactants that produces the maximum amount of precipitate. Be sure to balance the equations.

   a. $AgNO_3(aq) + NaCl(aq) \longrightarrow AgCl(s) + NaNO_3(aq)$

   b. $Cu(NO_3)_2(aq) + K_2CO_3(aq) \longrightarrow CuCO_3(s) + KNO_3(aq)$

   c. $ZnCl_2(aq) + NaOH(aq) \longrightarrow Zn(OH)_2(s) + NaCl(aq)$

   d. $CaCl_2(aq) + Na_2C_2O_4(aq) \longrightarrow CaC_2O_4(s) + NaCl(aq)$

4. **Lab Report** Write a lab report for the experiment you did in which you found the mole ratio by precipitating solids.

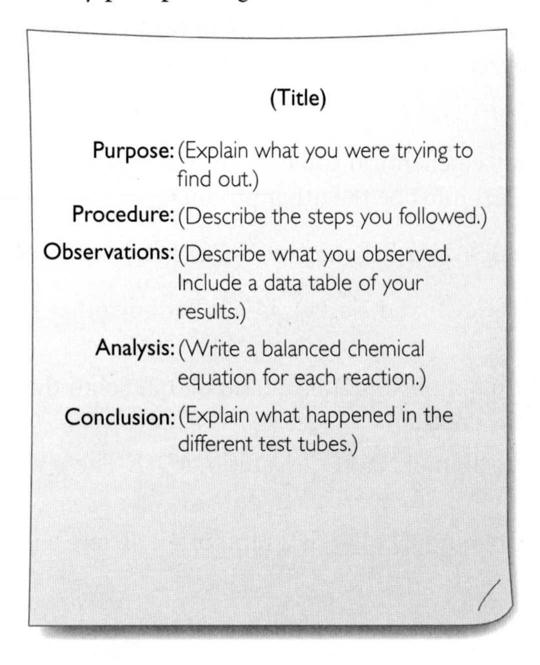

5. Aqueous silver nitrate reacts with aqueous sodium chloride producing a precipitate of silver chloride.

$$AgNO_3(aq) + NaCl(aq) \longrightarrow AgCl(s) + NaNO_3(aq)$$

   a. How many moles of NaCl do you need to react with 0.10 mol of $AgNO_3$?

   b. How many grams of NaCl does this represent?

6. Imagine you have 500 mL of 0.20 M silver nitrate solution. How many milliliters of 0.50 M sodium chloride solution do you need to add to remove all the silver ions from the solution?

LESSON

25 Mole Tunnel

Stoichiometry

Think About It

Sales of medications bring in billions of dollars per year. Most of these medications are produced through chemical reactions that are performed in a laboratory. In manufacturing the medications, it is important to know exactly how much reactant you need to prepare a specified mass of product. The balanced chemical equation for the reaction serves as a guide. However, the calculation requires conversion from mass to moles and back to mass.

How do you convert between grams and moles to determine the mass of product?

To answer this question, you will explore

1. Gram-Mole Conversions
2. Solving Stoichiometry Problems

Exploring the Topic

1 Gram-Mole Conversions

One type of medication available is an antacid made of aluminum hydroxide, $Al(OH)_3$. It neutralizes excess acid in the stomach. In order to make aluminum hydroxide, many manufacturers start with aluminum chloride, which is commercially available in large quantities. When solutions of aluminum chloride are mixed with sodium hydroxide, a precipitate of aluminum hydroxide is formed. The balanced chemical equation for the reaction is

$$AlCl_3(aq) + 3NaOH(aq) \longrightarrow Al(OH)_3(s) + 3NaCl(aq)$$

The coefficients in the equation indicate how many moles to mix together. But in the laboratory, reactants are measured in grams, kilograms, or pounds.

Going Through the Mole Tunnel

A chemical equation is like a recipe for making a product. It indicates how much of the reactants to mix. If you want to make aluminum hydroxide, $Al(OH)_3$, you need one formula unit of aluminum chloride, $AlCl_3$, for every three formula units of sodium hydroxide, NaOH. Before the recipe can be put into practice, it must be converted to mass. This is because in the laboratory substances are weighed, not counted.

$$1\text{ mol } AlCl_3(aq) + 3\text{ mol } NaOH(aq) \longrightarrow 1\text{ mol } Al(OH)_3(s) + 3\text{ mol } NaCl(aq)$$

↓ ↓ ↓ ↓

$$\underline{\ ?\ }\text{ g } AlCl_3(aq) + \underline{\ ?\ }\text{ g } NaOH(aq) \longrightarrow \underline{\ ?\ }\text{ g } Al(OH)_3(s) + \underline{\ ?\ }\text{ g } NaCl(aq)$$

If you know the mass of reactants, you can determine the mass of products that can be made.

This requires a trip through the "mole tunnel." Starting with grams of reactant and the balanced chemical equation, there are three steps you must take to convert from grams of reactants to grams of products.

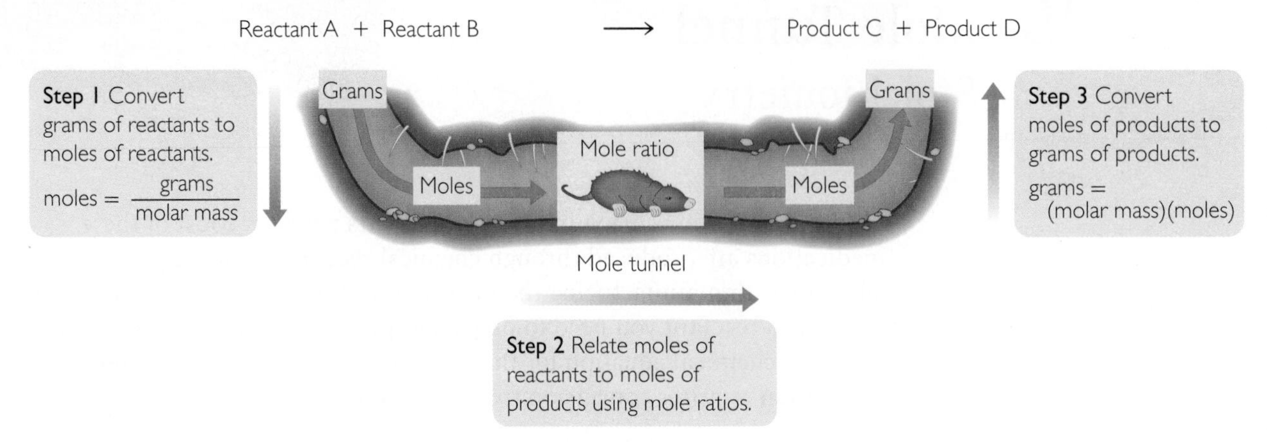

To go from one side of the equation to the other, you must go through moles, represented by the tunnel under the equation. The "mole tunnel" in the illustration is one way to remember these steps. These types of calculations, involving quantities of reactants and products, are referred to as gram-mole conversions, or **stoichiometry** calculations. The word *stoichiometry* comes from the Greek words *stoikheion,* meaning "element," and *metrein,* meaning "to measure."

The mole tunnel also works in reverse. If you want a certain mass of a product, you can determine the mass of reactants required.

2 Solving Stoichiometry Problems

The chemical equation for the formation of aluminum hydroxide, $Al(OH)_3$, shows the reactants combining in a mole ratio of 1:3. The chemical equation also indicates that the mole ratio of the $AlCl_3$ reactant to the $Al(OH)_3$ product is 1:1. So, for every mole of $AlCl_3$ that reacts, one mole of $Al(OH)_3$ is produced. If you want to make 100 moles of $Al(OH)_3$, you will need 100 moles of $AlCl_3$ and 300 moles of NaOH.

The molar masses on the periodic table allow you to convert between grams and moles.

Molar Masses

Reactants: $AlCl_3$ = 133.5 g/mol NaOH = 40.0 g/mol

Products: $Al(OH)_3$ = 78.0 g/mol NaCl = 58.5 g/mol

Next the mole amounts shown in the chemical equation can be converted to gram amounts. The table summarizes these calculations.

	Reactants		Products	
	$AlCl_3$	NaOH	$Al(OH)_3$	NaCl
Moles	1 mol	3 mol	1 mol	3 mol
Grams	133.5 g	120.0 g	78.0 g	175.5 g
Total mass	253.5 g		253.5 g	

CONSUMER CONNECTION

Aluminum chloride, $AlCl_3$, and other aluminum compounds are used in antiperspirants to control underarm wetness. When the aluminum ions enter the skin cells, they cause more water to pass into the cells. The cells swell and squeeze the sweat ducts closed, trapping the sweat inside the sweat glands.

Example 1

From Moles to Grams

Suppose you have 5.00 mol of $AlCl_3$ in solution and an unlimited supply of NaOH. How many grams of $Al(OH)_3$ and NaCl can you make?

$$AlCl_3(aq) + 3NaOH(aq) \longrightarrow Al(OH)_3(s) + 3NaCl(aq)$$

Solution

Use mole ratios to determine the moles of product you can make.

$$AlCl_3(aq) + 3NaOH(aq) \longrightarrow Al(OH)_3(s) + 3NaCl(aq)$$

1 : 1

1 : 3

So, with 5.00 mol of $AlCl_3$, you can make 5.00 mol of $Al(OH)_3$ and 15.0 mol of NaCl. Use molar mass to convert these quantities to grams.

Multiply the number of moles by the molar mass. $5.0 \text{ mol } Al(OH)_3 \cdot 75.0 \text{ g/mol} = 375 \text{ g}$

$15.0 \text{ mol NaCl} \cdot 58.5 \text{ g/mol} = 878 \text{ g}$

If you start with 5.0 mol $AlCl_3$, you can react it with 15.0 mol NaOH to prepare 375 g $Al(OH)_3$ and 878 g NaCl.

Example 2

From Grams to Grams

Suppose you have 500 g of $AlCl_3$ in solution and an unlimited supply of NaOH. How many grams of each product can you make?

Solution

You cannot convert directly from grams of reactant to grams of product. You must use the "mole tunnel."

Use molar mass to convert from grams to moles of reactants.

Divide by the molar mass. $\dfrac{500 \text{ g } AlCl_3}{133.5 \text{ g/mol}} = 3.75 \text{ mol } AlCl_3$

Use mole ratios to determine the moles of each product.

So 3.75 mol $AlCl_3$ will make 3.75 mol $Al(OH)_3$ and 11.25 mol NaCl.

Use molar mass to convert between moles and grams.

Multiply the number of moles by the molar mass. $3.75 \text{ mol } Al(OH)_3 \cdot 75.0 \text{ g/mol} = 281 \text{ g } Al(OH)_3$

$11.25 \text{ mol NaCl} \cdot 58.5 \text{ g/mol} = 658 \text{ g NaCl}$

So, 500 g $AlCl_3$ produces 281 g $Al(OH)_3$ and 658 g NaCl.

Key Term

stoichiometry

Lesson Summary

How do you convert between grams and moles to determine the mass of product?

In order to determine the mass of product produced by a certain mass of reactant (and vice versa), it is necessary to convert mass to moles and then back to mass. This is because the balanced chemical equation is written in moles. Typically, when you're working with chemical reactions, you know the grams of reactant you need to produce a certain number of grams of product. Calculations involving mole ratios and masses of reactants and products, are referred to as gram-mole conversions, or stoichiometry calculations.

EXERCISES

Reading Questions

1. Explain why you need to do gram-mole conversions when carrying out chemical reactions.

Reason and Apply

2. Consider this balanced chemical equation.

$$\underset{\text{copper}}{Cu(s)} + \underset{\text{silver nitrate}}{2AgNO_3(aq)} \longrightarrow \underset{\text{copper (II) nitrate}}{Cu(NO_3)_2(aq)} + \underset{\text{silver}}{2Ag(s)}$$

   a. Find the mole ratio of $AgNO_3$ to $Cu(NO_3)_2$.
   b. Find the mole ratio of $AgNO_3$ to Ag.
   c. Suppose you have 6.0 mol of Cu. How many moles of $AgNO_3$ are needed to react completely with 6.0 mol of Cu? How many mol of $Cu(NO_3)_2$ are produced?
   d. Suppose you want to make 30.0 g Ag. How many moles of Ag is that? How many moles of Cu do you need?

3. Consider this balanced chemical equation.

$$\underset{\text{sodium chloride}}{2NaCl(aq)} + \underset{\text{lead (II) nitrate}}{Pb(NO_3)_2(aq)} \longrightarrow \underset{\text{sodium nitrate}}{2NaNO_3(aq)} + \underset{\text{lead (II) chloride}}{PbCl_2(s)}$$

   a. Suppose you want to make 30.0 g $PbCl_2$. How many grams of NaCl do you need?
   b. Suppose you have 56 g $Pb(NO_3)_2$. How many grams of NaCl are needed to react completely with 56 g $Pb(NO_3)_2$? How many grams of $PbCl_2$ are produced?

4. An aqueous solution of sodium carbonate reacts with an aqueous solution of calcium sulfate to produce solid calcium carbonate and an aqueous solution of sodium sulfate.

$$\underset{\text{sodium carbonate}}{2Na_2CO_3(aq)} + \underset{\text{calcium sulfate}}{CaSO_4(aq)} \longrightarrow \underset{\text{sodium sulfate}}{Na_2SO_4(aq)} + \underset{\text{calcium carbonate}}{CaCO_3(s)}$$

   a. What type of reaction is this?
   b. How many grams of each reactant are needed to produce 4500 g of $CaCO_3$?

HISTORY CONNECTION

Early black-and-white photographs, called *calotypes,* were made on paper brushed with silver nitrate. The paper was loaded into a camera under near-total darkness, then exposed to a scene for 10–60 seconds to make a photo negative. Prints could then be produced from the negative. An early photograph of President Abraham Lincoln is shown.

LESSON

26 Get the Lead Out

Limiting Reactant and Percent Yield

Think About It

Imagine you want to make as many cheese sandwiches as possible from a loaf of bread and a large package of sliced cheese. The package of cheese contains 15 slices. The loaf of bread contains 24 slices. Which ingredient will you run out of first, and how many sandwiches can you make?

Which reactant determines how much product you can make?

To answer this question, you will explore

1. Limiting Reactants
2. Solving Limiting Reactant Problems
3. Percent Yield

Exploring the Topic

1 Limiting Reactants

Mixing reactants to form products is a little like making cheese sandwiches from specific quantities of cheese and bread. In the example above, you could make 12 sandwiches before running out of bread. You would have 3 slices of cheese left over. In the world of chemistry, substances rarely come together in the exact mole ratios specified by a chemical equation. One of the reactants usually runs out first. As you learned in Lesson 24: Mole to Mole, the reactant that runs out is called the limiting reactant.

Imagine you have a beaker with an aqueous solution of sulfuric acid containing a total of 4 mol of H_2SO_4. You add 112 g of solid potassium hydroxide, KOH. Will the addition of this amount of KOH(*s*) completely neutralize the $H_2SO_4(aq)$? Will either reactant be left over?

The balanced chemical equation for the reaction you are examining is

$$H_2SO_4(aq) + 2KOH(s) \longrightarrow K_2SO_4(aq) + 2H_2O(aq)$$

You must first find out how many moles are represented by 112 g of KOH. The molar mass of KOH is 56 g/mol, so 112 g KOH is equal to 2 mol of KOH.

The chemical equation shows that you will need 2 mol of KOH to neutralize 1 mol of H_2SO_4. There are 4 mol of H_2SO_4 in the aqueous solution, and you have only 2 mol of KOH. You will be able to neutralize only 1 mol of H_2SO_4. There will be 3 mol of H_2SO_4 left over.

The illustration represents what happens in your beaker. Each unit represents 1 mol.

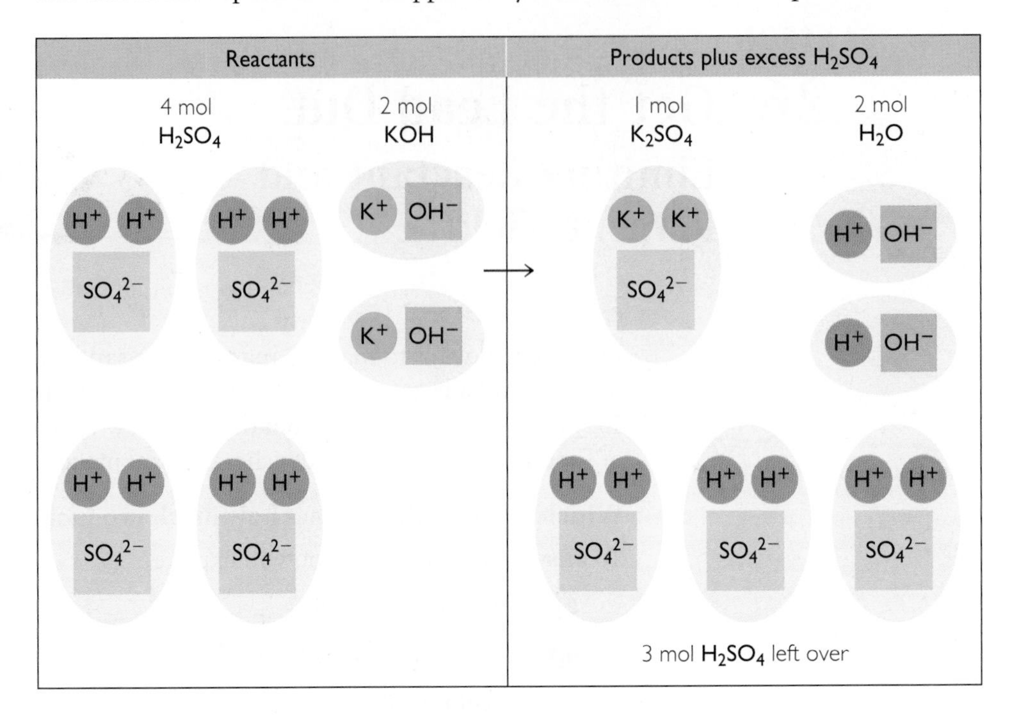

There is not enough KOH to react with all of the H_2SO_4. KOH is the limiting reactant, and at the end of the reaction there will be excess H_2SO_4.

2 Solving Limiting Reactant Problems

Knowing the identity of the limiting reactant allows you to figure out the maximum amount of product you can make with the substances you have. Mole ratios from the balanced chemical equation are the key to solving limiting reactant problems.

BIG IDEA The limiting reactant determines how much product you can make from a chemical reaction.

The steps to complete a limiting reactant problem are shown here.

Solving a Limiting Reactant Problem

Step 1: Write the balanced chemical equation.

Step 2: Determine the molar mass of each compound.

Step 3: Determine the number of moles of each reactant that you have.

Step 4: Use the mole ratio to identify the limiting reactant.

Step 5: Use the limiting reactant to determine the maximum amount of product.

CONSUMER CONNECTION

Magnesium chloride, $MgCl_2$, is used in the preparation of tofu from soy milk. When it is added to the soy milk, it causes the tofu to clot up and separate out.

Example 1

Preparing Magnesium Chloride

Adding hydrochloric acid, HCl, to magnesium hydroxide, $Mg(OH)_2$, produces magnesium chloride, $MgCl_2$, and water. Imagine you start with 10.0 g of $Mg(OH)_2$ and 100 mL of 4.0 M HCl. How much $MgCl_2$ will be produced?

Solution

Step 1: Write the balanced chemical equation.

$$Mg(OH)_2(s) + 2HCl(aq) \longrightarrow MgCl_2(aq) + 2H_2O(l)$$

Step 2: Determine the molar mass of each compound.

Reactants: $Mg(OH)_2 = 58.3$ g/mol   HCl = 36.5 g/mol

Products: $MgCl_2 = 95.3$ g/mol $H_2O = 18.0$ g/mol

Step 3: Determine the number of moles of each reactant that you have.

$$\text{Moles of } Mg(OH)_2 = \frac{10.0 \text{ g}}{58.3 \text{ g/mol}} = 0.17 \text{ mol}$$

$$\text{Moles of HCl} = (0.100 \text{ L})(4.0 \text{ mol/L}) = 0.40 \text{ mol}$$

Step 4: Use the mole ratio to identify the limiting reactant.

The reactants combine in a 1:2 ratio. So you need 0.17 mol · 2 = 0.34 mol of hydrochloric acid, HCl, to react with the 0.17 mol of magnesium hydroxide, $Mg(OH)_2$.

You have 0.40 mol of HCl, so the limiting reactant is $Mg(OH)_2$. When the reaction is complete, there will be 0.06 mol of HCl left over.

Step 5: Use the limiting reactant to determine the maximum amount of product.

For each mole of $Mg(OH)_2$, 1 mol of $MgCl_2$ is produced. So, 0.17 mol of $MgCl_2$ is produced. Use the molar mass to determine the mass of $MgCl_2$.

$$(0.17 \text{ mol})\ (95.3 \text{ g/mol}) = 16.2 \text{ g } MgCl_2$$

So starting with 10.0 g $Mg(OH)_2$ and 100 mL of 4.0 M HCl, you can make 16.2 g $MgCl_2$.

Important to Know The reactant present in the smallest amount is not necessarily the limiting reactant. The mole ratio specified by the coefficients of the balanced equation must be taken into consideration. ◀

3 Percent Yield

When you use the limiting reactant to calculate the amount of a product produced in a chemical reaction, you are calculating the **theoretical yield** for that product or, what you should be able to produce in theory. For a variety of reasons, including experimental error, reactions rarely produce the predicted theoretical yield when run in the laboratory. The amount of a product produced when the reaction is actually run is called the **actual yield** for that product

INDUSTRY CONNECTION

Pharmaceutical companies can calculate percent yield for each chemical reaction involved in creating a medicine. A low percent yield can be an indication of possible problems with the manufacturing process. Correcting these problems helps the manufacturer to make a safer, better product and to make the most of it at the lowest cost of raw materials.

and is usually a bit less than the theoretical yield. When you compare the actual yield with the theoretical yield and express it as a percent, it is called the **percent yield.**

$$\text{Percent yield} = \frac{\text{actual yield}}{\text{theoretical yield}} \cdot 100\%$$

The percent yield tells you how successful or efficient your procedure was in the laboratory. The closer your percent yield is to 100%, the more efficient your procedure was.

Example 2

Percent Yield of $MgCl_2$

Suppose you ran the reaction from Example 1 and 15.4 g of $MgCl_2$ are produced. What is the percent yield of your reaction?

Solution

Calculate the percent yield for the reaction you ran. The theoretical yield was 16.2 g $MgCl_2$.

$$\text{Percent yield} = \frac{\text{actual yield}}{\text{theoretical yield}} \cdot 100\%$$

$$= \frac{15.4\text{ g}}{16.2\text{ g}} \cdot 100\%$$

$$= 95.1\%$$

Your percent yield was very close to the theoretical yield.

Key Terms

theoretical yield
actual yield
percent yield

Lesson Summary

Which reactant determines how much product you can make?

In the real world, substances rarely come together in the exact mole ratios specified by a chemical equation. Usually there is more of one reactant than is required and less of the other. If reactants are not present in the exact mole ratio, one reactant, the limiting reactant, will get used up, therefore limiting the amount of product that can be produced. There will be excess of the other reactant. The percent yield of product is the actual amount of product produced in a chemical reaction as compared with the theoretical yield expressed as a percent. The theoretical yield of a chemical reaction is determined by the limiting reagent and the molar ratios in the chemical equation.

EXERCISES

Reading Questions

1. How can you determine how much product you can make from two compounds?
2. What is percent yield?

Reason and Apply

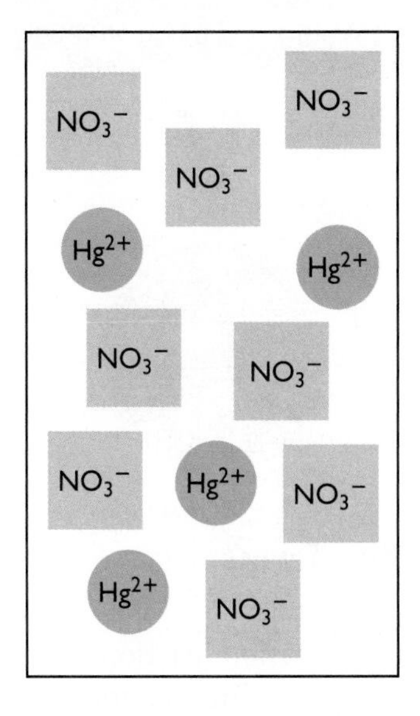

3. Consider the reaction to remove mercury from a water source through precipitation:

$$Hg(NO_3)_2(aq) + 2NaCl(aq) \longrightarrow HgCl_2(s) + 2NaNO_3(aq)$$

   a. On your paper, draw a diagram that shows the correct amount of sodium chloride needed to react with the amount of mercury (II) nitrate represented in the box on the right.

   b. What else is present in the beaker at the end of the reaction, besides the solid $HgCl_2$?

   c. Why aren't there any mercury ions left over at the end?

4. Suppose you were trying to remove 50.3 g of mercury (II) nitrate, $Hg(NO_3)_2$, from a water supply using the reaction described in Exercise 3. How many grams of sodium chloride would you need to add?

5. Silver nitrate, $AgNO_3$ reacts with sodium chloride, NaCl, in aqueous solution to form solid silver chloride, AgCl(*s*), and aqueous sodium nitrate, $NaNO_3(aq)$. Suppose you start with 6.3 g of $AgNO_3$ and 4.5 g of NaCl.

   a. Write a balanced chemical equation for this reaction.

   b. How many moles of each reactant are you starting with?

   c. What is the limiting reactant?

   d. How many grams of each product do you expect to produce?

   e. How many grams of excess reactant do you expect to have when the reaction is complete?

6. In the laboratory, you run a procedure for the reaction described in Exercise 5 and produce 2.9 g of silver chloride. What is the percent yield for your procedure?

SECTION

V

SUMMARY

Toxic Cleanup

Key Terms

precipitate
precipitation reaction
complete ionic equation
spectator ion
net ionic equation
mole ratio
limiting reactant (limiting reagent)
stoichiometry
theoretical yield
actual yield
percent yield

Toxins Update

One way to remove toxic substances from water samples is to add the appropriate ions to create a solid precipitate that can be filtered out. In other situations, the appropriate ions can be added to dissolve a solid into solution. A solubility table can help identify which ionic compounds can help remove a toxic ion from solution.

How do you know how much reactant you need? Sometimes it is necessary to translate between moles and mass. The relationship between mass and moles is a proportional one, connected by molar mass. When comparing quantities of reactants and products, it is necessary to consider the mole ratios.

Review Exercises

1. Write a complete ionic equation and a net ionic equation for three precipitation reactions that involve the formation of solid nickel compounds.
2. Calcium chloride reacts with sodium phosphate to produce calcium phosphate and sodium chloride. The unbalanced chemical equation is

$$\_\_\_\ CaCl_2(aq) + \_\_\_\ Na_3PO_4(aq) \longrightarrow \_\_\_\ Ca_3(PO_4)_2(s) + \_\_\_\ NaCl(aq)$$

   a. Balance the chemical equation for this reaction.
   b. If you start with 13.5 g of $CaCl_2$ and 9.80 g of Na_3PO_4, how many grams of calcium phosphate can you expect to produce?
   c. How many grams of excess reactant will be left over?
   d. In the laboratory, you run a procedure for this reaction, and you produce 8.20 g of calcium phosphate. What is the percent yield for your procedure?
3. Suppose you mix a slightly soluble compound with water.
   a. Draw a molecular view diagram of the resulting mixture.
   b. Would the resulting mixture be homogeneous or heterogeneous? Explain.

WEB  RESEARCH

PROJECT *Removing Toxins*

Choose a toxin (your teacher may assign you one). Find some information about your toxin. Your report should cover these topics.

- According to the Environmental Protection Agency, what is the Maximum Contaminant Level for your toxin in the drinking water supply? How is it reported?
- If your toxin dissolves in water, how would you remove it from water?
- If it does not dissolve in water, how would you remove it from soil?

UNIT 4

Toxins Review

Chemical changes make life possible on this planet. Through chemical reactions our bodies process the food we eat, make new cells and tissue, and eliminate waste products. Chemical equations allow us to track these changes on an atomic level.

However, some substances cause chemical changes that affect your body in negative ways, causing damage or harm. Acids and bases can react with your skin, causing burns and destroying tissue. Metal salts can interfere with ionic substances in the bloodstream, upsetting the natural balance. Precipitates can form in your body and block passages—examples are cholesterol in the bloodstream and kidney stones in the urinary tract.

But chemical reactions can be used to address toxic reactions. Acids and bases can be dealt with through neutralization reactions or by dilution with water. Toxic substances can often be safety removed from your body through precipitation reactions. Other times, the best approach is to dissolve solid toxic substances and then remove them from the body in liquid form. These same methods also have other applications, such as purifying the drinking water that comes to your home.

Every substance on the planet has the potential to be either toxic or therapeutic. The outcome depends on the dosage or amount and the way it is administered. Health officials test substances and set limits on exposure to all the different compounds that we might encounter in our tap water, medications, atmosphere, food, or the household products we buy.

Review Exercises

1. Balance the following chemical equations:
   a. $Zn(s) + HCl(aq) \longrightarrow ZnCl_2(aq) + H_2(g)$
   b. $KClO_3(s) \longrightarrow KCl(s) + O_2(g)$
   c. $S_8(s) + F_2(g) \longrightarrow SF_6(g)$
   d. $Fe(s) + O_2(g) \longrightarrow Fe_2O_3(s)$
2. For each of the four equations in Exercise 1, name the type of reaction it represents.
3. Using the same equations you balanced in Exercise 1, suppose you start with 1 g of each reactant. Identify the limiting reactant and determine the mass of each product that can be made.
4. Write a balanced chemical equation for these reactions:
   a. Calcium metal reacts with water to form calcium hydroxide and hydrogen gas.
   b. A solution of hydrochloric acid reacts with solid calcium bicarbonate to produce water, carbon dioxide, and calcium chloride.

5. What is a neutralization reaction?

6. Write out the complete and net ionic equations for reactions between these compounds in solution:

   a. magnesium nitrate and calcium chloride

   b. silver nitrate and potassium iodide

   c. sodium sulfate and barium chloride

7. Does a precipitate form in any of the reactions from Exercise 6? Circle them.

8. Describe the steps you would follow to make a 1.0 M solution of silver nitrate.

9. How many grams of solute are dissolved in a 500.0 mL sample of 0.50 M sodium sulfate?

10. What is the pH of a solution with $[H^+] = 2.0 \times 10^{-8}$? What is the $[OH^-]$ of the solution?

11. This reaction is completed in a factory to create potassium phosphate:

$$\_\_\_\ H_3PO_4(aq) + \_\_\_\ KOH(aq) \longrightarrow \_\_\_\ K_3PO_4(aq) + \_\_\_\ H_2O(l)$$

   a. What type of reaction is represented by the equation?

   b. Copy the equation and balance it.

   c. If you want to make 5.0 mol of potassium phosphate, how many moles of phosphoric acid, H_3PO_4, do you need?

   d. If 628.0 g of phosphoric acid are reacted with 1122.0 g of potassium hydroxide, what is the maximum amount of potassium phosphate that can be produced?

   e. Which is the limiting reactant in part d? How much (in moles and in grams) of the excess reactant is left over?

12. The reaction shown here is completed by a chemical company to create phosphoric acid, H_3PO_4.

$$\_\_\_\ K_3PO_4(aq) + \_\_\_\ H_2SO_4(aq) \longrightarrow \_\_\_\ H_3PO_4(aq) + \_\_\_\ K_2SO_3(aq)$$

   a. What type of reaction is represented by the equation?

   b. Copy the equation and balance it.

   c. How many moles of potassium phosphate, K_3PO_4, are required to produce 6 mol of phosphoric acid?

   d. How many moles of sulfuric acid, H_2SO_4, are required to produce 6 mol of phosphoric acid?

   e. If 750.0 g of potassium phosphate are reacted with 0.75 L of 18.4 M sulfuric acid, what is the maximum number of moles of phosphoric acid that can be produced?

   f. Which is the limiting reactant in part e? How much (in moles and in grams) of the excess reactant is left over?

UNIT

5 Fire

Fire is a fascinating phenomenon and a vital source of energy on our planet.

Why Fire?

Every change that happens to matter is accompanied by a change in energy. Fire is visible evidence of the energy associated with one particular type of chemical change. When a compound burns, it is broken down into smaller, less complex substances, and heat and light are released. When fire is uncontrolled, it can be destructive. However, this same chemical reaction can also provide heat, light, and mechanical or electrical energy. This unit explores how the energy from chemical and physical change can be observed, measured, understood, and controlled.

In this unit, you will learn

- the nature of heat, energy, and fire
- how to keep track of and measure changes in energy
- the source of energy in chemical changes
- how chemical energy is transformed into work
- about energy exchanges during reactions with metals and ionic compounds

SECTION

I

Observing Energy

The snow on the branch is changing from the solid to the liquid phase. Energy is involved in phase changes.

All around you is evidence of energy. Any time matter moves or changes, energy is involved. Some expressions of energy are familiar, like the heat from a campfire or the radiant energy from the Sun. But energy can be a tricky concept to define. This section of Unit 5: Fire introduces you to energy through observations of its effects on matter.

In this section, you will study

- energy exchange in chemical reactions
- heat, temperature, and thermal energy
- the effect of heat on different substances
- the energy of phase changes

LESSON

1 Fired Up!

Energy Changes

Think About It

Fire is central to our lives. The most familiar fires are wood, oil, and gas fires. When these substances are ignited, brilliant flames leap into the air and there is intense heat. The energy from fire is used to warm homes, to cook food, and to move cars. For millennia, fire has been a vital energy source for humankind.

ENVIRONMENTAL CONNECTION

Lightning strikes sometimes cause wildfires. It's probable that humans first discovered fire through lightning strikes.

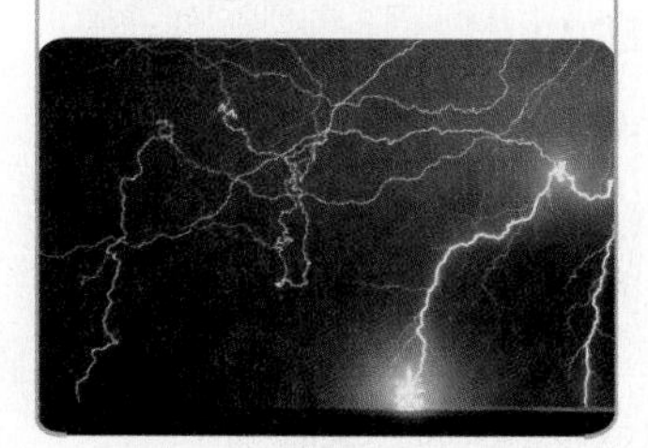

What reactions are sources of heat?

To answer this question, you will explore

1. Energy
2. Exothermic Processes

Exploring the Topic

1 Energy

The universe consists of matter and energy. As discussed in Unit 1: Alchemy, matter is made up of atoms of different elements. So, what is energy?

Because energy is not matter, it is not a substance. Energy does not have mass and it does not take up space. Energy is difficult to describe and define. Take a moment to examine how the word *energy* is used in the following sentences. What do these situations have in common?

- A plant needs energy from the Sun to grow.
- An athlete eats a snack bar for energy to continue running.
- A campfire provides energy to heat water.
- Energy from a waterfall makes the waterwheel rotate.
- Electrical energy causes the filament in the light bulb to glow.
- Pressure from steam provides the energy to move a locomotive.

You may notice that all of the statements are about energy causing something to happen to matter. Matter moves, falls, glows, melts, breaks apart, or burns. In each case, matter changes in some way and energy causes the change to happen. So, one definition of energy is that it is about change. **Energy** is a measure of the ability to cause change to occur. Even though you cannot see energy or hold it in your hand, you can measure amounts of energy transferred to a substance or from a substance.

② Exothermic Processes

The chemical reactions listed here all release energy. This is evident in the flames, the sparks, the light you see, the sounds you hear, and the heat you feel when you observe these reactions. A reaction that creates products that are hotter than the reactants is called **exothermic.** Take a moment to examine these exothermic reactions.

Heat, Light, Flames

The first two equations involve molecular covalent compounds. They react with oxygen, O_2. The products include carbon dioxide and water. Heat, light, and flames accompany these reactions.

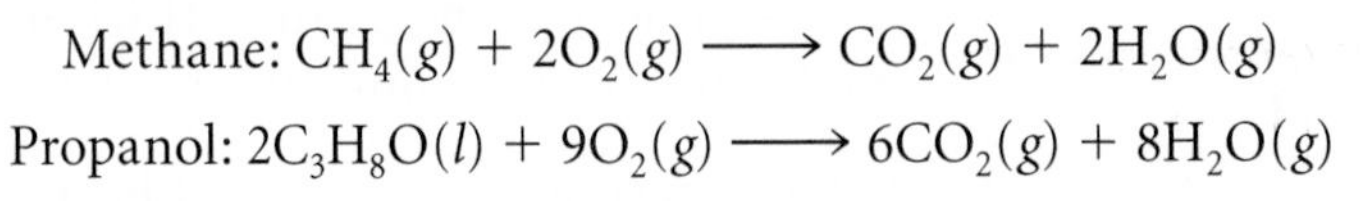

$$\text{Methane: } CH_4(g) + 2O_2(g) \longrightarrow CO_2(g) + 2H_2O(g)$$

$$\text{Propanol: } 2C_3H_8O(l) + 9O_2(g) \longrightarrow 6CO_2(g) + 8H_2O(g)$$

Heat, Light, Sparks

The second set of equations involves elemental substances. Elemental metals and many nonmetals also react with oxygen to produce heat and light. There are no flames associated with these reactions, but sometimes sparks are emitted.

$$\text{Phosphorus: } P_4(s) + 5O_2(g) \longrightarrow 2P_2O_5(s)$$

$$\text{Iron: } 4Fe(s) + 3O_2(g) \longrightarrow 2Fe_2O_3(s)$$

HISTORY CONNECTION

The Great Fire of London occurred in 1666. The fire lasted for four days, destroying 13,200 homes. Historians credit this fire with eradicating the rat population and helping to end the Great Bubonic Plague that had killed almost 20 percent of the city's populace.

In all of these exothermic reactions, the reactants change into products that are very hot. These hot products cool down over time as they transfer heat to the surroundings. Exothermic changes are associated with big changes to matter.

Fire

When you observe fire, matter is changing drastically. As a result of a fire, an entire forest can be reduced to smoke and ashes. During this change, the trees and brush are converted to carbon dioxide and water, which spread out in the atmosphere in gaseous form. Some ash and charred fuel remain after a fire. These are the result of reactants that did not react completely.

Flames are generally what distinguish fires from other exothermic reactions. Flames consist mainly of hot gases. The glowing yellow color of a wood fire is caused by small particles of carbon carried into the air by the gaseous products of this exothermic reaction.

BIG IDEA Changes in matter are accompanied by changes in energy.

Key Terms

energy
exothermic

Lesson Summary

What reactions are sources of heat?

Energy is not a substance, so it does not have mass and it does not take up space. However, any change in matter is accompanied by a change in energy. Many chemical reactions are exothermic, which means they result in products that are hotter than the reactants. Fire is an important example of an exothermic reaction that transfers energy in the form of light and heat. In a fire, reactants combine with oxygen to form carbon dioxide and water. The energy associated with fires has many uses such as heating your home, cooking your food, and moving cars.

EXERCISES

Reading Questions

1. In your own words, define the word *energy.*
2. Use your own words to write a short paragraph describing fire to someone who has never seen it before.

Reason and Apply

3. Write five sentences that use the word *energy.*
4. Describe three ways in which fire is central to life on our planet.
5. Overnight, the ashes of a fire cool down. What has happened to the products and the heat?
6. **WEB RESEARCH** Describe how fire is used in gas heaters to heat homes.
7. **WEB RESEARCH** Describe how fire is used to generate electricity in a coal-burning power plant.

LITERATURE CONNECTION

Many myths and legends have attempted to explain the origins of fire. According to Greek mythology, Prometheus stole fire from the gods on Mount Olympus and brought it to humankind. He was punished for it by Zeus.

LESSON

2 Not So Hot

Exothermic and Endothermic

Think About It

You just received a bad bruise on your leg playing sports. The bruise is swelling and is quite painful. Your coach pulls out a disposable cold pack from the first-aid kit. A quick twist of the package activates it, and the temperature of the cold pack suddenly decreases. It feels nice and cold on your injury. What is the source of this cold sensation?

In what direction is heat transferred during a chemical process?

To answer this question, you will explore

1. Heat
2. Exothermic and Endothermic Processes
3. Kinetic Energy

Exploring the Topic

1 Heat

Therapeutic hot and cold packs are fairly easy to make. In one type of hot pack, solid calcium chloride, $CaCl_2$, is separated from water in two different pouches. When the pack is twisted, the pouches break open and the solid $CaCl_2$ dissolves in the water, releasing heat. The temperature of the solution increases and the bag feels hot. In contrast, one type of cold pack has solid ammonium nitrate, NH_4NO_3, separated from water in two different pouches. When you twist the pouch and the NH_4NO_3 dissolves in the water, the temperature of the solution decreases and the bag feels cold. If you perform these two reactions in beakers, you can measure the temperature changes with a thermometer.

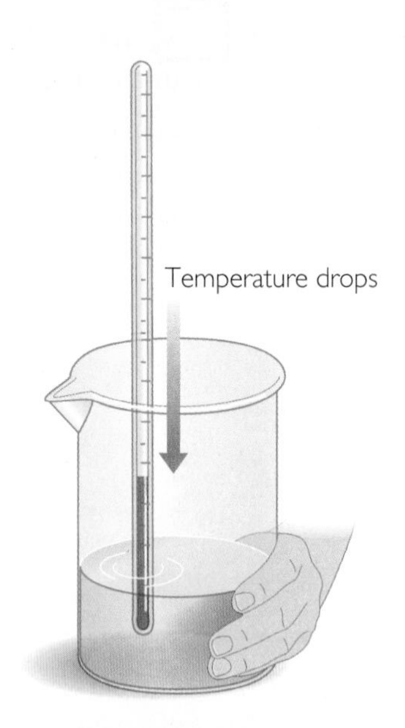

When you dissolve ammonium nitrate in water, the temperature decreases. The beaker feels cold.

What the thermometer records and your hand experiences is energy transferred into or out of the products of the chemical change. This process of energy transfer is called heat. **Heat** is a transfer of energy between two objects due to temperature differences. Heat always transfers from a higher temperature to a lower temperature. So, for instance, when the beaker feels cold to your hand, it is because heat transfers from your hand to the beaker.

System and Surroundings

In order to communicate clearly about heat transfer, it is necessary to specify where the heat is transferring to or from. The matter that you are focusing on is referred to as the **system.** Once you have defined the system, everything else is referred to as the **surroundings,** or environment. If the solution of ammonium nitrate and water is the system, then the beaker, the thermometer, the air around the beaker, your hand, and everything else in the universe constitutes the surroundings. Heat transfers between a system and its surroundings.

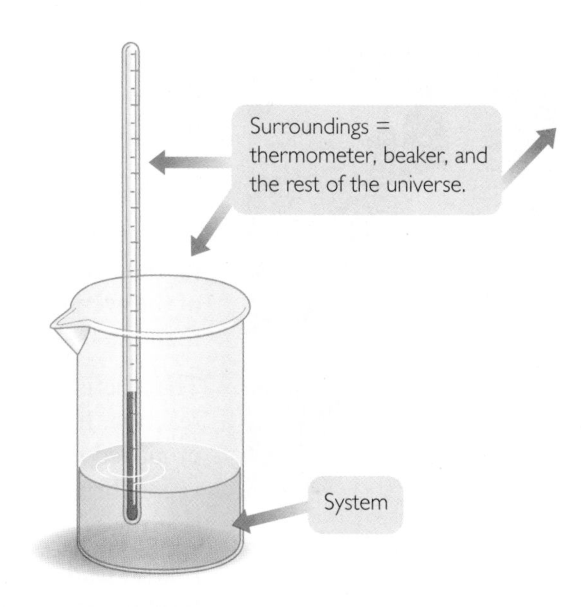

BIG IDEA Heat is a transfer of energy due to temperature differences.

❷ Exothermic and Endothermic Processes

Chemists categorize chemical changes according to the direction of heat transfer. As you learned in the previous lesson, when heat is transferred out of the system to the surroundings, the process is exothermic. When heat is transferred from the surroundings to the system, the process is **endothermic.** (In Latin, *exo* means outside and *endo* means inside.) Exothermic processes are experienced as warm by an observer. Endothermic processes are experienced as cold by an observer.

Some examples of exothermic and endothermic changes are listed below. The word *heat* is included in the chemical equation to highlight the direction of heat transfer.

Exothermic Processes

Burning methane gas

$$CH_4(g) + O_2(g) \longrightarrow CO_2(g) + H_2O(g) + \text{heat}$$

Dissolving calcium chloride in water

$$CaCl_2(s) \longrightarrow Ca^{2+}(aq) + 2Cl^-(aq) + \text{heat}$$

Neutralization of sodium hydroxide with hydrochloric acid

$$NaOH(aq) + HCl(aq) \longrightarrow H_2O(l) + NaCl(aq) + \text{heat}$$

Endothermic Processes

Dissolving ammonium nitrate in water

$$NH_4NO_3(s) + \text{heat} \longrightarrow NH_4^+(aq) + NO_3^-(aq)$$

Decomposition of mercury oxide

$$2HgO(l) + \text{heat} \longrightarrow 2Hg(l) + O_2(g)$$

Most household refrigerators contain a liquid called a ***refrigerant*** that circulates in a series of coiled tubes. As it evaporates, the refrigerant absorbs heat from inside of the refrigerator. The refrigerant is then condensed back into a liquid in tubes located in back of the refrigerator, releasing heat, and the process starts all over again.

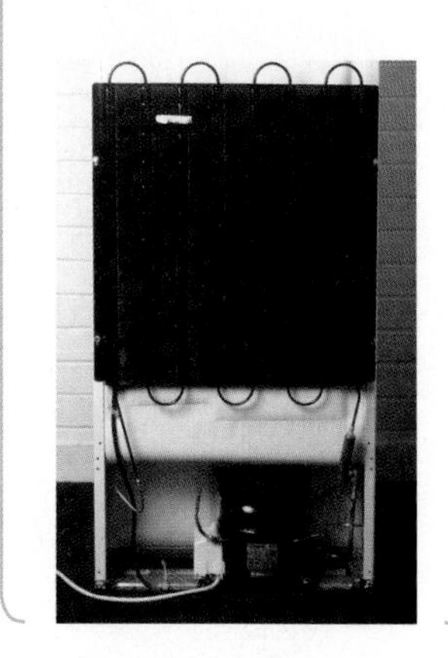

The shuttle launches due to the reaction of H_2 and O_2. The water vapor produced expands rapidly to thrust the shuttle upward.

Notice that heat transfer is not limited to chemical reactions. The process of dissolving is also exothermic or endothermic depending on the substance. This is because energy is involved every time atoms are rearranged.

3 Kinetic Energy

Changes in temperature due to a chemical reaction are associated with changes in motion. **Kinetic energy** is the energy of motion, and temperature is a reflection of the average kinetic energy of a sample. If the products of a reaction are hotter than the reactants, they must be moving faster.

The reaction between H_2 and O_2 to produce H_2O is an extremely exothermic reaction. It is the reaction used to launch the space shuttle. The water molecules produced in the reaction are so hot and they expand so rapidly that the water vapor thrusts the space shuttle up into the air. The photo here provides evidence that the product molecules are moving with explosive speed.

A great deal of the heat energy of this reaction is converted to kinetic energy in the form of rapidly moving water molecules. While the water molecules push the space shuttle up, they transfer some of their kinetic energy to the space shuttle. They also transfer some of their energy to the surroundings as heat. With less kinetic energy, the water molecules gradually become cooler and move more slowly.

Key Terms

heat
system
surroundings
endothermic
kinetic energy

Lesson Summary

In what direction is heat transferred during a chemical process?

Heat is a transfer of energy between two objects due to temperature differences. Heat transfer accompanies all chemical changes. The heat is transferred either into the system or out of the system. Exothermic reactions are chemical processes that result in the transfer of heat *from* the products of the reaction (the system) *to* the surroundings. These reactions feel hot to the observer. Endothermic reactions are chemical processes that result in the transfer of heat *from* the surroundings *to* the products of the reaction. These reactions feel cold to the observer. Because temperature is directly related to the motion of molecules, hotter products mean faster moving molecules. Colder products mean slower moving molecules.

EXERCISES

Reading Questions

1. In your own words, define exothermic and endothermic chemical changes.
2. Why does an endothermic reaction that takes place in a beaker cause the beaker to feel cold?

Reason and Apply

3. Methane, $CH_4(g)$, reacts with oxygen, $O_2(g)$, to produce carbon dioxide, $CO_2(g)$, and water, $H_2O(g)$. You observe flames.
   a. How does the average kinetic energy of the reactants differ from the average kinetic energy of the products?
   b. Describe what you would expect to see if you had a molecular view of the products and reactants.

4. You mix solid copper (II) sulfate, $CuSO_4(s)$, with a small amount of water, $H_2O(l)$ in a beaker. Hydrated copper (II) sulfate, $CuSO_4 \cdot 5H_2O(s)$ is produced. A thermometer inside the beaker indicates that the temperature has increased from 19 °C to 48 °C.

$$CuSO_4(s) + 5H_2O(l) \longrightarrow CuSO_4 \cdot 5H_2O(s)$$

   a. List all the substances that are part of the system.
   b. List at least four things that are part of the surroundings.
   c. Which is at a lower temperature: $CuSO_4(s)$ or $CuSO_4 \cdot 5H_2O(s)$? Explain your thinking.
   d. What will you feel if you touch the beaker?
   e. Is the process endothermic or exothermic? Explain your thinking.
   f. What evidence do you have that heat is transferred *to* the surroundings *from* the products of the reaction?

5. You mix solid hydrated barium hydroxide, $Ba(OH)_2 \cdot 8H_2O(s)$, and solid ammonium nitrate, $2NH_4NO_3(s)$, in a beaker. The reaction is shown here. A small pool of water in contact with the outside of the beaker freezes.

$$Ba(OH)_2 \cdot 8H_2O(s) + 2NH_4NO_3(s) \longrightarrow 2NH_3(g) + 10H_2O(l) + Ba(NO_3)_2(aq)$$

   a. List all the substances that are part of the system.
   b. List at least four objects that are part of the surroundings.
   c. Which is at a lower temperature: $NH_4NO_3(s)$ or $Ba(NO_3)_2(aq)$? Explain your thinking.
   d. What will you feel if you touch the beaker?
   e. Is the reaction endothermic or exothermic? Explain your thinking.
   f. What evidence do you have that heat is transferred *from* the surroundings *to* the products of the reaction?

LESSON

3 Point of View

First and Second Laws

Think About It

Suppose you have just come inside on a very cold day. There is a fire in the fireplace. If you go near the fire, you will feel the warmth. Heat is transferred to your body by the warm gases of the flames and the energy radiating from the burning wood.

What do temperature differences indicate about heat transfer?

To answer this question, you will explore

1. Heat Transfer
2. The First and Second Laws of Thermodynamics

Exploring the Topic

1 Heat Transfer

There are three main ways heat is transferred: conduction, convection, and radiation.

- Conduction takes place when a substance transfers heat to another substance or object it is in contact with. For example, the handle on a frying pan gets hot because the frying pan is hot.
- Convection takes place when a warm substance changes location, such as when warm air rises.
- Radiation takes place when electromagnetic waves carry energy from an energy source, such as the Sun.

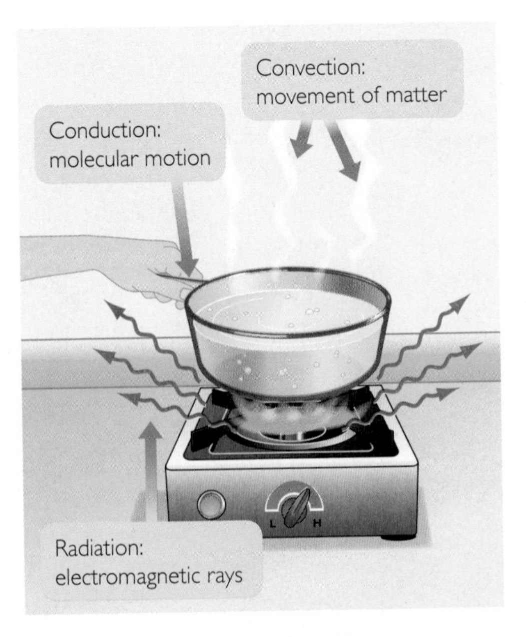

CONSUMER CONNECTION

Materials that do not conduct heat well are called *thermal insulators*. We use them in the walls and ceilings of our homes to slow down the transfer of heat from our homes in the wintertime and into our homes in the summertime. A thermos bottle is composed of several layers of insulating material, often with a vacuum in the middle, to limit heat transfer and to keep cold liquids cold and hot liquids hot.

System and Surroundings

Consider each of the three illustrations. Assume water is the system. Try to identify the direction of heat transfer, the type of heat transfer, and the surroundings.

Water boils on the stove, producing water vapor. When the water vapor comes in contact with the cold pane of glass it condenses and forms drops.

Water is placed in the freezer to make ice cubes.

An ice cube is on a table and melts.

BIOLOGY CONNECTION

Our bodies are little heaters. The internal temperature of the body is around 98.6 °F, or 37 °C. The temperature on the surface of our skin is usually right around 91 °F, or 33 °C. Our bodies regulate their own temperatures. The body cools itself by producing sweat that evaporates, and warms itself by shivering.

In the first illustration, water molecules are in the form of both water vapor and liquid water. For the water vapor, the surroundings include the stove, the pot, the window, and the air in the room. Heat transfers from the stove to the pot and then to the water by conduction, then to the air in the room by convection. Heat also transfers from the water vapor to the window by conduction.

In the second illustration, water takes the form of both liquid and solid. For the liquid water in the ice cube tray, the surroundings include the ice cube tray, the freezer, and the air in the freezer. Heat transfers from the water to the surroundings by conduction. When enough heat energy transfers from the liquid water to the interior of the freezer, the water freezes and becomes solid.

For the ice cube on the table, the surroundings include the table and the air in the room. Heat transfers to the ice cube by conduction. When enough energy is transferred into the ice cube on the table, it melts.

Hot and Cold

The sensation of hot or cold depends on the direction of heat transfer. For example, if you hold an ice cube, your hand will feel cold as heat transfers from your hand to the ice cube. If you hold a cup of hot tea, your hand will feel warm as heat transfers from the hot tea to the cup and then to your hand. However, "hot" and "cold" are relative.

Just because something is at a low temperature does not mean it will feel cold. For example, if your hands are cold because you have been out in the snow all day, even holding them under cold water can feel painfully hot because the water is at a higher temperature than your hands. The experience of hot or cold depends on whether energy is being transferred to or from the observer.

Important to Know There is no such thing as "cold transfer." Cold is the experience of heat transfer away from our bodies. ◄

Thermal Equilibrium

Heat transfer occurs between two substances that are in contact with each other until there is no longer a temperature difference between them. If you take a pot of hot soup off the stove and place it on the counter, it will transfer heat to the air and the countertop. The soup (and the metal pot) will continue to transfer heat until they are at the same temperature as the room.

If two objects at different temperatures are in contact with one another long enough, the two substances will reach the same temperature. When they reach the same temperature, they are at **thermal equilibrium.**

BIG IDEA The direction of heat transfer is always from a hotter substance or object to a colder one.

❷ The First and Second Laws of Thermodynamics

The study of heat transfer is called *thermodynamics.* The principles behind thermodynamics have been summarized into scientific laws that are consistent with observations of a large number of systems.

The **first law of thermodynamics** states that energy is always conserved. The heat transferred *from* a hotter object is equal to the heat transferred *to* the colder objects surrounding it. There is no loss of energy. Likewise, you cannot create energy. It can only be transferred from one place to another place, and it can change form, such as when solar energy is converted to electrical energy.

BIG IDEA Energy is conserved. It cannot be created or destroyed.

The **second law of thermodynamics** states that energy tends to disperse, or spread out. Energy transfers from hotter objects to colder objects until the temperature evens out. It does not transfer from a cold object to a hot object because this would concentrate the energy in one place rather than spread it out. This concept helps to explain why you can sit far away from a fire and become warmed by it. The energy from the fire disperses, or spreads out, from its source.

Another way to explain this concept is to say that the **entropy** of a system tends to increase over time. This means that energy and matter have a natural tendency to become more dispersed and disordered rather than more collected and ordered. For example, when you open a bottle of perfume, the molecules disperse throughout the room. The molecules will not naturally collect back into the bottle.

BIG IDEA Energy tends to disperse.

Key Terms

thermal equilibrium
first law of thermodynamics
second law of thermodynamics
entropy

Lesson Summary

What do temperature differences indicate about heat transfer?

The sensory experience of hot or cold depends on the direction of heat transfer. If your hand is at a higher temperature than what you are touching, the object will feel cold. If your hand is at a lower temperature, the object will feel warm. Heat is always transferred from a hotter object or to a colder one. Heat transfer is associated with phase changes. An ice cube melts because heat is transferred into it from the surroundings. If objects remain in contact with each other, heat is transferred until objects are at the same temperature. This balanced state is referred to as thermal equilibrium. Two laws of thermodynamics are related to these observations: The first law states that energy is conserved, and the second law states that energy tends to disperse, or spread out.

EXERCISES

Reading Questions

1. Why does an ice cube feel cold to your hand?
2. What is thermal equilibrium?

Reason and Apply

3. Is energy transferred to a liquid or from a liquid when it evaporates? When it solidifies?
4. Is energy transferred to a gas or from a gas when it condenses?
5. What is evaporation and why is it a cooling process? What cools during evaporation?
6. What determines the direction of heat transfer?
7. Do humans ever reach thermal equilibrium with the surrounding air? Explain.
8. Provide evidence to support the first law of thermodynamics.
9. Provide evidence to support the second law of thermodynamics.
10. Explain why a campfire burns out. Explain why the ashes eventually cool to the temperature of the surrounding air.
11. Could there ever be a situation in which an ice cube actually heats up something else? Explain.

RECREATION CONNECTION

Hypothermia occurs when the body loses heat faster than it can generate heat. A body temperature below 95 °F (35 °C) is considered dangerous. People who are outdoors in winter without enough protective clothing, or who accidentally fall into very cold water, may develop hypothermia. Hot liquids and external heat sources can help a person recover from hypothermia.

LAB Heat Transfer

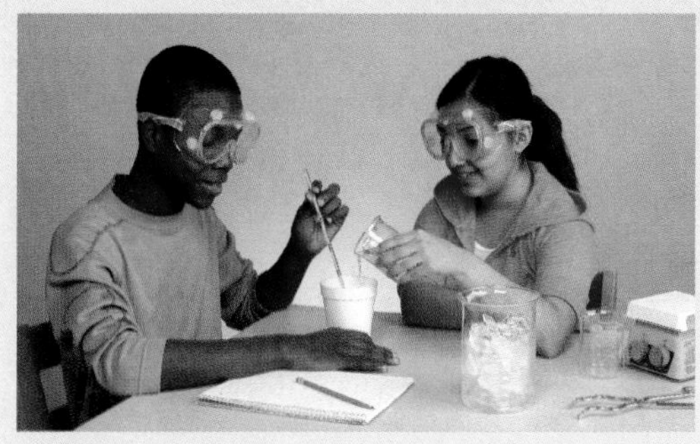

Purpose

To explore the factors that affect heat transfer between two samples of water.

Materials

- large foam cup
- 250 mL beakers (2)
- 100 mL graduated cylinder
- thermometer
- hot plate
- beaker tongs
- ice in a large beaker or plastic tub

Procedure

1. Measure and record a volume of water. Pour it into the foam cup and record the temperature.
2. Measure and record a second volume of water. Pour it into a beaker and either heat or cool the water using a hot plate or ice. Record the temperature.
3. Add the second sample to the first. Stir and watch the thermometer closely. Record the temperature after the temperature has stabilized.
4. Repeat the experiment using different volumes and temperatures. Create a data table with results from at least three trials.
5. Convert your volume measurements to mass. The density of water is 1 g/mL.

Analysis

1. Look for relationships in your data table. What factors affect the temperature change?
2. Chemists measure the amount of heat transfer from one sample to another in units of calories. A *calorie* is the amount of energy needed to raise the temperature of 1 g of water by 1 °C. How does this relate to your data?
3. Write an equation relating the masses of the two water samples, m_1 and m_2, and the temperature changes, ΔT_1 and ΔT_2, of these two masses of water.
4. **Making Sense** What happens to the motions of the water molecules if you mix hot and cold water? Use changes in molecular motion to explain how energy from the hot water transfers to the cold water.
5. **If You Finish Early** Predict the final temperature if you mix 15 g of water at 20 °C with 45 g of water at 80 °C. Show your work.

LESSON

4 Heat Versus Temperature

Heat Transfer

Think About It

Imagine you want two samples of really hot water. You want a large sample to make soup and a smaller sample for a cup of hot cocoa. Will it take the same amount of heat to get both samples to the same temperature?

What is the difference between temperature and heat?

To answer this question, you will explore

1. Particle View of Energy
2. Measuring Heat Transfer

Exploring the Topic

1 Particle View of Energy

Consider a sample of water molecules that is heated. What happens to the molecules as more energy is transferred to the sample?

As shown in the illustration, atoms and molecules move faster as they are heated. As the movement of the molecules increases, the temperature of the sample goes up.

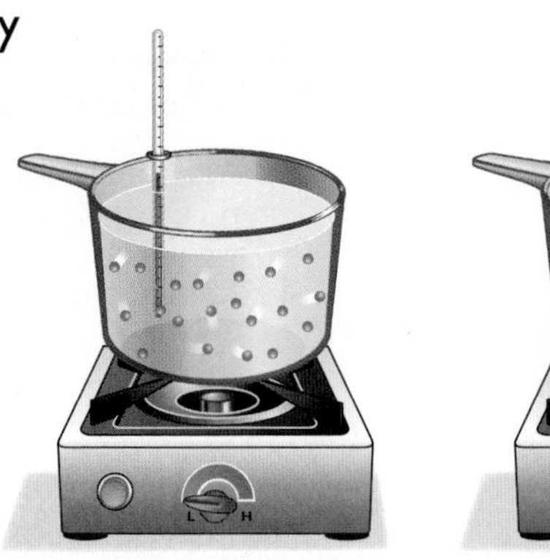

Same volume, different amounts of energy.

Now consider two samples of water that contain different numbers of molecules. While the temperatures might be identical, the amount of energy represented by the two samples is not the same. In one sample you have many quickly moving particles. In the other sample you have fewer particles, even though they are moving just as quickly. There will be more total energy in the larger sample.

If the water samples in the large pot and the small cup shown on the next page are both at 85 °C, you could say that they both have the same "degree of hotness," or temperature. However, the water in the large pot has a greater total "quantity of hotness" because it is a larger sample. This second concept is referred to as **thermal energy.** Thermal energy describes the total amount of energy in the particles of a sample. Because the temperature is the same, the gas molecules in both samples are moving with the same average kinetic energy. Because their volumes are different, the two samples have different amounts of thermal energy. So thermal energy is dependent on both the temperature of the sample *and* on the number of particles in the sample.

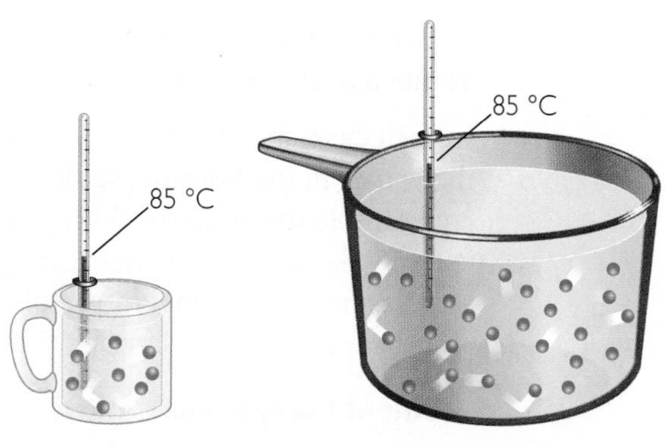

Same temperatures, different amounts of energy.

BIG IDEA Temperature depends on the average kinetic energy of matter. Thermal energy depends on the average kinetic energy and the mass of the sample.

❷ Measuring Heat Transfer

Unfortunately, you cannot measure the thermal energy of a sample of matter directly. A thermometer gives you only an indication of the average kinetic energy. However, changes in thermal energy can be determined using heat transfer.

Imagine that you have three samples of water at different temperatures: a pitcher of water at 80 °C, a beaker of water at 80 °C, and a pitcher of water at 50 °C. Your common sense tells you that the largest 80 °C sample is the one with the greatest thermal energy. But can that energy be measured or quantified?

You can calculate the amount of thermal energy in each sample by determining how much energy can be transferred from each one. One way to do this is to use the energy in the three samples to warm identical samples of cold water.

Warming 100 mL samples of water

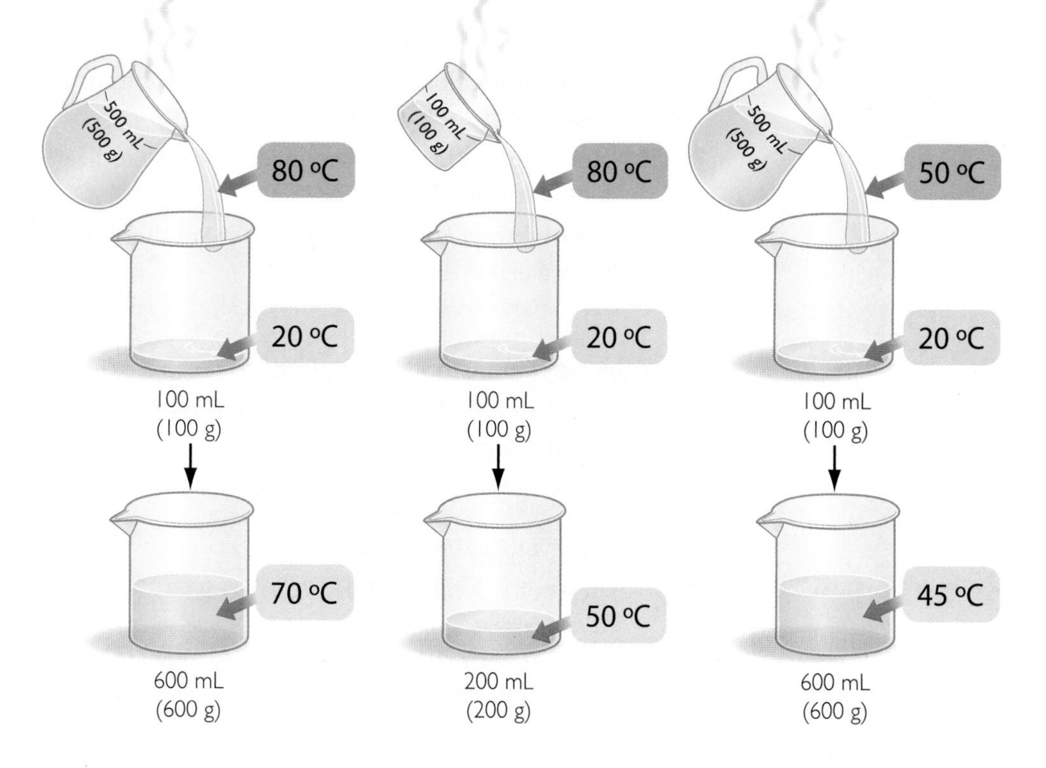

Notice that the large sample of water at 50 °C warms the cold water sample nearly as much as the small sample of water at 80 °C.

In each case, the mixture reaches thermal equilibrium. That is, the faster moving molecules in the hot sample slow down and the slower moving molecules in the cold sample speed up until they are all moving with the same average kinetic energy. The transfer of energy happens through collisions of the molecules as the samples mix. The temperature of the final mixture depends on the temperatures and the masses of the samples that were mixed.

A Unit of Energy: The Calorie

Scientists define a unit called a **calorie** to measure heat transfer. A calorie is the amount of energy needed to raise the temperature of 1 gram of water by 1 Celsius degree. So, this unit of energy is based on heat transfer to or from water. To determine the number of calories needed, you multiply the mass of the water by the temperature increase in Celsius degrees. This can be summarized by the mathematical equation for the heat transfer, q, in terms of the mass of water, m, and the temperature change, ΔT. The uppercase Greek letter delta, Δ, is used to represent a change in quantity.

$$q = m(1 \text{ cal/g} \cdot {}^\circ\text{C})\Delta T$$

(Remember, 1 g of water is equivalent to 1 mL of water at room temperature.)

So, for example, it would take 5000 calories of energy to raise the temperature of 500 grams of water by 10 °C from a temperature of 20 °C to a temperature of 30 °C.

[For a review of this math topic, see **MATH** ***Spotlight: Averages*** on page 623.]

Example 1

Calories

Suppose you need to heat some water.

a. How many calories of energy does it take to heat a 75 mL sample of water for cocoa from 25 °C to 38 °C?

b. How many calories of energy does it take to heat a 2 L sample of water for a footbath from 25 °C to 38 °C?

c. Which sample of water contains more thermal energy, the water for cocoa or the water for the footbath?

Solution

Convert volume to mass in grams. Determine the change in temperature ΔT for both situations. Then use the equation for heat transfer to solve for q.

a. 75 mL water = 75 g water

$$\Delta T = 38 - 25 = 13\ {}^\circ\text{C}$$

$$\begin{aligned} q &= m(1 \text{ cal/g} \cdot {}^\circ\text{C})\Delta T \\ &= 75 \text{ g}(1 \text{ cal/g} \cdot {}^\circ\text{C})13\ {}^\circ\text{C} \\ &= 975 \text{ calories} \end{aligned}$$

b. 2 L water = 2000 g water

$$\Delta T = 38 - 25 = 13\ °C$$

$$q = m(1\ \text{cal/g} \cdot °C)\Delta T$$
$$= 2000\ \text{g}(1\ \text{cal/g} \cdot °C)13\ °C$$
$$= 26{,}000\ \text{calories}$$

c. The water in the footbath represents a greater amount of thermal energy.

Example 2

Conservation of Energy

You mix 300 mL of water at 60 °C with 100 mL of water at 20 °C.

a. Find the final temperature of the mixture.

b. Show that the heat transferred *from* the hot water is equal to the heat transferred *to* the cold water.

Solution

a. The final temperature will be a weighted average of the two initial temperatures. (Each temperature is "weighted" by multiplying it by the mass of each sample that is mixed.) The masses of the samples are 300 g and 100 g.

$$\frac{300\ \text{g}(60\ °C) + 100\ \text{g}(20\ °C)}{400\ \text{g}} = 50\ °C$$

The final temperature is 50 °C. It makes sense that the final temperature is closer to 60 °C, since there was more of the 60 °C water.

b. The calories of energy transferred is given by this equation:

$$q = m(1\ \text{cal/g} \cdot °C)\Delta T$$

Remember, 300 mL of water = 300 g of water.

Hot water: $q = (300\ \text{g})(1\ \text{cal/g} \cdot °C)(50\ °C - 60\ °C) = -3000$ calories

Cold water: $q = (100\ \text{g})(1\ \text{cal/g} \cdot °C)(50\ °C - 20\ °C) = 3000$ calories

The negative sign means that energy is transferred *from* the hot water. The positive sign for the cold water indicates that energy is transferred *to* the cold water. Based on the calculated answers, energy is conserved.

Key Terms

thermal energy
calorie

Lesson Summary

What is the difference between temperature and heat?

Temperature and heat are different, although they are both related to the motion of atoms and molecules. The temperature of a substance is a measure of the average kinetic energy of its particles. Heat is a process of energy transfer to or from a sample of matter. Thermal energy is a term often used to describe the "energy content" of a sample. The thermal energy of a sample depends on the number of particles in a sample, whereas temperature is the same regardless of the size of the sample. A calorie is a unit of energy. It is defined as the energy required to raise the temperature of 1 g of water by 1 °C. The energy transferred is referred to as heat.

EXERCISES

Reading Questions

1. If you heat up a large and a small sample of cold tap water to the same temperatures, explain why it takes more energy to heat up the larger sample.
2. What is the difference between thermal energy and temperature?

Reason and Apply

3. How many calories of energy do you need to transfer for each of the following changes?
   a. Raise the temperature of 1 g of water by 5 °C
   b. Raise the temperature of 2 g of water by 5 °C
   c. Raise the temperature of 9 g of water by 35 °C
4. Which will warm a child's inflatable pool more: adding 500 g of water at 50 °C or 100 g at 95 °C? The temperature of the water in the pool is 20 °C. Explain your reasoning.

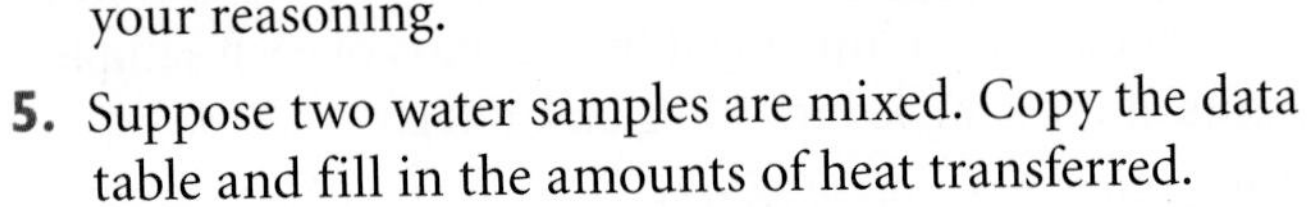

5. Suppose two water samples are mixed. Copy the data table and fill in the amounts of heat transferred.

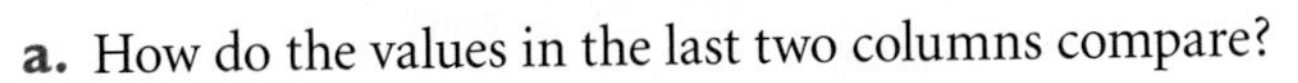

   a. How do the values in the last two columns compare?

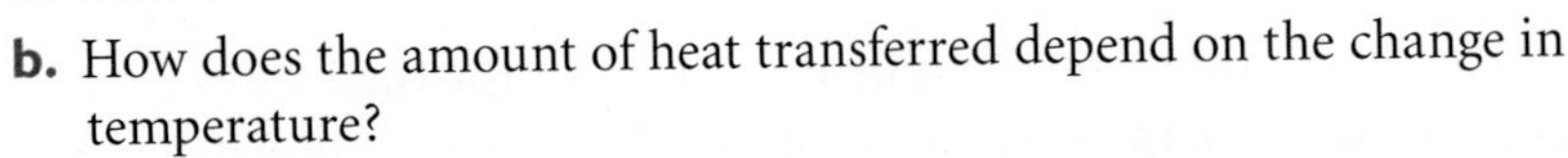

   b. How does the amount of heat transferred depend on the change in temperature?
   c. How does the amount of heat transferred depend on the mass of each sample?

	Sample 1 mass (g)	Sample 1 temperature (°C)	Sample 2 mass (g)	Sample 2 temperature (°C)	Final temperature (°C)	Heat transfer to sample 1 (cal)	Heat transfer from sample 2 (cal)
Trial 1	100 g	20 °C	100 g	80 °C	50 °C		
Trial 2	100 g	20 °C	100 g	70 °C	45 °C		
Trial 3	100 g	20 °C	300 g	80 °C	65 °C		
Trial 4	200 g	20 °C	100 g	80 °C	40 °C		

6. Suppose you mix two water samples: 300 g of water at 20 °C and 200 g of water at 50 °C. What do you expect the final temperature of the water to be?
7. From a molecular viewpoint, explain how thermal equilibrium is reached when hot and cold water mix.

LESSON

5 The Heat Is On

Specific Heat Capacity

Think About It

Imagine you put a metal pan and a pizza in a hot oven. Both the pizza and the pan start out at room temperature. You leave them in the oven for the same amount of time. Will they heat up to the same temperature? And once they are taken out of the oven, will they cool off differently?

How do different substances respond to heat?

To answer this question, you will explore

1. Specific Heat Capacity
2. Bonding, Numbers, and Heat

Exploring the Topic

1 Specific Heat Capacity

Every substance has a unique response to heat. Suppose you place an aluminum pot of water on the stove to heat. After a minute, the metal pot will be too hot to touch, but the water will still be cool. This is because different amounts of energy are needed to raise the temperature of each type of substance.

To compare heat transfer to different substances, you can measure the energy required to raise the temperature of the same mass of each substance by the same number of degrees. The amounts of energy needed to raise the temperature of 1 g of several sample substances by 1 °C are given in the table.

Sample	Mass (g)	ΔT (°C)	Energy required (cal)
water, $H_2O(l)$	1.0 g	1 °C	1.0 cal
methanol, $CH_3OH(l)$	1.0 g	1 °C	0.58 cal
aluminum, $Al(s)$	1.0 g	1 °C	0.21 cal
copper, $Cu(s)$	1.0 g	1 °C	0.09 cal

Notice that fewer calories of energy are required to raise the temperature of methanol, aluminum, and copper from 20 °C to 21 °C, compared with water. This is consistent with the observation that the aluminum pot heats up and cools off faster than the water in the pot. Less energy is involved in changing the temperature of the aluminum.

ENVIRONMENTAL CONNECTION

Deserts can be extremely hot at midday and extremely cold at night. The lack of water vapor in the air allows for greater temperature fluctuations. As a result, near the coast summers are not as hot and winters are not as cold as they are farther inland.

The heat required to raise the temperature of 1 g of a substance by 1 °C is called the **specific heat capacity.** Every substance has a specific heat capacity. Some values are shown in the table.

Substance	Specific heat capacity (cal/g · °C)
aluminum, $Al(s)$	0.21
water, $H_2O(l)$	1.00
copper, $Cu(s)$	0.09
iron, $Fe(s)$	0.11
wood (cellulose), $(C_6H_{10}O_5)_n(s)$	0.41
glass, $SiO_2(s)$	0.16
nitrogen, $N_2(g)$	0.24
ethanol, $CH_3CH_2OH(l)$	0.57
hydrogen, $H_2(g)$	3.34

The greater the specific heat capacity of a substance, the less its temperature will rise when it absorbs a given amount of energy. If a substance has a low specific heat capacity, it will change temperature easily, with a small transfer of energy. This applies whether the energy is being transferred into or out of the substance. This is why the aluminum pot heats up faster, and also cools down faster, than water. Water does not change temperature as easily as most substances.

BIG IDEA Substances with low specific heat capacities can heat up and cool down easily.

The specific heat capacity of a substance can be used to determine the amount of energy needed to raise the temperature of any mass by any number of degrees. The formula for heat transfer to any substance is a product of the mass, m, its specific heat capacity, C_p, and the temperature change, ΔT.

$$q = mC_p\Delta T$$

Example 1

Cooling

Suppose you have 15 g of methanol, CH_3OH, and 15 g of water, H_2O, both at 75 °C. You want to cool both samples to 20 °C. How much energy (in calories) do you need to remove from each sample?

Solution

The energy transferred is a product of the mass, the specific heat capacity, and the change in temperature.

$$q = mC_p\Delta T$$

The mass of each sample is 15 g, and the change in temperature is 20 °C minus 75 °C, or −55 °C. However, the specific heat capacities of water and methanol are different. The specific heat capacity of water is 1 cal/g · °C. The specific heat capacity of methanol is 0.58 cal/g · °C.

Water:

$$q = (15\text{ g})(1\text{ cal/g}\cdot{}^\circ\text{C})(-55\ {}^\circ\text{C}) = -830\text{ cal}$$

Methanol:

$$q = (15\text{ g})(0.58\text{ cal/g}\cdot{}^\circ\text{C})(-55\ {}^\circ\text{C}) = -480\text{ cal}$$

It takes a greater transfer of energy to cool water than to cool methanol. Because of this, water is a very good insulator and retains its temperature.

Example 2

Final Temperature

Imagine that you have 3.5 g of copper and 3.5 g of water. Both are at 25 °C. Which sample will be at a higher temperature if you transfer 150 cal to each? Show your work.

Solution

The energy transferred is a product of the mass, the specific heat capacity, and the change in temperature.

$$q = mC_p\Delta T$$

You know the value of q and the masses of the water and copper samples. You want to find the temperature change. Solve for ΔT. You can rearrange the equation:

$$\Delta T = \frac{q}{mC_p}$$

You can look up the specific heat capacities of water and copper in a table like the one on page 494. The specific heat capacity of water is 1 cal/g · °C. The specific heat capacity of copper is 0.09 cal/g · °C.

Water:

$$\Delta T = \frac{q}{mC_p} = \frac{150\text{ cal}}{(3.5\text{ g})(1\text{ cal/g}\cdot{}^\circ\text{C})} = 43\ {}^\circ\text{C}$$

Copper:

$$\Delta T = \frac{q}{mC_p} = \frac{150\text{ cal}}{(3.5\text{ g})(0.09\text{ cal/g}\cdot{}^\circ\text{C})} = 480\ {}^\circ\text{C}$$

The temperature of copper rises about ten times as much as the temperature of the same quantity of water when 150 calories of energy are transferred.

To find the final temperature, add the temperature change to the initial temperature.

Water: Final temperature = 25 °C + 43 °C = 68 °C

Copper: Final temperature = 25 °C + 480 °C = 505 °C

For the same amount of heat transfer, the copper is much hotter.

CONSUMER CONNECTION

In order to make ice cream, the ingredients must be mixed at below freezing temperature. This is why you add salt to the ice in the outermost chamber of an ice-cream maker. Salt causes the ice to melt because salt water has a lower freezing point than water. However, it requires energy to melt ice, so heat is transferred from the ice cream to the salt-ice mixture, lowering the temperature to below 0 °C.

❷ Bonding, Numbers, and Heat

It may seem strange that different amounts of energy are needed to raise the temperature of different substances. Why would you need more energy for water, compared with the same mass of methanol, aluminum, or copper to heat them all to the same temperature?

The answer is partially related to molar mass. Look back at the table of specific heat capacities. While the molar mass of aluminum is 27.0 g/mol, the molar mass of copper is 63.6 g/mol. So, there are more atoms in 1 g of aluminum than in 1 g of copper. It makes sense that more energy is required to cause a larger number of atoms to move faster.

Bonding also has an effect on the specific heat capacity of a substance. Substances that are polar tend to have high specific heat capacities. This means that molecules will have higher specific heat capacities than metals.

Finally, complicated molecules with large numbers of atoms have a variety of internal vibrational motions. The atoms within the molecule can vibrate back and forth like balls on a spring. And the entire molecule can rotate. All of these motions need to be increased in order to raise the temperature. So, substantially more heat is required to increase the average kinetic energy of molecules compared with individual atoms.

High Specific Heat Capacity of Water

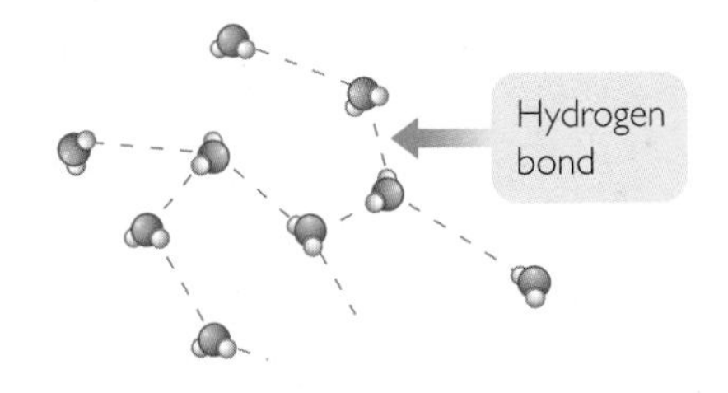

Water has a particularly high specific heat capacity for several reasons. First, the molar mass of water, 18.0 g/mol, is quite small. So there are a lot of water molecules to get moving in a 1 g sample. Second, because water is molecular, it has complex internal movements. Third, water consists of H_2O molecules with strong intermolecular attractions called **hydrogen bonds.** A hydrogen bond is the attraction between the positive and negative dipoles of different water molecules. The hydrogen bonding restricts the motions of the molecules. Extra heat needs to be added in order to overcome these attractions and raise the kinetic energy. This is why the water in the metal pot heats up so much more slowly than the metal pot itself.

Key Terms

specific heat capacity
hydrogen bond

Lesson Summary

How do different substances respond to heat?

In order to increase the temperature of a sample of matter, you must supply energy to it. But the addition of a given amount of energy does not always result in the same rise in temperature. Substances differ in their responses to heat. Some substances, like metals, change temperature more than others in response to the transfer of the same amount of energy. This is because they have a lower specific heat capacity. Specific heat capacity is the amount of energy needed to raise the temperature of 1 g of a substance by 1 °C. The specific heat capacity of a substance depends on a number of factors, including the number of atoms or molecules in a sample and the type of bonding.

EXERCISES

Reading Questions

1. What is specific heat capacity? Give the specific heat capacity of a substance, and explain what that means in terms of that substance.
2. Why is more energy required to raise the temperature of 1.0 g of water compared with 1.0 g of aluminum?

Reason and Apply

3. Which substance in each pair has the higher specific heat capacity? Justify your choice with an atomic view.

   a. Al or Pb b. H_2 or Ar c. F_2 or Cl_2
4. Imagine that you transfer the same amount of energy to all of the substances in the table of specific heat capacities on page 494. Which one would be hottest to the touch? Coolest? Explain your reasoning.
5. Use specific heat capacity to explain why metal keys left in the sun are hotter than a plastic pair of sunglasses left in the sun.
6. How much energy is required to raise the temperature of 50 g of methanol from 20 °C to 70 °C?
7. If you place 20 g of aluminum and 20 g of copper, both at 20 °C, into ice water, which will cool faster? Explain.
8. Imagine you have 50 g of water and 50 g of methanol, each in a 100 mL beaker. You place both on a hot plate on low heat.
   a. Which sample will be at a higher temperature after 5 min? Explain your thinking.
   b. If the initial temperature of each liquid is 23 °C, what is the temperature of each after 25 cal of energy are transferred from the hot plate to each sample.
9. Would water make a good thermal insulator for a home? Explain your reasoning.
10. Why do farmers spray water on oranges to protect them from frost? Give two reasons why the water on the orange might help.

LESSON

6 Where's the Heat?

Heat and Phase Changes

Think About It

Water boils at 100 °C, at 1 atmosphere pressure. Once it reaches that temperature, liquid water does not get any hotter, no matter how big the flames are under its container or how much you heat it. If the temperature of the boiling water is not changing, where is all that heat going?

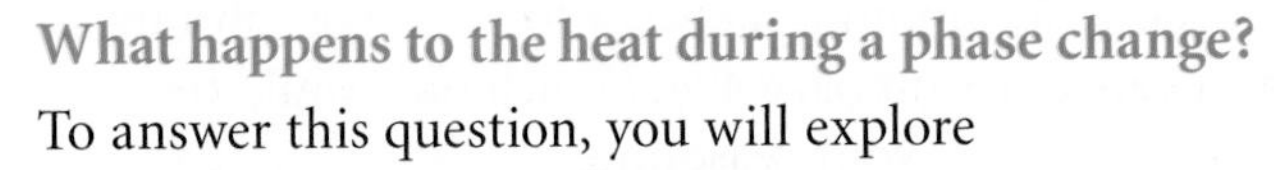

What happens to the heat during a phase change?

To answer this question, you will explore

1. Heating Curve of Water
2. Heat and Phase Changes

Exploring the Topic

1 Heating Curve of Water

Imagine you have an ice cube that you have just removed from the freezer. The temperature of the freezer is below zero. The graph shows the changes in temperature as you heat the ice cube from below 0 °C to above 100 °C. The *x*-axis represents time. Because the water is being heated steadily, as time goes by, more and more heat is transferred to the water. This graph is the heating curve of water.

Heating Curve of Water

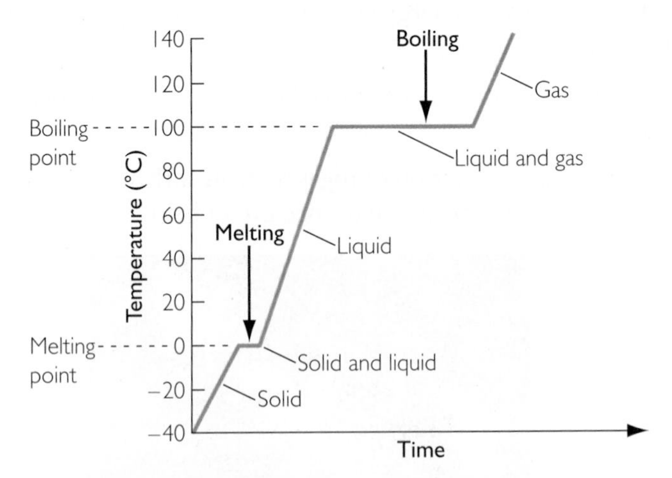

As you can see, the heating curve of water is a series of steep inclines with flat horizontal parts in between. Along the horizontal parts, energy continues to be transferred into the water, but the temperature does not change. The flat lines occur when melting or boiling is taking place, at 0 °C or 100 °C.

Note that whenever the temperature is not changing, two phases are present. As soon as only one phase—solid, liquid, or gas—is present, the temperature is able to change again. Something is happening to the energy going into the water during phase change.

2 Heat and Phase Changes

During melting and boiling, there is no doubt that energy is going into the substance. However, this energy is doing something besides making the molecules move faster. Remember, temperature is a measure of the average kinetic energy of the molecules. If the temperature is not going up, then the molecules are not moving any faster on average. Where is the energy going?

The answer lies in the intermolecular forces. These attractive forces hold particles in a solid or a liquid close to each other. When heat is transferred into a substance, like ice, the molecules in the ice move faster and the temperature rises. Once the temperature rises to 0 °C, the ice begins melting. When heat is transferred into melting ice, the average kinetic energy does not change, and the temperature does not rise. Instead, the energy transferred weakens the attractive forces between the molecules so that they can wiggle and rotate. Once the ice has melted completely, the energy transferred causes the molecules in the liquid to move faster, and the temperature rises once again.

During the boiling process, all of the energy transferred is used to completely break the intermolecular attractions so that molecules can move freely as a gas. As a result, no temperature change is observed during boiling.

BIG IDEA Heat transfer does not always result in a temperature change.

When a solid substance changes phase from solid to liquid or from liquid to gas, a certain amount of energy must be supplied in order to overcome the molecular or ionic attractions between the particles. This is the energy that is changing the solid into the liquid or the liquid into a gas, without changing the temperature. For example, the heat required to change 1 gram of a substance from a liquid into a gas is called the **heat of vaporization.** The heat required to change 1 gram of a substance from a solid to a liquid is called the **heat of fusion.**

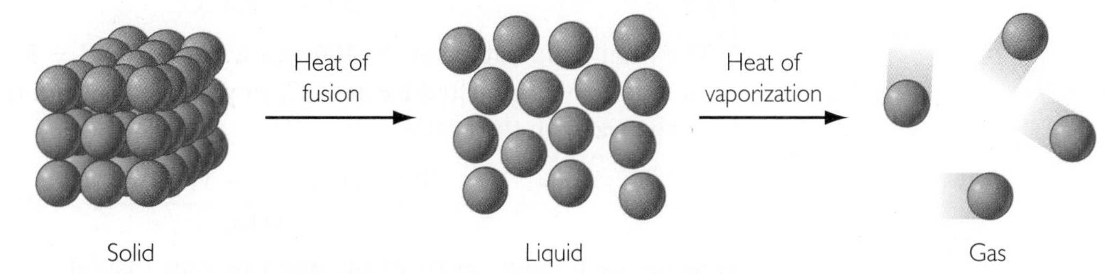

When a substance changes phase from a solid to a liquid, or from a liquid to a gas, energy is transferred *to* the substance from the surroundings. When a substance changes phase from a gas to a liquid, or from a liquid to a solid, energy is transferred *from* the substance to the surroundings.

The table on the next page shows some phase change data for five common substances. Notice that it takes very little energy per gram to melt mercury (2.7 calories per gram), whereas it takes a great deal of energy to melt aluminum (94.5 calories per gram). Substances also require significantly more energy to change into gases than to melt.

Phase Change Data

	Water	Ethanol	Mercury	Aluminum	Gold
Heat of fusion	79.9 cal/g	24.9 cal/g	2.7 cal/g	94.5 cal/g	15.4 cal/g
Melting point	0 °C	−117 °C	−39 °C	659 °C	1063 °C
Heat of vaporization	539 cal/g	205 cal/g	70 cal/g	2500 cal/g	377 cal/g
Boiling point	100 °C	78 °C	357 °C	2467 °C	2660 °C

Example

Heat of Fusion

You place 11 g of ice at 0 °C into 100 g of liquid water at 20 °C in an insulated water bottle. After a few minutes, the ice melts completely. Calculate the final temperature of the water in the bottle.

Solution

The heat of fusion of water given in the table is 79.9 cal/g. This means that, to melt 11 g of ice, 79.9 cal/g · 11 g, or 879 calories are transferred. This energy is transferred from the 100 g of liquid water in the bottle. Use the equation for heat transfer to determine how much this amount of energy changes the temperature.

$$q = mC_p\Delta T$$

$$-879 \text{ cal} = (100 \text{ g})\ (1 \text{ cal/g} \cdot {}^\circ\text{C})\ \Delta T$$

$$\Delta T = -8.8\ {}^\circ\text{C}$$

The final temperature of the 100 g of water is 20 °C − 8.8 °C, or 11.2 °C. But there is also 11 g melted ice at 0 °C, so you need to find the weighted average of all the water in the bottle.

$$\frac{100 \text{ g } (11.2\ {}^\circ\text{C}) + 11 \text{ g } (0\ {}^\circ\text{C})}{111 \text{ g}} = 10\ {}^\circ\text{C}$$

So the final temperature of the water in your insulated water bottle will be around 10 °C.

Key Terms

heat of vaporization
heat of fusion

Lesson Summary

What happens to the heat during a phase change?

Heat transfer does not always result in a temperature increase. During a phase change, the transfer of energy into a substance does not cause any rise in temperature. Instead, this energy is used to overcome the attractions between atoms and molecules.

The heat of fusion is the energy required to change a substance from solid to liquid. The heat of vaporization is the energy required to change a substance from liquid to gas.

EXERCISES

Reading Questions

1. Why doesn't the temperature of water change during boiling?
2. What is heat of vaporization? Use a substance from the phase change data table on page 500 as an example.

Reason and Apply

3. Use the table of phase change data on page 500 to sketch a heating curve for mercury.
4. A jeweler pours liquid gold into a mold to make a 6 g wedding ring. How much energy is transferred from the ring when it solidifies? What temperature is the gold at the instant all of its atoms become solid?
5. Which of the following processes are exothermic? Explain your thinking.
   A. boiling ethanol
   B. freezing liquid mercury
   C. subliming carbon dioxide
6. How much energy is required to melt these quantities of ice?
   a. 56 g
   b. 56 mol
7. If you transfer 5000 cal, how many grams of ethanol can you vaporize?
8. **a.** How much energy do you need to transfer in order to raise the temperature of 150 g of aluminum from 20 °C to its melting point?
   b. How much energy do you need to transfer in order to melt 150 g of aluminum?
9. You place 25.0 g of ice at 0 °C into 100 g of liquid water at 45 °C. The final temperature is 20 °C. Show that energy is conserved.

ENVIRONMENTAL CONNECTION

Iron melts at over 1500 °C and boils at nearly 2800 °C. Industrial furnaces that are used to melt iron and make steel operate at about 1800 °C.

SECTION

I

Key Terms

energy
exothermic
heat
system
surroundings
endothermic
kinetic energy
thermal equilibrium
first law of
thermodynamics
second law of
thermodynamics
entropy
thermal energy
calorie
specific heat capacity
hydrogen bond
heat of vaporization
heat of fusion

SUMMARY

Observing Energy

Fire Update

Energy and heat can be difficult to define. However, one thing is certain: Whenever matter changes in any way, energy is involved. Here are some generalizations you may have made from your observations of heat:

- Fire is an exothermic chemical change. Fire releases energy in the form of heat and light.
- Heat is a transfer of thermal energy. Heat always transfers from a system at a high temperature to a system at a lower temperature until both systems are at thermal equilibrium.
- In an endothermic reaction, energy is transferred from the surroundings to the system. An ice pack feels cold because your body transfers heat to the pack.
- The energy transferred by heating does not always result in temperature change. Some of that energy goes into breaking intermolecular attractions when a substance goes through a phase change.
- A metal pot will heat up and cool off more quickly than the water inside the pot. This is because different substances have different specific heat capacities.

Review Exercises

1. Provide evidence to support the first and second laws of thermodynamics.
2. Why is it possible to touch a piece of aluminum foil shortly after it has been removed from a hot oven, but a piece of hot cheese pizza can burn you long after it has been removed?
3. Thermal energy is transferred to raise the temperature of several substances as specified here. List the substances in order of increasing amount of energy that needs to be transferred. Show your work by determining the number of calories of energy transferred.
   a. Raise the temperature of 20 g of water by 25 °C.
   b. Raise the temperature of 200 g of copper by 50 °C.
   c. Raise the temperature of 20 g of methanol by 40 °C.
4. An ice cube tray can make 16 ice cubes. Each mold is filled with 25 mL of water, and the ice cube tray is placed in the freezer.
   a. As the water is freezing, is energy transferred from the freezer to the water, or from the water to the freezer? Draw a diagram of this scenario with an arrow showing the direction of heat transfer.

b. Would the freezing of the water in the tray be considered endothermic or exothermic for the ice cube?

c. Based upon your answer to part a, calculate the amount of heat transfer needed in order to freeze one ice cube.

d. Based upon your answer to part a, calculate the amount of heat transfer needed in order to freeze all the ice cubes in the tray.

5. A 5 kg iron pot is filled with 500 mL of water, placed on the stove, and the unit is turned on. After three minutes have passed, the temperature of the iron pot has increased from 25 °C to 150 °C.

   a. How much heat was transferred to the pot?

   b. After three minutes, would you expect the water in the pot to be the same temperature, hotter, or colder than the iron pot? Why?

6. A 30.4 g copper coin is heated to 400 °C with a Bunsen burner, and then dropped into a beaker containing 100 mL of 25 °C, room temperature water. The water absorbs 1000 cal of heat when this happens.

   a. What is the final temperature of the water in the beaker?

   b. Based upon your answer to part a, what will be the final temperature of the coin?

   c. What is the specific heat of the copper coin?

7. If 70 g of hot water at 85 °C are poured into 50 g of cold water at 18 °C, what is the temperature of the water mixture after it has come to thermal equilibrium?

8. A 50.0 g sample of pure water is heated from 10 °C to 120 °C.

   a. Draw and label a heating curve that represents the different stages of this process.

   b. If the specific heat capacity is 0.49 cal/g · °C for ice, 1.0 cal/g · °C for liquid water, and 0.50 cal/g · °C for steam, how much energy is added to a 50.0 g sample of water to raise its temperature from 10 °C to 120 °C.

   c. Use the heating curve and your calculations to determine which part of the process requires the most heat transfer. Explain your reasoning.

PROJECT *Uses of Fire*

Pick an object that you might purchase, and consider different ways that fire might have been used to either make it, harvest it, transport it, and so on. Describe step-by-step how fire may be involved in getting that object from its natural beginnings to you.

SECTION

II

Measuring Energy

When charcoal burns in a barbecue grill, energy is released, allowing us to cook a tasty meal.

One way to make use of the energy from chemical change is to burn a fuel. In homes, methane gas is burned in hot water heaters and kitchen stoves. In most automobiles, gasoline or some other fuel is burned. However, not all substances burn, and not all fuels are equally productive from an energy standpoint. This section of Unit 5: Fire introduces you to ways that energy from chemical change can be observed and measured.

In this section, you will study

- substances that will and will not burn
- experimental methods for measuring heat
- how to compare different fuels
- quantifying energy

LESSON

7 You're Fired!

Combustion

Think About It

An automobile engine burns gasoline or ethanol. In a fireplace, you burn paper and wood. In a barbecue grill, you burn charcoal, lighter fluid, and an occasional marshmallow. While many substances burn, others—like sodium chloride, table salt, and water—do not.

What types of substances burn?

To answer this question, you will explore

1. Substances That Burn
2. Combustion Reactions

Exploring the Topic

1 Substances That Burn

A wide variety of substances can burn, or combust. Solids, such as wood and coal, liquids, such as methanol or kerosene, and gases, such as methane or hydrogen, all combust. Something besides phase must determine whether a material can burn.

Use the table to compare substances that combust with those that do not combust.

Combustion and Bond Type

Substance	Chemical formula	Type of bond	Does it combust?
sodium chloride	$NaCl(s)$	ionic	no
iron (III) oxide	$Fe_2O_3(s)$	ionic	no
nickel	$Ni(s)$	metallic	yes
magnesium	$Mg(s)$	metallic	yes
water	$H_2O(l)$	molecular covalent	no
cellulose (paper)	$C_6H_{12}O_6(s)$	molecular covalent	yes
octane	$CH_3(CH_2)_6CH_3(l)$	molecular covalent	yes
carbon dioxide	$CO_2(g)$	molecular covalent	no
hydrogen	$H_2(g)$	molecular covalent	yes
silicon dioxide (sand)	$SiO_2(s)$	network covalent	no

HISTORY CONNECTION

From the 1880s until the 1930s, photographers would ignite a sample of magnesium to produce the flash needed to take a picture. For safety and simplicity, flash bulbs, which contained magnesium wire or thin sheets of magnesium foil, were then invented. In the 1980s, these bulbs were replaced with electronic flashes.

INDUSTRY CONNECTION

When pure alcohols, such as ethanol and methanol, combust, the flame is almost colorless and therefore dangerous because it is invisible. In 2005, the fuel for the cars racing in the Indianapolis 500 was changed from methanol to ethanol to promote safety. The ethanol now used contains 2% gasoline, which creates a visible yellow-orange flame when it is burning.

Using the information in the table, it is possible to make some generalizations about whether a substance will combust.

- Most ionic compounds do not combust.
- Most molecular covalent compounds do combust, especially those that contain carbon and hydrogen.
- Most elemental metals combust.
- Water and carbon dioxide do not combust.

2 Combustion Reactions

Combustion is a chemical reaction in which an element or a compound reacts with oxygen and releases energy in the form of heat and light. When molecular covalent substances containing mainly carbon and hydrogen burn, they produce carbon dioxide and water. When metallic substances combust, they produce a metal oxide.

These are the chemical equations for the combustion of methane and of magnesium.

$$CH_4(g) + 2O_2(g) \longrightarrow CO_2(g) + 2H_2O(g) + \text{heat} + \text{light}$$

$$2Mg(s) + O_2(g) \longrightarrow 2MgO(s) + \text{heat} + \text{light}$$

Oxygen is always a reactant in a combustion reaction. The other reactant in a combustion reaction is usually called a fuel. In these reactions, methane and magnesium are considered the fuels.

Notice that both reactions have heat and light on the product side of the equation. Remember, heat is not a substance, but chemists sometimes include it in the equation to keep track of energy changes.

If a compound is already combined with oxygen, it is not likely to combust and combine with *more* oxygen. This is why some molecular covalent compounds like water, H_2O, and carbon dioxide, CO_2, do not burn. Other examples include sulfur trioxide, $SO_3(g)$, and nitrogen dioxide, $NO_2(g)$.

Topic: Combustion
Visit: www.SciLinks.org
Web code: KEY-501

Fire—A Special Type of Combustion

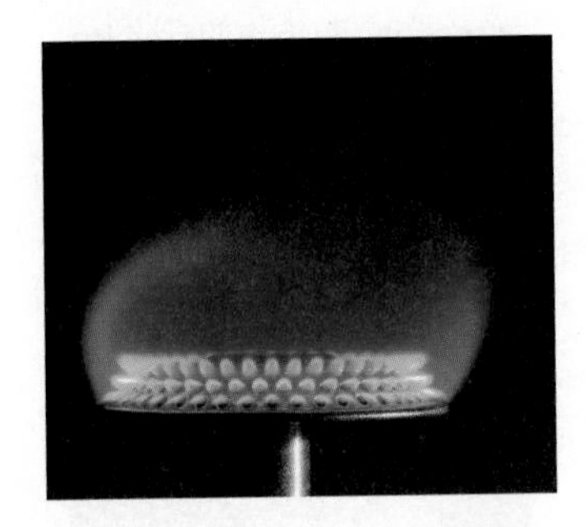

The photo shows the flame produced when methane burns. When we call something a fire, it is because a flame is present during the combustion reaction. The combustion of methane produces a fire.

Not all combustion reactions produce fire. The burning of magnesium produces a bright light but no flame. The light produced is so bright that it can damage your eyes. Most elemental metals are combustible but do not produce fire.

Ash is often left over after a fire. For molecular compounds, ash is composed of unburned or partially burned reactants. After combustion of metallic substances, the product may look like ash, but it is actually the metal oxide product of the reaction.

Key Term

combustion

Lesson Summary

What types of substances burn?

The process of burning is also referred to as combustion. Combustion is the combining of compounds or elements with oxygen through a chemical reaction that produces energy in the form of heat and light. Most ionic compounds do not combust. Most molecular covalent compounds are combustible. When molecular substances that are made mostly of hydrogen and carbon combust, they produce carbon dioxide and water. When metals combust, they form metal oxides, producing heat and light but no flame.

EXERCISES

Reading Questions

1. What should you look for in a chemical formula to decide if a compound or element will not combust?
2. Name three things that are true of every combustion reaction.

Reason and Apply

3. Which of the substances listed below will combust? Explain your reasoning.
   - **a.** $Na_2SO_4(s)$ **b.** $Cu(s)$ **c.** $MgO(s)$
   - **d.** $Na(s)$ **e.** $C_2H_6O(l)$ **f.** $Ar(g)$
4. Balance the following equations for combustion reactions. Circle the fuel for each reaction.
   - **a.** $C_2H_6(g) + O_2(g) \longrightarrow CO_2(g) + H_2O(l)$
   - **b.** $C_6H_{12}O_6(s) + O_2(g) \longrightarrow CO_2(g) + H_2O(l)$
   - **c.** $C_2H_5OH(l) + O_2(g) \longrightarrow CO_2(g) + H_2O(l)$
   - **d.** $C_{21}H_{24}N_2O_4(s) + O_2(g) \longrightarrow CO_2(g) + H_2O(l) + NO_2(g)$
   - **e.** $C_2H_5SH(l) + O_2(g) \longrightarrow CO_2(g) + H_2O(l) + SO_2(g)$
5. What are the products of these combustion reactions? Write balanced chemical equations for each reaction.
   - **a.** $C_7H_6O(l) + O_2(g) \longrightarrow$ **b.** $CH_3COCH_3(l) + O_2(g) \longrightarrow$
   - **c.** $H_2C_2O_4(s) + O_2(g) \longrightarrow$ **d.** $Ca(s) + O_2(g) \longrightarrow$
   - **e.** $Li(s) + O_2(g) \longrightarrow$ **f.** $Si(s) + O_2(g) \longrightarrow$

HISTORY CONNECTION

Irish physicist Robert Boyle created the first match in the 17th century when he coated a small piece of wood with sulfur and struck it against a phosphorus-coated piece of paper. Joshua Pusey, a Pennsylvania lawyer, invented book matches in 1889.

LAB Calorimetry

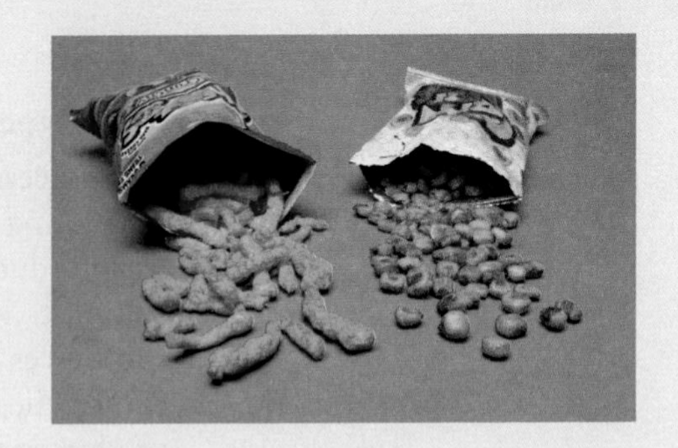

Purpose

To determine which snack food, cheese puffs or toasted corn snacks, transfers the most heat in a combustion reaction.

Materials

- cheese puffs, toasted corn snacks
- ring stand with ring and clamp
- Bunsen burner
- wire mesh, paper clips, safety pins, corks, modeling clay, rubber bands, tape
- matches
- crucible
- thermometer
- 150 mL beaker
- balance
- water
- empty tuna fish can, empty soft drink cans with tabs, aluminum foil

Safety Instructions

 Safety goggles must be worn at all times. Be sure to know the location of the fire blanket and fire extinguisher.

Procedure

1. Design an experiment using any of the materials above to determine which snack food will transfer more energy when burned. You must provide measurements and calculations to support your conclusion.
2. Write a detailed materials list of what you will use and a step-by-step procedure. You must get teacher approval before beginning your test.
3. When your teacher approves your procedure, conduct your experiment and record your observations and data.

Analysis

1. List all the types of data your team collected.
2. How did you decide what data were important for your experiment?
3. Why do you need to know the mass of the snack food burned?
4. **Making Sense** Which snack food supplies more energy to your body, cheese puffs or toasted corn snacks? What evidence do you have to support your answer?
5. **If You Finish Early** Repeat the experiment to determine whether the volume of water or the mass of fuel used makes a difference in the outcome.

LESSON 8 Now We're Cooking

Calorimetry

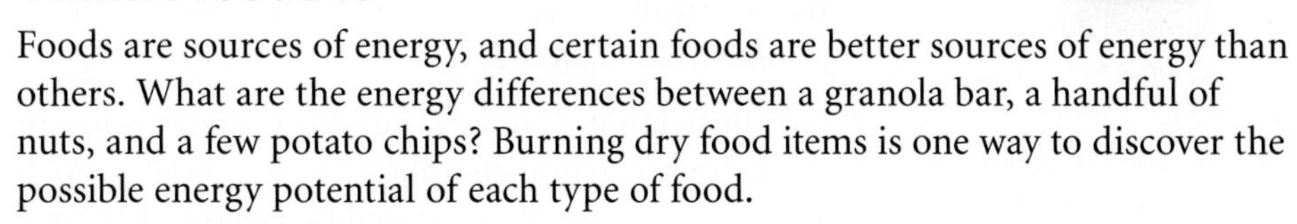

Think About It

Foods are sources of energy, and certain foods are better sources of energy than others. What are the energy differences between a granola bar, a handful of nuts, and a few potato chips? Burning dry food items is one way to discover the possible energy potential of each type of food.

How are food Calories measured?

To answer this question, you will explore

1. Foods As Fuel
2. Calorimetry

Exploring the Topic

1 Foods As Fuel

The foods you eat are sources of energy. There are no fires in your stomach, but your body processes the food you eat to generate energy. Burning two foods and comparing the energy transfer that results is a way to measure the energy of each food.

Food Calories

Food labels contain a wide variety of nutritional data, including Calorie content, which is a measure of energy. Food Calories, or Calories with a capital C, are actually kilocalories of energy, abbreviated kcal or Cal. This product label tells you that 1 ounce of potato chips has 160 Cal. In comparison, 1 ounce of walnuts has 190 Cal.

Nutrition Facts

Serving Size 1 oz.

Amount Per Serving	
Calories 160	Calories from fat 80
	% Daily Value*
Total Fat 10g	16%
Saturated Fat 1g	5%
Polyunsaturated Fat 3g	
Monounsaturated Fat 6g	
Trans Fat 0g	
Cholesterol 0mg	0%
Sodium 160mg	7%
Potassium 340mg	10%
Total Carbohydrate 14g	5%
Dietary Fiber 1g	4%
Sugars 0g	
Protein 2g	

Vitamin A 0%	Vitamin C 10%
Calcium 0%	Iron 0%
Vitamin E 8%	Thiamin 2%
Niacin 4%	Vitamin B_6 6%
Phosphorus 4%	

* Percent Daily Values are based on a 2,000 calorie diet. Your daily values may be higher or lower depending on your calorie needs:

	Calories	2,000	2,500
Total Fat	Less than	65g	80g
Sat Fat	Less than	20g	25g
Cholesterol	Less than	300mg	300mg
Sodium	Less than	2,400mg	2,400mg
Potassium		3,500mg	3,500mg
Total Carbohydrate		300g	375g
Dietary Fiber		25g	30g

Calories per gram:
Fat g • Carbohydrate 4 • Protein 4

> **Example**
>
> **Comparing Calories per Gram**
>
> Are there more Calories in 1.0 g of potato chips or 1.0 g of walnuts? (28.3 g = 1.0 oz)
>
> ***Solution***
>
> There are 160 Cal in 1.0 oz of potato chips. There are 190 Cal in 1.0 oz of walnuts. An ounce is the same as 28.3 g.
>
> $$\text{Potato chips: } 160 \text{ Cal}/28.3 \text{ g} = 5.6 \text{ Cal/g}$$
>
> $$\text{Walnuts: } 190 \text{ Cal}/28.3 \text{ g} = 6.7 \text{ Cal/g}$$
>
> So, there are more Calories available in 1.0 g of walnuts than 1.0 g of potato chips.

According to these calculations, an ounce of walnuts contains more energy than an ounce of potato chips. However, the number of Calories per gram is fairly close. In contrast, raisins contain 3.0 Cal per gram, and raw carrots contain 0.4 Cal per gram.

② Calorimetry

To measure the Calories in food, you need to find out how much energy the food can release when it reacts with oxygen. One way to determine the amount of heat transfer during a combustion reaction is to use the fuel to heat a measured amount of some other substance, such as water. You can then use a thermometer to measure the change in temperature of the water. The science of measuring the energy released or absorbed in a chemical reaction or physical change is called **calorimetry.** (In Latin, *calor* means heat and *meter* means measure.) Chemists and nutritionists alike use a bomb calorimeter to figure out the calorie content of combustible substances.

HEALTH CONNECTION

A healthy diet for the average adult consists of an intake of about 2000 food Calories per day. One fast-food double cheeseburger contains about 750 Calories, a medium soda is about 210 Calories, and a medium order of French fries is 450 Calories. If you ate this for lunch, you would have only 600 Calories left for the rest of the day.

The illustration shows a bomb calorimeter. The combustion reaction takes place in an inner reaction chamber called a bomb, which is completely surrounded by water. The sample is ignited by electrical energy, and the energy, due to the burning of the fuel, is transferred to the water. A stirring rod makes sure the heat is uniform in the water. If you determine the change in temperature of the water and you know the quantity of water in the calorimeter, you can determine the heat transferred during the combustion reaction.

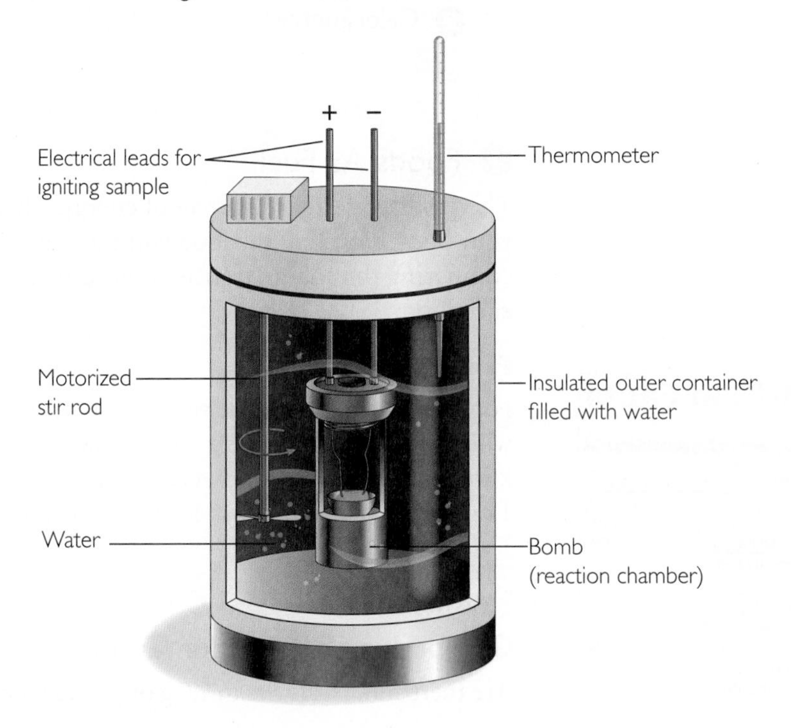

Key Terms
calorimetry

Lesson Summary

How are food Calories measured?

There are a number of different ways to compare the energy of foods. Some foods can be burned and then their energies can be measured and compared. Calorimetry is the science of measuring the energy released or absorbed by a chemical or physical change. You can use a bomb calorimeter to measure the amount of heat transferred to the surroundings in a combustion reaction.

EXERCISES

Reading Questions

1. What do Calories have to do with combustion?
2. Why does a bomb calorimeter have an inner chamber and an outer chamber?

Reason and Apply

3. Can you measure the heat energy of a combustion reaction by placing a thermometer directly into the flames? Explain.
4. What is calorimetry?
5. Most school laboratories do not have bomb calorimeters.
   a. Design an experiment that would allow you to compare whether a cheese puff snack or a roasted corn snack transfers more energy when burned.
   b. In your experiment, does it matter whether you burn the same mass of cheese puff and roasted corn snack? Why or why not?
   c. Why is it important to measure the mass of each snack food at the end of your experiment?
   d. What are some possible sources of error that could occur in your experiment?
6. **Lab Report** Write a lab report for the Lab: Calorimetry. In the procedure section, describe your procedure in detail and include a sketch of your setup. In your conclusion, include a discussion of how well your setup worked and whether you would change anything.

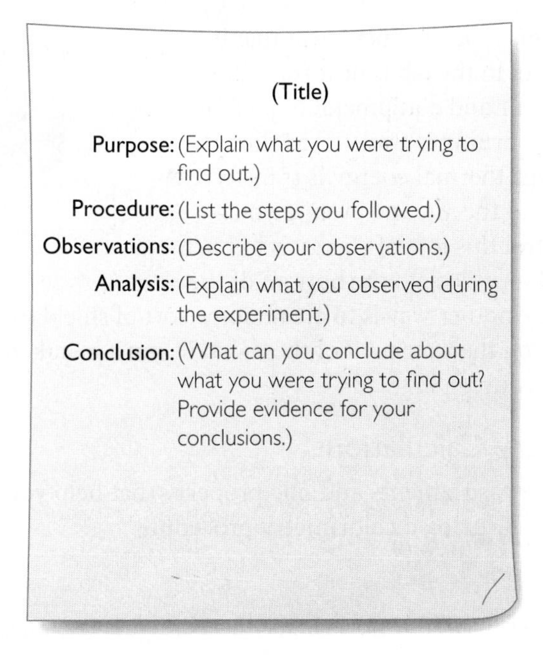
(Title)

Purpose: (Explain what you were trying to find out.)

Procedure: (List the steps you followed.)

Observations: (Describe your observations.)

Analysis: (Explain what you observed during the experiment.)

Conclusion: (What can you conclude about what you were trying to find out? Provide evidence for your conclusions.)

LESSON

9 Counting Calories

Calorimetry Calculations

Think About It

Heating a beaker of water with a burning potato chip seems like an unusual way to determine the number of Calories in this snack food. However, this is very close to the actual procedure used by nutritionists.

How does a calorimetry experiment translate into Calories?

To answer this question, you will explore

1. Calorimetry in the Lab
2. Calorimetry Calculations

Exploring the Topic

1 Calorimetry in the Lab

Suppose you wanted to measure the amount of energy transfer from combustion of a cashew nut. Because the digestive processes of the human body cannot be duplicated in your classroom, the next best approach is to determine the energy that can be transferred when food reacts directly with oxygen. To do this, you need to burn the food item and transfer the energy released due to combustion to another substance. Many substances can be used, but the most common one to use is water.

HEALTH CONNECTION

The actual amount of energy a human obtains after the digestive processes are completed is about 85% of the Calorie content listed on nutrition labels. Fats have high food energy densities, around 9 Cal/g. Sugars and proteins are around 4 Cal/g.

One feature of this type of experiment that is difficult to control in the lab is heat transfer to the surrounding air and equipment. Not all of the energy from burning is transferred directly to the water. Some thermal energy is transferred to the container, to the air, and even to you. One way to control this loss of energy is to burn the food sample very close to the bottom of the water container. Another way is to create some sort of shield to keep the energy transfer focused on the water. All of these details must be taken into account when designing a procedure.

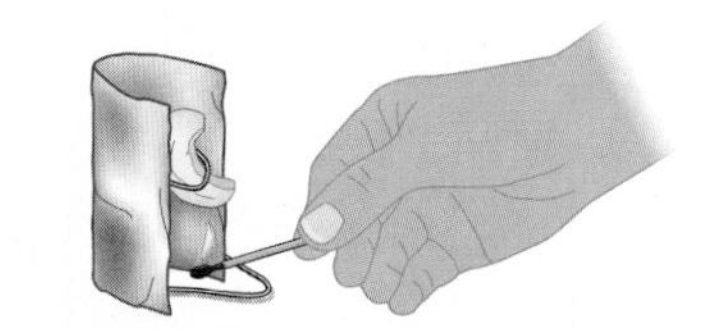

Building a shield of some sort helps to direct the heat from burning.

2 Calorimetry Calculations

There are three measurements and one property that help you to determine the energy transferred during a calorimetry procedure.

Data needed:

1. Mass of water heated in grams; 1 mL = 1 g
2. Temperature change of the water in degrees Celsius, ΔT (delta T)
3. Mass of fuel burned in grams
4. Specific heat capacity of water = 1.00 cal/g °C at 25 °C and 1 atm pressure

With the first two pieces of data, you can calculate how many calories of thermal energy are transferred to the water. Recall that the equation for heat transfer is

$$q = mC_p\Delta T$$

where C_p is the specific heat capacity of the substance being heated.

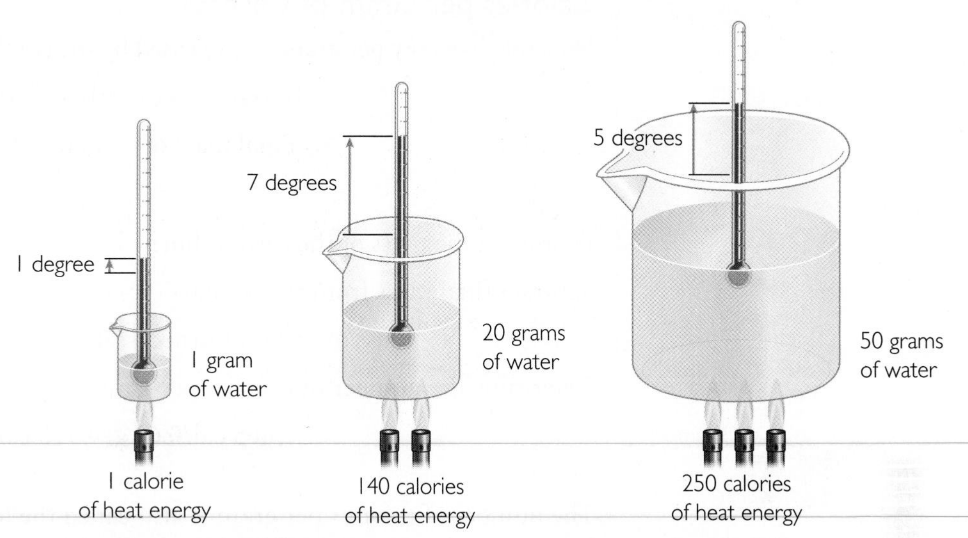

The number of calories of energy transferred is dependent on the mass of the water and the temperature change.

The illustration shows that if you multiply the change in temperature by the mass of the water, you can calculate the number of calories of heat transferred to the water. This is because the specific heat capacity of water is 1.00 cal/g · °C.

Example 1

Calories from a Cashew

These data were collected from the burning of one cashew. How much thermal energy was transferred?

Initial T of water = 19.0 °C

Final T of water = 34.5 °C

Volume of water = 30.0 mL

Solution

First, figure out the temperature change.

Subtract the initial temperature from the final temperature.

$$\Delta T = 34.5\ °C - 19.0\ °C = 15.5\ °C$$

Since 1 mL of water is equal to 1.0 g of water, the mass of the water heated is 30.0 g.

Use the equation for heat transfer.

Substitute values and solve.

$$q = mC_p\Delta T = 30.0\ g(1.00\ cal/g \cdot °C)15.5\ °C = 465\ cal$$

One cashew transferred 435 cal of energy to the water. In order to compare different foods, it is necessary to figure out the calories *per gram* of food burned.

Example 2

Calories per Gram of Cashew

How much energy per gram was released by the combustion of the cashew?

Initial mass of cashew = 0.66 g

Final mass of cashew = 0.06 g

Solution

Figure out the mass of the cashew burned.

Subtract final mass from initial mass.

$$\text{Mass of cashew burned} = 0.66\text{ g} - 0.06\text{ g} = 0.60\text{ g}$$

Determine the number of calories per gram.

$$465\text{ cal}/0.60\text{ g} = 775\text{ cal/g}$$

The number of calories per gram is also called the food energy density. Cashews have an energy density of 775 cal/g of cashew burned. Remember, these are chemist's calories. To convert them to food Calories, you must divide by 1000 to get 0.775 Cal/g.

Comparing Foods

In the real world, nutritionists use calorimeters to figure out the Calorie content of foods. The word *content* is misleading. The food does not actually contain the Calories. The Calorie content represents the energy transferred specifically by *combustion* of these compounds.

Bomb calorimeters minimize the heat transfer to the surroundings, and the results are much more precise than our lab experiments. The method of calculation is very similar to the one used here. If a bomb calorimeter is used, a cashew would be found to have closer to 5500 chemists' calories per gram, or 5.5 Cal/g. Our experimental results came up with 775 cal/g. There is a lot of error in the classroom procedure.

Lesson Summary

How does a calorimetry experiment translate into Calories?

Different fuels transfer different amounts of energy during a combustion reaction. The temperature change, fuel mass, and water volume data from a calorimetry procedure can be converted into Calories. Using the specific heat capacity of the substance that was heated by the reaction allows you to calculate the exact number of calories transferred by the combustion of the fuel. This provides you with a measure of the amount of thermal energy "contained" in a fuel. The number of calories transferred from a substance that burns depends on the identity of the substance and its mass.

EXERCISES

Reading Questions

1. Name three possible sources of error in a calorimetry experiment.
2. Why is it important to know the specific heat of the substance being heated by the combustion of a fuel?

Reason and Apply

3. Could you use ethanol instead of water in a calorimetry experiment? Explain.
4. The experimental value for the Calorie content of a cashew is 0.775 Cal/g. The actual value provided on the package label is 5.5 Cal/g. Determine the percent error.
5. A cereal flake is burned under a beaker containing 25 mL of water. If the water temperature goes up 6 °C, how many calories of energy were transferred to the water?
6. Fuel pellets are used in modern energy-saving wood stoves. If the pellets used for these stoves release 742 cal/g, how many calories of energy will be released by combustion of an entire 40 lb sack of pellets?
7. Examine the illustration on page 513. Use a kinetic molecular view to explain why it takes more calories of heat to raise the water temperature in the third beaker by 5 °C than it does to raise the temperature of the water in the second beaker by 7 °C.
8. The calorie content of a peanut is measured by burning it beneath a can of water and measuring the temperature change of the water. Which of these is a possible source of error?

   A. The initial mass of the peanut is measured incorrectly.

   B. Some of the heat of combustion is transferred to the air.

   C. Some unburned remnants of the peanut are lost before finding the final mass.

   D. All of the above.

   E. None of the above.

HISTORY CONNECTION

In the late 18th century, the French scientists Antoine Lavoisier and Pierre Laplace developed an ice calorimeter. With this apparatus, the amount of ice melted by a reaction was measured to determine the energy transferred. A century later, around 1860, the French chemist Pierre Bertholet constructed the first modern calorimeter.

LESSON

10 Fuelish Choices

Heat of Combustion

Think About It

Combustion reactions are used to power vehicles, heat water and homes, and create electricity. But not all combustion reactions provide the same amount of energy. Some fuels heat water or power furnaces better than others.

How do different fuels compare?

To answer this question, you will explore

1. Burning Fuels
2. Heat of Reaction
3. Comparing Fuels

Exploring the Topic

1 Burning Fuels

There are a number of different choices of fuels to power race cars. Ideally, you want to choose a fuel that transfers the most energy per gram. That way, the fuel does not unnecessarily add to the weight of the car. You can compare the energy from combustion of a variety of fuels using calorimetry to see which one would be best for a race car.

Ethanol Versus Methanol

Both ethanol, C_2H_6O, and methanol, CH_4O, have been used to power race cars. Suppose you put the two fuels into alcohol burners and use them to heat two identical 50 mL samples of water by 30 °C.

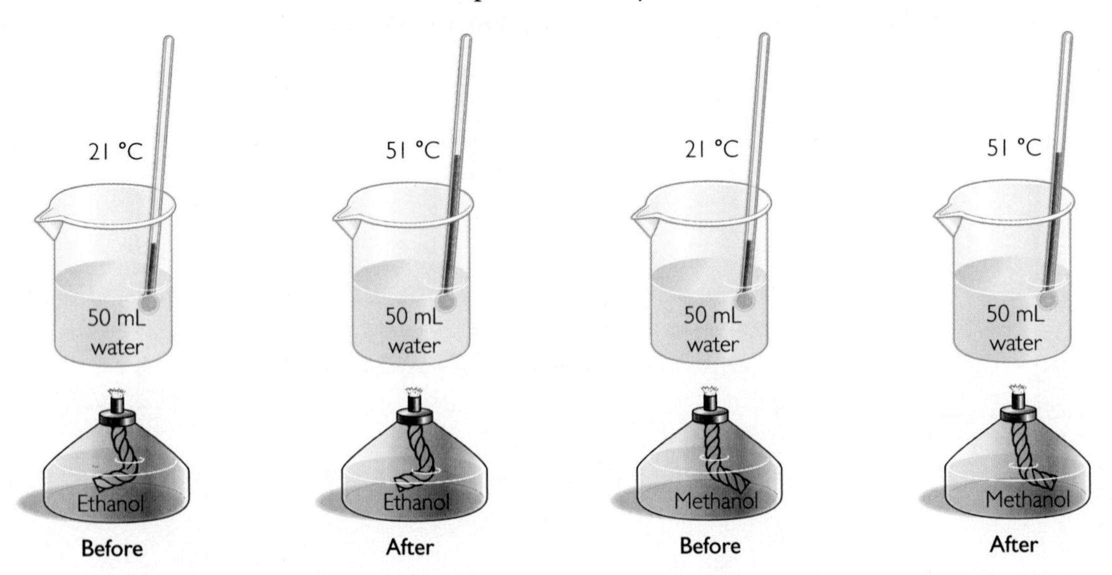

Burner plus ethanol = 128.0 g   Burner plus ethanol = 126.8 g   Burner plus methanol = 128.0 g   Burner plus methanol = 119.4 g

ENVIRONMENTAL CONNECTION

A biofuel is any plant substance that can be used as a fuel. For example, some people have customized their cars to burn vegetable oils. These people can collect used French fry oil from restaurants for use in their fuel tanks.

The chemical equations for the combustion of each type of fuel are shown here.

$$\text{Combustion of ethanol: } 2C_2H_6O(l) + 6O_2(g) \longrightarrow 4CO_2(g) + 6H_2O(l)$$

$$\text{Combustion of methanol: } 2CH_4O(l) + 3O_2(g) \longrightarrow 2CO_2(g) + 4H_2O(l)$$

	Mass of fuel burned (g)	Mass of water heated (g)	ΔT for the water (°C)	Energy transferred to water (cal)	Energy per gram of fuel (cal/g)
Ethanol	1.2 g	50 g	30 °C	1500 cal	1250 cal/g
Methanol	1.6 g	50 g	30 °C	1500 cal	938 cal/g

The experimental results show that the combustion of ethanol delivers more energy per gram of fuel consumed, compared with the combustion of methanol. So you would want to use ethanol, not methanol, in a race car if weight is a concern.

2 Heat of Reaction

The amount of energy transferred during a chemical reaction is called the **heat of reaction.** It is often designated with the symbol ΔH. The heat of reaction for a combustion reaction is usually referred to as the **heat of combustion.** Heat of reaction, ΔH, is expressed as a negative number when the reaction is exothermic. This is because energy is transferred from the system. When a reaction is endothermic, the system gains energy. Heat of reaction, ΔH, is expressed as a positive number when the reaction is endothermic.

Scientists have done carefully controlled experiments to measure the heat of combustion for many different fuels. Accurate measurements for the combustion of methanol result in a heat of reaction that is much larger in magnitude than the experimental value reported earlier.

$$\text{Combustion of methanol: } 2CH_4O + 3O_2 \longrightarrow 2CO_2 + 4H_2O$$

$$\Delta H = -5400 \text{ cal/g}$$

$$\text{Combustion of ethanol: } C_2H_6O(l) + 6O_2(g) \longrightarrow 4CO_2(g) + 6H_2O(l)$$

$$\Delta H = -7100 \text{ cal/g}$$

Important to Know The heat of combustion is reported with a negative sign to indicate that energy is released by the reaction. ◂

Units

The metric unit used by scientists to express energy is the joule, J. A **joule** is a unit of energy that is 4.184 times larger than a calorie. The heat of reaction can be expressed in kilocalories per gram, kilocalories per mole, kilojoules per gram, or kilojoules per mole.

$$1 \text{ joule} = 0.239 \text{ calorie}$$

$$1 \text{ calorie} = 4.184 \text{ joules}$$

❸ Comparing Fuels

One way to compare fuels is to calculate the number of kilojoules of energy per mole (kJ/mol) of fuel. Simply multiply the kilojoules per gram (kJ/g) by the molar mass (g/mol) of each fuel.

This table provides the heat of combustion for several fuels in both kilojoules per gram and kilojoules per mole.

Heats of Combustion

	Fuel	Chemical formula	ΔH (kJ/g)	ΔH (kJ/mol)
elements	hydrogen	$H_2(g)$	−121.3	−242
	carbon (coal)	$C(s)$	−32.6	−393
alkanes	methane	$CH_4(g)$	−55.6	−891
	ethane	$C_2H_6(g)$	−51.9	−1560
	butane	$C_4H_{10}(l)$	−49.6	−2882
	hexane	$C_6H_{14}(l)$	−48.1	−4163
	octane	$C_8H_{18}(l)$	−47.7	−5508
alcohols	methanol	$CH_4O(l)$	−22.6	−724
	ethanol	$C_2H_6O(l)$	−29.7	−1368
	butanol	$C_4H_{10}O(l)$	−36.0	−2669
sugar	glucose	$C_6H_{12}O_6(s)$	−15.9	−2828

Notice that when the number of carbon atoms increases, the energy per mole increases. Also, the energy transferred per mole is smaller for the alcohols compared with the alkanes. This makes sense because combustion is a reaction in which a substance combines with oxygen, and the alcohols already contain an oxygen atom. Also notice that elemental hydrogen releases the highest energy per gram. This is why hydrogen is used as the fuel to launch the space shuttle. You get a lot of energy without adding much weight.

Key Terms

heat of reaction
heat of combustion
joule

Lesson Summary

How do different fuels compare?

One way to compare different fuels is to measure the amount of energy they transfer as a result of combustion. Calorimetry is the most common method of measuring this energy. The energy transferred can be expressed in kcal/mol, kcal/g, kJ/mol, or kJ/g of substance. The amount of energy transferred per gram or per mole during a chemical change is commonly called the heat of reaction, ΔH. For combustion reactions, the heat of reaction is negative and is also referred to as the heat of combustion.

EXERCISES

Reading Questions

1. How you can use water to compare two fuels?
2. What is heat of combustion?
3. Explain why heat of combustion is expressed as a negative number.

Reason and Apply

4. The combustion of 4.0 g of propanol transfers energy to 150 mL of water. The temperature of the water rises from 20 °C to 45 °C.
   a. How many calories of energy are transferred to the water?
   b. How many kilocalories of energy are transferred to the water?
   c. How many kilojoules of energy are transferred to the water?
   d. What temperature increase do you expect for 300 mL of water?
   e. What temperature increase do you expect for 2.0 g of propanol and 150 mL of water?
5. Write balanced chemical equations for the combustion of the fuels in the heats of combustion table on page 518. What generalization can you make about the amount of oxygen required for combustion of hydrocarbons?
6. Explain why the combustion of glucose provides a large energy per mole, but a rather small energy per gram.
7. **WEB RESEARCH** In the heats of combustion table, hydrogen has the highest energy per gram of fuel that is combusted. Give three reasons why it is challenging to use hydrogen as a fuel for your automobile.

SECTION

II

Key Terms

combustion
calorimetry
heat of combustion
heat of reaction
joule

SUMMARY

Measuring Energy

Fire Update

Combustion can be defined as a reaction in which a substance combines with oxygen, releasing energy. The process of combustion is commonly referred to as burning, and the substance that burns is called a fuel. Some fuels are better sources of energy than others. To measure the energy from a fire, or any chemical reaction, you must measure it indirectly by measuring its effect on matter. Energy is usually measured in units of calories or joules.

The procedure used to measure the transfer of energy is called calorimetry. The amount of energy transferred during a chemical reaction is called the heat of reaction, ΔH. The energy released by the combustion of a fuel is called the heat of combustion.

Review Exercises

1. Describe an experimental setup that would help you compare two fuels. What would you measure?
2. A chemist wants to find out how many Calories are in a bag of peanuts. He sets up a calorimeter, and burns a peanut underneath it. This is his data table.

Initial mass of peanut (g)	Mass of water (g)	ΔT of water (°C)	Final mass of peanut (g)
3.75 g	100 g	10 °C	1.20 g

   a. Calculate the amount of energy, in calories, that can be transferred from the combustion of one peanut.
   b. Calculate the amount of calories per gram of peanut.
   c. If a bag of peanuts is 500 g, how many food Calories does it contain?
3. If iron nails are left outside, they will rust according to the reaction: $4Fe(s) + 3O_2(g) \longrightarrow 2Fe_2O_3(s)$. Why do you think this reaction does not appear to release light and heat?
4. Propane, $C_3H_8(g)$, is used as fuel in small gas stoves and barbecue grills.
   a. Write the balanced equation for the combustion of propane.
   b. The heat of combustion of propane is −2220 kJ/mol. How many kJ/g would combusting propane release?
   c. According to the table on page 518, is the heat of combustion per gram of propane greater than the heat of combustion per gram of methane? Per mole? Can you explain why?

SECTION

III

Understanding Energy

Early trains were powered by steam engines. A steam engine burns coal or wood to boil water, then uses the pressure from the water vapor to drive the engine. Modern trains run on diesel or electricity instead.

When chemical change occurs, new substances with new properties are produced. This means some chemical bonds must be broken and new chemical bonds must be formed. The energy transferred during chemical change is related to bond making and bond breaking. People have figured out how to make productive use of the energy that accompanies chemical changes. This section of Unit 5: Fire focuses on the source of energy in chemical change. It also explores how easily or how quickly a reaction might happen.

In this section, you will study

- bond energies and heat of reaction
- reversing chemical reactions
- rates of chemical reactions
- how energy is converted to work

ACTIVITY Make It or Break It

Purpose

To explore bond energies and calculate the net energy exchange for several reactions.

Materials

- ball-and-stick model kit

Questions

1. List at least two patterns you see in the table of average bond energies.

Average Bond Energies (per mole of bonds)

Bond	H–H	C–H	C–C	C=C	O–H	C–O	C=O	O=O	O–O
Bond energy (kJ/mol)	432	413	347	614	467	358	799	495	146

2. Create the reactants for this reaction with the ball-and-stick models.

   Burning methane: $CH_4(l) + 2O_2(g) \longrightarrow CO_2(g) + 2H_2O(g)$

3. Rearrange the reactant models into carbon dioxide and water. Count how many of each type of bond you must break when this reaction takes place. Record the information in a table. Use the average bond energies to determine the amount of energy transferred to break all the bonds.
4. Count how many of each type of bond you must make in order to form the products. Record this information in your table. Use the average bond energies to determine the amount of energy transferred out when the new bonds are formed.
5. What is the net energy you expect to be transferred to the surroundings by this reaction?
6. The energy of reaction for exothermic reactions is normally expressed as a negative number. Why do you think this is?
7. **Making Sense** When a substance combusts, energy is transferred to the surroundings as heat and light. Where does that energy come from?
8. **If You Finish Early** What is the energy of reaction for 1 g of methane burned, in kilocalories per gram?

LESSON

11 Make It or Break It

Bond Energy

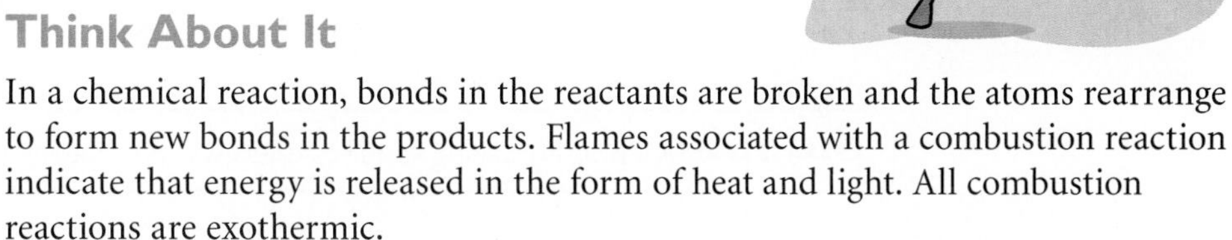

Think About It

In a chemical reaction, bonds in the reactants are broken and the atoms rearrange to form new bonds in the products. Flames associated with a combustion reaction indicate that energy is released in the form of heat and light. All combustion reactions are exothermic.

Where does the energy from an exothermic reaction come from?

To answer this question, you will explore

1. Bond Breaking and Bond Making
2. Energy of Change
3. Energy Exchange Diagrams

Exploring the Topic

1 Bond Breaking and Bond Making

In chemical compounds, atoms are bonded together. When the compound methane burns, it combines with oxygen to make new products. In order to make carbon dioxide and water from molecules of methane and oxygen, the atoms must rearrange as shown. This means that covalent bonds in the methane and oxygen molecules must be broken and new covalent bonds must be formed.

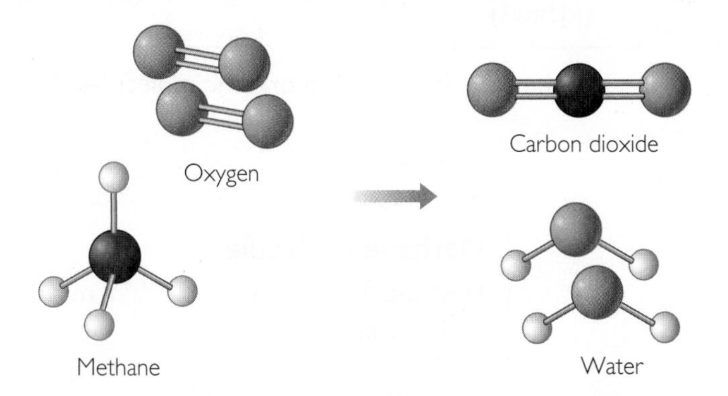

Energy is required to break bonds. Atoms that are bonded together are attracted to each other, so it takes effort to get these atoms apart. So bond breaking is an endothermic process.

In contrast, when new bonds are formed, energy is released. In terms of energy, bond making is the opposite of bond breaking. Because energy is conserved, the amount of energy released when a bond is made is assumed to be equal to the amount of energy required to break that same bond.

During a combustion reaction, energy is required to break the bonds in the reactant molecules and energy is released when the bonds form in the product molecules. The net energy released is observed as heat and light.

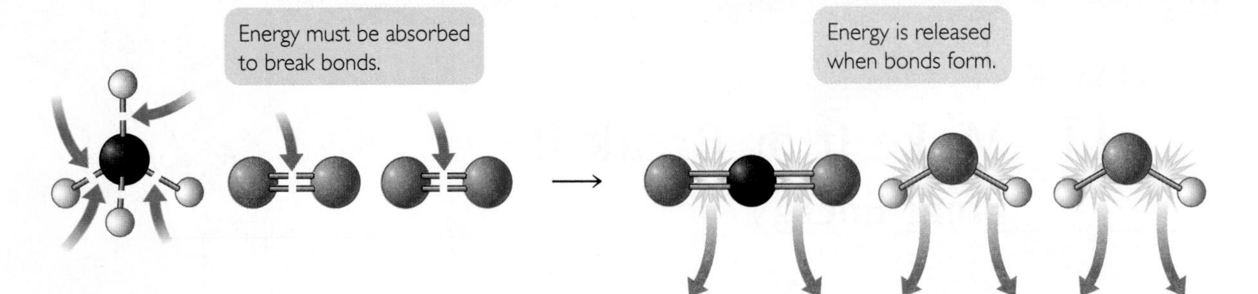

BIG IDEA Bond breaking requires energy. Bond making releases energy.

Bond Energy

The amount of energy it takes to break a specific bond, such as a carbon-hydrogen bond, is called the **bond energy.** Bond energy is considered a measure of the strength of that bond. However, not all C—H bonds have the same bond energy. For example, it takes 435 kJ/mol to break the first C—H bond in a methane molecule, as compared to 339 kJ/mol to break the last C—H bond.

For this reason, bond energies are reported as averages. The bond energy of an average carbon-hydrogen bond is approximately 413 kJ/mol (or 99 kcal/mol). This means that it takes 413 kilojoules of energy to break one mole of carbon-hydrogen bonds.

This table shows the bond energies of some common bonds.

Average Bond Energies (per mole of bonds)

Bond	H—H	C—H	C—C	C=C	C≡C	O—H	C—O	C=O	O=O	O—O
Bond energy (kJ/mol)	432	413	347	614	811	467	358	799*	495	146

*C=O in CO_2: 799, C=O in organic molecules: 745

Example

Methane Molecule

How much energy would it take to break all the bonds in 1 mol of methane molecules?

Solution

Each molecule of methane has four C—H bonds. Therefore, each mole of methane molecules has 4 mol of carbon-hydrogen bonds.

$$413 \text{ kJ/mol} \cdot 4 \text{ mol C—H bonds} = 1652 \text{ kJ}$$

In reality, all of the bonds of the reactants do not necessarily break in order to form the products in a chemical reaction. However, this is a useful model to explain energy exchanges related to a chemical reaction.

❷ Energy of Change

To estimate the energy of an entire chemical reaction, you can consider the reaction as if it takes place in two parts–*energy in* for bond breaking and *energy out* for bond making.

Combustion of Methane

Consider the reaction for the combustion of 1 mol of methane molecules with 2 mol of oxygen molecules. What is the net energy exchange?

$$CH_4 + 2O_2 \longrightarrow CO_2 + 2H_2O$$

First, figure out how much energy is required to break all the bonds in the reactants.

Bond Breaking—Energy In

413 kJ 413 kJ 495 kJ 495 kJ

413 kJ 413 kJ

1 mol of CH_4 2 mol of O_2

Total energy input = 4(413 kJ) + 2(495 kJ) = 2642 kJ

Notice that the energy required to break the bonds of the reactants is positive because energy is added to the system from the surroundings. Now figure out how much energy is released by the formation of new bonds.

Bond Making—Energy Out

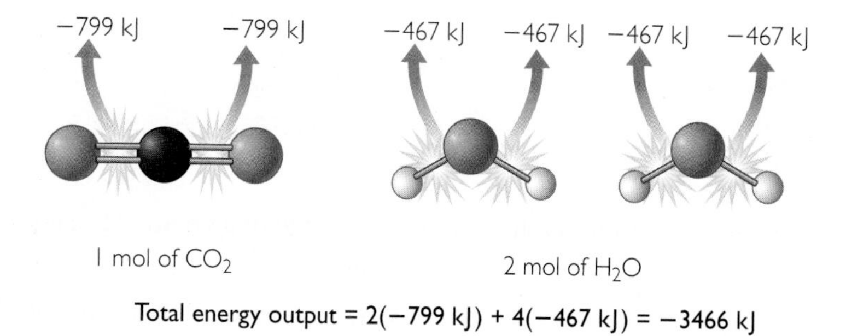

Total energy output = 2(−799 kJ) + 4(−467 kJ) = −3466 kJ

Notice that the energy required to make the bonds of the reactants is negative because energy leaves the system and goes to the surroundings. The net energy exchange is equal to the sum of the energy input and the energy output.

$$\text{Net energy exchange} = (2642 \text{ kJ}) + (-3466 \text{ kJ})$$
$$= -824 \text{ kJ}$$

The magnitude of energy out is greater than the magnitude of energy in, so this reaction is exothermic. Net energy is expressed as a negative number, because the system loses energy to the surroundings. This value is very close to the actual value for the heat of combustion, ΔH, for this reaction.

❸ Energy Exchange Diagrams

An energy exchange diagram is a way to keep track of the energy changes during a chemical reaction. The arrows pointing up represent the energy going into the system to break the bonds. The arrows pointing down represent the energy released when new bonds are made. The net energy change determines whether a reaction is exothermic or endothermic.

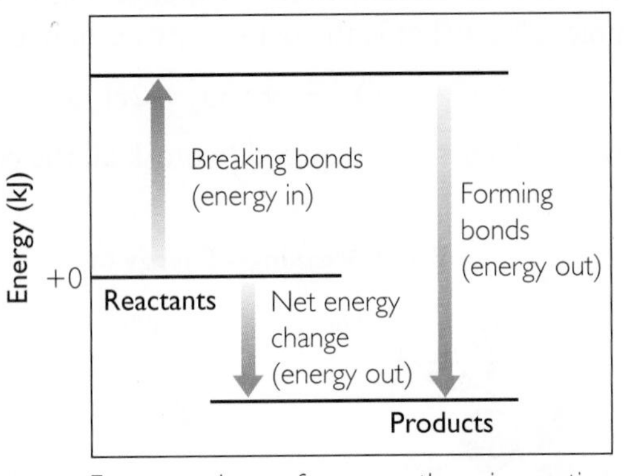

Energy exchange for an exothermic reaction.

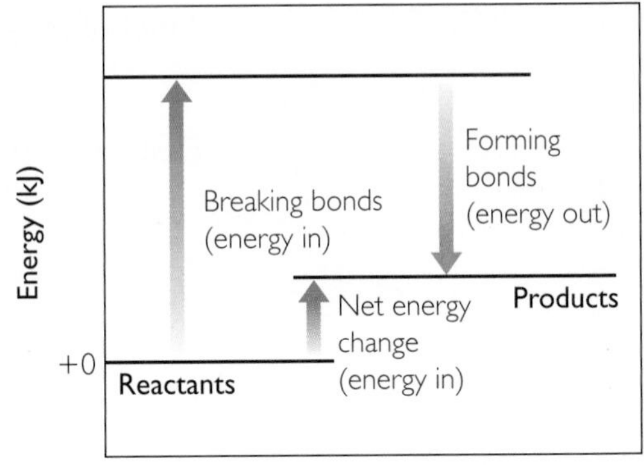

Energy exchange for an endothermic reaction.

Key Term

bond energy

Lesson Summary

Where does the energy from an exothermic reaction come from?

When reactions take place, bonds between atoms are broken and new bonds are formed. Energy must be transferred *to* a reactant to break a bond. The average energy required to break a certain type of bond is called its bond energy. Conversely, when new bonds are made, energy is released. The heat transferred during a chemical reaction is equal to the difference in energy between the bond-breaking process and the bond-making process. An energy diagram can be used to keep track of energy exchanges in a chemical reaction.

EXERCISES

Reading Questions

1. How can you determine the net energy exchange of a chemical reaction using average bond energies?
2. Use the energies involved in bond breaking and bond making to explain why combustion reactions are exothermic.

Reason and Apply

3. Place these bonds in order of increasing energy required to break the bond: H–H, C–C, C=C, O–O, O=O.
4. Place these bonds in order of increasing energy released when the bond is formed: H–H, C–H, O–H, C=O.
5. What is the total energy required to break all the bonds in 1 mol of ethanol, C_2H_6O?

6. Consider the combustion of hydrogen. Use the table of average bond energies to answer the questions.

$$2H_2(g) + O_2(g) \longrightarrow 2H_2O(l)$$

a. What is the total energy required to break the bonds in the reactant molecules?

b. What is the total energy released by the formation of the bonds in the product molecules?

c. Use the values from (a) and (b) to determine if this reaction is exothermic or endothermic.

7. Consider the combustion of methanol and butanol. Use the table of average bond energies to answer these questions.

$$\text{Methanol: } 2CH_4O(l) + 3O_2(g) \longrightarrow 2CO_2(g) + 4H_2O(l)$$

$$\text{Butanol: } C_4H_{10}O(l) + 6O_2(g) \longrightarrow 4CO_2(g) + 5H_2O(l)$$

a. Calculate the net energy exchange for each reaction.

b. Compare your answers to the heat of combustion values for methanol and butanol in Lesson 10: Fuelish Choices. What did you discover?

8. Calculate the net energy change for this reaction:

$$C_2H_6(g) \longrightarrow C_2H_4(g) + H_2(g)$$

9. In exothermic reactions,

A. It takes more energy to make all the new bonds in the products than to break all the bonds in the reactants.

B. There are more bonds made than broken.

C. There are more bonds broken than made.

D. It takes more energy to break all the bonds in the reactants than to make all the new bonds in the products.

LESSON

12 Over the Hill

Reversing Reactions

Think About It

So far you have been shown chemical reactions that proceed in one direction. This means that reactants are converted to products until the reaction is complete. Some reactions can be reversed. If a reaction is run in the reverse direction, the products of the reaction are converted back into reactants.

What is the energy associated with reversing reactions?

To answer this question, you will explore

1. Reversing Reactions
2. Kinetic and Potential Energy
3. Activation Energy

Exploring the Topic

1 Reversing Reactions

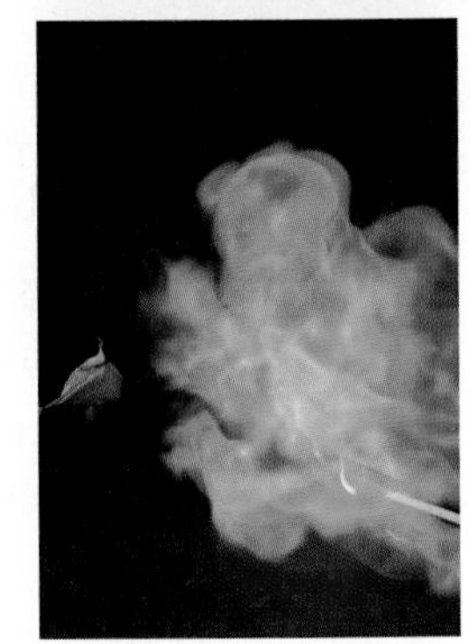

If you place a lit candle under a balloon filled with hydrogen, you can observe the formation of water. When hydrogen combines with oxygen to form water there is a large "bang."

$$2H_2(g) + O_2(g) \longrightarrow 2H_2O(g) \qquad \Delta H = -482 \text{ kJ}$$

The energy released by this reaction is the heat of reaction, −482 kJ. This energy corresponds to 2 mol of $H_2(g)$ reacting with 1 mol of $O_2(g)$ to form 2 mol of $H_2O(g)$. This converts to −241 kJ per mole of $H_2(g)$, the value given in the heat of combustion table in Lesson 10: Fuelish Choices.

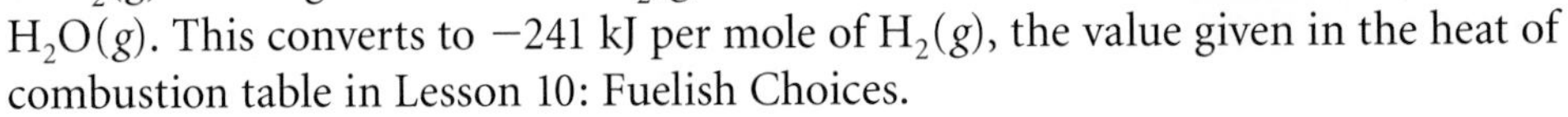

Important to Know Heat of reaction values are determined experimentally by calorimetry. You can use average bond energies to estimate the heat of reaction. ◀

Notice that the decomposition of water into molecules of hydrogen and oxygen is the reverse of the combustion of hydrogen to form water. What is the heat of reaction for this change?

$$2H_2O(g) \longrightarrow 2H_2(g) + O_2(g) \qquad \Delta H = +482 \text{ kJ}$$

The first law of thermodynamics states that energy is conserved. So the energy exchange of the reaction in the forward direction is equal and opposite to the energy exchange required for the reverse of the reaction. Notice that when the forward reaction is exothermic, the reverse reaction must be endothermic.

Diagrams showing the energy exchange for the formation of water and the decomposition of water are provided here. These energy diagrams are similar to the ones in Lesson 11: Make It or Break It.

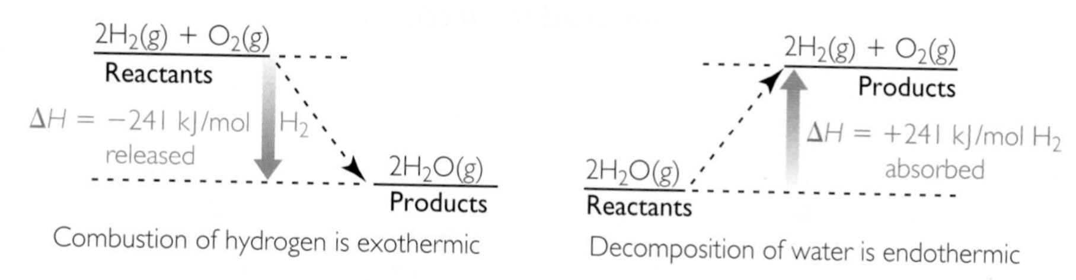

❷ Kinetic and Potential Energy

The energy in a system is a combination of kinetic energy and potential energy. Kinetic energy is the energy of motion. **Potential energy** is energy that is stored within a physical system that can be converted into another type of energy, such as kinetic energy.

Potential energy can also be defined as the energy associated with the composition of a substance or the position or location of an object in space. For example, a ball at the top of a hill has high potential energy. This potential energy is converted into kinetic energy if the ball rolls down the hill. Likewise, some molecules have a high potential energy. For an exothermic reaction, this potential energy is converted into kinetic energy by converting reactants into products.

Conserving Energy

These diagrams show the changes in potential energy for an exothermic reaction and an endothermic reaction.

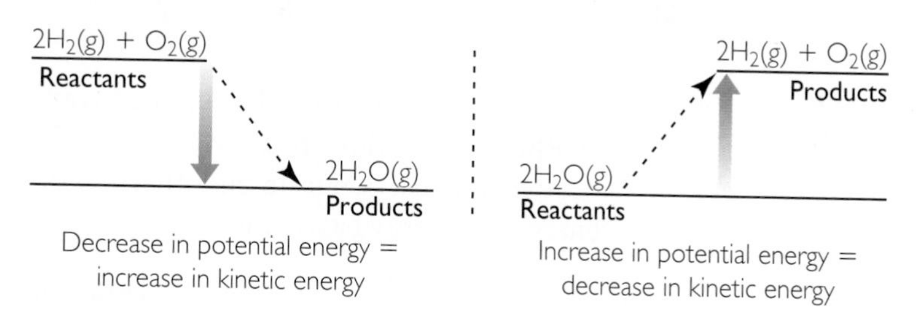

For an exothermic reaction, the potential energy of the system decreases in converting from reactants to products. Because energy is conserved, the sum of the kinetic energy and the potential energy must remain the same, so the kinetic energy increases. The products are hotter than the reactants, because they have a higher average kinetic energy. To reach thermal equilibrium with the surroundings, the hot products transfer energy to the surroundings.

For an endothermic reaction, the potential energy of the system increases. This means that kinetic energy must be converted into potential energy in order to conserve the total energy. The products are colder than the reactants because they have a lower kinetic energy. To reach thermal equilibrium with the surroundings, energy is transferred from the surroundings to the cold products.

❸ Activation Energy

A fire can be started with a match. The combustion reaction in an automobile engine is started with a spark plug. Every chemical reaction, not just combustion reactions, requires some energy input to get it started. This is called the **activation energy.**

GEOLOGY CONNECTION

The hydrocarbons that make up Earth's petroleum reserves are often referred to as "fossil fuels." This is because much of the oil buried beneath Earth's surface was formed from decayed plant and animal remains that were buried deep underground millions of years ago.

Examine the energy diagram for an exothermic reaction. The potential energy of the products is lower than that of the reactants. However, notice that the potential energy first increases on the pathway from reactants to products before it is converted to kinetic energy.

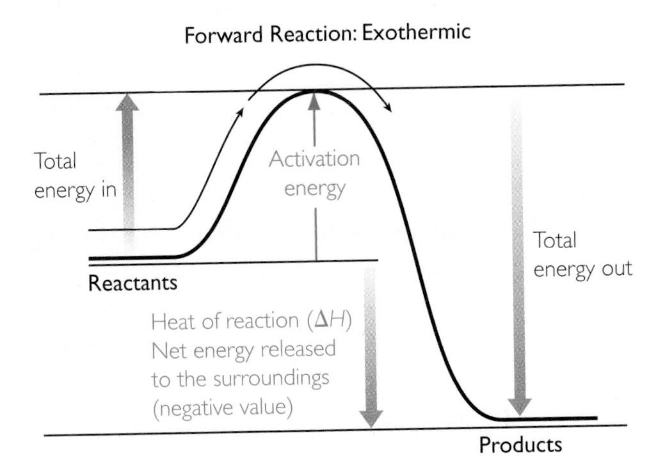

The potential energy between the reactants and products for this reaction is called the activation energy. Energy must be supplied to the reactants to get over this energy hill barrier. This is why reactions often require a spark to get started. Once an exothermic reaction is started, the reaction itself can provide the necessary energy for more molecules to react. This is why a single spark can cause an entire forest to burn down.

BIG IDEA Some reactions are energetically favored over other reactions.

In contrast, energy must be supplied continuously to cause the reverse of an exothermic reaction. You can see from the diagram that the reverse of the exothermic reaction is an endothermic reaction. Notice also that the activation energy, or energy barrier, is much greater in the reverse direction.

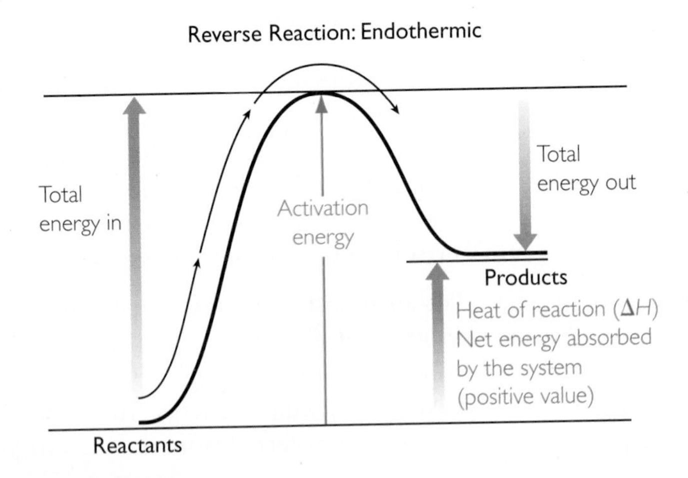

The net energy change from reactants to products of this reaction does not provide the necessary energy to activate more molecules to react. This is why you need to continuously supply energy to decompose water to hydrogen and oxygen. Endothermic reactions *generally* are less likely to occur than exothermic reactions.

Key Terms

potential energy
activation energy

Lesson Summary

What is the energy associated with reversing reactions?

Energy is conserved in chemical processes. This means that the net energy exchange in a forward process is equal and opposite to the net energy exchange in the reverse process. If a forward reaction is exothermic, the reverse reaction is endothermic. Energy in a chemical system is in the form of either kinetic energy or potential energy. When substances burn, a great deal of potential energy is converted into kinetic energy. Reactions with lower activation energies are easier to get started.

EXERCISES

Reading Questions

1. What happens to the potential energy and the kinetic energy of a chemical system during an exothermic reaction?
2. Explain why you do not need to keep sparking a fire once it is started.

Reason and Apply

3. Consider the combustion of ethane, $C_2H_6(g)$.
   a. Write a balanced equation for the reaction.
   b. Draw an energy diagram for the forward reaction. Label the reactants and the products. Use the table of heats of combustion from Lesson 10: Fuelish Choices to label the heat of reaction.
   c. What is the net energy exchange of the reverse reaction? Explain.
4. Consider the energy diagrams at right for the combustion of one mole of methane and the reverse reaction.

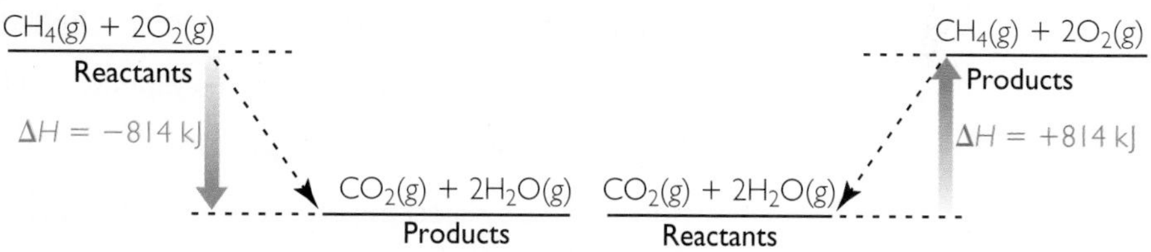

   a. Which diagram represents an endothermic reaction?
   b. Which substances have the lowest potential energy?
   c. When methane reacts with oxygen to form carbon dioxide and water, the potential energy decreases. What happens to the kinetic energy?
   d. Explain why the reverse reaction requires a constant input of energy.
   e. Use the table of average bond energies from Lesson 11: Make It or Break It, to estimate the heat of reaction. How does your value compare to the heat of reaction given in the diagram?
5. Copy the three energy diagrams shown here.

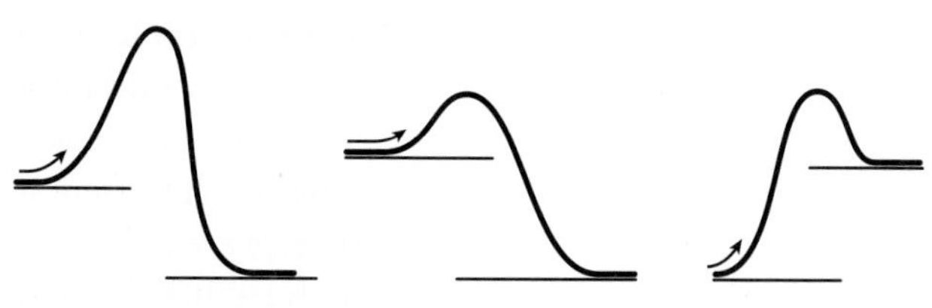

   a. Label the heat of reaction on each.
   b. Label the activation energy on each.
   c. Which diagram represents the most exothermic reaction?
   d. Which reactions require energy to get started?

LESSON

13 Speed Things Up

Rate of Reaction

Think About It

The Golden Gate Bridge in San Francisco is a national landmark. However, it is slowly eroding because of a chemical reaction in which iron in the steel beams combines with oxygen in the air to create iron oxide. This iron oxide falls away from the metal underneath. Is there a way to slow down the rate of this reaction?

How can you control the speed of a reaction?

To answer this question, you will explore

1. Reaction Rates
2. Catalysts

Exploring the Topic

1 Reaction Rates

Some reactions proceed quickly and others proceed slowly. When a balloon full of hydrogen is ignited, it reacts rapidly, exploding into flame. The formation of iron oxide on steel bridges is a slower process called rusting.

Combustion of hydrogen occurs rapidly. $2H_2(g) + O_2(g) \longrightarrow 2H_2O(g)$

Rusting of iron occurs slowly. $4Fe(s) + 3O_2(g) \longrightarrow 2Fe_2O_3(s)$

Both of these reactions are exothermic, yet the rates of each reaction are very different.

You might notice that the reactants are both gases for the quicker reaction. Gases mix on a molecular scale, so there is a lot of contact between the reactants. In contrast, the slow reaction involves a solid. The reactants are in contact with one another only at the surface of the solid. There are many more iron atoms on the interior of the solid that are not exposed to oxygen and therefore do not react. When a layer of the iron oxide, rust, flakes off, the underlying iron is then exposed to oxygen and will react.

Collision Theory

For a reaction to occur, the reactant particles must collide with enough energy to cause reactant bonds to break and product bonds to form. This energy is the activation energy of the reaction. In addition, the reactant particles must collide in the correct orientation with one another. When this happens the reactant particles are able to form a temporary arrangement called an *activated complex*, which facilitates bond breaking and bond making.

Enough energy but incorrect orientation:	Correct orientation but not enough energy:	Enough energy and correct orientation:
AA + BB → AABB → AA + BB	AA + BB → AA/BB → AA + BB	AA + BB → AA/BB → AB + AB
No products form.	No products form.	Products form.

So, for a reaction to occur, the reactant particles must collide in the correct orientation and with enough energy. If they do not, they will bounce apart and products will not be formed. Various factors can increase the frequency of collisions between particles and the energy of the collisions. These factors will speed up a reaction.

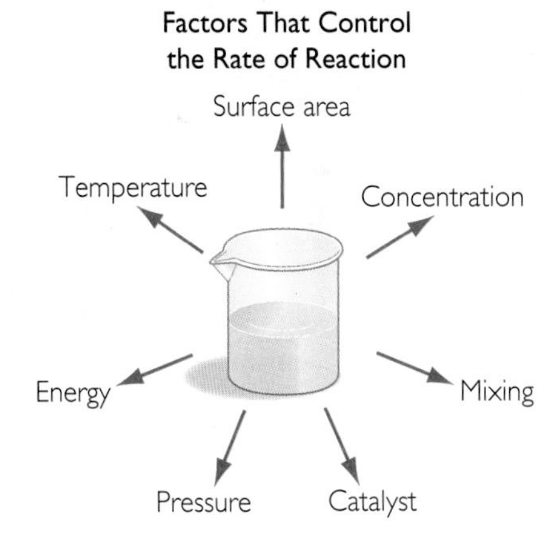

- **Surface area:** The exposed surface area of the reactants can affect the rate of a reaction. For example, in combustion reactions, fuels with larger exposed surface area, like sawdust or twigs, burn more rapidly than fuels with smaller exposed surface area, like an entire tree or a wooden table.
- **Mixing:** When reactants are mixed well, the reaction rate increases. This is related to the increased probability that the appropriate reactants will collide.
- **Temperature:** When the temperature is raised, the average kinetic energy of the reactant particles is also raised. The particles move more rapidly and more collisions between particles are possible.
- **Concentration or pressure:** Reactions usually proceed faster as the concentration of the reactants in solution increases or as the pressure of gases increases. Again, more reactants in a small volume results in more collisions.
- **Energy:** Adding energy, such as microwave radiation or light, can increase the rate of a reaction.

Topic: Reaction Rates
Visit: www.SciLinks.org
Web code: KEY-513

❷ Catalysts

Most of the methods of increasing the rate of reaction involve an increase in the number of collisions of the molecules or an increase in the speeds of the collisions. Another way to increase the rate of reaction is to add a catalyst to the reactants.

CONSUMER CONNECTION

The opposite of a catalyst is an *inhibitor,* a substance added to slow down a reaction. Sodium nitrite, ascorbic acid, and cinnamaldehyde are used to slow down rusting reactions. Inhibitors such as lemon juice, which contains ascorbic acid or 1-methylcyclopropene, can be added to fruit to slow down the reactions that cause browning.

A catalyst is a substance that assists a reaction, but is neither consumed nor permanently changed by the reaction. A catalyst lowers the energy of activation for the reaction. In Unit 2: Smells, sulfuric acid was used to catalyze the reaction that produced sweet-smelling ester molecules.

Chemical reactions may go through a series of intermediate steps to get from reactants to products. Some catalysts work by changing the sequence of steps to shorten the reaction path. Catalysts may undergo chemical transformations during a reaction, but they always return to their original form at the end of the reaction.

Another way a catalyst works is to provide a surface on which the reaction takes place. A catalytic converter in a car provides a platinum surface where pollutants can react to form harmless gases.

The energy diagram shows the same reaction with and without a catalyst. The catalyst lowers the energy barrier for the reaction.

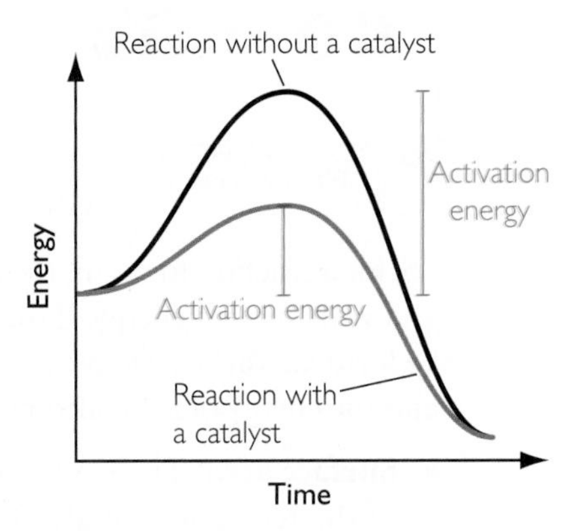

Lesson Summary

How can you control the speed of a reaction?

In order for molecules to react with each other, they must collide with a sufficient amount of energy for bond breaking or bond making. There are a number of ways to increase the rate of a reaction, such as increasing the temperature of the system or changing the concentration of the reactants. These methods increase collisions between reactants. One way catalysts increase the rate of a reaction is by lowering the activation energy to get the reaction started. Catalysts are substances that speed up a reaction, but are not permanently changed or consumed by the reaction.

EXERCISES

Reading Questions

1. Name three things you could do to increase the rate of a reaction between two liquids.
2. Name three things you could do to increase the rate of a reaction between two solids.
3. Name two ways you could decrease the rate of a reaction.

Reason and Apply

4. What could you do to speed up the reactions listed here?
   a. Lighting charcoal for a barbecue
   b. Baking cupcakes
   c. Dissolving sugar in water
   d. Removing a stain from clothing
5. Which will burn faster, wood chips or a tree? Explain the difference in the rates of reaction.
6. Which will dissolve faster, a lollipop or powdered sugar? Explain the difference in the rates of reaction.
7. Which will react faster, a steel beam or powdered iron? Explain the difference in the rates of reaction.
8. Why do you think you can extinguish a flame with a carbon dioxide fire extinguisher?
9. What makes an explosion happen so quickly?
10. **WEB RESEARCH** Research a particular enzyme in your body. Explain how the enzyme catalyzes a reaction in your body.
11. The rate of a reaction can be affected by
    A. The phase of the reactants.
    B. The temperature of the reactants.
    C. The presence of a catalyst.
    D. All of the above.
    E. None of the above.

BIOLOGY CONNECTION

Enzymes could be considered nature's catalysts. They are biomolecules that speed up the rates of chemical reactions in living things. Enzymes are usually protein molecules. For example, the enzyme amylase in your saliva helps to break down sugar molecules. The enzyme lactase in the small intestine breaks down lactose. People with insufficient lactase have trouble digesting dairy products, which contain lactose.

LESSON

14 Make It Work

Work

Think About It

All around us are machines doing work for us. They transport us from place to place, they help us to manufacture products, and they allow us to listen to music, wash our clothes, and watch television. The source of energy for most of these machines is chemical change.

How can a chemical reaction be used to do work?

To answer this question, you will explore

1. Work
2. Converting Chemical Energy to Work

Exploring the Topic

1 Work

In order to move anything, whether it is a truck or an atom, from one place to another, work must be done. **Work,** like heat, is a transfer of energy. Specifically, work is the result of a force acting through a distance. A force is a push or a pull upon an object as a result of an interaction with another object. Force is measured in newtons, N. Scientists define mechanical work as the product of force and distance.

$$\text{Work} = \text{force} \cdot \text{distance}$$

$$W = Fd$$

INDUSTRIAL CONNECTION

The electricity that runs through the wires in buildings comes from a variety of sources, including hydroelectric, nuclear, and geothermal power plants. However, the majority of electricity used in the United States is generated from power plants that burn coal, oil, or natural gas to heat water and make steam. This pressurized steam turns giant turbines that generate electricity.

Lifting an Object

Imagine you pick up your baby sister, who weighs 22 pounds, or 10 kilograms, and lift her 1.2 meters. To lift her, you need to exert 100 newtons of force. This force is equal to the force of gravity acting on your sister, or her weight if it were expressed in newtons.

The work done picking up your 22-pound sister is equal to the force of her weight in newtons multiplied by the distance she was lifted.

$$\text{Work} = 100\ \text{N} \cdot 1.2\ \text{m} = 120\ \text{N} \cdot \text{m}$$

Work is expressed in units of newton-meters, N · m. Work and energy are related. One newton-meter is equal to one joule.

Machines (or tools) help us to do work. Levers, ramps, pulleys, wheels, and wedges are all examples of simple machines. They give you a mechanical advantage to perform work. Consider how easy it is to lift a great deal of weight using a seesaw. A claw hammer acts as both a wedge and a lever, allowing you to pull a nail out of a board, something that would be nearly impossible with your bare hands.

A hammer acts as a lever when you use it to pull out a nail.

ENVIRONMENTAL CONNECTION

Hydroelectric power plants, windmills, and solar panels do not use chemical reactions to produce electricity. These power sources depend on physical change alone. They are good alternatives to using fossil fuels to produce electricity.

Energy from Chemical Reactions

Chemical reactions are sources of energy. Matter has the potential to transfer energy when it is transformed into new compounds. You have seen that burning different fuels can transfer large quantities of energy in the form of heat and light. But heat is not the only useful source of energy from chemical reactions. Consider the products of a combustion reaction. Both the gases formed and the heat produced are potential sources of work.

$$CH_4(g) + 2O_2(g) \longrightarrow CO_2(g) + 2H_2O(g) + \text{heat}$$

Two potential sources of work from a chemical reaction

Recall from Unit 3: Weather that the pressure from expanding gases can be a powerful force. In the case of a combustion reaction, the products are in the form of heated gases that can expand and push on objects. The work done by expanding gases is often referred to as *PV* work, where *P* and *V* stand for pressure and volume. *PV* work is equal to pressure times the change in volume.

$$W = P\Delta V$$

If pressure is expressed in atmospheres and the volume is expressed in liters, *PV* work is expressed in liter-atmospheres. Liter-atmospheres can also be converted to units of energy in joules.

$$1 \text{ liter-atmosphere} = 101 \text{ joules}$$

Important to Know A system cannot contain or store either heat or work. So, heat and work both represent energy in transition. ◀

2 Converting Chemical Energy to Work

Chemical systems are often paired up with machines to transform chemical change into useful work. The heat transferred, *q*, from a reaction or process can be utilized in some way, or the chemical products themselves can be used.

Steam Engine

A steam engine uses a combustion reaction to heat water in a cylinder with a piston. When the heat is transferred to the water, the liquid changes phase to a gas that expands and moves the piston.

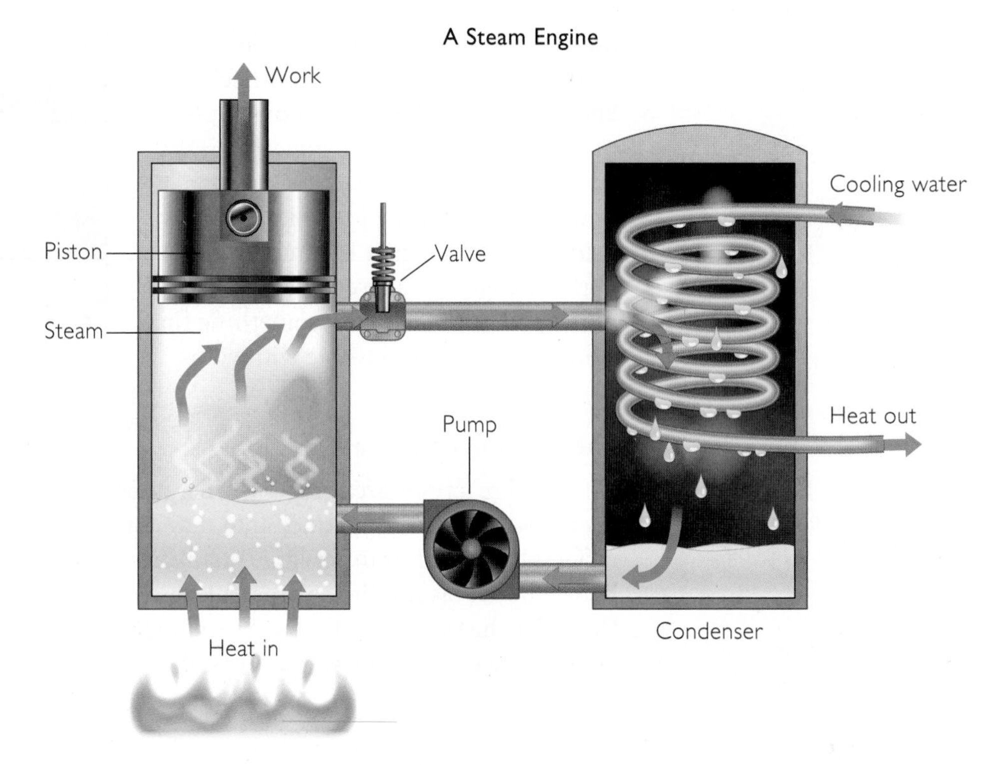

HISTORY CONNECTION

The earliest recorded steam engine was built by an ancient Greek mathematician named Hero of Alexandria. The steam came out through two pipes, making a metal arm spin around. There is no evidence that this engine was used for anything more than amusement.

Notice that this is a two-step process as the heat from a chemical change is used to cause a physical change. The expanding gas (water vapor) does the work in this type of engine. Early steam engines were used to pump water and to power steamboats, trains, farm machinery, factories, and the first automobiles.

Internal Combustion Engine

Internal combustion engines like those in cars also make use of expanding gases. However, this type of engine uses the gases produced in the combustion reaction, carbon dioxide and water vapor, to do the work.

The illustration shows an internal combustion engine going through a four-step process.

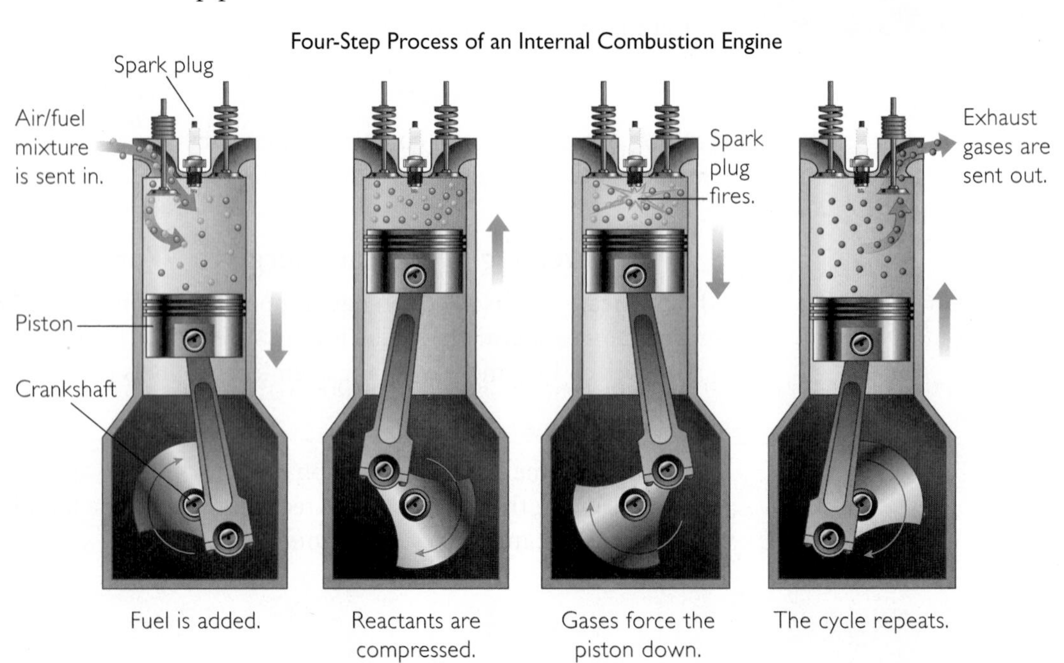

BIOLOGY CONNECTION

The plant life on our planet converts solar energy to chemical energy through photosynthesis. Without photosynthesis, the petroleum products we use now as fuel would not exist.

This four-step process is repeated at an extremely rapid rate to run the engine. The expanding carbon dioxide and water vapor from the combustion reaction move the pistons up and down, turning a large crank that ultimately makes tires, wheels, propellers, or turbines go around.

Entropy

Internal combustion and steam engines convert thermal energy to work. As required by the first law of thermodynamics, the total energy is conserved in the process. However, it is impossible to completely convert all that energy into work. Some of the energy is lost to the surroundings as heat.

This loss of heat to the surroundings is related to entropy and the second law of thermodynamics. As you recall, the second law states that energy and matter tend to disperse or become more disordered. Entropy is the measure of the unavailability of a system's energy to do work due to this dispersal. So when designing an efficient machine, scientists and engineers must account for entropy.

BIG IDEA Energy disperses; it does not collect.

Key Term

work

Lesson Summary

How can a chemical reaction be used to do work?

Work is force acting upon an object causing it to move in the direction of the force. So work is energy in motion. The heat transferred, q, from chemical reactions can be used to do work. The work done by expanding gases is called PV work. The combustion of fuels is one of the most common ways we transform chemical energy into work. However, some energy is always lost, due to entropy.

EXERCISES

Reading Questions

1. What is the scientific meaning of the word *work*?
2. Give an example of how a chemical reaction can be used to do work.

Reason and Apply

3. The expansion of gases pushes a piston. The volume of the gases increases from 0.5 L to 12.5 L. If the piston is pushing against a pressure of 1.2 atmospheres, how much work is done by the gases?
4. If it takes 45 N of force to lift a box 3.8 m, how much work is done to the box?
5. **WEB RESEARCH** Describe three ways that a rocket engine is different from an internal combustion engine. What extra item must a rocket engine carry with it into space?

SECTION

SUMMARY

Understanding Energy

Fire Update

All chemical reactions involve the breaking of bonds and the making of bonds. The net energy difference between these two processes determines if a reaction is exothermic or endothermic. This is because bond breaking requires energy and bond making releases energy. By using average bond energies, you can closely approximate the heat of reaction, ΔH.

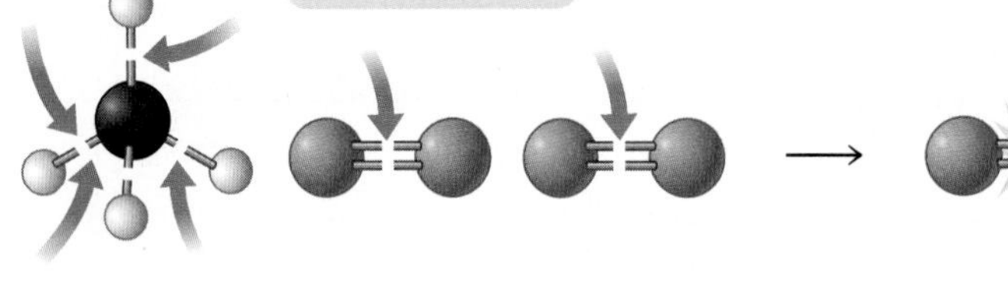

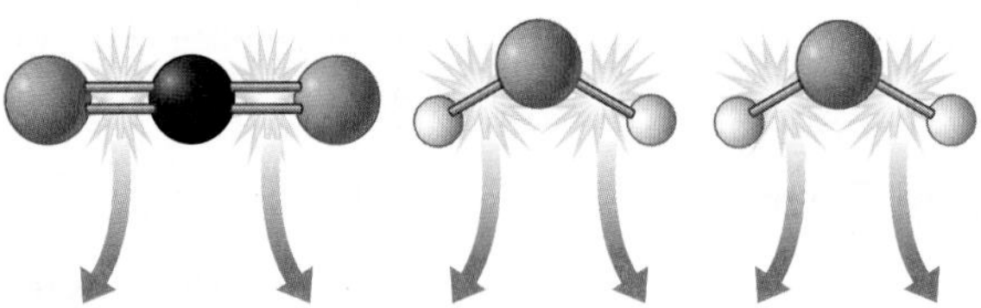

Key Terms

bond energy
potential energy
activation energy
work

Each chemical system expresses energy in the motions of its molecules or particles. This is known as kinetic energy. In addition, each chemical system has potential, or the potential for further energy in its chemical bonds. When an exothermic reaction occurs, potential energy of the system decreases and kinetic energy of the system increases. The reverse is true in an endothermic reaction.

It generally takes an input of energy called the activation energy to get a reaction started. This is why a fire requires a match or a spark. The overall conditions and the form the reactants are in can affect the rate of a reaction. For example, raising the temperature, reducing the surfaces of reactants, or adding a catalyst can increase the rate of a reaction.

Chemical reactions can be used to do work. For example, the internal combustion engine uses the gaseous products formed in a combustion reaction to do work on pistons.

Review Exercises

1. List seven ways you could speed up a chemical reaction. Explain how each method can increase the rate of reaction.
2. You are walking by a classroom, and you overhear two students having a discussion about boiling water. Whom do you agree with and why?

   Trey: When you're boiling water, you're breaking the bonds between the hydrogen and oxygen atoms in water. The steam you see above the boiling water is hydrogen and oxygen gas. Heating the water provides the energy to break these bonds.

Paige: Boiling water is not providing nearly enough energy to decompose water. The steam you see above the boiling water is still water; you have not broken any bonds to make it. Heating the water provides the energy to vaporize the water.

3. You have a cup of iced coffee that you are going to sweeten with a sugar cube. Name three ways that you could speed up the rate of dissolving sugar in the coffee. Explain why each of your methods works according to collision theory.
4. The chemical formula for heptane is C_7H_{16}.
   a. Draw the molecular structure for heptane.
   b. Write the balanced equation for the combustion of heptane.
   c. Using the table of average bond energies (located on page 524) and your answers to parts a and b, calculate the heat of reaction for heptane.
5. Methane, CH_4, is a combustible gas used in high school and college laboratories to burn in Bunsen burners. Carbon dioxide, CO_2, is an incombustible gas used in fire extinguishers.
   a. Draw the molecular structures for methane and carbon dioxide.
   b. Using the table of average bond energies on page 524, calculate the amount of energy that would be needed to break the bonds in a mole of methane and a mole of carbon dioxide.
   c. Why do you think that methane combusts, but carbon dioxide doesn't?
6. A balloon is placed over a flask with water in it. As the flask is heated on a hot plate, the balloon expands from a volume of 0.1 L to 1.6 L at a constant 1.1 atm.
   a. Calculate the amount of work that the water molecules did on the balloon.
   b. Is there anything else that needs to be taken into account when calculating the amount of work that the water vapor did?

RESEARCH

PROJECT ***Uses of Fire***

Write a report describing where coal comes from and how the energy from coal is converted into electricity in a powerplant. Include a diagram of the process in your report.

SECTION

IV Controlling Energy

The battery in your car converts chemical reactions into electricity.

Fires are not the only source of energy from chemical change. Energy from all types of reactions can be made useful. Reactions involving metals and metal compounds are particularly valuable sources of energy. Using certain techniques, these reactions can be made portable, controllable, and safe. This final section of Unit 5: Fire focuses on ways in which the energy from chemical change is contained and made easily available.

In this section, you will study

- exothermic reactions of metals and metal compounds
- the energy it takes to extract pure metals from compounds
- how chemical change is converted into electrical energy
- combinations of reactants that make useful batteries

LESSON

15 Metal Magic

Oxidation

Think About It

If you examine a handful of pennies, chances are some of them will be a bright coppery color and others will be discolored and dull. These tarnished pennies have reacted with the oxygen in the air, just like a fuel does when it burns. But is this a form of combustion?

What happens when metals react with oxygen?

To answer this question, you will explore

1. Oxidation
2. Electron Transfer

Exploring the Topic

1 Oxidation

No penny stays bright and shiny forever. Eventually they all become dark and dull when the copper metal on the surface forms copper (II) oxide and other copper compounds. This reaction happens very slowly at room temperature. The reaction is also exothermic, but it happens so slowly the heat is difficult to detect.

The reaction of copper with oxygen is described by this equation.

$$2Cu(s) + O_2(g) \longrightarrow 2CuO(s)$$

On the other hand, if magnesium is held to a flame it reacts rapidly and violently with oxygen, producing an almost blinding light. This metal is also combining with oxygen, but it is a much more exothermic reaction and happens quite rapidly once it gets started.

$$2Mg(s) + O_2(g) \longrightarrow 2MgO(s)$$

Copper and magnesium are just two examples of metals reacting with oxygen. Under the right conditions, *all* metals will react with oxygen. This process is called oxidation. So, both combustion of molecules and tarnishing of metals are forms of oxidation. Compare the oxidation of two metals with a common combustion reaction:

$$\text{Oxidation of copper: } 2Cu(s) + O_2(g) \longrightarrow 2CuO(s)$$

$$\text{Oxidation of magnesium: } 2Mg(s) + O_2(g) \longrightarrow 2MgO(s)$$

$$\text{Combustion of methane: } CH_4(g) + 2O_2(g) \longrightarrow CO_2(g) + 2H_2O(g)$$

The reaction of oxygen with a metal produces a solid, whereas the reaction of oxygen with a molecular substance produces two gases. But the reactions

HEALTH CONNECTION

Antioxidants are molecules that can slow or prevent the oxidation of other molecules. Antioxidants found in fruits and vegetables can cancel out the cell-damaging effects of free radicals. This is one of the reasons why eating fruits and vegetables can lower a person's risk for heart disease, neurological diseases, and some forms of cancer. Berries, especially acai berries, are high in antioxidants.

CONSUMER CONNECTION

Over time, silver oxidizes and becomes dull. One way to make it shiny again is to place it in a solution of sodium bicarbonate (baking soda) with aluminum foil. Within a few minutes, the silver becomes shiny again and the aluminum oxidizes instead.

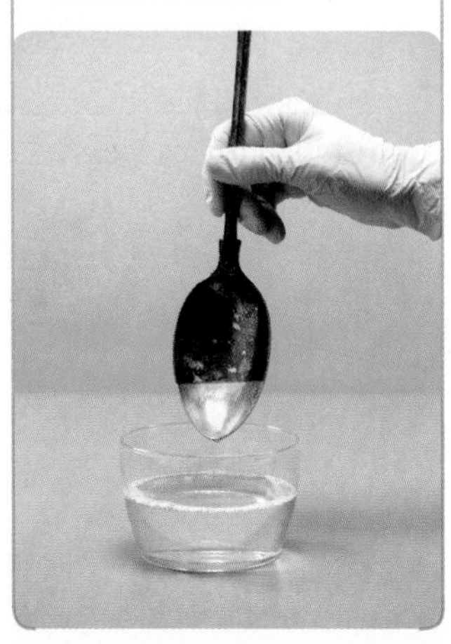

have several characteristics in common. Here is a comparison of the two types of reactions:

	Combustion of molecules	Oxidation of metals
combines with	oxygen	oxygen
releases heat?	yes	yes
releases light?	yes	sometimes
produces flames?	yes	no
product	gaseous molecules	solid metal oxides

Oxidation of metals transfers energy to the surroundings, just like combustion of molecular covalent compounds. This energy is in the form of heat and sometimes a bright light. However, you can see that the products of oxidation of metals and combustion of molecular compounds are different.

Flame Versus Glow

Flames are composed largely of superheated gases. When metals oxidize, they might glow with a very bright light, but there is no flame. The bright light is due to the excitation of electrons. Recall that the colored flame tests in Unit 1: Alchemy were also the result of excited electrons changing energy levels. The bright light from the oxidation of metals may resemble a flame, but it is not a flame.

2 Electron Transfer

All of the oxidation reactions discussed so far involve oxygen. However, **oxidation** is defined more broadly as any reaction in which an atom transfers electrons to oxygen or another atom. When magnesium oxide is formed, the magnesium atoms transfer electrons to the oxygen atom. When this happens, we say that the magnesium has been oxidized.

BIG IDEA When an atom or ion is oxidized, it loses electrons.

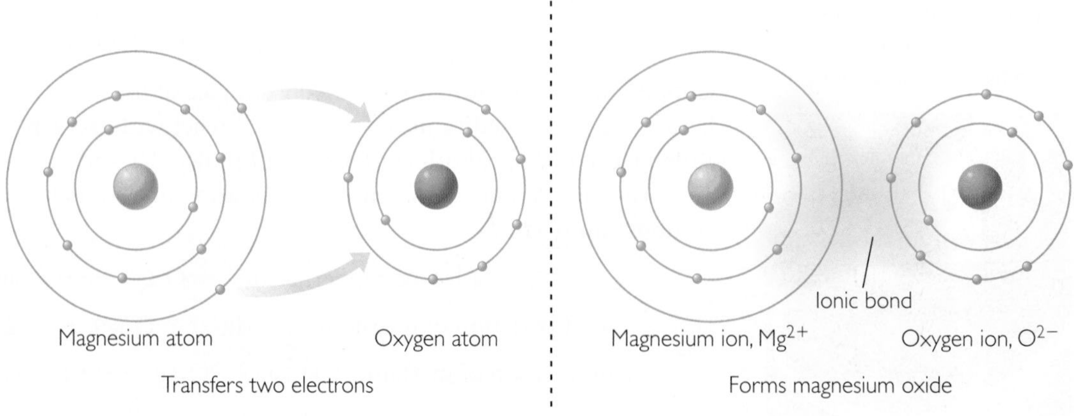

Oxidation of magnesium

Topic: Oxidation and Reduction

Visit: www.SciLinks.org

Web code: KEY-515

The shell models show a magnesium atom being oxidized. When magnesium is oxidized, it forms Mg^{2+} ions.

Recall that metal elements generally have one, two, or three electrons in their valence electron shells. Having such a small number of electrons in the valence shell makes it more likely that metal atoms will lose electrons, forming cations that have a charge of +1, +2, or +3. Through electron transfer, metal atoms form ionic bonds with nonmetal atoms.

Molecular covalent compounds also transfer electrons to oxygen when they react. Therefore, combustion reactions are considered oxidation reactions.

Key Term

oxidation

Lesson Summary

What happens when metals react with oxygen?

Metal atoms react with oxygen to form metal oxides. When metal oxides form, heat is released, and sometimes an intense bright light is emitted. This process is called oxidation. Molecular substances can also be oxidized as is the case in combustion. However, no flames are present when metals oxidize. This is partly what differentiates the oxidation of metals from the oxidation of molecular substances. Oxidation occurs whenever an atom, ion, or molecule transfers electrons to another atom, ion, or molecule.

EXERCISES

Reading Questions

1. How does oxidation of a metal compare to combustion of a covalent compound?
2. Explain what happens when magnesium metal is oxidized.

Reason and Apply

3. **a.** Write balanced equations for the reactions of calcium, chromium, and lithium with oxygen. Identify the charge on the cation and the charge on the anion for each product.

   $Ca(s) + O_2(g) \longrightarrow CaO(s)$

   $Cr(s) + O_2(g) \longrightarrow Cr_2O_3(s)$

   $Li(s) + O_2(g) \longrightarrow Li_2O(s)$

   b. Circle the elements that are oxidized in the chemical equations in part a.
4. Draw shell models for lithium and oxygen atoms to show the transfer of electrons during oxidation.
5. Draw shell models for silicon and oxygen atoms to show the transfer of electrons during oxidation.

LESSON

16 Pumping Iron

Heat of Formation

Think About It

With all the heat and light it gives off, fire is an obvious source of energy. Less obvious is the possible use of other exothermic chemical reactions as energy sources. Because the oxidation of metals is exothermic, it may be possible to use these reactions to generate power.

How much energy is transferred during oxidation of metals?

To answer this question, you will explore

1. Heat of Formation
2. Reversing Oxidation

Exploring the Topic

1 Heat of Formation

The oxidation of most metals is exothermic, just like the combustion of fuels. So, oxidation reactions are a potential source of energy for our daily lives. To explore the usefulness of metal oxidation as an energy source, we must examine and compare the amount of thermal energy transferred during these reactions.

Malachite, $Cu_2CO_3(OH)_2$, a green stone, was probably the first source of copper metal that was extracted through heating. Malachite may have been used to color a piece of pottery that was then fired in a very hot oven, leaving behind globs of copper.

Copper Versus Zinc

Copper metal and zinc metal react with oxygen to form copper (II) oxide, CuO, and zinc oxide, ZnO.

$$2Cu(s) + O_2(g) \longrightarrow 2CuO(s) \qquad \Delta H = -314 \text{ kJ}$$

$$2Zn(s) + O_2(g) \longrightarrow 2ZnO(s) \qquad \Delta H = -696 \text{ kJ}$$

Notice that ΔH is negative for both reactions. As you recall, this indicates that the reactions are exothermic and heat is transferred to the surroundings. Also notice that in both cases the heat of reaction is associated with *two* moles of metal oxide being produced. So, the formation of *one* mole of each oxide results in the release of half as much energy.

The energy released when one mole of a compound is formed from elements is called the **heat of formation,** or ΔH_f. Compare the heats of formation of copper (II) oxide, CuO, and zinc oxide, ZnO:

$$\text{Heat of formation of copper (II) oxide: } \Delta H_f(CuO) = -157 \text{ kJ/mol}$$

$$\text{Heat of formation of zinc oxide: } \Delta H_f(ZnO) = -348 \text{ kJ/mol}$$

These values indicate how much heat transfer to expect when one mole of each compound is formed from its elements. The formation of 1 mol of zinc oxide releases over twice as much energy as the formation of 1 mol of copper (II) oxide. You can find heats of formation values of some metal oxides in the table on the next page. These values are determined under standard conditions, at

Heats of Formation

Compound	ΔH_f (kJ/mol)
$Fe_2O_3(s)$	−826
$Al_2O_3(s)$	−1676
$PbO(s)$	−218
$MgO(s)$	−602
$CuO(s)$	−157
$Au_2O_3(s)$	+81

1 atm and 25 °C. Note that standard conditions are different from STP, standard temperature and pressure, which is 1 atm and 0 °C.

Example 1

Oxidation of Mg and Fe

Which reaction will release more energy?

$$2Mg(s) + O_2(g) \longrightarrow 2MgO(s)$$
$$4Fe(s) + 3O_2(g) \longrightarrow 2Fe_2O_3(s)$$

Solution

The heat of formation per mole of product can be found in the table. Multiply ΔH_f by the number of moles of each product to get the heat of the reaction, ΔH:

$$(-602 \text{ kJ/mol})(2 \text{ mol MgO}) = -1204 \text{ kJ}$$
$$(-826 \text{ kJ/mol})(2 \text{ mol } Fe_2O_3) = -1652 \text{ kJ}$$

The oxidation of iron, or the formation of iron (III) oxide, Fe_2O_3, produces more energy than the oxidation of magnesium. Heats of formation are not limited to metal oxides. For example, the heat of formation of magnesium chloride, $MgCl_2$, is −2686 kJ/mol.

2 Reversing Oxidation

When an oxidation reaction for a metal is reversed, the resulting products are a pure metal and oxygen gas. This reaction is a decomposition reaction. The metal oxide decomposes and one of the products is a metal element. The decomposition of a metal oxide is one way to extract pure metals from metal compounds. But how difficult is it to accomplish energetically? Consider the decomposition of aluminum oxide, Al_2O_3:

$$2Al_2O_3(s) \longrightarrow 4Al(s) + 3O_2(g)$$

Energy diagrams for the oxidation of aluminum and for the decomposition of aluminum oxide are shown here. Take a moment to examine the energy changes.

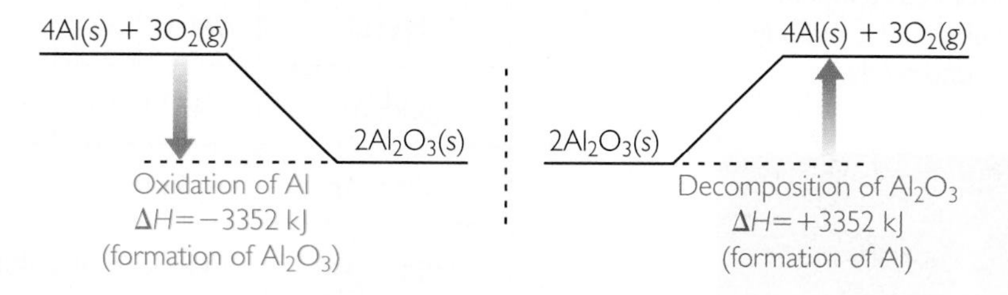

The oxidation of aluminum to form aluminum oxide is very exothermic. The energy for the reverse reaction is equal and opposite. So the decomposition of aluminum oxide is highly endothermic. This means that a large amount of energy is required to extract pure aluminum from aluminum oxide.

Indeed, aluminum was not discovered and brought into use until about 200 years ago, partly because so much energy is required to extract the pure aluminum metal from aluminum ores. The simple heating of aluminum ores does not provide enough energy to decompose the aluminum compounds and extract aluminum metal.

Example 2

Extracting Al

How much energy is required to extract a mole of aluminum metal from aluminum oxide, Al_2O_3?

Solution

The equation for the decomposition of aluminum oxide is

$$2Al_2O_3 \longrightarrow 4Al(s) + 3O_2(g)$$

This process is the reverse of the formation of $Al_2O_3(s)$, for which $\Delta H_f = -1676$ kJ/mol. (See the table on page 547.)

Because 2 mol of $Al_2O_3(s)$ are decomposed in the equation, the heat of reaction, ΔH, is 2(1676 kJ/mol), or 3352 kJ/mol. Note that the sign is positive since the reaction is reversed.

Because 4 mol of aluminum are produced, the energy required to extract 1 mol of aluminum metal is one-fourth of the heat of reaction.

$$3352 \text{ kJ/mol}/4 = 838 \text{ kJ/mol Al}$$

The History of Metal Use

It is apparent from archeological sites that gold was in common use thousands of years before iron. This makes sense from the energy data for these metals. The metals that have been in use the longest are the ones that are easiest to extract from their compounds.

This table shows the energy data for the decomposition of metal oxides to metals. In order to extract the metals, people needed to construct furnaces that transferred a sufficient amount of heat.

INDUSTRY CONNECTION

The process of removing metals from their ores is called *smelting*. It is usually necessary to add carbon to assist in the process of removing metals from ores. Smelting is often performed at high temperatures so that the metal obtained is already molten and ready to be cast.

Extraction of Metals from Metal Oxides

Metal	ΔH kJ/mol	Date
gold, Au	−41 kJ/mol	6000 B.C.E.
silver, Ag	+15 kJ/mol	4000 B.C.E.
copper, Cu	+156 kJ/mol	4000 B.C.E.
lead, Pb	+218 kJ/mol	3500 B.C.E.
iron, Fe	+413 kJ/mol	2500 B.C.E.
tin, Sn	+581 kJ/mol	1800 B.C.E.
magnesium, Mg	+602 kJ/mol	1755 C.E.
aluminum, Al	+838 kJ/mol	1825 C.E.

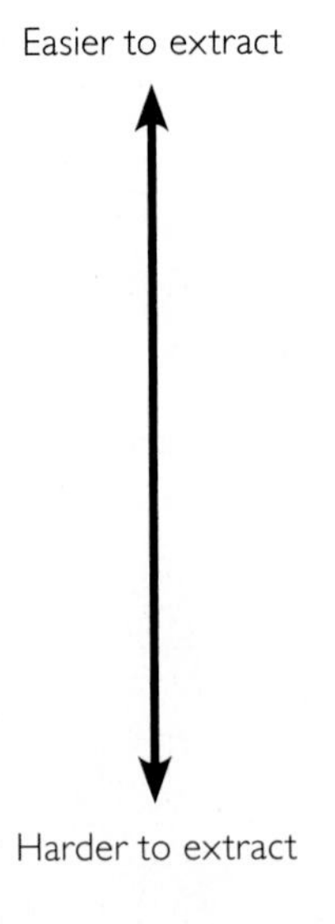

The difficulty of extraction of various metals from metal compounds is reflected in the chronological "discovery" and use of these metals through history. The more energy it takes to extract a metal, the later the metal was put into practical use.

Metals that are easily oxidized are hard to extract. In other words, the oxidation reaction is harder to reverse. These are also the metals that tarnish and rust easily. Gold is often found in its pure form in nature. This is because gold is not easily oxidized and is fairly unreactive.

Key Term

heat of formation

Lesson Summary

How much energy is transferred during oxidation of metals?

The formation of metal oxides is usually an exothermic reaction. This means oxidation reactions may be valuable sources of energy. The heat of formation of a compound is the thermal energy change associated with the formation of *one mole* of that compound from its constituent elements. The decomposition of metal oxides into pure metals and oxygen is the reverse of an oxidation reaction. These reactions are generally endothermic. This means it requires an input of energy to extract a metal from a metal compound. Metals that are easily oxidized (with high heats of formation) are more difficult to extract from their metal compounds than metals that are less easily oxidized (with low heats of formation).

EXERCISES

Reading Questions

1. Of all the possible reactions in this lesson, which ones have the potential of being the best sources of energy? Explain your answer.
2. What does the heat of formation indicate about a metal in terms of its practical use as a metal?

Reason and Apply

3. How much energy is released when 10 mol of lead (II) oxide, PbO, are formed?
4. Explain why metals that are easily oxidized are not found in nature in their pure forms. Name two of these metals.
5. Draw the energy diagrams for the oxidation of copper and its reverse reaction. Include the energy data.
6. How many kilojoules of energy does it take to extract 6 mol of tin from tin (II) oxide, SnO?
7. How many kilojoules of energy does it take to extract 10 g of copper from copper (II) oxide, CuO?
8. Which would be a source of more energy, the formation of 6 mol of magnesium oxide, MgO, or the formation of 2 mol of aluminum oxide, Al_2O_3? Explain.
9. **WEB RESEARCH** Research how one of the metals of antiquity was extracted from metal compounds dug out of the earth. Describe the process and draw a sketch.

LESSON

17 Electron Cravings

Oxidation-Reduction

Think About It

A silvery or reddish mineral called hematite, Fe_2O_3, is the main source of iron metal on Earth. When this iron (III) oxide is decomposed to iron and oxygen, the oxygen atoms lose electrons. This means the iron atoms gain electrons.

What happens to electrons during oxidation?

To answer this question, you will explore

1. Oxidation-Reduction
2. Redox Reactions

Exploring the Topic

1 Oxidation-Reduction

When iron oxides are decomposed to iron metal, the iron cations gain electrons and the oxygen anions lose electrons.

$$2Fe_2O_3(s) \longrightarrow 4Fe(s) + 3O_2(g)$$

Sample of hematite, Fe_2O_3

The oxygen anions in $Fe_2O_3(s)$ have a charge of -2. In the decomposition of $Fe_2O_3(s)$, each O^{2-} anion loses two electrons to form neutral oxygen atoms. The oxygen atoms bond in pairs to form elemental oxygen, O_2 (to satisfy the octet rule). In this process, the oxygen anions, O^{2-}, are oxidized to elemental oxygen, O_2.

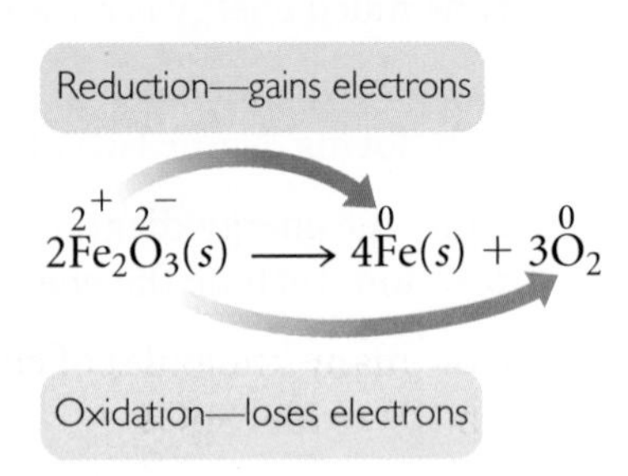

The iron cations in $Fe_2O_3(s)$ have a charge of $+3$. The iron atoms in $Fe(s)$ have a charge of 0. In the decomposition of $Fe_2O_3(s)$, each Fe^{3+} cation gains three electrons to form neutral iron atoms in the elemental metal. Just as there is a term for losing electrons (oxidation), there is a term for gaining electrons: **reduction.** In this process, the iron atoms, Fe^{3+}, are reduced to elemental iron, Fe.

Whenever metals are extracted from their compounds, they are reduced. While the term *reduction* refers to the process of gaining electrons, its origins have to do with the concept of reducing metal ores to their simplest elements. Whenever you see a metal element as the product of a reaction, you can be fairly certain that it has been reduced.

Every time a substance loses electrons, another substance gains electrons and vice versa. These two processes always take place together: oxidation and reduction. **Oxidation-reduction** reactions are often referred to as **redox** reactions.

INDUSTRY CONNECTION

Barium is a metal that is easily oxidized in air to form BaO. So, it is difficult to obtain this metal in its pure form. Because barium reacts so easily with oxygen, it is sometimes used to create vacuum tubes, by actually using it to remove the last of the oxygen.

Reaction of Zinc and Copper Sulfate

A strip of zinc metal is placed in a light blue solution of copper (II) sulfate, $CuSO_4$, as shown in the illustration. Soon, a reddish coating begins to form on the zinc strip. The blue color of the solution begins to fade. What is going on?

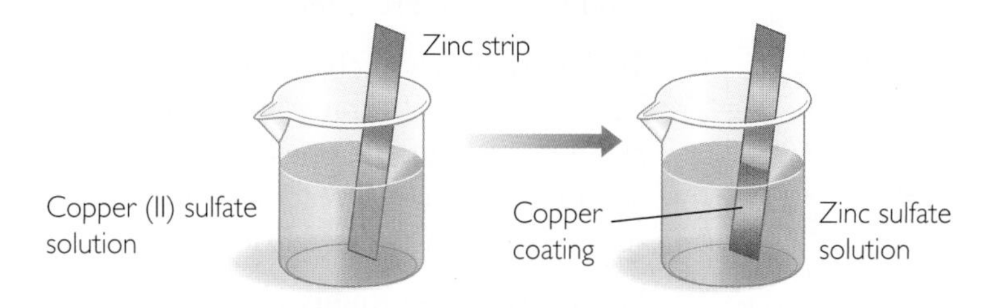

Start with the balanced chemical equation for this reaction:

$$Zn(s) + CuSO_4(aq) \longrightarrow ZnSO_4(aq) + Cu(s)$$

You know from Unit 1: Alchemy that when salts dissolve in water, the compound dissociates into ions. Often chemical equations are written in ionic form to specify the charges of the ions in solution as shown here.

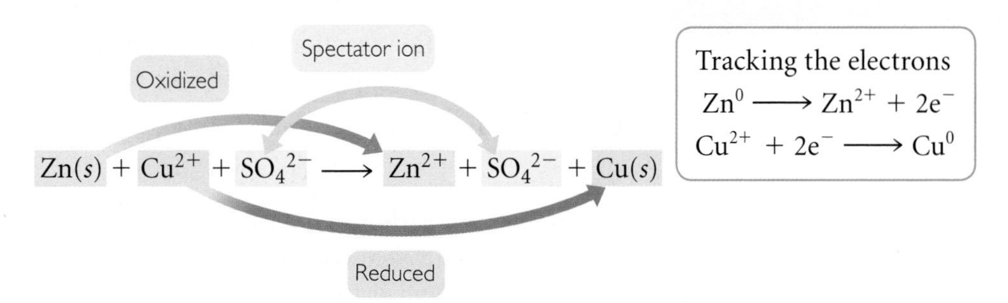

Notice that the sulfate ion is a spectator ion. The net ionic equation is therefore

$$Zn(s) + Cu^{2+} \longrightarrow Zn^{2+} + Cu(s)$$

The next illustration shows this reaction from a particle view. Take a moment to follow the transfer of the electrons.

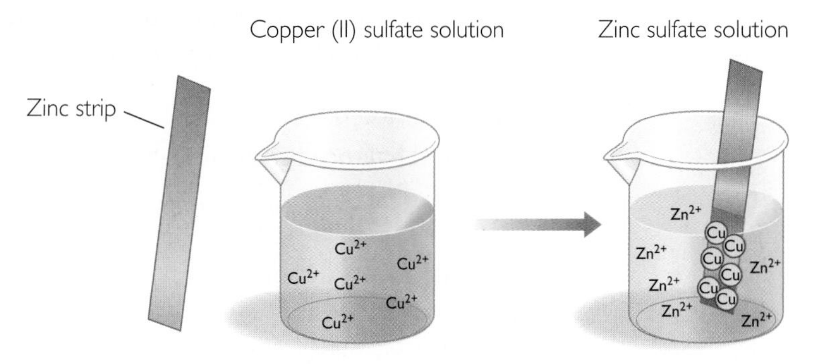

The copper ions, Cu^{2+}, in solution are reduced to copper metal, Cu^0. The zinc metal atoms on the strip, Zn^0, are oxidized to zinc ions, Zn^{2+}, and enter the solution.

Important to Know The atom that is being reduced gains electrons. The atom that is oxidized loses electrons. ◀

CONSUMER CONNECTION

When silverware made from pure silver comes into contact with certain foods, it is oxidized and becomes tarnished. Eggs are an example of a food item that will tarnish silver. Eggs are rich in sulfur compounds. The compound that forms as a result of this oxidation is silver sulfide, $Ag_2S(s)$.

2 Redox Reactions

Many chemical reactions are oxidation-reduction reactions. In fact, any reaction in which electrons are transferred from one reactant to another is a redox reaction. This includes most reactions that have an element as a reactant or product, which are often single exchange reactions.

Typically, double exchange reactions are *not* redox reactions. For example, in acid-base reactions, H^+ trades places with another cation. In precipitation reactions, an aqueous ion from one compound combines with an aqueous ion from another. So, double exchange reactions often involve transfer of ions, while single exchange reactions involve transfer of electrons.

Single exchange: Redox reaction

$$Mg(s) + 2HCl(aq) \longrightarrow H_2(g) + MgCl_2(aq)$$

Double exchange: Acid-base reaction

$$NaOH(aq) + HCl(aq) \longrightarrow H_2O(l) + NaCl(aq)$$

Double exchange: Precipitation reaction

$$AgNO_3(aq) + KCl(aq) \longrightarrow AgCl(s) + KNO_3(aq)$$

Oxidation-reduction reactions are not limited to metals and ionic compounds. Molecular covalent compounds also take part in oxidation-reduction reactions.

Combustion: Redox reaction

$$CH_4 + 2O_2 \longrightarrow CO_2 + 2H_2O$$

In this reaction, carbon is oxidized. Electron transfer is trickier to identify in reactions with covalent compounds. But we already know that combustion is a form of oxidation. For molecules, oxidation often involves a loss of hydrogen and a gain of oxygen, and reduction often involves a gain of hydrogen.

Example

Oxidation-Reduction

Solid magnesium, Mg, is placed in a solution of iron (III) nitrate, $Fe(NO_3)_3$. Magnesium nitrate and solid iron are formed. Show the ionic equation and determine what is oxidized and what is reduced.

Solution

Write and balance the chemical equation for this reaction:

$$3Mg(s) + 2Fe(NO_3)_3(aq) \longrightarrow 3Mg(NO_3)_2(aq) + 2Fe(s)$$

Translate this equation into an ionic equation. Ignore the spectator ions, in this case, the NO_3^- ions.

$$3Mg(s) + 2Fe^{3+}(aq) \longrightarrow 3Mg^{2+}(aq) + 2Fe(s)$$

The metal losing electrons is oxidized. The metal gaining electrons is reduced.

Each magnesium atom loses two electrons, so magnesium is oxidized. Two iron ions accept three electrons each, so iron is reduced.

Tracking the electrons

$3Mg^0 \longrightarrow 3Mg^{2+} + 6e^-$

$2Fe^{3+} + 6e^- \longrightarrow 2Fe^0$

Key Terms

reduction
redox reactions
(oxidation-reduction)

Lesson Summary

What happens to electrons during oxidation?

Whenever an atom or ion loses electrons, another atom or ion gains electrons. The losing of electrons is called oxidation, and the gaining of electrons is called reduction. Oxidation and reduction always occur together. This process is known as oxidation-reduction, or redox. When metals react with oxygen to form metal oxides, the metal atoms are oxidized and the oxygen is reduced. To extract elemental metals, the reverse reaction must occur and metal ores must be reduced.

EXERCISES

Reading Questions

1. Explain what is oxidized and what is reduced when copper reacts with oxygen to form copper (II) oxide, CuO.
2. Explain what is oxidized and what is reduced when copper (II) oxide is decomposed to copper and oxygen.

Reason and Apply

3. Describe how you might extract a metal from a metal salt solution.
4. Determine what is oxidized and what is reduced in these reactions:
   a. $CO_2(g) + H_2(g) \longrightarrow CO(g) + H_2O(g)$
   b. $SF_4(g) + F_2(g) \longrightarrow SF_6(g)$
   c. $4Ag(s) + 2H_2S(g) + O_2(g) \longrightarrow 2Ag_2S(s) + 2H_2O(g)$
   d. $C_6H_{12}O_6(aq) + 6O_2(g) \longrightarrow 6CO_2(g) + 6H_2O(l)$
5. Write the net ionic equations for these reactions:
   a. $Zn(s) + CuCl_2(aq) \longrightarrow Cu(s) + ZnCl_2(aq)$
   b. $Ba(NO_3)_2(aq) + CuSO_4(aq) \longrightarrow BaSO_4(s) + Cu(NO_3)_2(aq)$
6. Determine which of the reactions in Exercise 5 is not a redox reaction. Explain your reasoning.
7. Which of these reactions from Unit 4: Toxins can be classified as redox reactions? For each redox reaction, show what is oxidized, what is reduced, and the total number of electrons transferred.
   a. $Tl_2O(s) + 2HCl(l) \longrightarrow 2TlCl(aq) + H_2O(l)$
   b. $HgS(s) + 2HCl(aq) \longrightarrow HgCl_2(s) + H_2S(aq)$
   c. $PbCO_3(s) + 2HCl(aq) \longrightarrow PbCl_2(aq) + H_2CO_3(aq)$
   d. $Na_2C_2O_4(aq) + CaCl_2(aq) \longrightarrow CaC_2O_4(s) + 2NaCl(aq)$
   e. $Pb(s) + 2HCl(aq) \longrightarrow PbCl_2(aq) + H_2(g)$
   f. $2As(s) + 6HCl(aq) \longrightarrow 2AsCl_3(aq) + 3H_2(g)$

LESSON

18 The Active Life

Activity of Metals

Think About It

Imagine you set up three experiments. In one, a strip of copper is placed in a solution of silver nitrate. In the second, a strip of zinc is placed in a solution of copper (II) nitrate. In the third, a strip of silver is placed in a solution of zinc nitrate. Only two of these setups will result in reactions. How can you predict which reactions will occur?

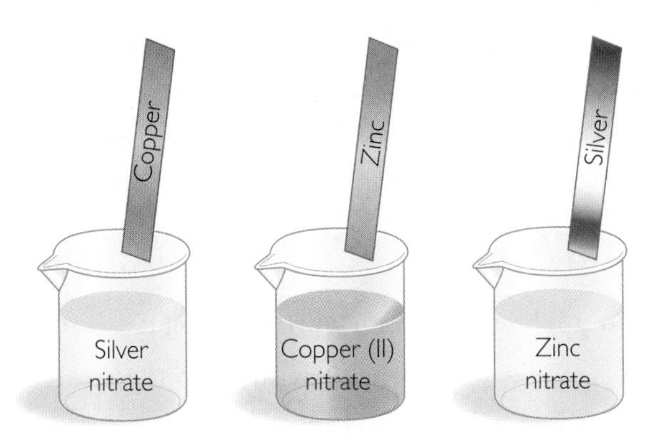

Which metal atoms are most easily oxidized?

To answer this question, you will explore

1. Comparing Metals
2. Activity Series

INDUSTRY CONNECTION

Redox reactions have been used for decades to coat metals with chromium. Chrome plating improves appearance and durability. Chromium sulfate, or chromium chloride, are often used in industrial plating solutions.

Exploring the Topic

1 Comparing Metals

When metal atoms are combined with metal ions in solution, you may or may not end up with a reaction. Some combinations of metals and metal salts result in redox reactions and others don't. It all depends on which combinations of metals are used.

Consider the three solutions shown on the next page. Each solution contains metal cations and a strip of elemental metal. If the metal atoms on the strip transfer electrons to the metal cations in solution, a reaction will occur.

In the first beaker, the copper atoms transfer electrons to the silver ions. Copper atoms are oxidized and silver cations are reduced. As a result, silver metal can be seen as a coating on the copper strip. In the second beaker, zinc atoms transfer electrons to the copper ions in solution. Zinc atoms are oxidized and copper is reduced. Copper metal coats the zinc strip. However, nothing happens in the third beaker.

Beaker 1: Copper in silver nitrate solution

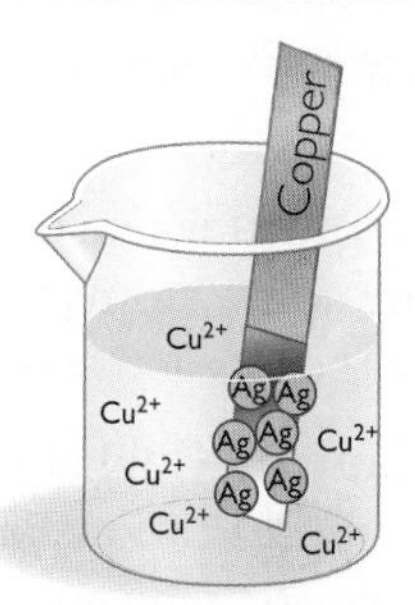

Silver: reduced
Copper: oxidized

Beaker 2: Zinc in copper (II) nitrate solution

Copper: reduced
Zinc: oxidized

Beaker 3: Silver in zinc nitrate solution

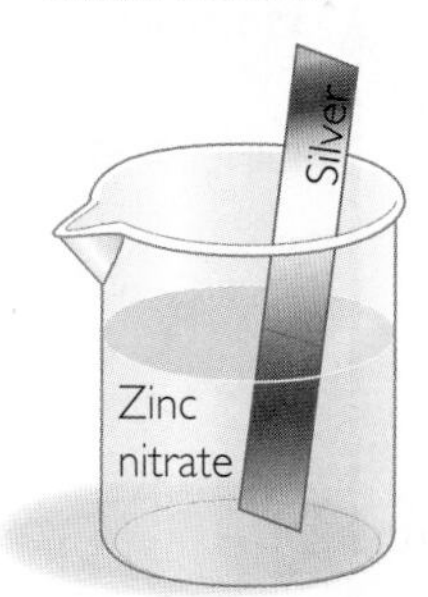

No reaction

$$\text{Beaker 1: } Cu(s) + 2AgNO_3(aq) \longrightarrow Cu(NO_3)_2(aq) + 2Ag(s)$$

$$Cu(s) + 2Ag^+(aq) \longrightarrow Cu^{2+}(aq) + 2Ag(s)$$

$$\text{Beaker 2: } Zn(s) + Cu(NO_3)_2(aq) \longrightarrow Zn(NO_3)_2(aq) + Cu(s)$$

$$Zn(s) + Cu^{2+}(aq) \longrightarrow Zn^{2+}(aq) + Cu(s)$$

$$\text{Beaker 3: } Ag(s) + Zn(NO_3)_2(aq) \longrightarrow \text{no reaction}$$

In the third beaker there is no reaction. Silver does not give up its electrons in the presence of zinc ions. Chemists say that silver is not as active as zinc.

2 Activity Series

By combining different metals and metal ions, you can determine experimentally which metals are more active than other metals. The result is called an **activity series.** Working with the data we have so far, zinc, copper, and silver can be placed in order of their activity.

Zinc
gives up electrons to
Copper
which gives up electrons to
Silver

More active ↕ Less active

Where would magnesium fit on this list? In the classroom you combined magnesium with zinc nitrate. The magnesium was oxidized and formed Mg^{2+} ions.

$$Mg(s) + Zn(NO_3)_2(aq) \longrightarrow Mg(NO_3)_2(aq) + Zn(s)$$

So, magnesium belongs above zinc on the list.

Magnesium
gives up electrons to
Zinc
which gives up electrons to
Copper
which gives up electrons to
Silver

More active ↕ Less active

CONSUMER CONNECTION

Lithium is the most active metal on the periodic table and would potentially be listed above potassium on the activity series table. Your cell phone may contain a lithium battery. A redox reaction in the battery supplies electrical energy to power your phone.

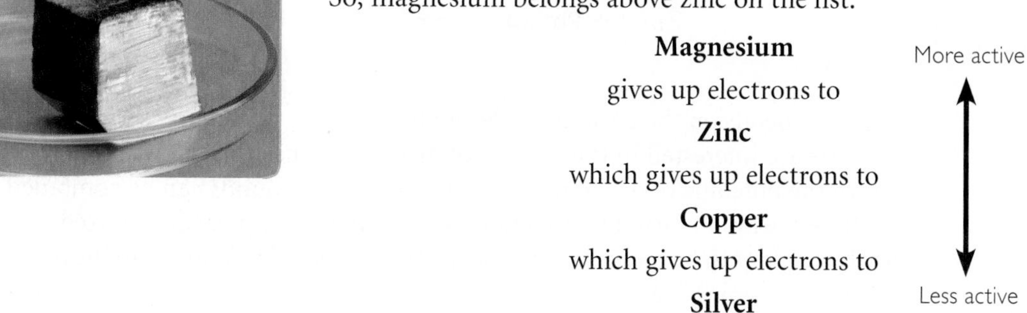

Activity Series

Metal	
Potassium	More active Easily oxidized
Barium	
Calcium	
Sodium	
Magnesium	
Aluminum	
Zinc	
Chromium	
Iron	
Nickel	
Copper	
Silver	
Mercury	Less active
Gold	Easily reduced

You may have noticed that all the reactions in this section have been single exchange reactions where the more easily oxidized metal displaces the other metal. In this way, the more active metal forms cations while the less active metal is reduced to a solid.

A series of experiments can reveal where other metals belong on the list. The more active atoms at the top of the list will always displace the less active ones below.

On this list, gold is at the bottom. Gold atoms are very difficult to oxidize, and gold cations are very easy to reduce. This is one reason why gold is rarely found in compounds combined with other atoms.

Because silver is easily reduced, it is relatively easy to plate metals with silver for ornamentation.

A metal high in the activity series

- reacts vigorously and quickly with compounds
- readily gives up electrons in reactions to form positive ions
- is corroded easily

A metal low in the activity series

- does not react vigorously and quickly with chemicals
- does not readily give up electrons in reactions to form positive ions
- is not corroded easily

Example

Activity of Lead

Lead, Pb, will give up electrons to copper cations, Cu^{2+}, but not to zinc cations, Zn^{2+}.

a. Write the net ionic equation for the reaction between Pb and Cu^{2+}.

b. Do you expect that Zn will transfer electrons to lead cations, Pb^{2+}? Explain your reasoning.

Solution

a. $Pb(s) + Cu^{2+}(aq) \longrightarrow Pb^{2+}(aq) + Cu(s)$

b. Pb does not give up electrons to Zn^{2+}. This means that the reverse reaction *will* happen. Zn will give up electrons to Pb^{2+}. Zn is more active than Pb, so it displaces Pb in the ionic compound.

$$Zn(s) + Pb^{2+}(aq) \longrightarrow Zn^{2+}(aq) + Pb(s)$$

All of the metals on the periodic table can be placed in an activity series table. Chemists are interested in the activity of metals, because this allows them to control the outcome of different reactions. Metal compounds can be combined in different ways to form other compounds. Also, the most easily oxidized metals turn out to be a good source of electrons. You'll learn more about this in Lesson 19: Current Events.

Key Term

activity series

Lesson Summary

Which metal atoms are most easily oxidized?

Metals that are easily oxidized give up their electrons easily. They are considered more active than other metals. Comparing metals experimentally allows you to rank the metals in terms of their ease of oxidation. The activity of a metal is demonstrated when it is combined with another metal cation. The more active metal will lose electrons, displacing the other metal in the compound. The displaced metal is reduced. The ranking of metals in order of their activity is referred to as an activity series.

EXERCISES

Reading Questions

1. Explain how you would figure out where tin should be listed in the activity series.
2. Why are metals that are more active also considered less stable?

Reason and Apply

3. If zinc nitrate, $Zn(NO_3)_2$, is paired with solid iron, what would you expect to observe? Explain your reasoning.
4. If iron (III) nitrate, $Fe(NO_3)_3$, were combined with solid zinc, what would you expect to observe? Explain your reasoning
5. Consider any reactions that occur in Exercises 3 and 4.
   a. Write chemical equations.
   b. Write the net ionic equation.
   c. Identify the more active metal.
   d. Which metal is oxidized? Which metal is reduced?
6. Platinum is more active than gold but less active than silver. Where does it belong in the activity series?
7. If cobalt ions are replaced by iron to form iron ions and solid cobalt, is cobalt above or below iron in the activity series? Explain.
8. Write net ionic equations for two reactions between magnesium and less active metals.
9. Name three combinations of ionic compounds and metals that will not react.

LAB Electrochemical Cell

Purpose

To explore how to get electrical energy from a redox reaction.

Materials

- zinc strip
- copper strip
- 250 mL beakers (2)
- 500 mL beaker
- 2 connecting wires with alligator clips
- tiny LED light bulb
- 1.0 M $CuSO_4$, 100 mL
- 1.0 M zinc sulfate solution, $ZnSO_4$, 100 mL
- saturated potassium nitrate (KNO_3) solution, 20 mL—for salt bridge
- filter paper approximately 1 in. wide and 6 in. long—for salt bridge
- gloves for handling salt bridge

Procedure

1. Make a salt bridge. Place a piece of folded up filter paper in the bottom of an empty 500 mL beaker. Soak it thoroughly with KNO_3. Set aside for later.
2. Carefully pour 100 mL of $CuSO_4$ into one beaker and 100 mL of $ZnSO_4$ into another beaker.
3. Set up the zinc strip in the $ZnSO_4$ solution and the copper strip in the $CuSO_4$ solution. Use a wire with alligator clips to connect the two metal strips. The clips should not touch the solutions.
4. Use gloves to place the salt bridge between the two beakers as shown.

Observation and Analysis

1. What do you observe?
2. Why is it necessary to have an ionic solution in the salt bridge?
3. Which substance is being oxidized? Which substance is being reduced? How do you know?
4. Connect the tiny LED light into your circuit using both sets of alligator clips. What do you observe? What does that prove?
5. Explain how you might reverse this reaction.
6. **Making Sense** Explain where the electricity in the electrochemical cell is coming from.

LESSON

19 Current Events

Electrochemical Cell

Think About It

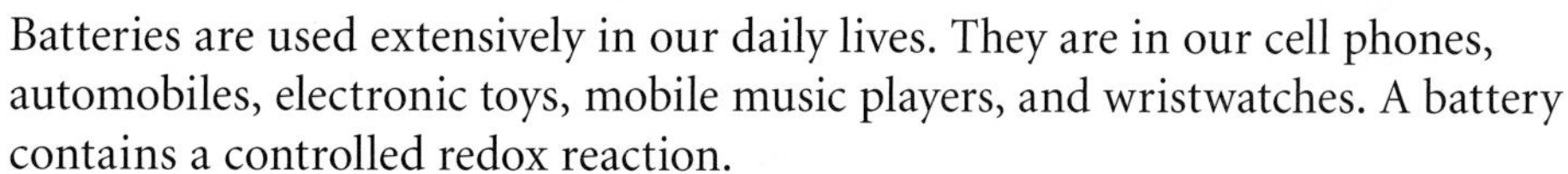

Batteries are used extensively in our daily lives. They are in our cell phones, automobiles, electronic toys, mobile music players, and wristwatches. A battery contains a controlled redox reaction.

How can you use a redox reaction as an energy source?

To answer this question, you will explore

1. Electrochemical Cells
2. Maximizing the Potential
3. Batteries

Exploring the Topic

1 Electrochemical Cells

The combustion of gasoline powers cars. The combustion of methane is used to cook food and heat homes. But it would be hard to imagine a cell phone or a toy running off of a combustion reaction.

There is another way, besides burning, to make use of the energy of a reaction. In the case of many oxidation-reduction reactions, the transfer of electrons can be converted into electrical energy using an apparatus called an *electrochemical cell.*

Controlling Redox Reactions

The first step to controlling a redox reaction is to separate the oxidation and reduction reactions. One possible way to do this is shown. Two strips of metal called *electrodes* are placed in two beakers. Both beakers are filled with aqueous salt solutions called **electrolytes.** Electrons move through a wire connecting the two metal strips, and ions move through a salt bridge connecting the two electrolyte solutions.

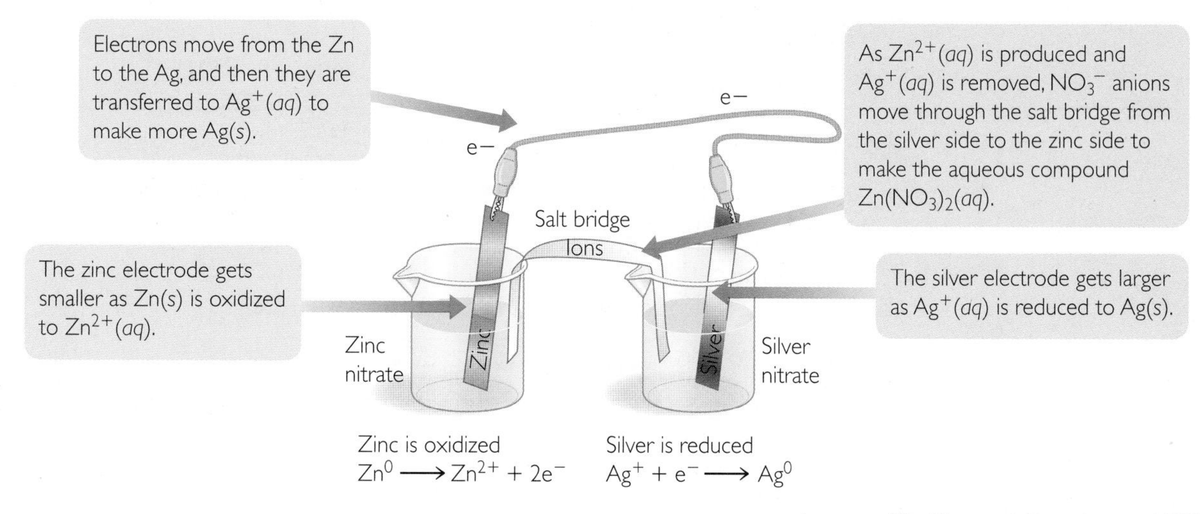

CONSUMER CONNECTION

The forward reaction of a car battery is used to start the engine, but once the engine is running, electrical energy causes the reverse reaction, recharging the battery. When you jump-start a car, you leave the engine running for a while to recharge the battery.

You can use the activity series table below to help you predict what will happen in the **electrochemical cell.** Because zinc is higher on the chart than silver, zinc is the more active metal and you can expect it to be oxidized. Likewise, because silver is not a very active metal, you can expect it to be reduced.

$$Zn(s) + 2Ag^{+}(aq) \longrightarrow Zn^{2+}(aq) + 2Ag(s)$$

If you place $Zn(s)$ and $Ag^{+}(aq)$ in the same beaker, the reaction will occur with direct transfer of electrons from the $Zn(s)$ to the $Ag^{+}(aq)$. The reaction just goes until one or both reactants are used up.

Separating the reactants allows you to control the reaction and to create an electrical current. Each half of an electrochemical cell is called a **half-cell.** The reaction in the electrochemical cell can be stopped and started when desired. It will proceed only when the circuit is closed and the two halves of the reaction are connected. The reaction that takes place in each half-cell is called a **half-reaction.**

2 Maximizing the Potential

Not every combination of metals and metal salts results in the same output of energy. Generally, the farther away two metals are from each other on the activities series list, the more energy will be produced by a reaction between them. For example, the energy released when calcium and silver are paired is greater than the energy released when iron and lead are paired.

Activity Series

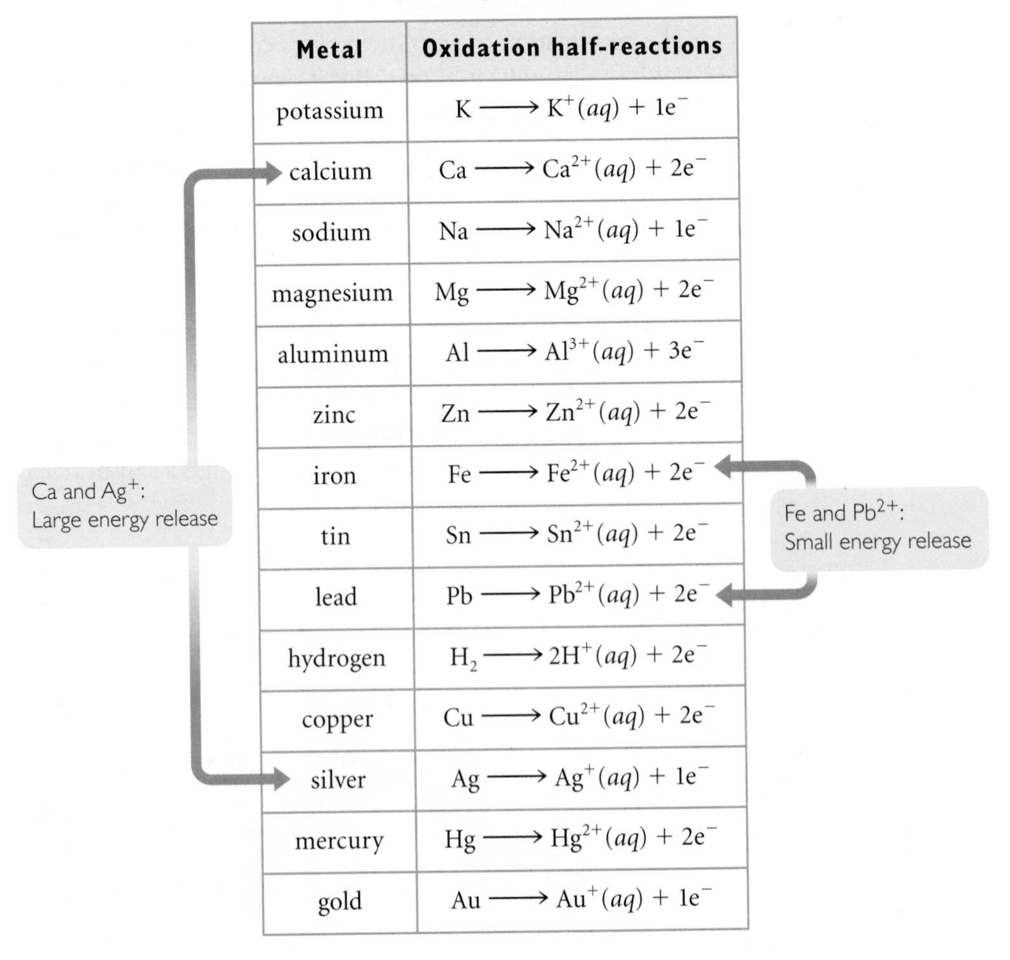

Metal	Oxidation half-reactions
potassium	$K \longrightarrow K^{+}(aq) + 1e^{-}$
calcium	$Ca \longrightarrow Ca^{2+}(aq) + 2e^{-}$
sodium	$Na \longrightarrow Na^{2+}(aq) + 1e^{-}$
magnesium	$Mg \longrightarrow Mg^{2+}(aq) + 2e^{-}$
aluminum	$Al \longrightarrow Al^{3+}(aq) + 3e^{-}$
zinc	$Zn \longrightarrow Zn^{2+}(aq) + 2e^{-}$
iron	$Fe \longrightarrow Fe^{2+}(aq) + 2e^{-}$
tin	$Sn \longrightarrow Sn^{2+}(aq) + 2e^{-}$
lead	$Pb \longrightarrow Pb^{2+}(aq) + 2e^{-}$
hydrogen	$H_2 \longrightarrow 2H^{+}(aq) + 2e^{-}$
copper	$Cu \longrightarrow Cu^{2+}(aq) + 2e^{-}$
silver	$Ag \longrightarrow Ag^{+}(aq) + 1e^{-}$
mercury	$Hg \longrightarrow Hg^{2+}(aq) + 2e^{-}$
gold	$Au \longrightarrow Au^{+}(aq) + 1e^{-}$

Example 1

Magnesium-Tin Cell

Imagine you want to create an electrochemical cell using magnesium, Mg, and tin, Sn.

a. Which metal would be oxidized and which would be reduced?

b. What ion will form and what metal will be plated out of solution as a solid?

c. Write the half-reactions and the net ionic equation.

Solution

Magnesium is higher in the activity series.

a. Magnesium will be oxidized. Tin will be reduced.

b. Tin will be plated out onto the tin strip. Magnesium ions, $Mg^{2+}(aq)$, will form.

c. The half-reactions are $Mg^0 \longrightarrow Mg^{2+} + 2e^-$ and $Sn^{2+} + 2e^- \longrightarrow Sn^0$. Adding these, the net reaction is $Mg(s) + Sn^{2+}(aq) \longrightarrow Mg^{2+}(aq) + Sn(s)$.

The concentrations of the reactants, as well as the identities of the metals and metal ions used, can affect the amount of energy that is produced.

Voltage

Topic: Electrochemical Cells

Visit: www.SciLinks.org

Web code: KEY-519

The energy that comes out of an electrochemical cell is expressed in volts. The **voltage** of a battery lets you know how much energy is "stored" in it, or how much potential it has to produce electricity.

Chemists can predict the voltage of an electrochemical cell by examining the two half-reactions involved. The voltage of half-cells is measured by chemists under uniform conditions. These voltages are normally expressed as *reduction* half-reactions and are listed in a standard reduction potentials table.

Standard Reduction Potentials at 25 °C, 1 atm, and 1 M Ion Concentration

Reduction Reaction	Half-cell Potential
$Li^+ + e^- \longrightarrow Li$	−3.05 volts
$Mg^{2+} + 2e^- \longrightarrow Mg$	−2.37 volts
$Zn^{2+} + 2e^- \longrightarrow Zn$	−0.76 volt
$Ni^{2+} + 2e^- \longrightarrow Ni$	−0.26 volt
$Pb^{2+} + 2e^- \longrightarrow Pb$	−0.13 volt
$Sn^{2+} + 2e^- \longrightarrow Sn$	0.15 volt
$Cu^{2+} + 2e^- \longrightarrow Cu$	0.34 volt
$Ag^+ + e^- \longrightarrow Ag$	0.80 volt
$Au^+ + e^- \longrightarrow Au$	1.69 volts

The overall voltage of an electrochemical cell can be calculated by finding the difference between the potentials of the half-reactions. This means subtracting the voltage of the oxidation process from the voltage of the reduction process.

Example 2

Silver-Zinc Electrochemical Cell

Determine the voltage of a silver-zinc electrochemical cell.

Solution

Write the half-reactions for silver and zinc from the standard reduction potentials table on page 561. The half-reaction that is more positive will proceed as a reduction, and the half-reaction that is more negative will proceed as an oxidation (in the reverse direction).

This half-reaction proceeds as reduction.	$Ag^+ + e^- \longrightarrow Ag$	0.80 volt
This half-reaction proceeds in the opposite direction as oxidation.	$Zn^{2+} + 2e^- \longrightarrow Zn$	−0.76 volt

Next, subtract the reduction potential of the oxidation half-reaction from the reduction potential of the reduction half-reaction.

$$0.80 - (-0.76) = 0.80 + 0.76 = +1.56 \text{ volts}$$

This cell has a voltage of +1.56 volts.

ENVIRONMENTAL CONNECTION

Batteries often contain substances that are potentially hazardous to living things. It is important to dispose of them properly so that they do not contaminate the water supply or the soil. This means they should be taken to a recycling center, not put in the trash.

3 Batteries

Electrochemical cells have been developed into batteries, which are basically containers of chemicals just waiting to be converted to electrical energy. The chemical reaction within a battery does not take place until the two reactions are connected to one another. When you flip the switch on a flashlight, you are connecting the halves of the battery inside. If you forget and leave the flashlight on, the chemical reaction inside will continue until the battery is dead.

Some batteries are rechargeable. This means that once the battery has gone dead, an electric current can be run through it in the opposite direction in order to push the reverse reaction to occur. This restores the original reactants and the battery is good once again.

A battery that is not connected to anything still has a voltage, even though it is not yet doing any work. Most of the batteries in your household can produce voltages from about 1.25 volts to 9 volts.

Key Terms

electrolyte
electrochemical cell
half-cell
half-reaction
voltage

Lesson Summary

How can you use a redox reaction as an energy source?

You can convert the transfer of electrons that occurs during oxidation-reduction into electrical energy. By separating the reactants into half-cells, you can control and essentially "store" electrical energy. The activity series allows you to choose electrodes and electrolytes that will produce electricity. The voltage of an electrochemical cell can be calculated from the half-cell potentials.

EXERCISES

Reading Questions

1. Explain how to determine the direction of electron transfer in an electrochemical cell.
2. Explain how you might choose which reactants to put together in an electrochemical cell.

Reason and Apply

3. Suppose you make an electrochemical cell. One beaker has a calcium metal electrode and a solution of aqueous calcium ions, $Ca^{2+}(aq)$. The other beaker has an iron metal electrode and a solution of aqueous iron ions, $Fe^{3+}(aq)$.
   a. Write the half-reactions for each beaker.
   b. Use the activity series table on page 560 to decide which metal is oxidized and which metal cation is reduced.
   c. Make a sketch of the electrochemical cell. Show the reaction that occurs in each beaker.
   d. Write a net ionic equation for the reaction that takes place to generate electricity.
4. Repeat Exercise 3 with the pairs of metals here.
   a. copper and tin
   b. silver and aluminum
   c. copper and lead
5. Predict which combination of metals in Exercises 3 and 4 will result in the greatest voltage. Explain your prediction. Then calculate the voltages of the redox reactions. Was your prediction correct?
6. What are the pros and cons of using combustion reactions as a source of energy?
7. What are the pros and cons of using electrochemistry as a source of energy?
8. **WEB RESEARCH** Look up common batteries, such as AA or D cell batteries, to find out the chemistry behind them. List the metals and compounds commonly used in their construction.
9. **WEB RESEARCH** Car batteries are usually 12-volt. Explain how such a high voltage is achieved.
10. **WEB RESEARCH** When car lights are left on for a long time, a battery is considered "dead" and will require a jump start from another car battery. In chemical terms, what does it mean to have a "dead" car battery? What does "jump-starting" a car battery mean in terms of the overall redox reaction?

SECTION

IV

SUMMARY

Controlling Energy

Key Terms

oxidation
heat of formation
reduction
redox reactions
(oxidation-reduction)
activity series
electrolyte
electrochemical cell
half-cell
half-reaction
voltage

Fire Update

Every chemical change is accompanied by an energy transfer. The chemical energy from reactions between metals and metal compounds can be converted to electrical energy with the appropriate apparatus.

Oxidation-reduction reactions are a particularly useful source of energy. Oxidation is the loss of electrons by an atom, ion, or molecule during a chemical change. Its counterpart, reduction, is the gaining of electrons by an atom, ion, or molecule during a chemical change. In fact, combustion is one type of oxidation-reduction reaction in which the fuel is oxidized and the oxygen is reduced.

Metals that are easily oxidized give up their electrons easily. These metals are considered more active.

Review Exercises

1. Explain how an electrochemical cell works to convert energy from chemical reactions into electrical energy.
2. Why are some metal elements easier to extract from metal ores than others?
3. Based upon the heat of formation value in the table on page 547, determine the heat of reaction for the decomposition of rust into iron and oxygen gas. Is this reaction endothermic or exothermic? Explain.

$$2Fe_2O_3 \longrightarrow 4Fe(s) + 3O_2(g)$$

4. If you place a piece of aluminum foil in a beaker of 1 M $AgNO_3$, would a single exchange reaction occur? What if you place a piece of silver foil in a beaker of 1 M $Al(NO_3)_3$? How can you explain these two different results?
5. An electrochemical cell is set up with 1 M $Pb(NO_3)_2$ on one side, and 1 M $AgNO_3$ on the other.
   a. Which metal will be oxidized? Which will be reduced? Write out balanced half-reactions for these reactions.
   b. Write out the balanced overall reaction for this electrochemical cell.
   c. How many moles of electrons will flow from the anode to the cathode?
   d. Calculate the voltage for this electrochemical cell.

PROJECT *Alternative Energy*

Investigate an energy source that is an alternative to fossil fuels. Create a poster promoting the benefits of the energy source and explaining how it works. Write a short paper on the pros and cons of your energy source.

UNIT

5

Fire Review

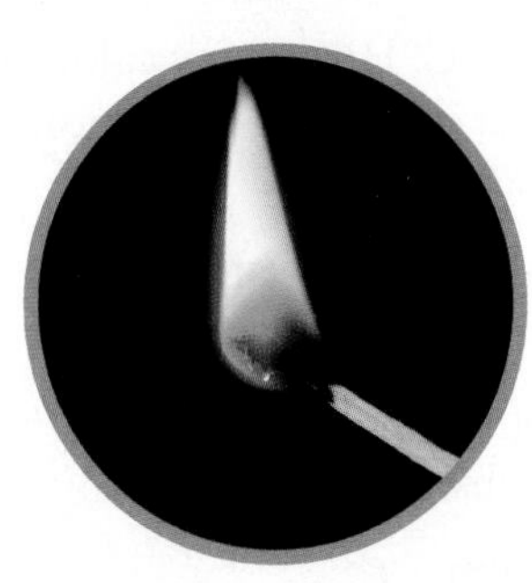

Combustion reactions are extremely useful in our daily lives. Even the digestion of our food could be considered a very controlled combustion reaction. Calorimetry is one technique used to figure out how much energy is released during combustion reactions.

Once the data from calorimetry and other experimental procedures are collected, they can be used to compare and classify reactions. Exothermic reactions release energy to the surroundings. Endothermic reactions require energy. Endothermic reactions are often sensed as "cold" by the observer. The heat of a reaction comes from bond breaking and bond making. It takes energy to break the bonds in a compound. In contrast, bond making is a process that releases energy. An energy diagram can be used to show the energetic process of a reaction.

When the energy of making the new bonds in the products is greater than the energy that it takes to break the bonds in the reactants, energy is released and the reaction is exothermic. When it takes more energy to break the bonds, the reaction is endothermic.

The net energy of a reaction is referred to as the heat of the reaction, or ΔH. In the case of an exothermic reaction, ΔH is negative, because energy is lost by the system. In the case of an endothermic reaction, ΔH is positive, because energy is absorbed by the system.

The energy associated with the formation of any compound from its elements is referred to as the heat of formation, or ΔH_f. Heat of formation is expressed in kilojoules per mole of product formed.

There is also an energy associated with getting a reaction started, called the activation energy. For example, it takes a match or a spark to get a fire started. Combustion is an oxidation-reduction reaction. Oxidation is defined as the loss of electrons by an atom, an ion, or a molecule during a chemical change. Reduction is when an atom, an ion, or a molecule gains electrons in a chemical change.

Different substances respond differently to heating. Some substances change temperature easily, with very little heat transfer. These substances, such as most metals, have low specific heat capacities.

Review Exercises

1. What have you learned about fire in this unit?
2. Generally describe how our energy needs are satisfied by chemical changes.

3. Where does the thermal energy from an exothermic reaction come from?

4. What do the first and second laws of thermodynamics indicate about energy?

5. How many calories of energy are transferred to
   a. raise the temperature of 50 grams of water from 25 °C to boiling?
   b. lower the temperature of 100 grams of water from boiling to 25 °C?

6. The specific heat capacity of aluminum is 0.21 cal/g · °C. How many calories of thermal energy must be transferred in order to heat up a 100 g aluminum pot from 20 °C to 80 °C?

7. Which takes less energy to heat up, aluminum metal or liquid water? Explain.

8. The heat of fusion for water is 79.9 cal/g. The heat of fusion for ethanol is 24.9 cal/g. What does this tell you about these two substances?

9. Name three ways you can increase the rate of a reaction.

10. Define work. How many joules of work is done if a 500 lb piano is lifted by a pulley to a height of 20 m?

11. Write and balance the equation for the oxidation of lead, Pb.

12. Write and balance a chemical equation for a redox reaction.

13. Use net ionic equations to show which reactant atoms are oxidized and which cations are reduced in the following equations.

$$Zn(s) + 2MnO_2(s) \longrightarrow ZnO(s) + Mn_2O_3(s)$$

$$3CaO(s) + 2Al(s) \longrightarrow Al_2O_3(s) + 3Ca(s)$$

14. Explain how you know which atoms are oxidized and which are reduced in a redox reaction.

15. All of the thermal energy produced in an internal combustion engine cannot be converted into work because
   A. combustion is an endothermic process, so heat is absorbed by the system.
   B. unlike matter, energy can be created or destroyed.
   C. combustion is an exothermic process, so heat is absorbed by the system.
   D. some of the energy is lost to the surroundings as heat.

UNIT

6 Showtime

Salt continuously dissolves in and precipitates out of the water in the ocean, maintaining a balance in the salt concentration.

Why Showtime?

Sometimes chemical reactions are used simply to entertain. After all, what is the purpose of a fireworks display except to delight and astonish the audience? In the chemistry classroom, chemical reactions can demonstrate new chemistry concepts in an intriguing way. Chemical demonstrations often involve reversible reactions. These reactions can be fine-tuned and controlled when you understand the balance between products and reactants in a reversible chemical system. This balance is referred to as equilibrium. Unit 6: Showtime explores chemical equilibrium and how it can be used in chemical demonstrations.

In this unit, you will learn

- what happens in a chemical system at equilibrium
- how the balance in a reversible reaction is maintained
- the mathematical relationship between reactants and products in a system at equilibrium
- about variables that affect reversible reactions

SECTION

I

Chemical Equilibrium

The extremely high levels of salt found in Mono Lake lead to the formation of calcium carbonate deposits called tufa towers.

Many chemical demonstrations feature reversible reactions. Unlike most combustion reactions, reversible reactions do not go all the way to completion. In reversible chemical systems both the forward and reverse reactions occur simultaneously, so products and reactants exist in the same reaction mixture. When two opposite processes in a system are in balance, the system has reached what is called a state of equilibrium. To the eye, a system at equilibrium may appear unchanging, but this is not the case. Changes are continuously happening on the particle level in systems at equilibrium.

In this section, you will learn

- to describe conditions that can affect chemical systems
- about reversible chemical and physical processes
- to explain the dynamic nature of chemical equilibrium
- the mathematical relationship between products and reactants at equilibrium

LESSON

1 How Awesome

Chemical Demonstrations

Think About It

Magicians perform "magic" tricks, like making fire appear or turning a liquid that looks like milk into a liquid that looks like grape juice. Tricks like these are often nothing more than glorified chemistry demonstrations. Chemistry teachers also conduct demonstrations that intrigue and amaze their students. What is the chemistry behind all these special effects?

What makes a good chemistry demonstration?

To answer this question, you will explore

1. The Chemistry of Demonstrations
2. Changing Conditions

Exploring the Topic

1 The Chemistry of Demonstrations

Magic acts often contain a lot of good chemical demonstrations. What makes a demonstration powerful is usually a combination of visual effects and surprise. You might see, hear, or experience something that you didn't expect. You may be puzzled by the outcome and want to see the demonstration again. Chemical demonstrations feature color changes, puffs of smoke, flames, and other surprises. To produce a good chemical demonstration, the conditions affecting the reaction must be carefully controlled.

Transforming a Liquid

One fun chemical demonstration involves a series of changes in liquids. The point of this demonstration is that a liquid that appears to be fruit punch is turned into a liquid that appears to be milk or ginger ale.

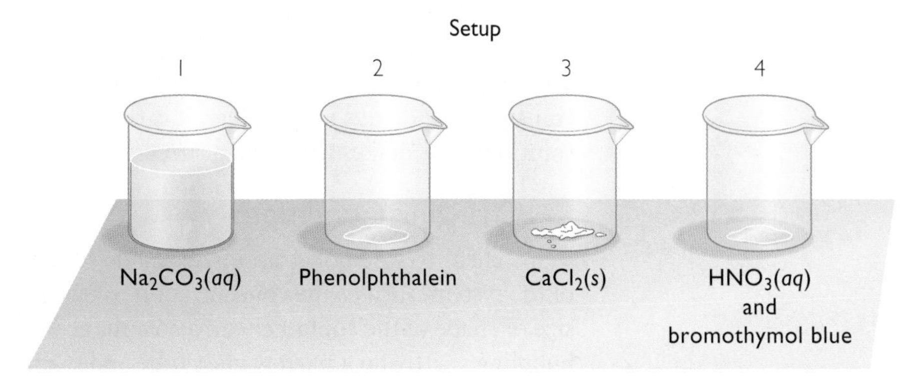

To the audience, the first beaker appears to be water, while the other beakers look empty. The demonstrator pours the contents of beaker 1 into beaker 2, and a pink solution that looks like fruit punch appears. Then the demonstrator pours

RECREATION CONNECTION

"Magic sand" is a special type of hydrophobic sand created by mixing ordinary beach sand with tiny particles of pure silica and trimethylsilanol $(CH_3)_3SiOH$ vapor. When poured into water, the grains of the sand will adhere to each other and form columns. As soon as the sand is taken out of the water, it is completely dry and flows freely.

this pink solution into beaker 3, where it instantly forms a cloudy pink solution that appears to be strawberry milk. Finally, the demonstrator pours the contents of this beaker into beaker 4, resulting in a clear yellow solution with bubbles that looks like ginger ale.

The Chemistry Behind the Transformations

The colorful liquids from this demonstration are shown in this illustration, along with the chemical equations for the reactions that cause them.

"Water"

The "water" in beaker 1 is really sodium carbonate solution, $Na_2CO_3(aq)$, which is basic.

$$Na_2CO_3(s) \longrightarrow Na^{2+}(aq) + CO_3^{2-}(aq)$$

Recall that under the Brønsted-Lowry definition of a base, sodium carbonate "adds" OH^- to aqueous solutions by removing H^+ from water.

$$CO_3^{2-}(aq) + H_2O(l) \longrightarrow HCO_3^-(aq) + OH^-(aq)$$

"Fruit punch"

When the basic solution in beaker 1 is poured into beaker 2, the phenolphthalein indicator in beaker 2 turns from clear to pink. The demonstrator has purposefully used pH to get the desired effect when the contents of beakers 1 and 2 are combined.

"Strawberry milk"

Beaker 3 contains sodium calcium chloride, $CaCl_2(s)$. When the contents in beaker 2 are poured into beaker 3, the calcium chloride dissolves and dissociates. The calcium ions combine with the carbonate ions to form a white precipitate of calcium carbonate suspended in a pink solution.

$$Ca^{2+}(aq) + CO_3^{2-}(aq) \longrightarrow CaCO_3(s)$$

"Ginger ale"

Beaker 4 contains nitric acid, $HNO_3(aq)$, and the indicator bromothymol blue, which is pale yellow in the presence of an acid. As soon as the contents of beaker 3 are poured into beaker 4, the white precipitate of calcium carbonate dissolves in the nitric acid and carbonic acid, $H_2CO_3(aq)$, forms.

$$CaCO_3(s) + 2H^+(aq) \longrightarrow H_2CO_3(aq) + Ca^{2+}(aq)$$

The carbonic acid decomposes into water and carbon dioxide gas.

$$H_2CO_3(aq) \longrightarrow CO_2(g) + H_2O(l)$$

The combination of the yellow of the bromothymol blue indicator and the carbon dioxide gas bubbles makes this solution look like ginger ale. The illusion of the transforming drinks is complete. It is the careful control of the pH of these solutions that makes this trick a success.

2 Changing Conditions

Many chemistry demonstrations are made spectacular by fine-tuning the conditions of the system, such as the concentrations of the reactants or the temperature. The size or shape of the container can also enhance the effects. For example, a reaction bubbling swiftly up a narrow glass tube and overflowing is visually entertaining.

If you think about the chemistry you have learned, you will find evidence that you can change the outcome of a chemical reaction by manipulating the conditions. In Unit 1: Alchemy, you turned copper into colorful compounds, and then obtained the copper back again. In Unit 3: Weather, you learned that the evaporation and condensation of water is key in causing changes in weather. A slight change in temperature can mean the difference between a dry day, rain showers, or a hurricane forming. Unit 4: Toxins was about reversing the effects of toxins. You were able to manipulate the acidity and basicity of solutions from one end of the pH scale to the other. You were also able to dissolve solid substances in solution and then precipitate them from the solution.

You examined chemical kinetics in Unit 5: Fire, and learned how chemical reactions can be affected by changing conditions. For example, when more oxygen is mixed with a fuel, a combustion reaction proceeds more quickly. Changing conditions can also force a reaction to occur even though it is not favored under standard conditions.

In Unit 6: Showtime, the focus moves to the balance between products and reactants in a reaction. In some chemical demonstrations you want more products to appear, and in others you want to see more reactants. In this unit, you will examine how changes in temperature, concentration, and pressure drive reactions in a forward direction toward products or in a reverse direction toward reactants.

Lesson Summary

What makes a good chemistry demonstration?

Chemical demonstrations fascinate and amaze audiences with color changes, puffs of smoke, flames, and other unexpected outcomes. For you to get the most out of a chemical demonstration, the conditions affecting the reaction must be carefully controlled. You can change certain conditions to affect the rate of a reaction. You can also change conditions to affect the amounts of products or reactants that are present at any point in time.

EXERCISES

Reading Questions

1. What are some characteristics of a good chemical demonstration?
2. What conditions can be changed to alter a chemical demonstration?

Reason and Apply

3. Write net ionic equations for the reactions on page 570.
4. **WEB RESEARCH** Explore the Web to find out how magician's flash paper works. What is the chemistry involved?

LAB Reversible Reactions

Purpose

To observe and explore reversible reactions.

Materials

- dropper bottle of 0.1 M cobalt (II) chloride, $CoCl_2$
- dropper bottle of saturated (~6 M) calcium chloride, $CaCl_2$
- 2 small test tubes
- test tube rack
- wash bottle
- water

Safety Instructions

Wear safety goggles at all times. Wash hands thoroughly before leaving the laboratory. Do not pour solutions in the sink; use the waste container.

Procedure and Analysis

1. Explore the reaction between a pink cobalt solution and a colorless calcium chloride solution. Begin with 20 drops of the pink cobalt solution in a test tube. What happens as you add drops of the calcium chloride solution? Record your observations.
2. The net ionic equation here describes the reaction.

$$Co(H_2O)_6^{2+}(aq) + 4Cl^-(aq) \longrightarrow CoCl_4^{2-}(aq) + 6H_2O(l)$$

Use the equation to explain what happens as you add the calcium chloride solution to the pink cobalt solution.

3. Add drops of water to the blue solution in the test tube. Record your observations.
4. Write a net ionic equation to explain what happens as you add water to the blue solution.
5. This reaction is often written with a double arrow. What do you think the double arrow indicates?

$$\underset{\text{pink}}{Co(H_2O)_6^{2+}(aq)} + 4Cl^-(aq) \rightleftharpoons \underset{\text{blue}}{CoCl_4^{2-}(aq)} + 6H_2O$$

6. Use your solutions to make a purple solution in the second test tube. Explain your procedure. Why is the solution purple? What ions are present?
7. **Making Sense** What evidence do you have that some reactions are reversible?

LESSON

2 How Backward

Reversible Reactions

Think About It

Imagine you put a bottle of water that is half full outside in the sun. The cap is on tight. After a short period of time, drops of water are visible on the inside walls of the bottle. This indicates that some of the water in the bottle turned into water vapor and then turned back into liquid water when it came in contact with the sides of the bottle. So, a phase change and its reverse happened in the same bottle.

What is a reversible process?

To answer this question, you will explore

1. Reversible Processes
2. The Double Arrow
3. Examples of Reversible Processes

HISTORY CONNECTION

The concept of a reversible reaction was first introduced around 1800. A French chemist named Claude Berthollet noticed that sodium carbonate crystals had formed at the edge of a salt lake. He recognized this reaction as being the reverse of a common reaction he had observed in the laboratory.

Exploring the Topic

1 Reversible Processes

So far, you might have the impression that reactions proceed only in the direction in which they are written, from reactants to products. When gasoline burns, the reaction goes in the forward direction, from reactants to products. This reaction is said to go to completion.

$$2C_8H_{18}(l) + 25O_2(g) \longrightarrow 16CO_2(g) + 18H_2O(g)$$

Though it *is* possible for gasoline to be created through chemical reactions, it takes many chemical steps and requires a great deal of energy. So this reaction, as it is written, is not reversible. Once you burn the gasoline, there's no going back.

BIG IDEA Many reactions are reversible.

Reversible reactions are chemical processes that proceed freely in the forward and reverse directions. Some physical processes are also reversible, such as the condensation and evaporation of water in a closed water bottle.

Extracting Mercury Metal (Reduction of Mercury)

Consider a chemical reaction in which mercury metal is extracted from mercury oxide. This is an oxidation-reduction reaction that can be reversed.

If you heat mercury (II) oxide, $HgO(s)$, to a very high temperature, you can obtain mercury metal and oxygen gas. If you flow an unreactive gas, such as argon, Ar, over the HgO, the oxygen gas that forms is swept away and the reaction keeps going.

$$2HgO(s) \longrightarrow 2Hg(l) + O_2(g)$$

Hg(*l*) forms at the surface.

Ar(g)

The $O_2(g)$ formed is swept away by the Ar(g).

HgO(s)

Heat

Reversing the Reaction (Oxidation of Mercury)

Under different conditions, the products of the reaction will also combine. If you flow oxygen gas over mercury metal at a high temperature, mercury (II) oxide is produced.

$$2Hg(l) + O_2(g) \longrightarrow 2HgO(s)$$

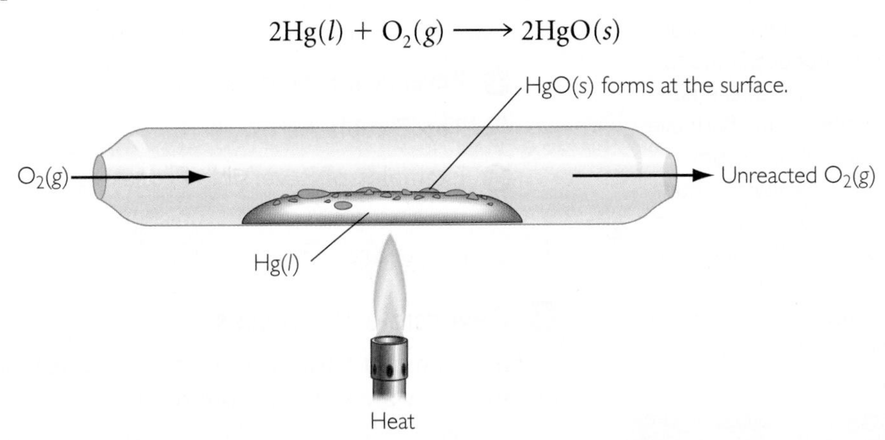

CONSUMER CONNECTION

Rechargeable batteries depend on a reversible reaction. Once the reactants are used up in a rechargeable battery, it no longer works. The rechargeable battery is put into an electronic device that provides enough energy to reverse the reaction and restore the original reactants.

So depending on the conditions, this reaction can go in the forward or reverse directions. But can you set up a situation in which the reaction goes in *both* directions in the same container?

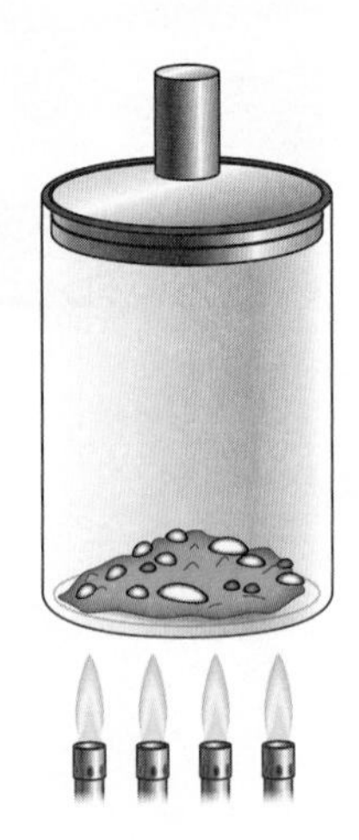

Suppose mercury (II) oxide is heated in a closed container. As heat is added to the closed system, mercury metal is produced. Oxygen gas is also produced as a product of the reaction but it cannot escape.

As the mercury metal and oxygen gas form, they react to form mercury (II) oxide. The reaction is proceeding in both directions. The reaction will continue to go in both directions simultaneously because the system is closed and the oxygen gas cannot escape. If you examined the contents of the container, you would find a mixture of the products *and* the reactants.

BIG IDEA In reversible systems, the forward and reverse reactions can occur at the same time.

BIOLOGY CONNECTION

Cobalt chloride paper can be used in an experiment to prove that plants lose water from the undersides of their leaves. If blue cobalt chloride paper is placed on the top and underside of a leaf and left for a while, the paper on the underside of the leaf will turn pink. This happens because plants release water from openings on the undersides of their leaves.

❷ The Double Arrow

A double arrow is used in the chemical equation to show that a reaction is reversible. The redox reaction described previously would be written as

$$HgO(s) \rightleftharpoons 2Hg(l) + O_2(g)$$

You could also write this equation as

$$2Hg(l) + O_2(g) \rightleftharpoons HgO(s)$$

Cobalt Chloride Reaction

Reversible reactions do not go to completion. The forward and reverse processes occur in the system at the same time. This means that reactants and products are both present in the system at the same time. This is easy to observe in the cobalt chloride reaction you completed in class. The net ionic equation below describes this reaction. The polyatomic ion on the left is referred to as the hexaaquocobalt ion, and the polyatomic ion on the right is called the tetrachlorocobalt ion. In the hexaaquocobalt polyatomic ion, six water molecules are organized around each cobalt cation.

$$\underset{\text{pink}}{Co(H_2O)_6^{2+}(aq)} + 4Cl^-(aq) \rightleftharpoons \underset{\text{blue}}{CoCl_4^{2-}(aq)} + 6H_2O(l)$$

When mostly $Co(H_2O)_6^{2+}$ ions are present in the test tube, the solution appears pink. When mostly $CoCl_4^{2-}$ ions are present in the test tube, the solution appears blue. When there is a mixture of $Co(H_2O)_6^{2+}$ and $CoCl_4^{2-}$ ions in the test tube, the solution is between pink and blue, a purplish color.

Cobalt chloride reaction

Mostly reactants | Half reactants/ half products | Mostly products

In class you added chloride ions to a sample of the pink solution, creating $CoCl_4^{2-}$ ions, which turned the test solution blue. In this reaction chloride ions replace the water molecules organized around the metal cation. If enough chloride ions are added, the solution becomes quite blue.

To reverse this reaction, you added more water to the system. This forced the formation of the original reactant, pink hexaaquocobalt ion, $Co(H_2O)_6^{2+}$.

❸ Examples of Reversible Processes

Common examples of reversible processes include redox reactions, phase changes, dissolving and precipitation, and acid-base reactions.

Phase Changes

Consider water in a drinking glass and water in a closed bottle on a hot day. In both cases, the water is evaporating. In the glass, eventually all of the liquid water turns to water vapor and escapes into the air. However, in the closed bottle, the water

Evaporation: $H_2O(l) \longrightarrow H_2O(g)$

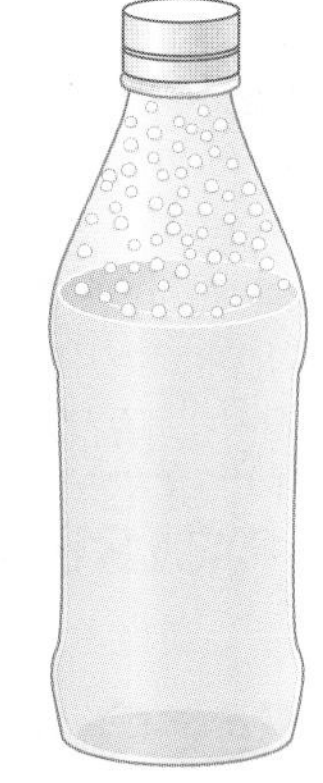

Evaporation and condensation: $H_2O(l) \rightleftharpoons H_2O(g)$

Water drops are evidence of the reverse reaction.

NATURE CONNECTION

Pointed stalactites and stalagmites form when limestone alternately dissolves in and precipitates out of the dripping water in a cave. These calcite structures build slowly, at the rate of about one quarter inch to one inch per century.

vapor cannot escape. Some of the water vapor condenses inside the bottle to re-form liquid water. Evaporation and condensation happen at the same time in the closed container.

Dissolving and Precipitation

Suppose you dissolve a small amount of salt in water. All of the salt dissolves. However, if you dissolve a large enough amount of salt in the water, a saturated solution forms. In a saturated solution, excess salt forms solid precipitate at the bottom of the container. While it might appear to the eye that all change has stopped, salt is still dissolving and precipitating, back and forth.

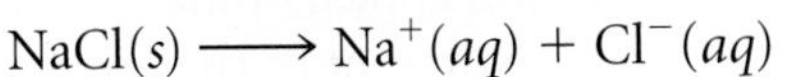

$NaCl(s) \longrightarrow Na^+(aq) + Cl^-(aq)$

$NaCl(s) \rightleftharpoons Na^+(aq) + Cl^-(aq)$

Acid-Base Reactions

When a strong acid such as hydrochloric acid, HCl, is dissolved in water, the process goes to completion. All of the acid molecules break apart into ions.

$$HCl(g) \longrightarrow H^+(aq) + Cl^-(aq)$$

In contrast, a weak acid such as hydrogen cyanide, HCN, does not dissociate completely. This is because the reverse reaction is also occurring.

$$HCN(g) \rightleftharpoons H^+(aq) + CN^-(aq)$$

Key Terms

reversible reaction

Lesson Summary

What is a reversible process?

A reversible process can proceed freely in both the forward and reverse directions. In reversible systems, the forward and reverse reactions can occur at the same time. In a reversible chemical reaction system, both products and reactants are present. A double arrow is used in chemical equations to indicate reversible reactions. Examples of reversible systems are phase changes in a closed container, saturated solutions, reactions of weak acids, and redox reactions.

EXERCISES

Reading Questions

1. What evidence is there that some reactions are reversible?
2. What does a double arrow in a chemical equation tell you?

Reason and Apply

3. Write chemical equations for three chemical processes that are reversible. Choose examples that are different from the ones in this lesson.
4. Write chemical equations for three chemical processes that are not easy to reverse. Choose examples that are different from the ones in this lesson.
5. Is it always possible to see that reactions are reversing? Why or why not?
6. This net ionic chemical equation describes the reaction between aqueous hexaaquonickel cations and aqueous calcium chloride solution.

$$\underset{\text{green}}{Ni(H_2O)_6^{2+}(aq)} + 4Cl^-(aq) \rightleftharpoons \underset{\text{yellow}}{NiCl_4^{2-}(aq)} + 6H_2O(l)$$

   a. Describe what happens to the color of the solutions as you add chloride to a solution of hexaaquonickel.
   b. Describe what happens to the color as you add water to a tetrachloronickel solution.
   c. Only those ions that participate in the reaction are included in the net ionic equation. What else is in the solution that is not included in the equation?
   d. What does the double arrow indicate?
7. Imagine that you heat lead oxide, $PbO(s)$, to extract elemental lead, $Pb(s)$.
   a. Write a balanced equation for the reaction.
   b. Describe what you would do in order to convert all the $PbO(s)$ to $Pb(s)$.
   c. Describe what you would do to reverse the reaction.
   d. Explain why $Pb(s)$ kept for a long time in a jar with air no longer looks shiny.

MUSIC CONNECTION

Compounds such as calcium sulfate, $CaSO_4$, absorb water from the air and become hydrate gels. For example, each $CaSO_4$ molecule bonds with two water molecules to become $CaSO_4 \cdot 2H_2O$. This property is useful in humidity control. For example, during shipping, the dry compound can be placed inside a musical instrument or pair of shoes to absorb excess moisture. If placed in the sun it will dry out and can be reused.

LESSON

3 How Dynamic

Dynamic Equilibrium

Think About It

A tightrope walker moves carefully and purposefully, keeping his balance as he goes. He tips back and forth while he walks, sometimes using an outstretched arm or a leg to counterbalance any movement one way or the other. He is maintaining his balance but not keeping still. Chemical systems that are at equilibrium maintain a similar type of dynamic balance.

What is equilibrium?

To answer this question, you will explore

1. Models of Equilibrium
2. Chemical Equilibrium

BIOLOGY CONNECTION

The word *equilibrium* also refers to the sense of balance that keeps people and animals from falling over. This sense of equilibrium is a combination of input from the inner ear, eyes, and the cerebellum in the brain. When your sense of equilibrium is impaired, you can experience motion sickness.

Exploring the Topic

1 Models of Equilibrium

A Dynamic Balance

If you put enough solid sodium chloride, NaCl, in water, the solution will be saturated. Addition of any more salt causes a precipitate to appear. To an observer, the system consisting of the salt solution and solid precipitate doesn't appear to change. However, there is a forward and a reverse process occurring in the solution. On a particle level, the dissolving and precipitation processes are still happening. This system is said to be at **equilibrium.** Equilibrium is a dynamic state in which opposing processes of change occur at the same time. At equilibrium, the rate of the forward process is equal to the rate of the reverse process. Because these rates are equal, no change is visible.

Fish Tank Model

To get a visual sense of dynamic equilibrium, imagine two aquariums. One aquarium contains orange fish and the other contains green fish. The aquariums are connected by a tunnel that contains a sliding door.

Initial conditions

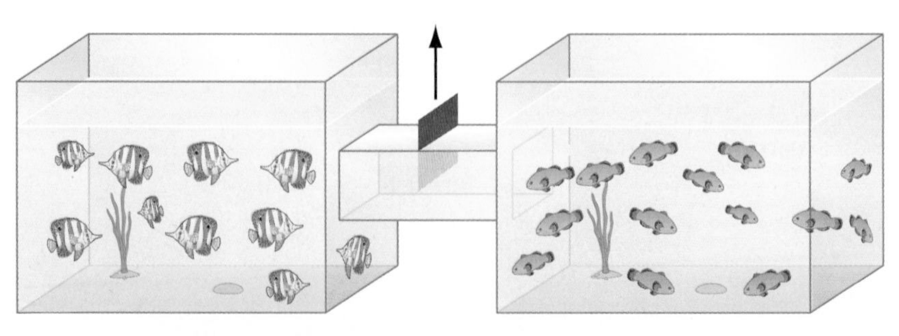

Not at equilibrium

Imagine that the door between the two tanks is opened. The fish are then free to swim back and forth. The green fish and orange fish begin to mix.

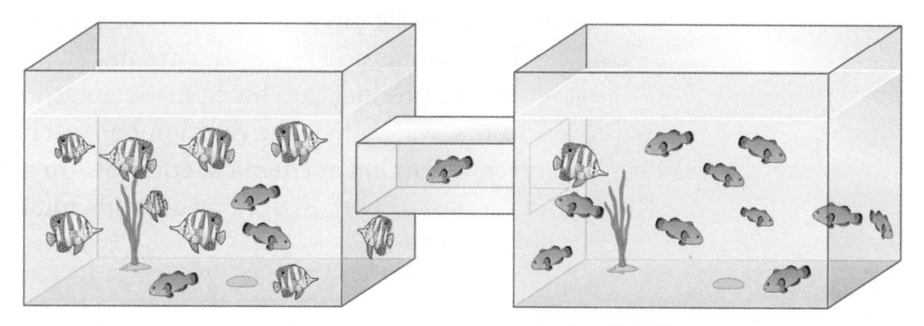

Moving toward equilibrium

Eventually, a balance is reached: The fish continue to swim back and forth, but the average number on each side remains the same. The system has reached dynamic equilibrium.

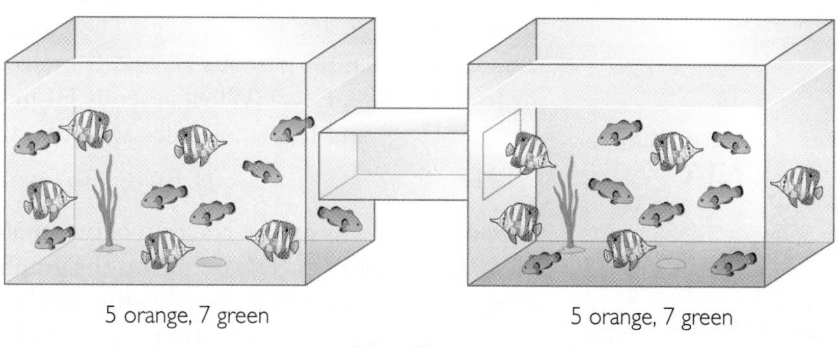

At equilibrium

The fish tank example is a reasonable analogy for what happens in a closed container of water at standard temperatures. Within a closed container, water evaporates and condenses. At equilibrium, the amount of water in the liquid and gaseous phases remains stable. However, the forward process of evaporation and the reverse process of condensation continue to occur. The equilibrium between gas and liquid phases in the bottle is dynamic.

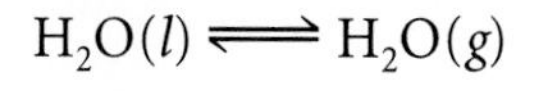

$$H_2O(l) \rightleftharpoons H_2O(g)$$

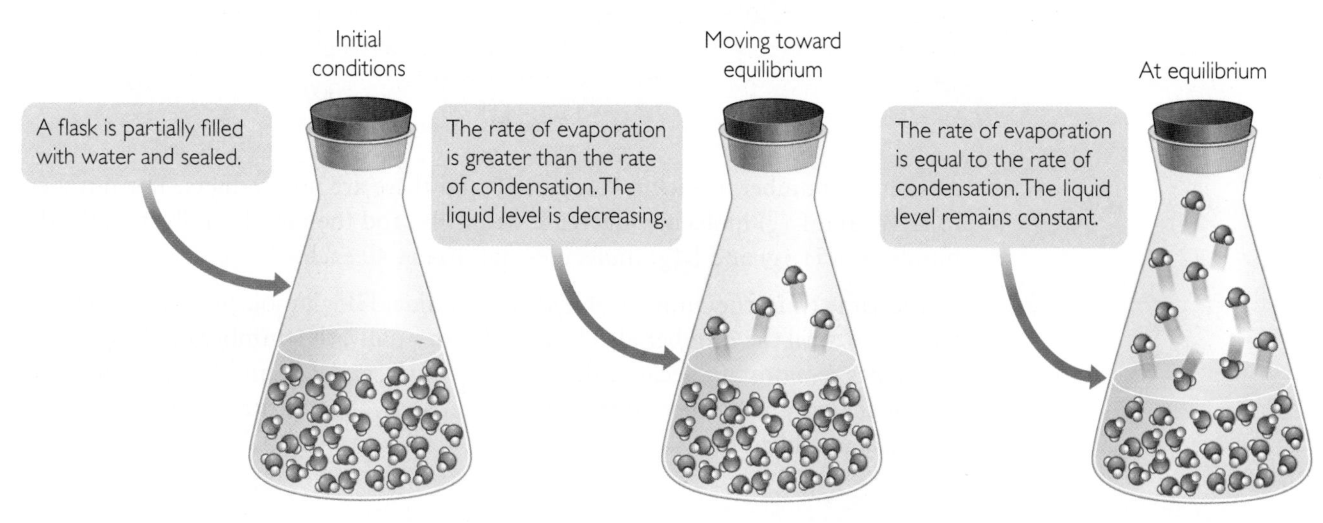

2 Chemical Equilibrium

The fish tank example illustrates the dynamic nature of equilibrium, but it only works to describe physical changes because no new "product" fish are created. Recall that atoms and molecules are in constant motion. At some point, the rates at which the products are being made and the reactants re-form are equal. When this occurs, we say that the reaction has reached **chemical equilibrium.** To an observer, a reaction at chemical equilibrium appears to no longer be changing. However, just as with dissolving and precipitation, the reaction is still proceeding in both directions.

BIG IDEA At equilibrium, the rate of the forward process is equal to the rate of the reverse process.

Starting with Reactants

Consider a reaction in which you begin with two types of gaseous molecules, hydrogen, $H_2(g)$, and iodine, $I_2(g)$. These reactants are combined in a sealed container. When the gaseous H_2 and I_2 molecules collide, they combine to form hydrogen iodide, $HI(g)$. When gaseous HI molecules collide, they can break apart to form $H_2(g)$ and $I_2(g)$ molecules again. So the reaction is reversible:

Topic: Chemical Equilibrium
Visit: www.SciLinks.org
Web code: KEY-603

$$H_2(g) + I_2(g) \rightleftharpoons 2HI(g)$$

Suppose you focus on the reaction between 50 molecules of $H_2(g)$ and 50 molecules of $I_2(g)$. The points on the graph represent the number of molecules of $H_2(g)$, $I_2(g)$, and $HI(g)$ shortly after mixing. Take a moment to examine the graph. (Because $H_2(g)$ and $I_2(g)$ are present in equal numbers and the mole ratio for the reaction is 1:1, both molecules are represented by the same line.)

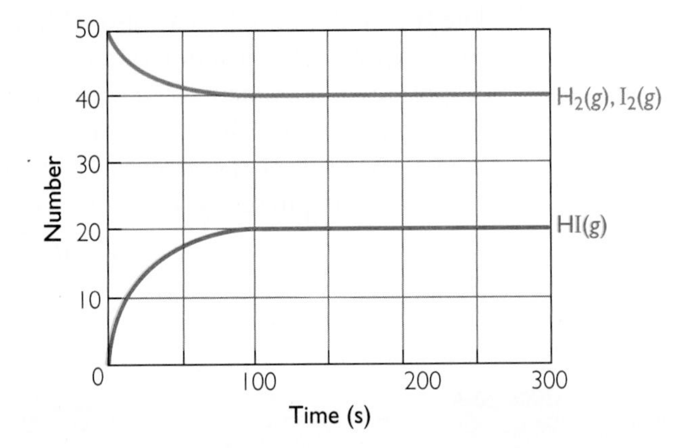

At first the number of each reactant is 50 and there are no products. The numbers of $H_2(g)$ and $I_2(g)$ molecules decrease with time and then level off. Eventually, the numbers of $H_2(g)$ and $I_2(g)$ molecules stabilize at 40 each.

At the same time, the number of hydrogen iodide, $HI(g)$, molecules increases with time until the number also levels off. Eventually, the number of $HI(g)$ molecules remains 20. Because the numbers of $H_2(g)$, $I_2(g)$, and $HI(g)$ molecules do not change any further after about 100 seconds, the mixture has reached chemical equilibrium.

Notice that not all of the $H_2(g)$ and $I_2(g)$ molecules have been converted to hydrogen iodide, $HI(g)$, molecules. This reaction does not go to completion because it is a reversible reaction. At equilibrium, molecules form and fall apart at equal rates. The straightening out of the lines on the graph indicates that the mixture has reached chemical equilibrium.

Important to Know Chemical equilibrium means equal rates of forward and reverse reactions, *not* equal amounts of reactants and products. ◄

Starting with Products

Imagine you repeat this experiment, but start with 50 hydrogen iodide, $HI(g)$, molecules in a closed container. How will this situation compare with the preceding one, in which you began with 50 hydrogen, $H_2(g)$, molecules and 50 iodine, $I_2(g)$, molecules?

The graph shows the progress of the reaction as it goes to equilibrium.

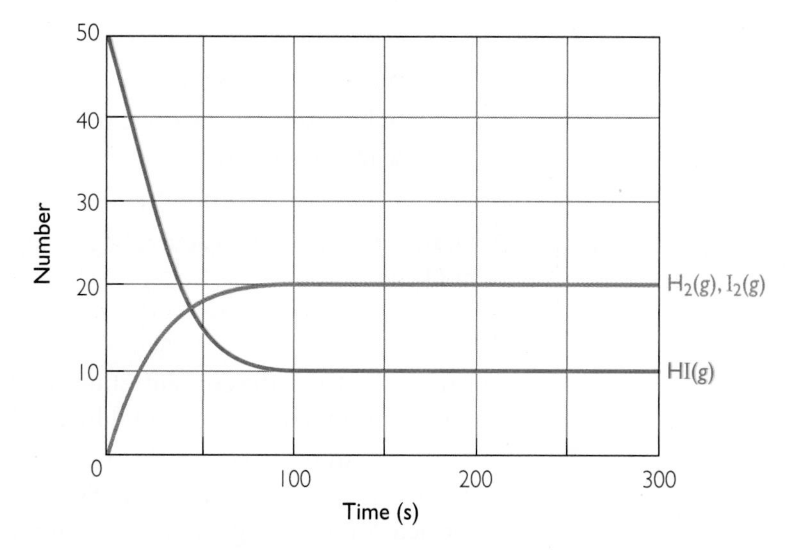

Notice this time that it is the reactants that start at 0 while the product starts at 50. The number of hydgrogen iodide, $HI(g)$, molecules decreases with time before leveling off at 10 molecules. At the same time, the numbers of hydrogen, H_2, and iodine, $I_2(g)$, molecules increase with time before leveling off at 20 molecules each. The equilibrium mixture consists of 10 hydrogen iodide, $HI(g)$, 20 hydrogen, $H_2(g)$, and 20 iodine, $I_2(g)$, molecules.

Although the final numbers of $H_2(g)$, $I_2(g)$, and $HI(g)$ molecules are different from the previous example, the ratios are the same. It does not matter whether you start the reaction with all reactants, with all products, or with a mixture. The ratios in the equilibrium mixture for this reaction under these conditions is the same.

Example

Formation of Sulfur Trioxide

Consider the reversible reaction in which sulfur dioxide, $SO_2(g)$, reacts with oxygen, $O_2(g)$, to form sulfur trioxide, $SO_3(g)$. The graph on the top of the next page shows the progress of the reaction when you start with the reactants.

The balanced chemical equation is

$$2SO_2(g) + O_2(g) \rightleftharpoons 2SO_3(g)$$

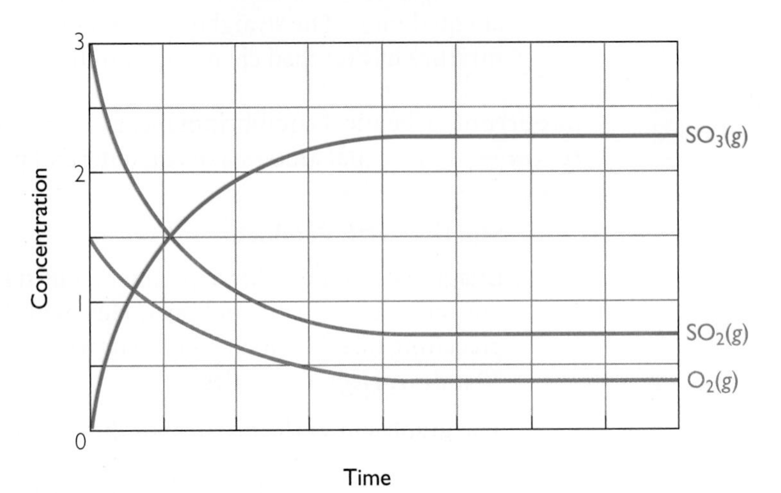

a. Explain how you know that the reaction has reached equilibrium.

b. The graph shows that the decrease in concentration for $SO_2(g)$ is equal to the increase in concentration for $SO_3(g)$. Explain why.

c. Why does the $O_2(g)$ concentration decrease less than the $SO_2(g)$ concentration?

d. Draw a graph for the reaction beginning with $SO_3(g)$ and with no $SO_2(g)$ or $O_2(g)$.

Solution

a. After some time, the concentrations of the molecules no longer change with time. (The graph is a horizontal line.) The mixture has reached equilibrium.

b. According to the balanced chemical equation, the mole ratio for SO_2:SO_3 is 2:2 or 1:1. For every molecule of $SO_2(g)$ that reacts, one molecule of $SO_3(g)$ is formed. So the changes in concentration are equal and opposite.

c. According to the balanced chemical equation, the mole ratio for O_2:SO_2 is 1:2. For every molecule of $O_2(g)$ that reacts, two molecules of $SO_2(g)$ react. So the changes in concentration for SO_2 are twice as large.

d.

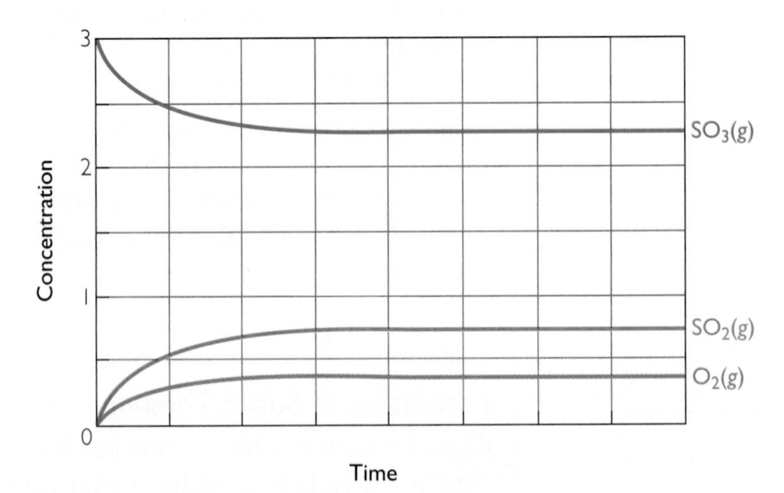

The graph levels off at the same quantities as the initial reaction.

Key Terms

equilibrium
chemical equilibrium

Lesson Summary

What is equilibrium?

Equilibrium is a dynamic process. The word *dynamic* means that things are constantly changing. Equilibrium is a balance that is reached between the forward and reverse processes in a reversible change. To an observer, it seems that things are no longer changing. In reality, however, changes are continuously occurring on a particle level but the changes balance each other out. When a system is at chemical equilibrium, the rate of the forward reaction is equal to the rate of the reverse reaction.

EXERCISES

Reading Questions

1. What does it mean when a system is in a state of dynamic equilibrium?
2. What processes are balanced when a system has reached chemical equilibrium?

Reason and Apply

3. The reversible reaction to form phosphorus pentachloride, $PCl_5(g)$, is carried out, and the concentrations of the reactants and products are carefully monitored. Use the data in the table to create a graph showing the concentrations of the reactants and products over time.

$$Cl_2(g) + PCl_3(g) \rightleftharpoons PCl_5(g)$$

Time (s)	$[Cl_2]$ (mol/L)	$[PCl_3]$ (mol/L)	$[PCl_5]$ (mol/L)
0	1.00	1.00	0.00
20	0.90	0.90	0.10
40	0.80	0.80	0.20
60	0.75	0.75	0.25
80	0.71	0.71	0.29
100	0.71	0.71	0.29
120	0.71	0.71	0.29

a. What are the equilibrium concentrations of the reactants and the products?

b. At what time did the mixture reach equilibrium? How do you know?

c. At 20 seconds, which reaction is faster, the forward or the reverse reaction? At 100 seconds?

d. Are the concentrations of reactants and products equal at equilibrium? Support your answer with evidence.

4. The reversible reaction to form ammonia, $NH_3(g)$, is performed, and data for concentrations of reactants and products are gathered. The data are incomplete. Copy this table and fill in estimates for the missing data.

$$N_2(g) + 3H_2(g) \rightleftharpoons 2NH_3(g)$$

Time (s)	$[N_2]$ (mol/L)	$[H_2]$ (mol/L)	$[NH_3]$ (mol/L)
0	1.00	1.00	0.00
1.0	0.90	0.70	0.20
2.0	0.80	?	?
3.0	0.75	0.25	0.50
4.0	0.72	?	?
5.0	0.70	?	?
6.0	?	?	0.60

a. What are the equilibrium concentrations of these compounds?

b. How did you determine the concentrations that belonged in each empty square of the data table?

5. Copy this graph and use it to answer the questions.

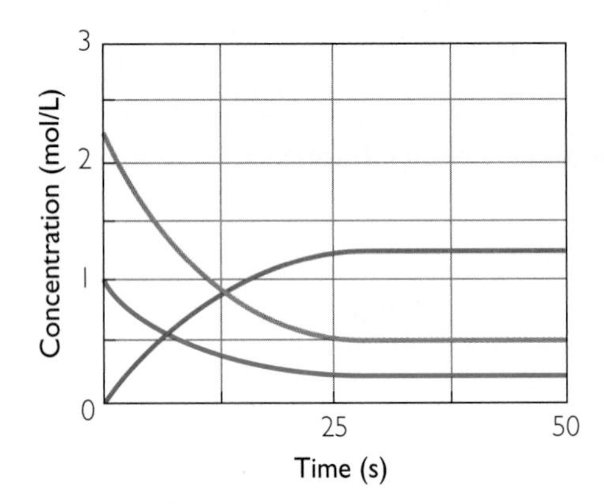

a. Label the reactant and product graphs.

b. Read the equilibrium concentrations of each reactant and product.

c. How long did it take for this reaction to reach equilibrium?

d. What is "equal" at equilibrium?

6. Describe how you could use the couples dancing at a prom versus those sitting and talking as a model for equilibrium.

LESSON

4 How Favorable

Equilibrium Constant K

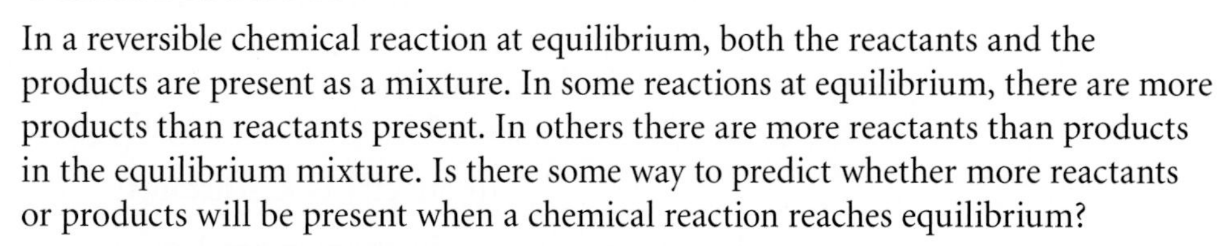

Think About It

In a reversible chemical reaction at equilibrium, both the reactants and the products are present as a mixture. In some reactions at equilibrium, there are more products than reactants present. In others there are more reactants than products in the equilibrium mixture. Is there some way to predict whether more reactants or products will be present when a chemical reaction reaches equilibrium?

How can you predict if products are favored in a reversible reaction?

To answer this question, you will explore

1. Products Versus Reactants
2. The Equilibrium Constant K
3. The Magnitude of K

Exploring the Topic

1 Products Versus Reactants

It is possible to simulate reversible reactions at equilibrium using household items such as pop beads or paper clips. Imagine that you are putting pop beads of different colors together, and your partner is taking them apart. After doing this for a while, you will have an equilibrium mixture. Now do the same thing with paper clips. How will the equilibrium mixtures of pop beads and paper clips compare?

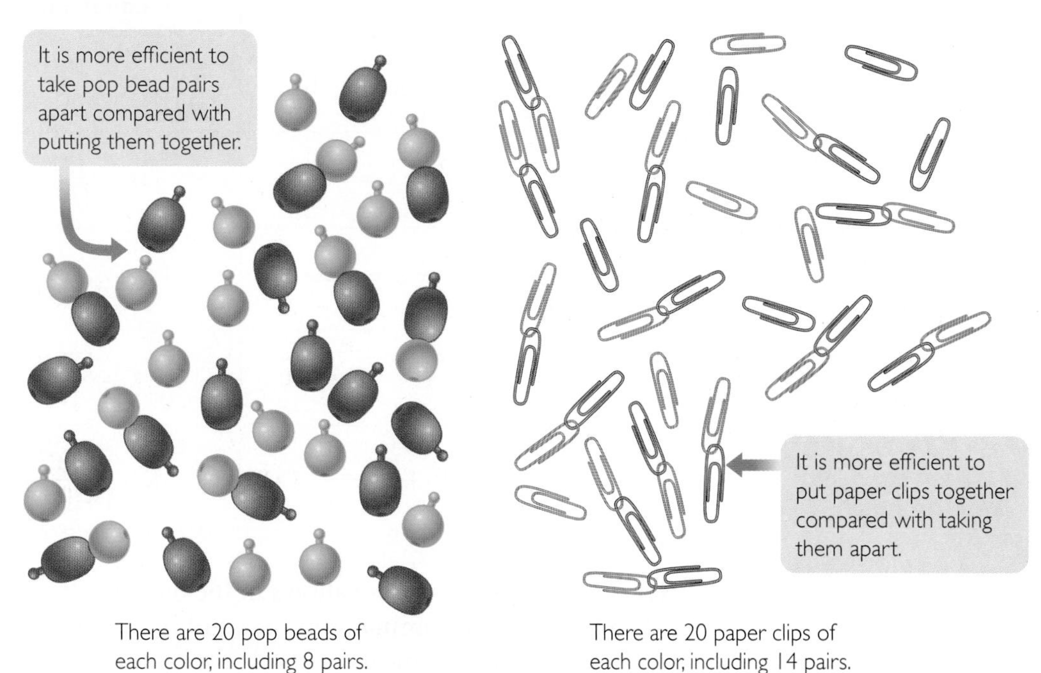

There are 20 pop beads of each color, including 8 pairs.

There are 20 paper clips of each color, including 14 pairs.

So, why are there more pairs of paper clips compared with single paper clips in the equilibrium mixture? This is because the rates of making pairs and breaking pairs depend on both the *efficiency* of your actions and the *opportunity* you have to find what you need in the mixture. Paper clips attach to each other easily, but after a while, the concentration of single paper clips is low enough that it is hard to find two to put together.

2 The Equilibrium Constant *K*

Reversible processes result in a mixture of reactants and products at equilibrium. There is a mathematical relationship between the concentrations of reactants and products in an equilibrium mixture. For the hypothetical reversible process $A \rightleftharpoons B$, the relationship is expressed by the general equation

$$[\text{products}] = K[\text{reactants}] \quad \text{or} \quad [B] = K[A]$$

The reactants and products are in brackets, indicating concentration in moles/liter. The number relating reactants and products is referred to as the **equilibrium constant** and designated with a *K*. Chemists use *K* to keep track of the extent to which products are favored in an equilibrium mixture. You can rearrange the equation above to solve for the value of *K*.

$$K = \frac{[\text{products}]}{[\text{reactants}]} = \frac{[B]}{[A]}$$

Each chemical process has a specific equilibrium constant *K* associated with it.

How to Write an Equilibrium-Constant Equation

The general equation above does not indicate how to deal with processes that have more than one reactant or product. The table shows how the equilibrium equation changes with different types of chemical processes. Examine the table to look for patterns.

Reaction	Equilibrium-constant equation
$A \rightleftharpoons B$	$K = \frac{[B]}{[A]}$
$2A \rightleftharpoons B$	$K = \frac{[B]}{[A]^2}$
$A + B \rightleftharpoons C$	$K = \frac{[C]}{[A][B]}$
$C \rightleftharpoons A + B$	$K = \frac{[A][B]}{[C]}$
$A_2 + B_2 \rightleftharpoons 2AB$	$K = \frac{[AB]^2}{[A_2][B_2]}$

Notice that the products are always in the numerator and the reactants are always in the denominator in the equilibrium-constant equation. The coefficients in the chemical equation show up in the equilibrium-constant equation as exponents.

These patterns can be generalized in what is called the **equilibrium-constant equation.**

$$aA + bB \rightleftharpoons cC + dD$$

$$K = \frac{[C]^c[D]^d}{[A]^a[B]^b}$$

Pure Solids and Liquids

When equilibrium-constant equations are written, pure solids and pure liquids do not appear in them. They are assigned a value of 1 because they don't have a "concentration" that changes. For example, in the reaction of the decomposition of water, energy is added to liquid water to form oxygen gas and hydrogen gas.

$$H_2O(l) \longrightarrow 2H_2(g) + O_2(g)$$

The equilibrium-constant equation does not include the liquid water.

$$K = \frac{[H_2]^2[O_2]}{1} = [H_2]^2[O_2]$$

When carbon dioxide gas, $CO_2(g)$, reacts with solid magnesium oxide, $MgO(s)$, solid magnesium carbonate, $MgCO_3(s)$, is produced.

$$CO_2(g) + MgO(s) \rightleftharpoons MgCO_3(s)$$

The equilibrium-constant equation for this reaction does not include the solids.

$$K = \frac{1}{[CO_2] \cdot 1} = \frac{1}{[CO_2]}$$

Example 1

Sulfur Oxides

Sulfur dioxide gas, $SO_2(g)$, reacts with oxygen gas, $O_2(g)$, to form sulfur trioxide gas, $SO_3(g)$. Write the equilibrium-constant equation for the reaction.

ENVIRONMENTAL CONNECTION

Sulfur trioxide is a primary agent in acid rain.

Solution

First, write the balanced chemical equation. $2SO_2(g) + O_2(g) \rightleftharpoons 2SO_3(g)$

All reactants and products are gases, so they appear in the equilibrium-constant equation. The coefficients become exponents. $K = \frac{[SO_3]^2}{[SO_2]^2[O_2]}$

To obtain a numerical value for K, insert the concentrations of the reactants and products into the equilibrium-constant equation and solve. The constant K is unusual in that it is dimensionless, or has no units associated with it. This is one of the few instances in chemistry where you do not need to include units in the answer.

Example 2

Determining K

If the concentrations listed here are measured at equilibrium for the reaction in Example 1, what is the equilibrium constant?

$$[SO_2] = 1.0 \times 10^{-3}\,M \quad [O_2] = 2.0 \times 10^{-3}\,M \quad [SO_3] = 3.0 \times 10^{-3}\,M$$

Solution

Recall that units are not used to calculate the equilibrium constant.

Use the equilibrium-constant equation for this reaction.

$$K = \frac{[SO_3]^2}{[SO_2]^2[O_2]}$$

Substitute the concentrations measured at equilibrium, without the units.

$$= \frac{[3.0 \times 10^{-3}]^2}{[1.0 \times 10^{-3}][2.0 \times 10^{-3}]}$$

Calculate the equilibrium constant.

$$= 1.8 \times 10^{-3}$$

❸ The Magnitude of *K*

The value of *K* can give an indication of whether there are more products or more reactants present in an equilibrium mixture.

Products are in the numerator.
Reactants are in the denominator.

$$K = \frac{[C]^c[D]^d}{[A]^a[B]^b}$$

If the value of the numerator is much larger than the denominator, the value for *K* is large. So, a large *K* means a lot of product. Likewise, if the value of the denominator is much larger than the value of the numerator, the value of *K* is very small. So, a *K* value less than one usually means reactants are favored. A really small *K* usually means lots of reactants. In the pop bead and paper clip models, the pop bead model favored reactants, so it has a small *K*. The paper clip model favored products, so it has a large *K*.

Example 3

Products or Reactants

Write the equilibrium-constant equation for each reaction. Do the reactions favor products or reactants?

a. $H_2(g) + Cl_2(g) \rightleftharpoons 2HCl(g) \quad K = 2.5 \times 10^4$

b. $N_2(g) + O_2(g) \rightleftharpoons 2NO(g) \quad K = 4.1 \times 10^{-4}$

Solution

a. $K = \frac{[HCl]^2}{[H_2][Cl_2]}$ *K* is large, so products are favored in this reaction.

b. $K = \frac{[NO]^2}{[N_2][O_2]}$ *K* is small, so reactants are favored in this reaction.

[For a review of this math topic, see **MATH** ***Spotlight: Scientific Notation*** on page 631.]

Key Terms

equilibrium constant K
equilibrium-constant equation

Lesson Summary

How can you predict if products are favored in a reversible reaction?

There is a mathematical relationship between the concentrations of the reactants and products in an equilibrium mixture. Generally speaking, the equilibrium constant K is a measure of the ratio of the concentrations of the products divided by the concentrations of the reactants. When a reaction is at equilibrium, the concentrations of reactants and products are not necessarily equal.

Some chemical processes favor reactants and others favor products. A large equilibrium constant indicates that an equilibrium mixture mainly consists of products. A small equilibrium constant usually indicates the equilibrium mixture is mostly reactants.

EXERCISES

Reading Questions

1. Why don't systems at equilibrium contain the same amounts of products and reactants in the mixture?
2. How can you tell whether reactants or products are favored in an equilibrium mixture?

Reason and Apply

3. Write an equilibrium-constant equation for each of these reversible processes.
   a. $2CO(g) + O_2(g) \rightleftharpoons 2CO_2(g)$
   b. $N_2(g) + 3H_2(g) \rightleftharpoons 2NH_3(g)$
   c. $H_2(g) + I_2(g) \rightleftharpoons 2HI(g))$
   d. $PCl_5(g) \rightleftharpoons PCl_3(g) + Cl_2(g)$
4. Use the data provided to calculate the equilibrium constant for the reactions from Exercise 3. Which reactions do you expect to clearly favor products? Explain.
   a. $[CO] = 2.5 \times 10^{-3}$ $[O_2] = 1.6 \times 10^{-3}$ $[CO_2] = 3.2 \times 10^{-2}$
   b. $[N_2] = 4.4 \times 10^{-2}$ $[H_2] = 1.2 \times 10^{-1}$ $[NH_3] = 3.4 \times 10^{-3}$
5. Nitrogen gas combines with oxygen gas to form nitrous oxide gas.
   a. Write the balanced chemical equation for this reaction.
   b. Write the equilibrium-constant equation for this reaction.
   c. Find $[O_2]$ if $K = 45.0$ $[N_2] = 1.00$ $[N_2O] = 1.00$
   d. What does the value of K tell you about the equilibrium mixture?

LESSON

5 How Balanced
Equilibrium Calculations

Think About It

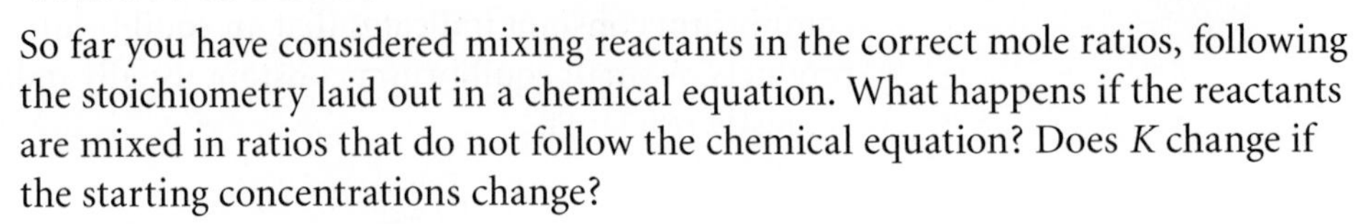

So far you have considered mixing reactants in the correct mole ratios, following the stoichiometry laid out in a chemical equation. What happens if the reactants are mixed in ratios that do not follow the chemical equation? Does *K* change if the starting concentrations change?

Why is *K* called a "constant"?

To answer this question, you will explore

1. Macroscopic Views
2. Molecular Views
3. Mathematical Relationships

Exploring the Topic

1 Macroscopic Views

Imagine that you have three test tubes all at standard conditions of 25 °C and 1 atm pressure. These solutions are all at equilibrium and all contain the same ions, but the concentrations of these ions differ. Are the concentrations of reactants and products still related to one another by an equilibrium constant *K*?

These three test tubes were prepared by dissolving iron (III) chloride, $FeCl_3(s)$ and sodium thiocyanate, $NaSCN(s)$, in water. The starting concentrations of the reactants are shown.

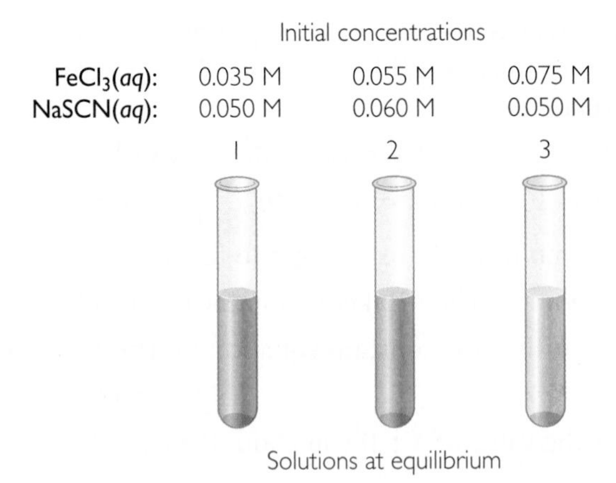

All three solutions are at equilibrium, so their colors are not changing over time. What causes the differences in color?

The net ionic equation is the same for the solutions in all three test tubes.

$$Fe^{3+}(aq) + SCN^{-}(aq) \rightleftharpoons FeSCN^{2+}(aq)$$

yellow colorless red

The Fe^{3+} ion is pale yellow and the $FeSCN^{2+}$ ion is red. Because the solution in test tube 3 is orange, it must have the highest ratio of Fe^{3+} ions to $FeSCN^{2+}$ ions.

2 Molecular Views

Molecular views are shown for each of the three solutions. Take a moment to examine the illustrations. How are the images different? Are the molecular views consistent with the color of the solution?

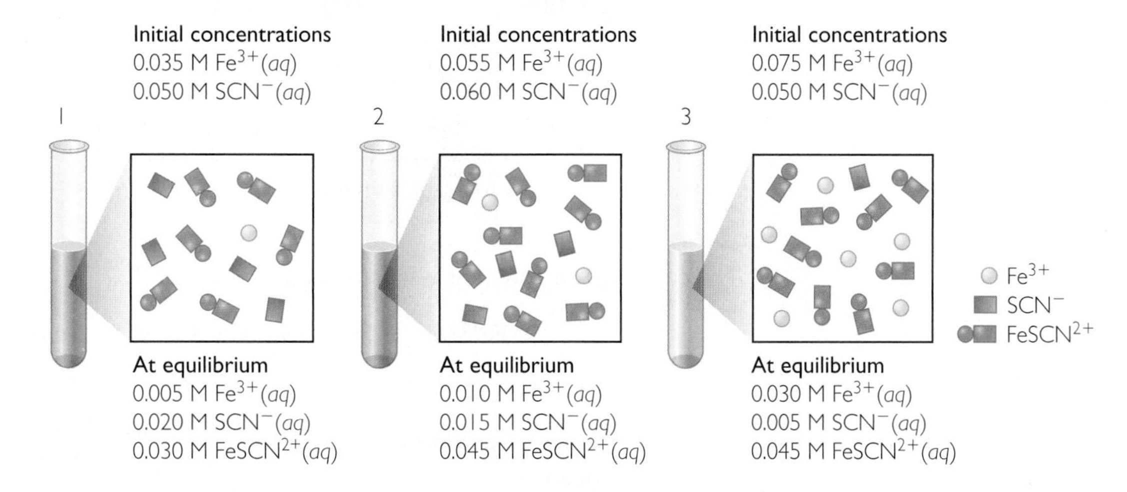

NATURE CONNECTION

Leaves turn different colors based on glucose content. Red and purple leaves have the highest glucose content.

If you were to track the number of moles of iron atoms in the solution from the initial conditions to equilibrium, you would see that they are conserved. Originally, all of the iron atoms in the solution came from solid iron (III) chloride, $FeCl_3(s)$, which formed iron and chloride ions when it dissolved. When sodium thiocyanate, NaSCN, was added, some of the Fe^{3+} ions combined with the SCN^- ions to form iron thiocyanate, $FeSCN^{2+}$.

Calculations confirm that the amount of iron atoms does not change.

For 10 mL (0.010 L) of solution in test tube 1:

Initial conditions	**At equilibrium**
Moles of Fe	Moles of Fe
= (0.035 M)(0.010 L)	= moles of Fe^{3+} + moles of $FeSCN^{2+}$
= 0.00035 mol	= (0.005 M)(0.010 L) + (0.030 M)(0.010 L)
	= 0.00035 mol

Iron atoms are conserved.

You might also notice how the equilibrium concentrations vary from test tube 1 to 3. The concentration of Fe^{3+} ions increases from test tube 1 to 3. The concentration of SCN^- ions decreases. The concentrations of $FeSCN^{2+}$ ions are the same in test tubes 2 and 3, and slightly less in test tube 1. This is consistent with the red color of the solutions in test tubes 1 and 2 and with the orange color of the solution in test tube 3. The color of the solutions at equilibrium depends on the ratio of Fe^{3+} ions to $FeSCN^{2+}$ ions.

③ Mathematical Relationships

The solutions in all three test tubes are at equilibrium, so what is the value of K for each one? The concentrations of the various species, shown in the table, can be inserted into the equilibrium equation.

Test tube	Color	$[Fe^{3+}]$ (M)	$[SCN^-]$ (M)	$[FeSCN^{2+}]$ (M)	K
1	red	0.005	0.020	0.030	300
2	red	0.010	0.015	0.045	300
3	orange	0.030	0.005	0.045	300

The equilibrium constant is calculated for test tube 1.

$$Fe^{3+}(aq) + SCN^-(aq) \rightleftharpoons FeSCN^{2+}(aq)$$

yellow red

Write the equilibrium-constant equation. $K = \dfrac{[FeSCN^{2+}]}{[Fe^{3+}][SCN^-]}$

Substitute values for test tube 1. $= \dfrac{(0.030\ M)}{(0.005\ M)(0.020\ M)}$

Evaluate. $= 300$

Notice that the value of K is the same for all three solutions in the test tubes. However, the concentration of the yellow reactant ion, Fe^{3+}, relative to the red iron (III) thiocyanate product ion, $FeSCN^{2+}$, depends on the colorless thiocyanate, SCN^-, ion concentration.

As a specific example, consider why the equilibrium concentrations of $FeSCN^{2+}$ ion in test tubes 2 and 3 are the same. Notice that in test tube 2 you get an equivalent amount of $FeSCN^{2+}$ ion with fewer Fe^{3+} ions by increasing the initial concentration of SCN^- ions.

Test tube 2 Test tube 3

$$K = \frac{[FeSCN^{2+}]}{[Fe^{3+}][SCN^-]} = \frac{(0.045\ M)}{(0.010\ M)(0.015\ M)} = \frac{(0.045\ M)}{(0.030\ M)(0.005\ M)} = 300$$

So the equilibrium mixtures in all three test tubes result in the same value for K even though the solutions have different concentrations of iron ions and different colors. Note that values of K are normally reported for measurements made at standard conditions of 25 °C and 1 atm. Under different conditions, the value of K may be different. If you were actually measuring equilibrium concentrations in the laboratory, you would very likely obtain values for K that were not identical, but rather showed some experimental variation.

Example

Formation of HI

Hydrogen gas, H_2, and iodine gas, I_2, are mixed in a closed container. The equilibrium mixture contains H_2 and I_2, each with a concentration of 0.033 mol/L. There is also hydrogen iodide gas, HI, in the equilibrium mixture. Given that $K = 50.3$, what is the concentration of the hydrogen iodide gas?

Solution

Begin by writing a balanced chemical equation and use it to write the equilibrium equation.

$$H_2(g) + I_2(g) \rightleftharpoons 2HI(g)$$

$$K = \frac{[HI]^2}{[H_2][I_2]}$$

Insert the values into the equation. The unknown concentration, [HI], can be represented as x.

$$50.3 = \frac{x^2}{[0.033][0.033]}$$

Solve for x^2, then find the square root.

$$x^2 = 0.055$$
$$x = 0.23$$

The concentration of HI is 0.23 mol/L.

Lesson Summary

Why is *K* called a "constant"?

The equilibrium constant K is called a constant because its value does not change when the concentrations of the reactants or products are changed. The concentrations of the ions in a system at equilibrium will reach a similar mathematical balance between products and reactants, no matter what concentrations you start with. Values for K are determined experimentally and can be used in calculations involving systems at equilibrium.

EXERCISES

Reading Questions

1. Explain why different solutions consisting of Fe^{3+} ions, SCN^- ions, and $FeSCN^{2+}$ ions can have different colors.
2. Explain how the solutions with different colors in Question 1 can all be at equilibrium.

Reason and Apply

3. Consider the reversible reaction: $H_2(g) + Cl_2(g) \rightleftharpoons 2HCl(g)$.
   a. Write out the equilibrium equation for this reaction.
   b. The equilibrium constant is $K = 4.05 \times 10^{31}$ for this reaction. If you mix $H_2(g)$ and $Cl_2(g)$, will you find more products or more reactants in the equilibrium mixture? Explain your thinking.

4. Consider the reversible reaction: $2NO_2(g) \rightleftharpoons N_2O_4(g)$. The data in the table show the equilibrium concentrations of reactants and products for four different mixtures. The color of the equilibrium mixture is given in the last column.

Mixture	Equilibrium concentrations (M)		Color
	$[NO_2]$	$[N_2O_4]$	
1	0.5400	0.643	orange
2	0.0451	0.448	light orange
3	0.0519	0.594	orange
4	0.0200	0.088	light orange

   a. Write out the equilibrium equation for this reaction.
   b. Determine a numerical value for the equilibrium constant for the mixture in each test tube.
   c. How do you explain the different colors in the four test tubes?
   d. How can the colors in the tubes be different if the value of *K* is the same for each situation?
   e. Determine the equilibrium concentration of $N_2O_4(g)$ if the equilibrium concentration of NO_2 is 0.124 M.
   f. Determine the equilibrium concentration of $NO_2(g)$ if the equilibrium concentration of N_2O_4 is 0.0324 M.

5. Give two possible sets of equilibrium concentrations of Fe^{3+} ions and SCN^- ions in a solution with 0.030 M $FeSCN^{2+}$ ions.

SECTION

I

SUMMARY

Chemical Equilibrium

Showtime Update

A colorful chemical demonstration in which a reaction proceeds in the forward and then in the reverse direction can baffle and impress an audience. A mixture of compounds begins with one color then shifts to a new color and then back to the original. This type of reaction is called reversible. Unlike combustion reactions, which go to completion, in reversible chemical processes both the forward and reverse reactions occur simultaneously. Reversible reactions are written with a double arrow to show that both the forward and reverse processes occur at the same time.

When both the forward and reverse processes in a chemical change are in balance, these systems are at equilibrium. Equilibrium systems can be physical, like the back and forth evaporation of water in a closed container, or chemical, such as the reaction of hydrogen and iodine gases to form hydrogen iodide:

$$H_2(g) + I_2(g) \rightleftharpoons 2HI(g)$$

At equilibrium, the rate of the forward process is equal to the rate of the reverse process, though this does not mean that the amounts of products and reactants are equal.

There is a mathematical relationship between the amount of products and the amount of reactants in an equilibrium mixture. This relationship can be described by the equilibrium-constant equation. The equilibrium constant K is the same for a specific chemical reaction at a given temperature and pressure.

Key Terms

reversible reaction
equilibrium
chemical equilibrium
equilibrium constant K
equilibrium-constant equation

Review Exercises

1. Suppose that you are studying hydrazine, N_2H_4. You determine that a 0.1 M solution of hydrazine has a pH of 10. The ionization equation for hydrazine is $N_2H_4(aq) + H_2O(l) \rightleftharpoons N_2H_5^+(aq) + OH^-(aq)$.
   a. Is hydrazine a strong acid, strong base, weak acid, or weak base? Explain how you know.
   b. List all the compounds and ions that would be present in a 0.1 M solution of hydrazine.
   c. Write the equilibrium equation for hydrazine ($K = 8.7 \times 10^{-7}$).
2. Write out the generic equilibrium equations for these chemical reactions. Remember to set the equation equal to K.
   a. $A(aq) + B(aq) \rightleftharpoons C(aq) + D(aq)$
   b. $CO(g) + Cl_2(g) \rightleftharpoons COCl_2(g)$
   c. $HCOOH(g) \rightleftharpoons CO(g) + H_2O(g)$ (*Hint:* Water *is included* in the equilibrium equation for this reaction. Why do you think this reaction *does* include water in the equation?)

3. For the reaction $N_2H_4(aq) + H_2O(l) \rightleftharpoons N_2H_5^+(aq) + OH^-(aq)$ at equilibrium $K = 8.7 \times 10^{-7}$.
   a. If $[N_2H_5^+] = [OH^-] = 3.1 \times 10^{-4}$, what is $[N_2H_4]$ at equilibrium?
   b. Would you expect the equilibrium mixture to be acidic, basic, or neutral? Explain your reasoning.
4. For the reaction $N_2(g) + 3H_2(g) \rightleftharpoons 2NH_3(g)$
   a. Write the equilibrium equation.
   b. Solve for K if the equilibrium concentrations are: $[N_2] = 0.1$ M, $[H_2] = 0.2$ M, $[NH_3] = 0.1$ M.
   c. Based upon your answer to part b, do you think that this reaction produces a reaction mixture that is product-favored, reactant-favored, or about equal?
5. For the reaction $H_2(g) + I_2(g) \rightleftharpoons 2HI(g)$
   a. Write the equilibrium equation.
   b. If $K = 55$, $[H_2] = 0.1$ M, and $[I_2] = 0.1$ M, what is $[HI]$?
   c. If you were to observe an equilibrium mixture of these compounds, would you expect to see more reactants, products, or equal amounts of both? Explain your choice.

WEB RESEARCH

PROJECT *Chemistry and Magic*

Look on the Web to find three chemistry demonstrations and three magic tricks that use chemistry. (The tricks do not need to be related to the demonstrations.) What makes each demonstration or trick exciting or intriguing? What do they have in common? How do they differ? Write a report. Your report should include

- descriptions of three chemistry demonstrations and three magic tricks involving chemistry, with illustrations.
- your explanation of what makes each one intriguing or exciting.
- a comparison of the different demonstrations and tricks, describing things they have in common and ways in which they are different.

SECTION

II

Changing Conditions at Equilibrium

If reactants or products are added to a system at equilibrium, the system adjusts to maintain equilibrium. Sometimes this shift is visible as a color change.

When a system is at equilibrium, there is a dynamic balance between the forward process and the reverse process. For a chemical reaction at equilibrium, this means that the rate of the forward reaction is equal to the rate of the reverse reaction. Conditions do not always stay the same for systems at equilibrium. A reversible chemical system at equilibrium will adjust in a systematic way to changes that occur so that equilibrium can be maintained.

In this section, you will learn

- about changing conditions at equilibrium
- how to use Le Châtelier's principle to explain how reactions respond to stresses
- about equilibrium constants for different types of reactions
- about the chemical equilibrium of acid-base indicators

LESSON

6 How Pushy

Le Châtelier's Principle

Think About It

Human blood has a certain pH. When people exercise they produce a lot of carbon dioxide, which in turn builds up carbonic acid in their bodies. Carbonic acid alters the pH of the blood, which puts stress on the body. How does the body return to a healthy equilibrium?

What happens to a system at equilibrium when conditions change?

To answer this question, you will explore

1. Changing Conditions
2. Le Châtelier's Principle

Exploring the Topic

1 Changing Conditions

When a system is at equilibrium the reaction is dynamically balanced, with the reaction going backward and forward at equal rates. However, conditions do not always stay the same for systems at equilibrium. The temperature may change or something new might be added to the mixture. A reversible chemical system at equilibrium will adjust systematically to any changes that occur, in order to reestablish equilibrium. The type of adjustment a system makes is very predictable and depends on the type of change that occurs.

Here are three types of changes that can occur in an equilibrium system.

- A change in the concentrations of the reactants or products
- A change in temperature
- A change in pressure—this applies mainly to reactions involving gases

Changing the conditions for a system at equilibrium will stress the system. Some examples will demonstrate how each of the above changes (or stresses) affects a reversible reaction at equilibrium.

Changes in Concentration

The reaction between hexaaquocobalt ions, $Co(H_2O)_6^{2+}$, and chloride ions, Cl^-, is especially useful in demonstrating changes in equilibrium because it involves a striking color change. When there are more reactants present in the solution, the solution is pink. When there are more products, the solution is blue. A mixture of reactants and products results in some shade of purple. By systematically changing the conditions and observing the color of the mixture after each change, you can track the effect of each change.

$$\underset{\text{pink}}{Co(H_2O)_6^{2+}(aq)} + 4Cl^-(aq) \rightleftharpoons \underset{\text{blue}}{CoCl_4^{2-}(aq)} + 6H_2O(l)$$

Imagine you have a pink sample of this solution. The system is at equilibrium and the equilibrium constant K is 2.5×10^{-4}.

If you change the conditions by adding some more chloride ions to the mix, the sample turns blue. So when more reactants are added, the result is the formation of more products. If you take this blue sample and add water to it, the mixture changes back to pink. So when more products are added, the result is the formation of more reactants. The illustration shows the results of these two changes.

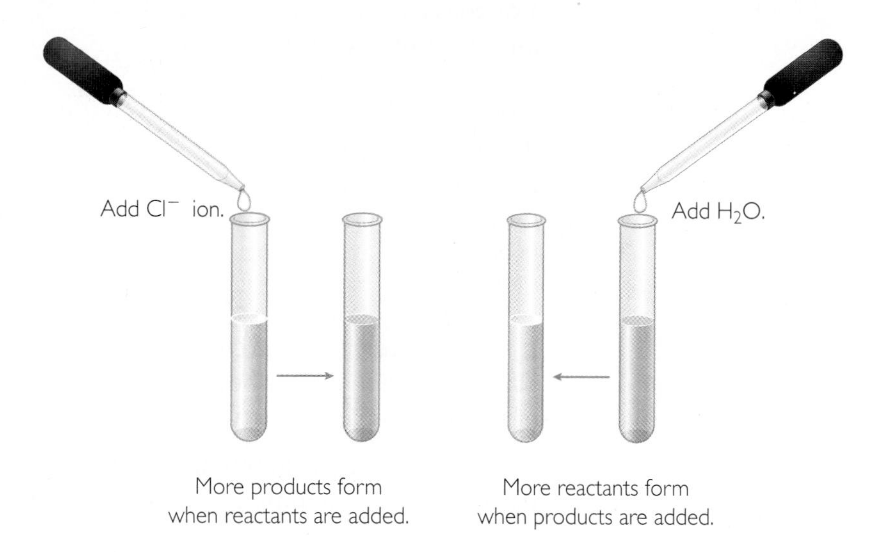

More products form when reactants are added.

More reactants form when products are added.

So the addition of one of the reactants or products to an equilibrium mixture has the effect of changing the final concentrations. The change depends on which side of the reversible reaction gets the stress.

Changes in Temperature

This same reversible reaction can be used to test the effects of changing temperatures on a mixture at equilibrium. If a pink starting sample of cobalt solution is heated, it will turn blue. If a blue sample of this mixture is cooled, it turns back to pink or pinkish-purple. In this particular reversible reaction, heating favors the products.

Heating does not always favor products. The direction of the change depends on whether the reaction is exothermic or endothermic. For an endothermic reaction, heating favors products. For an exothermic reaction, cooling favors products. The cobalt reaction is endothermic as written, so heating causes the formation of more reactants.

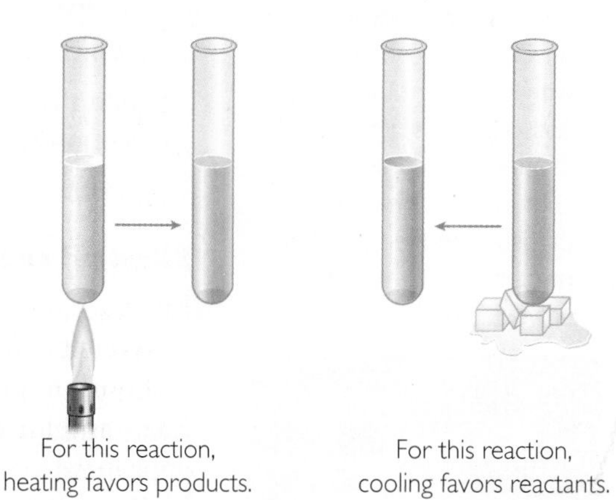

For this reaction, heating favors products.

For this reaction, cooling favors reactants.

BIG IDEA A system at equilibrium will try to minimize any changes that occur.

Changes in Pressure

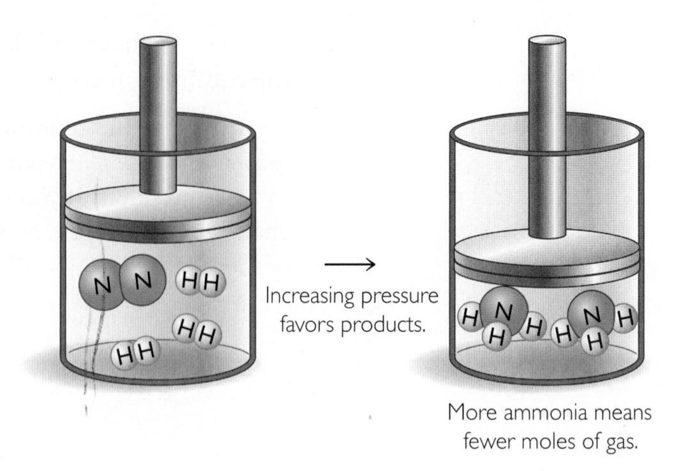

Pressure changes are particularly important when considering equilibrium mixtures that are in the gas phase. If the pressure on a gaseous equilibrium mixture is increased, a change will occur in the direction that decreases the number of moles of gas and relieves some of the pressure. For example, consider the reaction for the formation of ammonia gas from nitrogen and hydrogen gases.

$$N_2(g) + 3H_2(g) \rightleftharpoons 2NH_3(g)$$

If the pressure on this mixture is increased, the system will change so that more ammonia forms. This is because there are four moles of gas on the reactant side of the equation and only two moles of gas on the product side.

The creation of more ammonia relieves the stress of the added pressure by reducing the total number of moles of gas in the mixture.

Important to Know Changes in temperature actually change the value of *K*. Other changes, such as changes in pressure or concentration, stress the system but they do not change the value of *K*. ◄

AGRICULTURE CONNECTION

Ammonia is widely used in agriculture to produce fertilizers that contain nitrates. Under normal conditions, nitrogen is not very reactive, so it is difficult to produce ammonia by combining it with hydrogen. The Haber-Bosch process, developed in the early twentieth century, uses an optimal pressure and temperature and an iron catalyst to solve this problem.

2 Le Châtelier's Principle

These very predictable changes in equilibrium systems were first noted around 1885 by a French chemist named Henri Le Châtelier. He noted that within an equilibrium system, when conditions are modified, the system reacts in such a way as to relieve the stress. This idea is known as **Le Châtelier's principle.**

Le Châtelier's Principle

When there is a stress on a system at equilibrium, the system will react in the direction that serves to relieve the stress.

Blood pH and Exercise

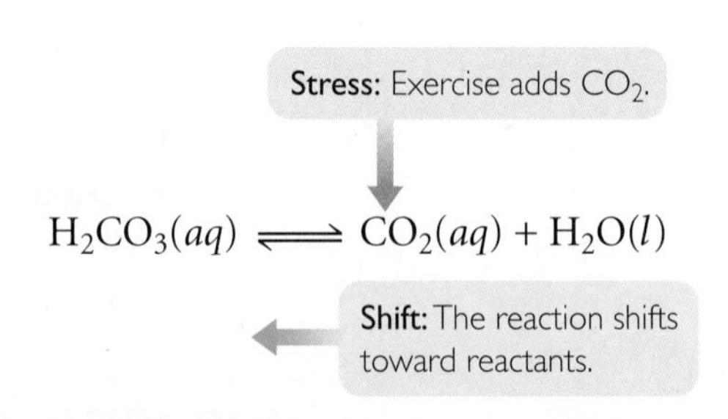

The human body is a large collection of reversible reactions. In order for reactions to proceed at the appropriate rates, and for substances to stay in the appropriate concentrations within our bodies, a balance must be maintained. For example, the body's blood pH is normally around 7.4, slightly basic. When you exercise, carbon dioxide is produced in the bloodstream. This is a natural waste product that you normally exhale from the lungs. However, a lot of carbon dioxide in the bloodstream leads to the formation of carbonic acid, H_2CO_3. This increases the acidity of the blood and lowers the pH.

One of the ways you can deal with this unhealthy shift in pH is to exhale more rapidly, removing CO_2 from the bloodstream and shifting the reaction back toward the right.

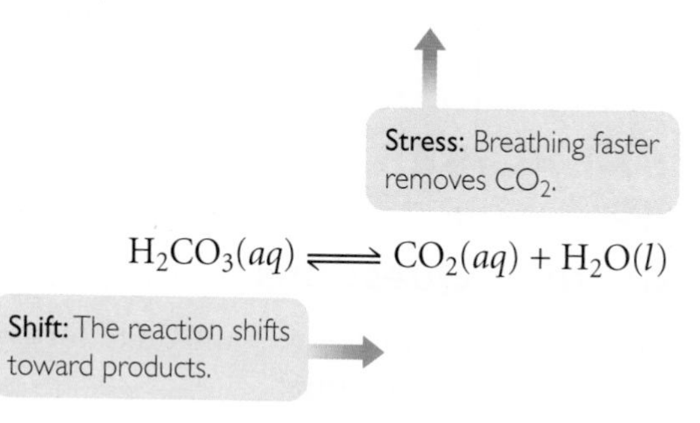

This is one reason why we breathe more heavily when exercising—to rid the body of excess carbon dioxide and increase the blood's pH.

Example

Decomposition of SO_3

Gaseous sulfur trioxide, SO_3, decomposes to form sulfur dioxide, SO_2, and oxygen, O_2; the reaction is endothermic. Predict the effect of the changes here on the reaction

$$2SO_3(g) \rightleftharpoons 2SO_2(g) + O_2(g)$$

a. Increasing the pressure on the system
b. Adding more O_2
c. Removing O_2 from the system
d. Adding heat

Solution

a. In the forward reaction, two moles of gas decompose to form three moles of gas. The system can minimize the effect of a pressure increase by converting some of the products back into SO_3 to decrease the number of moles of gas molecules present. This will push the reaction more toward the reactant side.
b. Adding more O_2 to the system will cause more reactant to form. By making more reactant some of the excess O_2 is removed, relieving some of the stress on the system.
c. Removing O_2 from the system has the opposite effect; it causes the formation of more products. This change has the effect of restoring some of the O_2 that was removed.
d. Because energy is absorbed in the forward reaction, heating will favor products.

Key Term
Le Châtelier's principle

Lesson Summary

What happens to a system at equilibrium when conditions change?

When conditions are changed for a system at equilibrium, the system will respond by reducing the effect of the change. This is known as Le Châtelier's principle. Changing conditions (or stresses) may be changes to temperature, concentration, or pressure. Equilibrium is regained by rebalancing the concentrations of reactants and products. The exact nature of the change is predictable and depends on the type of change encountered by the system. A temperature change affects the value of *K*. A pressure change or a change in concentration stress the system, but does not change the value of *K*.

EXERCISES

Reading Questions

1. Name two things you could do to form more reactants in an equilibrium mixture.
2. Explain how you can tell that the cobalt reaction on page 598 is endothermic as written from the evidence provided in this lesson.

Reason and Apply

3. If you add more ammonia, $NH_3(aq)$, to the equilibrium mixture described by the equation here, what will happen? Explain your thinking.

$$Cu(H_2O)_6{}^{2+}(aq) + 6NH_3(aq) \rightleftharpoons Cu(NH_3)_6{}^{2+}(aq) + 6H_2O(aq)$$

4. If you remove ammonia, $NH_3(g)$, from the equilibrium mixture described by the equation here, what will happen? Explain your thinking.

$$N_2(g) + 3H_2(g) \rightleftharpoons 2NH_3(g)$$

5. If you heat the equilibrium mixture described by this endothermic equation, what will happen? Explain your thinking.

$$2NiO(s) \rightleftharpoons 2Ni(s) + O_2(g)$$

6. Suppose you want to dissolve more copper (II) hydroxide, $Cu(OH)_2(s)$ in an aqueous solution. Would you add acid or base? Explain your thinking.

$$Cu(OH)_2(s) \rightleftharpoons Cu^{2+}(aq) + 2OH^-(aq)$$

7. For each reversible reaction listed, describe *two* things you can do for each system to change the concentration of one of the products.
   a. $2NO_2(g) \rightleftharpoons N_2O_4(g)$
   b. $HF(aq) \rightleftharpoons H^+(aq) + F^-(aq)$
   c. $PbCl_2(s) \rightleftharpoons Pb^{2+}(aq) + 2Cl^-(aq)$
   d. $CaCO_3(s) \rightleftharpoons CaO(s) + CO_2(g)$

LAB Applying Le Châtelier's Principle

Purpose

To explore equilibrium with acid-base indicators.

Materials

- bromothymol blue powder, 0.1 g
- methyl orange powder, 0.1 g
- alizarin yellow powder, 0.1 g
- scoopula
- weigh boats
- 3 test tubes
- test tube rack or beaker
- distilled water
- wash bottle
- universal indicator

Safety Instructions

Wear safety goggles and wash your hands before leaving the lab.

Acid-base indicators are generally weak acids or bases. The general reversible reaction for an indicator that is a weak acid can be written as

$$\mathrm{H}In(aq) \rightleftharpoons \mathrm{H}^{+}(aq) + In^{-}(aq)$$

The In^{-} stands for whatever indicator anion is present.

General information for three acid-base indicators is given in this table.

Indicator	H*In* color	*In*⁻ color	K_a
bromothymol blue	yellow	blue	1.0×10^{-7}
methyl orange	red	yellow	3.4×10^{-4}
alizarin yellow	yellow	red	1.0×10^{-11}

Procedure

Put a tiny amount of each powdered indicator into three different test tubes. Add a few mL of distilled water to each. Record the colors of the solutions.

Analysis

1. A weak acid dissociates only partially, whereas a strong acid dissociates completely. Use the equilibrium equations and the equilibrium constants in the table to decide if indicators are weakly dissociated or strongly dissociated. Explain your thinking.
2. How can you use indicators to get solutions of different colors for demonstrations? Describe how to obtain a blue solution. Describe how "lemonade" could appear to change into "grape juice."
3. **Making Sense** Explain how acid-base indicators are used in determining the acidity of a solution.

LESSON

7 How Colorful

Applying Le Châtelier's Principle

Think About It

Red cabbage juice changes color from red to purple to green as the H^+ concentration in a solution decreases. As the H^+ concentration increases, a green solution containing cabbage juice turns back to purple and then red. The color changes of an acid-base indicator are reversible, so equilibrium considerations apply to these systems.

How do acid-base indicators work?

To answer this question, you will explore

1. Types of Equilibrium Constants
2. Indicators

Exploring the Topic

1 Types of Equilibrium Constants

The equilibrium constant K can be used to solve a variety of equilibrium problems. In every case the equilibrium-constant equation can be used to figure out the concentration of one of the compounds in the equilibrium mixture. The table shows some specific types of chemical equilibrium constants. Each of these is designated by a unique subscript.

Equilibrium constant	Chemical equation	Equilibrium-constant equation
Solubility product constant, K_{sp}	$MX(s) \rightleftharpoons M^+(aq) + X^-(aq)$	$K_{sp} = [M^+][X^-]$
Acid dissociation constant, K_a	$HA(aq) \rightleftharpoons H^+(aq) + A^-(aq)$	$K_a = \frac{[H^+][A^-]}{[HA]}$
Base dissociation constant, K_b	$B(aq) + H_2O(l) \rightleftharpoons BH^+(aq) + OH^-(aq)$	$K_b = \frac{[BH^+][OH^-]}{[B]}$
Ionization of water constant, K_w	$H_2O(l) \rightleftharpoons H^+(aq) + OH^-(aq)$	$K_w = [H^+][OH^-] = 10^{-14}$

Notice that the equations are the same ones that you might have written, except that there is a subscript attached to the symbol K. As discussed previously,

HEALTH CONNECTION

Barium sulfate is often used as a tool to help doctors examine a patient's digestive tract. After ingesting a barium sulfate mixture, any abnormalities or blockages in the patient's stomach and intestines can be seen clearly in an x-ray. Since barium sulfate has a low solubility, it passes out of the body without doing any harm.

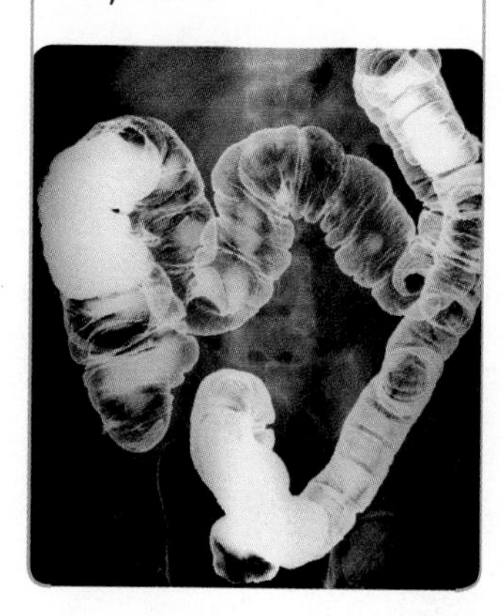

liquids and solids are not included in the equilibrium-constant equation because their concentrations do not change. So equilibrium-constant equations can be used to determine the concentrations of molecules and ions in solution and the concentrations of gaseous molecules in systems at equilibrium.

The equilibrium constant K_{sp} can be used to compare the solubilities of different substances. In these systems the value of the equilibrium constant K is referred to as the solubility product constant K_{sp}.

Example 1

Comparing Solubility

The solubility product constants K_{sp} for several alkaline earth metal sulfates at 25 °C are given in the table.

Metal sulfate	Solubility product constant K_{sp}
magnesium sulfate, $MgSO_4(s)$	4.7
calcium sulfate, $CaSO_4(s)$	4.9×10^{-5}
barium sulfate, $BaSO_4(s)$	1.1×10^{-10}

a. Put the three salts in order of increasing solubility—from least soluble to most soluble. Explain your thinking.

b. Use the periodic table to predict whether strontium sulfate, $SrSO_4(s)$, is more soluble than calcium sulfate, $CaSO_4(s)$. Explain your thinking.

c. Suppose you add sodium sulfate, $Na_2SO_4(s)$, to a saturated solution of calcium sulfate, $CaSO_4(aq)$. Use Le Châtelier's principle to explain why a small amount of $CaSO_4(s)$ precipitates from the solution.

Solution

a. The chemical equation for the solubility of a metal sulfate, $MSO_4(s)$, is

$$MSO_4(s) \rightleftharpoons M^{2+}(aq) + SO_4^{2-}(aq) \qquad K_{sp} = [M^{2+}][SO_4^{2-}]$$

If the solubility product constant K_{sp} is large, products are favored (the salt is more soluble). Therefore, the solubility from least soluble to most soluble is $BaSO_4 < CaSO_4 < MgSO_4$.

b. All of the metals are in Group 2A, the alkaline earth metals. The solubility decreases as you go down the group. So, you expect strontium sulfate, $SrSO_4(s)$, will be less soluble than calcium sulfate, $CaSO_4(s)$, but more soluble than $BaSO_4(s)$.

c. Adding sodium sulfate, $Na_2SO_4(s)$, adds more sulfate ions. Because $SO_4^{2-}(aq)$ is a product of dissolving calcium sulfate, $CaSO_4(aq)$, the system will reduce the stress by forming more reactants. This is why some $CaSO_4(s)$ precipitates.

The equilibrium constant K_a can be used to determine the strength of an acid or a base.

Example 2

Strength of an Acid

Imagine you prepare a 0.10 M solution of HCl and a 0.10 M solution of HOCl. The concentrations of the ions in equilibrium are given in this table.

General equation for acid dissociation: $HA(aq) \rightleftharpoons H^+(aq) + A^-(aq)$

Acid	[HA]	$[H^+]$	$[A^-]$
hydrochloric acid, $HCl(aq)$	~0	0.10 M	0.10 M
hypochlorous acid, $HOCl(aq)$	0.099994 M	6.0×10^{-5} M	6.0×10^{-5} M

a. Determine K_a for hypochlorous acid, $HOCl(aq)$.

b. Show that K_a for $HCl(aq)$ is a very large number.

c. Which acid is stronger? Explain your thinking.

Solution

a. $HOCl(aq) \rightleftharpoons H^+(aq) + OCl^-(aq)$

$$K_a = \frac{[H^+][OCl^-]}{[HOCl]}$$

$$= \frac{(6.0 \times 10^{-5})(6.0 \times 10^{-5})}{0.099994}$$

$$= 3.6 \times 10^{-8}$$

b. $HCl(aq) \rightleftharpoons H^+(aq) + Cl^-(aq)$

$$K_a = \frac{[H^+][Cl^-]}{[HCl]}$$

$$= \frac{(0.10)(0.10)}{0}$$

$$= \infty$$

So K_a for this acid is infinitely large.

c. The very large K_a for HCl indicates that HCl fully dissociates in water. The tiny K_a for HOCl indicates that HOCl does not dissociate as much as HCl. So the HCl is a stronger acid than the HOCl.

2 Indicators

Acid-base indicators are weak acids or bases, because they do not dissociate fully in solution. Most often they are molecules that break apart into H^+ and an anion in aqueous solutions. Like other acids and bases, the dissociation process is reversible for indicators.

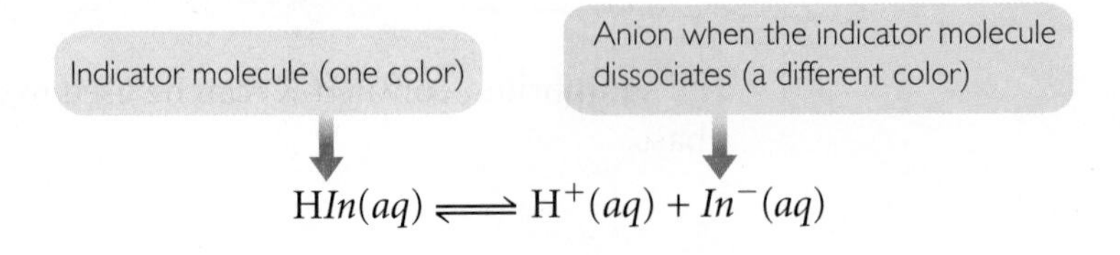

HISTORY CONNECTION

Litmus is a mixture of water-soluble organic compounds that are extracted from certain lichens. Often this mixture is infused into filter paper to make litmus paper. Lichens themselves are a form of fungus that exists in a close symbiotic relationship with a photosynthetic alga.

Indicator molecules are often written with the symbol H*In*. This is because indicator molecules are large molecules with complex structures. In^- is the general symbol for the anion part of the indicator when the indicator dissociates.

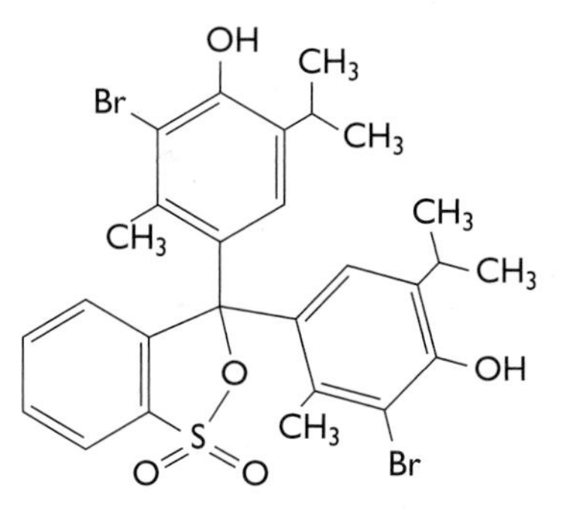

Bromothymol blue

Most indicator molecules, H*In*, and their anions have bright colors in solution. For example, the H*In* molecules of the indicator bromothymol blue are yellow, whereas the In^- anions that are produced when the indicator dissociates are blue.

Colors in Pure Water

The value of the equilibrium constant K_a for indicators is a measure of the degree to which an acid-base indicator dissociates in solution.

$$K_a = \frac{[H^+][In^-]}{[HIn]}$$

A large numerator indicates more dissociation. K_a is larger.

A large denominator indicates less dissociation. K_a is smaller.

Information on the colors for three common acid-base indicators is given in the table here. How does the color relate to the magnitude of the equilibrium constant?

Indicator	H*In*	In^-	K_a	Color in pure water
methyl orange	orange	yellow	3.4×10^{-4}	yellow
litmus	red	blue	3.2×10^{-7}	purple
phenolphthalein	colorless	red	5.0×10^{-10}	colorless

The equilibrium constants (also called dissociation constants) are all much less than 1, indicating that the concentrations of the products are quite small. This is because these indicators are very weak acids that don't dissociate very much. The value of K_a for methyl orange is the largest, so it dissociates the most of these three indicators, while phenolphthalein, with the smallest K_a value, dissociates the least. The value of K_a for litmus is in between the values of methyl orange and phenolphthalein.

The colors in the table support these conclusions. In pure water, at a pH of 7, the methyl orange is yellow, showing that it favors the In^- ion; the phenolphthalein is colorless, showing that it favors H*In;* and the litmus, being purple, appears to have a balance of In^- ions and H*In*.

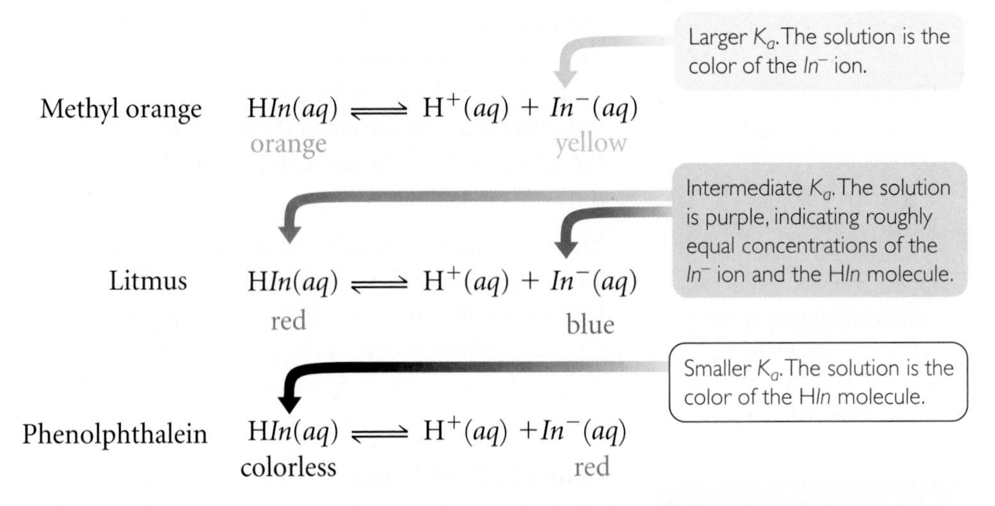

The next illustration shows what happens to the colors of the three indicators if you place them in solutions of different acidities.

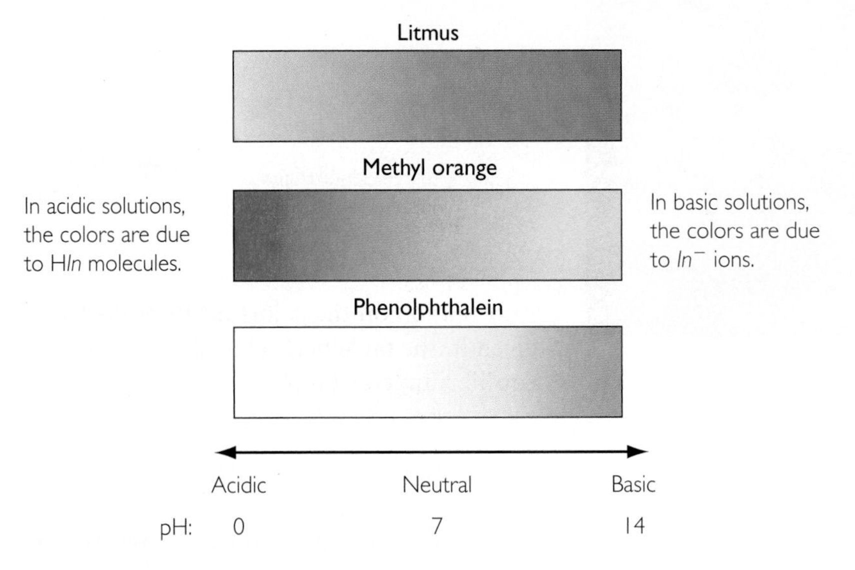

Notice that the colors of the indicators in basic solution (high pH) are the colors of the In^- ions for all three indicators. The colors in acidic solutions are those of the HIn molecules.

Applying Le Châtelier's principle

One way to understand the differences in color for indicators in acidic and basic solution is to apply Le Châtelier's principle.

$$HIn(aq) \rightleftharpoons H^+(aq) + In^-(aq)$$

When an indicator is added to an acidic solution, $[H^+]$ increases. This stresses the equilibrium system of the indicator and the system responds to minimize the change.

Stress: Add H^+ ions.

$$HIn \rightleftharpoons H^+(aq) + In^-(aq)$$

Shift: The reaction shifts toward the reactants.

NATURE CONNECTION

Gardeners can manipulate the color of the flowers on their hydrangeas by changing the pH of the soil. Adding aluminum to soil will increase the pH and make the flowers turn blue; adding lime juice to the soil will lower the pH and make them turn pink.

Alternatively, if an indicator is added to a base, $[H^+]$ decreases. Again, the system is stressed and responds to minimize the change.

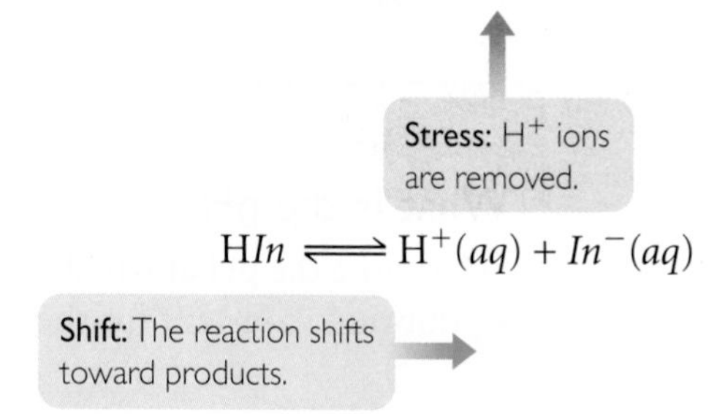

$$HIn \rightleftharpoons H^+(aq) + In^-(aq)$$

The color change of the indicator signals the shift that has occurred.

Another way to look at it is to consider the equilibrium-constant equation.

$$K_a = \frac{[H^+][In^-]}{[HIn]}$$

Since K_a is a constant, it does not change. If any variable in the equation changes, the other variables must adjust. For example, if $[H^+]$ increases, $[In^-]$ must decrease and $[HIn]$ must increase to keep K_a the same.

pH Range for Color Change

Take a moment to examine the information for the color change for several indicators in the table. You might notice that the color change occurs at different pH values for different indicators.

Indicator	HIn color	In^- color	K_a	$[H^+]$ when $[HIn] = [In^-]$	pH range for color change
methyl orange	red	yellow	3.4×10^{-4}	3.4×10^{-4}	3.2–4.4
bromocresol green	yellow	blue	2.0×10^{-5}	2.0×10^{-5}	3.8–5.4
methyl red	yellow	red	7.9×10^{-6}	7.9×10^{-6}	4.8–6.0
bromothymol blue	yellow	blue	1.0×10^{-7}	1.0×10^{-7}	6.0–7.6
phenol red	yellow	red	1.3×10^{-8}	1.3×10^{-8}	6.8–8.4
phenolphthalein	colorless	pink	4.0×10^{-10}	4.0×10^{-10}	8.2–10.0
alizarin yellow	yellow	lilac	1.0×10^{-11}	1.0×10^{-11}	10.0–12.0

Notice that the color change for an indicator occurs when the H^+ ion concentration is equal to K_a. This happens when the concentration of HIn molecules is equal to the concentration of the In^- ions. Rearranging the equilibrium-constant equation proves this to be true.

$$K_a = \frac{[H^+][In^-]}{[HIn]} = [H^+]\frac{[In^-]}{[HIn]} = [H^+]$$

When $[H^+] > K_a$, then $[HIn] > [In^-]$. The solution has the color of H*In* molecules. When $[H^+] < K_a$, then $[HIn] < [In^-]$. The solution has the color of In^- ions.

Example 3

What Is the pH?

Determine the pH at which the indicator congo red changes color. The equilibrium constant for the dissociation of the indicator is $K_a = 10^{-4}$.

Solution

The color change happens when the concentration of H*In* equals the concentration of In^-. If $[HIn] = [In^-]$, then $K_a = [H^+] = 10^{-4}$.

$$\begin{aligned} pH &= -\log[H^+] \\ &= -\log[10^{-4}] \\ &= 4 \end{aligned}$$

So the indicator changes color at pH = 4.

Lesson Summary

How do acid-base indicators work?

There are several specific types of equilibrium constants that apply to different processes. Acid-base indicators are weak acids or bases that change color depending on pH. The concentrations of an indicator molecule and its anion depend on the concentration of H^+ ion in solution. Therefore the color of an indicator is a measure of the acidity of solution it is placed in. The color of an indicator changes when the concentrations of H*In* molecule and In^- anion are identical. This occurs when $K_a = [H^+]$.

EXERCISES

Reading Questions

1. Explain how an acid-base indicator works to show the pH of a solution.
2. Does the value of K_a for an indicator change when the indicator is added to an acid or a base? Explain.

Reason and Apply

3. The solubility product constants K_{sp} at 25 °C for several chlorides are provided in the table on the top of the next page.

Metal chloride	Solubility product constant K_{sp}
thallium (I) chloride, $TlCl(s)$	1.86×10^{-4}
copper (I) chloride, $CuCl(s)$	1.72×10^{-7}
silver (I) chloride, $AgCl(s)$	1.77×10^{-10}

a. Put the three salts in order of increasing solubility—from least soluble to most soluble. Explain your thinking.

b. Suppose you add sodium chloride, $NaCl(s)$, to a saturated solution of copper chloride, $CuCl(aq)$. Use Le Châtelier's principle to explain what happens.

4. Imagine you prepare a 0.10 M solution of HNO_3 and a 0.10 M solution of HF. The concentrations of the ions in equilibrium are given in the table below.

General equation for acid dissociation: $HA(aq) \rightleftharpoons H^+(aq) + A^-(aq)$

Acid	[HA]	$[H^+]$	$[A^-]$
nitric acid, $HNO_3(aq)$	~0	0.10 M	0.10 M
hydrofluoric acid, $HF(aq)$	0.0974 M	2.6×10^{-3} M	2.6×10^{-3} M

a. Determine K_a for hydrofluoric acid, HF.

b. Show that K_a for HNO_3 is a very large number.

c. Which acid is stronger? Explain your thinking.

d. Which solution has the higher concentration of $H^+(aq)$, 0.10 M HOCl or 0.10 M HF? Explain your thinking.

5. Use the table on page 609 to predict the color of these indicators if you dissolve them in water. Explain your thinking.

a. alizarin yellow

b. bromocresol green

c. methyl red

d. phenol red

6. Use Le Châtelier's principle to explain why bromocresol green is blue if you dissolve it in a base, such as 0.10 M sodium hydroxide, NaOH.

7. Determine the pH at which thymolphthalein changes color. The equilibrium constant for the dissociation of the indicator is $K_a = 10^{-10}$.

8. Methyl violet changes color from the yellow H*In* molecule to the blue-violet In^- ion at a pH of 1. What is the equilibrium constant for the dissociation of methyl violet?

9. WEB RESEARCH Look up the source, structure, and properties of a particular indicator. Write a short description of the molecule.

SECTION

II

Key Term

Le Châtelier's principle

SUMMARY

Changing Conditions at Equilibrium

Showtime Update

Le Châtelier's principle states that when a stress is placed on a system at equilibrium, the system will react in the direction that serves to relieve that stress. Changes in reactant or product concentrations place a stress on a system at equilibrium. If you add reactants to a system at equilibrium, the reaction will shift in the direction of forming products. If you remove reactants from a system at equilibrium, the system relieves this stress by shifting in the direction of forming reactants. The same is true for adding or removing products. Changing temperature is another way to place a stress on a system at equilibrium. If the reaction is exothermic, heating will shift the reaction in the direction of forming reactants. For an endothermic reaction, heating will shift the reaction in the direction of forming products. For a system where the reactants and products are in the gas phase, changing the pressure places a stress on the system at equilibrium. If the pressure on a gaseous mixture at equilibrium is increased, a change will occur in the direction that decreases the total number of moles of gas and relieves some of the pressure.

There are several types of equilibrium constants. The solubility product constant, K_{sp}, is the equilibrium constant for a system of dissolving and precipitating. K_a and K_b are the acid and base dissociation constants. The strength of an acid or a base depends on how many acid or base molecules have dissociated in solution.

Review Exercises

1. Suppose this chemical reaction is in a state of dynamic equilibrium in a sealed gas container:

$$COCl_2(g) \rightleftharpoons CO(g) + Cl_2(g) \quad K = 4.63 \times 10^{-3}$$

a. Write the equilibrium-constant equation for this reaction.

b. At equilibrium, what would $[COCl_2]$ be if $[Cl_2] = [CO] = 2.0 \times 10^{-2}$ M?

c. Which direction would the equilibrium shift if some additional $COCl_2$ were added to the system? Explain why.

d. Suppose that the $COCl_2$ added in part c brought $[COCl_2]$ to 4.0 M at equilibrium. What would $[Cl_2]$ and $[CO]$ be?

2. At room temperature, $K = 3.75$ for this reaction:

$$SO_2(g) + NO_2(g) \rightleftharpoons SO_3(g) + NO(g)$$

If the reaction were in a state of dynamic equilibrium initially, predict the effect of these actions. Explain your choices.

a. Additional $SO_2(g)$ is added.

b. Additional $NO(g)$ is added.

c. Some $SO_3(g)$ is removed.

d. Equal amounts of $SO_3(g)$ and $SO_2(g)$ are added.

3. Hydrofluoric acid is a weak acid with $K_a = 7.2 \times 10^{-4}$. In water, it reacts according to this reaction:

$$HF(aq) \rightleftharpoons H^+(aq) + F^-(aq)$$

a. Write the equilibrium-constant equation for the reaction.

b. At equilibrium, is $[H^+]$ greater than, less than, or equal to that of $[F^-]$? Explain your answer.

c. At equilibrium, is $[H^+]$ greater than, less than, or equal to that of [HF]? Explain your answer.

d. Which direction would the equilibrium shift if more H^+ were added to the system?

e. In part d, would [HF] increase, decrease, or stay the same as the system proceeded back to equilibrium? Explain your answer.

4. Suppose this reaction is at equilibrium in a flexible container:

$$2NO(g) + Br_2(g) \rightleftharpoons 2NOBr(g)$$

a. Write the equilibrium-constant equation for the reaction.

b. What would happen to the mixture if some additional NOBr(*g*) were added?

c. Which direction would the reaction shift if the pressure inside the container were increased?

d. What would happen to the equilibrium if the volume of the container were increased?

5. This exothermic reaction is at equilibrium. In which direction will the equilibrium position shift with these changes? Explain your answers.

$$3NO(g) \rightleftharpoons N_2O(g) + NO_2(g)$$

a. Adding more N_2O

b. Removing some N_2O

c. Raising the temperature

d. Lowering the temperature

PROJECT *It's Showtime!*

Your teacher will assign you a chemistry demonstration to present to an audience. After presenting the demonstration once, you must change at least two variables and present the demonstration again, once for each change. Your demonstration should include

- a poster showing how your demonstration works.
- a written procedure for the two modifications to your demonstration plus a description of the outcomes.
- an explanation applying equilibrium or Le Châtelier's principle to your demonstration.
- an explanation of the chemistry involved in your demonstration appropriate for the particular audience.
- a creative story to engage the audience.

UNIT

6

Showtime Review

Many chemical demonstrations feature reversible reactions. These reactions do not go all the way to completion. In reversible chemical processes the forward and reverse reactions continue, with both products and reactants in the mixture. When the forward and reverse reactions are in balance, these systems are said to be at equilibrium. Reversible reactions are written with a double arrow to show that the forward and reverse reactions occur at the same time.

Equilibrium systems may be physical or chemical. For example, the back and forth evaporation and condensation of water on the planet is a form of phase equilibrium. Acid-base systems and redox reactions are chemical equilibrium systems.

The formation of sodium carbonate crystals on the edge of a salt lake is an example of a reversible dissolving/precipitation equilibrium process. Reversible processes are often written as net ionic equations. For example,

$$Na_2CO_3(s) \rightleftharpoons 2Na^+(aq) + CO_3^-(aq)$$

Equilibrium is a dynamic balance between the reactants and products in a reaction or process. When a system is at equilibrium, the rate of the forward reaction and the rate of the reverse reaction are equal. However, this does not mean that the amounts of reactants and products are equal.

There is a mathematical relationship between the products and reactants in an equilibrium system. The equilibrium constant K and the equilibrium-constant equation can be used to solve problems involving systems at equilibrium.

$$aA + bB \rightleftharpoons cC + dD$$

$$K = \frac{[C]^c[D]^d}{[A]^a[B]^b}$$

In general, a high value for K indicates that products are favored in a reaction, and a low value for K indicates that reactants are favored.

Le Châtelier's principle states that when conditions are changed for a system at equilibrium, the system will respond by reducing the effect of the change. When conditions such as temperature, pressure, and concentration are changed in a system, equilibrium is regained by rebalancing the concentrations of reactants and products.

Review Exercises

1. Write the reversible reactions for these processes.
   a. Chlorine gas becomes liquid chlorine.
   b. Nitrogen dioxide becomes dinitrogen tetroxide.
   c. Silver chloride precipitates from aqueous solution.
   d. Water evaporates.
2. Write the equilibrium-constant equations for the processes in Exercise 1.
3. Write two statements that are always true of a system at equilibrium.
4. When a dissolving and precipitation process is at equilibrium, the solution is saturated. Explain why this is true.
5. What is the concentration of copper ion, Cu^{2+}, in an aqueous solution of copper (II) carbonate, $CuCO_3$, if $K_{sp} = 2.5 \times 10^{-10}$ at 25 °C?
6. What does an acid-base indicator have to do with equilibrium?
7. The chemical equation for the dissolution of lead chloride, $PbCl_2(s)$, in water is given here.

   $$PbCl_2(s) \rightleftharpoons Pb^{2+}(aq) + 2Cl^-(aq)$$

   Determine the concentration of Pb^{2+} ions in a saturated $PbCl_2$ solution if $K_{sp} = 1.78 \times 10^{-5}$.
8. Solid lead (II) sulfate, $PbSO_4$, is dissolved in water to form a solution with excess undissolved solute at the bottom of the container. The value of K_{sp} for lead sulfate is 1.6×10^{-8}.
   a. How can you tell that this solution is saturated?
   b. Is the solution at equilibrium? How do you know?
   c. Write the chemical equation for the reversible change that is taking place.
   d. What is the equilibrium-constant equation for this equilibrium?
   e. What is the equilibrium concentration of each ion?
   f. Dissolving lead (II) sulfate is an endothermic change. Predict what will happen to the value of $[Pb^{2+}]$ and $[SO_4^{2-}]$ if the solution is heated.
   g. What would happen if you added more sulfate ions to the solution?
9. The value of K_{sp} for cadmium sulfide, CdS, is 8.0×10^{-27}. Compare this value with the K_{sp} for lead (II) sulfate given in Exercise 8.
   a. Which compound will have a higher concentration of ions in solution at equilibrium? Explain your reasoning.
   b. Calculate the concentration of cadmium ions in solution at equilibrium.
10. For the acid-base indicator litmus, the molecule, $HIn(aq)$, is red and the anion, $In^-(aq)$, is blue.
    a. Write a reaction for the dissociation of the HIn molecule in solution.
    b. Why is litmus purple when you dissolve it in water? Explain your reasoning.
    c. What color do you think litmus is in a basic solution? Use Le Châtelier's principle to explain your thinking.

MATH *Spotlights*

SI Units of Measure

Scientists rely on repeatable measurements as they study the physical world. It is important that they use consistent units of measure worldwide. In 1960 an international council standardized the metric system, creating the *Système International d'Unités* (International System of Units), abbreviated as SI.

These are the basic SI units used in chemistry. Other units such as density and volume are combinations of these.

Quantity	Unit (abbreviation)
length	meter (m)
mass	kilogram (kg)
time	second (s)
temperature	Kelvin (K)
amount of substance	mole (mol)

SI units are based on powers of 10. Larger and smaller units get their names by combining standard prefixes with these basic units. For example, the word *centimeter* is a combination of *centi-* and *meter.* A centimeter is one one-hundredth of a meter. These are the prefixes used in the SI, along with their abbreviations and their meanings. (A kilogram is the only basic SI unit that has a prefix as part of its name.)

Prefix	Multiple	Scientific Notation	Prefix	Multiple	Scientific Notation
tera- (T-)	1,000,000,000,000	10^{12}	pico- (p-)	0.000 000 000 001	10^{-12}
giga- (G-)	1,000,000,000	10^{9}	nano- (n-)	0.000 000 001	10^{-9}
mega- (M-)	1,000,000	10^{6}	micro- (μ-)	0.000 001	10^{-6}
kilo- (k-)	1,000	10^{3}	milli- (m-)	0.001	10^{-3}
hecto- (h-)	100	10^{2}	centi- (c-)	0.01	10^{-2}
deka- (da-)	10	10^{1}	deci- (d-)	0.1	10^{-1}

Example 1

Length Conversions

How many meters does each of these lengths represent?

a. 562 centimeters b. 2.5 kilometers

Solution

a. $562\ \text{cm} \cdot \dfrac{1\ \text{m}}{100\ \text{cm}} = 5.62\ \text{m}$

b. $2.5\ \text{km} \cdot \dfrac{1000\ \text{m}}{1\ \text{km}} = 2500\ \text{m}$

Example 2

The Mass of One Liter

One milliliter, or cubic centimeter, of water at 4 °C weighs 1 g. How much does 1 L of water at 4 °C weigh?

Solution

One milliliter is one one-thousandth of a liter; multiply the mass of 1 mL by 1000 to get the mass of 1 L.

$$1000 \cdot 1\ \text{g} = 1000\ \text{g} = 1\ \text{kg}$$

Practice Exercises

Convert these measurements to the indicated units.

1. 7 m = ______ cm
2. 3200 mL = ______ L
3. 20,012 cm = ______ km
4. 0.003 kg = ______ g
5. $16\ \text{m}^2$ = ______ cm^2
6. $2\ \text{m}^3$ = ______ dm^3

Answers

1. 700 cm 2. 3.2 L 3. 0.20012 km 4. 3 g 5. $160{,}000\ \text{cm}^2$ 6. $8000\ \text{dm}^3$

Accuracy, Precision, and Significant Digits

There are two kinds of numbers in the world—exact and inexact. For instance, counting is exact because you can safely say that there are exactly 12 eggs in a dozen. However, no measurement with a ruler, a balance, or a graduated cylinder is ever exact. So, when a measurement or the average of several measurements comes out extremely close to the actual true value, we say that the measurement or average is *accurate.*

If you measure something several times and get very similar answers each time, your measurements are *precise.*

The ability to make precise measurements depends partly on the equipment used. For example, a graduated cylinder is more precise than a beaker. Precision also depends on how carefully a measurement was made. For example, a measurement of 23.76 mL is more precise than a measurement of 24 mL.

To understand the difference between precision and accuracy, imagine a lab experiment to measure the boiling point of water at sea level. Several readings are taken: 97.2 °C, 97.0 °C, and 97.1 °C. The measurements are close to each other; repeating the experiment would likely give similar results, so they are precise. However, they are not accurate; the boiling point of water at sea level is known to be 100 °C. Perhaps the thermometer was faulty, or the person taking the measurements consistently read the thermometer incorrectly.

Using significant digits in a measurement allows you to indicate the degree of certainty in the measurements. In general, the last digit of any measurement is uncertain. For example, suppose you use a meterstick, marked in millimeters, to measure the length of an object. If you record a measurement of 24.33 cm, the last digit is an estimate based on the closest millimeter markings and is not certain. Another person might measure the length as 24.34 cm or 24.32 cm. However, you will both agree on the 24.3 because you can read the meterstick accurately to the millimeter, or 0.1 cm. The measurement 24.33 has four significant digits.

The rules for determining the number of significant digits are complicated. Nonzero digits always count. Zeros sometimes count depending on where they are. Zeros might be leading, trapped, or trailing.

- *Leading zeros,* such as those in 0.004728, never count as significant digits.
- *Trapped zeros,* as in 1.08, always count as significant digits.
- *Trailing zeros,* or those at the end of a number, count only when there is a decimal point. In the numbers 20.0, 300.00, and even 50., the zeros count as significant digits. If there is no decimal point, as in 500, then it isn't possible to tell whether the zeros are significant. Sometimes you can deduce that zeros are significant based on the instrumentation. For example, a thermometer generally measures to the nearest degree. You can use scientific notation to avoid ambiguity. The measurement 5.00×10^2 has three significant digits, whereas the measurement 5×10^2 has only one significant digit.

As you do calculations involving numbers with different numbers of significant digits, follow these two rules.

- Adding or subtracting: Your final answer will have only as many *decimal places* as the measurement with the fewest decimal places.
- Multiplying or dividing: The result can have only as many significant digits as the number with the fewest significant digits.

As you add, multiply, or combine measurements in other ways, you will often have to round the result of your calculation to obtain the correct number of significant digits.

Example 1

Rounding the Sum or Difference

Calculate each sum or difference and round the answer to the appropriate number of significant digits.

a. $2.24 + 3.4 + 5.231$ b. $10.5 \text{ cm} - 3.36 \text{ cm}$

Solution

a. First add the numbers to get the sum 10.871. Then consider the decimal places to arrive at the final answer.

2 decimal places	2.24
1 decimal place	3.4
3 decimal places	+ 5.231
	10.871

Because 3.4 has only one decimal place, the sum should be rounded to 10.9. Always complete the calculation before rounding.

b. First subtract the numbers.

1 decimal place	10.5 cm
2 decimal places	− 3.36 cm
	7.14 cm

Because 10.5 has only one decimal place, round the final answer to one decimal place: 7.1 cm.

Example 2

Rounding a Product

Calculate each product and round to the appropriate number of significant digits.

a. $12.34 \cdot 1.6$ b. $4.71 \text{ m} \cdot 5.28 \text{ m}$

Solution

a. First multiply the two decimals.

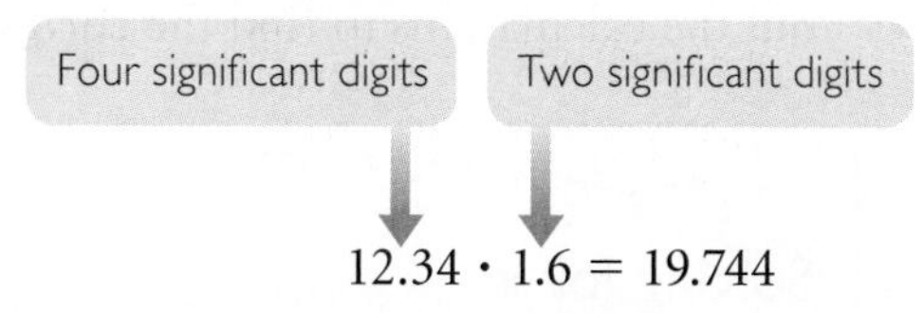

$$12.34 \cdot 1.6 = 19.744$$

Because there are only two significant digits in 1.6, you can have only two significant digits in your answer. You must round the product to 20. You might write this answer as $2.0 \cdot 10$ to make it clear that both digits are significant.

b. First multiply: $4.71 \text{ m} \cdot 5.28 \text{ m} = 24.8688 \text{ m}^2$.

Both factors have three significant digits, so the answer must be rounded to three significant digits: 24.9 m^2.

Practice Exercises

1. How many significant digits are in each of these numbers?
   a. 20.1 **b.** 300.0 **c.** 0.0031 **d.** 0.03010
2. Complete these calculations.
   a. $25.14 + 3.4 + 15.031$ **b.** $100.04 \text{ cm} - 7.362 \text{ cm}$ **c.** $3005 \cdot 45.20$
3. What is the volume of a box 14.5 cm by 15.9 cm by 21.1 cm?
4. The density of copper is 8.92 g/cm^3. What is the mass of 24 cm^3 of copper?

Answers

1a. 3 **1b.** 4 **1c.** 2 **1d.** 4 **2a.** 43.5 **2b.** 92.68 **2c.** 135,800
3. 4860 cm^3 **4.** 210 g

Solving Equations

Solving chemistry problems sometimes involves solving a math equation. When the quantity you are looking for is isolated, or alone on one side of the equation, all you need to do is complete the calculations. A simple example is finding the Fahrenheit equivalent of 28 °C using the equation

$$F = \frac{9}{5}C + 32°$$

Sometimes the quantity that answers your question is not alone on one side of the equation. To solve these problems, isolate the variable you are looking for on one side of the equation, so that calculations that will lead to your answer are on the other side of the equation. For example, to find a temperature in degrees Celsius, rearrange the equation in terms of *C*.

Start with the known relationship.	$F = \frac{9}{5}C + 32°$
Subtract 32° from both sides.	$F - 32° = \frac{9}{5}C$
Multiply both sides by $\frac{5}{9}$.	$\frac{5}{9}(F - 32°) = C$

Now that *C* is isolated, you can substitute any Fahrenheit temperature and carry out the calculations to find the corresponding Celsius temperature.

Example 1

Solving for *x*

Solve these equations for *x*.

a. $0.1x + 12 = 2.2$ b. $\frac{12 + 3.12x}{3} = -100$

Solution

a.

Original equation.	$0.1x + 12 = 2.2$
Subtract 12 from both sides.	$0.1x + 12 - 12 = 2.2 - 12$
	$0.1x = -9.8$
Divide both sides by 0.1.	$x = -98$

b.

Original equation.	$\frac{12 + 3.12x}{3} = -100$
Multiply both sides by 3.	$12 + 3.12x = -300$
Subtract 12 from both sides.	$-12 + 12 + 3.12x = -300 - 12$
	$3.12x = -312$
Divide both sides by 3.12.	$x = -100$

Example 2

Solving for V_2

Solve this equation for V_2.

$$\frac{V_1P_1}{T_1} = \frac{V_2P_2}{T_2}, \text{ for } V_2$$

Solution

Original equation.

$$\frac{V_1P_1}{T_1} = \frac{V_2P_2}{T_2}$$

Multiply both sides by T_2; remove the factor of 1.

$$\frac{T_2V_1P_1}{T_1} = \frac{T_2V_2P_2}{T_2}$$

Divide both sides by P_2; remove the factor of 1.

$$\frac{T_2V_1P_1}{T_1P_2} = \frac{V_2P_2}{P_2}$$

$$V_2 = \frac{T_2V_1P_1}{T_1P_2}$$

The equation is now expressed in terms of V_2. Substituting in values for the other five variables will give a value for V_2.

Practice Exercises

1. Solve these equations. Indicate the action you take at each stage.

a. $144x + 33 = 45$ **b.** $\frac{1}{6}x + 2 = 8$ **c.** $5(x - 7) = 15 + 5^2$

2. Solve these equations for the variable indicated.

a. $d = rt$, for t **b.** $P = 2(l + w)$, for w **c.** $A = \frac{1}{2}h(a + b)$, for h

Answers

1a. $x = \frac{1}{12}$ (On both sides, subtract 33 and divide by 144.)

1b. $x = 36$ (On both sides, subtract 2 then multiply by 6.)

1c. $x = 15$ (Combine 15 and 25 on the right side, divide both sides by 5, and add 7 to both.)

2a. $t = \frac{d}{r}$ **2b.** $l = \frac{P}{2} - w$ **2c.** $h = \frac{2A}{a} + b$

Order of Operations

Math expressions and equations often involve several operations. For example, to convert from Celsius to Fahrenheit, you first multiply the number of Celsius degrees by the fraction $\frac{9}{5}$ and then add 32 degrees. To convert from Fahrenheit to Celsius, you subtract 32 degrees from the number of Fahrenheit degrees and then multiply the result by $\frac{5}{9}$. A rule called the *order of operations* is used to write math expressions clearly so that anyone seeing the formula or equation would know whether multiplication was the first step or the second step.

Order of Operations

1. Evaluate all expressions within parentheses.
2. Evaluate all terms with exponents.
3. Multiply and divide from left to right.
4. Add and subtract from left to right.

Example 1

Temperature Conversions

Convert these temperatures.

a. 37 °C to degrees Fahrenheit

b. 48 °F to degrees Celsius

Solution

Substitute the known value into each equation and then solve using the order of operations.

a.	Substitute 37° into the equation.	$F = \frac{9}{5} \cdot 37° + 32°$
	Multiply.	$F = 66° + 32°$
	Add.	$F = 98$ °F
b.	Substitute 48° into the equation.	$C = \frac{5}{9}(48° - 32°)$
	Subtract.	$C = \frac{5}{9}(16°)$
	Multiply.	$C = 9$ °C

Example 2

Parentheses, Exponents, and Fractions

Evaluate these expressions.

a. $\frac{5}{9}(96 - 15)$

b. $\frac{(70 - 64)^2}{2 \cdot 5} + 12$

Solution

Evaluate the expression in the parentheses first.

a.	Original expression.	$\frac{5}{9}(96 - 15)$
	Subtract the numbers within the parentheses.	$= \frac{5}{9}(81)$
	Multiply $\frac{5}{9}$ by 81.	$= 45$

b. The fraction line acts like parentheses. In fact, when the expression is entered into a calculator, parentheses are required around the $2 \cdot 5$.

Original expression.	$\frac{(70 - 64)^2}{2 \cdot 5} + 12$
Evaluate the expressions above and below the fraction line.	$= \frac{6^2}{10} + 12$
Divide 36 by 10 and then add 12.	$= 15.6$

Practice Exercises

1. Evaluate these expressions.

a. $3 \cdot 24 \div 8$ **b.** $3 + 24 \cdot 8$ **c.** $3 - 24 + 8$

d. $(3 + 24) \cdot 8$ **e.** $(3 + 21) \div 8$ **f.** $3 - (24 + 8)$

2. Calculate the value of each expression.

a. $-2 + 5 - (-8)$ **b.** $(-5^2) - (-3)^2$ **c.** $-0.3 \cdot 20 + 15$

3. Insert parentheses as needed to make each equation true.

a. $15 \div 3 + 7 - 4 = -48$ **b.** $15 \div 3 + 7 - 4 = 8$ **c.** $-4^2 + -3^2 = -7$

Answers

1a. 9 **1b.** 195 **1c.** −13 **1d.** 216 **1e.** 3 **1f.** −29

2a. 11 **2b.** −34 **2c.** 9

3a. $(15 \div 3 + 7)(-4) = -48$ **3b.** $15 \div 3 + 7 - 4 = 8$ **3c.** $-4^2 + (-3)^2 = -7$

Averages

Scientists are often interested in the typical result of a repeated experiment. The average, also called the *mean,* is one way to determine a typical value. The average is calculated by totaling the data values and dividing by the number of values.

Example 1

The Mean

Find the average of these numbers: 14, 23, 10, 21, 7, 80, 32, 30, 92, 14, 26, 21, 38, 20, 35, 21.

Solution

Sum of the data values

Mean

$$\frac{14 + 23 + 10 + 21 + 7 + 80 + 32 + 30 + 92 + 14 + 26 + 21 + 38 + 20 + 35 + 21}{16} = 30.25$$

Number of data values

The average is 30.25.

Sometimes the data values are not of equal importance in contributing to the average. You need to weight the values differently.

Example 2

The Weighted Average

On a chemistry quiz, 1 student got 100%, 7 students got 95%, 12 students got 90%, 1 student got 85%, 5 students got 80%, 3 students got 75%, and 1 student got 70%. What was their average (mean) score?

Solution

One solution would be to change the list so that 95% appeared 7 times, 90% appeared 12 times, and so on. However, it is more efficient to use a weighted average as shown here.

$$(1)(100) + (7)(95) + (12)(90) + (1)(85) + (5)(80) + (3)(75) + (1)(70) = 2625$$

Thirty students took the quiz, so divide 2625 by 30.

The average score was 87.5%.

Example 3

Average Atomic Mass

The element silver, Ag, has two naturally occurring isotopes. Approximately 52% of all silver consists of atoms with 60 neutrons, and 48% consists of atoms with 62 neutrons. Calculate the average atomic mass of silver atoms.

Solution

Silver atoms have 47 protons. The atoms with 60 neutrons have masses of $47 + 60 = 107$ amu. The atoms with 62 neutrons have masses of $47 + 62 = 109$ amu. If you have a sample of 100 silver atoms, 52 atoms will have a mass of 109 amu and 48 atoms will have a mass of 109 amu. Use a weighted average. The total mass is 52(107 amu) + 48(109 amu) = 10,796 amu. The average mass is 10,796 amu/100 = 107.96. This is close to the atomic mass of 107.9 amu shown on the periodic table.

Practice Exercises

1. Find the average of these numbers: 52.3, 18.91, 35.66, 4.35.
2. A student had these scores on chapter tests: 87%, 90%, 95%, 92%. He got 92% on one unit test and 86% on the other. His final exam score was 91%. Unit tests count twice as much as chapter tests, and the final exam counts four times as much as a chapter test. What is his average score?
3. About 76% of all chlorine atoms have an atomic mass of 35.00 amu; about 24% have an atomic mass of 37.00 amu. Use this information to calculate the average atomic mass of 100 chlorine atoms.

Answers

1. 27.8 **2.** 90% **3.** 35.48 amu

Graphing

Many chemical experiments involve changing one variable and seeing how another variable changes as a result. The results of experimental procedures can be listed in a table, but it is often helpful to look for any trends in the data on a graph.

Data consisting of two variables, such as temperature and volume measurements, can be graphed using a coordinate plane. Each point on the coordinate plane can be identified by a pair of numbers (x, y) called coordinates. The first number, the x-coordinate, describes how far the point is to the right or left of the origin; the second number, the y-coordinate, describes how far up or down the point is.

In the example below, a coordinate pair represents one (temperature, volume) data point. The quantity that the experimenter is changing, in this case temperature, is called the *independent variable.* It is graphed on the horizontal axis, the x-axis. The resulting measurement, or the *dependent variable,* is graphed on the vertical axis, or y-axis. The dependent variable in this example is the volume. If the experiment had been set up to measure changes in temperature as a result of changes in volume, the independent variable would have been volume, and temperature would have been the dependent variable.

Normally, you graph the dependent variable versus the independent variable. (*Versus* means "compared with" and is abbreviated "vs.")

Example 1

Graphing Coordinate Pairs

Graph these results for the heating of a gas:

Temperature (°C)	Volume (mL)	Temperature (°C)	Volume (mL)
0	465	15	491
5	475	20	498
10	481	25	509

Solution

The temperature is being changed, so temperature is the independent variable and should be graphed on the x-axis. The volume changes as a result, so it is the dependent variable and should be graphed on the y-axis. To graph the data, you need to decide on the scale for each axis. The x-axis can start at 0 °C and go up to 30 °C. To save space, you don't need to show the y-axis starting at 0; it can go from 450 to 520. Label the axes "Temperature (°C)" and "Volume (mL)." Graph the coordinate pairs. Title the graph. In the title, the dependent variable is usually named first.

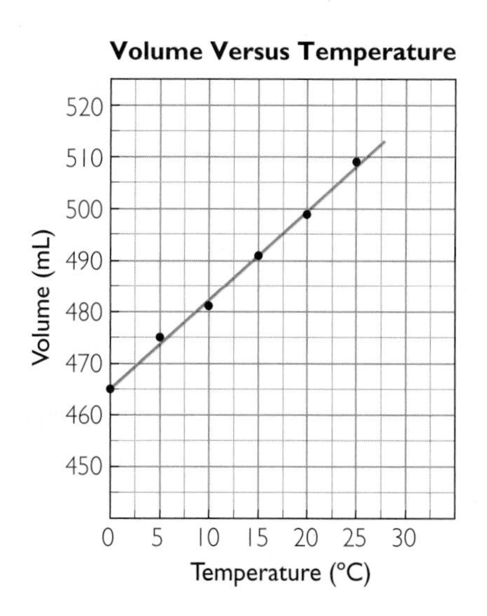

Notice that the points in the graph of volume versus temperature lie nearly in a straight line. Small errors in measurement account for the variation. The points are close enough to a line to indicate that the relationship between temperature and volume is linear. As one quantity increases, the other increases at a constant rate. So the quantities vary directly, or are *directly proportional.*

Example 2

Volume Versus Pressure

Graph the results of an experiment in which a gas's volume is measured as pressure is applied. The temperature of the gas is kept from changing. This table shows the data collected.

Pressure (atm)	Volume (L)	Pressure (atm)	Volume (L)
1.0 atm	10.0 L	3.0 atm	3.3 L
1.5 atm	6.7 L	3.5 atm	2.9 L
2.0 atm	5.0 L	4.0 atm	2.5 L
2.5 atm	4.0 L		

Solution

An appropriate scale for the independent variable on the *x*-axis is 0 atm to 4.5 atm. The scale for the dependent variable on the *y*-axis can go from 0 L to 12 L. Label the axes and indicate the units: "Pressure (atm)" and "Volume (L)." The coordinate pairs to graph are (1.0, 10), (1.5, 6.7), (2.0, 5.0), and so on. Title the graph *Volume Versus Pressure.*

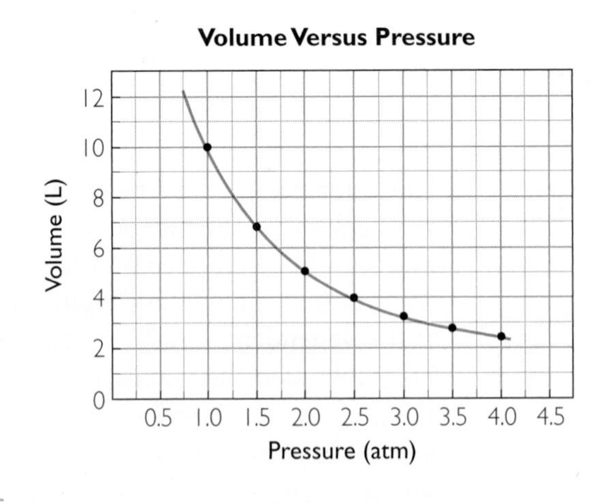

The points in the graph of volume versus pressure do not lie in a line, but you can draw a smooth curve through the points to show the trend. The shape of the graph is typical of relationships that are inversely proportional.

Practice Exercises

1. The examples all used coordinate pairs with positive values. But points can also have negative coordinates. The coordinates of point *A* are (−7, −4). Name the (*x*, *y*) coordinates of each point pictured.

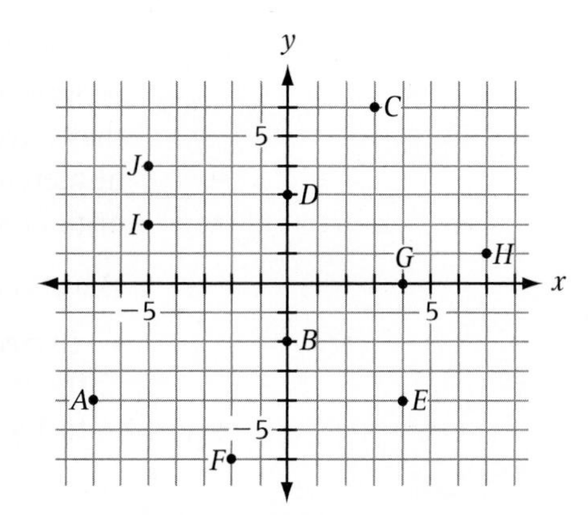

2. A sealed rigid container is heated and the pressure in the container is measured. Describe the graph of the data (your scales, axes titles, graph title, plotted points, and shape of the curve through the points).

Temperature (°C)	Pressure (atm)
22.0 °C	3.20 atm
35.0 °C	3.34 atm
45.0 °C	3.45 atm
60.0 °C	3.61 atm

3. These coordinate pairs represent measurements of volume and mass of copper: (1.1, 9.8), (2.3, 20.5), (2.7, 24.1), (4.5, 40.1). If these are graphed, what will be the shape of the graph? Are mass and volume directly proportional or inversely proportional?

Answers

1. *B* (0, −2), *C* (3, 6), *D* (0, 3), *E* (4, −4), *F* (−2, −6), *G* (4, 0), *H* (7, 1), *I* (−5, 2), *J* (−5, 4)

2. Sample answer: The *x*-axis is "Temperature (°C)" with a scale from 20 to 65. The *y*-axis is "Pressure (atm)" with a scale of 3.10 to 3.70. Plotted points are (22, 3.20), (35, 3.34), (45, 3.45), (60, 3.61). The graph title might be *Pressure Versus Temperature.* The points lie in a straight line; the quantities are directly proportional.

3. A straight line through the origin; directly proportional.

Ratios and Proportions

As chemists study matter, they are often working with ratios. The four statements here all express some sort of ratio.

Density is the ratio of mass to volume: $D = m/V$.

The units of molar mass are g/mol.

The mole ratio of H_2 to O_2 is 2:1.

The reaction gave a 78% yield.

Ratios can be written as fractions. For instance, 78% is $\frac{78}{100}$. The ratio 2:1 can be written $\frac{2}{1}$. Ratios can be reduced like fractions, or they can be added, once they have a common denominator. Unit fractions, which equal 1 because the numerator is equivalent to the denominator, can be used to convert between units or to create equivalent fractions.

Whenever you are working with ratios, use the rules for working with fractions.

- *To create equivalent fractions,* multiply a fraction by a unit fraction. In a unit fraction, the numerator is equivalent to the denominator. A unit fraction is equal to 1 and can be used to convert between units or create equivalent fractions.

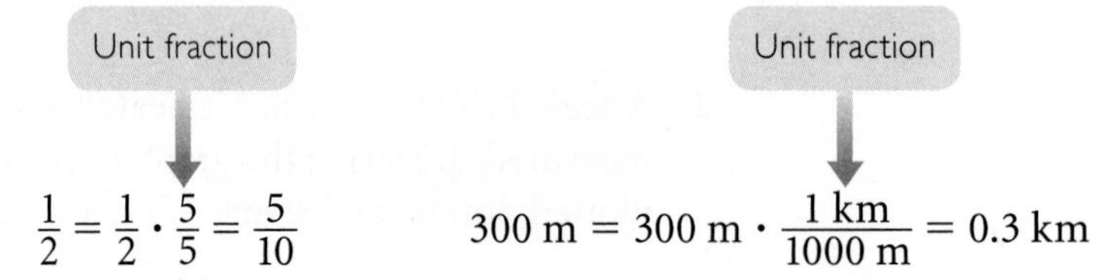

$$\frac{1}{2} = \frac{1}{2} \cdot \frac{5}{5} = \frac{5}{10} \qquad 300 \text{ m} = 300 \text{ m} \cdot \frac{1 \text{ km}}{1000 \text{ m}} = 0.3 \text{ km}$$

- *To add or subtract fractions,* find equivalent fractions with a common denominator, then add or subtract the numerators and put that number over the common denominator.

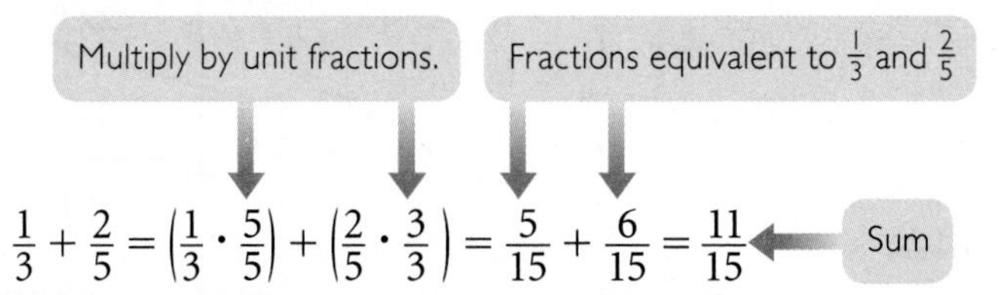

$$\frac{1}{3} + \frac{2}{5} = \left(\frac{1}{3} \cdot \frac{5}{5}\right) + \left(\frac{2}{5} \cdot \frac{3}{3}\right) = \frac{5}{15} + \frac{6}{15} = \frac{11}{15}$$

- *To multiply fractions,* multiply the two numerators, and multiply the two denominators. You may want to reduce the fraction so that there are no common factors between the numerator and the denominator.

$$\frac{1}{3} \cdot \frac{3}{5} = \frac{1 \cdot 3}{3 \cdot 5} = \frac{3}{15} = \frac{1}{5}$$

- *To divide fractions,* multiply by the reciprocal of the divisor.

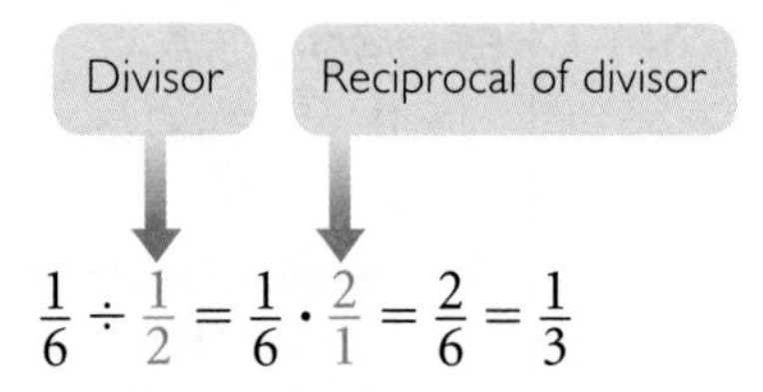

$$\frac{1}{6} \div \frac{1}{2} = \frac{1}{6} \cdot \frac{2}{1} = \frac{2}{6} = \frac{1}{3}$$

- *To reduce a fraction,* divide the numerator and denominator by any factor they have in common; this is often referred to as *canceling.*

$$\frac{6}{15} = \frac{2 \cdot \cancel{3}}{\cancel{3} \cdot 5} = \frac{2}{5} \qquad \frac{300 \cancel{\text{m}} \cdot 1 \text{ km}}{1000 \cancel{\text{m}}} = 0.3 \text{ km}$$

Proportions are equations that state two ratios are equal. Each of the equations shown here is a proportion.

$$\frac{P_1V_1}{T_1} = \frac{P_2V_2}{T_2} \qquad \frac{3.20 \text{ atm}}{22 \text{ °C}} = \frac{P_2}{50 \text{ °C}}$$

$$\frac{1 \text{ mol } H_2O}{18.0 \text{ g } H_2O} = \frac{x \text{ mol } H_2O}{51.2 \text{ g } H_2O}$$

Proportions can be solved by multiplying (or dividing) both sides of the equation by the same value. As you solve chemistry problems, you will often use ratios and proportions.

Example 1

Limiting Reactant

Imagine you start with 10.0 g of $Mg(OH)_2$ and 100.0 mL of 4.0 M HCl. How much $MgCl_2$ will be produced?

Solution

Write the balanced chemical equation.

$$Mg(OH)_2(s) + 2HCl(aq) \longrightarrow MgCl_2(aq) + 2H_2O(l)$$

Determine the molar masses of each compound.

Reactants: $Mg(OH)_2 = 58.3$ g/mol   HCl = 36.5 g/mol

Products: $MgCl_2 = 95.2$ g/mol $H_2O = 18.0$ g/mol

Determine the number of moles of each reactant that you have.

$$\text{Moles of } Mg(OH)_2 = \frac{10.0 \text{ g}}{58.3 \text{ g/mol}} = 0.17 \text{ mol}$$

$$\text{Moles of HCl} = (0.100 \text{ L})(4.0 \text{ mol/L}) = 0.40 \text{ mol}$$

Use the mole ratio to identify the limiting reactant.

The reactants combine in a 1:2 ratio. So you need 0.17 mol · 2 = 0.34 mol of hydrochloric acid, HCl, to react with the 0.17 mol of magnesium hydroxide, $Mg(OH)_2$.

You have 0.40 mol of HCl, which is plenty, so the limiting reactant is $Mg(OH)_2$. When the reaction is complete, there will be 0.06 mol of HCl left over.

Use the limiting reactant to determine the maximum amount of product.

For every 1 mol of $Mg(OH)_2$, 1 mol of $MgCl_2$ is produced.

So 0.17 mol of $MgCl_2$ is produced. Use the molar mass of $MgCl_2$ to determine the mass of $MgCl_2$.

$$\frac{95.2 \text{ g } MgCl_2}{1 \text{ mol}} = \frac{x \text{ g } MgCl_2}{0.17 \text{ mol}}$$

$$0.17 \text{ mol} \cdot \frac{95.2 \text{ g}}{1 \text{ mol}} = x$$

$$x = 16 \text{ g } MgCl_2$$

Example 2

Percent Yield

Suppose you ran the reaction from Example 1 again, this time starting with 35.0 g $Mg(OH)_2$ and the same amount of HCl as before. In the laboratory, 15.4 g of $MgCl_2$ are produced. What is the percent yield of your reaction?

Solution

First, determine the number of moles of each reactant you have. This will make it possible to identify the limiting reactant.

$$\frac{58.3 \text{ g}}{1 \text{ mol Mg(OH)}_2} = \frac{35.0 \text{ g}}{x \text{ mol Mg(OH)}_2}$$

$$x = 0.60 \text{ mol Mg(OH)}_2$$

$$\frac{y \text{ mol HCl}}{0.100 \text{ L}} = \frac{4.0 \text{ mol}}{1 \text{ L}}$$

$$y = 0.40 \text{ mol HCl}$$

The mole ratio of reactants is 1:2, so you need 1.2 mol of HCl to react with 0.60 mol of $Mg(OH)_2$. You have only 0.40 mol of HCl available, so this time HCl is the limiting reagent.

It takes 2 mol of HCl to produce each mole of $MgCl_2$, so you can expect to produce 0.20 mol of $MgCl_2$: $\left(\frac{2}{1} = \frac{0.40}{0.20}\right)$.

Next, use the molar mass of $MgCl_2$ to calculate the theoretical yield.

$$\frac{x \text{ g MgCl}_2}{0.20 \text{ mol}} = \frac{95.2 \text{ g}}{1 \text{ mol}}$$

$$x = \frac{95.2 \text{ g}}{1 \text{ mol}} \cdot 0.20 \text{ mol}$$

$$x = 19 \text{ g}$$

Now you can calculate the percent yield for the reaction you ran. Percent yield is yield per 100, or yield/100.

$$\frac{\text{yield}}{100} = \frac{\text{actual yield}}{\text{theoretical yield}}$$

$$\text{yield} = \frac{15.4 \text{ g}}{19 \text{ g}} \cdot 100$$

Percent yield = 81%

Your percent yield was significantly lower than 100%.

Practice Exercises

1. What ratios and proportions appear in Examples 1 and 2?

2. Write each expression as a fraction.

a. 68%

b. 3:4

c. the ratio of volume to temperature

d. grams per cubic centimeter

3. Complete these calculations.

a. $\frac{2}{3} + \frac{1}{5}$ **b.** $\frac{4}{5} - \frac{1}{6}$

c. $\frac{1}{3} \cdot \frac{9}{11}$ **d.** $\frac{3}{8} \div \frac{1}{4}$

Answers

1. $\frac{10\text{ g}}{58.3\text{ g/mol}}$, 1:2, $\frac{95.3\text{ g } MgCl_2}{1\text{ mol}} = \frac{x\text{ g } MgCl_2}{0.17\text{ mol}}$, $\frac{58.3\text{ g}}{1\text{ mol } Mg(OH)_2} = \frac{35.0\text{ g}}{x\text{ mol } Mg(OH)_2}$

$x = 0.60$ mol $Mg(OH)_2$, $\frac{15.4\text{ g}}{19.4\text{ g}}$

2a. $\frac{68}{100}$ **2b.** $\frac{3}{4}$ **2c.** volume/temperature, or $\frac{V}{T}$ **2d.** $\frac{\text{g}}{\text{cm}^3}$

3a. $\frac{13}{15}$ **3b.** $\frac{19}{30}$ **3c.** $\frac{3}{11}$ **3d.** $\frac{3}{2}$, or $1\frac{1}{2}$

Scientific Notation

The mass of a hydrogen atom is 0.0000000000000000000000014 g. The number of atoms in 1 mol of carbon is 602,000,000,000,000,000,000,000. To make these numbers easier to read, compare, and use in calculations, scientists use scientific notation. The long numbers above can be written as 1.4×10^{-24} and 6.02×10^{23}.

Each number is written as a decimal with one digit to the left of the decimal point times a power of 10. A number written in scientific notation has the form $a \times 10^n$, where $1 \leq a < 10$ or $-10 < a \leq -1$, and n is an integer. In other words, if a is a positive number, it is greater than or equal to 1 and less than 10. If a is a negative number, it is less than or equal to -1 and greater than -10.

Use the properties of exponents as you combine numbers written in scientific notation.

- For addition and subtraction, convert all the numbers so that the powers of 10 you are combining are the same. (The converted numbers might not be in scientific notation.)
- For multiplication, add the exponents. For any values of b, m, and n, $b^m \cdot b^n = b^{m+n}$.
- For division, subtract the exponents. For any nonzero value of b, and any values of m and n, $\frac{b^m}{b^n} = b^{m-n}$.

Example 1

Billions

Write these numbers in scientific notation.

a. 2,110,000,000

b. 0.0074

Solution

a. Move the decimal point so that there is only one digit to the left of it, in this case 2. The decimal point moved nine places. The number in scientific notation is 2.11×10^9. This means that to get the number in standard notation you start with 2.11 and move the decimal nine places to the right.

b. Move the decimal after the 7 to get 7.4. From here, the decimal will need to move four places to the left, so the number in scientific notation is 7.4×10^{-4}.

Example 2

Operations with Scientific Notation

Calculate these values.

a. $(3.2 \times 10^5)(4.0 \times 10^{-8})$

b. $\dfrac{1.4 \times 10^3}{2.8 \times 10^5}$

c. $(3.2 \times 10^5) + (4.0 \times 10^4)$

Solutions

a. 1.28×10^{-2}. Multiplying the decimals gives 12.8. The product of the powers of 10 is 10^{-3}. To put this into scientific notation, divide 12.8 by 10; to keep things balanced, increase the power of 10 by a factor of 10 to -2.

b. 5.0×10^7. The division of the numbers gives 0.50. To divide the powers of 10, subtract the exponents: $3 - (-5)$, or 8. To rewrite in scientific notation, multiply 0.50 by 10, so subtract 1 from the power of 10.

c. 3.6×10^5. Change the second number to 0.4×10^5 before you add.

Practice Exercises

1. Use scientific notation to write the speed of light, 299,792,458 meters per second, accurate to five digits.
2. How many zeros would follow the final 2 if 6.022×10^{23} were written without scientific notation?
3. Calculate these values.

a. $(3.0 \times 10^{14})(4.0 \times 10^{-4})$ b. $\dfrac{2.8 \times 10^{-31}}{7 \times 10^{-28}}$

c. $(1.21 \times 10^{-4})(4.18 \times 10^4)$ d. $(3.61 \times 10^7) - (2.5 \times 10^6)$

Answers

1. 2.9979×10^8 **2.** 20

3a. 1.2×10^{11} **3b.** (4×10^{-4}) **3c.** 5.06×10^0, or 5.06   **3d.** (3.36×10^7)

Dimensional Analysis

As you work on problems that involve numbers with units of measurement, it is convenient to consider the units (or dimensions) as factors. For example, you might want to change 0.002 kilometer to millimeters. You can always multiply by 1 without changing a value, so to convert between units you can multiply by a unit fraction. A unit fraction's value is 1 because its numerator and denominator are equivalent. Some unit fractions are shown here.

$$\frac{1000 \text{ m}}{1 \text{ km}} = 1 \qquad \frac{1000 \text{ mm}}{1 \text{ m}} = 1 \qquad \frac{1 \text{ m}}{1000 \text{ mm}} = 1$$

To convert from kilometers to millimeters, use the unit fraction with meters in the numerator and the equivalent kilometers in the denominator and a unit fraction with millimeters in the numerator and equivalent meters in the denominator.

$$0.002 \text{ km} \cdot \frac{1000 \text{ m}}{1 \text{ km}} \cdot \frac{1000 \text{ mm}}{1 \text{ m}} = 2000 \text{ mm}$$

You can see that the unit fractions have been chosen so that most of the units cancel out. The answer is then in millimeters as they are the only remaining units.

Dimensional analysis is also helpful when you need to convert between two different systems of measurement. For example, 1 inch is equivalent to 2.54 centimeters. To convert from centimeters to inches, use the unit fraction $\frac{1 \text{ in.}}{2.54 \text{ cm}}$. To convert from inches to centimeters, use the unit fraction $\frac{2.54 \text{ cm}}{1 \text{ in}}$.

Example 1

Calculating Volume

What is the volume of a box 15 cm by 10 cm by 5.12 in.?

Solution

The volume is the product of the three dimensions, but you need to include a factor to convert the inches to centimeters.

$$15 \text{ cm} \cdot 10 \text{ cm} \cdot 5.12 \text{ in.} \cdot \frac{2.54 \text{ cm}}{1 \text{ in.}} \approx 1950 \text{ cm}^3$$

If the problem specified the answer in cubic inches, you would have used this product.

$$15 \text{ cm} \frac{1 \text{ in.}}{2.54 \text{ cm}} \cdot 10 \text{ cm} \cdot \frac{1 \text{ in.}}{2.54 \text{ cm}} \cdot 5.12 \text{ in} \approx 119 \text{ in}^3$$

Example 2

Speed

A radio-controlled car travels 30 feet across the room in 1.6 seconds. How fast is it traveling in miles per hour?

Solution

From the information, the rate of the car is $\frac{30 \text{ ft}}{1.6 \text{ s}}$. Multiply by unit fractions such as $\frac{60 \text{ s}}{1 \text{ min}}$ to go from seconds to minutes to hours and from feet to miles. These fractions are chosen so that the final units are miles per hour.

$$\frac{30 \text{ ft}}{1.6 \text{ s}} \cdot \frac{60 \text{ s}}{1 \text{ min}} \cdot \frac{60 \text{ min}}{1 \text{ h}} \cdot \frac{1 \text{ mi}}{5{,}280 \text{ ft}} = \frac{108{,}000 \text{ mi}}{8{,}448 \text{ mi}} \approx \frac{12.8 \text{ mi}}{1 \text{ h}}$$

or 12.8 miles per hour

Practice Exercises

1. Show how you would use dimensional analysis to convert 6 cm/s to km/h.

2. Use dimensional analysis to change

   a. 50 meters per second to kilometers per hour

   b. 0.025 day to seconds

   c. the speed 60 miles per hour to kilometers per hour? (1609 meters = 1 mile)

3. The equation for the universal gas law is $PV = nRT$. Suppose the pressure P is in pascals (Pa), the volume V is in cubic meters (m^3), the amount of substance n is in moles (mol), and the temperature T is in kelvins (K). What must be the units of the universal gas law constant R?

Answers

1. $\frac{6\ \text{cm}}{\text{s}} \cdot \frac{1\ \text{m}}{100\ \text{cm}} \cdot \frac{1\ \text{km}}{1000\ \text{m}} \cdot \frac{60\ \text{s}}{1\ \text{min}} \cdot \frac{60\ \text{min}}{1\ \text{hr}} = 0.216\ \text{km/hr}$

2a. 180 km/h **2b.** 2160 s **2c.** 97 km/h **3.** $\frac{\text{m}^3 \cdot \text{Pa}}{\text{mol} \cdot \text{K}}$

Logarithms

Logarithms can be used to solve problems involving exponential functions, such as the half-life of a radioactive substance or the pH of a solution.

A *logarithm* is an exponent and is abbreviated *log.* For example, $\log_{10}x$ is the exponent you put on 10 to get x. So, if $x = 100$, then $\log_{10}x = 2$ because $10^2 = 100$. In the same way, $\log_{10}1000 = 3$, $\log_{10}10 = 1$, $\log_{10}1 = 0$, $\log_{10}\frac{1}{10} = -1$. Ten is a common base for logarithms, so $\log x$ is called a *common logarithm* and is shorthand for $\log_{10}x$. Check out these log equivalents on your calculator, and use your calculator to determine the logs of numbers that are not multiples of 10.

Because logarithms are exponents, they follow the properties of exponents.

Product property	$a^m \cdot a^n = a^{m+n}$	$\log xy = \log x + \log y$
Quotient property	$a^m/a^n = a^{m-n}$	$\log x/y = \log x - \log y$
Power property	$\log x^n = n \log x$	

In some problems you know what the logarithm is, but you want to find the number that has that logarithm. What you are looking for is called the *antilogarithm*, abbreviated antilog. Use the 10^x key on your calculator to find an antilog.

Example 1

Comparing Logarithms

Find the values a–f. Which values are equal?

$a = \log 18$ $b = \log 71$ $c = a + b$ $d = \text{antilog } c$ $e = 18 \cdot 71$ $f = \log(18 \cdot 71)$

Solution

$a \approx 1.255$; $b \approx 1.851$; $c \approx 3.106$; $d = 1278$; $e = 1278$; $f \approx 3.106$

$d = e$, or antilog $(\log 18 + \log 71) = 18 \cdot 71$

$c = f$, or $\log 18 + \log 71 = \log(18 \cdot 71)$

Example 2

pH

What is the pH of a solution in which $[H^+]$ is 5.3×10^{-4}?

Solution

Start with the definition of pH.	$pH = -\log [H^+]$
Substitute the value of $[H^+]$.	$= -\log [5.3 \times 10^{-4}]$
Find the log, then change the sign.	$= 3.35$

Practice Exercises

1. Find the log.

a. 456

b. $\frac{x}{15}$

c. 3.4×10^{-6}

2. Find the antilog

a. 2 **b.** -3 **c.** 3.4

3. The pH of a solution is 8.25. What is the concentration of hydrogen ions in the solution?

Answers

1a. 2.659 **1b.** $\log x - 1.176$ **1c.** 5.469

2a. 100 **2b.** $\frac{1}{1000}$ **2c.** 2512 **3.** 5.62×10^{-9}

Answers to Selected Exercises

Unit 1: Alchemy

Lesson 1

1. Possible answer: We use glass because substances in a glass container are visible and glass containers are relatively easy to clean and reuse. Tempered glass containers can be heated over flames without shattering.
3. Possible answers: Know the location of safety equipment. Read lab instructions carefully. Check to be sure that you are using the right chemicals and equipment.
5. Possible answers: Put all equipment in its proper place. Clean your work area. Make sure all bottles and containers holding chemicals are closed and stored properly.

Lesson 2

1. Alchemists developed some of the first laboratory tools and chemistry techniques. They classified substances into categories and experimented with mixing and heating different substances to create something new.
5. Observable changes that involve chemistry often involve an alteration in the appearance of matter. Examples include metal rusting, cookies baking, and ice cubes forming. Changes that do not involve chemistry only involve matter moving to a different location. Examples include the Sun going down, objects falling, and hands moving on a clock.

Lesson 3

1. Mass is the amount of material in an object. Volume is the amount of space the object takes up.
5. Possible answer: Examples of things that are matter: a car, a tree, a person, water. Examples of things that are not matter: sound, movement, feelings, energy.

Lesson 4

1. Measure the dimensions of the object and calculate its volume using a geometric formula or, if it does not float or dissolve, measure the amount of liquid that it displaces when it is submerged.
3. Yes, the volume of an object is the amount of space it fills. You can usually see how much space an object fills, and estimate its volume based on its dimensions. However, for objects that have an irregular shape, are very thin, or have a surface with lots of holes or pits, determining the volume of the object by sight may be difficult.
5. Yes, the mass of the rubber band is the same because only the shape of the rubber band changes, not the amount of matter in it.
11. Possible answer: The balance or scale that was used to measure the two different cubes may not be able to measure accurately to a hundredth of a gram. The two cubes could have exactly the same mass because the measurement error is greater than 0.03 g.

Lesson 5

1. Possible answer: Density is the mass of an object divided by its volume.
3. The density of aluminum is less than the density of gold. More matter is present in a given volume of gold than in the same volume of aluminum.
5. Possible answer: The object that has a density of 2.7 g/cm^3 has the larger volume. The two objects have the same mass, but the mass is packed into a smaller space in the denser object.
7. **a.** 2.87 g/cm^3 **b.** 111 g, 38.7 cm^3, 2.87 g/cm^3
   c. The density is the same.

Section I Review Questions

1. Possible answer: Determine the volume of a powdered solid or of a liquid by pouring the substance into a graduated cylinder or beaker and reading the markings on the side. Determine the volume of a rock by submerging the rock in a graduated cylinder partially filled with water and then reading how much the water level changes.
3. Density is no help in determining which object will displace more water. A large object will displace more water than a small object no matter how dense the two objects are.

Lesson 6

1. Elements are the building materials of all matter. A compound is matter that is made up of two or more elements combined in a specific ratio. An element cannot be broken down into simpler substances by chemical means, but a compound can be broken down into elements.

3. Sodium nitrate contains three elements: sodium, nitrogen, and oxygen.
5. The chemical formula for cubic zirconia is ZiO_2. The chemical formula for a diamond is C. The stone cannot be a diamond because it has a different chemical formula than a diamond.

Lesson 7

1. Possible answer: A chemical reaction is a change leading to the final substance or substances being different from the original substance or substances. Some of the signs that a new substance has formed include color changes, formation of a new solid or gas, and the release of energy as heat or light.
3. Possible answer: A chemical reaction combined the zinc with part of the dissolved copper compound. The resulting zinc compound was also dissolved in water. The zinc was still present, but not as a pure element.
5. a. The baking soda is a solid, the vinegar is a liquid, the clear colorless liquid is a liquid, and the CO_2 is a gas.
   b. Yes, the production of carbon dioxide gas is evidence that a chemical change has occurred after the original liquid and solid substances were mixed.
   c. Before the change, the sodium is in the solid baking soda. After the change, the sodium must be in the clear colorless liquid.

Lesson 8

1. Possible answer: The chemical names and symbols indicate what compounds were combined in each step. Because matter is conserved, the products must contain the same elements as the original compounds, which enables you to make a reasonable guess about the products. For example, when sulfuric acid combines with copper oxide, it is likely that copper sulfate and water are the products.
5. The solution would be yellow because the combination of nickel, Ni, and hydrochloric acid, HCl, can only produce a solution containing compounds with nickel, hydrogen, and chlorine. Nickel chloride, $NiCl_2$, is a possible product. Nickel sulfate, $NiSO_4$ is not.

Lesson 9

1. Three useful properties for sorting elements are reactivity, formulas of their compounds, and atomic mass.
5. a. CaS
   b. The compound with sulfur will have more mass for a given amount of calcium. The two compounds have the same number of atoms, but the atomic mass of sulfur is 32, and the atomic mass of oxygen is only 16.

Lesson 10

1. Within Group 1A, the elements tend to get more reactive as you move from the top of the column to the bottom.
5. b. titanium c. lead e. potassium f. silicon
7. Elements c, copper, and f, mercury, are the least reactive. On the periodic table the least reactive elements (aside from the noble gases) are the transition metals that are located in the center of the table. The other elements listed are from more reactive groups near the edge of the table: alkali metals (potassium and rubidium), alkaline earth metals (barium), and halogens (chlorine).

Section II Review Questions

1. Gold, represented by the symbol Au, is a transition metal that is a solid at room temperature. It has an atomic number equal to 79 and an average atomic mass of 196.97. Gold is nonreactive, a good conductor of heat, and a good conductor of electricity. It has properties similar to those of copper and silver.
3. Possible answer: A chemical formula is a symbol that represents a compound. The chemical formula shows what elements are in the compound and the ratio in which the elements combined. It can also show what physical form the compound is in.

Lesson 11

1. When Thomson zapped atoms with electricity, he found that a negatively charged particle was removed. Because the solid sphere model does not allow for particles splitting off atoms, he created the plum pudding model.
3. Bohr revised the nuclear model of the atom when he noticed different atoms giving off different colors of light when exposed to flame or electric fields. Because the nuclear model fails to account for this process, he created the solar system model.
5. Possible answer: The two types of atoms could have different amounts of positive fluid and different numbers of electrons. Each atom would still have a net charge of zero.
11. No, the Greeks were not correct. It is now known that atoms are made up of smaller particles such as electrons, protons, and neutrons. Each of these particles has a mass and a volume, so they are matter.

Lesson 12

1. The atomic number indicates the number of protons in the nucleus of an atom.
3. magnesium
5. The atomic mass is the sum of the number of protons and the number of neutrons in the nucleus of an atom.

Although boron and carbon each have six neutrons, carbon has six protons while boron has only five protons.

Lesson 13

1. Atomic number refers to the number of protons in the nucleus of an atom. Atomic mass, when expressed in amu, is the sum of the number of protons and the number of neutrons.
3. Possible answer: The isotopes differ from one another in the number of neutrons in their nuclei. The isotopes are $^{39}_{19}K$, $^{40}_{19}K$, $^{41}_{19}K$.
5. **a.** 58 amu **b.** $^{58}_{26}Fe$
7. 35.48 amu
9. **B.** Because nitrogen has an atomic number of 7 on the periodic table, any possible isotopes must have an atomic number of 7.

Lesson 14

1. Possible answer: An element is a fundamental building block of matter. An atom is the smallest possible unit of an element. An atom is the smallest unit of an element that still has the same characteristics as the element.
3. Oxygen has three stable isotopes. Neodymium has five stable isotopes. Copper has two stable isotopes. Tin has ten stable isotopes.
5. The diagonal line on the graph represents isotopes that have equal numbers of protons and neutrons, because the line passes through points that have the same *x*- and *y*-coordinates.
7. No, because no isotope is indicated on the graph of isotopes of the first 95 elements that has 31 protons and 31 neutrons.
11. no

Lesson 15

1. A nuclear reaction is a change in the nucleus of an atom.
3. Possible answer: Gamma radiation is the most harmful because it has the most power to penetrate living tissues and cause damage.
5. The mass number of the atom does not change during beta decay because the nucleus loses one electron, which has a mass that is only a tiny fraction of the total mass of the nucleus.
7. **a.** calcium-42, $^{42}_{20}Ca$✓ **b.** xenon-131, $^{131}_{54}Xe$✓ **c.** magnesium-24, $^{24}_{12}Mg$✓ **d.** cobalt-52, $^{52}_{27}Co$

Lesson 16

1. Possible answers: In alpha decay, a nucleus emits a particle consisting of two protons and two neutrons. The atomic number decreases by 2 and the atomic mass decreases by 4. In beta decay, a nucleus emits an electron. The atomic number increases by 1 as one of the neutrons becomes a proton. The atomic mass does not change. In nuclear fission, a nucleus splits apart to form the nuclei of two or more lighter elements. In nuclear fusion, two nuclei combine to form the nucleus of a heavier element.
3. $^{141}_{58}Ce \longrightarrow {}^{0}_{-1}e + {}^{141}_{59}Pr$
5. Possible answer: $^{50}_{24}Cr + {}^{4}_{2}He \longrightarrow {}^{54}_{26}Fe$

Section III Review Questions

1. Possible answer: An atom changes identity when the number of protons in its nucleus changes. Processes in which this can occur are radioactive decay, fission, and fusion. In radioactive decay, emission of an alpha particle decreases the atomic number by 2, while emission of a beta particle increases the atomic number by 1. Nuclear fission is the process in which a nucleus breaks apart, forming two or more smaller nuclei. In nuclear fusion, two nuclei join to form one larger nucleus.

Lesson 17

1. The color of the flame produced during a flame test is a characteristic of particular metallic elements. When a compound containing one of the metallic elements is heated, its atoms emit light of a specific color.
5. Possible answer: Red fireworks could contain lithium chloride, and purple fireworks could contain a mixture of lithium chloride and copper chloride. Red fireworks could contain lithium sulfate, and purple fireworks could contain a mixture of lithium sulfate and copper sulfate.
7. No, the flame color of each nitrate compound is different and matches the flame color of the metal in the compound. This indicates that the nitrate is not responsible for the color of the flame.
9. **a.** yellow-orange **b.** green **c.** pink-lilac **d.** pink-lilac **e.** green

Lesson 18

1. For main group elements, the number of shells containing electrons is equal to the period number. All of the shells, except the highest, are completely filled. The group number indicates the number of electrons in the outermost shell.
3. Possible answer: Beryllium, magnesium, and calcium are all alkaline earth metals. They are located in Group 2A, so their atoms all have two valence electrons.
5. The number of core electrons does not change across a period.
7. **a.** Element number 17 is chlorine. It has the chemical symbol Cl and is located in Group 7A. This

information comes directly from square 17 on the periodic table.

b. The nucleus contains 17 protons. The number of protons is equal to the atomic number.

c. Possible answer: The nucleus can contain either 18 or 20 neutrons. This information is given in the graph of isotopes of the first 95 elements.

d. The number of electrons in a neutral atom of chlorine is 17. In a neutral atom, the number of electrons equals the number of protons.

e. Chlorine has 7 valence electrons. The number of valence electrons is equal to the group number.

f. Chlorine has 10 core electrons. The number of core electrons equals the difference between the total number of electrons and the number of valence electrons, or 17 − 7.

g. Possible answer (any 3 elements): Fluorine, bromine, iodine, and astatine all have the same number of valence electrons. All of these elements are in the same Group, 7A, as chlorine.

Lesson 19

1. A cation is an ion that has a positive charge. An anion is an ion that has a negative charge.

3. 2 electrons, 3 protons, and either 3 or 4 neutrons

7. The noble gas closest to sulfur is argon. A sulfur atom gains two electrons to have an electron arrangement similar to that of argon.

9. Possible answers: S^{2-}, Cl^{-}, K^{+}, Ca^{2+}

11. Ti^{4+}

13. Elements on the right side of the table gain electrons to have a noble gas arrangement. They do not tend to lose electrons because the charge would be too large.

Lesson 20

1. The number of valence electrons can predict whether an atom will form a cation or an anion, as well as the size of the charge on the ion. Ionic compounds form between cations and anions in a ratio so that the charges are balanced.

3. a. +1 **b.** −3

c. Lithium nitride has three lithium ions with charge +1 and one nitride ion with charge −3. $3(+1) + (-3) = 3 + (-3) = 0$

d. 8 valence electrons

5. a. KBr has one potassium ion with charge +1 and one bromide ion with charge −1.
$+1 + (-1) = 0$

b. CaO has one calcium ion with charge +2 and one oxide ion with charge −2.
$+2 + (-2) = 0$

c. Li_2O has two lithium ions with charge +1 and one oxide ion with charge −2.
$2(+1) + (-2) = 2 + (-2) = 0$

d. $CaCl_2$ has one calcium ion with charge +2 and two chloride ions with charge −1.
$+2 + 2(-1) = 2 + (-2) = 0$

e. $AlCl_3$ has one aluminum ion with charge +3 and three chloride ions with charge −1.
$+3 + 3(-1) = 3 + (-3) = 0$

7. a. $NaCl_2$ does not form because it has a net charge of −1. The sodium ion has a charge of +1 and each chloride ion has a charge of −1.

b. CaCl does not form because it has a net charge of +1. The calcium ion has a charge of +2 and the chloride ion has a charge of −1.

c. AlO does not form because it has a net charge of +1. The aluminum ion has a charge of +3 and the oxide ion has a charge of −2.

Lesson 21

1. For main group elements, the group number shows the number of valence electrons. Metal atoms lose all of their valence electrons when they form an ion, adding a positive charge for each electron lost. Nonmetal atoms gain enough electrons to have eight valence electrons, adding one negative charge for each electron gained.

3. a. LiCl is possible because the total of the charges on the ions equals zero. The lithium ion has a charge of +1 and the chloride ion has a charge of −1.
b. $LiCl_2$ is not possible because the total of the charges on the ions equals −1. **c.** MgCl is not possible because the total of the charges on the ions equals +1. Magnesium ions have a charge of +2. **d.** $MgCl_2$ is possible because the total of the charges on the ions equals zero. **e.** $AlCl_3$ is possible because the total of the charges on the ions equals zero. Aluminum ions have a charge of +3.

7. a. $AlBr_3$, aluminum bromide
b. Al_2S_3, aluminum sulfide
c. AlAs, aluminum arsenide
d. Na_2S, sodium sulfide
e. CaS, calcium sulfide
f. Ga_2S_3, gallium sulfide

Lesson 22

1. A polyatomic ion is an ion that consists of two or more elements.

3. a. ammonium chloride **b.** potassium sulfate
c. aluminum hydroxide **d.** magnesium carbonate

5. −1

Lesson 23

1. The Roman numeral indicates the charge on the transition metal cation in the compound.

3. a. +2, mercury (II) sulfide **b.** +2, copper (II) carbonate **c.** +2, nickel (II) chloride **d.** +3, cobalt (III) nitrate **e.** +2, copper (II) hydroxide **f.** +2 iron (II) sulfate

5. $Co_3(PO_4)_2$

Lesson 24

1. Electron subshells are divisions within a specific electron shell of an atom.

3. As the number of electrons in an element increases, they are added in a specific sequence that is illustrated by the position of the element on the periodic table. Each section of the table corresponds to a particular subshell of electrons.

5. 15 subshells

7. 4p

9. a. $1s^22s^22p^63s^23p^1$ **b.** Element 13 has 3 valence electrons because it has three electrons in the outer shell, $n = 3$. **c.** Element 13 has 10 core electrons because it has two shells filled completely: $n = 1$ with 2 electrons and $n = 2$ with 8 electrons.

11. a. $1s^22s^22p^4$, [He] $2s^22p^4$
b. $1s^22s^22p^63s^23p^5$, [Ne] $3s^23p^5$
c. $1s^22s^22p^63s^23p^64s^23d^6$, [Ar] $4s^23d^6$
d. $1s^22s^22p^63s^23p^64s^2$, [Ar] $4s^2$
e. $1s^22s^22p^63s^2$, [Ne] $3s^2$
f. $1s^22s^22p^63s^23p^64s^23d^{10}4p^65s^24d^9$, [Kr] $5s^24d^9$
g. $1s^22s^22p^63s^23p^2$, [Ne] $3s^23p^2$
h. $1s^22s^22p^63s^23p^64s^23d^{10}4p^65s^24d^{10}5p^66s^24f^{14}5d^{10}$, [Xe] $6s^24f^{14}5d^{10}$

13. a. chromium **b.** silicon **c.** nitrogen **d.** cesium **e.** lead **f.** silver

Section IV Review Questions

1. Valence electrons are important because they are in the outermost electron shell of an atom and will interact with other atoms. This interaction is what determines the properties of an element.

3. As the number of electrons in an element increases, they are added in a specific sequence that is illustrated by the position of the element on the periodic table. Each section of the table corresponds to a particular subshell of electrons.

5. a. The anion is Mg^{2+} and has a charge of +2. The cation is Cl^- and has a charge of −1.
b. The anion is Ca^{2+} and has a charge of +2. The cation is NO_2^- and has a charge of −1.

Lesson 25

1. A substance is insoluble if it fails to dissolve in a particular solvent.

3. Possible answer: The substance is most likely an ionic compound. Many compounds dissolve in water but electrical conductivity is a characteristic of compounds that separate into ions, such as ionic compounds.

5. No, though ionic compounds generally do not conduct electricity as solids, they do conduct electricity as aqueous solutions.

Lesson 26

1. The atoms that make up substances are held together by chemical bonds. The bond is an attraction between the positively charged nuclei of atoms and the valence electrons of other atoms.

3. a. metallic **b.** molecular covalent **c.** ionic

5. NO_2 is a gas that is made up of nonmetal atoms, so it has molecular covalent bonds.

7. Possible answer: Drop the mixture into a beaker of water. The sodium chloride will dissolve in the water, but the carbon will not.

9. Carbon is a solid because many carbon atoms are held together in a large array by network covalent bonds.

Lesson 27

1. Possible answers: finding pure metals in nature, heating ionic compounds to separate the metal, extracting metals with electricity

3. Attach the coated object to the positive terminal of a battery in an electroplating circuit.

5. The copper sulfate solution is composed of cations (Cu^{2+}) and anions (SO_4^-) dissolved in water. The only thing that is added to the solution during the experiment is a stream of electrons. When the electrons are added to the copper ions, copper atoms (Cu) are formed on the metal strip. This indicates that the Cu^{2+} ions are simply copper ions that are missing electrons.

7. Possible answer: Nickel cannot change into copper unless the number of protons in the nucleus changes. The plating apparatus only adds electrons to the nickel strip, causing the plating to occur. Adding electrons does not change the nucleus, so it cannot change the identity of the atoms.

Section V Review Questions

1. Possible answer: While it is not practical to try to make gold, many substances that are quite valuable can be made through chemistry.

Unit I Review Exercises

1. Possible answer: An element is the basic building block of compounds. An element has only one type of atom, while a compound has at least two types of atoms held together by chemical bonds.

3. a. Lithium, Li, has atomic number 2 and is in Group 2A. It has 2 protons and 2 electrons. **b.** Bromine, Br, has atomic number 35 and is in Group 7A. It has 35 protons and 35 electrons. **c.** Zinc, Zn, has atomic number 30 and is in Group 2B. It has 30 protons and 30 electrons. **d.** Sulfur, S, has atomic number 16 and is in Group 6A. It has 16 protons and 16 electrons. **e.** Barium, Ba, has atomic number 56 and is in Group 2A. It has 56 protons and 56 electrons. **f.** Carbon, C, has atomic number 6 and is in Group 4A. It has 6 protons and 6 electrons.

5. An isotope is an atom of an element with a specific number of neutrons in its nucleus. Predict the most common isotope by rounding the average atomic mass of the element to the nearest whole number.

7. Cations are ions that have lost electrons, causing them to have a positive charge. Anions are ions that have gained electrons, causing them to have a negative charge.

9. a. Aluminum chloride has ionic bonding and conducts electricity in solution only. **b.** Oxygen has molecular covalent bonding and does not conduct electricity. **c.** Silver (I) hydroxide has ionic bonding and conducts electricity in solution only. **d.** Platinum has metallic bonding and conducts electricity.

11. A material that does not dissolve in water and does not conduct electricity is held together by network covalent bonds or molecular covalent bonds.

Unit 2: Smells

Lesson 1

1. Possible answer: Scientists classify smells by placing similar types of smells in a category. This allows scientists to talk about smells in a consistent way.

7. a. Methyl octenoate probably smells sweet because it has two oxygen atoms and twice as many hydrogen atoms as carbon atoms in each molecule. The chemical name also ends in "-ate." **b.** Monoethylamine probably smells fishy because it has one nitrogen atom and no oxygen atoms in each molecule. **c.** Ethyl acetate probably smells sweet because it has two oxygen atoms and twice as many hydrogen atoms as carbon atoms in each molecule. The chemical name also ends in "-ate."

Lesson 2

1. Structural formulas show what atoms are present in a molecule and how the atoms are bonded to one another.

3. Yes, the structural formula shows all of the atoms in a molecule, enabling use of the structural formula to determine the molecular formula.

5. a. $C_3H_6O_2$ **b.** $C_5H_{10}O$ **c.** $C_5H_{10}O$ **d.** $C_4H_8O_2$ **e.** $C_4H_{11}N$ **f.** $C_2H_4O_2$

Lesson 3

1. The HONC 1234 rule describes the bonding patterns in molecules. Hydrogen forms one bond with other atoms. Oxygen forms two bonds with other atoms. Nitrogen forms three bonds with other atoms. Carbon forms four bonds with other atoms.

3. Possible answers:

```
             H   H   H
             |   |   |
a. H—O—C—C—C—O—H
             |   |   |
             H   H   H

             H     H
              \   /
                N
        H   H   |   H
        |   |   |   |
b. H—C—C—C—C—H
        |   |   |   |
        H   H   H   H

        H   H   H   H
        |   |   |   |
c. H—C—C—C—C—H
        |   |   |   |
        H   H   H   H

        H   H   H   H   H
        |   |   |   |   |
d. H—C—C—C—C—C—O—H
        |   |   |   |   |
        H   H   H   H   H
```

5. Possible answer: Atoms on the molecules might react with smell receptors inside the nose. Different receptors may detect different types of molecular structures, leading to different sensations of smell.

Lesson 4

1. Possible answer: In an ionic bond, a valence electron is transferred from one metal atom to a nonmetal atom. In a covalent bond, two nonmetal atoms share a pair of electrons. Ionic and covalent bonds are similar because both involve the valence electrons of the atoms that are bonded together. The result of the bond is that the two atoms have an outer shell that is filled. In both types of bonds, electrical forces hold the atoms together.

3. a.

	K·	In·	·Pb·	·Bi·	·Te:	:I:
Group	1	3	4	5	6	7

b. two covalent bonds: Te, one covalent bond: I, does not form covalent bonds: K, In, Pb, Bi

5.

```
      ..              ..        ..
a.   :Cl         b. :I:H   c. :Br:
      ..  ..         ..        ..  ..
    :Te:Cl:                  :As:Br:
      ..  ..                   ..  ..
                              :Br:
                               ..
       ..
      :F:
   ..  ..  ..          ..  ..
d. :F:Si:F:       e. :F:F:
   ..  ..  ..          ..  ..
      :F:
       ..
```

Lesson 5 Exercises

1. Possible answer: Nitrogen has three unpaired electrons, as shown in the Lewis dot structure. Hydrogen atoms have one unpaired electron, so three hydrogen atoms can form bonds with one nitrogen, giving the nitrogen an octet of electrons and each hydrogen two electrons. In NH_2, nitrogen would have only seven electrons, and in NH_4, an extra hydrogen atom is left over after all the unpaired electrons in the nitrogen atom have formed bonds.

```
                                  H
                                  ..
 ..      ..       ..       ..     ..
·N·     ·N:H     H:N:H    H:N:H·
 ·       ..       ..       ..
         H        H        H
```

3. Possible answer: hydrogen, chlorine, or another fluorine

```
5. a.     ..           b.    H          c.      H
         :F:                 ..     ..     ..   ..   ..
       .. .. ..            H:C:Cl:       :Cl:Si:Cl:
      :F:C:F:                ..  ..        ..   ..   ..
       .. .. ..              H                  H
         :F:
          ..                      ..  ..
   d.    H               e. H:O:Cl:      f.    H H
         ..  ..                ..  ..          .. ..
       H:C:O:H                               H:C:N:
         ..  ..                                .. ..
         H                                     H H
```

9. CH_4 would form a stable compound because all of the atoms in the molecule are surrounded by the most stable number of valence electrons—eight for carbon, two for hydrogen. In CH_3, the carbon atom would have only seven electrons in its outer shell.

Lesson 6

1. A functional group is a portion of a molecular structure that is the same in all molecules of a certain type.

3. C_2H_4 has fewer hydrogen atoms than C_2H_6 because the two carbon atoms are held together by a double bond. That means there are only two unpaired electrons in each carbon atom available to form bonds with hydrogen.

5. a. $C_6H_{14}O$

```
        H   H   H   H   H   H
        |   |   |   |   |   |
b. H—C—C—C—C—C—C—O—H
        |   |   |   |   |   |
        H   H   H   H   H   H
```

c. The structural formula is more useful because it shows the –OH functional group that is characteristic of alcohols, and alcohols have a characteristic smell.

11. a. $C_3H_6O_2$ **b.** $C_5H_{10}O$ **c.** $C_5H_{10}O$ **d.** $C_4H_8O_2$
e. $C_4H_{11}N$ **f.** $C_2H_4O_2$

Lesson 7

1. Combine the alcohol and acid with a strong acid and heat the mixture.

3. Possible answer: When butyric acid is heated with methanol, the two compounds react to form a new compound, which is a sweet-smelling ester. The butyric acid, which has a foul smell, is no longer present.

5. Possible answer: Combine the acetic acid with an alcohol and sulfuric acid and then heat the mixture. The hydrogen atom of the acid is replaced by the part of the alcohol molecule that is attached to the –OH functional group.

Lesson 8

1. Possible answer: Converting a carboxylic acid into an ester is a way to make a new compound with more desirable properties. For example, eliminating a putrid odor is one possible goal of converting a carboxylic acid into an ester.

3. A catalyst is a chemical that is added to a reaction mixture to help get the reaction started, but is not consumed by the reaction.

5. a. $C_8H_{16}O_2$ **b.** $C_7H_{14}O_2$ **c.** $C_9H_{18}O_2$ **d.** $C_6H_{12}O_2$

Section I Review Exercises

1. a. ester **b.** ketone **c.** carboxyl **d.** amine

3. a. 4 lone pairs **b.** 12 lone pairs **c.** no lone pairs

5. Possible answer: The smell of a compound is strongly related to the functional group of the molecule.

Lesson 9

1. Possible answer: A structural formula shows the arrangement of atoms and the types of bonds within a molecule in a two-dimensional form. A ball-and-stick model adds information about the arrangement of the atoms in three dimensions.

3. 10 black balls (carbon), 18 white balls (hydrogen), 1 red ball (oxygen), 30 connectors (chemical bonds)

7. Possible answer: The three molecules geraniol, menthol, and fenchol each have ester functional groups in their structural formulas. However, each of the three molecules has a distinctive smell.

Lesson 10

1. A tetrahedral shape has four single bonds spaced equally around one central atom.

5. Possible answer: CH_4, $SiCl_4$, CF_4

7. AsH_3 will have a pyramidal shape because the three bonded pairs and the one lone pair on the arsenic atom form a tetrahedron. Therefore, the three single bonds form a pyramid shape with the arsenic atom at its top.

9.

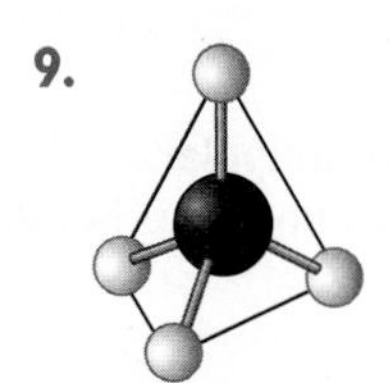
Methane

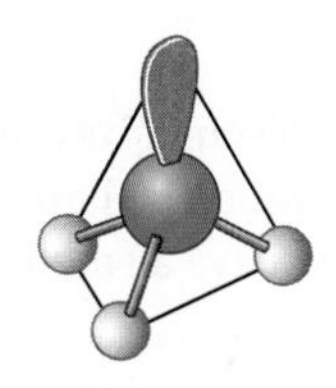
Ammonia

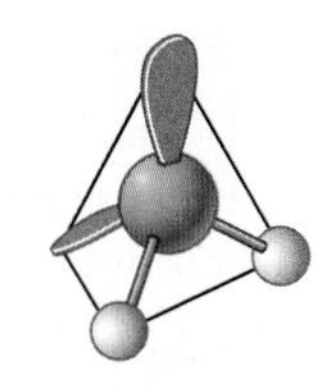
Water

Lesson 11

1. If a molecule has three electron domains, its shape will be trigonal planar.

3. Cl_2 and CO_2 are linear and H_2O is bent.

5. A molecule with two atoms will be linear. A molecule with three atoms can be either linear or bent. A molecule with four atoms can be linear, bent, trigonal planar, or pyramid-shaped. A molecule with five atoms can be linear, bent, tetrahedral, or trigonal planar or pyramidal with the fifth molecule attached to one of the triangle's or pyramid's vertices.

7. a.

```
   H H H H
   ‥ ‥ ‥ ‥
 H:C:C:C:C:H
   ‥ ‥ ‥ ‥
   H H H H
```

b. 13

c. Possible answer: a crooked, zigzag shape

d. Each carbon atom is surrounded by four electron domains. Therefore, the shape of the molecule is a series of linked tetrahedra.

Lesson 12

1. A space-filling model is a more accurate model of how the atoms of the molecule are arranged in space. Space-filling models give a better picture of the shape of the molecule than ball-and-stick models.

3. Possible answer:

```
  H   H   H   H   H   H   H   H   H
  |   |   |   |   |   |   |   |   |
H—C — C — C — C — C — C — C — C — C—O—H
  |   |   |   |   |   |   |   |   |
  H   H   H   H   H   H   H   H   H
```

$C_9H_{20}O$ is long and stringy, with a zigzag shape.

5. Possible answer:

```
      H   H
       \ /
  H     C
   \  /   \
    C       C=O
    ‖       |
    C       C—H
   / \     / \
  H    C      H   H   H
      / \         |   |
     H    H—C — C — C—H
              /    |   |
         H—C—H    H   H
             |
             H
```

$C_{10}H_{16}O$ has a ring-shaped structure of carbon atoms with a "handle." The overall shape of the molecule resembles a frying pan.

Lesson 13

1. The molecular formula of a compound can provide some information about its smell but is not useful in most cases because it fails to give information about functional groups. For alkanes and amines, the chemical formula is a good indicator of smell.

3. Possible answer: The minimum information required to determine that a molecule smells sweet is the shape and the functional group or, if it is an ester, just the functional group.

Lesson 14

1. According to the receptor site theory, the nose is lined with sites that match the shapes of molecules that have smells. When a molecule fits into this site, it stimulates nerves to send a message to the brain. The brain interprets the message as a smell.

3. Possible answer: The compound would most likely have a minty smell. The nose detects the minty smell when the molecule shown fits into a receptor site in the nose that detects minty smells. In order for this to happen, some of the rub must change phase and become a gas.

Section II Review Exercises

1. a. $C_5H_{10}O_2$

b.

```
                 ‥
   H H H        :O H
   ‥ ‥ ‥   ‥    ‥  ‥
 H:C:C:C:O:C:C:H
   ‥ ‥ ‥   ‥       ‥
   H H H           H
```

c. ester

d. The molecule probably has a sweet smell.

3. The shape of a molecule is determined by the number of electron domains around its atoms. The electron domains are all negatively charged, so they are positioned as far apart as possible.

5. Possible answer: Although the functional group appears to be a main factor that determines the smell of a molecule, its shape also affects the smell because it helps determine which receptors will be stimulated to send a signal to the brain.

Lesson 15

1. Possible answer: A polar molecule is a molecule in which the charge is not evenly distributed around the molecule. This means that different portions of the molecule will have partial electric charges.

5. Hexane would not be expected to dissolve in water because the information given indicates that it is a nonpolar compound. Compounds with nonpolar molecules do not tend to dissolve in water.

Lesson 16

1. Possible answer: In a polar covalent bond, the electron is attracted more by one atom than by the other, so one of the atoms has a partial negative charge and the other has a partial positive charge. In a nonpolar covalent bond, in which the two atoms share the electrons equally, there are no partial charges.
3. Although the two carbon-oxygen bonds of carbon dioxide are polar, the molecule itself has a linear shape, so the partial negative charges are on opposite sides of the molecule. These charges balance one another, so there is no dipole.

Lesson 17

1. Electronegativity values help determine the polarity of the bond between two atoms because they can be used to determine the tendency of an electron to be attracted to one atom rather than to another atom. The greater the difference in the electronegativities of the two atoms, the more polar the bond that forms between the atoms.
3. a. Li–F, Na–F, K–F, and Rb–F (same polarity), Cs–F; For an alkali metal bonding with fluorine, the polarity increases from the top of the group to the bottom.
   b. P–S, N–F, Al–N, Mg–O, K–Cl; The polarity of the molecules increases as the distance between the atoms on the periodic table increases.
5. No, hydrogen will have a partial positive charge only when it is bonded to an atom that has a greater electronegativity. If the hydrogen atom is bonded to an atom with a smaller electronegativity, such as boron, then the hydrogen atom will have a partial negative charge. If a hydrogen atom is bonded with another hydrogen atom, the bond will be nonpolar.
7. Possible answer: O–F, 0.54; C–H, 0.45; S–F, 1.40
9. To say that bonding is on a continuum means that the type of bonding changes gradually as the difference in electronegativity between atoms increases. There is no sharp distinction between polar covalent and ionic bonds.

Lesson 18

1. To determine whether a molecule is polar, use a Lewis dot structure or structural diagram to figure out the shape of the molecule. Then look at whether the individual bonds are polar. If the molecule is asymmetrical and has polar bonds, then it will be a polar molecule.
3. a.
```
  H
  ••
: Se : H
  ••
```
   H_2Se is a bent molecule. Because its bonds are polar and it is asymmetrical, it is likely to have a smell.

   b. H:H

   H_2 is a linear molecule. Its bond is nonpolar. Because it is symmetrical and has a nonpolar bond, it is not likely to have a smell.

   c.
```
 ••
:Ar:
 ••
```
   Because Ar consists of a single atom, it is not polar and is not likely to have a smell.

   d.
```
 ••
: F :
 ••
:O:H
 ••
```
   HOF is a bent molecule. Its bonds are polar and it is asymmetrical, so it is likely to have a smell.

   e.
```
        H
 ••  ••  ••
: F : C :Cl:
 ••  ••  ••
     : F :
      ••
```
   $CHClF_2$ is a tetrahedral molecule and it has polar bonds. Its shape is symmetrical, but because the atoms at each point in the tetrahedron are different, the bonds are not symmetrical. It is likely to have a smell.

   f.
```
H
 :C::O:
H
```
   H_2CO is a trigonal planar molecule. Its bonds are polar and it is asymmetrical, so it is likely to have a smell.
5. Possible answer: Water does not have a smell because there are no receptors in the nose that are sensitive to water molecules. If the nose had receptor sites that were sensitive to water, they would always be filled by the water in the mucous membrane.

Lesson 19

3. Decanol has a smell because it is a medium-sized molecular compound and because it is polar. Lead does not have a smell because it is a metal. Iron oxide does not have a smell because it is an ionic compound. Potassium chloride does not have a smell because it is an ionic compound.
5. Possible answer: If a substance can become a gas under normal conditions, you should be able to smell it as long as receptor sites in the nose can detect it. In general, small nonpolar molecules are odorless.
7. Possible answer: When the T-shirt comes out of the clothes dryer, it is so warm that molecules from the detergent and fabric softener are likely to have changed phase and become gases. When they reach your nose, you are able to smell them.

Section III Review Exercises

1. a. Br b. Li c. Au

3. a. 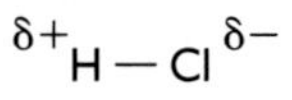

b. 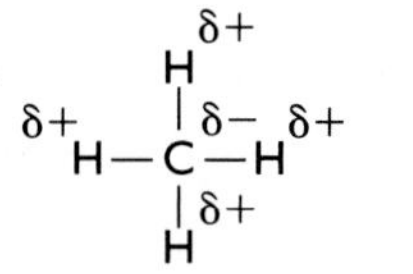

c. 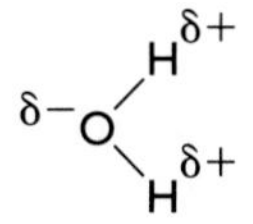

HCl and H_2O are polar molecules because they are asymmetrical and have polar bonds. CH_4 is not a polar molecule because its polar bonds are symmetrically arranged around the carbon atom.

Lesson 20

1. The mirror image of the letter "D" can be superimposed on the original as long as the mirror is placed horizontally with respect to the letter. Then the mirror image and the original image are identical.
3. Examples of objects that look different in a mirror include a glove, a written sentence, a pair of scissors. Examples of objects that look the same in a mirror include a spoon, an empty glass, and a pencil with no writing on it.
9. a. citronellol: $C_{10}H_{20}O$; geraniol: $C_{10}H_{18}O$
   b. Citronellol has two distinct smells because the structural formulas have mirror-image isomers that interact with different receptors in the nose. Citronellol has mirror-image isomers because the third carbon atom from the right in the structural formula is bonded to four different groups.
   c. The geraniol molecule has only one smell because the molecule and its mirror image are identical. None of the carbon atoms in the structure are bonded to four different groups.

Lesson 21

1. Amino acids are the building materials for many of the structures in the body, especially those related to the functioning of cells. An amino acid has a carbon backbone and two functional groups, an amine group and a carboxylic acid group.

Section IV Review Exercises

1. Possible answer: Two molecular models represent mirror-image isomers if they are mirror images of each other and one cannot be rotated in space to be superimposed on top of the other. When the models have a carbon atom that is attached to four different groups, then they will be mirror-image isomers.
3. Possible answer: A molecule fits into its receptor site only when the receptor has the same "handedness" as the molecule. This is similar to the way that each of your feet will fit into only one shoe of a pair of shoes, even though the shoes are mirror images of one another.

Unit 2 Review Exercises

1. Isomers are molecules that have the same molecular formula but different structural formulas.
3. A molecular formula provides the elements and number of atoms per element in a compound. A structural formula shows the way in which these atoms are bound to each other. A ball-and-stick model provides a three-dimensional image of the bonds. A space-filling model shows the actual amount of space each atom takes up, with atoms that are sharing electrons overlapped in space.
5. a. yes b. $C_8H_8O_3$
   c. The molecule has a flat section with the ester functional group attached to its side, similar to the shape of a frying pan.
   d. The bond is bent because there are four electron domains. The oxygen atom has two bonded pairs and two lone pairs. The shape that separates the four domains as much as possible is tetrahedral, which causes the two bonds to be bent.

Unit 3: Weather

Lesson 1

1. Possible answer: To have weather, a planet must have a layer of gases surrounding its surface.
3. A physical change is a change in the form of a substance that does not change the identity of the substance.

Lesson 2

1. Possible answer: One way to determine the volume of 3 cm of water in the rain gauge is to use the graph. Find 3 cm on the *x*-axis, move up to the line on the graph, and then read the corresponding *y*-value. A second way is to use a proportionality constant. Examining the data in the table shows that the value of the volume is always twice the value of the height.
3. Possible answer: Both the volume and height of water in the rain gauge will keep track of the increase because they both increase by the same proportion as additional rain falls into the gauge.
5. Possible answer: The three containers would have different volumes of water in them after the storm because their bases have different areas. The greatest volume of water will be in the washtub and the least will be in the graduated cylinder. However, the height of water in each container will be the same.

Lesson 3

1. You can convert the volume of snow in a snowpack to a volume of water by using the mathematical equation

for density, $D = \frac{m}{V}$. If you know the volume and density of snow, you can determine the mass of the snow in the snowpack. If the snow melts, the mass of the water will be the same as the mass of the snow. Then you can calculate the volume of water by dividing the mass by the density of water, which is equal to 1 g/mL.

7. a. 16.1 g **b.** 17.5 g

9. a. 4.3 g of iron occupies a greater volume because volume increases as density decreases, and iron is less dense than lead.

b. 2.6 mL of lead has a larger mass because mass increases as density increases, and lead is more dense than iron.

Lesson 4

1. Possible answer: Liquids usually expand when they are heated and contract when they are cooled. At any given temperature the volume of a fixed amount of liquid will always be the same. If the liquid is in a tube, the height of the column of liquid will change in a predictable way with changes in temperature.

5. −40 °F

7. Normal body temperature is about 98.6 °F. A body temperature of 40 °C is the same as 104 °F, which indicates fever and probable illness.

9. a.

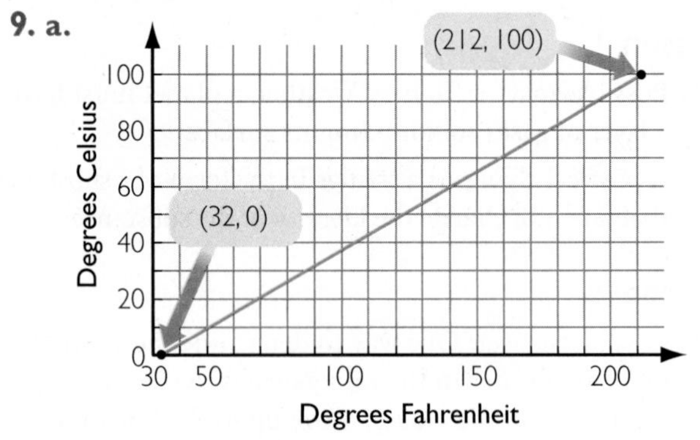

b. 50 °F **c.** 13 °C

Lesson 5

1. Possible answer: Absolute zero is the temperature at which the volume of a gas would be equal to zero. It is considered a theoretical temperature because a real gas would condense into a liquid and then a solid before reaching absolute zero, making a volume of zero impossible.

3. Possible answer: The kinetic theory of gases defines the temperature of a gas as the average kinetic energy of the particles of the gas.

5. The smallest unit is 1 °F because there are 180 Fahrenheit degrees between the freezing temperature of water and the boiling temperature of water, while there are only 100 Celsius degrees or kelvins between the freezing temperature of water and the boiling temperature of water.

7. a. −173 °C **b.** 333 K **c.** −23 °C **d.** 298 K **e.** 27 °C **f.** 173 K **g.** 127 °C

9. a. 308 K **b.** 450 K **c.** 258 K

11. B

Lesson 6

1. Determine the proportionality constant, k, for a gas by dividing the volume of the gas by its temperature measured on the Kelvin temperature scale.

3. 689 mL

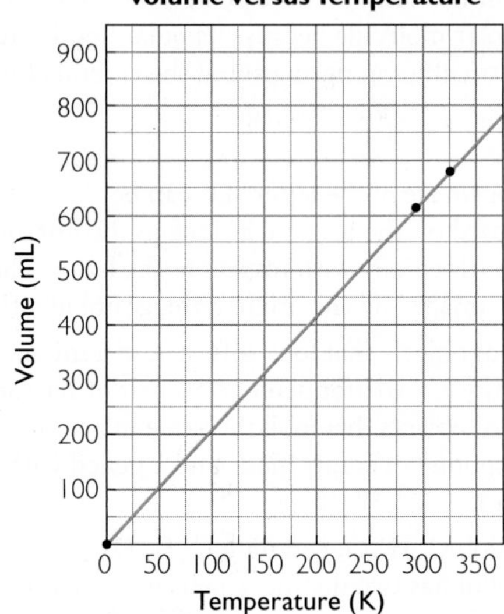

The answer is reasonable because 325 K corresponds to about 700 mL.

5. 186 K

Lesson 7

1. When air gets warmer, its volume increases and its density decreases. Air that is less dense than the surrounding air will rise.

3. A

Section I Review Exercises

1. 283 K

3. 760 mL

5. a. 3.3 mL **b.** 2700 mL. The volume of gas can be calculated using the density equation. The mass of the carbon dioxide does not change when it changes phase and becomes a gas.

Lesson 8

1. The density of a gas is much lower than the density of a solid.

3. Possible answer: Sublimation occurs when a substance changes from the solid phase directly to the gas phase without becoming a liquid. Evaporation occurs when a substance changes from the liquid phase to the gas phase.

5. Possible answer: Ice floats in water so it must have a slightly lower density than water.

7. 4.1 mL

9. 12 g

Lesson 9

3. Because the air inside the balloon exerts pressure on the inside wall but there is no pressure balancing it on the outside wall, the balloon will expand and most likely pop.

Lesson 10

1. As you push down on the plunger of the syringe, the gas pressure inside the syringe increases and the pressure on the scale increases. The pressure inside the syringe is equal to the pressure you are exerting on the plunger, which can be read from the scale. To calculate the pressure of the gas, divide the weight on the scale by the surface area of the plunger. Then add atmospheric pressure.

5. Possible answer: The relationship between gas pressure and gas volume is different from the relationship between gas volume and gas temperature because an increase in pressure causes the volume to decrease, while an increase in temperature causes the volume to increase.

Lesson 11

1. Possible answer: You can change the pressure, the volume, or the temperature of a gas sample.

3. 720 L

5. **a.** The volume of air inside the bottle stays the same because the bottle is a rigid container.
   b. Yes, the air inside the bottle will lose energy until it has the same temperature as its surroundings because glass does not insulate against temperature changes.
   c. Pressure and temperature are directly proportional. The pressure of the air inside the bottle will decrease as its temperature decreases.
   d. 0.93 atm

Lesson 12

1. According to the kinetic theory of gases, increasing the gas volume decreases the gas pressure because the molecules of gas have more room in which to travel and therefore they strike the walls of the container less often.

3. Possible answer: An increase in temperature probably caused the change in pressure because pressure and temperature are proportional and the volume of a gas cylinder doesn't change.

5. The volume of a gas cannot decrease to zero because the particles of the gas themselves occupy space.

7. **a.** The volume of your lungs at 3.5 atm is much lower than the volume at sea level.
   b. Holding your breath when you ascend quickly is dangerous because the pressure decreases rapidly as you rise. If you hold your breath, your lungs will act as sealed containers, making the pressure inside the lungs much greater than the pressure outside the lungs, possibly causing your lungs to rupture.

Lesson 13

1. The combined gas law is a mathematical equation that relates gas pressure, temperature, and volume. This law applies when all three variables can change at the same time.

3. **a.** The volume of the gas will decrease because pressure and volume are inversely proportional when the temperature and the amount of gas remain unchanged.
   b. $PV = k$ **c.** 0.49 L

5. **a.** The volume of the balloon will increase because, based on the values given, the change in pressure has a greater effect on the final volume than the change in temperature.
   b. $\frac{PV}{T} = k$ **c.** 1.6 L

Lesson 14

1. Possible answer: High-pressure areas are generally associated with clear skies. Low-pressure areas are generally associated with cloudy weather and precipitation.

Section II Review Exercises

1. The pressure of a gas is proportional to the temperature of the gas. As a gas becomes warmer, its molecules move faster and collide with each other and other objects more frequently.

3. 0.80 atm

5. 2.0 atm

Lesson 15

1. Possible answer: Air pressure decreases with increasing altitude because the number of molecules of air in a given volume decreases. This means that there are fewer collisions of air molecules with one another and with other objects, and therefore the air exerts less pressure.

5. Possible answers: Increase the amount of air in the tire. Increase the temperature of the air in the tire. Press on the outside of the tire to decrease its volume.

7. When a high-pressure system moves into a region, the air pressure increases. When the air pressure on the

open end of the barometer increases, it pushes on the mercury harder than it did before. Mercury is pushed higher into the closed end of the barometer until the downward pressure caused by the weight of the column of mercury is equal to the atmospheric pressure.

Lesson 16

1. Possible answer: Chemists invented the mole so as to have a sufficiently large unit with which to count the enormous number of particles in a sample of gas.

3. **a.** Each gas sample contains 1 mol of atoms. **b.** 0.0446 mol/L **c.** The xenon sample has the largest mass because the mass of each xenon atom is greater than the mass of each neon atom or argon atom. **d.** the xenon sample

5. 8.0 g of helium

7. The two balloons hold the same volume.

Lesson 17

1. The ideal gas law is an equation that relates the variables pressure, volume, number of moles, and temperature for any sample of any gas. The equation is $PV = nRT$.

3. 0.13 mol

5. 3.90 atm

7. Possible answers: Reduce the number of particles to 0.5 mol. Reduce the temperature to 137 K. Increase the volume to 44.8 L.

Lesson 18

1. Humidity is a measure of the number density of molecules of water vapor in the air.

3. about 1.7 mol per 1000 L

5. 2.2 mol per 1000 L

7. 0.43 mol per 1000 L

11. **a.** No, if the amount of water vapor in the air remains constant, it will still be less than the maximum possible vapor.
b. Yes, the amount of water vapor in the air cannot remain constant, because the maximum possible vapor at 20 °C is less than the original amount of water vapor. Therefore, some water vapor will condense and form fog.

Lesson 19

1. Possible answer: For a hurricane to form, the temperature at the ocean's surface must be at least 80 °F and the air must contain a lot of moisture.

Section III Review Exercises

3. D

5. 68%

Unit 3 Review Exercises

1. The ice has a lower mass because it has a lower density than liquid water. The density of water is 1.0 g/mL. Mass can be calculated by multiplying the density by the volume, and because the samples have the same volume, the ice has less mass.

3. According to the kinetic theory of gases, gas particles move faster when they have more energy. When a gas is heated, its particles gain energy, move faster, and scatter farther apart, causing the gas to expand. When the gas is cooled, its particles lose energy, causing them to move more slowly and remain closer together, so the gas contracts.

7. 2.05 atm

9. **a.** 273 K, 1.0 atm **b.** 1 mol **c.** 2 mol **d.** 0.0446 mol/L

11. 12 mol

Unit 4: Toxins

Lesson 1

1. Reactants are the substances that are present before a chemical reaction, and products are the substances that are formed during the reaction.

3. Possible answer: Toxic substances cause harm to living organisms.

5. **a.** Solid mercury chloride is added to a solution of EDTA to produce a solution containing a compound of mercury with EDTA and hydrochloric acid.
b. If the two reactants are mixed, the solid will disappear and a mixture of two liquids remains.

Lesson 2

1. Possible answer: When sugar melts, it changes from the solid phase to the liquid phase. When sugar dissolves in water, the sugar molecules spread throughout the water.

3. **a.** The solid magnesium will react and disappear into an aqueous compound, and gas bubbles will appear.
b. Gas bubbles will appear in the solution.
c. A solid will form when the two solutions are mixed.

5. In both cases, the mixture of two clear solutions produces gas bubbling out of the resulting solution.

Lesson 3

1. During a physical change, substances change form but new chemical substances are not made. During a chemical change, chemical substances are changed into other chemical substances.

3. In each case, the material changes its appearance or breaks apart, but it does not change identity. Examples include melting ice, breaking glass, and grinding pepper.

5. 1: chemical change; 2: physical change; 3: chemical change; 4: chemical change; 5: physical change; 6: chemical change; 7: chemical change

Lesson 4

1. Possible answer: The law of conservation of mass states that matter is not created or destroyed during a chemical reaction. The atoms involved in the reaction are rearranged but are not created or destroyed.

Lesson 5

1. Possible answer: You need to balance chemical equations to show how many molecules, atoms, or formula units of each substance take part in the reaction or are produced by the reaction.

Lesson 6

1. Combination reactions and decomposition reactions are opposites because combination reactions put two or more atoms or compounds together to make a new compound, while decomposition reactions break a compound into two or more atoms or compounds.

3. **a.** double exchange reaction
   $NaOH(aq) + HNO_3(aq) \longrightarrow NaNO_3(aq) + H_2O(l)$
   b. combination reaction
   $C_2H_4(g) + Cl_2(g) \longrightarrow C_2H_4Cl_2(g)$
   c. single exchange reaction
   $Cl_2(g) + MgBr_2(s) \longrightarrow Br_2(s) + MgCl_2(s)$

5. $SO_3(g) + H_2O(l) \longrightarrow H_2SO_4(aq)$
   Because sulfur trioxide is a gas, it could enter the body through the nose and mouth and travel to the lungs. Sulfuric acid is an aqueous liquid that could enter the body by ingestion or through the skin.

Section I Review Exercises

1. Possible answer: Identifying physical or chemical changes through observations alone is difficult because often no change occurs that can be detected by sight or smell.

3. **a.** Each reactant has ionic bonds because they both consist of metal atoms bonded to nonmetal atoms or polyatomic ions. **b.** double exchange reaction **c.** chemical change **d.** A balanced reaction shows the matter is conserved because all of the atoms in the reactants are accounted for in the products. **e.** calcium phosphate

Lesson 7

1. Possible answer: Toxicities of most substances are measured by exposing laboratory animals to the substance in different dosages.

5. **a.** 90,600 mg **b.** 227,000 tablets

Lesson 8

1. Possible answer: If you know the mass of an individual object in a large group of objects with the same mass, you can use it and the total mass to calculate the total number of objects.

5. 740 marbles will have a greater mass than 740 tiny plastic beads because the mass of a single marble is greater than the mass of a single bead.

7. 333 g of red jelly beans

9. 0.0000000000000000000000327 g

Lesson 9

1. The average mass of atoms of an element is given on the periodic table in units of amu.

5. **a.** 1 mol carbon **b.** 1 mol iron **c.** 1 mol gold **d.** 1 mol gold

7. **a.** 10 g beryllium **b.** 2 mol arsenic **c.** 500 g lead

Lesson 10

1. To determine the molar mass of sodium chloride, add the atomic mass of sodium and the atomic mass of chlorine to determine the mass of one unit of sodium chloride. The atomic mass in amu is the same as the molar mass in moles.

3. $Ne(g)$: 20.2 g/mol, $Ca(s)$: 40.1 g/mol, $CO_2(g)$: 44.0 g/mol, $CaCO_3(s)$: 100.1 g/mol, $CH_4O(l)$: 32.0 g/mol, $C_2H_6O(l)$: 46.1 g/mol, $Fe_2O_3(s)$: 159.7 g/mol

5. **a.** 10.0 g calcium, Ca
   b. 5.0 g sodium fluoride, NaF
   c. 2.0 g iron oxide, FeO

Lesson 11

1. To convert between moles of a substance and grams of the substance, use the formula: mass (g) = molar mass · moles.

3. 12.0 mol

5. 169 g BaO_2

7. **a.** 1.0 mol Cu_2O
   b. 1.0 mol Cu_2O
   c. Cu_2O represents the best deal for the company because you get twice as much copper for the same price.

Lesson 12

1. Possible answer: A comparison of the amount of each compound needed to make a soft drink taste sweet provides evidence that aspartame is sweeter than fructose. It takes many more molecules of fructose than aspartame to sweeten the drink.
3. **a.** about 360 cans **b.** 0.00072 mol/kg

Section II Review Exercises

3. Possible answer: Scientific notation is a useful tool for chemists because it is often necessary to use very large or very small numbers in calculations. Scientific notation simplifies these calculations and also makes comparison of the values easier.
5. AgCl, NaCl, LiCl

Lesson 13

1. Possible answer: "Uniform throughout" means that the components of the solution are so well mixed that two samples taken from anywhere in the solution will be identical.
5. In order of increasing molarity, the solutions are c, a, b.
7. In order of increasing molarity, the solutions are a, c, and b.

Lesson 14

1. Concentration refers to the amount of a substance divided by the volume in which the substance is dissolved. Two solutions of different volumes can have the same concentration if the proportion of the substance to the volume is the same.
5. **a.** The first two solutions have the same number of molecules. Each of these solutions has more molecules than the third solution. The number of molecules in each sample is the product of the molarity and the volume. The first two solutions have 1 mol of molecules, and the third solution has 0.5 mol of molecules. **b.** All three solutions have the same concentration, 1.0 M. Molarity is a measure of concentration. **c.** The 1.0 L sucrose solution has the greatest mass because each sucrose molecule has more mass than each glucose molecule. Therefore, one mole of sucrose has more mass than one mole of glucose. Also, one mole of sucrose has more mass than half a mole of sucrose.
9. **a.** 0.25 L **b.** 1.5 L **c.** 6.0 L **d.** 300 L

Lesson 15

1. Possible answer: To prepare a 0.25 molar solution of sugar, you would calculate the mass of 0.25 mol of sugar, then measure the amount of sugar and dissolve it in enough water to make 1 L of solution.
3. **a.** 29 g **b.** 220 g **c.** 60 g
5. **a.** 10 L **b.** 0.33 L **c.** 0.18 L
7. 1.0-liter solution with 20 g of glucose

Lesson 16

1. A substance dissolved in the water adds mass to the solution. A sample of contaminated water should have a greater density than a sample of pure water. If this difference can be measured, the greater density indicates that the water is not pure.
3. The 0.10 M copper chloride ($CuCl_2$) solution has a greater density than the 0.10 M potassium chloride (KCl) solution. Copper chloride has a molar mass of 134.5 g and potassium chloride has a molar mass of 75.6 g. So if you have equal volumes of each solution there would be more mass in the copper chloride solution. And if the mass is greater, the density is greater.
7. The sodium bromide (NaBr) solution would weigh the most because it has more solute added than the sodium hydroxide (NaOH) solution. The KCl weighs about half as much as the other two samples because a 500 mL aqueous solution will always weigh less than a 1 L aqueous solution.

Section III Review Exercises

1. Possible answer: Knowing exactly how many moles or grams of contaminant are in a water sample determines whether the sample is toxic because toxicity is based on dosage.
3. **a.** 0.00002 mol Pb^{2+} **b.** 0.0066 g $Pb(NO_3)_2$ **c.** 0.0041 g Pb^{2+}

Lesson 17

1. Possible answer: Acids have a sour taste and they can burn the skin in concentrated form. Bases have a bitter taste and a slippery feel. Bases can also cause skin burns in concentrated form. Most acidic and basic solutions are colorless and odorless.

Lesson 18

1. According to the Arrhenius theory, an acid is a molecule that dissociates to form hydrogen ions in solution and a base is a substance that dissociates to form hydroxide ions in solution.
3. In a strong acid, all of the molecules of the solute dissociate into ions, while in a weak acid only some of the molecules dissociate.
7. Washing soda, Na_2CO_3, forms a basic solution because it accepts a proton from water, thereby forming a hydroxide ion in solution.

Lesson 19

1. The pH of a solution is equal to the negative logarithm of the H^+ concentration. Therefore, the pH increases by 1 as the hydrogen ion concentration decreases by a factor of 10.

7. pH $= -0.40$

Lesson 20

1. When you add water to an acid solution, the pH of the solution increases to a maximum of 7.

5. Because the concentration of hydrogen ions contributed by the acid is less than the concentration of hydrogen ions in pure water, the addition of 1×10^{-9} moles of HCl does not affect the pH of pure water.

7. a. 0.075 mol **b.** 0.068 M **c.** pH $= 1.2$

Lesson 21

1. Two ways to make a strong acid solution safer are dilution with water and neutralization with a base.

5. Each mole of sulfuric acid forms two moles of hydrogen ions in solution, while each mole of sodium hydroxide forms one mole of hydroxide ions. Therefore, the final solution has more hydrogen ions than hydroxide ions, so it is acidic and the pH is less than 7.

Lesson 22

3. 1000 mL

5. a. The final solution is acidic because the solution has more H^+ ions than OH^- ions. **b.** The final solution is neutral because the solution has equal numbers of H^+ ions and OH^- ions. **c.** The final solution is basic because the solution has fewer H^+ ions than OH^- ions.

Section IV Review Exercises

1. According to the Arrhenius theory, an acid is a substance that dissociates to form hydrogen ions in solution and a base is a substance that dissociates to form hydroxide ions in solution. According to the Brønsted-Lowry theory, an acid is a substance that donates protons in solution while a base is a substance that accepts protons. This is similar to the Arrhenius theory, but it accounts for bases that do not dissociate to form hydroxide ions.

5. a. Sodium bicarbonate can be classified as a base because it accepts a proton from water, forming hydroxide ions. **b.** Sodium bicarbonate was available in case sulfuric acid spilled. The sodium bicarbonate could have been used to neutralize any acid that was spilled.

Lesson 23

1. A precipitate is an insoluble compound that forms as a product of a chemical reaction in solution.

3. Possible answer: A spectator ion is an ion that is present in both the reactants and the products of a chemical reaction. Spectator ions do not participate in the reaction.

Lesson 24

1. Possible answer: Coefficients are important in chemical equations because they represent the proportion in which reactants combine to form products.

5. a. 0.10 mol **b.** 5.8 g

Lesson 25

1. Possible answer: You need to convert between grams and moles to use mole ratios to find the right mass proportions of reactants.

3. a. 12.6 g **b.** 19.8 g NaCl, 47.0 g $PbCl_2$

Lesson 26

5. a. $AgNO_3(aq) + NaCl(aq) \longrightarrow AgCl(s) + NaNO_3(aq)$

b. 0.037 mol $AgNO_3$; 0.077 mol NaCl

c. silver nitrate, $AgNO_3$

d. 5.3 g AgCl; 3.1 g $NaNO_3$

e. 2.3 g NaCl

Section V Review Exercises

1. Possible answer:

$2NaOH(aq) + NiCl_2(aq) \longrightarrow 2NaCl(aq) + Ni(OH)_2(s)$

$2Na^+(aq) + 2OH^-(aq) + Ni^{2+}(aq) + 2Cl^-(aq) \longrightarrow 2Na^+(aq) + 2Cl^-(aq) + Ni(OH)_2(s)$

$2OH^-(aq) + Ni^{2+}(aq) \longrightarrow Ni(OH)_2(s)$

$K_2CO_3(aq) + NiSO_4(aq) \longrightarrow K_2SO_4(aq) + NiCO_2(s)$

$2K^+(aq) + CO_3^{2}(aq) + Ni^{2+}(aq) + SO_4^{2+}(aq) \longrightarrow 2K^+(aq) + SO_4^{2+}(aq) + NiCO_3(s)$

$CO_3^{2+}(aq) + Ni^{2+}(aq) \longrightarrow NiCO_3(s)$

$2Li_3PO_4(aq) + 3NiCl_2(aq) \longrightarrow 6LiCl(aq) + Ni_3(PO_4)_2(s)$

$6Li^+(aq) + PO_4^{3+}(aq) + 3Ni^{2+}(aq) + 6Cl^-(aq) \longrightarrow 6Li^+(aq) + 6Cl^-(aq) + Ni_3(PO_4)_2(s)$

$2PO_4^{3+}(aq) + 3Ni^{2+}(aq) \longrightarrow Ni_3(PO_4)_2(s)$

Unit 4 Review Exercises

1. a. $Zn(s) + 2HCl(aq) \longrightarrow ZnCl_2(aq) + H_2(g)$

b. $2KClO_3(s) \longrightarrow 2KCl(s) + 3O_2(g)$

c. $S_8(s) + 24F_2(g) \longrightarrow 8SF_6(g)$

d. $4Fe(s) + 3O_2(g) \longrightarrow 2Fe_2O_3(s)$

3. a. limiting reactant: HCl; 1.87 g $ZnCl_2$ made
   b. limiting reactant: $KClO_3$; 0.608 g KCl made
   c. limiting reactant: F_2; 1.28 g SF_6
   d. limiting reactant: Fe; 1.43 g Fe_2O_3
5. A neutralization reaction is a chemical reaction in which an acid and a base combine to form a salt and water.
9. 36 g

Unit 5: Fire

Lesson 1

1. Possible answer: Energy is a measure of the ability to cause a change in matter.
5. Possible answer: Most of a fire is hot gases that become part of the atmosphere. Ashes are leftover material that was not converted to light and heat by a fire. The heat from the ashes transfers into the air and other materials around the site of the fire as it cools.

Lesson 2

1. Possible answer: Exothermic chemical changes are changes that transfer heat into the environment, and endothermic chemical changes are changes that absorb heat from the environment.
3. a. The average kinetic energy of the products is greater than the average kinetic energy of the reactants because the reaction is exothermic.
   b. In a molecular view, the gas particles in the reactant would be moving more slowly than the particles in the products.
5. a. barium hydroxide, $Ba(OH)_2 \cdot 8H_2O$, ammonium nitrate, NH_4NO_3, ammonia, NH_3, water, H_2O, barium nitrate, $Ba(NO_3)_2$
   b. Possible answer: the beaker, the pool of water outside the beaker, air, my hand
   c. The $Ba(NO_3)_2$ is at a lower temperature than the NH_4NO_3 because the water freezing outside the beaker shows the temperature decreased after the reactants were mixed.
   d. The beaker will feel cold.
   e. The process is endothermic because the temperature decreases during the reaction. Energy was transferred from the surrounding environment into the reaction.
   f. A chemical reaction must have occurred for a change in energy to occur. The reaction must have been the cause of the water outside the beaker turning into ice.

Lesson 3

1. The ice cube feels cold to your hand because heat is being transferred from your hand to the ice cube, which is at a lower temperature.
3. When a liquid evaporates energy is transferred to it, and when it solidifies it transfers energy to the surroundings.
5. Evaporation is the process of a liquid changing into a gas. It is a cooling process because the substance absorbs energy during the phase change, so that energy must be transferred from the surrounding environment. Objects and substances in the surroundings of the evaporating liquid are cooled.
9. Possible answer: When an object is warmer than its environment, the heat spreads into the environment. For example, a fire will eventually cause air in all parts of a room to feel warmer.
11. Yes, if the ice cube is warmer than the air around it or an object that is touching it, heat will transfer from the ice cube to its surroundings or into the other object.

Lesson 4

1. Because the larger sample has more particles, it takes more energy to increase the average kinetic energy of that sample by the same amount as the smaller sample.
3. a. 5 cal b. 10 cal c. 315 cal

Lesson 5

1. Possible answer: The specific heat capacity of a substance is the amount of thermal energy needed to raise the temperature of 1 g of the substance by 1 °C. For example, the specific heat capacity of aluminum is 0.21 cal/g °C, which means that 0.21 cal of thermal energy is needed to increase the temperature of 1 g of aluminum by 1 °C.
5. Possible answer: Because metals have a lower specific heat capacity than plastic, they experience a greater change in temperature when they absorb a specific amount of energy from sunlight. Plastics are composed of large molecules that require more energy to cause a rise in temperature than a metal does.
7. The copper will cool faster because it has a lower specific heat capacity. Copper does not need to transfer as much thermal energy as aluminum does to cool to the temperature of the water.

Lesson 6

1. Water temperature doesn't change during boiling because the energy added to the water doesn't increase the kinetic energy of molecules.
3.

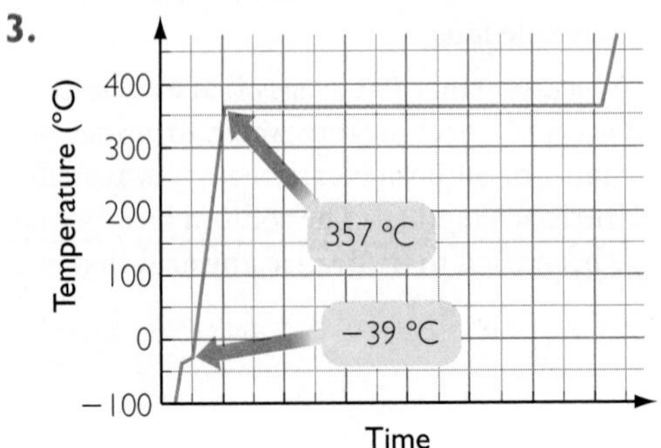

5. B.
7. 24.4 g

Section I Review Exercises

3. methanol (460 cal), water (500 cal), copper (900 cal)
7. 57 °C

Lesson 7

1. If a compound is ionic, or a substance contains water or carbon dioxide, it is unlikely to combust.
3. a. Na_2SO_4 will not combust because it is an ionic compound.
   b. Cu will combust because it is a metallic element.
   c. MgO will not combust because it is an ionic compound.
   d. Na will combust because it is a metallic element.
   e. C_2H_6O will combust because it is a molecular substance containing carbon and hydrogen.
   f. Ar will not combust because it is a noble gas.

Lesson 8

1. Possible answer: Calories are a measure of heat. The heat that is produced by the reaction of the substance with oxygen during combustion can be measured in calories.
3. No, placing a thermometer in the flames measures the temperature of the combustion reaction, but it cannot measure the total heat energy given off by the reaction.

Lesson 9

5. 150 cal
7. Temperature is a measure of the average kinetic energy of the particles of water in the beakers. It takes more energy to increase the temperature of the third beaker because there are many more particles of water, making the amount of thermal energy needed to increase the temperature greater even though there is less temperature change.

Lesson 10

3. Heat of combustion is expressed as a negative number because energy transfers out of the system and into the environment.

Section II Review Exercises

3. Possible answer: The reaction does not appear to release light and heat because it occurs slowly, so the energy is lost to the environment without being noticed.

Lesson 11

3. O–O, C–C, H–H, O=O, C=C
5. a. 3237 kJ
7. a. methanol: −660 kJ/mol; butanol: −2509 kJ/mol
   b. The calculated values are slightly higher (slightly less negative) than the values given in the table.

Lesson 12

1. During an exothermic reaction, the potential energy of the system decreases and the kinetic energy of the system increases. Some of the potential energy in the system is converted into kinetic energy.
3. a. $2C_2H_6(g) + 7O_2(g) \longrightarrow 4CO_2(g) + 6H_2O(g)$
   b.

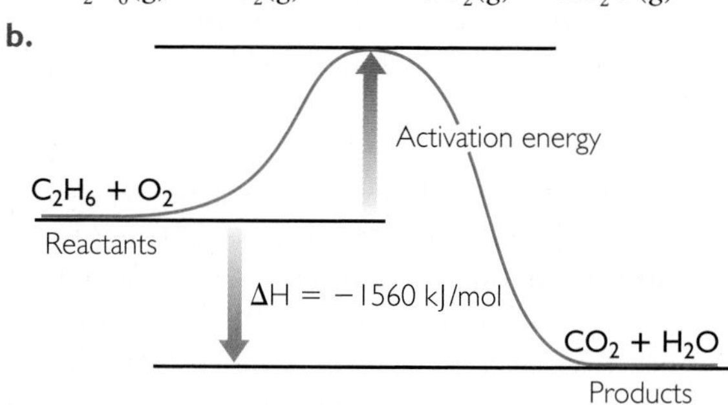

   c. The net energy exchange of the reverse reaction is 1560 kJ/mol. Conservation of energy requires that the amount of energy used to make ethane and oxygen from carbon dioxide and water is equal to the amount of energy released when ethane and oxygen are converted into carbon dioxide and water.
5. a. and b.

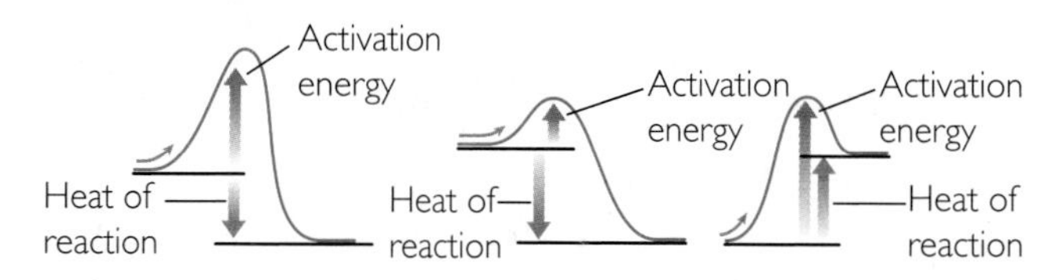

   c. the second diagram
   d. All of the reactions require energy to get started.

Lesson 13

5. Wood chips will burn faster because they have a much greater surface area than a tree, allowing the molecules in the wood chips more contact with the oxygen in the air.
7. The powdered iron oxidizes faster than the steel beam because the small particles have much more total surface area than the beam. Because a greater percentage of the iron is exposed to the oxygen in the air, the reaction is faster.
9. During an explosion, a large amount of energy is released very quickly from chemical bonds in molecules. This energy acts as activation energy to other molecules and causes the explosion to occur rapidly.

Lesson 14

1. Work is the result of a force acting through a distance. It is a transfer of energy, like heat.

3. 14 atm · L

Section III Review Exercises

3. Paige's answer is correct.

Lesson 15

3. a. $2Ca(s) + O_2(g) \longrightarrow 2CaO(s)$; Ca^{2+}, O^{2-}
   $4Cr(s) + 3O_2(g) \longrightarrow 2Cr_2O_3(s)$; Cr^{3+}, O^{2-}
   $4Li(s) + O_2(g) \longrightarrow 2Li_2O(s)$; Li^+, O^{2-}

   b. Calcium, Ca, is oxidized. Chromium, Cr, is oxidized. Lithium, Li, is oxidized.

Lesson 16

1. Possible answer: The oxidation reactions are better sources of energy than the reverse reactions because they are almost always exothermic. The metals that are the hardest to extract have the lowest (most negative) heats of formation, which means that they release the most energy into the environment. The oxidation of aluminum is the best source of energy listed in the lesson.
3. 2180 kJ
5. 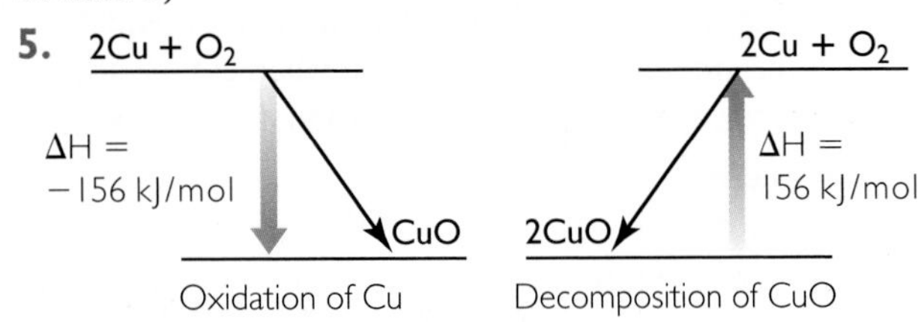

7. 24.5 kJ

Lesson 17

1. In this reaction, copper is oxidized because the neutral copper atom loses the electrons and becomes a positively charged copper ion. The oxygen atom is reduced because it gains electrons and becomes a negatively charged oxygen ion.
5. $Zn(s) + Cu^{2+}(aq) \longrightarrow Cu(s) + Zn^{2+}(aq)$
   $Ba^{2+}(aq) + SO_4^{2-}(aq) \longrightarrow BaSO_4(s)$
7. a. not a redox reaction b. not a redox reaction c. not a redox reaction d. not a redox reaction e. redox reaction, lead is oxidized and hydrogen is reduced, 2 electrons are transferred f. redox reaction, arsenic is oxidized and hydrogen is reduced, 6 electrons are transferred

Lesson 18

3. No change would be observed because iron is less active than zinc. Therefore, iron cannot be oxidized by the zinc nitrate solution and the iron strip would not be affected.
5. a. No reaction occurs in Exercise 3. The chemical equation for the reaction in Exercise 4 is
   $2Fe(NO_3)_2(aq) + 3Zn(s) \longrightarrow 2Fe(s) + 3Zn(NO_3)_2(aq)$

   b. $2Fe^{3+}(aq) + 3Zn(s) \longrightarrow 2Fe(s) + 3Zn^{2+}(aq)$

   c. zinc

   d. Zinc is oxidized, and iron is reduced.
7. Cobalt is below iron in the activity series because the iron is oxidized by the cobalt ions, making iron the more active metal.

Lesson 19

1. In an electrochemical cell, electrons are transferred from the more active metal to the less active metal.
3. a. $Ca \longrightarrow Ca^{2+} + 2e^-$
   $Fe \longrightarrow Fe^{3+} + 3e^-$

   b. Calcium is oxidized and iron is reduced.

   c. 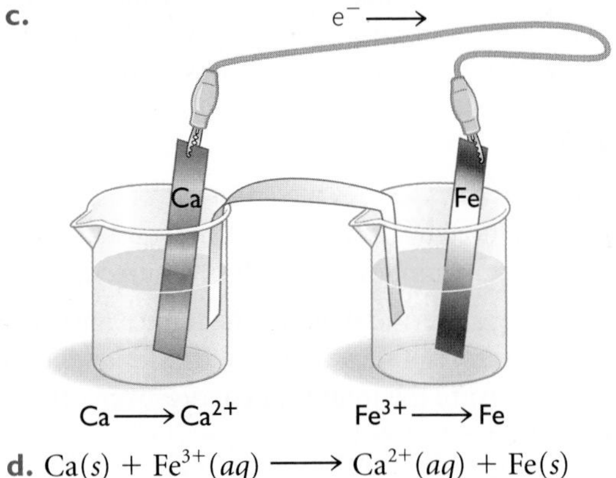

   d. $Ca(s) + Fe^{3+}(aq) \longrightarrow Ca^{2+}(aq) + Fe(s)$

Section IV Review Exercises

1. Possible answer: An electrochemical cell works by separating the two parts of a redox reaction. As electrons are transferred from one metal to the other, they travel through a wire connecting the two half-reactions. The electrical energy is the stream of moving electrons. The chemical energy is the voltage potential of the redox reaction.
3. In the extraction of metals from metal oxides table, the heat of reaction of iron (III) oxide decomposing into pure iron and oxygen gas is listed as +413 kJ/mol. Because this value is positive, the reaction is endothermic. Energy must be continuously added into the system for the reaction to occur.

Unit 5 Review Exercises

3. The thermal energy of an exothermic reaction is the result of bond making releasing more energy than is required for bond breaking for the reaction.
5. a. 3800 cal

   b. −7500 cal

7. More energy is required to raise the temperature of water than to raise the temperature of aluminum by the same amount because water has a higher specific heat capacity than aluminum.

13. $Zn(s) + 2Mn^{4+}(aq) \longrightarrow Zn^{2+}(aq) + 2Mn^{3+}(aq)$
Zinc loses electrons, so it is oxidized. Magnesium gains electrons, so magnesium ions are reduced.
$2Al(s) + 3Ca^{2+}(aq) \longrightarrow 2Al^{3+}(aq) + 3Ca(s)$
Aluminum loses electrons, so it is oxidized. Calcium gains electrons, so calcium ions are reduced.

Unit 6: Showtime

Lesson 1

1. Possible answer: A good chemical demonstration includes a combination of visual effects and surprising results. Chemical demonstrations could feature color changes, puffs of smoke, or flames.

3. $Na_2CO_3(s) + 2H_2O(l) \longrightarrow 2H^+(aq) + CO_3^{2-}(aq) + 2Na^+(aq) + 2OH^-(aq)$
$CaCl_2(s) + CO_3^{2-}(aq) \longrightarrow CaCO_3(s) + 2Cl^-(aq)$
$CaCO_3(s) \longrightarrow Ca^{2+}(aq) + CO_3^{2-}(aq)$
$2H^+(aq) + CO_3^{2-}(aq) \longrightarrow CO_2(g) + 2H_2O(l)$

Lesson 2

3. Possible answers:
$H_2SO_4(aq) \rightleftharpoons 2H^+(aq) + SO_4^{2-}(aq)$
$Fe(s) + 2AgNO_3(aq) \rightleftharpoons Fe(NO_3)_2(aq) + 2Ag(s)$
$CH_3OH(l) \rightleftharpoons CH_3OH(g)$

5. Reactions that are reversing are not always visible because many chemical reactions do not have a noticeable change. If one of the directions of the reaction does not produce a noticeable color change, bubbling gas, or other visual effect, then the reaction may occur without being detectable by sight.

7. a. $2PbO(s) \longrightarrow 2Pb(s) + O_2(g)$
b. Remove the oxygen so that the reaction cannot go in the opposite direction.
c. Possible answer: Increase the amount of oxygen that comes in contact with lead. One way to do this would be by grinding the lead into a powder to increase its surface area.
d. The outer surface of the lead becomes coated with lead oxide, which is not shiny.

Lesson 3

1. Possible answer: Dynamic equilibrium means that the rate of a forward process and the rate of its reverse process are equal.

3.

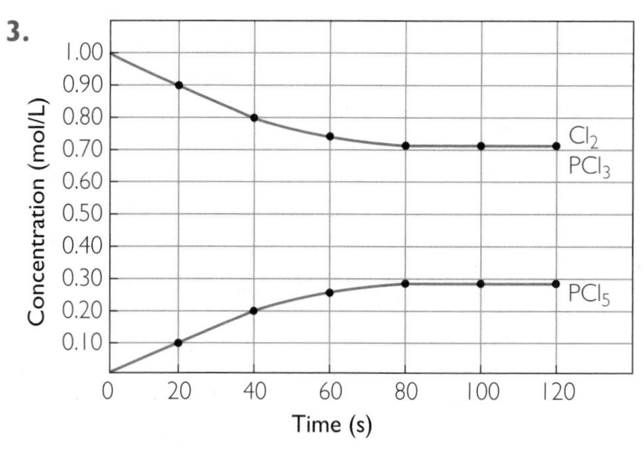

a. 0.71 mol/L Cl_2, 0.71 mol/L PCl_3, 0.29 mol/L PCl_5
b. At 80 seconds, when the concentrations of the reactants and the products stopped changing, the reaction was at chemical equilibrium.
c. At 20 seconds, the forward reaction is faster. At 100 seconds, both reactions occur at the same rate.
d. No, according to the table and the graph, the concentration of reactants is higher than the concentration of products at equilibrium.

Lesson 4

1. The ratio of the products to reactants depends on the rate of the reaction in each direction.

3. a. $K = \frac{[CO_2]^2}{[CO]^2[O_2]}$ **b.** $K = \frac{[NH_3]^2}{[N_2][H_2]^3}$
c. $K = \frac{[HI]^2}{[H_2][I_2]}$ **d.** $K = \frac{[PCl_3][Cl_2]}{[PCl_5]}$

5. a. $2N_2(g) + O_2(g) \longrightarrow 2N_2O(g)$
b. $K = \frac{[N_2O]^2}{[N_2]^2[O_2]}$
c. 2.22×10^{-2}
d. The reaction favors the products at equilibrium.

Lesson 5

1. The color of the solution depends on the relative concentrations of the Fe^{3+} and $FeSCN^{2+}$ ions. As the concentration of the Fe^{3+} ions increases, the solution becomes more yellow. As the concentration of the $FeSCN^{2+}$ ions increases, the solution becomes more red.

3. a. $K = \frac{[HCl]^2}{[H_2][Cl_2]}$
b. The equilibrium mixture will have more of the product because the equilibrium constant is much greater than 1.

Section I Review Exercises

1. Possible answers:
   a. Hydrazine is a weak base. Hydrazine is a base because the ionization equation includes a hydroxide ion as a product. Hydrazine is a weak base because a 0.1 M solution has a pH of 10. A strong base has a pH of 14 at 1.0 M and 13 at 0.1 M.
   b. $N_2H_4(aq)$, $H_2O(l)$, $N_2H_5^+(aq)$, $OH^-(aq)$
   c. $8 \times 10^{-7} = \frac{[N_2H_5^+][OH^-]}{[N_2H_4]}$

3. a. 0.11 M b. The solution is basic because the concentration of hydroxide ions is greater than 1×10^{-7}. Also, the hydroxide ion is present in the chemical equation and the hydrogen ion is not.

5. a. $K = \frac{[HI]^2}{[H_2][I_2]}$
   b. 0.74 M
   c. K is greater than 1, so the product is favored over the reactants.

Lesson 6

3. Adding ammonia will cause the equilibrium to change in favor of the products because the equilibrium constant does not change. An increase in the concentration of a reactant means that more reactants are available for the reaction. Therefore, the equilibrium conditions adjust so that the product concentration is increased as well.

5. This is a decomposition reaction that is the reverse reaction of nickel combustion. Because nickel combustion is exothermic, nickel oxide decomposition is endothermic. Because the reaction is endothermic, adding heat will shift the equilibrium toward the products.

7. Possible answers: **a.** increase temperature, increase pressure, increase $[NO_2]$, decrease $[N_2O_4]$ **b.** increase [HF], decrease $[F^-]$, add base **c.** decrease $[Pb^+]$, increase $[Cl^-]$, increase $[PbCl_2]$ **d.** decrease pressure, decrease $[CO_2]$, decrease [CaO], increase $[CaCO_3]$

Lesson 7

1. An indicator is a weak acid or a weak base that has a different color in the molecular form than in the ionized form. A change in pH shifts the equilibrium between the two forms. When the equilibrium favors the molecular form, the indicator has one color, and when it favors the ionized form, it has the other color.

3. a. AgCl, CuCl, TlCl
   The chemical equation for the solubility of a metal chloride is $MCl(s) \rightleftharpoons M^+(aq) + Cl^-(aq)$. Therefore, the solubility product is given by $K_{sp} = [M^+][Cl^-]$. If the solubility product constant is large, products are favored. The products are the dissolved ions, so a larger solubility product constant means a higher solubility.
   b. If NaCl is added to the solution, copper chloride will precipitate from the solution. When NaCl is added to the solution, some of the NaCl will dissolve, increasing the concentration of Cl^- ions. This will add stress to the product side of the reaction. According to Le Châtelier's principle, the equilibrium will shift to relieve stress, so the shift will favor the reactants, meaning more copper chloride will form.

5. When the indicator is dissolved in pure water, the water has a pH of 7. So, $[H^+]$ is 1.0×10^{-7}. The formula for the acid disassociation constant of an indicator is
   $$K_a = \frac{[H^+][In^-]}{[HIn]}, \text{ so}$$
   $$\frac{[HIn]}{[In^-]} = \frac{[H^+]}{K_a}$$
   If the ratio is greater than 1, then $[HIn] > [In^-]$ and the solution has the color of HIn molecules. The ratio is greater than 1 if K_a is less than 1.0×10^{-7}.
   If the ratio is less than 1, then $[HIn] < [In^-]$ and the solution has the color of In^- ions. The ratio is less than 1 if K_a is greater than 1.0×10^{-7}.
   a. The solution is yellow because K_a for alazarin yellow is 1.0×10^{-11}, and yellow is the color of the HIn molecules. **b.** The solution is blue because K_a for bromocresol green is 2.0×10^{-5}, and blue is the color of the In^- ions. **c.** The solution is red because K_a for methyl red is 7.9×10^{-6}, and red is the color of the In^- ions. **d.** The solution is yellow because K_a for phenol red is 1.3×10^{-8}, and yellow is the color of the HIn molecules.

7. 10

Section II Review Exercises

1. a. $4.63 \times 10^{-3} = \frac{[CO][Cl_2]}{[COCl_2]}$
   b. 8.6×10^{-2} M
   c. The equilibrium would shift toward the products. The additional stress would be on the reactant side of the equation. According to Le Châtelier's principle, the equilibrium will shift to relieve stress, so the shift will favor the product.
   d. 1.34×10^{-1} M

3. a. $7.2 \times 10^{-4} = \frac{[H^+][F^-]}{[HF]}$
   b. At equilibrium, the hydrogen ion and fluorine ions are at the same concentration because the mole

ratio of the ions in the reaction is 1:1. **c.** $[H^+]$ is less than [HF] because the equilibrium constant is much less than 1. The denominator in the ratio, [HF], is much greater than the numerator. **d.** toward the reactants **e.** [HF] increases because more HF molecules are formed to relieve the stress of the excess hydrogen ions.

5. a. Adding N_2O would shift the equilibrium toward the reactants to relieve the stress of having a higher concentration of products. **b.** Removing N_2O would shift the equilibrium toward the products to relieve the stress of having a higher concentration of reactants. **c.** Raising the temperature of the mixture would shift the reaction toward the reactants because raising the temperature is the same as adding heat to the mixture. **d.** Lowering the temperature would shift the reaction toward the products because that direction is exothermic.

Unit 6 Review Exercises

1. a. $Cl_2(g) \rightleftharpoons Cl_2(l)$
b. $2NO_2(g) \rightleftharpoons N_2O_4(g)$
c. $Ag^+(aq) + Cl^-(aq) \rightleftharpoons AgCl(s)$
d. $H_2O(l) \rightleftharpoons H_2O(g)$

3. Possible answers: The reaction is reversible. The equilibrium constant does not change as the concentrations of products and reactants change.

5. 1.6×10^{-5} M

Reference Tables

Table of Average Atomic Masses

Element and symbol	Atomic number	Average atomic mass (amu)
Actinium, Ac	89	(227)
Aluminum, Al	13	26.98
Americium, Am	95	(243)
Antimony, Sb	51	121.8
Argon, Ar	18	39.95
Arsenic, As	33	74.92
Astatine, At	85	(210)
Barium, Ba	56	137.3
Berkelium, Bk	97	(247)
Beryllium, Be	4	9.012
Bismuth, Bi	83	209.0
Bohrium, Bh	107	(272)
Boron, B	5	10.81
Bromine, Br	35	79.90
Cadmium, Cd	48	112.4
Calcium, Ca	20	40.08
Californium, Cf	98	(251)
Carbon, C	6	12.01
Cerium, Ce	58	140.1
Cesium, Cs	55	132.9
Chlorine, Cl	17	35.45
Chromium, Cr	24	52.00
Cobalt, Co	27	58.93
Copper, Cu	29	63.55
Curium, Cm	96	(247)
Darmstadtium, Ds	110	(281)

Element and symbol	Atomic number	Average atomic mass (amu)
Dubnium, Db	105	(268)
Dysprosium, Dy	66	162.5
Einsteinium, Es	99	(252)
Erbium, Er	68	167.3
Europium, Eu	63	152.0
Fermium, Fm	100	(257)
Fluorine, F	9	19.00
Francium, Fr	87	(223)
Gadolinium, Gd	64	157.3
Gallium, Ga	31	69.72
Germanium, Ge	32	72.64
Gold, Au	79	197.0
Hafnium, Hf	72	178.5
Hassium, Hs	108	(277)
Helium, He	2	4.003
Holmium, Ho	67	164.9
Hydrogen, H	1	1.008
Indium, In	49	114.8
Iodine, I	53	126.9
Iridium, Ir	77	192.2
Iron, Fe	26	55.85
Krypton, Kr	36	83.80
Lanthanum, La	57	138.9
Lawrencium, Lr	103	(262)
Lead, Pb	82	207.2
Lithium, Li	3	6.941

The values in parentheses are the average atomic mass of the longest lasting isotope of the element at the time of writing.

Element and symbol	Atomic number	Average atomic mass (amu)
Lutetium, Lu	71	175.0
Magnesium, Mg	12	24.31
Manganese, Mn	25	54.94
Meitnerium, Mt	109	(276)
Mendelevium, Md	101	(258)
Mercury, Hg	80	200.6
Molybdenum, Mo	42	95.94
Neodymium, Nd	60	144.2
Neon, Ne	10	20.18
Neptunium, Np	93	(237)
Nickel, Ni	28	58.69
Niobium, Nb	41	92.91
Nitrogen, N	7	14.01
Nobelium, No	102	(259)
Osmium, Os	76	190.2
Oxygen, O	8	16.00
Palladium, Pd	46	106.4
Phosphorus, P	15	30.97
Platinum, Pt	78	195.1
Plutonium, Pu	94	(244)
Polonium, Po	84	(209)
Potassium, K	19	39.10
Praseodymium, Pr	59	140.9
Promethium, Pm	61	(145)
Protactinium, Pa	91	231.0
Radium, Ra	88	(226)
Radon, Rn	86	(222)
Rhenium, Re	75	186.2
Rhodium, Rh	45	102.9
Roentgenium, Rg	111	(280)

Element and symbol	Atomic number	Average atomic mass (amu)
Rubidium, Rb	37	85.47
Ruthenium, Ru	44	101.1
Rutherfordium, Rf	104	(267)
Samarium, Sm	62	150.4
Scandium, Sc	21	44.96
Seaborgium, Sg	106	(271)
Selenium, Se	34	78.96
Silicon, Si	14	28.09
Silver, Ag	47	107.9
Sodium, Na	11	22.99
Strontium, Sr	38	87.62
Sulfur, S	16	32.07
Tantalum, Ta	73	180.9
Technetium, Tc	43	(98)
Tellurium, Te	52	127.6
Terbium, Tb	65	158.9
Thallium, Tl	81	204.4
Thorium, Th	90	232.0
Thulium, Tm	69	168.9
Tin, Sn	50	118.7
Titanium, Ti	22	47.87
Tungsten, W	74	183.8
Ununbium, Uub	112	(277)
Uranium, U	92	238.0
Vanadium, V	23	50.94
Xenon, Xe	54	131.3
Ytterbium, Yb	70	173.0
Yttrium, Y	39	88.91
Zinc, Zn	30	65.39
Zirconium, Zr	40	91.22

Table of Electronegativities

H 2.10																	He
Li 0.98	Be 1.57											B 2.04	C 2.55	N 3.04	O 3.44	F 3.98	Ne
Na 0.93	Mg 1.31											Al 1.61	Si 1.90	P 2.19	S 2.58	Cl 3.16	Ar
K 0.82	Ca 1.00	Sc 1.36	Ti 1.54	V 1.63	Cr 1.66	Mn 1.55	Fe 1.83	Co 1.88	Ni 1.91	Cu 1.90	Zn 1.65	Ga 1.81	Ge 2.01	As 2.18	Se 2.55	Br 2.96	Kr
Rb 0.82	Sr 0.95	Y 1.22	Zr 1.33	Nb 1.60	Mo 2.16	Tc 1.90	Ru 2.2	Rh 2.28	Pd 2.20	Ag 1.93	Cd 1.69	In 1.78	Sn 1.96	Sb 2.05	Te 2.1	I 2.66	Xe
Cs 0.79	Ba 0.89	La* 1.10	Hf 1.30	Ta 1.50	W 2.36	Re 1.90	Os 2.20	Ir 2.20	Pt 2.28	Au 2.54	Hg 2.00	Tl 1.62	Pb 2.33	Bi 2.02	Po 2.00	At 2.20	Rn
Fr 0.70	Ra 0.89	Ac* 1.10															

* Electronegativity values for the lanthanides and actinides range from about 1.10 to 1.50.

Bonding Continuum

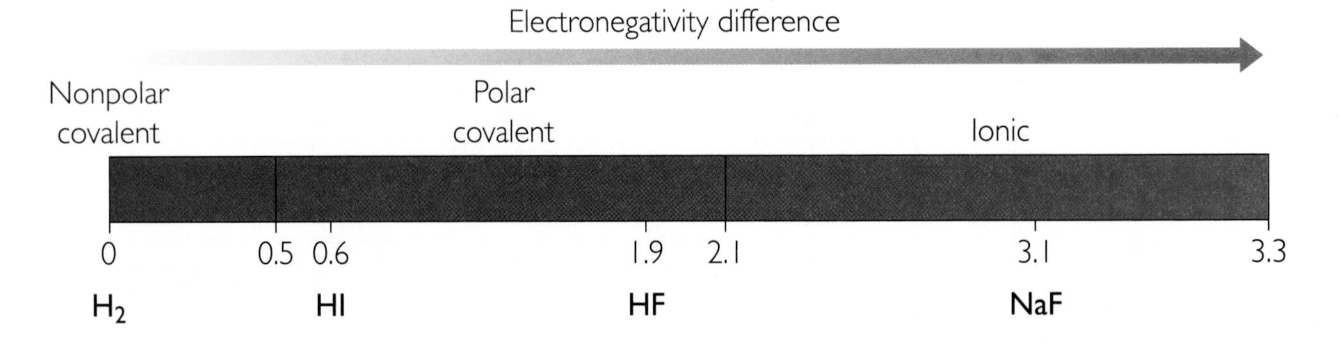

Average Bond Energies (per mole of bonds)

Bond	H–H	C–H	C–C	C=C	C≡C	O–H	C–O	C=O	O=O	O–O
Bond energy (kJ/mol)	432	413	347	614	811	467	358	799*	495	146

*C=O in CO_2: 799, C=O in organic molecules: 745

Heats of Combustion

	Fuel	Chemical formula	ΔH (kJ/mol)	ΔH (kJ/g)
Elements	hydrogen	$H_2(g)$	−243	−121.3
	carbon (coal)	$C(s)$	−393	−32.6
Alkanes	methane	$CH_4(g)$	−891	−55.6
	ethane	$C_2H_6(g)$	−1560	−51.9
	butane	$C_4H_{10}(l)$	−2882	−49.6
	hexane	$C_6H_{14}(l)$	−4163	−48.1
	octane	$C_8H_{18}(l)$	−5508	−47.7
Alcohols	methanol	$CH_4O(l)$	−724	−22.6
	ethanol	$C_2H_6O(l)$	−1368	−29.7
	butanol	$C_4H_{10}O(l)$	−2669	−36.0
Sugar	glucose	$C_6H_{12}O_6(s)$	−2828	−15.9

Solubility Trends

		Anions						
		NO_3^-	Cl^-	OH^-	SO_4^{2-}	CO_3^{2-}	$C_2O_4^{2-}$	PO_4^{3-}
Cations	Most alkali metals, such as Li^+, Na^+, K^+, NH_4^+	S	S	S	S	S	S	S
	Most alkaline earth metals, such as Mg^{2+}, Ca^{2+}, Sr^{2+}	S	S	N	S	N	N	N
	Some Period 4 transition metals, such as Fe^{3+}, Co^{3+}, Ni^{2+}, Cu^{2+}, Zn^{2+}	S	S	N	S	N	N	N
	Other transition metals, such as Ag^+, Pb^{2+}, Hg^2	S	N	N	N	N	N	N

S = very soluble, N = not very soluble or insoluble

Polyatomic Ions

Anions

Charge: 1⁻

Name	Formula
acetate	$CH_3CO_2^-$
chlorate	ClO_3^-
chlorite	ClO_2^-
hypochlorite	ClO^-
perchlorate	ClO_4^-
cyanide	CN^-
formate	HCO_2^-
hydroxide	OH^-
nitrate	NO_3^-
nitrite	NO_2^-
permanganate	MnO_4^-
thiocyanate	SCN^-

Charge: 2⁻

Name	Formula
oxalate	$C_2O_4^{2-}$
carbonate	CO_3^{2-}
chromate	CrO_4^{2-}
sulfate	SO_4^{2-}
sulfite	SO_3^{2-}

Charge: 3⁻

Name	Formula
phosphite	PO_3^{3-}
phosphate	PO_4^{3-}

Cations

Charge: 1⁺

Name	Formula
ammonium	NH_4^+

Charge: 2⁺

Name	Formula
hexaaquocobalt	$Co(H_2O)_6^{2+}$

Specific Heat Capacities at 20 °C

Substance	Specific heat capacity (cal/g °C)	Specific heat capacity (J/g °C)
aluminum	0.21	0.88
water	1.00	4.18
brass	0.09	0.38
iron	0.11	0.46
rubber	0.41	1.72
glass	0.16	0.67
sand	0.19	0.95
mercury	0.03	0.13
hydrogen	3.34	13.97
copper	0.09	0.38
wood	0.41	1.72
silver	0.05	0.21
concrete	0.20	0.84
air (dry, sea level)	0.24	1.00
zinc	0.09	0.38
ethanol	0.57	2.38
lead	0.03	0.13
helium	1.24	5.19
silver	0.05	0.23
tin	0.05	0.21
gold	0.03	0.13

SI Units and Conversion Factors

Length

SI unit: meter (m)
1 meter = 100 cm = 1.0936 yards
1 inch = 2.54 centimeters (exactly)
1 kilometer = 1000 m = 0.62137 mile

Mass

SI unit: kilogram (kg)
1 kilogram = 1000 grams = 2.2046 pounds
1 pound = 453.59 grams = 0.45359 kilogram = 16 ounces
1 atomic mass unit = 1.66054×10^{-27} kilogram

Volume

SI unit: cubic meter (m^3)
1 liter = 1000 milliliters = 0.001 cubic meter
1 cubic centimeter = 1 milliliter

Pressure

SI unit: pascal (Pa)
1 Pa = 1 newton/m^2 = 0.01 millibar
1 atmosphere = 101.325 kilopascals
= 760 torr (mm Hg)
= 14.70 pounds per square inch
= 1.01 bar

Energy

SI unit: joule (J)
1 joule = 0.23901 calorie
1 calorie = 4.184 joules

Physical Constants

Constant	Symbol	Value
atomic mass unit	amu	1.66054×10^{-24} g
Avogadro's number	N_A	6.02214×10^{23} particles/mole
charge of an electron	e	-1.60218×10^{-19} coulomb
universal gas constant	R	8.31451 J/K · mol $0.08206 \frac{L \cdot atm}{mol \cdot K}$
mass of an electron	m_e	9.10938×10^{-28} g 5.48580×10^{-4} amu
mass of a neutron	m_n	1.67493×10^{-24} g 1.00866 amu
mass of a proton	m_p	1.67262×10^{-24} g 1.00728 amu
density of water (at 4 °C)	—	1.0000 g/cm^3
mass of carbon-12 atom	—	12 amu (exactly)
ideal gas molar volume at STP	V_m	$22.4 \frac{L}{mol}$

Symbols and Abbreviations

m	mass
V	volume
D	density
α	alpha ray
β	beta ray
γ	gamma ray
δ^+ δ^-	partial ionic charge
n	number of moles
T	temperature
ΔT	change in temperature
P	pressure
STP	standard temperature and pressure
R	universal gas constant
F	force
A	area
h	height
LD_{50}	lethal dose
[ ]	concentration in moles per liter
q	heat
E	energy
C_p	specific heat capacity
ΔH	heat of reaction (change in enthalpy)
w	work
K_{eq}	equilibrium constant
K_{sp}	solubility product constant
K_a, K_b	acid/base dissociation constant

Proportional Relationships

Mathematical relationship	Variables	Proportionality constant
$m = D \cdot V$ $D = \frac{m}{V}$	mass, m volume, V	density, D
$n = M \cdot V$ $M = \frac{n}{V}$	moles, n volume, V	molarity, M
$m = \text{molar mass} \cdot n$ $\text{molar mass} = \frac{m}{n}$	mass, m moles, n	molar mass (in g/mol)
$F = P \cdot A$ $P = \frac{F}{A}$	force, F area, A	pressure, P
$V = A \cdot h$ $A = \frac{V}{h}$	volume, V height, h	area, A

Proportional Relationships in the Gas Laws

Law	Variables	Mathematical relationships	Variables held constant
Charles's law	volume, V temperature, T	$V = k \cdot T$ $k = \frac{V}{T}$	pressure, P number of moles, n
Gay-Lussac's law	pressure, P temperature, T	$P = k \cdot T$ $k = \frac{P}{T}$	volume, V number of moles, n
Boyle's law	pressure, P volume, V	$k = P \cdot V$ $P = \frac{k}{V}$	temperature, T number of moles, n
combined gas law	pressure, P volume, V temperature, T	$PV = kT$ $k = \frac{PV}{T}$	number of moles, n
ideal gas law	pressure, P volume, V temperature, T moles, n	$PV = nRT$ $R = \frac{PV}{nT}$	none

Other Important References

Table	Page number
Isotopes of the Elements	p. 68, Unit 1: Alchemy, Lesson 14
Four Types of Bonding	p. 131, Unit 1: Alchemy, Lesson 26
Water Vapor Density Versus Temperature	p. 336, Unit 3: Weather, Lesson 18
Lethal Doses	p. 376, Unit 4: Toxins, Lesson 7
Activity Series	p. 560, Unit 5: Fire, Lesson 19
Standard Reduction Potentials	p. 561, Unit 5: Fire, Lesson 19

Some Key Formulas

Density	$D = \frac{m}{V}$
Ideal gas law	$PV = nRT$
Heat	$q = C_p m \Delta T$
Acidity	$\mathrm{pH} = -\log [\mathrm{H}^+]$
Basicity	$\mathrm{pOH} = -\log [\mathrm{OH}^-]$

Glossary

English/Inglés	Spanish/Español

A

absolute zero The temperature defined as 0 K on the Kelvin scale and −273.15 °C on the Celsius scale. Considered to be the lowest possible temperature that matter can reach. (p. 274)

cero absoluto La temperatura definida como 0 K en la escala Kelvin y −273.15 °C en la escala Celsius. Se considera como la temperatura más baja que puede alcanzar la materia. (p. 274)

acid A substance that adds hydrogen ions, H^+, to an aqueous solution; a substance that donates a proton to another substance in solution. (p. 424)

ácido Sustancia que cede iones de hidrógeno, H^+, a una solución acuosa; una sustancia que dona un protón a otra sustancia en la solución. (p. 424)

actinides A series of elements that follow actinium in Period 7 of the periodic table and that are typically placed separately at the bottom of the periodic table. (p. 46)

actínidos Serie de elementos que están después del actinio en el séptimo período de la tabla periódica y que usualmente, aparecen en la parte de abajo de la tabla. (p. 46)

activation energy The minimum amount of energy required to initiate a chemical process or reaction. (p. 529)

energía de activación La mínima cantidad de energía que se necesita para iniciar una reacción o un proceso químico. (p. 529)

activity series A table showing elements in order of their chemical activity, with the most easily oxidized at the top of the list. (p. 555)

serie de actividad Una tabla que muestra elementos ordenados de acuerdo con la actividad química de cada uno de dichos elementos, empezando por aquellos que se oxidan con mayor facilidad. (p. 555)

actual yield The amount of a product obtained when a reaction is run (as opposed to the theoretical yield). (p. 467)

rendimiento rea La cantidad de un producto que se obtiene cuando se ejecuta una reacción (al contrario del rendimiento teórico). (p. 467)

air mass A large volume of air that has consistent temperature and water content. (p. 284)

masa de aire Gran volumen de aire cuya temperatura y contenido de agua son constantes. (p. 284)

alkali metals The elements in Group 1A on the periodic table, except for hydrogen. (p. 46)

metales alcalinos Los elementos del grupo 1A de la tabla periódica, con excepción del hidrógeno. (p. 46)

alkaline earth metals The elements in Group 2A on the periodic table. (p. 46)

metales alcalinotérreos Los elementos del grupo 2A de la tabla periódica. (p. 46)

alpha decay A nuclear reaction in which an atom emits an alpha particle consisting of two protons and two neutrons. Alpha decay decreases the atomic number of an atom by 2 and the mass number by 4. (p. 74)

desintegración alfa Reacción nuclear en la que un átomo emite partículas alfa que contienen dos protones y dos neutrones. La desintegración alfa disminuye el número atómico de un átomo en 2 y el número de masa en 4. (p. 74)

alpha particle A particle made of two protons and two neutrons, equivalent to the nucleus of a helium atom. (p. 74)

partícula alfa Partícula que se compone de dos protones y dos neutrones, lo que equivale al núcleo de un átomo de helio. (p. 74)

amino acid A molecule that contains both an amine ($-NH_2$) and a carboxylic acid ($-COOH$) functional group. (p. 248)

aminoácido Molécula que contiene un grupo funcional amino ($-NH_2$) y un ácido carboxílico ($-COOH$). (p. 248)

anion An ion that has a negative charge. (p. 99)

anión Ion que tiene una carga negativa. (p. 99)

aqueous solution A solution in which water is the dissolving medium or solvent. (p. 25)

solución acuosa Solución en la que el agua es el medio de disolución o el disolvente. (p. 25)

Arrhenius theory of acids and bases An acid-base theory that defines an acid as a substance that adds hydrogen ions, H^+, to an aqueous solution and a base as a substance that adds hydroxide ions, OH^-, to an aqueous solution. (p. 428)

teoría de ácidos y bases de Arrhenius Teoría ácido-base que define un ácido como una sustancia que cede iones de hidrógeno, H^+, a una solución acuosa; y una base como una sustancia que cede iones de hidróxido, OH^-, a una solución acuosa. (p. 428)

atmosphere (atm) A unit of measurement for gas pressure. One atmosphere is equivalent to 14.7 pounds of pressure per square inch or a barometric reading of 760 millimeters of mercury. (p. 296)

atmósfera estándar (atm) Unidad de medida de la presión de los gases. Una atmósfera estándar equivale a 14.7 libras de presión por pulgada cuadrada o a la medida barométrica de 760 milimetros de mercurio. (p. 296)

atmospheric pressure Pressure exerted by the weight, or force, of the air pressing down on a surface. Air pressure is a result of air molecules colliding with the surfaces of objects on Earth. At sea level, atmospheric pressure equals approximately 1 atm or 14.7 lb/in^2. (p. 296)

presión atmosférica La presión ejercida por el peso o fuerza del aire sobre una superficie. La presión del aire es el resultado del choque de las moléculas de aire con la superficie de los objetos que están sobre la Tierra. Al nivel del mar, la presión atmosférica es igual a 1 atm o 14.7 $lb/pulg^2$, aproximadamente. (p. 296)

atom The smallest unit of an element that retains the chemical properties of that element and can exist as a separate particle. (p. 53)

átomo La unidad más pequeña de un elemento que mantiene sus propiedades químicas y que puede existir como una partícula independiente. (p. 53)

atomic mass The mass of a single atom (or isotope) of an element. (p. 40)

masa atómica La masa de un solo átomo (o isótopo) de un elemento. (p. 40)

atomic mass unit (amu) The unit used for expressing atomic mass. 1 amu = 1.66×10^{-24} g, the mass of one hydrogen atom. This is 1/12 the mass of a carbon-12 atom. (p. 40)

unidad de masa atómica (uma) La unidad que se usa para expresar la masa atómica. 1 uma = 1.66×10^{-24} g, la masa de un átomo de hidrógeno. Esto equivale a 1/12 de la masa de un átomo de carbono-12. (p. 40)

atomic number The consecutive whole numbers associated with the elements on the periodic table. The atomic number is equal to the number of protons in the atomic nucleus of an element. (p. 43)

número atómico Los números enteros consecutivos asociados con los elementos de la tabla periódica. El número atómico es igual a la cantidad de protones que hay en el núcleo de un elemento. (p. 43)

atomic theory A theory that states that all matter is made up of individual particles called atoms. (p. 54)

teoría atómica Teoría que afirma que toda la materia se compone de partículas individuales llamadas átomos. (p. 54)

average atomic mass The weighted average of the mass of the isotopes of an element. (p. 64)

masa atómica promedio El promedio ponderado de la masa de los isótopos de un elemento. (p. 64)

Avogadro's law A scientific law stating that equal volumes of gases at the same temperature and pressure contain equal numbers of particles. (p. 329)

ley de Avogadro Ley científica que establece que volúmenes iguales de gases que tienen la misma temperatura y la misma presión contienen la misma cantidad de partículas. (p. 329)

Avogadro's number A number equal to 6.02×10^{23}. It is the number of particles (atoms, ions, molecules, or formula units) present in one mole of a substance. (p. 328)

número de Avogadro Número igual a 6.02×10^{23}. Es el número de partículas (átomos, iones, moléculas o unidades de fórmula) que existen en un mol de una sustancia. (p. 328)

B

ball-and-stick model A three-dimensional representation of a molecule that uses color-coded balls to represent atoms and sticks to represent bonds. (p. 187)

modelo de esferas y varillas Representación tridimensional de una molécula que usa esferas de colores específicos para representar átomos y varillas para representar enlaces. (p. 187)

base A substance that adds hydroxide ions OH^-, to an aqueous solution or a substance that accepts protons, H^+ ions, from another substance in solution. (p. 424)

base Sustancia que cede iones de hidróxido, OH^-, a una solución acuosa o una sustancia que recibe protones, iones H^+, de otra sustancia en una solución. (p. 424)

bent shape The nonlinear shape around a bonded atom with two lone pairs of electrons. (p. 194)

forma curva La forma curvilínea alrededor de un átomo enlazado con dos pares aislados de electrones. (p. 194)

beta decay A nuclear reaction in which a neutron changes into a proton and an electron, and the atom emits a beta particle, which is the electron. Beta decay increases the atomic number of the atom without changing the mass. (p. 75)

desintegración beta Reacción nuclear en la cual un neutrón pasa a ser un protón y un electrón, y el átomo emite una partícula beta, que es el electrón. La desintegración beta aumenta el número atómico de un átomo sin alterar su masa. (p. 75)

beta particle An electron emitted from the nucleus of an atom during beta decay. (p. 75)

partícula beta Electrón emitido por el núcleo de un átomo durante una desintegración beta. (p. 75)

boiling point (boiling temperature) The temperature at which both liquid and gas phases of a single substance are present and in equilibrium; the temperature at which equilibrium is established between a liquid and its vapor at a pressure of 1 atm. (p. 271)

punto de ebullición (temperatura de ebullición) La temperatura a la cual los estados líquido y gaseoso de una misma sustancia están presentes y en equilibrio; la temperatura a la cual se establece el equilibrio entre un líquido y su vapor a una presión de 1 atm. (p. 271)

bond energy The amount of energy that is required to break a specific chemical bond. (p. 524)

energía de enlace La cantidad de energía necesaria para romper un enlace químico. (p. 524)

bonded pair A pair of electrons that are shared in a covalent bond between two atoms. (p. 161)

par enlazado Par de electrones que están siendo compartidos en un enlace covalente entre dos átomos. (p. 161)

Boyle's law The scientific law that states that the volume of a given sample of gas at a given temperature is inversely proportional to its pressure. (p. 301)

ley de Boyle La ley científica que establece que el volumen de una cantidad dada de un gas a una temperatura dada es inversamente proporcional a la presión. (p. 301)

Brønsted-Lowry theory of acids and bases An acid-base theory that defines an acid as a substance that donates a proton to another substance, and a base as a substance that accepts a proton from another substance. (p. 428)

teoría de ácidos y bases de Brønsted-Lowry Teoría ácido-base que define un ácido como una sustancia que dona protones a otra, y una base como una sustancia que recibe un protón de otra. (p. 428)

C

calorie (cal) A unit of measurement for thermal energy. The amount of heat required to raise the temperature of 1 gram of water by 1 Celsius degree. A kilocalorie, 1000 calories, is equal to one food Calorie. (p. 490)

caloría (cal) Unidad de medida de la energía térmica. La cantidad de calor que se necesita para elevar la temperatura de 1 gramo de agua en 1 grado centígrado. Una kilocaloría, 1000 calorías, es igual a una caloría alimenticia. (p. 490)

calorimetry A procedure used to measure the heat transfer that occurs as a result of chemical reactions or physical changes. (p. 510)

calorimetría Procedimiento usado para medir la transferencia de calor que se produce como resultado de las reacciones químicas o de los cambios físicos. (p. 510)

catalyst A substance that accelerates a chemical reaction but is itself not permanently consumed or altered by the reaction. (p. 182)

catalizador Sustancia que acelera una reacción química pero que no se consume o se transforma de manera permanente por dicha reacción. (p. 182)

cation An ion that has a positive charge. (p. 99)

catión Ion que tiene una carga positiva. (p. 99)

Charles's law The law that states that the volume of a given sample of gas is proportional to its Kelvin temperature if the pressure is unchanged. (p. 280)

ley de Charles La ley que establece que el volumen de una cantidad dada de un gas es proporcional a su temperatura en grados Kelvin si la presión no cambia. (p. 280)

chemical bond An attraction between atoms that holds them together. (p. 130)

enlace químico Fuerza de atracción entre átomos que los mantiene unidos. (p. 130)

chemical change See **chemical reaction.**

cambio químico Ver **reacción química.**

chemical equation A representation of a chemical reaction written with chemical symbols and formulas. (p. 180)

ecuación química Representación de una reacción química escrita con símbolos químicos y fórmulas. (p. 180)

chemical equilibrium The dynamic state of a chemical system in which the rate of the forward reaction is equal to the rate of the reverse reaction. At equilibrium, there is no net change in the concentrations of the products and reactants in the system. (p. 580)

equilibrio químico El estado dinámico de un sistema químico en el que la velocidad de una reacción es igual a la velocidad de una reacción inversa. En estado de equilibrio, no hay un cambio neto en la concentración de los productos y los reactantes de un sistema. (p. 580)

chemical formula A combination of element symbols and numbers representing the composition of a chemical compound. (p. 24)

chemical reaction (chemical change) A transformation that alters the composition of one or more substances such that one or more new substances with new properties are produced. (p. 29)

chemical symbol A one- or two-letter representation of an element. The first letter is always uppercase. If there is a second letter, it is lowercase. (p. 24)

chemistry The study of substances, their properties, and how they can be transformed; the study of matter and how it can be changed. (p. 8)

coefficients The numbers in front of the chemical formulas of the reactants and products in a balanced chemical equation. They indicate the correct ratio in which the reactants combine to form the products. (p. 366)

combination reaction A reaction in which two or more reactants combine to form a single product. (p. 369)

combined gas law The law that describes the proportional relationship among the pressure, temperature, and volume of a gas. It states that the value of PV/T will be constant for a given sample of any gas. (p. 313)

combustion An exothermic chemical reaction between a fuel and oxygen, often producing flames; burning. (p. 506)

complete ionic equation A chemical equation that shows all of the soluble ionic compounds as independent ions. (p. 453)

compound A pure substance that is a chemical combination of two or more elements in a specific ratio. (p. 24)

concentration A measure of the amount of solute dissolved per unit of volume of solution, often expressed as moles of solute per liter of solution, mol/L. (p. 403)

fórmula química Combinación de símbolos de elementos y números que representan la composición de un compuesto químico. (p. 24)

reacción química (cambio químico) Transformación que altera la composición de una o más sustancias de tal manera que se producen una o más sustancias nuevas con propiedades nuevas. (p. 29)

símbolo químico Representación de un elemento por medio de una o dos letras. La primera letra es siempre mayúscula y la segunda, si la hay, es minúscula. (p. 24)

química El estudio de las sustancias, sus propiedades y cómo se pueden transformar; el estudio de la materia y cómo se puede transformar. (p. 8)

coeficientes Los números delante de las fórmulas químicas de los reactantes y los productos en una ecuación química balanceada. Indican la velocidad de reacción exacta a la que los reactantes se combinan para formar productos. (p. 366)

reacción de combinación Reacción en la que dos o más reactantes se combinan para formar un solo producto. (p. 369)

ley combinada de los gases La ley que describe la relación proporcional entre la presión, la temperatura y el volumen de un gas. Establece que el valor de PV/T será constante para una cantidad dada de un gas. (p. 313)

combustión Reacción química exotérmica entre un combustible y oxígeno que suele producir llamas; fuego. (p. 506)

ecuación iónica completa Ecuación química que muestra todos los compuestos iónicos solubles como iones independientes. (p. 453)

compuesto Sustancia pura que es una combinación química de dos o más elementos en una proporción determinada. (p. 24)

concentración Medida de la cantidad de soluto diluido por unidad de volumen de solución, usualmente expresada en moles de soluto por litro de solución, mol/L. (p. 403)

conductivity A property that describes how well a substance transmits electricity, heat, or sound. (p. 126)

conductividad Propiedad que describe la capacidad de una sustancia de transmitir electricidad, calor o sonido. (p. 126)

core electrons All electrons in an atom that are not valence electrons. (p. 93)

electrones internos Todos los electrones de un átomo que no son electrones de valencia. (p. 93)

covalent bonding A type of chemical bonding in which one or more pairs of valence electrons are shared between the atoms. (p. 131)

enlace covalente Enlace químico entre átomos en el que uno o más pares de electrones de valencia son compartidos por dichos átomos. (p. 131)

D

daughter isotope An isotope that is formed as a result of a nuclear reaction. (p. 81)

isótopo descendiente Isótopo que se forma como resultado de una reacción nuclear. (p. 81)

decomposition reaction A chemical change in which a single substance is broken down into two or more simpler substances. (p. 370)

reacción de descomposición Cambio químico en el que una sola sustancia se descompone en dos o más sustancias más simples. (p. 370)

density The measure of the mass of a substance per unit of volume, often expressed as grams per milliliter, g/mL, or grams per cubic centimeter, g/cm^3. (p. 17)

densidad La medida de la masa de una sustancia por unidad de volumen, la cual se expresa usualmente en gramos por mililitro, g/mL, o gramos por centímetro cúbico, g/cm^3. (p. 17)

diatomic molecule A molecule consisting of two atoms. (p. 230)

molécula diatómica Molécula compuesta por dos átomos. (p. 230)

dilution The process of adding solvent to a solution to lower the concentration of solute. (p. 436)

dilución El proceso de añadir solvente a una solución para bajar la concentración del soluto. (p. 436)

dipole A molecule or covalent bond with a nonsymmetrical distribution of electrical charge that makes the molecule or bond polar. (p. 226)

dipolo Molécula o enlace covalente con una distribución asimétrica de la carga eléctrica que hace que dicha molécula o enlace sea polar. (p. 226)

dissociate To break apart to form ions in solution. (p. 427)

disociar Se romper para formar iones en solución. (p. 427)

dissolve To disperse a substance homogeneously into another substance at the molecular, ionic, or atomic level. (p. 126)

disolver Dispersar un sustancia de manera homogénea en otra sustancia a nivel molecular, iónico o atómico. (p. 126)

double bond A covalent bond where four electrons are shared between two atoms. (p. 165)

doble enlace Enlace covalente en el que dos átomos comparten cuatro electrones. (p. 165)

double exchange reaction (double displacement reaction) A chemical change in which both reactants break apart and then recombine to form two new products; chemical change where there is an exchange of ions between reactants to form new products. (p. 371)

reacción de doble intercambio (reacción de doble desplazamiento) Cambio químico en el que los dos reactantes se rompen y se vuelven a combinar para formar dos nuevos productos; cambio químico en el que hay un intercambio de iones entre los reactantes para formar productos nuevos. (p. 371)

E

electrochemical cell A device used for generating electrical energy from chemical reactions. The electrical current is caused by substances releasing and accepting electrons in oxidation-reduction reactions. (p. 560)

celda electroquímica Dispositivo usado para generar energía con reacciones químicas. La corriente eléctrica se genera mediante sustancias que ceden y reciben electrones en reacciones de oxidación-reducción. (p. 560)

electrolyte solution A solution that contains ions and can conduct electricity. (p. 559)

solución de electrolitos Solución que contiene iones y que es conductora de la electricidad. (p. 559)

electron An elementary particle with a negative charge that is located outside of the nucleus of an atom. It has a mass of about 1/1838 amu. (p. 56)

electrón Partícula elemental con carga negativa que está fuera del núcleo de un átomo. Tiene una masa de 1/1838 uma, aproximadamente. (p. 56)

electron configuration A notation for keeping track of where the electrons in an atom are distributed among the shells and subshells in an atom. (p. 118)

configuración electrónica Notación que sirve para llevar la cuenta de cómo están distribuidos los electrones de un átomo en las capas y las subcapas electrónicas. (p. 118)

electron domain The space occupied by bonded pairs or lone pairs of valence electrons in a molecule. Electron domains affect the overall shape of a molecule. (p. 193)

dominio del electrón Espacio que ocupan los pares de electrones de valencia enlazados o aislados en una molécula. El dominio de los electrones afecta la forma de una molécula. (p. 193)

electron domain theory A scientific theory that states that every electron domain is located as far away as possible from every other electron domain in a molecule. (p. 194)

teoría del dominio del electrón Teoría científica que establece que el dominio de cada electrón se encuentra lo más lejos posible del dominio de otro electrón en una molécula. (p. 194)

electronegativity A measure of the ability of an atom in a molecular substance to attract electrons to itself. (p. 225)

electronegatividad Medida de la capacidad de atraer electrones que tiene un átomo en una sustancia molecular. (p. 225)

electroplating The process by which a material is coated with a thin layer of metal using electricity passed through a suitable ionic solution. (p. 137)

galvanostegia Proceso mediante el cual se cubre un material con una capa delgada de metal haciendo pasar electricidad a través de una solución iónica especial. (p. 137)

element A unique substance that cannot be broken down into simpler substances through physical or chemical processes. Elements serve as the building materials of all matter. (p. 24)

elemento Sustancia única que no se puede descomponer en sustancias más simples a través de procesos físicos o químicos. Los elementos son los componentes que constituyen toda la materia. (p. 24)

endothermic Describes a process in which heat transfers from the surroundings to the system; heat-absorbing. (p. 478)

endotérmico Describe un proceso en el que se transfiere calor del entorno al interior del sistema; absorción de calor. (p. 478)

energy A measure of the capacity of an object or system to do work or produce heat. (p. 475)

energía Medida de la capacidad que tiene un objeto o un sistema de hacer trabajo o producir energía. (p. 475)

entropy The tendency of energy or matter in a system to disperse; the energy of a system that cannot be used for external work. (p. 485)

entropía La tendencia de la energía o la materia de un sistema a dispersarse; la energía de un sistema que no puede usarse para trabajo externo. (p. 485)

equilibrium A dynamic condition where opposing processes of change occur simultaneously and at the same rate. (p. 578)

equilibrio Condición dinámica en la que procesos opuestos de transformación suceden simultáneamente y en la misma proporción. (p. 578)

equilibrium constant *K* The value obtained when the concentrations of the reactants and products are substituted into the equilibrium equation for a chemical system at equilibrium. The value of K for a reaction is constant at a given temperature. (p. 586)

constante de equilibrio *K* El valor que se obtiene cuando la concentración de los reactantes y productos en la ecuación de equilibrio se sustituye por un sistema químico en equilibrio. El valor de K para una reacción es constante a una temperatura dada. (p. 586)

equilibrium equation A mathematical equation that relates the concentrations of reactants and products in a chemical system at equilibrium. (p. 587)

ecuación de equilibrio Ecuación matemática que relaciona la concentración de reactantes y productos en un sistema químico en equilibrio. (p. 587)

equivalence point The point in an acid-base titration when the acid and base have completely neutralized each other. (p. 445)

punto de equivalencia Momento, en una valoración química ácido-base, en que el ácido y la base se han neutralizado totalmente entre sí. (p. 445)

evaporation The phase change from a liquid to a gas. (p. 292)

evaporación El cambio de estado de líquido a gas. (p. 292)

exothermic Describes a process in which heat is transferred from a system to the surroundings; heat-releasing. (p. 476)

exotérmico Describe un proceso en el que el calor se transfiere del interior del sistema al entorno; que emite calor. (p. 476)

extensive property A characteristic, such as volume or mass, that is specific to the amount of matter and therefore changes if the quantity of the substance changes. (p. 19)

propiedad extensiva Característica, como el volumen o la masa, que es específica a una cantidad de materia y por tanto, cambia si la cantidad de la sustancia cambia. (p. 19)

F

first law of thermodynamics The scientific law that states that energy is conserved, therefore it cannot be created or destroyed. (p. 484)

primera ley de la termodinámica La ley científica que establece que la energía se conserva y por tanto, no se puede crear ni destruir. (p. 484)

fission (nuclear) The splitting apart of an atomic nucleus into two smaller nuclei, accompanied by a release of energy. (p. 76)

fisión (nuclear) La ruptura del núcleo de un átomo en dos núcleos más pequeños, que viene acompañada de liberación de energía. (p. 76)

flame test A laboratory procedure used to determine the presence of certain metal atoms in a chemical sample by heating the sample in a flame and observing the resulting color of the flame. (p. 88)

prueba a la llama Procedimiento de laboratorio que se usa para determinar la presencia de átomos de un determinado metal en una muestra química, calentando la muestra con un mechero Bunsen y observando la coloración que adquiere la llama. (p. 88)

formula unit The simplest chemical formula that can be used to represent network covalent or ionic compounds that shows the elements present in the smallest whole number ratio. (p. 366)

unidad de fórmula La fórmula química más simple que se puede usar para representar una red de compuestos iónicos o covalentes, y que muestra la proporción simplificada de los elementos presentes con números enteros. (p. 366)

functional group A structural feature of a molecule consisting of a specific arrangement of atoms, responsible for certain properties of the compound. (p. 169)

grupo funcional Característica estructural de una molécula que consiste en una distribución específica de los átomos, y a la que se deben ciertas propiedades de un compuesto. (p. 169)

fusion (nuclear) The joining of two atomic nuclei to form a larger nucleus, accompanied by a release of energy. (p. 77)

fusión (nuclear) La unión de dos núcleos atómicos para formar un núcleo más grande, que viene acompañada de liberación de energía. (p. 77)

G

gamma ray A form of high-energy electromagnetic radiation emitted during nuclear reactions. (p. 76)

rayo gamma Forma de radiación electromagnética de alta energía emitida durante reacciones nucleares. (p. 76)

Gay-Lussac's law The scientific law that states that the pressure of a given amount of gas is directly proportional to temperature, if the gas volume does not change. (p. 305)

ley de Gay-Lussac La ley científica que establece que la presión de una cantidad dada de un gas es directamente proporcional a la temperatura, si su volumen no cambia. (p. 305)

group A vertical column on the periodic table, also called a family. Elements in a group have similar properties. (p. 46)

grupo Columna vertical en la tabla periódica, también denominada familia. Los elementos de un grupo tienen propiedades similares. (p. 46)

H

half-cell One half of an electrochemical cell, the site of either oxidation or reduction. (p. 560)

semicelda La mitad de una celda electroquímica, en la que puede producirse oxidación o reducción. (p. 560)

half-life The amount of time required for one half of the radioactive atoms in a sample to decay. (p. 75)

vida media El tiempo que se necesita para que la mitad de los átomos radiactivos de una muestra se desintegren. (p. 75)

half-reaction A chemical equation that represents either the oxidation or the reduction part of an oxidation-reduction reaction. (p. 560)

semirreacción Ecuación química que puede representar la oxidación o la reducción de una reacción de oxidación-reducción. (p. 560)

halogens The elements in Group 7A on the periodic table. (p. 46)

halógenos Los elementos del grupo 7A de la tabla periódica. (p. 46)

heat A transfer of energy between two substances, from a hotter body to a colder body. (p. 478)

calor Una transferencia de energía entre dos sustancias, de un cuerpo más caliente a uno más frío. (p. 478)

heat of combustion The energy released as heat when a compound undergoes complete combustion with oxygen, usually expressed per mole of the fuel. (p. 517)

calor de combustión La energía liberada en forma de calor cuando un compuesto pasa por una combustión completa con oxígeno. Usualmente se expresa por mol de combustible. (p. 517)

heat of formation The energy transferred as heat during a chemical reaction when a compound is formed from its constituent elements, expressed per mole of product. (p. 546)

calor de formación La energía que se transfiere en forma de calor durante una reacción química cuando se produce un compuesto a partir de sus elementos constituyentes. Se expresa por mol de producto. (p. 546)

heat of fusion The amount of heat transfer required to change a substance from a solid to a liquid, usually expressed per mole or gram. (p. 499)

calor de fusión La cantidad de calor que debe ser transferida para que una sustancia pase de ser un sólido a ser un líquido. Usualmente se expresa por mol o gramo. (p. 499)

heat of reaction The amount of energy transferred as heat during a chemical reaction. (p. 517)

calor de reacción La cantidad de energía que se transfiere en forma de calor durante una reacción química. (p. 517)

heat of vaporization The amount of heat transfer required to change a substance from a liquid to a gas, usually expressed per mole or gram. (p. 499)

calor de vaporización La cantidad de calor que debe ser transferida para que una sustancia pase de ser un líquido a ser un gas. Usualmente se expresa por mol o gramo. (p. 499)

heterogeneous mixture A mixture whose composition is not uniform throughout. (p. 403)

mezcla heterogénea Mezcla cuya composición no es uniforme. (p. 403)

homogeneous mixture A mixture whose composition is uniform throughout.(p. 403)

mezcla homogénea Mezcla cuya composición es uniforme. (p. 403)

HONC 1234 rule A rule that states that in most molecules, hydrogen makes 1 bond, oxygen makes 2 bonds, nitrogen makes 3 bonds, and carbon makes 4 bonds. (p. 154)

regla de HONC 1234 Regla que establece que en la mayoría de moléculas, el hidrógeno hace 1 enlace, el oxígeno hace 2 enlaces, el nitrógeno hace 3 enlaces y el carbono hace 4 enlaces. (p. 154)

humidity The concentration of the water vapor in the air at any given time. (p. 336)

humedad La concentración de vapor de agua en el aire en un momento determinado. (p. 336)

hydrogen bond An intermolecular attraction between a hydrogen atom in a molecule and an electronegative atom in another molecule (especially nitrogen, oxygen, or fluorine). The hydrogen atoms in water molecules form hydrogen bonds with the oxygen atoms in other water molecules. (p. 496)

enlace de hidrógeno Atracción intermolecular entre un átomo de hidrógeno en una molécula y un átomo electronegativo en otra (especialmente de nitrógeno, oxígeno o flúor). Los átomos de hidrógeno en las moléculas de agua forma enlaces con átomos de oxígeno en otras moléculas de agua. (p. 496)

hypothesis A proposed explanation for an observation or scientific problem, which can be tested by further investigation. (p. 7)

hipótesis Una explicación dada a una observación o a un problema científico, la cual se puede probar con más investigación. (p. 7)

I

ideal gas law The scientific law that relates volume, pressure, temperature, and the number of moles of a gas sample: $PV = nRT$, where R is the universal gas constant. (p. 332)

ley de los gases ideales La ley científica que relaciona el volumen, la presión, la temperatura y el número de moles de una muestra de gas: $PV = nRT$, cuando R es la constante universal de los gases ideales. (p. 332)

indicator (acid-base) A chemical compound that indicates the relative acidity or basicity of a solution through its characteristic color changes. (p. 423)

indicador (ácido-base) Compuesto químico que sirve para mostrar la acidez o la basicidad relativa de una solución a través de cambios de color característicos. (p. 423)

insoluble Unable to be dissolved in another substance. (p. 126)

insoluble Que no se puede disolver en otra sustancia. (p. 126)

intensive property A characteristic, such as boiling point or density, that does not depend on the size or amount of matter and can be used to identify matter. (p. 19)

propiedad intensiva Característica como el punto de ebullición o la densidad que no depende del tamaño o la cantidad de materia y que puede ser usada para identificar dicha materia. (p. 19)

intermolecular force A force of attraction that occurs between molecules. (p. 218)

fuerza intermolecular Fuerza de atracción que se produce entre moléculas. (p. 218)

inversely proportional Related in such a way that when one quantity increases, the other decreases in a mathematically predictable way. The variables x and y are inversely proportional to each other if $y = k/x$, where k is the proportionality constant. (p. 301)

inversamente proporcional Relacionadas de tal manera que cuando una variable aumenta, la otra disminuye de forma que se puede predecir matemáticamente. Las variables x y y son inversamente proporcionales si $y = k/x$, donde k es la constante de proporcionalidad. (p. 301)

ion An atom or group of bonded atoms that has a positive or negative charge. (p. 97)

ion Átomo, o grupo de átomos enlazados, que tiene carga positiva o negativa. (p. 97)

ionic bonding A type of chemical bonding that is the result of the transfer of electrons from one atom to another, typically between metal and nonmetal atoms. (p. 131)

enlace iónico Atracción entre átomos que es el resultado de la transferencia de electrones de un átomo a otro, y que es común entre átomos de metales y de no metales. (p. 131)

ionic compound A compound that consists of positively charged metal cations and negatively charged nonmetal anions formed when valence electrons are transferred. (p. 101)

compuesto iónico Compuesto que contiene cationes de metales con carga positiva y aniones de no metales con carga negativa que se forman al transferirse los electrones de valencia. (p. 101)

isomers Compounds with the same molecular formula but different structural formulas. Isomers differ in molecular structure and in chemical and physical properties. (p. 150)

isómeros Compuestos que tienen una misma fórmula molecular pero fórmulas estructurales diferentes. Los isómeros son distintos en su estructura molecular y en sus propiedades químicas y físicas. (p. 150)

isotopes Atoms of the same element that have different numbers of neutrons. These atoms have the same atomic number but different mass numbers. (p. 64)

isótopos Átomos de un mismo elemento que tienen distinta cantidad de neutrones. Estos átomos tienen un mismo número atómico pero diferente número de masa. (p. 64)

J

joule (J) A unit of measurement of energy. One calorie is equal to 4.184 joules. (p. 517)

julio (J) Unidad de medida de la energía. Una caloría es igual a 4.184 julios. (p. 517)

K

Kelvin scale A temperature scale with units in kelvins, K, that sets the zero point at −273.15 °C, which is also known as absolute zero. Kelvin units are equivalent in scale to Celsius units. (p. 275)

escala Kelvin Escala de temperatura cuyas unidades son los kelvins, K, con el punto cero en −273.15 °C, temperatura también conocida como cero absoluto. Las unidades Kelvin son equivalentes en escala a las unidades Celsius. (p. 275)

kinetic energy The energy of motion. (p. 480)

energía cinética La energía del movimiento. (p. 480)

kinetic theory of gases The scientific theory that states that gases are composed of tiny particles in continuous, random, straight-line motion and collide with each other and the walls of the container. (p. 275)

teoría cinética de los gases La teoría científica que establece que los gases están compuestos por pequeñas partículas en movimiento continuo, aleatorio y en línea recta, y que se chocan entre sí y con las paredes que las contienen. (p. 275)

L

lanthanides A series elements that follow lanthanum in Period 6 of the periodic table and that are typically placed separately at the bottom of the periodic table. (p. 46)

lantánidos Conjunto de elementos que están después del lantano en el sexto período de la tabla periódica y que usualmente aparecen en la parte de abajo de la tabla. (p. 46)

law of conservation of mass The scientific law that states that mass cannot be gained or lost in a chemical reaction and that matter cannot be created or destroyed. (p. 36)

ley de conservación de la masa La ley científica que establece que en una reacción química no se gana ni se pierde masa y que la materia no se puede crear ni destruir. (p. 36)

Le Châtelier's principle A scientific principle that states that when a stress is put on a system at equilibrium, the system will react in a way to relieve the stress. (p. 600)

principio de Le Châtelier Principio científico que establece que cuando se aplica tensión sobre un sistema en equilibrio, el sistema reaccionará de forma que se alivie la tensión. (p. 600)

Lewis dot structure A diagram of a molecule's structure that uses dots to represent the valence electrons. (p. 161)

estructura de puntos de Lewis Diagrama de la estructura de una molécula en el que se usan puntos para representar los electrones de valencia. (p. 161)

Lewis dot symbol A diagram that uses dots to represent the valence electrons of a single atom. (p. 161)

símbolo de puntos de Lewis Diagrama en el que se usan puntos para representar los electrones de valencia de un solo átomo. (p. 161)

limiting reactant The reactant that runs out first in a chemical reaction. It is the reactant that limits the amount of product that can be produced in the reaction. (p. 458)

reactante limitador El reactante que se acaba primero en una reacción química. Es el que limita la cantidad de producto que se puede producir en una reacción. (p. 458)

linear shape A straight-line shape found in small molecules. (p. 198)

forma recta Forma de línea recta que aparece en moléculas pequeñas. (p. 198)

lone pair A pair of unshared valence electrons that are not involved in bonding in a molecule. (p. 162)

par aislado Par de electrones de valencia que no están siendo compartidos y que no están involucrados en el enlace de una molécula. (p. 162)

M

main group elements The elements in Groups 1A to 7A on the periodic table. (p. 46)

elementos del grupo principal Los elementos de los grupos 1A a 7A de la tabla periódica. (p. 46)

mass A measure of the quantity of matter in an object. (p. 9)

masa Medida de la cantidad de materia en un objeto. (p. 9)

mass number The sum of the number of protons and neutrons in the nucleus of an atom. (p. 64)

número de masa La suma del número de protones y neutrones en el núcleo de un átomo. (p. 64)

matter Anything that has substance and takes up space; anything that has mass and volume. (p. 9)

materia Todo lo que tiene sustancia y ocupa un lugar en el espacio; todo lo que tiene masa y volumen. (p. 9)

melting point (melting temperature) The temperature at which both solid and liquid phases of a single substance can be present and in equilibrium. (p. 271)

punto de fusión (temperatura de fusión) La temperatura a la cual los estados sólido y líquido de un misma sustancia están presentes y en equilibrio. (p. 271)

meniscus The curvature of the top of a liquid in a container, which is the result of intermolecular attractions between the liquid and the container. (p. 11)

menisco La curvatura que se forma en la parte superior de un líquido que está dentro de un envase y que es el resultado de las atracciones intermoleculares entre el líquido y el envase. (p. 11)

metal An element that is generally shiny and malleable and an excellent conductor of heat and electricity. Metals are located to the left of the stair-step line on the periodic table. (p. 47)

metal Elemento que generalmente es brillante y maleable y un excelente conductor del calor y de la electricidad. Los metales están a la izquierda de la línea escalonada en la tabla periódica. (p. 47)

metallic bonding A type of bonding between metal atoms in which the valence electrons are free to move throughout the substance. (p. 131)

enlace metálico Enlace entre átomos de un metal en el que los electrones de valencia se mueven libremente por la sustancia. (p. 131)

metalloid An element that has properties of both metals and nonmetals. Metalloids are located along the stair-step line of the periodic table. (p. 48)

metaloide Elemento que tiene propiedades tanto de los metales como de los no metales. Los metaloides están sobre la línea escalonada de la tabla periódica. (p. 48)

mirror-image isomer Molecules whose structures are mirror images of each other and cannot be superimposed on one another. (p. 244)

isómero especular Moléculas cuyas estructuras se reflejan entre sí y no se pueden superponer. (p. 244)

mixture A blend of two or more substances that are not chemically combined. (p. 25)

mezcla Combinación de dos o más sustancias que no están combinadas químicamente. (p. 25)

model A simplified representation of a real object or process that facilitates understanding or explanation of that object or process. (p. 54)

modelo Representación simplificada de un objeto o proceso real que facilita la comprensión o explicación de dicho objeto o proceso. (p. 54)

molarity The concentration of dissolved substances in a solution, expressed in moles of solute per liter of solution. (p. 403)

molaridad La concentración de sustancias disueltas en una solución. Se expresa en moles de soluto por litro de solución. (p. 403)

molar mass The mass in grams of one mole of a substance. (p. 385)

masa molar La masa en gramos de un mol de una sustancia. (p. 385)

mole A counting unit used to keep track of large numbers of particles. One mole represents 6.02×10^{23} items. (p. 328)

mol Unidad de conteo usada para llevar la cuenta de grandes cantidades de partículas. Un mol representa 6.02×10^{23} partículas. (p. 328)

mole ratio The ratio of the moles of one reactant or product to the moles of another reactant or product in a balanced chemical equation. (p. 457)

razón molar La proporción entre los moles de un reactante o producto y los moles de otro reactante o producto en una ecuación química balanceada. (p. 457)

molecular covalent bonding A type of chemical bonding characterized by the sharing of valence electrons between atoms, resulting in individual units called molecules. (p. 131)

enlace covalente molecular Tipo de enlace químico caracterizado por átomos que comparten electrones de valencia y forman unidades individuales llamadas moléculas. (p. 131)

molecular formula The chemical formula of a molecular substance, showing the identity of the atoms in each molecule and the ratios of those atoms to one another. (p. 146)

fórmula molecular La fórmula química de una sustancia molecular que muestra la identidad de los átomos de cada molécula y la proporción de dichos átomos con respecto a los demás. (p. 146)

molecule A group of atoms that are covalently bonded together. (p. 132)

molécula Grupo de átomos que están unidos por enlaces covalentes. (p. 132)

monatomic ion An ion that consists of only one atom. (p. 111)

ion monoatómico Ion que contiene un solo átomo. (p. 111)

N

net ionic equation A chemical equation that is written without including spectator ions. (p. 453)

ecuación iónica total Ecuación química en la que no se incluyen los iones espectadores. (p. 453)

network covalent bonding A type of chemical bonding characterized by the sharing of valence electrons throughout the entire solid sample. (p. 131)

enlace covalente encadenado Tipo de enlace químico caracterizado porque se comparten los electrones de valencia en toda la muestra sólida. (p. 131)

neutralization reaction A chemical reaction in which an acid and base react to form a salt and water. (p. 442)

reacción de neutralización Reacción química en la que un ácido y una base reaccionan para formar sal y agua. (p. 442)

neutron A particle that is located in the nucleus of an atom and does not have an electric charge. The mass of a neutron is almost exactly equal to that of a proton, about 1 amu. (p. 56)

neutrón Partícula que está en el núcleo de un átomo y que no tiene carga eléctrica. La masa de un neutrón es casi igual a la de un protón, 1 uma, aproximadamente. (p. 56)

noble gases The elements in Group 8A on the periodic table. Noble gases are known for not being reactive. (p. 46)

gases nobles Los elementos del grupo 8A de la tabla periódica. Los gases nobles se conocen por no ser reactivos. (p. 46)

nonmetal An element that does not exhibit metallic properties. Nonmetals are often gases or brittle solids at room temperature. Nonmetals are poor conductors of heat and electricity and are located to the right of the stair-step line on the periodic table. (p. 47)

no metal Elemento que no tiene propiedades metálicas. Los no metales suelen ser gases o sólidos quebradizos cuando están a temperatura ambiente. Los no metales no son buenos conductores del calor ni de la electricidad, y están a la derecha de la línea escalonada en la tabla periódica. (p. 47)

nonpolar molecule A molecule that is not attracted to an electrical charge. A molecule is nonpolar if each atom shares electrons equally or there is no net dipole in the molecule. (p. 217)

molécula apolar Molécula que no es atraída por una carga eléctrica. Una molécula es apolar si cada átomo comparte electrones equitativamente o si no hay bipolaridad neta en la molécula. (p. 217)

nuclear chain reaction A nuclear reaction in which neutrons emitted from the nucleus during fission strike surrounding nuclei, causing them to split apart as well. (p. 164)

reacción nuclear en cadena Reacción nuclear en la que los neutrones emitidos desde el núcleo durante la fisión chocan con los núcleos alrededor haciendo que estos también se dividan. (p. 164)

nuclear equation A representation of a nuclear reaction written with isotope symbols. (p. 81)

ecuación nuclear Representación de una reacción nuclear que se expresa con símbolos de isótopos. (p. 81)

nuclear reaction A process that changes the energy, composition, or structure of an atom's nucleus. (p. 73)

reacción nuclear Proceso que transforma la energía, la composición o la estructura del núcleo de un átomo. (p. 73)

nucleus The dense, positively charged structure composed of protons and neutrons that is found in the center of an atom. (p. 56)

núcleo Estructura densa de carga positiva que está compuesta de protones y neutrones, y que se encuentra en el centro de un átomo. (p. 56)

number density The number of gas particles per unit volume usually expressed in moles per liter or moles per cubic centimeter. (p. 324)

densidad numérica El número de partículas de gas por unidad de volumen que usualmente se expresa en moles por litro o moles por centímetro cúbico. (p. 324)

O

octet rule Nonmetal atoms combine by sharing electrons so that each atom has a total of eight valence electrons. After bonding, each atom resembles a noble gas in its electron arrangements. (p. 164)

regla del octeto Los átomos de no metales se combinan con otros átomos hasta que cada uno queda rodeado por ocho electrones de valencia. Después de enlazarse, cada átomo se asemeja a un gas noble en la distribución de sus electrones. (p. 164)

oxidation A chemical reaction in which an atom or ion loses electrons. Oxidation is always accompanied by reduction. (p. 544)

oxidación Reacción química en la que un átomo o un ion pierde electrones. La oxidación siempre viene acompañada de reducción. (p. 544)

oxidation-reduction reaction A chemical reaction that involves electron transfer. (p. 550)

reacción de oxidación-reducción Reacción química que involucra transferencia de electrones. (p. 550)

P

parent isotope A radioactive isotope that undergoes decay. (p. 81)

isótopo precursor Isótopo radiactivo que sufre desintegración. (p. 81)

partial charge A less than full charge on part of a molecule, created by the unequal sharing of electrons. Partial charges are represented with the symbol delta ($\delta+$ for partial positive charge and $\delta-$ for partial negative charge). (p. 217)

carga parcial Una carga incompleta en un sector de una molécula que se crea porque se comparten electrones de manera desigual. Las cargas parciales se representan con el símbolo delta ($\delta+$ para cargas positivas y $\delta-$ para cargas negativas). (p. 217)

partial pressure The pressure exerted by one gas in a mixture of nonreacting gases. The partial pressures of all the gases add up to the total pressure exerted by that mixture of gases. (p. 336)

presión parcial La presión ejercida por un gas en una mezcla de gases no reactivos. La suma de la presión parcial de todos los gases es igual a la presión total ejercida por esa mezcla de gases. (p. 336)

peptide bond The bond between two amino acids; also called an amide bond. (p. 250)

enlace péptido El enlace entre dos aminoácidos; también se llama enlace amino. (p. 250)

percent error A calculation used to find the accuracy of a measurement. The lower the percent error, the more accurate the measurement. (p. 382)

error porcentual Cálculo que se usa para hallar la precisión de una medición. Entre más bajo es el error porcentual, más precisa es la medición. (p. 382)

percent yield A calculation that expresses the success of a chemical process in terms of product yield; the ratio of the actual yield to the theoretical yield expressed as a percentage. (p. 468)

rendimiento porcentual Cálculo que expresa el éxito de un proceso químico en términos del rendimiento del producto; la proporción entre el rendimiento real y el rendimiento teórico, expresada como un porcentaje. (p. 468)

period The elements in a horizontal row on the periodic table. (p. 46)

período Los elementos en una línea horizontal de la tabla periódica. (p. 46)

periodic table of the elements A table with elements organized in order of increasing atomic number, and grouped such that elements with similar properties are in vertical columns. (p. 41)

tabla periódica de los elementos Tabla con todos los elementos organizados en orden ascendente de acuerdo a su número atómico y agrupados de tal manera que los elementos con propiedades similares están en columnas verticales. (p. 41)

phase The physical form of matter such as the solid, liquid, or gaseous state. (p. 25)

fase La forma física de la materia como el estado sólido, líquido o gaseoso. (p. 25)

phase change A transition between solid, liquid, or gaseous states of matter. (p. 258)

cambio de fase Transición entre los estados sólido, líquido y gaseoso de la materia. (p. 258)

pH scale A logarithmic scale describing the concentration of hydrogen ions, H^+, in solution. $pH = -\log [H^+]$. (p. 424)

escala de pH Escala logarítmica que describe la concentración de iones de hidrógeno, H^+, en una solución. $pH = -\log [H^+]$. (p. 424)

physical change A change that alters the form of a substance but does not change the chemical identity of a substance. (p. 257)

cambio físico Cambio que altera la forma de una sustancia pero que no transforma la identidad química de la sustancia. (p. 257)

polar molecule A molecule that has a negatively charged end and a positively charged end due to electronegativity differences between the atoms and/or the asymmetry of its structure. (p. 217)

molécula polar Molécula que tiene un lado con carga negativa y otro con carga positiva debido a las diferencias de electronegatividad entre los átomos y/o a que su estructura es asimétrica. (p. 217)

polyatomic ion An ion that consists of two or more atoms covalently bonded. (p. 111)

ion poliatómico Ion que contiene dos o más átomos con enlaces covalentes. (p. 111)

potential energy Energy that is stored within a system and can be converted into other types of energy. This energy may be associated with the composition of a substance or its location. (p. 529)

energía potencial La energía que está almacenada en un sistema y que puede convertirse en otros tipos de energía. Esta energía se puede asociar con la composición de la sustancia o con su ubicación. (p. 529)

precipitate A solid produced in a chemical reaction between two solutions. (p. 451)

precipitado Sólido que se produce como resultado de una reacción química entre dos soluciones. (p. 451)

precipitation reaction A chemical reaction that results in the formation of a solid substance (a precipitate) that separates out of a solution because it is not very soluble. (p. 451)

reacción de precipitación Reacción química que resulta en la formación de una sustancia sólida (un precipitado) que se separa de la solución porque no es muy soluble. (p. 451)

pressure Force applied over a specific area. Force per unit area. (p. 294)

presión Fuerza aplicada sobre un área específica. Fuerza por unidad de área. (p. 294)

product A substance produced as the result of a chemical reaction. (p. 181)

producto Sustancia que se produce como resultado de una reacción química. (p. 181)

property A characteristic or quality of a substance. (p. 7)

propiedad Característica o cualidad de una sustancia. (p. 7)

proportional Related such that when one quantity increases, the other also increases. Two variables are proportional when you can multiply one variable by a constant to obtain the other. The variable y is proportional to the variable x if $y = kx$ where k is the proportionality constant. (p. 263)

proporcional Relacionadas de tal manera que cuando una cantidad aumenta, la otra también lo hace. Dos variables son proporcionales cuando una se puede multiplicar por una constante para obtener la otra. La variable y es proporcional a la variable x si $y = kx$ donde k es la constante de proporcionalidad. (p. 263)

proportionality constant The number that relates two variables that are proportional to one another. It is often represented by k. (p. 263)

constante de proporcionalidad El número que relaciona dos variables que son proporcionales entre sí. Generalmente se representa con la letra k. (p. 263)

protein A large molecule made up of chains of amino acids bonded together. Typical protein molecules consist of more than 100 amino acids. (p. 250)

proteína Molécula grande hecha de cadenas de aminoácidos enlazados. Las moléculas de proteínas suelen tener más de 100 aminoácidos. (p. 250)

proton A positively charged particle located in the nucleus of an atom. The mass of a proton is almost exactly equal to that of a neutron, about 1 amu. (p. 56)

protón Partícula de carga positiva localizada en el núcleo de un átomo. La masa de un protón es casi igual a la de un neutrón, 1 uma, aproximadamente. (p. 56)

pyramidal shape The shape assumed by other bonded atoms around an atom with one lone pair of electrons. (p. 194)

forma piramidal Forma que adquieren otros átomos enlazados alrededor de un átomo con un par aislado de electrones. (p. 194)

R

radiation Energy emitted by matter in the form of waves or particles. (p. 76)

radiación Energía emitida por la materia en forma de ondas o partículas. (p. 76)

radioactive decay Spontaneous disintegration of an atomic nucleus accompanied by the emission of particles and radiation. A radioactive substance will decay with a specific half-life. (p. 73)

desintegración radiactiva Desintegración espontánea del núcleo de un átomo que viene acompañada por la emisión de partículas y radiación. Una sustancia radiactiva se desintegrará en una vida media específica. (p. 73)

radioactive isotope Any isotope that has an unstable nucleus and decays over time. (p. 69)

isótopo radiactivo Cualquier isótopo que tiene un núcleo inestable y que se desintegra con el tiempo. (p. 69)

reactants The starting materials in a chemical reaction that are transformed into products during the reaction. (p. 181)

reactantes Los materiales iniciales de una reacción química que se transforman en productos durante dicha reacción. (p. 181)

reactivity The tendency of an element or compound to combine chemically with other substances, as well as the ease or speed of the reaction. (p. 39)

reactividad La tendencia de un elemento o compuesto a combinarse químicamente con otras sustancias, así como la facilidad o la velocidad de la reacción. (p. 39)

receptor site theory The currently accepted model explaining how specific molecules are detected by the nose. Molecules fit into receptor sites that correspond to the overall shape of the molecule. This stimulates a response in the body. (p. 208)

teoría de los sitios receptores El modelo aceptado en la actualidad que explica cómo detectar moléculas específicas mediante la nariz. Las moléculas caben dentro de sitios receptores que corresponden a la forma general de cada molécula. Esto estimula una respuesta en el cuerpo. (p. 208)

redox reaction Another name for an *oxidation-reduction* reaction. (p. 550)

reacción redox Otro nombre para la reacción de *oxidación-reducción* (p. 550)

reduction A chemical reaction in which an atom or ion gains electrons. Reduction is always accompanied by oxidation. (p. 550)

reducción Reacción química en la que un átomo o un ion gana electrones. La reducción siempre viene acompañada de oxidación. (p. 550)

relative humidity The amount of water vapor in the air compared to the maximum amount of water vapor possible for a specific temperature, expressed as a percentage. (p. 337)

humedad relativa La cantidad de vapor de agua en el aire comparada con la máxima cantidad de vapor de agua posible para una temperatura específica. Se expresa como un porcentaje. (p. 337)

reversible reaction A reaction that can proceed freely in both the forward and reverse directions. (p. 573)

reacción reversible Reacción que puede suceder libremente tanto en una dirección como en la dirección contraria. (p. 573)

rule of zero charge The rule that states that in an ionic compound, the positive charges on the metal cations and the negative charges on the nonmetal anions add up to zero. (p. 102)

regla de la carga cero La regla que establece que en un compuesto iónico, la suma total de las cargas positivas de los cationes de metales y de las cargas negativas de los aniones de no metales, es cero. (p. 102)

S

saturated solution A solution that contains the maximum amount of solute that can be dissolved in a given amount of solvent at a particular temperature. (p. 404)

solución saturada Solución que contiene la cantidad máxima de soluto que se puede disolver en una cantidad dada de solvente a un temperatura determinada. (p. 404)

scientific notation A shorthand notation used for writing numbers that are very large or very small. In this notation, the number is expressed as a decimal number with one digit to the left of the decimal point, multiplied by an integer power of 10. For example, the number 890,000 is written as 8.9×10^5 in scientific notation. (p. 385)

notación científica Notación abreviada que se usa para escribir números muy grandes o muy pequeños. En esta notación, el número se expresa como un número decimal con un dígito a la izquierda del punto decimal, multiplicado por una potencia entera de 10. Por ejemplo, el número 890,000 se escribe 8.9×10^5 en notación científica. (p. 385)

second law of thermodynamics The scientific law that states that energy tends to disperse or spread out. Thermal energy is always spontaneously transferred from a hotter object to a cooler object. (p. 484)

segunda ley de la termodinámica La ley científica que establece que la energía tiende a dispersarse o diseminarse. La energía térmica siempre se transfiere espontáneamente de un objeto más caliente a uno más frío. (p. 484)

single bond A covalent bond where two electrons are shared between two atoms. (p. 154)

enlace sencillo Enlace covalente en el que dos átomos comparten dos electrones. (p. 154)

single exchange reaction (single displacement reaction) A chemical change in which an element is displaced from a compound by a more reactive element. (p. 371)

reacción de intercambio simple (reacción de desplazamiento simple) Cambio químico en el que un elemento es desplazado de un compuesto por un elemento más reactivo. (p. 371)

soluble Capable of being dissolved into another substance. (p. 126)

soluble Que tiene la capacidad de disolverse en otra sustancia. (p. 126)

solute A substance dissolved in a solvent to form a solution. (p. 402)

soluto Sustancia disuelta en un solvente para formar una solución. (p. 402)

solution A homogeneous mixture of two or more substances. (p. 402)

solución Mezcla homogénea de dos o más sustancias. (p. 402)

solvent A substance in which another substance is dissolved, forming a solution. (p. 402)

disolvente Sustancia en la que otra sustancia se disuelve, para formar una solución. (p. 402)

space-filling model A three-dimensional representation of a molecule with no space between bonded atoms, as distinct from a ball-and-stick model. (p. 201)

modelo de espacio relleno Representación tridimensional de una molécula en la que no hay espacio entre los átomos enlazados, a diferencia del modelo de esferas y varillas. (p. 201)

specific heat capacity The amount of heat required to raise the temperature of 1 gram of a substance by 1 °C. (p. 494)

capacidad calorífica específica La cantidad de calor que se necesita para aumentar la temperatura de 1 gramo de una sustancia en 1 °C. (p. 494)

spectator ion An ion that does not directly participate in a chemical reaction. Spectator ions appear on both sides of a complete ionic equation. (p. 453)

ion espectador Ion que no participa directamente en una reacción química. Los iones espectadores aparecen a ambos lados de una ecuación iónica completa. (p. 453)

standard temperature and pressure (STP) A standard set of conditions at which gases can be measured and compared. Standard pressure is 1 atm and standard temperature is 273 K (0 °C). (p. 328)

temperatura y presión estándar (TPE) Conjunto estándar de condiciones en las que los gases pueden ser medidos y comparados. La presión estándar es 1 atm y la temperatura estándar es 273 K (0 °C). (p. 328)

stoichiometry The quantitative relationship between amounts (usually moles) of reactants and products in a chemical reaction. (p. 462)

estequiometría La relación cuantitativa entre la cantidad (usualmente en moles) de reactantes y productos en una reacción química. (p. 462)

strong acid (or base) An acid (or a base) that dissociates completely in solution. (p. 429)

ácido (o base) fuerte Ácido (o base) que se disocia completamente en una solución. (p. 429)

structural formula A two-dimensional drawing or diagram that shows how the atoms in a molecule are connected. Each line represents a covalent bond. (p. 150)

fórmula estructural Dibujo o diagrama de dos dimensiones que muestra cómo están conectados los átomos en una molécula. Cada línea representa un enlace covalente. (p. 150)

sublimation The process of changing phase from a solid to a gas without passing through the liquid phase. (p. 290)

sublimación El proceso de pasar de una fase sólida a una fase gaseosa sin pasar por la fase líquida. (p. 290)

surroundings Everything in the universe outside the system being investigated. (p. 478)

entorno Todo el universo que está por fuera del sistema que se está investigando. (p. 478)

synthesis The creation of specific compounds by chemists, through controlled chemical reactions. (p. 177)

síntesis La creación de compuestos específicos por parte de científicos, a través de reacciones químicas controladas. (p. 177)

system The part of the universe being investigated. (p. 479)

sistema La parte del universo que se está investigando. (p. 479)

T

temperature A measure of the average kinetic energy of the atoms and molecules in a sample of matter. (p. 276)

temperatura Medida del promedio de energía cinética de los átomos y las moléculas en una muestra de materia. (p. 276)

tetrahedral shape The shape defined by the symmetrical distribution of four bonded pairs of electrons around a central atom. (p. 193)

forma tetraédrica La forma definida por la distribución simétrica de cuatro pares de electrones enlazados alrededor de un átomo central. (p. 193)

theoretical yield The maximum amount of product that could be produced in a chemical reaction when a limiting reactant is entirely consumed. The value is calculated based on a balanced chemical equation. (p. 467)

rendimiento teórico La máxima cantidad de producto que puede producirse en una reacción química cuando el reactante limitador se ha consumido completamente. El valor se calcula basándose en una ecuación química balanceada. (p. 467)

thermal energy The total kinetic energy associated with the mass and motions of the particles in a sample of matter measured as heat energy. (p. 488)

energía térmica La energía cinética total asociada con la masa y el movimiento de partículas en una muestra de materia medida como energía calorífica. (p. 488)

thermal equilibrium When two objects in contact with one another reach the same temperature, they are in thermal equilibrium. (p. 484)

equilibrio térmico Cuando dos objetos que están en contacto alcanzan la misma temperatura, están en equilibrio térmico. (p. 484)

titration An analytical procedure used to determine the concentration of an acid or a base. A measured volume of an acid or a base of known concentration is reacted with a sample in the presence of an indicator. (p. 445)

valoración química Procedimiento analítico usado para determinar la concentración de un ácido o de una base. En presencia de un indicador, se hace reaccionar un ácido o una base de volumen y concentración conocidos, con una muestra de la sustancia que se quiere analizar. (p. 445)

toxicity The degree to which a substance can harm an organism. Toxicity depends on the toxin and the dose in which it is received. (p. 375)

toxicidad El grado de daño que una sustancia puede hacerle a un organismo. La toxicidad depende de la toxina y de la dosis suministrada. (p. 375)

transition elements The elements in Groups 1B to 8B on the periodic table. (p. 46)

elementos de transición Los elementos de los grupos 1B a 8B de la tabla periódica. (p. 46)

trigonal planar shape A flat triangular shape that is found in small molecules with three electron domains surrounding a central atom. (p. 197)

forma plana triangular Forma plana triangular que aparece en las moléculas pequeñas con dominio de tres electrones alrededor del átomo central. (p. 197)

triple bond A covalent bond in which three electron pairs are shared between two atoms. (p. 166)

enlace triple Enlace covalente en el que dos átomos comparten tres pares de electrones. (p. 166)

U

universal gas constant, *R* A number that relates the volume, temperature, pressure and number of moles of gas in the ideal gas law. The value of R is dependent on the units used. One value of R is 0.08206 L · atm/mol · K. (p. 332)

constante universal de los gases ideales, *R* Número que relaciona el volumen, la temperatura, la presión y el número de moles de un gas en la ley de los gases ideales. El valor de R depende de las unidades de medida. Un valor de R es 0.08206 L · atm/mol · K. (p. 332)

V

valence electrons The electrons located in the outermost electron shell of an atom, which participate in chemical bonding. (p. 93)

electrón de valencia Los electrones ubicados en la capa electrónica exterior de un átomo, que participan de un enlace químico. (p. 93)

valence shell The outermost electron shell in an atom. (p. 93)

capa de valencia La capa electrónica exterior de un átomo. (p. 93)

voltage The electrical potential of an electrochemical cell expressed in volts. (p. 561)

voltaje El potencial eléctrico de una celda electroquímica expresado en voltios. (p. 561)

volume The amount of space a sample of matter occupies. (p. 9)

volumen La cantidad de espacio que ocupa una muestra de materia. (p. 9)

W

water displacement (method) A method for measuring the volume of a solid object by immersing it in water. The volume of the object is equal to the amount of water displaced by the object when fully submerged. (p. 14)

desplazamiento de agua (método) Método para medir el volumen de un objeto sólido sumergiéndolo en agua. El volumen del objeto es igual a la cantidad de agua desplazada cuando el objeto se sumerge completamente. (p. 14)

weak acid (or base) An acid (or a base) that does not dissociate completely in solution. (p. 429)

ácido (o base) débil Ácido (o base) que no se disocia completamente en una solución. (p. 429)

weather The day-to-day atmospheric conditions such as temperature, cloudiness, and rainfall, affecting a specific place. (p. 257)

tiempo atmosférico Las condiciones atmosféricas diarias como la temperatura, la nubosidad y la precipitación, que afectan un lugar específico. (p. 257)

work Describes what happens when a force causes an object to move in the direction of the force applied. Work can be calculated as force · distance (p. 536)

trabajo Describe qué pasa cuando una fuerza hace que un objeto se mueva en la dirección en la que se aplica dicha fuerza. El trabajo se puede calcular como fuerza · distancia (p. 536)

Index

B

C

Index

Index

H

I

J

K

L

M

N

O

P

Index

Q

R

S

T

U

V

W

X

Z

Index

Photo Credits

Abbreviations: top (**t**), middle (**m**), bottom (**b**), left (**l**), right (**r**)

Unit 1

v, 1 (l): Dirk Wiersma/Science Source; **1 (r):** Jean-Loup Charmet/Science Photo Library; **2:** Gianni Cigolini/Getty Images; **3 (t):** Ken Karp Photography; **3 (bl):** Diane Hirsch/Fundamental Photos; **3 (bm):** Ken Karp Photography; **3 (br):** Richard Megna/Fundamental Photos; **4 (tl):** Richard Megna/Fundamental Photos; **4 (tm, tr):** Ken Karp Photography; **4 (m):** Shutterstock; **4 (bl, br):** Ken Karp Photography; **5 (t):** Lew Robertson/Getty Images; **5 (b):** Paul Rapson/Science Photo Library; **7:** Fotosearch/StockByte; **8 (t):** Shutterstock; **8 (b):** Sheila Terry/Science Photo Library; **10:** Clouds Hill Imaging Ltd./Corbis; **11 (l):** NASA/Science Photo Library; **11 (c, r):** Ken Karp Photography; **15:** Influx Productions/Getty Images; **17 (t):** image100/Corbis; **17 (b):** Joel Sartore/Getty Images; **18:** iStockPhoto; **19:** The Bridgeman Art Library/Getty Images; **20:** U.S. Mint; **22:** W. H. Freeman & Company/Worth Publishers; **24 (tl):** iStockPhoto; **24 Table: (mercury):** Harry Taylor/Getty Images/Dorling Kindersley; **24 (copper):** iStockphoto/Thinkstock; **24 (gold):** Dirk Wiersma/Science Source; **24 (sodium):** Andrew Lambert Photography/Photo Researchers; **24 (iron):** iStockPhoto; **24 (carbon):** Library of Congress; **24 (chlorine):** Charles D. Winters/Science Photo Library; **24 (iodine):** Charles D. Winters/Science Photo Library; **24 (phosphorus):** Charles D. Winters/Science Photo Library; **24 (bl):** Science Photo Library; **24 (ml):** Library of Congress; **24 (mr):** DEA/A. Rizzi/Getty; **24 (br):** Maurice Nimmo/Frank Lane Picture Agency/Corbis; **25 (t):** Virginija Valatkiene/Shutterstock; **25 (m, b):** Key Curriculum Press; **29 (t):** Reinhard Dirscherl/Getty Images; **29 (b-all):** Ken Karp Photography; **30 (l):** iStockPhoto; **30 (r):** Shutterstock; **31 (l):** Tony Freeman/PhotoEdit; **31 (r):** Blend Images/Thinkstock; **32 (m):** iStockPhoto; **32 (t):** Clive Streeter/Getty Images; **34:** Conor Caffrey/Science Photo Library; **36:** U.S. Mint; **39 (l):** Nation Wong/zefa/Corbis; **39 (m):** Martyn F. Chillmaid/Science Photo Library; **39 (r):** iStockPhoto; **40:** Hoberman Collection/Corbis; **46:** Stockbyte/Getty Images; **47:** PhotosIndia/Getty Images; **52:** CERN/Science Photo Library; **53:** Joel Sartore/Getty Images; **56 (t):** Andrew Dunn/Alamy. **56 (b):** Joseph Sohm; Visions of America/Corbis; **57:** Arthur S. Aubry/Getty Images; **59:** iStockPhoto; **60:** Jan Mika/New Zealand photos/Alamy; **64 (l):** Andrew Brookes National Physical Laboratory/Science Photo Library/Science Source; **64 (r):** Adam Buchanan/Danita Delimont/Alamy Images; **65:** Noel Hendrickson/Getty Images; **69:** Roger Ressmeyer/Corbis; **73:** iStockPhoto; **74:** Hulton-Deutsch Collection/Corbis; **75:** moodboard/Corbis; **76:** Shutterstock; **81:** NASA/JPL-Caltech/Corbis; **82:** Vladimir Repik/Reuters/Corbis; **85:** Lester V. Bergman/Corbis; **87 (t):** Tom Grill/Getty Images; **87 (b-all):** Lester V. Bergman/Corbis; **89:** Dr. Juerg Alean/Photo Researchers; **92:** Shutterstock; **97:** Shutterstock; **98:** iStockPhoto; **102:** iStockPhoto; **103:** Tobias Titz/Getty Images; **108:** Manchan/Getty Images; **109 (t):** Vincon/Klein/plainpicture/Corbis; **109 (b):** iStockPhoto; **113:** Condé Nast Archive/Corbis; **114:** The Gallery Collection/Corbis; **115:** Werner Forman/Art Resource, NY; **116:** iStockPhoto; **123:** iStockPhoto; **124:** David Bishop/Jupiter Images; **125:** Ken Karp Photography; **126:** iStockPhoto; **128:** Oleg Dolenko/Shutterstock; **129:** Jeffrey Hamilton/Getty Images; **130:** Michael Branscom/Lemelson-MIT Program; **133 (t):** Masakazu Watanabe/Aflo/Getty Images; **133 (b):** Stockbyte/Getty Images; **136 (l):** The Bridgeman Art Library/Getty Images; **136 (r):** iStockPhoto; **138 (l):** iStockPhoto; **138 (r):** Réunion des Musées Nationaux/Art Resource, NY

Unit 2

vi, 143 (l): Alex Cao/Getty Images; **143 (r):** Dan Suzio/Getty Images; **144:** Tim Pannell/Corbis; **145 (l):** iStockPhoto; **145 (m):** J.Garcia/photocuisine/Corbis; **145 (r):** Getty Images; **146:** Purestock/Getty Images; **147:** Exactostock/SuperStock; **148 (t):** David P. Hall/Corbis; **148 (b):** Thomas Northcut/Getty Images; **149 (l):** Shutterstock; **149 (r):** iStockPhoto; **150:** W. Perry Conway/Corbis; **155:** NASA; **156:** Alamy; **162:** Jonathon Nourok/PhotoEdit; **165:** iStockPhoto; **170 (t):** iStockphoto/Thinkstock; **170 (b):** Lew Robertson/Corbis; **171 (t):** Art Directors & TRIP/Alamy; **171 (b):** Felicia Martinez/

PhotoEdit; **173:** Ken Karp Photography; **176:** Corbis; **177:** SuperStock; **178 (t):** Alison Wright/Corbis; **179 (b):** iStockPhoto; **180:** Adalberto Rios Szalay/Sexto Sol/Getty Images; **181:** Shutterstock; **182:** Clive Streeter/Getty Images; **183:** Eric Risberg/Associated Press; **185:** Ken Karp Photography; **186 (l):** iStockPhoto; **186 (m):** Helen Sessions/Alamy; **186 (r):** iStockPhoto; **187 (l):** iStockPhoto; **187 (r):** Ken Karp Photography; **188:** Everett Collection/SuperStock; **191 (both):** Ken Karp Photography; **195:** Richard Hamilton Smith/Corbis; **196:** iStockPhoto; **198:** Getty Images; **199:** iStockPhoto; **201:** Sherry Yates Young/Shutterstock; **202:** iStockPhoto; **207:** Todd Gipstein/Getty Images; **208:** iStockPhoto; **210 (both):** iStockPhoto; **211:** Mark Karrass/Corbis; **213:** Creativa/Shutterstock; **214:** Ken Karp Photography; **217 (t):** Buck Forester/Getty Images; **217 (b):** Martyn F. Chillmaid/Science Photo Library/Science Source; **218 (m-both):** Ken Karp Photography; **218 (b):** Charles D. Winters/Science Photo Library; **220:** iStockPhoto; **228 (l):** iStockPhoto; **228 (r):** Comstock Images/Thinkstock; **230:** Getty Images; **233:** DAJ/Getty Images; **235:** iStockPhoto; **238:** Jim Craigmyle/Corbis; **239:** Getty Images; **242:** Leonard Lessin/Photo Researchers; **243 (tl):** Danilo Calilung/Corbis; **243 (tr):** iStockPhoto; **243 (b):** Ken Karp Photography; **244 (l):** iStockPhoto; **244 (r):** Shutterstock; **249:** Getty Images; **250 (l):** Comstock Select/Corbis; **250 (r):** Ken Karp Photography; **251 (b):** Getty Images/Thinkstock; **253 (l):** Jennifer Altman/epa/Corbis; **253 (r):** Dan Lamont/Corbis

Unit 3

vii, 255 (l): Stockbyte/Getty Images; **255 (r):** Grafton Marshall Smith/Corbis; **256:** NASA/Roger Ressmeyer/Corbis; **257 (l):** NASA; **257 (t):** Dex Image/Corbis; **257 (m):** Alley Cat Productions/Brand X/Corbis; **257 (b):** Lawrence Migdale/Science Photo Library; **259:** William Manning/Corbis; **260:** NASA; **263:** Mark Duffy/Alamy Images; **265:** Alamy; **266 (l):** Tom Stewart/Corbis; **266 (m):** USDA; **266 (r):** Philippe Psaila/Science Photo Library; **267:** Pierre Jacques/Hemis/Corbis; **268:** Ralph A. Clevenger/Corbis; **269:** Ken Karp Photography; **270 (l):** Spencer Grant/PhotoEdit; **270 (r):** Alamy; **274:** NASA; **275:** Corbis; **276:** NASA; **277:** Sheila Terry/Science Photo Library; **279:** Gianni Dagli Orti/Corbis; **280:** Alamy; **282 (t):** Shutterstock; **282 (b):** Bettmann/Corbis; **285 (m):** Alamy; **285 (b):** david tipling/Alamy Images; **286 (t):** Eric Nguyen/Corbis; **286 (b):** Nigel Cattlin/Photo Researchers; **288 (l):** Alamy; **288 (mt):** iStockPhoto; **288 (mb):** Image Bank/Getty Images; **288 (mr):** Alex Cao/Getty Images; **288 (r):** Alamy; **289:** Alamy; **290:** Charles D. Winters/Photo Researchers; **292 (t):** Manabu Ogasawara/Brand X/Corbis; **292 (b):** NASA/Roger Ressmeyer/Corbis; **293:** NASA; **294:** Ken Karp Photography; **295 (l):** Mark M. Lawrence/Corbis; **295 (r):** Ken Karp Photography; **296:** Macmillan Higher Education; **297:** Bettmann/Corbis; **298:** Ken Karp Photography; **300:** Duomo/Corbis; **301:** iStockPhoto; **303:** Robert Yin/Corbis; **304:** Randy Faris/Corbis; **306:** David Madison/Corbis; **307:** David Madison/Corbis; **308:** Robin Nelson/PhotoEdit; **310:** Paul Bradforth/Alamy Images; **311:** BSIP/Photo Researchers; **313:** SuperStock; **316:** Ken Karp Photography; **318:** Jeff Greenberg/PhotoEdit; **319:** Stefano Stefani/Getty Images; **320:** Chris Hellier/Alamy Images; **322:** Kent Wood/Photo Researchers; **324:** Science and Society/SuperStock; **325:** Photo Researchers; **326:** Paul Seheult; Eye Ubiquitous/Corbis; **328:** Frank Cezus/Getty Images; **334:** Davis Barber/PhotoEdit; **336 (l):** Juniors/SuperStock; **336 (r):** Studio Paggy/Getty Images; **337:** Randall Benton/Sacramento Bee/ZumaPress.Com/Alamy Images; **338:** Mark A. Johnson/Corbis; **339:** NOAA/Handout/Reuters/Corbis; **340:** NASA; **341 (t):** Dan Guravich/Corbis; **341 (bl):** Field, W. O. 1941. Muir Glacier. From the Glacier Photograph Collection. Boulder, Colorado USA: National Snow and Ice Data Center/World Data Center for Glaciology. Digital media; **341 (br):** USGS Photograph by Bruce F. Molnia; **343:** Carlos Barria/Reuters/Corbis; **345:** Dennis MacDonald/Alamy

Unit 4

viii, 347 (l): Charles D. Winters/Photo Researchers; **front cover, 347 (r):** Steve Cooper/Photo Researchers; **348:** Erik Reis/Alamy Images; **350:** Lisa Maree Williams/Getty Images; **354:** Philip Evans; **355:** iStockPhoto; **357 (l):** StockFood America; **357 (m):** Charles D. Winters/Photo Researchers; **357 (r):** Ken Karp Photography; **358 (l):** Andrew Lambert Photography/Science Photo Library; **358 (m):** Shutterstock; **358 (r):** Philip Evans; **360:** Martin Gray/Getty Images; **361:** Bryan Allen/Corbis; **362:** moodboard/Corbis; **363 (t):** Louie Psihoyos/Corbis; **363 (b):** ImageShop/Corbis; **364:** Joseph Sohm/Visions of America/Corbis; **365:** Corbis; **369:** Stephen Frink/zefa/Corbis; **371:** Roger Ressmeyer/Corbis; **372:** Alamy; **374:** Guy Cali/Corbis; **375:** Shutterstock; **376 (t):** Ted Kinsman/Photo Researchers; **376 (b):** Alejandro Ernesto/

epa/Corbis; **377:** Naturfoto Honal/Corbis; **378:** Michael Keller/Corbis; **379:** Studio Sato/amanaimages/Corbis; **382:** William West/AFP/Getty Images; **384:** Ken Karp Photography; **385:** NASA; **387:** iStockPhoto; **389:** Image Source/Corbis; **390:** iStockPhoto; **394:** Shutterstock; **395:** ImageShop/Corbis; **396:** Rachel Epstein/PhotoEdit; **397:** Food and Drug Administration; **400:** Jeremy Walker/Getty Images; **402:** Bill Aron/PhotoEdit; **403:** Tom Grill/PhotoEdit; **404:** Rob Melnychuk/Getty Images; **407:** Paul J. Sutton/PCN/Corbis; **408:** Shutterstock; **410:** Markus Moellenberg/zefa/Corbis; **413:** Tony Freeman/PhotoEdit; **415:** ASAP Agency/Photo Researchers; **416:** Bettmann/Corbis; **418:** Tony Wilson-Bligh; Papilio/Corbis; **420:** Nature Picture Library; **421:** Ken Karp Photography; **423 (both):** iStockPhoto; **424:** Gusto Productions/Photo Researchers; **426:** Nathan Hunsinger/Dallas Morning News/Corbis; **428:** Rachel Epstein/PhotoEdit; **429:** Kim Taylor/Nature Picture Library; **436:** Ken Karp Photography; **437:** Peter Byron/PhotoEdit; **440:** Ken Karp Photography; **442:** Christian Delbert/Shutterstock; **443:** David R. Frazier/David R. Frazier PhotoLibrary; **445:** Science Source; **446:** Will & Deni McIntyre/Corbis; **447:** Yoav Levy/Phototake/Newscom; **450:** Lawrence Migdale/Photo Researchers; **451:** Southern Illinois University/Science Source; **452:** David Taylor/Science Photo Library/Science Source; **453 (t):** Philip Schermeister/Getty Images; **453 (b):** Jose Luis Pelaez, Inc./Blend Images/Corbis; **455:** Ken Karp Photography; **456:** Will & Deni McIntyre/Corbis; **458:** Andrew Brookes/Corbis; **459:** Edward Kinsman/Photo Researchers; **463:** Shutterstock; **464:** Corbis; **467:** Envision/Corbis; **468:** Bernd Auers/Getty Images

Unit 5

ix, 473 (l): iStockPhoto; **473 (r):** Shutterstock; **474:** Shutterstock; **475 (t):** Blaine Franger/Getty Images; **475 (b-both):** iStockPhoto; **476 (tl-both):** iStockPhoto; **476 (m):** Michael Nicholson/Corbis; **476 (r):** Reuters/Corbis; **477:** The Bridgeman Art Library/Getty Images; **478:** Lon C. Diehl/PhotoEdit; **479:** sciencephotos/Alamy Images; **480:** Corbis; **482:** Ole Graf/zefa/Corbis; **486:** Melissa McManus/Getty Images; **487:** Ken Karp Photography; **493:** iStockPhoto; **494:** Kennan Harvey/Aurora Open/Corbis; **496:** Becky Luigart-Stayner/Corbis; **497:** Wayne Eastep/Getty Images; **498:** F. Hammond/photocuisine/Corbis; **500:** Steve Skjold/PhotoEdit; **502:** Shutterstock; **504:** iStockPhoto; **505:** Alamy; **506 (l):** Chris Howell/AP; **506 (r):** iStockPhoto; **507:** iStockPhoto; **508:** Ken Karp Photography; **510:** Burke/Triolo Productions/Brand X/Corbis; **515:** SSPL / Getty Images; **517:** Gary Houlder/Corbis; **521:** Colin Garratt; Milepost 92 1/2/Corbis; **522:** Ken Karp Photography; **528 (t):** Charles D. Winters/Photo Researchers; **528 (m):** Charles D. Winters/Photo Researchers; **528 (b):** Charles D. Winters/Photo Researchers; **530:** Photo Researchers; **532:** Roger Ressmeyer/Corbis; **534:** Ken Karp Photography; **535:** Tony Freeman/PhotoEdit; **536 (l):** Joseph Sohm/Visions of America/Corbis; **536 (r):** Tom Stewart/zefa/Corbis; **537 (t):** Corbis; **537 (b):** Guenter Rossenbach/zefa/Corbis; **538:** Science and Society/SuperStock; **539:** Image Werks Co.,Ltd./Alamy Images; **542:** iStockPhoto; **543 (t):** iStockPhoto; **543 (b):** SambaPhoto/Rogerio Assis/Getty Images; **544:** Ken Karp Photography; **546:** iStockPhoto; **548 (t):** The Bridgeman Art Library/Getty Images; **548 (b):** Craig Aurness/Corbis; **550:** Mark Schneider/Getty Images; **551:** Charles D. Winters/Photo Researchers; **552:** image100/Corbis; **554:** IPS Co., Ltd./Beateworks/Corbis; **555:** Martyn F. Chillmaid/Photo Researchers; **556:** Jim Cummins/Corbis; **560:** Tony Freeman/PhotoEdit; **562:** Richard T. Nowitz/Corbis

Unit 6

x, 567 (both): iStockPhoto; **568:** Jose Fuste Raga/Corbis; **569:** Blue Syndicate/Corbis; **570:** Educational Innovations Inc., www.teachersource.com; **571 (l):** Ken Karp Photography; **571 (m):** Stefano Stefani/Getty Images; **571 (r):** Lawrence Migdale/Science Photo Library; **572:** Ken Karp Photography; **573 (t):** iStockPhoto; **573 (b):** Richard T. Nowitz/Corbis; **575:** Ken Karp Photography; **576 (l):** Francesco Tomasinelli/Photo Researchers; **576 (m, r):** Ken Karp Photography; **577:** iStockPhoto; **578:** iStockPhoto; **591:** iStockPhoto; **597:** Achim Sass/Getty Images; **600:** Andia/Alamy Images; **601:** Ryan McVay/Getty Images; **605:** Photo Researchers; **607:** Michael P. Gadomski/Photo Researchers; **609:** iStockPhoto